SECOND EDITION

LINEAR ALGEBRA

with Applications

SECOND EDITION

LINEAR ALGEBRA

with Applications

Charles G. Cullen

University of Pittsburgh

ADDISON-WESLEY

An imprint of Addison Wesley Longman, Inc.

Reading, Massachusetts • Menlo Park, California • New York • Harlow, England
Don Mills, Ontario • Sydney • Mexico City • Madrid • Amsterdam

Sponsoring Editor: Kevin M. Connors
Project Coordination: Elm Street Publishing Services, Inc.
Art Development Editor: Vita Jay
Design Administrator: Jess Schaal
Text and Cover Design: Anderson Creative Services/Julie E. Anderson
Production Administrator: Randee Wire
Compositor: Interactive Composition Corporation
Printer and Binder: R. R. Donnelley & Sons Company
Cover Printer: The Lehigh Press, Inc.

Linear Algebra with Applications, Second Edition

Library of Congress Cataloging-in-Publication Data

Cullen, Charles G.
Linear algebra with applications/Charles G. Cullen.—2nd ed.
p. cm.
Includes index.
ISBN 0-673-99386-8
1. Algebras, Linear. I. Title.
QA184.C85 1996

512′.5—dc20 96-10857
CIP

96 97 98 99—DOC—9 8 7 6 5 4 3 2 1

Contents

Preface

This edition of *Linear Algebra with Applications* provides a carefully motivated introduction to linear algebra and a selection of its many applications. The text is suitable for introductory courses, which typically include a significant number of students who are not majoring in mathematics and may not have completed the full calculus sequence. Although some applications to calculus are included in the text, they are easily omitted so that the text is accessible to students who have had little or no calculus background. The book may be used at different levels by adjusting the sections covered and the exercises assigned.

The emphasis is on those topics from linear algebra that the students will encounter in their later work; extraneous material is carefully avoided. Interesting and accessible applications are included, in a nonintrusive way, throughout the text. The last two chapters are devoted to more involved applications. The importance of the computer to the development of linear algebra and to its practical applications is acknowledged and incorporated. Applications to computer science are included, with some pointers for implementing the techniques of linear algebra on a computer. Interactive software (for the IBM-PC and compatibles or the Apple Macintosh), which actually helps to teach the subject, is available from the publisher. This software can be most helpful in focusing attention on the interpretation of calculations rather than on the details of the calculations themselves. Hints on the use of the available software, as well as the MATLAB program and the TI-85 graphics calculator, are included at the end of selected sections. The final decision on how much to use software and calculators is left to the instructor.

The treatment is strongly computational, with many examples. The writing style is very student-oriented, with a careful balance between rigor and intuition. Proofs of theorems have been included when I felt that they either added to understanding or provided the basis for computational techniques; difficult, less relevant proofs have been omitted. The proof is occasionally presented first, with the formal statement of the related theorem acting as a summary statement. Where proofs have been omitted, the material is presented in a logical manner so that all the details can be added by the interested student. Some of the exercises provide the opportunity to complete or extend the theory; hints are frequently given. Because it is reasonable to expect some increase in mathematical maturity as the course evolves, the level of mathematical sophistication gradually increases as the book progresses. There is enough general vector-space theory to give the math majors a feel for the generality of the subject without overwhelming the nonmajors with abstract theory. There is an emphasis on reducing general vector-space questions to equivalent questions about systems of linear equations. In most cases the emphasis is on methods that involve first principles.

Because students learn best by doing, the exercises form a very important part of the text. There are many computational exercises, plus exercises designed to reinforce, supplement, and extend the development of the text. The use of motivating exercises (experimental evidence in support of general results to follow) is a key feature of this text. All exercises that are referred to in the main body of the text are flagged with a dagger. These exercises build intuition and provide motivation for topics to follow. Complete or partial answers to the odd-numbered exercises are included in the text. Detailed solutions of the odd-numbered problems are contained in the Student Solutions Manual, which is sold separately. The Instructor's Manual contains answers to the remainder of the computational problems and outlines of solutions for the theoretical problems. A set of review exercises is at the end of each of the first five chapters.

CHANGES IN THE SECOND EDITION

In preparing the second edition, I have been influenced both by the many helpful suggestions made by the prerevision reviewers and by the 1991 preliminary report of the AMS/AMA Linear Algebra Curriculum Study Group. I have tried to make it easier to focus on matrices and $\mathfrak{R}^n$ while still including enough general vector-space theory so that students can begin to see the power of the abstract approach. I have added many new examples and about 300 new exercises. I have also added chapter summaries at the end of each of the first five chapters. The software is a more integral part of the second edition than the first; MATMAN and MATALG are now available on-line from the publisher, and the help files are in an appendix to the text rather than in the Instructor's Manual. In many places, I have provided computational notes that, in parallel, describe how to do the relevant calculations using the MATMAN/MATALG programs, the MATLAB program, and the TI-85 graphics calculator. Use of the software or graphics calculators makes it possible to deal with a richer set of examples and to do some numerical experimentation.

TEXT

Chapter 1 begins with the linear equations problem and the elimination methods of solution. Matrix algebra is then developed and related to the elimination methods via row equivalence. There is a thorough discussion of systems with multiple right-hand sides that is used frequently in later chapters. Two optional sections discuss the LU factorization, machine arithmetic, and partial pivoting. By using these ideas, it is possible to describe a reasonably robust linear equations solver that has much in common with most production codes for solving the linear equations problem.

Chapter 2 develops the essential properties of the determinant function from an inductive point of view. Cramer's rule and the adjoint formula are developed and placed in their proper computational perspective. Some applications to geometry are included. Eigenvalues and eigenvectors for matrices are introduced in Section 2.4 as an application of the determinant function, but Section 2.4 can be postponed until the beginning of Chapter 5 if desired.

Chapter 3 deals with vector spaces. There is a brief geometric introduction for $\Re^2$ and $\Re^3$ and then a discussion of $\Re^n$. Most important vector-space concepts are introduced first for $\Re^n$ and then reexamined in the discussion of general real vector spaces that comes at the end of the chapter. This layering approach should aid students in understanding traditionally difficult abstract concepts. Inner products, vector norms, orthogonality, and the Gram-Schmidt process are treated only for $\Re^n$. The confusion that results from treating the elements of $\Re^n$ as either row or column matrices is carefully avoided; in this text the elements of $\Re^n$ are always column vectors.

Chapter 4 introduces linear transformations, their fundamental subspaces, and their matrix representatives. The algebra of linear transformations is developed, compared with the algebra of matrices, and used to interpret linear transformations on $\Re^n$ geometrically. The representation theory for linear operators, developed in Sections 4.5, 4.6, and 4.7, is the hardest theoretical material in the text. This material can be omitted if the students are nonmajors and the instructor is willing to give up the geometric motivation for the eigenvalue problem.

Chapter 5 introduces the matrix relation of similarity and completes the discussion of eigenvalues, eigenvectors, and diagonalizability introduced in Section 2.4. Orthogonal similarity and reduction to triangular form are discussed in Section 5.3; this important and useful topic is frequently omitted from books at this level. An optional section on the power method gives some insight into practical methods of computing eigenvalues. Instructions are given on how to compute eigenvalues and eigenvectors using the software or the TI-85 graphics calculator.

Chapter 6 on linear programming requires only Chapter 1 as background. The problem is first treated geometrically in the plane, and then the tableau version of the simplex method is introduced and used in several interesting applications. The two-phase method is introduced so that one is not restricted to problems in standard form. The MATMAN program is essential for this section if anything but the simplest applications are to be considered.

Chapter 7 contains four independent applications: graph theory, least squares approximations, quadratic forms, and linear economic models. These sections can be covered anytime after the indicated prerequisite sections have been covered.

Chapters 1–5 form the core of the text and provide a reasonable first course for math majors who have completed the calculus sequence. The prerequisites for the remainder of the material are indicated in the following diagram. I have recently taught a course for nonmajors that covered Chapters 1–3, Chapter 6, and Sections 7.1 and 7.4.

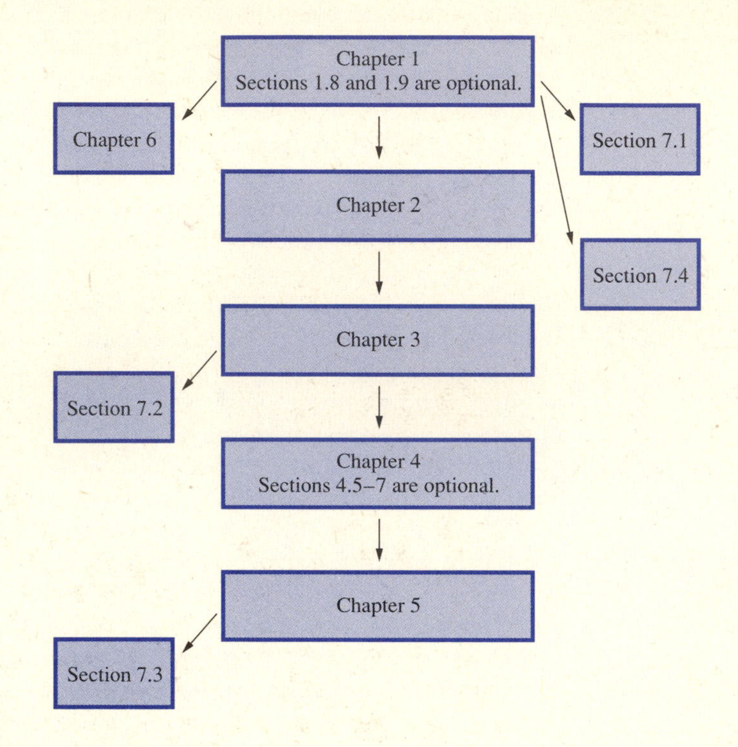

SOFTWARE

The software available to users of this text is very user-friendly and can be a valuable aid in teaching and learning the material. It consists of two main programs, MATMAN and MATALG. These programs are designed primarily to help teach linear algebra; they are not designed as production codes for large-scale problems. The programs are menu driven and require no programming knowledge; all that is required is knowing where to insert the program disk and how to turn on the machine. MATMAN (The MATrix MANipulator) is an interactive program that does row reductions. It can be used to solve linear systems, invert matrices, evaluate determinants, and solve linear programming problems using the tableau implementation of the simplex algorithm. In the primary mode, the user specifies the row operations to be performed, and the machine does the arithmetic quickly and accurately. The arithmetic can be done using either rational or real numbers. An automatic switch can be unlocked by the instructor after the student has progressed to the desired point. MATALG (The MATrix ALGebraist) is an interactive program for automating basic matrix operations. It can be used to save computation time with moderate-sized problems, to illustrate round-off error, and to provide experimental evidence in support of many of the theoretical observations in the text. These programs, which will run on IBM PCs and most IBM-PC compatibles or the Apple Macintosh, are available either electronically or on disk from the publisher. (See Appendix A for details on how to obtain these programs.) These programs were written in FORTRAN; the available files contain only the compiled versions of the programs. The available files and the appendices of this text contain full documentation for these programs. If the MATLAB program is available, it can be used instead of MATALG. MATLAB is a more powerful, more sophisticated, and costlier program than MATALG. Most of the calculational problems that require the software can also be solved on a variety of graphics calculators. Hints on using the TI-85 graphics calculator are included throughout the text.

SUPPLEMENTS

The Instructor's Manual, which is available to instructors considering this text, includes sample syllabi, answers to the even-numbered problems, and sample exams. The Student Solutions Manual, sold separately, contains solutions to the odd-numbered problems.

ACKNOWLEDGMENTS

The first edition of this text was started while I was on sabbatical leave at the University of California at Santa Barbara. I would like to thank the faculty for their hospitality and the use of their word-processing facilities during a most delightful winter term. I would also like to acknowledge the contribution of several classes of linear algebra students at the University of Pittsburgh who helped me class test and debug the original manuscript and its revisions. The software was written by John Burkardt, who took my rough ideas and turned them into accurate, efficient, and friendly courseware. I greatly admire his skill and his patience. The development of the courseware was partially supported by a grant from the College of General Studies of the University of Pittsburgh. I would also like to thank my colleagues at other schools who reviewed the manuscript and made many valuable suggestions. They include

Raymond Beasley, *University of Central Oklahoma*
Richard Blecksmith, *Northern Illinois University*
George Boros, *University of New Orleans*
Gordon Brown, *University of Colorado, Boulder*
John Bryant, *Florida State University*
Vincent F. Connolly, *Worcester Polytechnic Institute*
Daniel J. Curtin, *Northern Kentucky University*
George Feissner, *State University of New York at Cortland*
Roosevelt Gentry, *Jackson State University*

Noal Harbertson, *California State University, Fresno*
Diane Henderson, *Pennsylvania State University*
Matt Insall, *University of Missouri, Rolla*
Irving Katz, *George Washington University*
Magnhild Lien, *California State University, Northridge*
Philip Luft, *Salisbury State University*
Eugene Madison, *University of Iowa*
Mary Dowlen Pearce, *Greensboro College*
David Royster, *University of North Carolina, Charlotte*
Mary Salter, *Franciscan University*
James Snodgrass, *Xavier University*

Charles G. Cullen

1 Linear Systems and Matrices

In the first chapter of this text we will consider techniques for the efficient solution of systems of linear algebraic equations. Such systems arise frequently in all areas of investigation in which mathematical modeling is a useful tool. Frequently the number of unknowns and the number of equations is very large, and computational assistance, in the form of a computer, is needed to solve the system. In this chapter we will introduce matrices and matrix algebra in order to present efficient methods for solving linear systems.

1.1 GEOMETRIC VIEW OF LINEAR SYSTEMS

Let us begin by considering the following scheduling problem, which is a simplified version of a typical business application. A manufacturer of stuffed animals makes three products: Yogi Bear, Peter Rabbit, and Donald Duck. Each toy must pass through three separate departments: sewing, stuffing, and decorating. Yogi spends 24 minutes in the sewing department, 18 minutes in the stuffing department, and 9 minutes in decorating. Peter spends 16 minutes in sewing, 12 minutes in stuffing, and 8 minutes in decorating, while Donald requires 18 minutes of sewing, 9 minutes of stuffing, and 4 minutes of decorating. The sewing department can supply 50 hours of labor per day, the stuffing department 33 hours, and the decorating department 18 hours. How many of each animal should be produced in order to make full use of the available labor?

To analyze this situation we let

x = the number of bears produced;

y = the number of rabbits produced; and

z = the number of ducks produced.

The total usage of the sewing department is thus $24x + 16y + 18z$ minutes. Since the sewers are available for 50 hours or 3000 minutes we must have

$$24x + 16y + 18z = 3000.$$

The total usage of the stuffing department is $18x + 12y + 9z$ minutes and this department can supply 33 hours or 1980 minutes per day. Thus,

$$18x + 12y + 9z = 1980.$$

Similar analysis of the use of the decorators leads to the equation

$$9x + 8y + 4z = 1080.$$

Thus, the unknowns x, y, and z must satisfy each of the three equations

$$\begin{aligned} 24x + 16y + 18z &= 3000 \\ 18x + 12y + 9z &= 1980 \\ 9x + 8y + 4z &= 1080. \end{aligned}$$

This is an example of a system of linear equations. One major concern of this text will be with techniques for solving such systems; much of what we do will be motivated by the desire to deal with such systems efficiently and intelligently.

The general linear equation in three variables is an equation of the form $ax + by + cz = k$, where a, b, c, and k are constants. If the problem has more than three variables, we normally use subscripted variables like x_i, y_j, or z_k and denote the constants in the equation by a_i, b_j, or c_k. Thus, the **general linear equation in n unknowns** is an equation of the form

$$a_1x_1 + a_2x_2 + \cdots + a_nx_n = b,$$

where b and the a_i are constants. The **solution set** of this equation consists of all n-tuples of values of the variables that satisfy the equation. To find a particular solution of a linear equation you may randomly pick values for all but one of the unknowns and then use the equation to compute the appropriate value of the remaining unknown. For the linear equation $4x + 5y = 3$ we could choose x and then compute y, or we could first choose y and then compute x. Thus, if we choose $x = 1$ it follows that $5y = 3 - 4 = -1$ or $y = -\frac{1}{5}$; if we choose $y = 4$, then $4x = 3 - 20$ or $x = -\frac{17}{4}$. Thus, $(1, -\frac{1}{5})$ and $(-\frac{17}{4}, 4)$ are two members of the solution set of $4x + 5y = 3$. Can you find a solution for which $x = 5$? For the general case we make the following definition.

DEFINITION 1.1 An m by n **linear system** is a set of m linear equations in n unknowns. A **solution** of this linear system is an n-tuple of values for the variables that satisfy each equation in the system. □

Given such a system, we need to ask the following questions:

(1) Do solutions exist?
(2) How many solutions are there?
(3) How can we find all solutions?

We begin with a very simple case in which there are only two variables and only two equations.

$$\begin{aligned} 3x - 4y &= 12 \\ 3x + 4y &= 0 \end{aligned} \tag{1.1}$$

It is easy to check that (4, 0), that is, $x = 4$ and $y = 0$, is a solution of the first equation but not of the second, while $(4, -3)$ is a solution of the second equation but not of the first. You should check that $(2, -\frac{3}{2})$ is a solution of both equations and hence a solution of System (1.1). We will see one way to obtain this solution shortly.

In the case of two variables there is a familiar and useful geometric interpretation of linear equations:

> The equation $ax + by = c$ represents a straight line in the xy-plane.

The solutions of the linear equation $ax + by = c$ are just the coordinates of the points that lie on the line. The solution set is the set of all points that lie on the line. The solution set of the first equation of System (1.1) is easy to obtain. Since

$$3x - 4y = 12 \text{ is equivalent to } y = -3 + \frac{3x}{4},$$

we can choose for x any real number, say $x = c$, and then compute $y = -3 + (\frac{3}{4})c$. Thus $x = c$, $y = -3 + (\frac{3}{4})c$, is a solution of the equation for any choice of c ($c = 4$ gives the first solution). We can describe the solution set for this equation as

$$\left\{\left(c, -3 + \frac{3c}{4}\right) \middle| c \text{ real}\right\} \quad \text{or} \quad \left\{\left(x, -3 + \frac{3x}{4}\right), \middle| x \text{ real}\right\}. \tag{1.2}$$

Note that this amounts to giving a parametric description of the straight line whose equation is $3x - 4y = 12$; $x = c$ is the parameter.

The problem of solving System (1.1) amounts to finding the point of intersection of the two lines; see Figure 1.1.

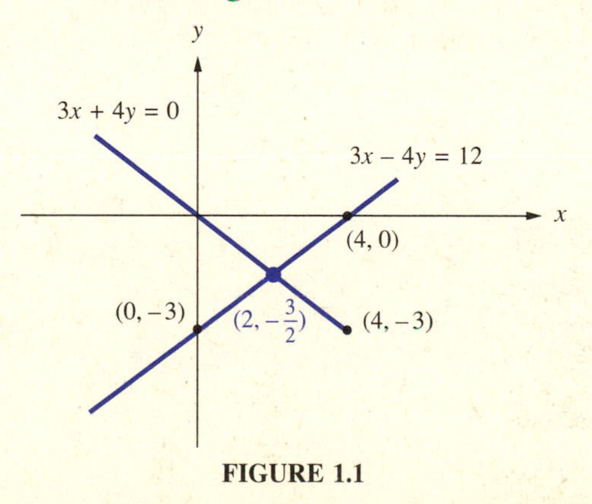

FIGURE 1.1

Finding the point of intersection of the two lines amounts to finding a value of x for which the two y values are the same. Thus we solve each equation for y, obtaining

$$y = -3 + \frac{3x}{4} \qquad \text{and} \qquad y = \frac{-3x}{4}.$$

These y values will be the same if

$$-3 + \frac{3x}{4} = \frac{-3x}{4} \qquad \text{or, equivalently,} \qquad \frac{3x}{4} + \frac{3x}{4} = \frac{6x}{4} = 3.$$

It follows that $x = 2$ and that $y = -\frac{3x}{4} = -\frac{3}{2}$. Thus, the point of intersection of the two lines has coordinates $(2, -\frac{3}{2})$.

Since solving two equations in two unknowns is equivalent to finding the point of intersection of two straight lines, it is clear that such a system may have either:

(1) no solutions, if the lines are parallel;
(2) exactly one solution, if the lines intersect; or
(3) an infinite number of solutions, if the lines are the same.

Figure 1.2 illustrates these three cases.

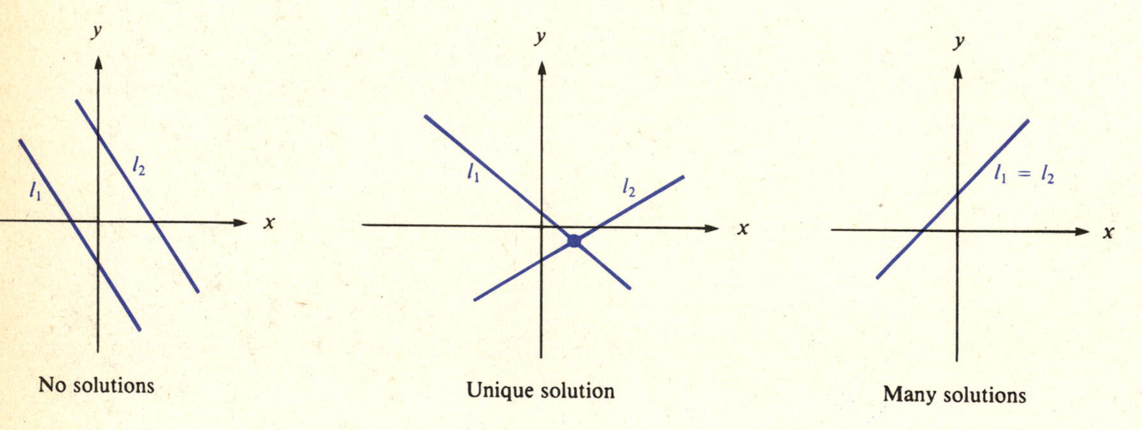

FIGURE 1.2

For n equations in two unknowns ($n > 2$), the same three possibilities exist, as illustrated for the case $n = 3$ in Figure 1.3. Note that in the first case none of the lines need be parallel. For $n > 2$ the first possibility (no solutions) is clearly the most likely, at least if the lines are chosen at random. If the equations describe some physical system, then it is more likely that a solution exists.

In general, *if there are more equations than unknowns, then it is a very real possibility that no solutions exist.*

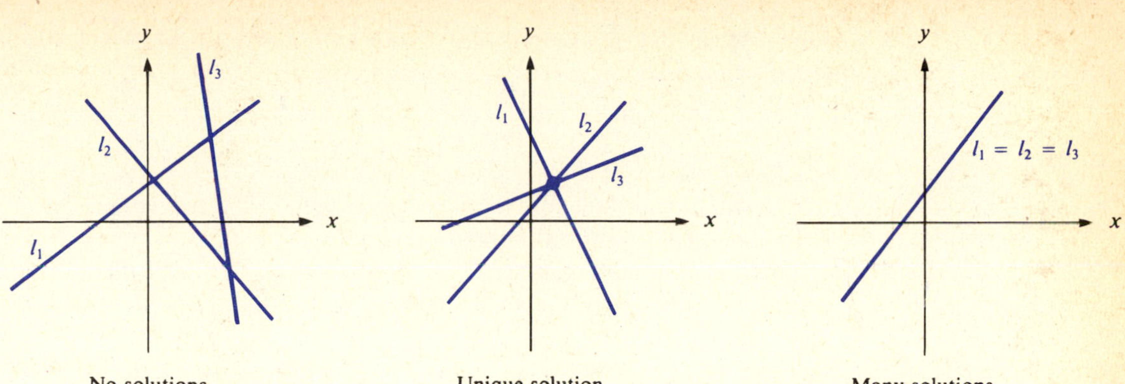

FIGURE 1.3

If you have studied three-dimensional analytic geometry, perhaps in the third term of calculus, then you know that a linear equation in three variables such as

$$ax + by + cz = d, \tag{1.3}$$

represents a plane in 3-space as shown in Figure 1.4.

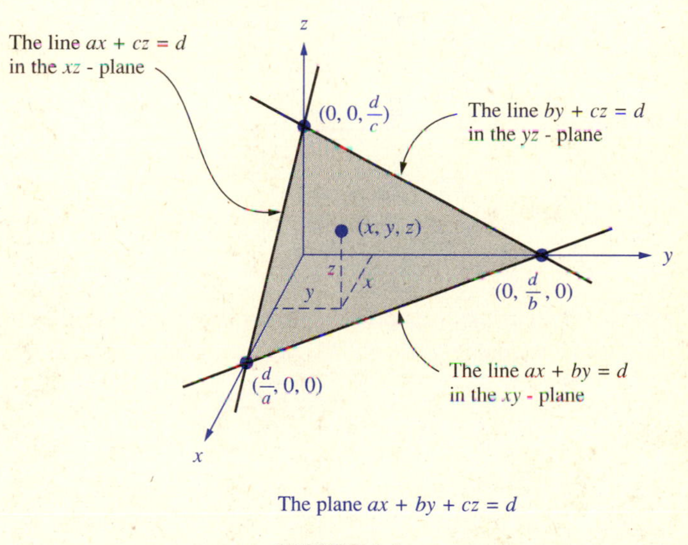

FIGURE 1.4

This observation allows us to give a geometric interpretation of the problem of solving n equations in three unknowns; in each case we are looking for a point common to several planes. If $n = 2$, then the two planes are either parallel or meet in a line; therefore there are either no solutions or an infinite number of solutions. Note that *we cannot have a unique solution when there are fewer equations than unknowns.* If $n = 3$, then either:

(1) There is no point of intersection; that is, the third plane is parallel to the line of intersection of the first two planes (like the ceiling, floor, and one side wall of your room);
(2) The three planes meet in a unique point (where the third plane meets the line of intersection of the first two); or
(3) There is an infinite number of solutions; that is, the three planes have at least one line in common (like the pages of a book have the binding in common).

Figure 1.5 illustrates these three cases. If $n > 3$, then the same three possibilities occur, the most likely (if the planes are chosen at random) being that there are no solutions.

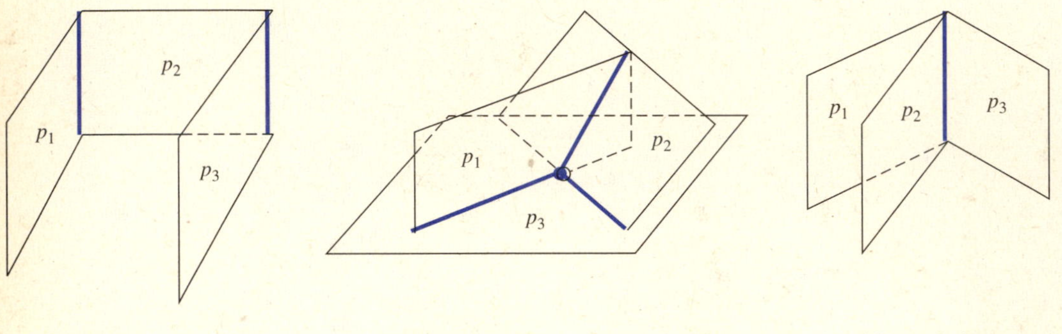

FIGURE 1.5

Note that in both cases considered there are always three possibilities: no solutions (in which case we say that the system is **inconsistent**), a unique solution, or an infinite number of solutions. From our examples it appears that the seemingly desirable case of a unique solution is most likely when the number of unknowns is equal to the number of equations. We will establish results later that will tell us precisely when a linear system has a unique solution.

EXAMPLE 1 Sketch of a Plane

Find the solution set of $3x + 2y + z = 6$ and sketch the plane.

In this case we will treat two of the variables as parameters: if we set $x = c_1$, and $y = c_2$, then we must have

$$z = 6 - 3x - 2y = 6 - 3c_1 - 2c_2.$$

Thus, we can describe the solution set of $3x + 2y + z = 6$ as

$$\{(c_1, c_2, 6 - 3c_1 - 2c_2) \mid c_1, c_2 \text{ real}\}$$

or, equivalently, as

$$\{(x, y, 6 - 3x - 2y) \mid x, y \text{ real}\}.$$

In order to sketch the plane we first locate the points of intersection of the plane with the coordinate axes. The point of intersection with the x-axis is obtained by setting $y = z = 0$ and computing $3x = 6$ or $x = 2$. Similarly, the point of intersection with the y-axis is $(0, 3, 0)$ and the intersection with the z-axis is $(0, 0, 6)$. The portion of the plane which lies in the first octant is shown in Figure 1.6. ■

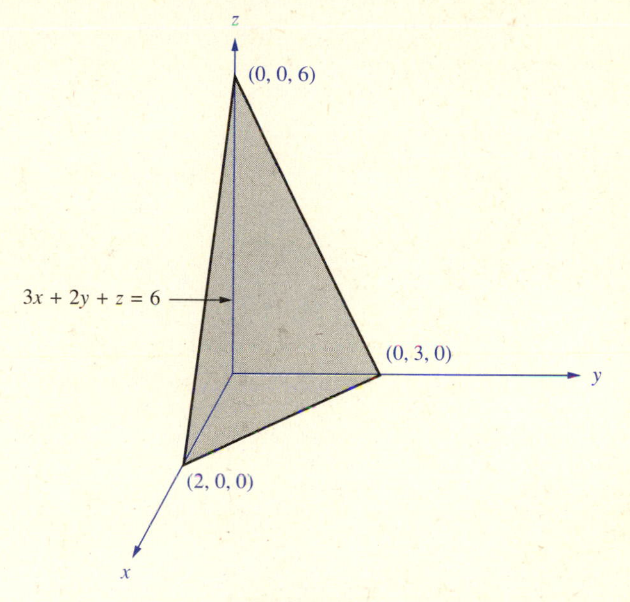

FIGURE 1.6

Our next example shows how linear systems arise in curve-fitting problems.

■ EXAMPLE 2 Find a Parabola Through Three Points

Determine constants a, b, and c so that the parabola $y = ax^2 + bx + c$ will pass through the points $(1, 3)$, $(0, 5)$, and $(3, 11)$.

Each point on the curve leads to a linear equation in a, b, and c. Thus, because the point $(1, 3)$ is on the curve we must have

$$a + b + c = 3.$$

Similarly, because $(0, 5)$ and $(3, 11)$ are to be on the curve, we must also have

$$0a + 0b + c = 5 \qquad \text{and} \qquad 9a + 3b + c = 11.$$

Thus a, b, and c must satisfy the 3 by 3 linear system

$$\begin{aligned} a + b + c &= 3 \\ c &= 5 \\ 9a + 3b + c &= 11. \end{aligned}$$

This system is easy to solve because the second equation tells us that $c = 5$. The other two equations then reduce to

$$\begin{aligned} a + b &= -2 \\ 9a + 3b &= 6. \end{aligned}$$

From the first equation we have $b = -2 - a$, which, when substituted into the second equation, yields $9a + 3(-2 - a) = 6$ or $6a = 12$. Thus, $a = 2$, $b = -4$, and the desired parabola is $y = 2x^2 - 4x + 5$ which is shown in Figure 1.7. ■

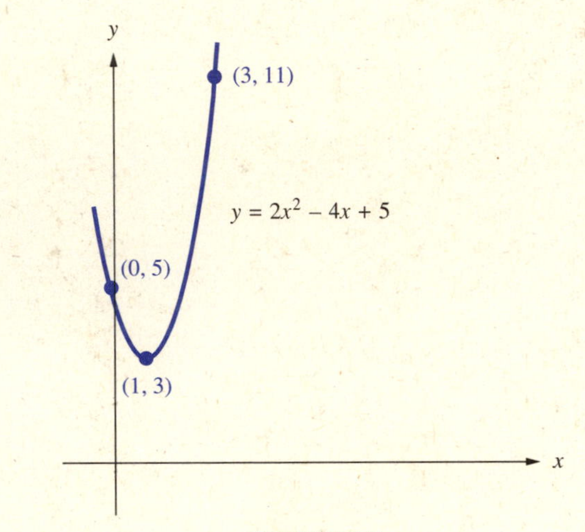

FIGURE 1.7

Let us now consider a general system of m equations in the n unknowns $x_1, x_2, \ldots, x_n$:

$$\begin{aligned} a_{11}x_1 + a_{12}x_2 + \cdots + a_{1n}x_n &= b_1 \\ a_{21}x_1 + a_{22}x_2 + \cdots + a_{2n}x_n &= b_2 \\ \vdots \qquad \vdots \qquad\quad \vdots \qquad &\\ a_{m1}x_1 + a_{m2}x_2 + \cdots + a_{mn}x_n &= b_m. \end{aligned} \tag{1.4}$$

Note carefully the notation used to describe the general system; *the coefficients are doubly subscripted with the first subscript indicating the equation and the second subscript indicating the variable;* for example a_{ij} is the coefficient of x_j in equation i. In the case $m = 4$, $n = 3$ System (1.4) becomes

$$\begin{aligned} a_{11}x_1 + a_{12}x_2 + a_{13}x_3 &= b_1 \\ a_{21}x_1 + a_{22}x_2 + a_{23}x_3 &= b_2 \\ a_{31}x_1 + a_{32}x_2 + a_{33}x_3 &= b_3 \\ a_{41}x_1 + a_{42}x_2 + a_{43}x_3 &= b_4. \end{aligned}$$

A **matrix** is a rectangular array of numbers; we will associate two important matrices with System (1.4).

The **coefficient matrix** of System (1.4) is the array

$$A = \begin{bmatrix} a_{11} & a_{12} & \cdots & a_{1n} \\ a_{21} & a_{22} & \cdots & a_{2n} \\ \vdots & \vdots & & \vdots \\ a_{m1} & a_{m2} & \cdots & a_{mn} \end{bmatrix}. \tag{1.5}$$

Note that the first subscript indicates the row in which the entry occurs while the second subscript determines the column. Thus, a_{21} is in row 2 and column 1, while a_{47} is in row 4 and column 7, and a_{rc} is the entry in row r and column c of the matrix A. We will say that the matrix A is m by n ($m \times n$) with the first number indicating the number of rows of A and the second indicating the number of columns of A.

The **augmented matrix** is the coefficient matrix with one extra column, that column being the matrix B whose entries are the constants from the right-hand side of the system. Thus, the augmented matrix of System (1.4) is

$$[A \mid B] = \begin{bmatrix} a_{11} & a_{12} & \cdots & a_{1n} & b_1 \\ a_{21} & a_{22} & \cdots & a_{2n} & b_2 \\ \vdots & \vdots & & \vdots & \vdots \\ a_{m1} & a_{m2} & \cdots & a_{mn} & b_m \end{bmatrix}, \tag{1.6}$$

$$\text{where } B = \begin{bmatrix} b_1 \\ b_2 \\ \vdots \\ b_m \end{bmatrix}.$$

As a general rule, if A is $m \times n$, then the augmented matrix is $m \times (n + 1)$.

There is an obvious one-to-one correspondence between systems and their augmented matrices. For example, the augmented matrix of the system

$$\begin{aligned} 3x - 4 &= 12 \\ 4x + 3y &= 0 \end{aligned} \quad \text{is} \quad \begin{bmatrix} 3 & -4 & 12 \\ 4 & 3 & 0 \end{bmatrix},$$

while the system with augmented matrix

$$\begin{bmatrix} 3 & 0 & 4 & 2 \\ 5 & 8 & -2 & 4 \\ 4 & 2 & 8 & 11 \end{bmatrix} \quad \text{is} \quad \begin{aligned} 3x_1 \phantom{{}+ 8x_2} + 4x_3 &= 2 \\ 5x_1 + 8x_2 - 2x_3 &= 4 \\ 4x_1 + 2x_2 + 8x_3 &= 11 \end{aligned}$$

Be careful to supply zeros in the appropriate places when going from the system to the augmented matrix. For example, the coefficient and augmented matrices of the system

$$\begin{aligned} 2x_1 - x_2 \qquad\qquad &= 5 \\ -x_1 + 2x_2 - x_3 \qquad &= 0 \\ - x_2 + 2x_3 - x_4 &= 0 \\ - x_3 + 2x_4 &= 9 \end{aligned}$$

are, respectively,

$$\begin{bmatrix} 2 & -1 & 0 & 0 \\ -1 & 2 & -1 & 0 \\ 0 & -1 & 2 & -1 \\ 0 & 0 & -1 & 2 \end{bmatrix} \quad \text{and} \quad \begin{bmatrix} 2 & -1 & 0 & 0 & 5 \\ -1 & 2 & -1 & 0 & 0 \\ 0 & -1 & 2 & -1 & 0 \\ 0 & 0 & -1 & 2 & 9 \end{bmatrix}.$$

The software programs MATMAN, MATALG, and MATLAB all have slightly different ways of entering a matrix. The following table lists the commands for entering the matrix $A = \begin{bmatrix} 1 & 2 & 3 & 4 \\ 5 & 6 & 7 & 8 \\ 9 & 10 & 11 & 12 \end{bmatrix}$.

MATMAN	MATALG	MATLAB	TI-85
E 3,4	E A 3,4	A=[1 2 3 4	[[1,2,3,4] [5,6,7,8]
1,2,3,4	1,2,3,4	5,6,7,8	[9,10,11,12]] STO A
5,6,7,8	5,6,7,8	9,10,11,12]	
9,10,11,12	9,10,11,12	or	
		A=[1 2 3 4; 5 6 7 8; 9 10 11 12]	

EXERCISES 1.1

1. For each of the following systems, graph the lines represented by the equations and, from the graphs, find the solution(s) of the systems if any exist.
   (a) $\begin{aligned} x + y &= 4 \\ 2x - 2y &= 8 \end{aligned}$
   (b) $\begin{aligned} x + y &= 4 \\ 2x + 2y &= 6 \end{aligned}$
   (c) $\begin{aligned} x + y &= 4 \\ x - y &= 8 \\ 2x + 3y &= 6 \end{aligned}$

2. Find the solution sets of each of the following linear equations. (Cf. Example 1.)
   (a) $3x - 5y = 15$
   (b) $4x_1 + 3x_2 = 9$
   (c) $3x + 5y - 7z = 10$
   (d) $4x_1 + 2x_2 - 6x_3 = 12$

3. Which of the following equations are linear in x, y, and z?
   (a) $x + 2y + z = 2$
   (b) $x + y + z = \sin 3$
   (c) $x + \frac{1}{y} + 3z^2 = 11$
   (d) $x + y + \sqrt{2}z = 4$
   (e) $3x + \sin y + z = 0$
   (f) $3x^2 + 5xy = 11$
   (g) $x^2 + y^2 = 25$
   (h) $2 \sin x + 3 \cos y = 4$

4. Find solutions of the systems in Exercise 1 using the substitution technique illustrated in the solution of System (1.1).

5. Find the coefficient matrix and the augmented matrix for the following systems.

(a) $$\begin{aligned} 2x - 3z &= 11 \\ 4x + 5y + z &= 6 \\ 4y + 12z &= 0 \end{aligned}$$

(b) $$\begin{aligned} 4x_1 - 3x_2 + 15x_3 &= 0 \\ 4x_1 + 17x_2 &= 14 \\ 3x_1 + 5x_2 + x_3 &= 4 \end{aligned}$$

(c) $$\begin{aligned} 7x + 8y + 8z &= -x \\ 9x - 16y - 18z &= -y \\ -5x + 11y + 13z &= -z \end{aligned}$$

(d) $$\begin{aligned} x_1 - x_3 &= 0 \\ x_2 + 3x_3 &= 1 \\ x_1 + x_2 + x_3 &= 5 \end{aligned}$$

(e) $4x_{i-1} + 6x_i - 5x_{i+1} = 10 + i, \quad i = 1, 2, \ldots, 5; x_0 = x_6 = 0$

(f) $\sum_{i=1}^{4} (i + j)x_i = j, \quad j = 1, 2, 3, 4$

(g) $\sum_{j=1}^{4} \frac{1}{i + j - 1} x_j = i, \quad i = 1, 2, 3.$

[†]**6.** Find the system whose augmented matrix is the following matrix.

(a) $$\begin{bmatrix} 2 & 1 & 5 & 1 & 5 \\ 1 & 1 & -3 & -4 & -1 \\ 3 & 6 & -2 & 1 & 8 \\ 2 & 2 & 2 & -3 & 2 \end{bmatrix}$$

(b) $$\begin{bmatrix} 3 & 0 & 0 & 0 & 9 \\ 2 & 1 & 0 & 0 & 0 \\ -2 & 3 & 1 & 0 & 4 \\ 0 & 3 & 2 & 1 & 7 \end{bmatrix}$$

(c) $$\begin{bmatrix} 4 & 1 & 0 & 0 & 5 \\ 1 & 4 & 1 & 0 & 0 \\ 0 & 1 & 4 & 1 & 0 \\ 0 & 0 & 1 & 4 & 7 \end{bmatrix}$$

(d) $$\begin{bmatrix} 1 & -3 & 0 & 3 \\ 0 & 1 & 1 & 1 \\ 0 & 0 & -6 & -7 \end{bmatrix}$$

(e) $$\begin{bmatrix} 2 & 0 & 0 & 5 \\ 0 & 1 & 0 & 3 \\ 0 & 0 & 7 & 2 \end{bmatrix}$$

(f) $$\begin{bmatrix} 1 & 2 & 4 & 8 \\ 0 & 0 & 3 & 9 \\ 0 & 0 & 0 & 5 \end{bmatrix}$$

Which, if any, of the above systems are easy to solve?

7. Solve the systems in exercises 5(a) and 5(d) using the substitution technique used to solve System (1.1).

[†]**8.** Show that if (x_0, y_0) is a solution of the system

$$\begin{aligned} ax + by &= c \\ dx + ey &= g, \end{aligned}$$

then (x_0, y_0) is also a solution of the equation

$$(d + ka)x + (e + kb)y = (g + kc),$$

obtained by adding k times the first equation to the second.

[†]These exercises are particularly important. They either provide motivation for results to follow or they have been referred to in the text.

9. Sketch the following sets in the xy-plane.

(a) $\{(x, y) \mid 2x + 3y = 6\}$
(b) $\{(x, y) \mid 2x + 3y < 6\}$
(c) $\{(x, y) \mid 2x + 3y > 6\}$

10. Show that if (a, b) and (c, d) are both in the solution set of $rx + sy = k$, then so is the point $(ta + (1 - t)c, tb + (1 - t)d)$ for any real number t. Interpret this result geometrically. What point corresponds to $t = \frac{1}{2}$?

11. A certain farm has only cows and chickens. It is observed that altogether there are 60 heads and 200 legs on the farm. How many of each kind of animal are on the farm?

12. Most computers do not understand subscript notation, so one uses $A(I, J)$ or $A[I, J]$ and $X(I)$ or $X[I]$ in place of a_{ij} and x_i. Rewrite System (1.4) and Matrices (1.5) and (1.6) using this notation.

13. A candy store wishes to make 100 lbs. of mixed nuts, including peanuts, cashews, almonds, and sunflower seeds. Peanuts cost \$1.25/lb., cashews are \$3.00/lb., almonds are \$2.50/lb., and sunflower seeds are \$1.75/lb. The mix is to contain twice as many peanuts as cashews, 3 times as many cashews as almonds, and is to be worth \$1.80/lb. How much of each raw ingredient is needed?

14. Show that it is *not* possible to choose 17 coins, with a total value of \$1.85, from a pile of quarters, dimes, and nickels in such a way that you have twice as many dimes as quarters.

15. Find the equation of the parabola $y = ax^2 + bx + c$ that passes through the points $(1, 1)$, $(2, 12)$, and $(-1, -9)$.

16. Find the equation of the circle $x^2 + y^2 + ax + by = c$ that passes through the points $(5, 5)$, $(4, 6)$, and $(5, -1)$.

17. In order to control a certain crop pest it is recommended that the crop be sprayed with a mixture of 6 units of chemical A, 10 units of chemical B, and 8 units of chemical C. These ingredients are available in three ready-mixed proportions: one barrel of spray $S1$ contains 1, 3, and 4 units of chemicals A, B, and C, respectively; one barrel of spray $S2$ contains 3 units of each chemical, and one barrel of spray $S3$ contains 2 units of A and 5 units of B. How much of each commercial spray should be used to obtain the exact amounts of chemical needed?

†18. Sketch the plane represented by each of the following linear equations:
(a) $2x + 3y + 4z = 12$ **(b)** $3x - 2y + 3z = 6$ **(c)** $2x + 5y = 10$

19. Find a, b, and c so that the plane $z = ax + by + c$ passes through the points $(1, 1, 3)$, $(2, 3, -1)$, and $(-1, 3, 7)$.

20. Consider an $n \times n$ matrix A.
(a) Where are the entries a_{ij} with $i = j$?
(b) Where are the entries a_{ij} with $i > j$?
(c) Where are the entries a_{ij} with $i \leq j$?

21. Find a, b, and c so that the parabola $y = ax^2 + bx + c$ contains the points $(1, 5)$, $(0, 4)$, and $(4, -4)$.

22. Enter the matrix

$$A = \begin{bmatrix} 2 & 3 & 0 & 5 & 4 \\ 6 & 3 & -\frac{5}{3} & 4 & 7 \\ -1 & 5 & 3 & 2 & 8 \\ 9 & -7 & 4 & 0 & 3 \\ 0 & 11 & \frac{3}{5} & 2 & -1 \end{bmatrix}$$

into your TI-85 calculator and, using the STO key, name it A. Use the edit command on the matrix menu to change a_{22} to -8 and a_{45} to 19. List the items under the MATH submenu of the MATRIX menu. Try the "T" command on the matrix A. What does it do? List the items available under the OPS submenu of the MATRIX menu. Try the "ref" command on the matrix A.

23. Enter the matrix

$$A = \begin{bmatrix} 2 & 3 & 0 & 5 & 4 \\ 6 & 3 & -\frac{5}{3} & 4 & 7 \\ -1 & 5 & 3 & 2 & 8 \\ 9 & -7 & 4 & 0 & 3 \\ 0 & 11 & \frac{3}{5} & 2 & -1 \end{bmatrix}$$

into the MATMAN program. By typing H, display the main command menu. Use the D command to open a diary file called "test" and add to it a note saying that this is your first experiment with MATMAN. Type H again so that the main menu will be included in your diary file. Use the C command to change a_{22} to -8 and a_{45} to 19. Display the new matrix using the TA command. Exit MATMAN using the Q command. Print the diary file you have just constructed.

24. Enter the matrix

$$A = \begin{bmatrix} 2 & 3 & 0 & 5 & 4 \\ 6 & 3 & -\frac{5}{3} & 4 & 7 \\ -1 & 5 & 3 & 2 & 8 \\ 9 & -7 & 4 & 0 & 3 \\ 0 & 11 & \frac{3}{5} & 2 & -1 \end{bmatrix}$$

into the MATLAB program. Look at some of the Demo files. Use the "diary" command to open a diary file called "test" and add to it a note saying that this is your first experiment with MATLAB. Change a_{22} to -8 and a_{45} to 19. Exit MATLAB using the exit command. Print the diary file you have just constructed.

1.2 GAUSSIAN AND GAUSS-JORDAN ELIMINATION

We will say that two linear systems are **equivalent** if they have precisely the same solution set. The best method of solving systems of linear equations is to replace the given system by an equivalent system that is easier to solve, as shown in Figure 1.8.

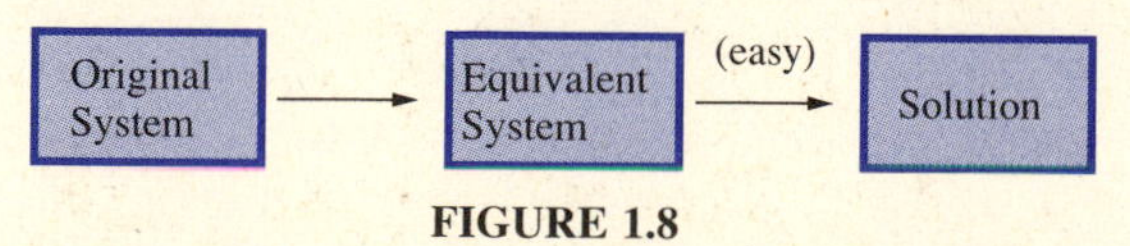

FIGURE 1.8

In order to implement this solution technique we need to answer the following two questions.

1. When is a linear system easy to solve?
2. What can we do to a linear system to change it into an equivalent system?

The exercises in Section 1.1 contained some hints to the answers to these questions. Certainly a system is easy to solve if the coefficient matrix is a **diagonal matrix;** that is, if $a_{ij} = 0$ for $i \neq j$. In the case of a system of four equations in four unknowns, the augmented matrix would look like

$$\begin{bmatrix} a_{11} & 0 & 0 & 0 & b_1 \\ 0 & a_{22} & 0 & 0 & b_2 \\ 0 & 0 & a_{33} & 0 & b_3 \\ 0 & 0 & 0 & a_{44} & b_4 \end{bmatrix}.$$

The corresponding system consists of the four equations

$$a_{ii}x_i = b_i, \qquad i = 1, 2, 3, 4$$

which are very easy to solve. In this case the equations are said to be **uncoupled.** Note that if any $a_{ii} = 0$, while $b_i \neq 0$, then this system has no solution.

A system is also easy to solve when its coefficient matrix is an **upper-triangular matrix,** that is, when $a_{ij} = 0$ for $j < i$. In this case, with four equations in four unknowns, the augmented matrix would have the following form:

$$\begin{bmatrix} a_{11} & a_{12} & a_{13} & a_{14} & b_1 \\ 0 & a_{22} & a_{23} & a_{24} & b_2 \\ 0 & 0 & a_{33} & a_{34} & b_3 \\ 0 & 0 & 0 & a_{44} & b_4 \end{bmatrix}.$$

A system like this can be solved by a technique called **back substitution.** We illustrate this technique for the 3-by-3 system

$$\begin{aligned} x + 2y + 3z &= 5 \\ y - 2z &= 6 \\ 2z &= 4 \end{aligned}$$

whose augmented matrix is

$$\begin{bmatrix} 1 & 2 & 3 & 5 \\ 0 & 1 & -2 & 6 \\ 0 & 0 & 2 & 4 \end{bmatrix}.$$

The last equation is certainly easy to solve for z; the solution is $z = 2$. We then substitute this value of z into the next-to-last equation, obtaining $y - 2(2) = 6$, or $y = 10$. Finally, we substitute the values of y and z, that we have already found, into the first equation to obtain $x + 2(10) + 3(2) = 5$, or $x = -21$. In either the diagonal or the upper triangular case, the solution process amounts to solving a series of equations of the form $ax = b$ which can fail to have a solution only if $a = 0$ when $b \neq 0$.

We now look at the second question raised at the beginning of this section. We can generate new systems, equivalent to our original system, by manipulating the equations of the system. Note that the following operations will lead to equivalent systems:

(1) multiply an equation by a nonzero constant;
(2) interchange the order of two equations; and
(3) replace a given equation by the sum of itself and a multiple of another equation (see Exercise 8, Section 1.1).

In Example 1 we show how to use these operations to solve a system of three equations in three unknowns. Each time we change the system we will also show the new augmented matrix. Note that we could go from one augmented matrix to the next by operating on the rows of the augmented matrix.

EXAMPLE 1 **Solve a 3 × 3 Linear System**

Solve the following system for x, y, and z.

System / Augmented Matrix

$$\begin{aligned} x + 2y + 3z &= 1 \\ 2x + 5y + 5z &= -3 \\ 3x + 5y + 11z &= 2 \end{aligned} \qquad \left[\begin{array}{ccc|c} 1 & 2 & 3 & 1 \\ 2 & 5 & 5 & -3 \\ 3 & 5 & 11 & 2 \end{array}\right]$$

We will eliminate x from the second equation by replacing the second equation by the sum of itself plus (-2) times the first equation. This operation yields

$$\begin{aligned} x + 2y + 3z &= 1 \\ y - z &= -5 \\ 3x + 5y + 11z &= 2 \end{aligned} \qquad \left[\begin{array}{ccc|c} 1 & 2 & 3 & 1 \\ 0 & 1 & -1 & -5 \\ 3 & 5 & 11 & 2 \end{array}\right].$$

In order to eliminate x from the third equation we replace the third equation by itself plus (-3) times the first equation to obtain

$$\begin{aligned} x + 2y + 3z &= 1 \\ y - z &= -5 \\ -y + 2z &= -1 \end{aligned} \qquad \left[\begin{array}{ccc|c} 1 & 2 & 3 & 1 \\ 0 & 1 & -1 & -5 \\ 0 & -1 & 2 & -1 \end{array}\right].$$

In the next step we eliminate y from the third equation by replacing the third equation by the sum of itself plus the second equation. This step yields

$$\begin{aligned} x + 2y + 3z &= 1 \\ y - z &= -5 \\ z &= -6 \end{aligned} \qquad \left[\begin{array}{ccc|c} 1 & 2 & 3 & 1 \\ 0 & 1 & -1 & -5 \\ 0 & 0 & 1 & -6 \end{array}\right].$$

At this point the system can be solved by the back-substitution technique, but we continue the reduction by adding the third equation to the second and then adding (-3) times the third equation to the first. These operations yield

$$\begin{aligned} x + 2y &= 19 \\ y &= -11 \\ z &= -6 \end{aligned} \qquad \left[\begin{array}{ccc|c} 1 & 2 & 0 & 19 \\ 0 & 1 & 0 & -11 \\ 0 & 0 & 1 & -6 \end{array}\right].$$

Finally we add (-2) times the second equation to the first to obtain the solution of the original system.

$$\begin{aligned} x &= 41 \\ y &= -11 \\ z &= -6 \end{aligned} \qquad \left[\begin{array}{ccc|c} 1 & 0 & 0 & 41 \\ 0 & 1 & 0 & -11 \\ 0 & 0 & 1 & -6 \end{array}\right]. \qquad \blacksquare$$

It is clearly possible, and somewhat more convenient, to work just with the augmented matrices; certainly this is the case with modern machine calculations.* The preceding operations on the equations of a system are equivalent to operations on the rows of the augmented matrix; we now define these operations formally.

DEFINITION 1.2 An **elementary row operation** on a matrix A is one of the following.

Type I: Multiply row j by $k \neq 0$ $\quad (R_j \leftarrow kR_j)$
Type II: Interchange row i and row j $\quad (R_i \leftrightarrow R_j)$
Type III: Replace row j by itself plus k times row i $\quad (R_j \leftarrow kR_i + R_j)$. $\square$

The abbreviations in parentheses will frequently be used to describe the details of a specific row reduction; the arrow indicates replacement, R_i represents the ith row of the matrix being reduced, and R_j indicates the jth row of the same matrix. The process of changing one matrix to another by elementary row operations is called **row reduction;** it is the most basic computational procedure of linear algebra. The schematic diagram of our solution process, given in Figure 1.8, can now be more accurately described in Figure 1.9. This solution

* There exist programs, like MAPLE and MATHEMATICA, that do symbolic algebra, but it is still more efficient to deal with just the augmented matrix.

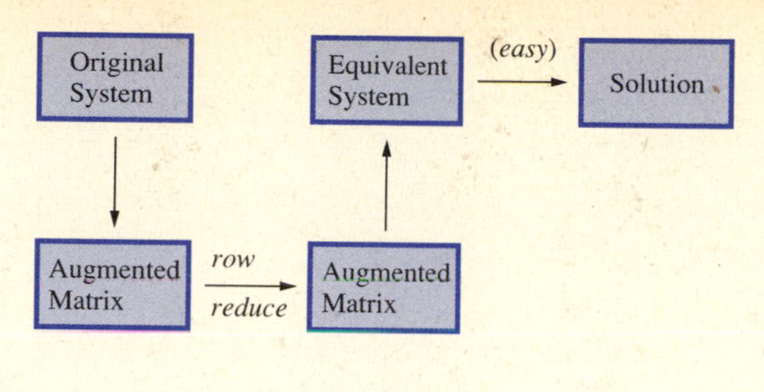

FIGURE 1.9

process, or any of a number of minor variants, is called **Gaussian elimination** if the reduction is stopped when the coefficient matrix first becomes upper triangular. In machine calculations we do not normally change the diagonal elements to 1, as is convenient with hand calculations. If, as in Example 1, the reduction is carried as far as possible toward a diagonal coefficient matrix, then the process is called **Gauss-Jordan elimination.**

EXAMPLE 2 Gaussian Elimination

Solve the following system by Gaussian elimination:

$$\begin{aligned} x + y + 2z &= 2 \\ x + y - 3z &= 2 \\ 2x + y + 5z &= 5. \end{aligned}$$

The augmented matrix is

$$\begin{bmatrix} 1 & 1 & 2 & 2 \\ 1 & 1 & -3 & 2 \\ 2 & 1 & 5 & 5 \end{bmatrix}$$

and the operations $(R_2 \leftarrow -R_1 + R_2)$ and $(R_3 \leftarrow -2R_1 + R_3)$ yield

$$\begin{bmatrix} 1 & 1 & 2 & 2 \\ 0 & 0 & -5 & 0 \\ 0 & -1 & 1 & 1 \end{bmatrix}.$$

Multiplying the third row by -1 $(R_3 \leftarrow -R_3)$, and then interchanging rows 2 and 3 $(R_2 \leftrightarrow R_3)$ leads to

$$\begin{bmatrix} 1 & 1 & 2 & 2 \\ 0 & 1 & -1 & -1 \\ 0 & 0 & -5 & 0 \end{bmatrix}$$

which is upper triangular. The corresponding system can now be solved by back substitution. Since the third equation of the system is $-5z = 0$ we have $z = 0$, and it follows directly that $y = -1$ and $x = 3$. ■

EXAMPLE 3 Gauss-Jordan Elimination

Solve the following system by Gauss-Jordan elimination.

$$\begin{aligned} x_1 + 2x_2 - x_3 &= 4 \\ 2x_1 + 4x_2 + 3x_3 &= 5 \\ -x_1 - 2x_2 + 6x_3 &= -7 \end{aligned}$$

The augmented matrix is

$$\begin{bmatrix} 1 & 2 & -1 & 4 \\ 2 & 4 & 3 & 5 \\ -1 & -2 & 6 & -7 \end{bmatrix},$$

and the operations $(R_2 \leftarrow -2R_1 + R_2)$ and $(R_3 \leftarrow R_1 + R_3)$ yield

$$\begin{bmatrix} 1 & 2 & -1 & 4 \\ 0 & 0 & 5 & -3 \\ 0 & 0 & 5 & -3 \end{bmatrix}.$$

The row operations $(R_3 \leftarrow -R_2 + R_3)$, $(R_1 \leftarrow \frac{1}{5}R_2 + R_1)$, and $(R_2 \leftarrow \frac{1}{5}R_2)$ now yield

$$\begin{bmatrix} 1 & 2 & 0 & \frac{17}{5} \\ 0 & 0 & 1 & -\frac{3}{5} \\ 0 & 0 & 0 & 0 \end{bmatrix}.$$

The system corresponding to this augmented matrix is

$$\begin{aligned} x_1 + 2x_2 &= \frac{17}{5} \\ x_3 &= -\frac{3}{5}. \end{aligned}$$

A different solution of this system can be obtained for each value of x_2. If we arbitrarily let $x_2 = c$, then the most general solution of the system is

$$\begin{aligned} x_1 &= \frac{17}{5} - 2c \\ x_2 &= c \\ x_3 &= -\frac{3}{5}. \end{aligned}$$

We do not obtain a unique solution because the final form of the system has fewer equations than unknowns. Geometrically, this is an example of three planes with a line in common. The line of intersection lies in the plane $x_3 = -\frac{3}{5}$.

Note carefully that it was absolutely critical that the entry in the (3, 4) position of the final matrix was zero. If that entry had been $h \neq 0$, then the last equation would have been

$$0x_1 + 0x_2 + 0x_3 = h \neq 0.$$

Since this equation has no solutions at all, neither would the original system; that is, the system would have been inconsistent. Geometrically this would mean that the three planes had no point in common. ■

EXAMPLE 4 **Heat Conduction**

Consider the insulated metal rod in Figure 1.10; the numbers represent the temperatures at the indicated points.

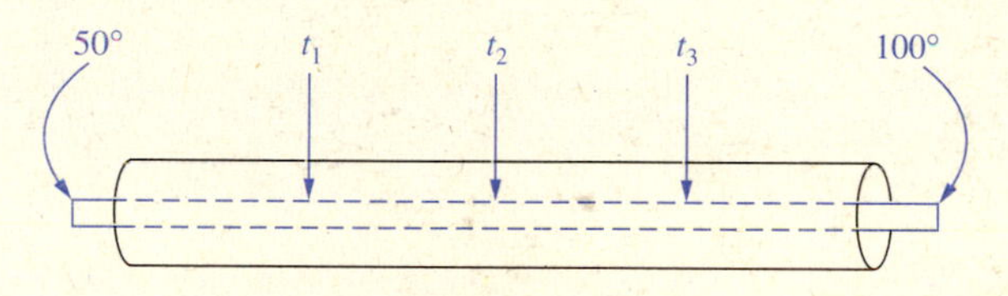

FIGURE 1.10

Find the three interior temperatures t_1, t_2, and t_3 if it is assumed that the temperature at each interior point is the average of the temperatures at the two adjacent points. The given assumption about the interior temperatures leads to the following system of three linear equations.

$$t_1 = \frac{50 + t_2}{2}$$

$$t_2 = \frac{t_1 + t_3}{2}$$

$$t_3 = \frac{t_2 + 100}{2}$$

If we rewrite these equations in the usual form we have the 3 by 3 linear system

$$\begin{aligned} 2t_1 - t_2 &= 50 \\ -t_1 + 2t_2 - t_3 &= 0 \\ -t_2 + 2t_3 &= 100. \end{aligned}$$

The augmented matrix of this system is

$$\begin{bmatrix} 2 & -1 & 0 & 50 \\ -1 & 2 & -1 & 0 \\ 0 & -1 & 2 & 100 \end{bmatrix},$$

which the sequence of row operations

$$(R_1 \leftrightarrow R_2), (R_1 \leftarrow -R_1), (R_2 \leftarrow -2R_1 + R_2), (R_2 \leftrightarrow R_3),$$
$$(R_3 \leftarrow -R_3), (R_3 \leftarrow -3R_2 + R_3)$$

reduces (verify this yourself) to

$$\begin{bmatrix} 1 & -2 & 1 & 0 \\ 0 & 1 & -2 & -100 \\ 0 & 0 & 4 & 350 \end{bmatrix}.$$

Thus, the original system is equivalent to the system

$$\begin{aligned} t_1 - 2t_2 + t_3 &= 0 \\ t_2 - 2t_3 &= -100 \\ 4t_3 &= 350. \end{aligned}$$

Solving this system by back substitution yields

$$t_3 = \frac{350}{4} = 87.5$$
$$t_2 = -100 + 2t_3 = -100 + 175 = 75$$
$$t_1 = 2t_2 - t_3 = 150 - 87.5 = 62.5.$$ ■

It is important to note that in each of the above reductions we have simplified the augmented matrix in a very orderly fashion—one column at a time, from left to right. Please do the same in the exercises that follow.

The MATMAN program, which is available to users of this text, is an interactive program for doing elementary row operations; you specify the operations and the machine does the arithmetic quickly and accurately. MATMAN will work with either decimal arithmetic or rational arithmetic (fractions). The MATLAB program can also be used to do elementary row operations. Some graphics calculators can also do elementary row operations. The table below lists the commands for MATMAN, MATLAB, and the TI-85 calculator.

ERO	MATMAN	MATLAB	TI-85
$R_j \leftarrow kR_j$	M, j, k or $Rj <= kRj$	$A(j, :) = k * A(j, :)$	multR(k, A, j) STO A
$R_i \leftrightarrow R_j$	I, i, j or $Ri <=> Rj$	$T = A(i, :), A(i, :) = A(j, :), A(j, :) = T$	rSwap(A, i, j) STO A
$R_j \leftarrow kR_i + R_j$	A, k, i, j or $Rj <= kRi + Rj$	$A(j, :) = k * A(i, :) + A(j, :)$	mRAdd(k, A, i, j) STO A

EXERCISES 1.2

1. Find the most general solutions of the systems whose augmented matrices are as follows.

(a) $\left[\begin{array}{ccc|c} 1 & 3 & -2 & 0 \\ 0 & 1 & 5 & 2 \\ 0 & 0 & 1 & -11 \end{array}\right]$

(b) $\left[\begin{array}{cccc|c} 1 & -3 & 4 & 2 & 0 \\ 0 & 1 & 2 & 0 & 0 \\ 0 & 0 & 0 & 1 & 5 \end{array}\right]$

(c) $\left[\begin{array}{ccc|c} 2 & 0 & 0 & 10 \\ 5 & 3 & 0 & 15 \\ 6 & 1 & -5 & 0 \end{array}\right]$

(d) $\left[\begin{array}{ccccccc|c} 1 & -2 & 0 & 5 & 0 & 0 & 0 & 3 \\ 0 & 0 & 1 & 3 & -1 & 0 & 0 & 2 \\ 0 & 0 & 0 & 0 & 0 & 1 & 0 & 3 \\ 0 & 0 & 0 & 0 & 0 & 0 & 1 & 0 \\ 0 & 0 & 0 & 0 & 0 & 0 & 0 & 0 \end{array}\right]$

(e) $\left[\begin{array}{cccc|c} 1 & 0 & 0 & 4 & 3 \\ 0 & 1 & 0 & -7 & 5 \\ 0 & 0 & 1 & 8 & 2 \\ 0 & 0 & 0 & 0 & 5 \end{array}\right]$

(f) $\left[\begin{array}{cccc|c} 1 & 2 & 0 & 0 & 5 \\ 2 & 0 & 1 & 0 & 3 \\ -3 & 0 & 2 & 1 & -2 \end{array}\right]$

Use Gaussian elimination to find all solutions of the following systems.

2. $\begin{aligned} x_1 + 2x_2 &= 7 \\ x_2 + 3x_3 &= 1 \\ x_1 - x_3 &= 2 \end{aligned}$

3. $\begin{aligned} 3x_1 - x_2 + 2x_3 &= -4 \\ 2x_1 + x_2 + x_3 &= -1 \\ x_1 + 3x_2 &= 2 \end{aligned}$

4. $\begin{aligned} x_1 - x_2 - x_3 - x_4 &= 5 \\ x_1 + 2x_2 + 3x_3 + x_4 &= -2 \\ 3x_1 + x_2 + 2x_4 &= 1 \\ 2x_1 + 2x_3 + 3x_4 &= 3 \end{aligned}$

5. $\begin{aligned} 2x_1 + 4x_2 + 3x_3 + 2x_4 &= 2 \\ 3x_1 + 6x_2 + 5x_3 + 2x_4 &= 2 \\ 2x_1 + 5x_2 + 2x_3 - 3x_4 &= 3 \\ 4x_1 + 5x_2 + 14x_3 + 14x_4 &= 11 \end{aligned}$

6. $\begin{aligned} 3x_1 - 7x_2 + 4x_3 &= 10 \\ -x_1 - 2x_2 + 3x_3 &= 1 \\ x_1 + x_2 + 2x_3 &= 8 \end{aligned}$

7. $\begin{aligned} x_1 + 3x_2 + 7x_3 + 2x_4 &= 2 \\ x_1 - 2x_2 + x_3 - 4x_4 &= 1 \\ x_1 - 12x_2 - 11x_3 - 16x_4 &= 5 \end{aligned}$

8. $\begin{aligned} x_1 + x_2 + 2x_3 &= 8 \\ - x_2 + 5x_3 &= 9 \\ 3x_1 - 7x_2 + 4x_3 &= 10 \end{aligned}$

9. $\begin{aligned} x_1 - x_2 + 2x_3 - x_4 &= -1 \\ 2x_1 + x_2 - 2x_3 - 2x_4 &= -2 \\ -x_1 + 2x_2 - 4x_3 + x_4 &= 1 \\ 2x_1 + 2x_2 - 4x_3 - 2x_4 &= -2 \end{aligned}$

10. $\begin{aligned} 3x + 2y - z &= -15 \\ 5x + 3y + 2z &= 0 \\ 3x + y + 3z &= 11 \\ 11x + 7y &= -30 \end{aligned}$

11. Solve each of the systems in Exercises 2 through 9 by Gauss-Jordan elimination.

12. Solve the following system by Gauss-Jordan elimination. After each row operation graph the equations of the new system.

$$\begin{aligned} 3x + 4y &= 12 \\ 2x + 3y &= 10 \end{aligned}$$

13. Find all values of h for which the following system is solvable.

$$\begin{aligned} x_1 + 2x_2 + 3x_3 + x_4 &= 3 \\ 3x_1 + 2x_2 + x_3 + 4x_4 &= 7 \\ 2x_2 + 4x_3 + x_4 &= 1 \\ x_1 + x_2 + x_3 + x_4 &= h \end{aligned}$$

14. Count the number of multiplications which would be required to solve a 3×3 system in the following ways:

(a) by a Gaussian elimination algorithm which does not change the diagonal elements to 1;

(b) by a Gaussian elimination algorithm which changes the diagonal elements to 1 before the column is reduced;

(c) by a Gauss-Jordan elimination algorithm in which the coefficient matrix is first reduced to triangular form, and then the columns are simplified from right to left; and

(d) by a Gauss-Jordan elimination in which each column is completely reduced before moving to the next column.

Note that exact 1*s* and exact 0*s* should be set, not computed; thus they require no computational effort.

15. For which values of t will the following system have a unique solution? No solutions? Many solutions?

$$\begin{aligned} x_1 + 2x_2 - 3x_3 &= 4 \\ 4x_1 + x_2 + 2x_3 &= 2 \\ 4x_1 + x_2 + (t^2 - 14)x_3 &= t - 2 \end{aligned}$$

16. Show that the first two elementary row operations in Example 1 are equivalent to solving the first equation for x in terms of y and z and then substituting this expression for x into the second and third equations.

17. Try to solve the general 2×2 system

$$\begin{aligned} a_{11}x_1 + a_{12}x_2 &= b_1 \\ a_{21}x_1 + a_{22}x_2 &= b_2 \end{aligned}$$

by Gaussian elimination. What condition on the coefficients will ensure that a unique solution exists?

18. Solve the following system of equations for $\sin\alpha$, $\cos\beta$, and $\tan\gamma$ and then determine the unknown angles α, β, and γ.

$$\begin{aligned} -2\sin\alpha - 3\cos\beta + 5\tan\gamma &= 1 \\ 4\sin\alpha + 2\cos\beta - 2\tan\gamma &= 2 \\ 2\sin\alpha - 5\cos\beta + 3\tan\gamma &= 7 \end{aligned}$$

How many solutions are there for $0 \le \alpha \le 2\pi$, $0 \le \beta \le 2\pi$, and $0 \le \gamma < \pi$?

19. Solve the following nonlinear system for x^2, y^2, and z^2 and then find all solutions for x, y, and z.

$$\begin{aligned} 3x^2 + y^2 + 4z^2 &= 6 \\ x^2 - y^2 + 2z^2 &= 2 \\ 2x^2 + y^2 - z^2 &= 3 \end{aligned}$$

20. Show that $\begin{bmatrix} a & b \\ c & d \end{bmatrix}$ can be row reduced to $\begin{bmatrix} 1 & 0 \\ 0 & 1 \end{bmatrix}$ if, and only if, $ad - bc \neq 0$.

21. Find conditions on the k_i so that the given system is consistent.

(a) $$\begin{aligned} 2x_1 + 4x_2 &= k_1 \\ 4x_1 + 8x_2 &= k_2 \end{aligned}$$

(b) $$\begin{aligned} x_1 - x_2 + 3x_3 + 2x_4 &= k_1 \\ -3x_1 + 2x_2 + 2x_3 - x_4 &= k_2 \\ 4x_1 - 3x_2 + x_3 + 3x_4 &= k_3 \\ 2x_1 - 2x_2 + 6x_3 + 4x_4 &= k_4 \end{aligned}$$

22. Give an example of an inconsistent system of 3 equations in 4 unknowns.

23. A certain company makes three models of its basic product: the regular, deluxe, and limited versions. Each unit requires processing by departments A, B, and C. The regular model requires 3 time units in department A, 7 in department B, and 5 in department C. The deluxe model requires 4 time units in department A, 8 in department B, and 7 in department C. The limited model requires 4 time units in department A, 9 in department B, and 6 in department C. The company has a special order which must be filled by overtime work. Department A can supply 129 units of overtime, B can supply 285 units, and C can supply 210 units. Management decides to use all the available overtime labor. How many of each model can be supplied?

24. A certain farm has 3 kinds of livestock: sheep, goats, and pigs. There are three food types available: A, B, and C. Each sheep consumes 1 unit of food A, 1 unit of B, and 2 units of C per month. Each pig consumes 3 units of A, 4 units of B, and 5 units of C per month, while each goat eats 2 units of A, 1 unit of B, and 5 units of C. How many animals of each type can the farm support if each month there are available 44 units of food A, 49 units of food B, and 83 units of food C, and all the available food is consumed?

25. Consider the insulated metal rod pictured below. The numbers represent the temperatures at the indicated points. Find the four interior temperatures. Make the same assumption as in Example 4 of this section.

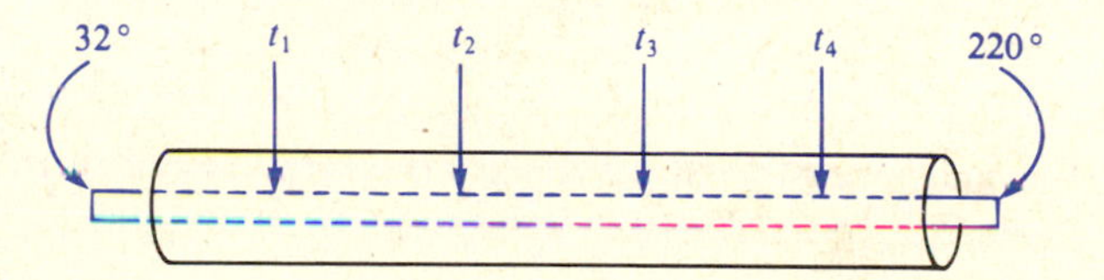

26. Consider the insulated metal plate pictured below. The numbers represent temperatures at the indicated points. Find the four interior temperatures t_1, t_2, t_3, and t_4 if it is assumed that each interior temperature is the average of the temperatures at the four neighboring grid points.

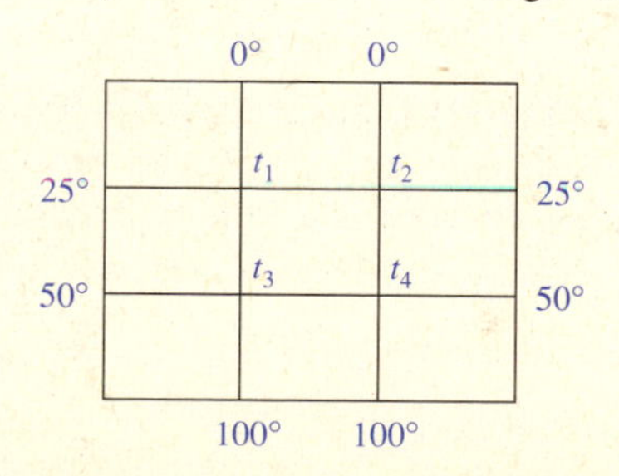

27. Use the SIMULT menu on the TI-85 calculator to enter and solve the linear systems of Examples 1 and 2.

28. Use the MATMAN program to solve the linear systems of Exercises 9 and 10.

1.3 ROW EQUIVALENCE AND ECHELON MATRICES

In the last section we saw how to solve linear systems by row reducing the augmented matrix. In each case we reduced the augmented matrix to the point where the system was easy to solve. In this section we will address the following question:

> Given the matrix A, what is the simplest matrix R that we can get from A using elementary row operations?

By *simple* we mean that the linear system corresponding to R is easy to solve; thus we want R to be as nearly diagonal as possible. The experience obtained in the last two sections should make it clear that we can reduce any matrix A to a matrix like

$$R = \begin{bmatrix} 1 & 2 & 0 & 4 & 0 & 0 & 7 \\ 0 & 0 & 1 & 2 & 0 & 0 & -2 \\ 0 & 0 & 0 & 0 & 1 & 0 & 4 \\ 0 & 0 & 0 & 0 & 0 & 1 & 3 \\ 0 & 0 & 0 & 0 & 0 & 0 & 0 \end{bmatrix},$$

which has the following properties:

(1) If a row of R contains at least one nonzero entry, then the first nonzero entry in that row (the **row leader** or **pivot**) is 1.
(2) The zero rows, if any, come last.
(3) In any two consecutive nonzero rows, the leading 1 in the lower row occurs farther to the right than the leading 1 in the upper row.
(4) Each column that contains a row leader has zeros everywhere else.

The bulk of the work in the reduction is aimed at satisfying Properties 3 and 4, and is accomplished mainly by Type III operations. Type II operations are used to get the rows in the proper order. Any zero rows which occur are moved to the bottom using Type II operations. The row leaders are changed to 1 by using Type I operations.

DEFINITION 1.3 A **reduced row-echelon matrix** is any matrix satisfying the four properties just listed. Such a matrix is said to be in **reduced row-echelon form.** A matrix satisfying the first three properties is said to be a **row-echelon matrix,** or to be in **row-echelon form.** □

The terminology in Definition 1.3 is suggested by the nontechnical meaning of the word *echelon* which is "step-like." The matrices

$$R_1 = \begin{bmatrix} 1 & 2 & 3 & 1 \\ 0 & 1 & 3 & 5 \\ 0 & 0 & 0 & 1 \end{bmatrix}, R_2 = \begin{bmatrix} 1 & 5 & 2 & 3 \\ 0 & 2 & 0 & 2 \\ 0 & 0 & 1 & 1 \end{bmatrix}, \text{ and } R_3 = \begin{bmatrix} 1 & 3 & 9 & 6 \\ 0 & 0 & 0 & 1 \\ 0 & 0 & 1 & 0 \end{bmatrix}$$

are close to being in reduced row-echelon form. Can you explain why they are not? Which are in row-echelon form? The matrices

$$R_4 = \begin{bmatrix} 1 & 0 & 0 & 3 \\ 0 & 1 & 0 & 5 \\ 0 & 0 & 1 & 7 \end{bmatrix}, R_5 = \begin{bmatrix} 0 & 1 & 3 & 0 \\ 0 & 0 & 0 & 1 \\ 0 & 0 & 0 & 0 \end{bmatrix}, \text{ and } R_6 = \begin{bmatrix} 1 & 2 & 0 & 2 & 4 \\ 0 & 0 & 1 & 3 & 5 \\ 0 & 0 & 0 & 0 & 0 \\ 0 & 0 & 0 & 0 & 0 \end{bmatrix}$$

are all in reduced row-echelon form.

Note that the Gaussian elimination process for solving a linear system amounts to reducing the augmented matrix to a row-echelon matrix form while the Gauss-Jordan elimination method amounts to reducing the augmented matrix to reduced row-echelon form. The Gauss-Jordan method replaces the back substitution of the Gaussian elimination method with the further reduction of the coefficient matrix. A careful accounting of the arithmetic required (one normally counts only the number of multiplications of two numbers required) shows that the Gaussian elimination and back-substitution combination, when properly implemented, is slightly more efficient than the Gauss-Jordan reduction; it is the method of choice for use on a digital computer. When working small problems by hand most students prefer Gauss-Jordan elimination. Both of these methods are quite easy to program for use on a modern digital computer. The Gauss-Jordan algorithm is described here in a way that is independent of machine or programming language. You can obtain the Gaussian elimination algorithm by omitting steps 3 and 6.

■ THE GAUSS-JORDAN ELIMINATION ALGORITHM ■

1. Locate the leftmost nonzero column.
2. By interchanging rows, if necessary, move a nonzero entry to the top position of the column in step 1.
3. Change the top entry in step 2 to 1 by multiplying the first row by a nonzero constant.
4. Add suitable multiples of the top row to the rows below so that all entries below the row leader become 0.
5. Remove the top row and the column just reduced from consideration and repeat steps 1–4 until the matrix is in row-echelon form (see Figure 1.11).
6. Beginning with the last row and working upward, add suitable multiples of each row to the rows above in order to introduce zeros above the row leaders.
7. Solve the reduced system.

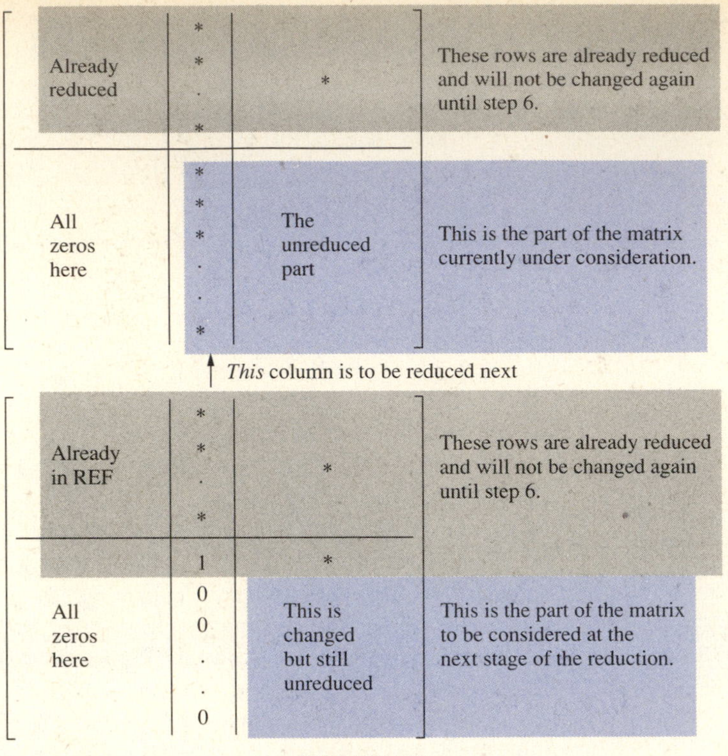

FIGURE 1.11

The Gauss-Jordan elimination algorithm will fail to produce a solution of the original linear system only if there is a nonzero row, in the reduced matrix, whose row leader is in the last column. Such a row would be associated with the equation

$$0x_1 + 0x_2 + \cdots + 0x_n = 1,$$

which clearly is inconsistent.

Any computer program that is designed to implement this elimination algorithm should make provisions to stop the computation and alert the user if the system is found to be inconsistent. (In practice, round-off error makes it unlikely that exact zeros will occur, but one should nevertheless provide for this case.) In Section 1.8 we will describe a better way of implementing this algorithm. If the reduced system is found to be consistent, then it has a unique solution if the number of nonzero equations is equal to the number of unknowns. If, in the reduced system, there are fewer equations than unknowns, then the solution is not unique. We summarize this discussion in the following theorem.

THEOREM 1.1 Consider a consistent system of m linear equations in n unknowns whose augmented matrix has been reduced to a reduced row-echelon matrix R that has r nonzero rows. The solution of the system is unique if $r = n$. If $r < n$, then the solution is not unique; the general solution involves expressing the r unknowns associated with the row leaders in terms of the remaining $n - r$ unknowns, the values of which are arbitrary. □

■ EXAMPLE 1 System with Many Solutions

Find all solutions of the system whose augmented matrix has been reduced to

$$\begin{bmatrix} 1 & 2 & 0 & 4 & 0 & 0 & 7 \\ 0 & 0 & 1 & 2 & 0 & 0 & -2 \\ 0 & 0 & 0 & 0 & 1 & 0 & 4 \\ 0 & 0 & 0 & 0 & 0 & 1 & 3 \\ 0 & 0 & 0 & 0 & 0 & 0 & k \end{bmatrix}.$$

If $k \neq 0$, then the system fails the consistency test and has no solutions. If $k = 0$, then the associated consistent system is

$$\begin{aligned} x_1 + 2x_2 \qquad + 4x_4 \qquad &= 7 \\ x_3 + 2x_4 \qquad &= -2 \\ x_5 &= 4 \\ x_6 &= 3. \end{aligned}$$

These four equations in six unknowns determine four of the unknowns (the ones associated with the row leaders) in terms of the other two (x_2 and x_4). These variables are called the **basic variables;** the others are called the **free variables.** If we set the free variables $x_2 = c_1$ and $x_4 = c_2$, then the most general solution of the system is

$$\begin{aligned} x_1 &= 7 - 2c_1 - 4c_2 \\ x_2 &= c_1 \\ x_3 &= -2 - 2c_2 \\ x_4 &= c_2 \\ x_5 &= 4 \\ x_6 &= 3 \quad . \end{aligned}$$

This system has two arbitrary constants in its general solution; such a system is said to have two *degrees of freedom.* ■

A system whose augmented matrix or coefficient matrix is in echelon form is certainly easy to solve, as we have seen in Example 1 and in the exercises and examples of the last section. In fact, the system may still be easy to solve even if properties 1 and 2 (on page 24) are not satisfied.

EXAMPLE 2 An Easy System

Solve the system whose augmented matrix has been reduced to

$$\left[\begin{array}{rrr|r} 2 & -1 & 3 & 0 \\ 0 & 0 & 3 & 6 \\ 0 & 5 & 1 & 17 \end{array}\right].$$

Allthough this matrix fails to be in row-echelon form, because the rows are out of order and the row leaders are not 1, the corresponding system is still easy to solve. The associated system is

$$\begin{aligned} 2x_1 - x_2 + 3x_3 &= 0 \\ 3x_3 &= 6 \\ 5x_2 + x_3 &= 17, \end{aligned}$$

and from the second equation we obtain $x_3 = 2$. Substituting this value into the third equation yields $5x_2 + 2 = 17$, so $x_2 = 3$. Finally, substituting these values into the first equation yields $x_1 = -\frac{3}{2}$. ■

In our next example we consider a curve-fitting problem which comes up in the computer-graphics problem of fitting a function with a cubic-spline approximation.

EXAMPLE 3 Fitting a Cubic Given Two Points and Two Slopes

Find a cubic polynomial $y = ax^3 + bx^2 + cx + d$ which passes through the point (1, 3) with slope -5 and through the point $(3, -7)$ with slope -1.

Since (1, 3) and $(3, -7)$ are to be on the curve we must have

$$a + b + c + d = 3 \qquad \text{and} \qquad 27a + 9b + 3c + d = -7.$$

Recall, from elementary calculus, that the slope at a point (x_1, y_1) on the curve is the value $y'(x_1)$ of the derivative at $x = x_1$. For $y = ax^3 + bx^2 + cx + d$ the derivative is $y' = 3ax^2 + 2bx + c$. Since the slope at (1, 3) is to be -5 we must have

$$3a + 2b + c = -5,$$

and since the slope at $(3, -7)$ is to be -1 we must also have

$$27a + 6b + c = -1.$$

Thus a, b, c, and d must satisfy the linear system

$$\begin{aligned} a + b + c + d &= 3 \\ 27a + 9b + 3c + d &= -7 \\ 3a + 2b + c &= -5 \\ 27a + 6b + c &= -1. \end{aligned}$$

The augmented matrix and the details of the row reduction are as follows.

$$\left[\begin{array}{rrrr|r} 1 & 1 & 1 & 1 & 3 \\ 27 & 9 & 3 & 1 & -7 \\ 3 & 2 & 1 & 0 & -5 \\ 27 & 6 & 1 & 0 & -1 \end{array}\right] \xrightarrow[\substack{R_2 \leftarrow -27R_1 + R_2 \\ R_3 \leftarrow -3R_1 + R_3 \\ R_4 \leftarrow -27R_1 + R_4}]{} \left[\begin{array}{rrrr|r} 1 & 1 & 1 & 1 & 3 \\ 0 & -18 & -24 & -26 & -88 \\ 0 & -1 & -2 & -3 & -14 \\ 0 & -21 & -26 & -27 & -82 \end{array}\right]$$

$$\xrightarrow[\substack{R_2 \leftrightarrow R_3 \\ R_3 \leftarrow -18R_2 + R_3 \\ R_4 \leftarrow -21R_2 + R_4}]{} \left[\begin{array}{rrrr|r} 1 & 1 & 1 & 1 & 3 \\ 0 & -1 & -2 & -3 & -14 \\ 0 & 0 & 12 & 28 & 164 \\ 0 & 0 & 16 & 36 & 212 \end{array}\right]$$

$$\xrightarrow[\substack{R_2 \leftarrow -R_2 \\ R_4 \leftarrow -\frac{4}{3} R_3 + R_4}]{} \left[\begin{array}{rrrr|r} 1 & 1 & 1 & 1 & 3 \\ 0 & 1 & 2 & 3 & 14 \\ 0 & 0 & 12 & 28 & 164 \\ 0 & 0 & 0 & -\frac{4}{3} & -\frac{20}{3} \end{array}\right]$$

$$\xrightarrow[\substack{R_4 \leftarrow -\frac{3}{4}R_4 \\ R_3 \leftarrow \frac{1}{12}R_3}]{} \left[\begin{array}{rrrr|r} 1 & 1 & 1 & 1 & 3 \\ 0 & 1 & 2 & 3 & 14 \\ 0 & 0 & 1 & \frac{7}{3} & \frac{41}{3} \\ 0 & 0 & 0 & 1 & 5 \end{array}\right]$$

The back substitution now yields $d = 5$, $c = \frac{41}{3} - \frac{35}{3} = 2$, $b = -5$, and $a = 1$. Thus the desired cubic polynomial is

$$y = x^3 - 5x^2 + 2x + 5. \quad ■$$

> Given the matrix A, the key to success in using the elimination methods to solve linear systems is to carry out the reduction in an orderly way. The matrix is reduced one column at a time from left to right.

Once you have mastered the basic technique you may wish to experiment with some modifications of the basic algorithm. In particular, if the problem involves only integers and you are working by hand, it is desirable to minimize work with fractions; we all are faster and more accurate with integer arithmetic than with fractions. For example, to produce a 0 in the (1, 3) position of

$$\begin{bmatrix} 1 & 5 & 6 & 3 & 8 \\ 0 & 0 & 7 & 4 & 11 \end{bmatrix}$$

the basic algorithm would use $(R_2 \leftarrow \frac{1}{7}R_2)$ and $(R_1 \leftarrow -6R_2 + R_1)$, yielding

$$\begin{bmatrix} 1 & 5 & 0 & -\frac{3}{7} & -\frac{10}{7} \\ 0 & 0 & 1 & \frac{4}{7} & \frac{11}{7} \end{bmatrix}.$$

Subsequent calculations using these fractions could become a burden. An alternate approach would be to sacrifice the leading 1 in the first row and use the operations $(R_1 \leftarrow 7R_1)$ and $(R_1 \leftarrow -6R_2 + R_1)$ to produce first

$$\begin{bmatrix} 7 & 35 & 42 & 21 & 56 \\ 0 & 0 & 7 & 4 & 11 \end{bmatrix} \quad \text{and then} \quad \begin{bmatrix} 7 & 35 & 0 & -3 & -10 \\ 0 & 0 & 7 & 4 & 11 \end{bmatrix}.$$

If such modifications are made to the basic process, and the scaling of the rows is delayed until the final step, then arithmetic with fractions can be avoided completely, provided that there are no fractional elements in the original matrix. We illustrate this procedure in our next example. Note that we do not depart from our basic scheme of reducing one column at a time from left to right.

EXAMPLE 4 Avoiding Fractions

Consider the system whose augmented matrix is

$$\left[\begin{array}{rrr|r} 3 & -1 & 2 & 1 \\ 2 & 1 & 1 & 1 \\ 1 & -3 & 0 & 2 \end{array}\right].$$

The normal first step would be to divide the first row by 3 to obtain a 1 in the (1, 1) position. However, two alternatives are available; we could interchange the first and third rows, or we could add (-1) times row 2 to row 1. We choose the first option and follow it with the operations $(R_2 \leftarrow -2R_1 + R_2)$ and $(R_3 \leftarrow -3R_1 + R_3)$ to complete the reduction of the first column. This yields the matrix

$$\left[\begin{array}{rrr|r} 1 & -3 & 0 & 2 \\ 0 & 7 & 1 & -3 \\ 0 & 8 & 2 & -5 \end{array}\right].$$

In order to obtain a 1 in the (2, 2) position we could divide the second row by 7; to avoid creating fractions, we choose instead to use the operations $(R_2 \leftarrow -R_3 + R_2)$ and $(R_2 \leftarrow -R_2)$ to produce

$$\left[\begin{array}{rrr|r} 1 & -3 & 0 & 2 \\ 0 & 1 & 1 & -2 \\ 0 & 8 & 2 & -5 \end{array}\right].$$

The operations $(R_1 \leftarrow 3R_2 + R_1)$ and $(R_3 \leftarrow -8R_2 + R_3)$ now reduce the second column and yield

$$\left[\begin{array}{rrr|r} 1 & 0 & 3 & -4 \\ 0 & 1 & 1 & -2 \\ 0 & 0 & -6 & 11 \end{array}\right].$$

To clear the third column without introducing fractions we will first make the elements in the (1, 3) and (2, 3) positions multiples of 6 by the operations $(R_1 \leftarrow 6R_1)$ and $(R_2 \leftarrow 6R_2)$. These operations, followed by $(R_3 \leftarrow -R_3)$, yield

$$\begin{bmatrix} 6 & 0 & 18 & -24 \\ 0 & 6 & 6 & -12 \\ 0 & 0 & 6 & -11 \end{bmatrix}.$$

The operations $(R_2 \leftarrow -R_3 + R_2)$ and $(R_1 \leftarrow -3R_3 + R_1)$ now lead to

$$\begin{bmatrix} 6 & 0 & 0 & 9 \\ 0 & 6 & 0 & -1 \\ 0 & 0 & 6 & -11 \end{bmatrix}$$

from which we obtain, by multiplying each row by $\frac{1}{6}$,

$$\begin{bmatrix} 1 & 0 & 0 & \frac{9}{6} \\ 0 & 1 & 0 & -\frac{1}{6} \\ 0 & 0 & 1 & -\frac{11}{6} \end{bmatrix}.$$

This final matrix is in reduced row-echelon form, and we have obtained it without doing any arithmetic with fractions. ■

In order to study elementary row operations more efficiently we make the following definition.

DEFINITION 1.4 The matrices A and B are said to be **row equivalent** $(A \underset{R}{\sim} B)$ if B can be obtained from A by a finite sequence of elementary row operations. □

Although row equivalence is a less restrictive relation on matrices than equality, it nevertheless satisfies three very important properties that are also satisfied by equality:

(1) **Reflexive** ($A \underset{R}{\sim} A$ for every matrix A);
(2) **Symmetric** (if $A \underset{R}{\sim} B$, then $B \underset{R}{\sim} A$); and
(3) **Transitive** (if $A \underset{R}{\sim} B$ and $B \underset{R}{\sim} C$, then $A \underset{R}{\sim} C$).

The first and third of these properties are essentially obvious; to verify the second we observe that *every elementary row operation can be reversed by an elementary row operation of the same type.*

(1) $(R_i \leftarrow \frac{1}{k}R_i)$ reverses $(R_i \leftarrow kR_i)$.
(2) $(R_i \leftrightarrow R_j)$ reverses itself.
(3) $(R_j \leftarrow -kR_i + R_j)$ reverses $(R_j \leftarrow kR_i + R_j)$.

For example, $(R_2 \leftarrow \frac{1}{2}R_2)$ reverses $(R_2 \leftarrow 2R_2)$, while $(R_4 \leftarrow 3R_2 + R_4)$ reverses $(R_4 \leftarrow -3R_2 + R_4)$, and $(R_2 \leftrightarrow R_3)$ reverses itself. It follows that if a sequence of elementary row operations reduces A to B, then we can get from B

back to A by using the sequence of operations that reverse the operations in the original sequence. The operations must be undone in the reverse order; that is, the last operation must be reversed first, the next to last operation must be reversed second, and so on.

If the matrix A is square, it is quite likely that the reduced row-echelon matrix associated with A is the diagonal matrix

$$I_n = \begin{bmatrix} 1 & 0 & 0 & \cdots & 0 \\ 0 & 1 & 0 & \cdots & 0 \\ \vdots & \vdots & \vdots & & \vdots \\ 0 & 0 & 0 & \cdots & 1 \end{bmatrix}.$$

This particular diagonal matrix is called an **identity matrix** for reasons that will become clear in later sections of this chapter. The subscript indicates that this matrix is $n \times n$; the subscript is frequently omitted when the size of the matrix is clear from the context.

The augmented matrix of a system of n equations in n unknowns is $n \times (n + 1)$. The reduced row-echelon matrix of such a system may very well be of the form

$$[I \mid B] = \begin{bmatrix} 1 & 0 & 0 & \cdots & 0 & b_1 \\ 0 & 1 & 0 & \cdots & 0 & b_2 \\ \vdots & \vdots & \vdots & & \vdots & \vdots \\ 0 & 0 & 0 & \cdots & 1 & b_n \end{bmatrix}.$$

This is the case if the solution of the original system is unique. **The only way that the reduced row-echelon matrix for such a system could fail to be of this form is if a zero row appeared in the augmented matrix during the reduction process.**

We conclude this discussion with an important theorem whose proof will be omitted.

THEOREM 1.2 Every matrix is row equivalent to a unique reduced row-echelon matrix. □

Theorem 1.2 assures us that we will always arrive at the same reduced row-echelon form even if we use different sequences of operations to do the reduction. Thus, the final solution to the original linear system will not be changed by the choice of the operations used for the reduction. The important thing is to be orderly and to do the arithmetic correctly; machines have a distinct advantage at this point.

We next consider a very simple example of a transportation problem. Such problems are of great importance in the military and in business.

EXAMPLE 5 A Transportation Problem

A manufacturer of small computers receives orders for 150 units from each of three stores located in Cleveland, St. Louis, and Dallas. There are 150 units in a Pittsburgh warehouse and 300 units in a Houston warehouse. The shipping costs from Pittsburgh to the three stores are \$5, \$15, and \$20 per unit, while the shipping costs from Houston to the three stores are \$20, \$10, and \$5 per unit. How should the order be filled in order to minimize shipping costs?

Let x, y, and z be the number of units shipped from Pittsburgh to Cleveland, St. Louis, and Dallas, respectively, and let u, v, and w be the number of units shipped from Houston to Cleveland, St. Louis, and Dallas, respectively. We describe the data in the problem with the data graph in Figure 1.12.

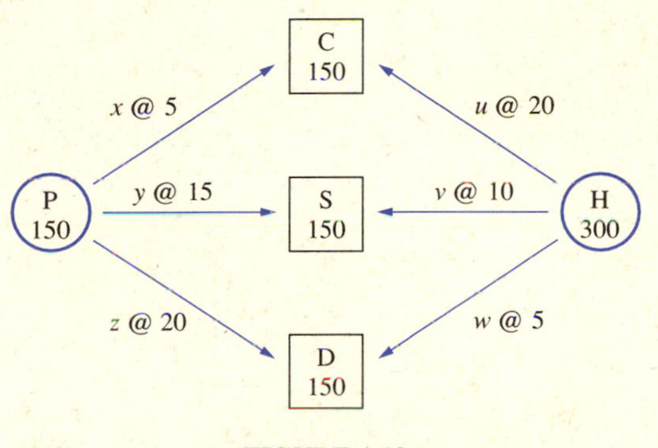

FIGURE 1.12

Since each store needs 150 units we must satisfy the demand conditions

$$\begin{aligned} x + u &= 150 \qquad \text{(Cleveland)} \\ y + v &= 150 \qquad \text{(St. Louis)} \\ z + w &= 150 \qquad \text{(Dallas).} \end{aligned}$$

Since each warehouse has limited stock we must also satisfy the supply conditions

$$\begin{aligned} x + y + z &= 150 \qquad \text{(Pittsburgh)} \\ u + v + w &= 300 \qquad \text{(Houston).} \end{aligned}$$

Thus, our six unknowns must satisfy a system of five linear equations. This system is frequently referred to as the constraint system. Because it has more unknowns than equations, this system, if it is consistent, will have many solutions. Of these solutions, we wish to choose a solution that will minimize

$$C = 5x + 15y + 20z + 20u + 10v + 5w,$$

where C is the shipping cost associated with filling the orders.

The augmented matrix of the system and its echelon form are

$$\begin{bmatrix} 1 & 0 & 0 & 1 & 0 & 0 & 150 \\ 0 & 1 & 0 & 0 & 1 & 0 & 150 \\ 0 & 0 & 1 & 0 & 0 & 1 & 150 \\ 1 & 1 & 1 & 0 & 0 & 0 & 150 \\ 0 & 0 & 0 & 1 & 1 & 1 & 300 \end{bmatrix} \rightarrow \begin{bmatrix} 1 & 0 & 0 & 0 & -1 & -1 & -150 \\ 0 & 1 & 0 & 0 & 1 & 0 & 150 \\ 0 & 0 & 1 & 0 & 0 & 1 & 150 \\ 0 & 0 & 0 & 1 & 1 & 1 & 300 \\ 0 & 0 & 0 & 0 & 0 & 0 & 0 \end{bmatrix}.$$

The general solution of the constraint system is thus

$$\begin{aligned} x &= -150 + v + w \\ y &= 150 - v \\ z &= 150 - w \\ u &= 300 - v - w, \end{aligned}$$

where v and w are arbitrary, except that $0 \leq v \leq 150$, and $0 \leq w \leq 150$. Note also that, because $x \geq 0$, we must have $150 \leq (v + w) \leq 300$. The cost function now becomes

$$\begin{aligned} C &= 5(-150 + v + w) + 15(150 - v) + 20(150 - w) \\ &\quad + 20(300 - v - w) + 10v + 5w \\ &= 10500 - 20v - 30w. \end{aligned}$$

Clearly, C will be smallest when v and w are as large as possible, that is, when $v = w = 150$. Then $x = 150$, $y = z = u = 0$, and the minimum cost is \$3000. ■

EXERCISES 1.3

1. Find a row-echelon form for each of the following matrices.

(a) $\begin{bmatrix} 3 & -1 & 2 & 1 & 0 & 0 \\ 2 & 1 & 1 & 0 & 1 & 0 \\ 1 & -3 & 0 & 0 & 0 & 1 \end{bmatrix}$ **(b)** $\begin{bmatrix} 1 & 2 & 3 & 1 & 1 & 1 \\ 2 & 4 & 5 & 1 & -3 & 2 \\ 3 & 5 & 6 & 1 & 2 & -2 \end{bmatrix}$

(c) $\begin{bmatrix} 1 & 1 & 2 & 6 \\ 1 & -1 & -4 & -8 \\ 3 & -2 & 5 & 11 \\ 2 & 5 & -2 & 3 \end{bmatrix}$ **(d)** $\begin{bmatrix} 1 & -1 & 2 & 3 & 1 & 0 & 0 & 0 \\ 2 & -1 & 0 & 2 & 0 & 1 & 0 & 0 \\ 4 & 1 & -11 & -1 & 0 & 0 & 1 & 0 \\ 1 & 2 & 3 & 83 & 0 & 0 & 0 & 1 \end{bmatrix}$

2. Find the reduced row-echelon form for each of the matrices in Exercise 1.

3. Find the general solution of the systems whose augmented matrices are the following.

(a) $\begin{bmatrix} 3 & 0 & 5 & 15 \\ 0 & 0 & 7 & 21 \\ 0 & 1 & 2 & 7 \end{bmatrix}$ **(b)** $\begin{bmatrix} 1 & 5 & 0 & 0 & 5 & -1 \\ 0 & 0 & 1 & 0 & 3 & 1 \\ 0 & 0 & 0 & 1 & 4 & 2 \\ 0 & 0 & 0 & 0 & 0 & 0 \end{bmatrix}$

(c) $\begin{bmatrix} 3 & 0 & 5 & 15 \\ 0 & 0 & 0 & 21 \\ 0 & 1 & 2 & 7 \end{bmatrix}$ **(d)** $\begin{bmatrix} 1 & 0 & 2 & 0 \\ 0 & 1 & 3 & 0 \\ 0 & 0 & 0 & 0 \end{bmatrix}$

4. From your work in either 1(a) or 2(a) find the solution of each of the following systems.

(a) $\begin{aligned} 3x - y + 2z &= 1 \\ 2x + y + z &= 0 \\ x - 3y &= 0 \end{aligned}$ **(b)** $\begin{aligned} 3x - y + 2z &= 0 \\ 2x + y + z &= 1 \\ x - 3y &= 0 \end{aligned}$ **(c)** $\begin{aligned} 3x - y + 2z &= 0 \\ 2x + y + z &= 0 \\ x - 3y &= 1 \end{aligned}$

5. From your work in either 1(b) or 2(b) find the solution of each of the three following systems.

(a) $\begin{aligned} x + 2y + 3z &= 1 \\ 2x + 4y + 5z &= 1 \\ 3x + 5y + 6z &= 1 \end{aligned}$ **(b)** $\begin{aligned} x + 2y + 3z &= 1 \\ 2x + 4y + 5z &= -3 \\ 3x + 5y + 6z &= 2 \end{aligned}$ **(c)** $\begin{aligned} x + 2y + 3z &= 1 \\ 2x + 4y + 5z &= 2 \\ 3x + 5y + 6z &= -2 \end{aligned}$

6. Theorem 1.2 states the reduced row-echelon form of any matrix A is unique. Show that the row-echelon form of A is not unique by giving an example of a matrix with a nonunique row-echelon form.

7. Reduce $A = \begin{bmatrix} 1 & 2 & 3 & 1 \\ 2 & 5 & 5 & -3 \\ 5 & 10 & 16 & -1 \end{bmatrix}$ to a reduced row-echelon matrix R, and then find a sequence of elementary row-operations which will change R back to A.

8. The following algorithm will reduce an $m \times n$ $(m < n)$ matrix A to a matrix which is nearly in row-echelon form.

```
For i = 1 to m − 1
    For k = i + 1 to m
        t = a_ki / a_ii
        For j = i to n
            a_kj = a_kj − t a_ij
        end
    end
end
```

(a) Describe carefully what these instructions will do to A provided that no $a_{ii} = 0$.

(b) Write a segment of computer code, in your favorite programming language, that will implement the above algorithm. Assume that the matrix A is already in the computer.

(c) Modify the code in part (b) so that the resulting matrix will be in row-echelon form.

(d) Further modify the code so that the resulting matrix will be in reduced row-echelon form.

(e) How would the code need to be changed to handle the case $a_{ii} = 0$?
(f) How would the code need to be changed if you wanted to save the original matrix?
(g) Provide proper input, output, and dimension statements for your code. Test your code on the matrix of Example 3.

9. What is the inverse of the composite row operation $(R_i \leftarrow aR_i + bR_j)$?

†10. In this problem you will need the following algebraic identities:

$$1 + 2 + 3 + \cdots + (k - 1) + k = \frac{k(k + 1)}{2}$$

$$1 + 2^2 + 3^2 + \cdots + (k - 1)^2 + k^2 = \frac{k(k + 1)(2k + 1)}{6}$$

Count the number of multiplications required for:
(a) the solution of n equations in n unknowns if the coefficient matrix is in row-echelon form with no zero rows;
(b) the reduction of an $n \times (n + 1)$ matrix to row-echelon form; and
(c) the reduction of an $n \times (n + 1)$ matrix to reduced row-echelon form.

11. Solve the system of equations (found on page 2) associated with the stuffed animal manufacturing problem at the beginning of Section 1.1

12. Find a cubic polynomial $y = ax^3 + bx^2 + cx + d$ which passes through the point (1, 2) with slope -1 and through the point (2, 5) with slope 8.

13. Find a cubic polynomial $y = ax^3 + bx^2 + cx + d$ which passes through (1, 1) with slope 3 and which also passes through (2, 11) and $(-1, -1)$.

14. Determine values of a, b, and c so that $x = 2$, $y = -2$, and $z = 4$ is a solution of

$$\begin{aligned} -2x - by + cz &= -2 \\ ax + 3y - cz &= -6 \\ ax + by - 3z &= -6. \end{aligned}$$

15. Solve the transportation problem whose data graph is shown below.

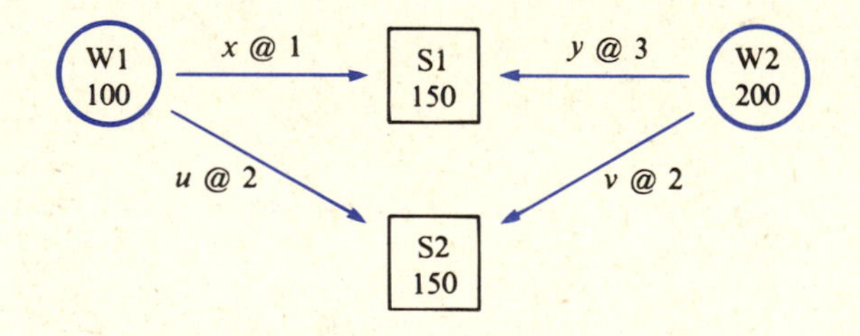

16. Solve the transportation problem of Example 5 if each of the two warehouses contains 225 units.

17. Solve the transportation problem whose data graph is shown below.

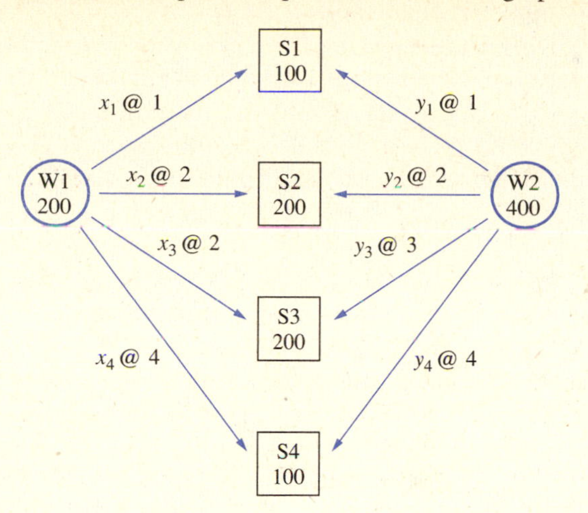

1.4 HOMOGENEOUS SYSTEMS

In this section we will treat two special cases of the linear equations problem which are important enough to deserve special emphasis: homogeneous systems and systems with multiple right-hand sides. The general system of linear equations

$$\begin{aligned} a_{11}x_1 + a_{12}x_2 + \cdots + a_{1n}x_n &= b_1 \\ a_{21}x_1 + a_{22}x_2 + \cdots + a_{2n}x_n &= b_2 \\ \vdots \qquad\quad \vdots \qquad\qquad\quad \vdots \quad &\quad \vdots \\ a_{m1}x_1 + a_{m2}x_2 + \cdots + a_{mn}x_n &= b_m \end{aligned} \tag{1.7}$$

is said to be **homogeneous** if all the constants b_i on the right-hand side of the equations are zero. Thus, the general homogeneous system looks like

$$\begin{aligned} a_{11}x_1 + a_{12}x_2 + \cdots + a_{1n}x_n &= 0 \\ a_{21}x_1 + a_{22}x_2 + \cdots + a_{2n}x_n &= 0 \\ \vdots \qquad\quad \vdots \qquad\qquad\quad \vdots \quad &\quad \vdots \\ a_{m1}x_1 + a_{m2}x_2 + \cdots + a_{mn}x_n &= 0. \end{aligned} \tag{1.8}$$

For linear systems in general we know that there are three possibilites: no solutions, a unique solution, or an infinite number of solutions. For a homogeneous system the first case can be ruled out by observing that a solution is obtained by setting all $x_i = 0$. (We will call this the **zero solution** or the **trivial solution.**) If the solution is unique, then it must be the zero solution. The only time there are nonzero solutions is when there are an infinite number of such solutions; homogeneous systems with an infinite number of solutions occur in many important applications.

Solving a homogeneous system is no different from solving any other system; the elimination methods do the job very nicely. It is worth noting that when we apply the Gaussian elimination method to a homogeneous system the last column of the augmented matrix is zero and stays zero throughout the reduction process. There is a temptation to omit this last column in the homogeneous case, but we will always include it to emphasize the homogeneous nature of the system.

It follows from Theorem 1.1 that a homogeneous system has a nonzero solution precisely when the number of nonzero rows, in the echelon form, is less than the number of unknowns. In particular, if the original homogeneous system has fewer equations than unknowns, then the system is certain to have nonzero solutions—an infinite number of them. We summarize these important facts about the solutions of homogeneous systems in Theorem 1.3.

THEOREM 1.3 The Homogeneous System (1.8) has an infinite number of nonzero solutions whenever there are fewer equations than unknowns, that is, if $m < n$. In general, System (1.8) will have an infinite number of nonzero solutions whenever the number of nonzero rows, in a row-echelon form for the augmented matrix of System (1.8), is less than the number of unknowns. □

EXAMPLE 1 A Homogeneous System with Only One Solution

Find all solutions of the homogeneous system

$$\begin{aligned} x + 2y + 3z &= 0 \\ 4x + 5y + 6z &= 0 \\ 7x + 8y \phantom{{}+6z} &= 0. \end{aligned}$$

The augmented matrix and the details of the row reduction are given below; the operations used are listed beneath the arrows.

$$\begin{bmatrix} 1 & 2 & 3 & 0 \\ 4 & 5 & 6 & 0 \\ 7 & 8 & 0 & 0 \end{bmatrix} \xrightarrow[R_3 \leftarrow -7R_1 + R_3]{R_2 \leftarrow -4R_1 + R_2} \begin{bmatrix} 1 & 2 & 3 & 0 \\ 0 & -3 & -6 & 0 \\ 0 & -6 & -21 & 0 \end{bmatrix}$$

$$\xrightarrow[R_3 \leftarrow 6R_2 + R_3]{R_2 \leftarrow -\frac{1}{3}R_2} \begin{bmatrix} 1 & 2 & 3 & 0 \\ 0 & 1 & 2 & 0 \\ 0 & 0 & -9 & 0 \end{bmatrix}$$

$$\xrightarrow[R_3 \leftarrow -\frac{1}{9}R_3]{} \begin{bmatrix} 1 & 2 & 3 & 0 \\ 0 & 1 & 2 & 0 \\ 0 & 0 & 1 & 0 \end{bmatrix}.$$

The final matrix is a row-echelon form for the augmented matrix of the original system. Since the number of nonzero rows (three) in the reduced matrix is equal to the number of unknowns, this system has a unique solution. The only solution of this system is $x = y = z = 0$. Each of the original equations represents a plane through the origin. In this case the planes have only the origin in common. ■

EXAMPLE 2 **An Underdetermined Homogeneous System**

Find all solutions of the homogeneous system

$$\begin{aligned} x_1 + x_2 + x_3 - 3x_4 &= 0 \\ 3x_1 - 2x_2 - 17x_3 + 16x_4 &= 0 \\ 3x_1 + 2x_2 - x_3 - 4x_4 &= 0. \end{aligned}$$

Since this homogeneous system has fewer equations (three) than unknowns (four), Theorem 1.3 tells us that there are certain to be an infinite number of solutions. To find these solutions we will use Gauss-Jordan elimination:

$$\begin{bmatrix} 1 & 1 & 1 & -3 & 0 \\ 3 & -2 & -17 & 16 & 0 \\ 3 & 2 & -1 & -4 & 0 \end{bmatrix} \xrightarrow[R_3 \leftarrow -3R_1 + R_3]{R_2 \leftarrow -3R_1 + R_2} \begin{bmatrix} 1 & 1 & 1 & -3 & 0 \\ 0 & -5 & -20 & 25 & 0 \\ 0 & -1 & -4 & 5 & 0 \end{bmatrix}$$

$$\xrightarrow[\substack{R_2 \leftarrow 5R_3 + R_2 \\ R_2 \leftrightarrow R_3 \\ R_1 \leftarrow -R_2 + R_1}]{R_3 \leftarrow -R_3} \begin{bmatrix} 1 & 0 & -3 & 2 & 0 \\ 0 & 1 & 4 & -5 & 0 \\ 0 & 0 & 0 & 0 & 0 \end{bmatrix}$$

The reduced system is

$$\begin{aligned} x_1 - 3x_3 + 2x_4 &= 0 \\ x_2 + 4x_3 - 5x_4 &= 0 \end{aligned} \qquad \text{or} \qquad \begin{aligned} x_1 &= 3x_3 - 2x_4 \\ x_2 &= -4x_3 + 5x_4 \end{aligned}$$

and the general solution is obtained by assigning arbitrary values to the free variables x_3 and x_4. If we set $x_3 = c_1$ and $x_4 = c_2$ we obtain, as the general solution,

$$x_1 = 3c_1 - 2c_2, \quad x_2 = -4c_1 + 5c_2, \quad x_3 = c_1, \quad x_4 = c_2. \quad ■$$

EXAMPLE 3 **Preview of Coming Attractions**

Find all values of λ such that the following system has nonzero solutions:

$$\begin{aligned} 2x + y &= \lambda x \\ 4x - y &= \lambda y. \end{aligned}$$

At first glance this system does not appear to be homogeneous, but when the terms on the right are moved to the left side we obtain the homogeneous system

$$\begin{aligned} (2 - \lambda)x + y &= 0 \\ 4x - (1 + \lambda)y &= 0. \end{aligned}$$

The elementary row operations $(R_2 \leftarrow \frac{1}{4}R_2)$, $(R_1 \leftrightarrow R_2)$, and $(R_2 \leftarrow -(2-\lambda)R_1 + R_2)$ reduce the augmented matrix from

$$\begin{bmatrix} (2-\lambda) & 1 & 0 \\ 4 & -(1+\lambda) & 0 \end{bmatrix} \quad \text{to} \quad \begin{bmatrix} 1 & \frac{-(1+\lambda)}{4} & 0 \\ 0 & 1 + \frac{(2-\lambda)(1+\lambda)}{4} & 0 \end{bmatrix}$$
$$= \begin{bmatrix} 1 & \frac{-(1+\lambda)}{4} & 0 \\ 0 & \frac{-(\lambda^2-\lambda-6)}{4} & 0 \end{bmatrix}.$$

This system will have a nonzero solution precisely when the second row of the last matrix is zero. This will occur when $\lambda^2 - \lambda - 6 = 0$. The solutions of this quadratic equation are easily seen to be $\lambda = 3$ and $\lambda = -2$. ■

The second topic we deal with in this section is the frequently occurring case in which there are several systems with the same coefficient matrix but different right-hand sides. Consider the case of three systems like (1.7) on page 37, each with the same coefficient matrix A, but with right-hand sides

$$B = \begin{bmatrix} b_1 \\ b_2 \\ \vdots \\ b_m \end{bmatrix}, C = \begin{bmatrix} c_1 \\ c_2 \\ \vdots \\ c_m \end{bmatrix}, \text{ and } D = \begin{bmatrix} d_1 \\ d_2 \\ \vdots \\ d_m \end{bmatrix}.$$

This situation occurs frequently in applications where the algebraic system models a physical system in which the constants on the right-hand sides of the equations represent inputs to the system and the unknowns represent the response of the system to those inputs. (See for example Exercises 9–12 of this section.) In this case our problem asks for the response of the system to several different inputs. We could, of course, solve each of the systems separately using the reduction techniques of this chapter. The bulk of the work in each case would be the reduction of the coefficient matrix A to echelon form. In each case the same operations would be used, so it would be inefficient to do this work three times. A better approach to this set of problems was suggested in Exercises 4 and 5 in Section 1.3, where we saw that if

$$[A|B|C|D] \xrightarrow[\text{reduce}]{\text{row}} [R|H|K|J],$$

then the system with right-hand side B is equivalent to the system with coefficient matrix R and right-hand side H, the system with right-hand side C is equivalent to the system with coefficient matrix R and right-hand side K, and so on. In particular, if $R = I$, then H, K, and J are the unique solutions of the three systems.

EXAMPLE 4 **A Set of Three Systems**

Solve the following three systems with a single row reduction.

$$\begin{aligned} x + 2y &= 4 \\ 2x + 3y &= 7 \\ x + 4y &= 6 \end{aligned} \qquad \begin{aligned} x + 2y &= 1 \\ 2x + 3y &= 1 \\ x + 4y &= 3 \end{aligned} \qquad \begin{aligned} x + 2y &= 2 \\ 2x + 3y &= 9 \\ x + 4y &= 5 \end{aligned}$$

The 3-fold augmented matrix and the routine row reduction are given below:

$$\begin{bmatrix} 1 & 2 & 4 & 1 & 2 \\ 2 & 3 & 7 & 1 & 9 \\ 1 & 4 & 6 & 3 & 5 \end{bmatrix} \to \begin{bmatrix} 1 & 2 & 4 & 1 & 2 \\ 0 & -1 & -1 & -1 & 5 \\ 0 & 2 & 2 & 2 & 3 \end{bmatrix}$$

$$\to \begin{bmatrix} 1 & 0 & 2 & -1 & 12 \\ 0 & 1 & 1 & 1 & -5 \\ 0 & 0 & 0 & 0 & 13 \end{bmatrix}.$$

From this calculation we see, using the third columns, that the first system has $x = 2$, $y = 1$ as its unique solution. From the fourth columns we see that the unique solution of the second system is $x = -1$, $y = 1$. From the fifth columns we see, because $13 \neq 0$, that the third system is inconsistent. ■

Our next example shows how homogeneous systems can be used in balancing the equations which describe chemical reactions.

EXAMPLE 5 Application to Chemistry

When sodium carbonate (Na_2CO_3) is combined with bromine vapor (Br_2) the by-products are sodium bromide ($NaBr$), sodium bromate ($NaBrO_3$), and carbon dioxide (CO_2). Balance the following equation that describes the reaction:

$$Na_2CO_3 + Br_2 \longrightarrow NaBr + NaBrO_3 + CO_2.$$

The above description of the reaction does not indicate the relative amounts of each chemical involved. In order to provide that information we need to balance the chemical equation. Thus we seek positive integers x_1, x_2, x_3, x_4, and x_5 so that

$$x_1 Na_2CO_3 + x_2 Br_2 = x_3 NaBr + x_4 NaBrO_3 + x_5 CO_2.$$

We can identify linear relations between the x_i by comparing the number of atoms of each of the chemical elements involved in the above equation. Thus, for sodium (Na) we must have

$$2x_1 = x_3 + x_4.$$

Similarly, we must have

$$\begin{aligned} x_1 &= x_5 && \text{for the carbon (C),} \\ 3x_1 &= 3x_4 + 2x_5 && \text{for the oxygen (O), and} \\ 2x_2 &= x_3 + x_4 && \text{for the bromine (Br).} \end{aligned}$$

Thus we see that the five x_i must satisfy a system of four homogeneous equations. Since there are fewer equations than unknowns we know that nonzero solutions exist. We will find these solutions by Gauss-Jordan elimination. The augmented matrix of the system and the reduction to echelon form are given below.

$$\begin{bmatrix} 2 & 0 & -1 & -1 & 0 & 0 \\ 1 & 0 & 0 & 0 & -1 & 0 \\ 3 & 0 & 0 & -3 & -2 & 0 \\ 0 & 2 & -1 & -1 & 0 & 0 \end{bmatrix} \to \begin{bmatrix} 1 & 0 & 0 & 0 & -1 & 0 \\ 0 & 0 & -1 & -1 & 2 & 0 \\ 0 & 0 & 0 & -3 & 1 & 0 \\ 0 & 1 & -\frac{1}{2} & -\frac{1}{2} & 0 & 0 \end{bmatrix}$$

$$\to \begin{bmatrix} 1 & 0 & 0 & 0 & -1 & 0 \\ 0 & 1 & -\frac{1}{2} & -\frac{1}{2} & 0 & 0 \\ 0 & 0 & 1 & 1 & -2 & 0 \\ 0 & 0 & 0 & 1 & -\frac{1}{3} & 0 \end{bmatrix} \to \begin{bmatrix} 1 & 0 & 0 & 0 & -1 & 0 \\ 0 & 1 & -\frac{1}{2} & 0 & -\frac{1}{6} & 0 \\ 0 & 0 & 1 & 0 & -\frac{5}{3} & 0 \\ 0 & 0 & 0 & 1 & -\frac{1}{3} & 0 \end{bmatrix}$$

$$\to \begin{bmatrix} 1 & 0 & 0 & 0 & -1 & 0 \\ 0 & 1 & 0 & 0 & -1 & 0 \\ 0 & 0 & 1 & 0 & -\frac{5}{3} & 0 \\ 0 & 0 & 0 & 1 & -\frac{1}{3} & 0 \end{bmatrix}$$

We obtain the general solution of this homogeneous system by letting x_5 be arbitrary, say $x_5 = c$. Then we have $x_1 = c, x_2 = c, x_3 = \frac{5c}{3}, x_4 = \frac{c}{3}$, and $x_5 = c$. Since we want the x_i to be positive integers we need to choose c to be some multiple of 3. If we choose $c = 3$, then the balanced chemical equation is

$$3Na_2CO_3 + 3Br_2 = 5NaBr + NaBrO_3 + 3CO_2. \quad ■$$

If a homogeneous system has integer coefficients, then, as illustrated in Example 5, it is always possible to find an integer solution. In some applications it is important that the solutions be integers.

EXERCISES 1.4

1. Find all solutions of the following homogeneous systems.

(a)
$$\begin{aligned} 3x_1 + 2x_2 + 16x_3 + 5x_4 &= 0 \\ 2x_2 + 10x_3 + 8x_4 &= 0 \\ x_1 + x_2 + 7x_3 + 3x_4 &= 0 \end{aligned}$$

(b)
$$\begin{aligned} 2x_1 + x_2 + 3x_3 &= 0 \\ x_1 + 2x_2 &= 0 \\ x_2 + x_3 &= 0 \end{aligned}$$

(c)
$$\begin{aligned} x_1 + 3x_2 + 5x_3 + x_4 &= 0 \\ 4x_1 - 7x_2 - 3x_3 - x_4 &= 0 \\ 3x_1 + 2x_2 + 7x_3 + 3x_4 &= 0 \end{aligned}$$

(d)
$$\begin{aligned} 7x_1 - 8x_2 - 8x_3 &= -x_1 \\ 9x_1 - 16x_2 - 18x_3 &= -x_2 \\ -5x_1 + 11x_2 + 13x_3 &= -x_3 \end{aligned}$$

(e)
$$\begin{aligned} x_1 - 4x_3 &= 3x_1 \\ 5x_2 + 4x_3 &= 3x_2 \\ -4x_1 + 4x_2 + 3x_3 &= 3x_3 \end{aligned}$$

(f)
$$\begin{aligned} x_1 + x_2 - 3x_3 + x_4 &= 0 \\ 2x_1 + x_3 - x_4 &= 0 \\ x_1 + 3x_2 - 10x_3 + 4x_4 &= 0 \end{aligned}$$

2. Consider the general homogeneous linear system (1.8), on page 37, with $m = n = 3$. Suppose we know that

$$x_1 = a, x_2 = b, x_3 = c \quad \text{and} \quad x_1 = a', x_2 = b', x_3 = c'$$

are two solutions of the system. Let h and k be arbitrary real constants and show that

$$x_1 = ha + ka', x_2 = hb + kb', x_3 = hc + kc'$$

is also a solution of the system. Is this true if the system is *not* homogeneous? What condition on h and k would make the statement true in the nonhomogeneous case?

3. Find all values λ such that the following systems have nontrivial solutions.

(a) $$\begin{aligned} 5x + y &= \lambda x \\ 4x + 8y &= \lambda y \end{aligned}$$

(b) $$\begin{aligned} 3x - 5y &= \lambda x \\ 5x + 3y &= \lambda y \end{aligned}$$

(c) $$\begin{aligned} 2x + y + z &= \lambda x \\ 2x + 3y + 2z &= \lambda y \\ x + y + 2z &= \lambda z \end{aligned}$$

(d) $$\begin{aligned} y &= \lambda x \\ z &= \lambda y \\ -2x + 3y &= \lambda z \end{aligned}$$

4. From your solutions of Exercises 2–10 in Section 1.2 find the general solutions of the associated homogeneous systems (change the right-hand side to zero in each case). No additional calculations should be necessary.

5. Solve the following set of systems of linear equations, each of which has coefficient matrix A and respective right-hand sides B_1, B_2, and B_3.

$$A = \begin{bmatrix} 1 & 0 & 1 \\ 2 & 1 & 0 \\ 0 & 0 & 1 \\ 1 & -2 & 0 \end{bmatrix} \quad B_1 = \begin{bmatrix} 0 \\ 1 \\ -1 \\ 3 \end{bmatrix} \quad B_2 = \begin{bmatrix} 0 \\ 0 \\ 1 \\ -5 \end{bmatrix} \quad B_3 = \begin{bmatrix} 0 \\ 0 \\ -3 \\ 5 \end{bmatrix}$$

6. Solve the following set of systems of linear equations, each of which has coefficient matrix A and respective right-hand sides B_1, B_2, and B_3.

$$A = \begin{bmatrix} 3 & -1 & 2 \\ 2 & 2 & 1 \\ 1 & -3 & 0 \end{bmatrix} \quad B_1 = \begin{bmatrix} 1 \\ 0 \\ 0 \end{bmatrix} \quad B_2 = \begin{bmatrix} 0 \\ 1 \\ 0 \end{bmatrix} \quad B_3 = \begin{bmatrix} 0 \\ 0 \\ 1 \end{bmatrix}$$

7. What condition on the coefficients of the general 2×2 homogeneous system

$$\begin{aligned} a_{11}x_1 + a_{12}x_2 &= 0 \\ a_{21}x_1 + a_{22}x_2 &= 0 \end{aligned}$$

will ensure that a nonzero solution exists?

8. Obtain balanced chemical equations for the following reactions.

(a) $N_2 + H_2 \rightarrow NH_3$

(b) $Cl_2 + KOH \rightarrow KCl + KClO_3 + H_2O$

(c) $C_3H_5(NO_3)_3 \rightarrow H_2O + CO + N_2 + O_2$

9. Consider the insulated rod pictured below. Recall Example 4 of Section 1.2 and then use the techniques of this section to estimate the internal temperatures t_1, t_2, t_3, and t_4 under the following circumstances.

(a) $t_0 = 0°$, $t_5 = 75°$ **(b)** $t_0 = 32°$, $t_5 = 212°$
(c) $t_0 = -20°$, $t_5 = 100°$ **(d)** $t_0 = 212°$, $t_5 = 0°$

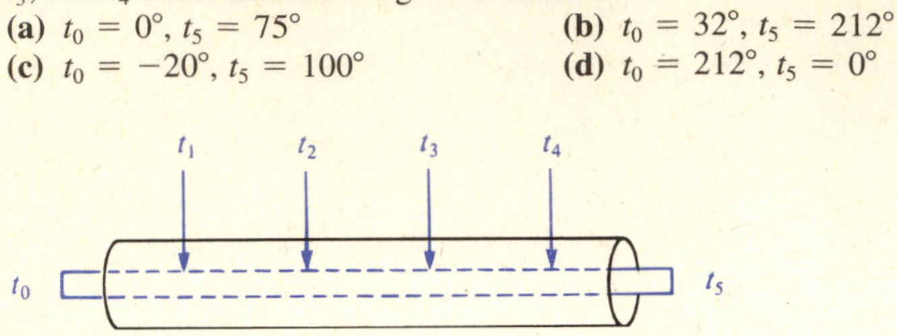

10. Describe how the application described in Exercise 17 in Section 1.1 could lead to a set of systems with the same coefficient matrix.

11. Consider the insulated metal plate pictured below. The numbers represent temperatures at the indicated points. Assume that each interior temperature is the average of the temperatures at the four neighboring grid points. Find the four interior temperatures t_1, t_2, t_3, and t_4 if the other temperatures are as listed below.

(a) $b_1 = b_2 = 50°$, $b_3 = b_4 = 100°$, $b_5 = b_6 = 200°$.
(b) $b_1 = b_3 = 0°$, $b_2 = b_4 = b_5 = b_6 = 100°$.
(c) $b_1 = 220°$, $b_2 = b_3 = b_4 = b_5 = b_6 = 0°$.

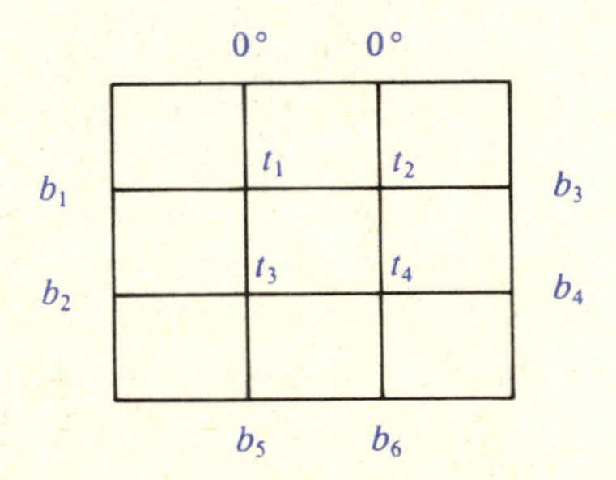

12. Describe how the curve-fitting problem of Example 3 in Section 1.3 could lead to a set of systems with the same coefficient matrix.

13. It is known from geometry that three points in the plane determine a circle. Find the circle through the points (4, 7), (−2, −3), and (−4, 5). (*Hint:* Assume the equation of the circle is $a + bx + cy + d(x^2 + y^2) = 0$.)

14. Let $A = \begin{bmatrix} 1 & 3 & 5 \\ -1 & -2 & 0 \\ 2 & 5 & 4 \end{bmatrix}$. Find solutions of the systems with coefficient matrix A and right-hand sides **(a)** $\begin{bmatrix} 0 \\ 1 \\ 1 \end{bmatrix}$, **(b)** $\begin{bmatrix} 1 \\ 0 \\ -11 \end{bmatrix}$, **(c)** $\begin{bmatrix} -1 \\ 1 \\ 2 \end{bmatrix}$.

1.5 MATRIX ALGEBRA

Matrices have already been encountered in our discussion of systems of linear equations. In this section we will develop an algebra for matrices that has many applications. In particular, the results of this section will make it much easier to deal with systems of linear equations. We begin by formally repeating the basic definitions and notational conventions used in dealing with matrices.

DEFINITION 1.5 A **matrix** is a rectangular array of numbers. The normal way of describing a matrix with m rows and n columns is

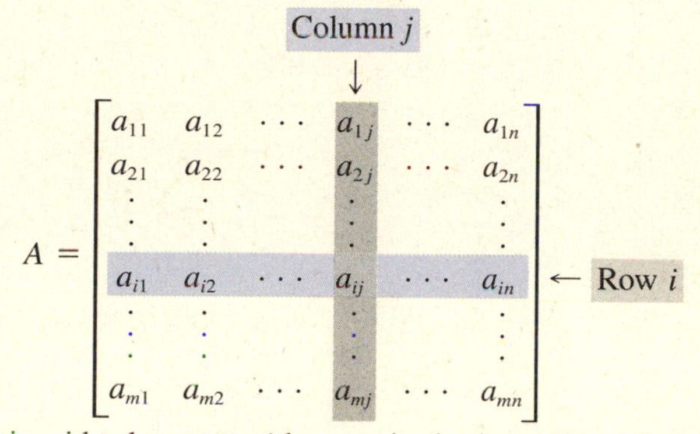

The matrix A is said to be $m \times n$ (the row size is always listed first) and the entry a_{ij} occurs in row i and column j. Two matrices are **equal** if they are identical in every way; that is, they are the same size and have the same entry in every position. □

The **diagonal entries** of A are $a_{11}, a_{22}, a_{33}, \ldots$, that is, those entries with equal row and column indices. The **upper-triangular entries** of A are those a_{ij} with $i < j$ while the **lower-triangular entries** of A are those a_{ij} with $j < i$. See Figure 1.13.

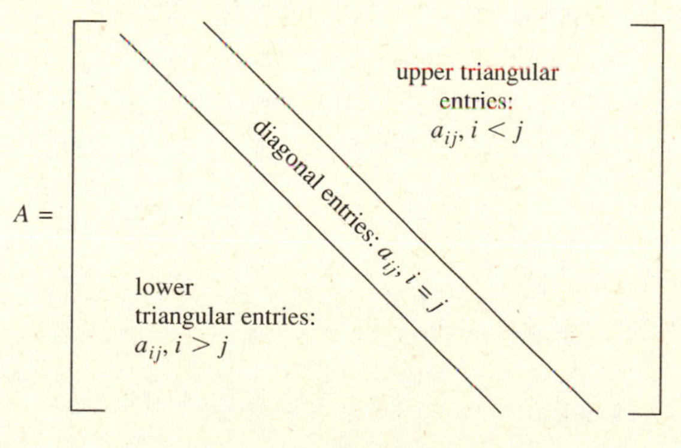

(A is $m \times n$, with $m < n$)

FIGURE 1.13

A matrix is said to be **upper triangular** if all its lower triangular entries are zero; that is, all the nonzero entries are diagonal or upper triangular entries. A matrix is said to be **lower triangular** if all its upper triangular entries are zero. A matrix is **diagonal** if it is square ($m = n$) and all the nondiagonal entries are zero, that is, $a_{ij} = 0$ if $i \neq j$. Of the matrices

$$D = \begin{bmatrix} 1 & 0 & 0 \\ 0 & 3 & 0 \\ 0 & 0 & 2 \end{bmatrix}, L = \begin{bmatrix} 1 & 0 & 0 \\ 2 & 0 & 0 \\ 3 & 3 & 5 \end{bmatrix}, U = \begin{bmatrix} 1 & 2 & 1 & 1 \\ 0 & 3 & 2 & 3 \\ 0 & 0 & 4 & 5 \end{bmatrix}, R = [1 \quad 2 \quad 3],$$

$$C = \begin{bmatrix} 5 \\ 6 \\ 3 \end{bmatrix}, A = \begin{bmatrix} 0 & 3 & 2 \\ 3 & 0 & 5 \\ 2 & 6 & 0 \end{bmatrix}, \text{ and } Z = \begin{bmatrix} 0 & 0 & 0 \\ 0 & 0 & 0 \\ 0 & 0 & 0 \end{bmatrix},$$

D and Z are diagonal, L, D, and Z are lower triangular, U, D, and Z are upper triangular, R is a row matrix ($m = 1$), C is a column matrix ($n = 1$), and D, L, A, and Z are square matrices while U, R, and C are not square.

We will sometimes wish to consider submatrices of a given matrix A. By a **submatrix** of A we mean any matrix obtained from A by deleting certain rows and columns of A. Of particular interest are submatrices obtained by partitioning the matrix into other submatrices. For example, the 5×5 matrix

$$A = \left[\begin{array}{ccc|cc} 1 & 3 & 7 & 4 & 2 \\ 3 & 4 & 6 & 5 & 0 \\ 8 & 0 & 2 & 1 & 2 \\ \hline 1 & 3 & 2 & 4 & 1 \\ 4 & 6 & 7 & 2 & 2 \end{array}\right] = \begin{bmatrix} A_{11} & A_{12} \\ A_{21} & A_{22} \end{bmatrix},$$

where $A_{11} = \begin{bmatrix} 1 & 3 & 7 \\ 3 & 4 & 6 \\ 8 & 0 & 2 \end{bmatrix}$ is a 3×3 submatrix of A and $A_{21} = \begin{bmatrix} 1 & 3 & 2 \\ 4 & 6 & 7 \end{bmatrix}$ is a 2×3 submatrix of A. The submatrix of A obtained by deleting row 3 and column 2 of A is

$$\begin{bmatrix} 1 & 7 & 4 & 2 \\ 3 & 6 & 5 & 0 \\ 1 & 2 & 4 & 1 \\ 4 & 7 & 2 & 2 \end{bmatrix}.$$

We already utilized this notation in the first section when we wrote the augmented matrix of the linear system (1.4) on page 9 as $[A \mid B]$. Of particular importance are the rows and columns of the matrix A; we will use the following notation for these submatrices:

$$\text{Row}_i(A) = [a_{i1} \quad a_{i2} \quad \cdots \quad a_{in}] \text{ is the ith row of } A$$

and

$$\mathrm{Col}_j(A) = \begin{bmatrix} a_{1j} \\ a_{2j} \\ \vdots \\ a_{mj} \end{bmatrix} \text{ is the jth column of } A.$$

We will, on occasion, wish to partition A into its rows.

$$A = \begin{bmatrix} \mathrm{Row}_1(A) \\ \mathrm{Row}_2(A) \\ \vdots \\ \mathrm{Row}_m(A) \end{bmatrix} = \begin{bmatrix} a_{11} & a_{12} & \cdots & a_{1n} \\ a_{21} & a_{22} & \cdots & a_{2n} \\ \vdots & \vdots & & \vdots \\ a_{m1} & a_{m2} & \cdots & a_{mn} \end{bmatrix}$$

or into its columns

$$A = [\,\mathrm{Col}_1(A) \quad \cdots \quad \mathrm{Col}_n(A)\,] = \begin{bmatrix} a_{11} & a_{12} & \cdots & a_{1n} \\ a_{21} & a_{22} & \cdots & a_{2n} \\ \vdots & \vdots & & \vdots \\ a_{m1} & a_{m2} & \cdots & a_{mn} \end{bmatrix}.$$

We begin our discussion of the algebra of matrices by defining the sum of two matrices.

DEFINITION 1.6 If A and B are matrices of the same size, then the **sum** $A + B$ is the matrix obtained by adding the corresponding entries of A and B; that is, if A and B are both $m \times n$ then $C = A + B$ is the $m \times n$ matrix whose entries satisfy:

$$c_{ij} = a_{ij} + b_{ij}; \qquad i = 1, 2, \ldots, m, \qquad j = 1, 2, \ldots, n.$$

If A and B are not the same size, then their sum is not defined. □

EXAMPLE 1 **Matrix Addition**

For the matrices

$$A = \begin{bmatrix} 2 & 0 \\ 4 & 3 \\ 6 & 7 \end{bmatrix}, B = \begin{bmatrix} 1 & 5 \\ 0 & 3 \\ 2 & 6 \end{bmatrix}, C = \begin{bmatrix} 1 & 2 & 3 \\ 4 & 5 & 6 \\ 7 & 8 & 9 \end{bmatrix}, \text{ and } D = \begin{bmatrix} 0 & 1 & 1 \\ 2 & 0 & 2 \\ 3 & 3 & 0 \end{bmatrix}$$

it follows from Definition 1.6 that

$$A + B = \begin{bmatrix} 3 & 5 \\ 4 & 6 \\ 8 & 13 \end{bmatrix} \quad \text{and} \quad C + D = \begin{bmatrix} 1 & 3 & 4 \\ 6 & 5 & 8 \\ 10 & 11 & 9 \end{bmatrix}.$$

Note that the sums $A + C$, $A + D$, $B + C$, and $B + D$ are not defined. ■

We next define the product of a scalar (real number) and a matrix.

DEFINITION 1.7 If A is any matrix and k is any number, then the **scalar multiple** kA is the matrix obtained by multiplying each entry of A by k. We define the **additive inverse** (negative) of A by $-A = (-1)A$. □

These definitions allow us to write $A + A + A = 3A$, and so on.

EXAMPLE 2 Scalar Multiplication

If A, B, C, and D are the matrices of Example 1, then

$$2A + 3B = 2\begin{bmatrix} 2 & 0 \\ 4 & 3 \\ 6 & 7 \end{bmatrix} + 3\begin{bmatrix} 1 & 5 \\ 0 & 3 \\ 2 & 6 \end{bmatrix}$$

$$= \begin{bmatrix} 4 & 0 \\ 8 & 6 \\ 12 & 14 \end{bmatrix} + \begin{bmatrix} 3 & 15 \\ 0 & 9 \\ 6 & 18 \end{bmatrix} = \begin{bmatrix} 7 & 15 \\ 8 & 15 \\ 18 & 32 \end{bmatrix},$$

$$5C - 4D = 5C + (-4)D = \begin{bmatrix} 5 & 10 & 15 \\ 20 & 25 & 30 \\ 35 & 40 & 45 \end{bmatrix} + \begin{bmatrix} 0 & -4 & -4 \\ -8 & 0 & -8 \\ -12 & -12 & 0 \end{bmatrix}$$

$$= \begin{bmatrix} 5 & 6 & 11 \\ 12 & 25 & 22 \\ 23 & 28 & 45 \end{bmatrix}, \quad \text{and}$$

$$C - C = C + (-1)C = \begin{bmatrix} 1 & 2 & 3 \\ 4 & 5 & 6 \\ 7 & 8 & 9 \end{bmatrix} + \begin{bmatrix} -1 & -2 & -3 \\ -4 & -5 & -6 \\ -7 & -8 & -9 \end{bmatrix}$$

$$= \begin{bmatrix} 0 & 0 & 0 \\ 0 & 0 & 0 \\ 0 & 0 & 0 \end{bmatrix} = Z.$$ ■

The matrix Z, all of whose entries are zero, is called the **zero matrix.** We will denote the zero matrix either by Z or 0; no confusion will result from using the same notation for the zero scalar and the zero matrix.

Another important algebraic operation on matrices is matrix multiplication. We first give the definition for an important special case.

DEFINITION 1.8 The **product** of a $1 \times n$ row matrix R and an $n \times 1$ column matrix C is defined to be

$$RC = [r_1 \quad r_2 \quad \cdots \quad r_n] \begin{bmatrix} c_1 \\ c_2 \\ \vdots \\ c_n \end{bmatrix}$$

$$= r_1c_1 + r_2c_2 + \cdots + r_nc_n = \sum_{i=1}^{n} r_ic_i. \quad \square$$

Note that we have defined the product of a $1 \times n$ matrix and an $n \times 1$ matrix to be a scalar. (We will not distinguish between scalars and 1×1 matrices.) This definition allows us to write the general linear equation

$$a_1x_1 + a_2x_2 + \cdots + a_nx_n = b$$

as the matrix equation

$$[a_1 \quad a_2 \quad \cdots \quad a_n] \begin{bmatrix} x_1 \\ x_2 \\ \vdots \\ x_n \end{bmatrix} = b.$$

We now use Definition 1.8 to give the general definition of matrix multiplication.

DEFINITION 1.9 If A is an $m \times r$ matrix and B is an $r \times n$ matrix, then the **product** AB is the $m \times n$ matrix C whose entries are given by

$$c_{ij} = \text{Row}_i(A)\ \text{Col}_j(B)$$

$$= a_{i1}b_{1j} + a_{i2}b_{2j} + \cdots + a_{ir}b_{rj} = \sum_{k=1}^{r} a_{ik}b_{kj}. \quad \square$$

The row-by-column characterization of matrix multiplication is clearly the way to remember Definition 1.9. Note that, in the expression for c_{ij}, each term has an a with first subscript i and a b with second subscript j. Moreover, in every term the second subscript on a and the first subscript on b are the same, as shown:

$$AB = \begin{bmatrix} a_{11} & a_{12} & \cdots & a_{1r} \\ \vdots & \vdots & & \vdots \\ a_{i1} & a_{i2} & \cdots & a_{ir} \\ \vdots & \vdots & & \vdots \\ a_{m1} & a_{m2} & \cdots & a_{mr} \end{bmatrix} \begin{bmatrix} b_{11} & \cdots & b_{1j} & \cdots & b_{1n} \\ \vdots & & \vdots & & \vdots \\ b_{i1} & \cdots & b_{ij} & \cdots & b_{in} \\ \vdots & & \vdots & & \vdots \\ b_{r1} & \cdots & b_{rj} & \cdots & b_{rn} \end{bmatrix}$$

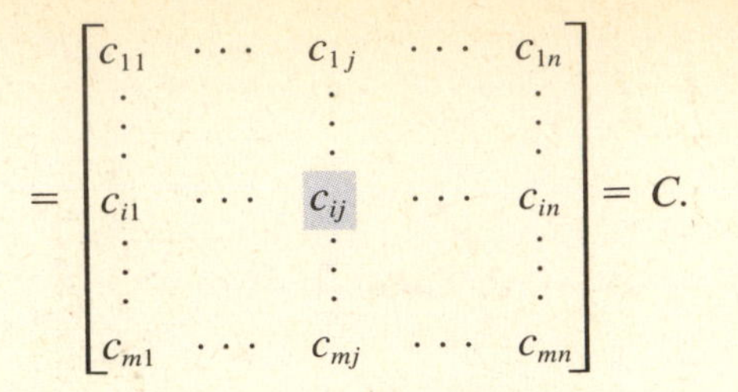

$$= \begin{bmatrix} c_{11} & \cdots & c_{1j} & \cdots & c_{1n} \\ \vdots & & \vdots & & \vdots \\ c_{i1} & \cdots & c_{ij} & \cdots & c_{in} \\ \vdots & & \vdots & & \vdots \\ c_{m1} & \cdots & c_{mj} & \cdots & c_{mn} \end{bmatrix} = C.$$

Note that the matrix product is defined only when the number of columns of the first factor is equal to the number of rows of the second factor. The following diagram is useful in remembering how the sizes of the factors affect the size of the product:

$$\underset{(m \times r)}{(A)}\ \underset{(r \times n)}{(B)} = \underset{(m \times n)}{(AB)}$$

EXAMPLE 3 Matrix Multiplication

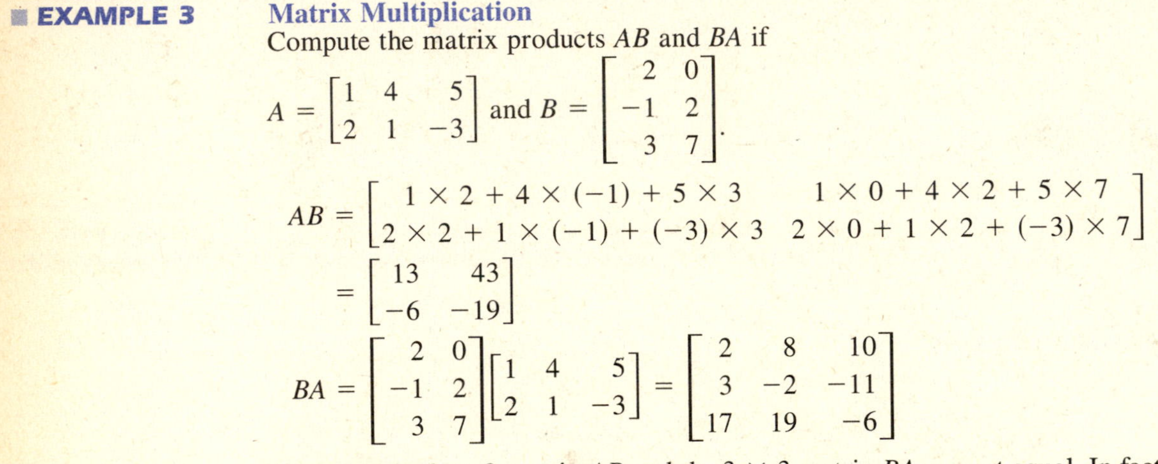

Compute the matrix products AB and BA if

$$A = \begin{bmatrix} 1 & 4 & 5 \\ 2 & 1 & -3 \end{bmatrix} \text{ and } B = \begin{bmatrix} 2 & 0 \\ -1 & 2 \\ 3 & 7 \end{bmatrix}.$$

$$AB = \begin{bmatrix} 1 \times 2 + 4 \times (-1) + 5 \times 3 & 1 \times 0 + 4 \times 2 + 5 \times 7 \\ 2 \times 2 + 1 \times (-1) + (-3) \times 3 & 2 \times 0 + 1 \times 2 + (-3) \times 7 \end{bmatrix}$$

$$= \begin{bmatrix} 13 & 43 \\ -6 & -19 \end{bmatrix}$$

$$BA = \begin{bmatrix} 2 & 0 \\ -1 & 2 \\ 3 & 7 \end{bmatrix} \begin{bmatrix} 1 & 4 & 5 \\ 2 & 1 & -3 \end{bmatrix} = \begin{bmatrix} 2 & 8 & 10 \\ 3 & -2 & -11 \\ 17 & 19 & -6 \end{bmatrix}$$

Note that the 2×2 matrix AB and the 3×3 matrix BA are not equal. In fact, they are not even the same size! ■

EXAMPLE 4 Noncommutativity

Compute CD and DC if

$$C = \begin{bmatrix} 1 & 2 \\ 2 & 4 \end{bmatrix} \quad \text{and} \quad D = \begin{bmatrix} 2 & -4 \\ -1 & 2 \end{bmatrix}.$$

Directly from Definition 1.9 we obtain

$$CD = \begin{bmatrix} 0 & 0 \\ 0 & 0 \end{bmatrix} \quad \text{while} \quad DC = \begin{bmatrix} -6 & -12 \\ 3 & 6 \end{bmatrix}.$$ ■

Again, note that CD and DC are not equal even though in this case they are the same size. Note also that for matrices A and C of Examples 3 and 4, AC is not defined although CA is a well-defined 2×3 matrix.

These last two examples point out some major differences between matrix multiplication and the multiplication of scalars.

1. The product of two matrices is not always defined.
2. Matrix multiplication is **noncommutative** (in general $AB \neq BA$).
3. The product of two nonzero matrices may be the zero matrix.

We will investigate the algebraic properties of matrix algebra in more detail in the next section.

It is very important to note that the general linear system

$$\begin{array}{ccccccc} a_{11}x_1 & + & a_{12}x_2 & + \cdots + & a_{1n}x_n & = & b_1 \\ a_{21}x_1 & + & a_{22}x_2 & + \cdots + & a_{2n}x_n & = & b_2 \\ \vdots & & \vdots & & \vdots & & \vdots \\ a_{m1}x_1 & + & a_{m2}x_2 & + \cdots + & a_{mn}x_n & = & b_m \end{array}$$

can be written as the single matrix equation

$$AX = B.$$

Here A is the coefficient matrix, X is the matrix of unknowns, and B is the matrix of constants from the right-hand sides of the equations:

$$AX = \begin{bmatrix} a_{11} & a_{12} & \cdots & a_{1n} \\ a_{21} & a_{22} & \cdots & a_{2n} \\ \vdots & \vdots & & \vdots \\ a_{m1} & a_{m2} & \cdots & a_{mn} \end{bmatrix} \begin{bmatrix} x_1 \\ x_2 \\ \vdots \\ x_n \end{bmatrix} = \begin{bmatrix} b_1 \\ b_2 \\ \vdots \\ b_m \end{bmatrix} = B.$$

In general, a **linear combination** of a set of matrices $C_1, C_2, \ldots, C_n$ is any sum of scalar multiples of the C_i; that is, any sum of the form

$$a_1C_1 + a_2C_2 + \cdots + a_nC_n.$$

We will see next that matrix multiplication can be profitably viewed as a process of forming linear combinations of either the rows or the columns of the matrices involved.

A careful examination of the definition of matrix multiplication shows that each of the elements in the jth column of the product matrix AB is computed by taking the product of a row of A and the jth column of B; that is, only the jth column of B is involved in computing the jth column of AB. In fact, if A is $m \times n$, we have

$$\mathrm{Col}_j(AB) = \begin{bmatrix} \mathrm{Row}_1(A)\ \mathrm{Col}_j(B) \\ \mathrm{Row}_2(A)\ \mathrm{Col}_j(B) \\ \vdots \\ \mathrm{Row}_m(A)\ \mathrm{Col}_j(B) \end{bmatrix} = \begin{bmatrix} a_{11}b_{1j} + a_{12}b_{2j} + \cdots + a_{1n}b_{nj} \\ a_{21}b_{1j} + a_{22}b_{2j} + \cdots + a_{2n}b_{nj} \\ \vdots \quad\quad \vdots \quad\quad\quad \vdots \\ a_{m1}b_{1j} + a_{m2}b_{2j} + \cdots + a_{mn}b_{nj} \end{bmatrix}$$
$$= A\mathrm{Col}_j(B) = \mathrm{Col}_1(A)b_{1j} + \mathrm{Col}_2(A)b_{2j} + \cdots + \mathrm{Col}_n(A)b_{nj}$$
$$= b_{1j}\mathrm{Col}_1(A) + b_{2j}\mathrm{Col}_2(A) + \cdots + b_{nj}\mathrm{Col}_n(A). \tag{1.9}$$

Equation (1.9) says that the jth column of AB is a linear combination of the columns of A, with coefficients from the jth column of B.

Since the single column of the product AX is a linear combination of the columns of A, we see that the linear system $AX = B$ has a solution precisely when B is expressible as a linear combination of the columns of the coefficient matrix A. Solving the system amounts to finding the coefficients which express B as a linear combination of the columns of A. That is, solving the linear system $AX = B$ amounts to finding scalars x_i such that

$$x_1\,\mathrm{Col}_1(A) + x_2\,\mathrm{Col}_2(A) + \cdots + x_n\,\mathrm{Col}_n(A) = B.$$

A useful expression equivalent to Equation (1.9) is

$$\begin{aligned} AB &= A[\mathrm{Col}_1(B)\quad \mathrm{Col}_2(B)\quad \cdots\quad \mathrm{Col}_n(B)] \\ &= [A\mathrm{Col}_1(B)\quad A\mathrm{Col}_2(B)\quad \cdots\quad A\mathrm{Col}_n(B)]. \end{aligned}$$

Similarly, one sees that

$$AB = \begin{bmatrix} \mathrm{Row}_1(A) \\ \mathrm{Row}_2(A) \\ \vdots \\ \mathrm{Row}_m(A) \end{bmatrix} B = \begin{bmatrix} \mathrm{Row}_1(A)B \\ \mathrm{Row}_2(A)B \\ \vdots \\ \mathrm{Row}_m(A)B \end{bmatrix}$$

where

$$\mathrm{Row}_i(AB) = \mathrm{Row}_i(A)B = a_{i1}\mathrm{Row}_1(B) + \cdots + a_{in}\mathrm{Row}_n(B). \tag{1.10}$$

Equation (1.10) says that the ith row of AB is a linear combination of the rows of B, with coefficients from the ith row of A. For the matrices A and B of Example 3 we see that

$$\mathrm{Col}_2(AB) = \begin{bmatrix} 43 \\ -19 \end{bmatrix} = A\begin{bmatrix} 0 \\ 2 \\ 7 \end{bmatrix} = 0\mathrm{Col}_1(A) + 2\mathrm{Col}_2(A) + 7\mathrm{Col}_3(A)$$

and

$$\begin{aligned} \mathrm{Row}_1(AB) &= [13\quad 43] = [1\quad 4\quad 5]B \\ &= \mathrm{Row}_1(B) + 4\mathrm{Row}_2(B) + 5\mathrm{Row}_3(B). \end{aligned}$$

These two results will be important in the development in Section 1.7.

Our next example shows how matrix multiplication is useful in studying population growth.

EXAMPLE 5 **Population Growth**

A new species of game fish is to be introduced into a small lake. The initial stocking consists of 1000 one-year-old fish, 100 two-year-old fish and 10 three-year-old fish. For this species it is estimated that 25 percent of the one-year-old fish and 50 percent of the two-year-old fish will survive the first year. The three-year-old fish will spawn and then die. The average number of eggs that hatch per three-year-old fish is 1000, only 2 percent of which will survive the first year. What will be the population of the species after four years? If the lake can support 2900 of this species, how long will it take to become overcrowded?

Let $f_1(k)$ be the population of 1-year-old fish, k years later,
$f_2(k)$ be the population of 2-year-old fish, k years later, and
$f_3(k)$ be the population of 3-year-old fish, k years later.

The above assumptions lead to the following equations:

$$\begin{aligned} f_2(k+1) &= .25 f_1(k) \\ f_3(k+1) &= .50 f_2(k) \\ f_1(k+1) &= 1000 f_3(k) \times .02 = 20 f_3(k). \end{aligned}$$

In matrix form these equations are

$$F(k+1) = \begin{bmatrix} f_1(k+1) \\ f_2(k+1) \\ f_3(k+1) \end{bmatrix} = \begin{bmatrix} 0 & 0 & 20 \\ .25 & 0 & 0 \\ 0 & .5 & 0 \end{bmatrix} F(k) = AF(k),$$

where the matrix A is called the **Leslie matrix** of this population. The population-distribution matrices $F(k)$ for this population are then

$$F(1) = \begin{bmatrix} 1000 \\ 100 \\ 10 \end{bmatrix}, \text{ which is the given initial distribution, and}$$

$$F(2) = AF(1) = A\begin{bmatrix} 1000 \\ 100 \\ 10 \end{bmatrix} = \begin{bmatrix} 200 \\ 250 \\ 50 \end{bmatrix},$$

$$F(3) = AF(2) = \begin{bmatrix} 1000 \\ 50 \\ 125 \end{bmatrix}, \quad F(4) = AF(3) = \begin{bmatrix} 2500 \\ 250 \\ 25 \end{bmatrix},$$

$$F(5) = AF(4) = \begin{bmatrix} 500 \\ 625 \\ 125 \end{bmatrix}, \quad \text{and} \quad F(6) = AF(5) = \begin{bmatrix} 2500 \\ 125 \\ 312.5 \end{bmatrix}.$$

From these calculations we see that the population at the end of the fourth year is 2775 and that the population at the end of the sixth year is 2937. Thus overcrowding becomes a problem after six years. ■

The problems in the following exercise set are designed not only to help you understand the definitions given above, but also to provide laboratory evidence for some of the general discussion of the next section. They require your diligent attention.

EXERCISES 1.5

1. Compute AB and BA if

$$A = \begin{bmatrix} 1 & 2 & 3 \\ 4 & 5 & 6 \end{bmatrix} \quad \text{and} \quad B = \begin{bmatrix} 1 & 0 \\ 2 & 3 \\ 0 & 4 \end{bmatrix}.$$

2. (a) Compute AB and BA if

$$A = \begin{bmatrix} 3 & -1 & 2 & 2 \\ 2 & 1 & 1 & 2 \\ 1 & -3 & 0 & -3 \end{bmatrix} \quad \text{and} \quad B = \begin{bmatrix} -\frac{3}{6} & \frac{6}{6} & \frac{3}{6} \\ -\frac{1}{6} & \frac{2}{6} & -\frac{1}{6} \\ \frac{7}{6} & -\frac{8}{6} & -\frac{5}{6} \\ \frac{6}{6} & -\frac{6}{6} & -\frac{6}{6} \end{bmatrix}.$$

(b) Compute I_3A, I_4B, and BI_3, where I_n is the $n \times n$ identity matrix.

3. (a) Compute AB, BA, IA, and BI if

$$A = \begin{bmatrix} 0 & 1 & 2 \\ -1 & 3 & 0 \\ 1 & -2 & 1 \end{bmatrix} \quad \text{and} \quad B = \begin{bmatrix} -3 & 5 & 6 \\ -1 & 2 & 2 \\ 1 & -1 & -1 \end{bmatrix}.$$

(b) Verify that $2(AB) = A(2B)$ and that $3(AB) = (3A)B$.

(c) Compute $2A + 3B$.

4. Compute AB and BA if

$$A = \begin{bmatrix} 2 & -1 & 1 \\ -1 & 2 & -1 \\ 1 & -1 & 2 \end{bmatrix} \quad \text{and} \quad B = \begin{bmatrix} 6 & -5 & 5 \\ -5 & 6 & -5 \\ 5 & -5 & 6 \end{bmatrix}.$$

5. (a) Compute $(AB)C$ and $A(BC)$ for

$$A = \begin{bmatrix} 2 & -1 & 5 \\ 3 & 2 & 4 \\ 8 & 0 & -2 \end{bmatrix}, \quad B = \begin{bmatrix} -3 & 5 & 6 \\ -1 & 2 & 2 \\ 1 & -1 & -1 \end{bmatrix},$$

$$\text{and} \quad C = \begin{bmatrix} 0 & 1 & 2 \\ -1 & 3 & 0 \\ 1 & -2 & 1 \end{bmatrix}.$$

(b) Compute $A(B + C)$ and $AB + AC$ for these matrices.

6. For the matrices

$$A = \begin{bmatrix} 1 & 2 & 0 \\ 5 & -3 & 4 \end{bmatrix}, \quad B = \begin{bmatrix} -2 & 5 & 6 \\ 2 & 0 & -1 \end{bmatrix}$$

$$\text{and} \quad C = \begin{bmatrix} 1 & 0 & 1 & 7 \\ 5 & 3 & 5 & -2 \\ -2 & 4 & 3 & 2 \end{bmatrix},$$

verify the right-handed distribution law; that is, show directly that

$$(A + B)C = AC + BC.$$

7. Solve the following matrix equations for x, y, u, and v;

(a) $\begin{bmatrix} x + 2u & y + 2v \\ 3x + 5u & 3y - 5v \end{bmatrix} = \begin{bmatrix} 10 & 3 \\ 26 & 7 \end{bmatrix}$ **(b)** $\begin{bmatrix} x & y \\ u & v \end{bmatrix} \begin{bmatrix} 1 & 2 \\ -1 & 4 \end{bmatrix} = \begin{bmatrix} 3 & 10 \\ 3 & 4 \end{bmatrix}$.

8. Let A, B, and C be the matrices of Exercise 5. Using as few multiplications as possible determine the entry in the (2, 3) position of $A(BC)$.

9. Let A be an $n \times n$ matrix and let D be an $n \times n$ diagonal matrix with diagonal entries $d_1, d_2, \ldots, d_n$. Compute the products AD and DA. What happens if all the $d_i = 1$?

10. **(a)** Find the most general matrix B which commutes with $A = \begin{bmatrix} 0 & 1 \\ -1 & 0 \end{bmatrix}$.

(*Hint:* Assume $B = \begin{bmatrix} a & b \\ c & d \end{bmatrix}$, compute AB and BA, set $AB = BA$, and then solve for the entries of B.)

(b) Show that any two matrices that commute with A commute with each other.

†**11.** Suppose $A = \begin{bmatrix} a & b \\ c & d \end{bmatrix}$ with $ad - bc \neq 0$.

Let $B = \dfrac{1}{ad - bc} \begin{bmatrix} d & -b \\ -c & a \end{bmatrix}$ and compute AB and BA.

†**12.** Compute LU and UL for

$$L = \begin{bmatrix} 1 & 0 & 0 \\ 3 & 2 & 0 \\ 0 & 1 & -1 \end{bmatrix} \quad U = \begin{bmatrix} 2 & 6 & 2 \\ 0 & -2 & 1 \\ 0 & 0 & 0 \end{bmatrix}.$$

†**13.** Compute the following matrix products.

(a) $\begin{bmatrix} 1 & 0 & 0 \\ 0 & 0 & 1 \\ 0 & 1 & 0 \end{bmatrix} \begin{bmatrix} a & b & c \\ d & e & f \\ g & h & i \end{bmatrix}$ **(b)** $\begin{bmatrix} 1 & 0 & 0 \\ 0 & k & 0 \\ 0 & 0 & 1 \end{bmatrix} \begin{bmatrix} a & b & c \\ d & e & f \\ g & h & i \end{bmatrix}$

(c) $\begin{bmatrix} 1 & 0 & 0 \\ 0 & 1 & 0 \\ k & 0 & 1 \end{bmatrix} \begin{bmatrix} a & b & c \\ d & e & f \\ g & h & i \end{bmatrix}$ **(d)** $\begin{bmatrix} a & b & c \\ d & e & f \\ g & h & i \end{bmatrix} \begin{bmatrix} 1 & 0 & 0 \\ 0 & 0 & 1 \\ 0 & 1 & 0 \end{bmatrix}$

(e) $\begin{bmatrix} a & b & c \\ d & e & f \\ g & h & i \end{bmatrix} \begin{bmatrix} 1 & 0 & 0 \\ 0 & k & 0 \\ 0 & 0 & 1 \end{bmatrix}$ **(f)** $\begin{bmatrix} a & b & c \\ d & e & f \\ g & h & i \end{bmatrix} \begin{bmatrix} 1 & 0 & 0 \\ 0 & 1 & 0 \\ k & 0 & 1 \end{bmatrix}$

14. Suppose A is $m \times r$ and B is $s \times t$.
(a) When are AB and BA both defined?
(b) When can $AB = BA$?
(c) When are AB and $A + B$ both defined?

15. Find the most general matrix B such that $AB = BA$ if $A = \begin{bmatrix} 2 & 1 \\ 0 & 2 \end{bmatrix}$.

16. Find the most general matrix B that commutes with $A = \begin{bmatrix} 2 & 1 \\ -1 & 2 \end{bmatrix}$. Show that any two such matrices commute.

17. Write the following system as a single matrix equation and then use matrix multiplication to make the substitution

$$\begin{aligned} x &= 5u + 7v - 2w \\ y &= 2u + w \\ z &= 3v + 7w \end{aligned}$$

in the linear system of Exercise 18(a).

18. Write the following systems of linear equations as matrix equations of the form $AX = B$.

(a)
$$\begin{aligned} 3x + 5y + 8z &= 12 \\ 4x - 2y - 4z &= 0 \\ 5x + 3y + 12z &= 13 \end{aligned}$$

(b)
$$\begin{aligned} 2x_1 + x_2 + 5x_3 + x_4 &= 5 \\ x_1 + x_2 - 3x_2 - 4x_4 &= -1 \\ 3x_1 + 6x_2 - 2x_3 + x_4 &= 8 \\ 2x_1 + 2x_2 + 2x_3 - 3x_4 &= 2 \end{aligned}$$

(c)
$$\begin{aligned} 2x_1 - x_2 + x_3 &= 4x_1 \\ -x_1 + 2x_2 - x_3 &= 4x_2 \\ x_1 - 2x_2 - x_3 &= 4x_3 \end{aligned}$$

†**19.** Let A, B, and C be arbitrary $n \times n$ matrices and h and k be real numbers. Use summation notation to write a formula for the element in the i, j position of
(a) $A + (B + C)$ **(b)** $(A + B) + C$ **(c)** $(h + k)A$
(d) $hA + kA$ **(e)** $A(B + C)$ **(f)** $AB + AC$
(g) $A(BC)$ **(h)** $(AB)C$

†**20.** Referring to your answers to Exercise 19, **(a)** write out in full and compare the answers to parts (e) and (f) for $n = 3$; and **(b)** write out in full and compare the answers to parts (g) and (h) for $n = 3$.

21. Show that if A has a zero row, then so does AB, and if A has two equal rows, then so does AB.

22. Count the number of multiplications required to compute the product AB if A is n by r and B is r by m. What does this count reduce to if A and B are both n by n?

23. **(a)** Use a calculator, MATALG, or MATLAB to compute the matrix product

$$AB = \begin{bmatrix} 3.12 & 4.00 & -5.07 \\ 5.23 & 7.00 & 3.89 \\ 11.40 & 0.00 & 5.27 \end{bmatrix} \begin{bmatrix} 0.35 & 4.01 & -1.00 \\ 1.00 & 3.00 & 4.30 \\ 1.87 & 6.37 & -9.09 \end{bmatrix}.$$

(b) Compute $3A^2B + 5B^3 - 7BA$ using MATALG or MATLAB.

[†]**24.** Show that it is not possible to find a, b, c, and d so that

$$A = \begin{bmatrix} 0 & 1 \\ 2 & 3 \end{bmatrix} = LU, \quad \text{where} \quad L = \begin{bmatrix} 1 & 0 \\ a & 1 \end{bmatrix} \quad \text{and} \quad U = \begin{bmatrix} b & d \\ 0 & c \end{bmatrix}.$$

25. **(a)** Compute the matrix product $[x \quad y \quad z] \begin{bmatrix} 4 & -1 & 2 \\ -1 & 0 & 1 \\ 2 & 1 & 0 \end{bmatrix} \begin{bmatrix} x \\ y \\ z \end{bmatrix}$.

(b) Write the quadratic polynomial

$$3x^2 - 8xy + 2y^2 + 6xz - 3z^2$$

as a matrix product like the one in part (a).

26. Show that for $A = \begin{bmatrix} 1 & 2 \\ 2 & 4 \end{bmatrix}$ it is not possible to find a 2×2 matrix B such that $BA = I$.

27. **(a)** Show that the following algorithm computes $C = AB$ if A is $m \times r$ and B is $r \times n$:

For $i = 1$ to m
  For $j = 1$ to n
    $c_{ij} = 0$
    For $k = 1$ to r
      $c_{ij} = c_{ij} + a_{ik}b_{kj}$.
    end
  end
end

(b) Write a computer program in your favorite programming language to compute $C = AB$.

28. Write a computer program to compute $C = aA + bB$ if A and B are $m \times n$.

29. In a certain small town there are 3000 families and three supermarkets. Let $N_i(k)$ be the number of families which shop at market i in the kth week and let p_{ji} be the fraction of those who shop in market i one week but will shop in market j the next week. Given the matrix

$$P = \begin{bmatrix} .5 & .2 & .4 \\ .3 & .6 & .1 \\ .2 & .2 & .5 \end{bmatrix},$$

and the fact that in the first week 1000 families shopped in each market, determine how many customers each market had in the fourth week.

30. Consider a population in which no one lives longer than 100 years. Let $n_1(k)$ be the number of people between 0 and 10 years old during the kth decade. Let $n_2(k)$ be the population in the 11–20 age bracket, and so on for each ten-year bracket. Let s_i be the survival rate of the ith group over a 10-year period and let b_i be the birthrate of the ith group over a 10-year period. Given that the birthrate matrix is

$$B = [0 \quad .15 \quad .50 \quad .40 \quad .05 \quad 0 \quad 0 \quad 0 \quad 0 \quad 0]$$

and that the survival-rate matrix is

$$S = [.95 \quad .95 \quad .90 \quad .90 \quad .85 \quad .75 \quad .60 \quad .40 \quad .30 \quad 0],$$

find the Leslie model for this population; that is, find a matrix A such that $N(k + 1) = AN(k)$.

1.6 PROPERTIES OF MATRIX OPERATIONS

What properties do the matrix operations introduced in the last section have in common with ordinary arithmetic?

In the last section we introduced algebraic operations for matrices and showed how a system of linear equations can be expressed as a single matrix equation. In order to exploit this point of view we need to explore some of the manipulative properties of matrix algebra. It would be particularly useful if these properties were like the properties of addition and multiplication of numbers, that is, the 1×1 case. In this section we will explore those properties, trying to point out carefully the areas of similarity and dissimilarity. A start has already been made on this exploration in the examples and exercises of the last section. It is very important to have worked those exercises before reading this section.

Since matrix addition and scalar multiplication are so like the addition and multiplication of numbers, it is not surprising that they behave very similarly. In the next theorem we list the important properties of matrix addition and scalar multiplication. All of these properties are familiar in the case of ordinary algebra (the 1×1 case). In fact, careful proofs of these properties rely heavily on the corresponding properties of ordinary algebra.

THEOREM 1.4 Let A, B, and C be any matrices of the same size and let h and k be any scalars. The following properties are then true:

(1) $A + B = B + A$ Matrix addition is commutative.
(2) $A + (B + C) = (A + B) + C$ Matrix addition is associative.
(3) $h(A + B) = hA + hB$ Scalar multiplication distributes over matrix addition.
(4) $(h + k)A = hA + kA$ Scalar multiplication distributes over scalar addition.
(5) $(hk)A = h(kA)$ Scalar multiplication is associative.
(6) There is a unique matrix Z such that for all matrices A, $A + Z = A$ (Z is called the **additive-identity element**).
(7) For each matrix A there is a unique matrix C such that $A + C = Z$ (C is called the **additive inverse of** A).

Proofs of several of these properties were outlined in Exercises 19 and 20 in Section 1.5. The proof of any one of these properties requires only an examination of a typical element of each of the two matrices. For example, the entry in the (i, j) position of $h(A + B)$ is $h(a_{ij} + b_{ij})$, while the entry in the (i, j) position of $hA + hB$ is $ha_{ij} + hb_{ij}$. Clearly these two numbers are the same, which establishes part 3 of the theorem. The additive-identity matrix of property 6 is the **zero matrix** of the same size as A:

$$Z = 0 = \begin{bmatrix} 0 & \cdots & 0 \\ \vdots & & \vdots \\ 0 & \cdots & 0 \end{bmatrix}.$$

The additive inverse of property 7 is the negative of the matrix A introduced in Definition 1.7; that is, $C = -A = (-1)A$. □

The only real difference between matrix addition and scalar addition is that matrix addition is not always defined. Matrix multiplication is quite another story. We have already seen that matrix multiplication is noncommutative (in general $AB \neq BA$) and that the product of two nonzero matrices can be the zero matrix.

In the next theorem we summarize the important properties of matrix multiplication. All of these properties have familiar analogs in the case of ordinary algebra.

THEOREM 1.5 Let A, B, and C be any three matrices. Then, provided only that the indicated operations are defined, we have the following properties:

(1) $A(BC) = (AB)C$ Matrix multiplication is associative.

(2) (a) $A(B + C) = AB + AC$ Matrix multiplication distributes over matrix addition.

(b) $(B + C)A = BA + CA$

(3) There is a unique $n \times n$ matrix I such that $IA = AI = A$ for every $n \times n$ matrix A.

(4) $(kA)B = k(AB) = A(kB)$ for any scalar k.

The general proofs of these properties are not particularly instructive, only messy, so we will not write them all down. The proof of the associative property was outlined in Exercises 19 and 20 of Section 1.5, and some experimental evidence for the other properties was given in the last exercise set. To verify the right-handed distributive law (property 2(b)) we assume that B and C are $m \times n$ and that A is $n \times t$. Then the entry in the (i, j) position of $(B + C)A$ is

$$\sum_{k=1}^{n} (b_{ik} + c_{ik})a_{kj} = \sum_{k=1}^{n} (b_{ik}a_{kj} + c_{ik}a_{kj}) = \sum_{k=1}^{n} b_{ik}a_{kj} + \sum_{k=1}^{n} c_{ik}a_{kj},$$

while the entry in the (i, j) position of $BA + CA$ is

$$\sum_{k=1}^{n} b_{ik}a_{kj} + \sum_{k=1}^{n} c_{ik}a_{kj}.$$

Since these two expressions are the same, and since $(B + C)A$ and $BA + CA$ are both $m \times t$, it follows that $(B + C)A = BA + CA$ as claimed.

The matrix I in the third property is called the **identity** matrix. The $n \times n$ matrix

$$I_n = \begin{bmatrix} 1 & 0 & \cdots & 0 \\ 0 & 1 & \cdots & 0 \\ \vdots & \vdots & & \vdots \\ 0 & 0 & \cdots & 1 \end{bmatrix}$$

first encountered in Section 1.3, is clearly what we need here; it is the matrix analog of the number 1. Note that if A is not square, say $m \times n$ with $m \neq n$, then I_nA is undefined. It is, however, true that $I_mA = AI_n = A$. □

To complete our comparison of matrix multiplication and multiplication of numbers we need a matrix analog of the reciprocal of a nonzero number. The **reciprocal,** or multiplicative inverse, of the number a, denoted by $\frac{1}{a}$ or a^{-1}, is the unique number x such that $ax = 1$.

DEFINITION 1.10 If A is a square matrix and B satisfies

$$AB = BA = I,$$

then B is a **multiplicative inverse** of A. □

Matrices that have inverses are said to be **nonsingular** or **invertible.** Only square matrices can have inverses. If a square matrix A does not have an inverse, it is said to be **singular** or **noninvertible.** Exercise 26 in Section 1.5 shows that not every nonzero square matrix has a multiplicative inverse.

EXAMPLE 1 A Matrix Inverse

The matrix $A = \begin{bmatrix} 1 & 3 \\ 1 & 2 \end{bmatrix}$ has $B = \begin{bmatrix} -2 & 3 \\ 1 & -1 \end{bmatrix}$ as a multiplicative inverse. To verify this we compute

$$AB = \begin{bmatrix} 1 & 3 \\ 1 & 2 \end{bmatrix}\begin{bmatrix} -2 & 3 \\ 1 & -1 \end{bmatrix} = \begin{bmatrix} 1 & 0 \\ 0 & 1 \end{bmatrix}$$

and

$$BA = \begin{bmatrix} -2 & 3 \\ 1 & -1 \end{bmatrix}\begin{bmatrix} 1 & 3 \\ 1 & 2 \end{bmatrix} = \begin{bmatrix} 1 & 0 \\ 0 & 1 \end{bmatrix}. \quad ■$$

It is easy to establish, from the associative property of matrix multiplication, that a matrix can have no more than one inverse. To see that this is so, suppose that B and C are both inverses of A; that is,

$$AB = BA = I \qquad \text{and} \qquad AC = CA = I.$$

It follows that

$$C(AB) = CI = C$$

and, from the associativity of matrix multiplication, that

$$C(AB) = (CA)B = IB = B.$$

Therefore $C = B$, as claimed.

DEFINITION 1.11 If the matrix A is invertible, then the unique multiplicative inverse of A will be denoted by A^{-1}. □

We avoid using the reciprocal notation $\frac{I}{A}$ because of the possible ambiguity of expressions like $\frac{B}{A}$. (Does it mean BA^{-1} or $A^{-1}B$? In general these two expressions will be unequal because of the noncommutativity of matrix multiplication.)

There is a simple formula for the inverse of a 2×2 matrix which is worth noting:

$$\begin{bmatrix} a & b \\ c & d \end{bmatrix}^{-1} = \frac{1}{ad - bc} \begin{bmatrix} d & -b \\ -c & a \end{bmatrix}, \text{ if } ad - bc \neq 0. \qquad \textbf{(1.11)}$$

You should do the multiplication necessary to check this last assertion (recall Exercise 11 in Section 1.5) and verify that this formula gives the correct answer for the matrix A of Example 1. For the matrix

$$C = \begin{bmatrix} 5 & -4 \\ 2 & 3 \end{bmatrix}, \text{ Equation (1.11) yields } C^{-1} = \frac{1}{23}\begin{bmatrix} 3 & 4 \\ -2 & 5 \end{bmatrix}.$$

In general it will not be practical to give such general formulas for A^{-1}; we will instead provide an algorithm for computing A^{-1}. Given a candidate for the inverse, one can always check to see whether or not it really is the inverse.

EXAMPLE 2 **Checking an Inverse**

$$\text{If } T = \begin{bmatrix} 1 & 0 & 0 \\ 1 & 2 & 0 \\ 2 & 1 & 3 \end{bmatrix}, \text{ and } L = \begin{bmatrix} 1 & 0 & 0 \\ -\frac{1}{2} & \frac{1}{2} & 0 \\ -\frac{1}{2} & -\frac{1}{6} & 1 \end{bmatrix} \text{ does } L = T^{-1}?$$

To answer this question we compute the product of T and L:

$$\begin{bmatrix} 1 & 0 & 0 \\ 1 & 2 & 0 \\ 2 & 1 & 3 \end{bmatrix}\begin{bmatrix} 1 & 0 & 0 \\ -\frac{1}{2} & \frac{1}{2} & 0 \\ -\frac{1}{2} & -\frac{1}{6} & 1 \end{bmatrix} = \begin{bmatrix} 1 & 0 & 0 \\ 0 & 1 & 0 \\ 0 & 0 & 3 \end{bmatrix} \neq I.$$

It follows that L is not T^{-1}. It appears to be close, but not quite right. We leave it to you to verify that replacing the $(3, 3)$ element of L by $\frac{1}{3}$ will yield the correct T^{-1}. ■

Some elementary properties of the matrix inverse can be established by direct multiplication.

THEOREM 1.6 If A and B are invertible n by n matrices, then the following assertions are true.

(1) A^{-1} is also invertible and $(A^{-1})^{-1} = A$.

(2) If $k \neq 0$, then kA is invertible and $(kA)^{-1} = \left(\frac{1}{k}\right)A^{-1}$.

(3) The product AB is invertible and $(AB)^{-1} = B^{-1}A^{-1}$, that is, the inverse of the product is the product of the inverses in the reverse order.

In order to verify the third assertion of the theorem we compute

$$(AB)(B^{-1}A^{-1}) = A(BB^{-1})A^{-1} = AIA^{-1} = I$$

and

$$(B^{-1}A^{-1})(AB) = B^{-1}(A^{-1}A)B = B^{-1}IB = I.$$

Therefore, because of the uniqueness of matrix inverses, $(AB)^{-1} = B^{-1}A^{-1}$ as claimed. The other proofs are left to the exercises. □

In the next example we will illustrate a brute-force method for finding the inverse of a matrix A. A more complete discussion of techniques for computing matrix inverses will be given in the next section.

EXAMPLE 3 **Brute-force Inverse Computation**

Find the inverse of $A = \begin{bmatrix} 1 & 2 \\ 3 & 5 \end{bmatrix}$.

Let us assume that $A^{-1} = B = \begin{bmatrix} b_{11} & b_{12} \\ b_{21} & b_{22} \end{bmatrix}$ and try to choose the b_{ij} so that $AB = I$. By direct multiplication

$$\begin{bmatrix} 1 & 2 \\ 3 & 5 \end{bmatrix}\begin{bmatrix} b_{11} & b_{12} \\ b_{21} & b_{22} \end{bmatrix} = \begin{bmatrix} b_{11} + 2b_{21} & b_{12} + 2b_{22} \\ 3b_{11} + 5b_{21} & 3b_{12} + 5b_{22} \end{bmatrix} = \begin{bmatrix} 1 & 0 \\ 0 & 1 \end{bmatrix}.$$

Thus we need to satisfy the four equations

$$\left.\begin{aligned} b_{11} + 2b_{21} &= 1 \\ 3b_{11} + 5b_{21} &= 0 \end{aligned}\right\} \quad \text{or} \quad A\begin{bmatrix} b_{11} \\ b_{21} \end{bmatrix} = \begin{bmatrix} 1 \\ 0 \end{bmatrix}$$

$$\left.\begin{aligned} b_{12} + 2b_{22} &= 0 \\ 3b_{12} + 5b_{22} &= 1 \end{aligned}\right\} \quad \text{or} \quad A\begin{bmatrix} b_{12} \\ b_{22} \end{bmatrix} = \begin{bmatrix} 0 \\ 1 \end{bmatrix}.$$

Note that these four equations really amount to two 2×2 systems with the same coefficient matrix. We can use the method suggested in Section 1.4 to efficiently find a solution of this pair of systems:

$$[A\,|\,I] = \begin{bmatrix} A & \begin{matrix} 1 & 0 \\ 0 & 1 \end{matrix} \end{bmatrix}$$

$$= \begin{bmatrix} 1 & 2 & 1 & 0 \\ 3 & 5 & 0 & 1 \end{bmatrix} \to \begin{bmatrix} 1 & 2 & 1 & 0 \\ 0 & -1 & -3 & 1 \end{bmatrix} \to \begin{bmatrix} 1 & 0 & -5 & 2 \\ 0 & 1 & 3 & -1 \end{bmatrix}$$

From the above row reduction we see that the solutions of the two systems are

$$\begin{bmatrix} b_{11} \\ b_{21} \end{bmatrix} = \begin{bmatrix} -5 \\ 3 \end{bmatrix} \quad \text{and} \quad \begin{bmatrix} b_{12} \\ b_{22} \end{bmatrix} = \begin{bmatrix} 2 \\ -1 \end{bmatrix}$$

so that $B = \begin{bmatrix} -5 & 2 \\ 3 & -1 \end{bmatrix}$ satisfies $AB = I$.

A simple calculation shows that $BA = I$ also; hence $B = A^{-1}$. ■

The MATLAB and MATLAG (available on-line from the publisher) programs as well as many graphics calculators have built in facilities for doing matrix calculations. The following table lists the appropriate commands.

MATRIX OPERATION	MATALG	MATLAB	TI-85
$A + B$	$= A + B$	$A + B$	$A + B$
$S = A + B$	$S = A + B$	$S = A + B$	$A + B$ [STO] S
AB	$= A * B$	$A * B$	$A * B$
$P = AB$	$P = A * B$	$P = A * B$	$A * B$ [STO] P
kA	$= k * A$	$k * A$	$k * A$
$M = kA$	$M = k * A$	$M = k * A$	$k * A$ [STO] M
I_n	$=$ id(n)	eye(n)	ident n
A^{-1}	$=$ inv(A)	inv(A)	A [x^{-1}]
$B = A^{-1}$	$B =$ inv(A)	$B =$ inv(A)	A [x^{-1}] [STO] B

EXERCISES 1.6

1. For $A = \begin{bmatrix} 1 & 1 \\ 0 & 1 \end{bmatrix}$ and $B = \begin{bmatrix} 2 & 1 \\ 5 & 3 \end{bmatrix}$ use Formula (1.11) to compute A^{-1}, B^{-1}, $A^{-1} + B^{-1}$, and $(A + B)^{-1}$.

2. Verify by direct multiplication that

$$A^{-1} = \begin{bmatrix} 0 & 1 & 2 \\ -1 & 3 & 0 \\ 1 & -2 & 1 \end{bmatrix} \quad \text{if} \quad A = \begin{bmatrix} -3 & 5 & 6 \\ -1 & 2 & 2 \\ 1 & -1 & -1 \end{bmatrix}$$

and that

$$B^{-1} = \frac{1}{6}\begin{bmatrix} -3 & 6 & 3 \\ -1 & 2 & -1 \\ 7 & -8 & -5 \end{bmatrix} \quad \text{for} \quad B = \begin{bmatrix} 3 & -1 & 2 \\ 2 & 1 & 1 \\ 1 & -3 & 0 \end{bmatrix}.$$

Find $(AB)^{-1}$, $(BA)^{-1}$, $(A^2)^{-1}$, and $(ABA)^{-1}$.

3. Let B be a nonsingular matrix for which $B^{-1} = \begin{bmatrix} 3 & 4 \\ 5 & 6 \end{bmatrix}$. Find B.

4. Let A and B be square matrices of the same size. Which of the following are valid equations?
(a) $(AB)^2 = A^2B^2$ **(b)** $(AB)^2 = A(BA)B$
(c) $(A + B)^2 = A^2 + 2AB + B^2$ **(d)** $(A - B)^2 = A^2 - AB - BA + B^2$

5. Find a formula for $(A + B)^3$. What does this reduce to if $AB = BA$?

6. Establish the following cancellation laws.
(a) If A^{-1} exists and $AB = AC$, then $B = C$.
(b) If A^{-1} exists and $BA = CA$, then $B = C$.

7. If $(5A)^{-1} = \begin{bmatrix} 4 & 5 \\ 2 & 3 \end{bmatrix}$, find A.

8. Let A and B be the matrices of Exercise 2. Show that

$$\text{Row}_2(AB) = -\text{Row}_1(B) + 2\text{Row}_2(B) + 2\text{Row}_3(B)$$

and that

$$\text{Col}_3(AB) = 2\text{Col}_1(A) + \text{Col}_2(A).$$

9. For $J = \begin{bmatrix} 3 & 1 & 0 \\ 0 & 3 & 0 \\ 0 & 0 & -1 \end{bmatrix}$ show that $J^3 - 5J^2 + 3J + 9I = 0$. Show that $J^{-1} = (\frac{1}{9})(-J^2 + 5J - 3I)$ and use this result to compute J^{-1}.

10. Let B be a square matrix such that $B^3 + 5B^2 - 7B + I = 0$. Show that $B^{-1} = 7I - 5B - B^2$.

11. Let A and B be $n \times n$. Show that if A has a zero row, then so does AB. Show also that if A has a zero row, then A must be singular.

12. Let A and B be $n \times n$. Show that if A has two equal rows, then so does AB and hence A must be singular.

13. Show that the product of two lower triangular $n \times n$ matrices must be lower triangular.

14. Let $AX = B$ be any consistent system of linear equations, and let X_p be a fixed solution. Show that every solution to the system can be written in the form $X = X_p + X_h$ where X_h is a solution of the homogeneous system $AX = 0$. Show also that every matrix of this form is a solution.

15. Let $M = \left[\begin{array}{ccc|c} 1 & -1 & 0 & 2 \\ 3 & 2 & 1 & 5 \\ \hline 4 & -2 & 1 & 2 \\ 3 & 5 & 0 & 1 \end{array}\right] = \begin{bmatrix} M_{11} & M_{12} \\ M_{21} & M_{22} \end{bmatrix}$

and $N = \left[\begin{array}{cc|cc} 5 & 0 & -3 & 1 \\ 2 & 2 & 0 & 1 \\ 3 & 0 & 2 & 1 \\ \hline 5 & 0 & 0 & 2 \end{array}\right] = \begin{bmatrix} N_{11} & N_{12} \\ N_{21} & N_{22} \end{bmatrix}.$

Verify by direct calculation that

$$MN = \begin{bmatrix} M_{11}N_{11} + M_{12}N_{21} & M_{11}N_{12} + M_{12}N_{22} \\ M_{21}N_{11} + M_{22}N_{21} & M_{21}N_{12} + M_{22}N_{22} \end{bmatrix}.$$

16. Let A, B, and C be $n \times n$ nonsingular matrices. Show that $(ABC)^{-1} = C^{-1}B^{-1}A^{-1}$. Can you generalize this result? If so, how?

17. What is the unique solution of the linear system $AX = K$ if the coefficient matrix A is invertible?

18. Show that if A^{-1} exists and $AB = I$, then $BA = I$.

19. Use the method of Example 3 to find inverses of the following matrices:

$$A = \begin{bmatrix} 4 & 3 \\ 3 & 3 \end{bmatrix}, \quad B = \begin{bmatrix} .5 & 5 & -3 \\ 2 & 2 & -1 \\ -1 & -1 & 1 \end{bmatrix} \quad \text{and} \quad C = \begin{bmatrix} 2 & -1 & 1 \\ -1 & 2 & -1 \\ 1 & -1 & 2 \end{bmatrix}.$$

20. Complete the proof of Theorem 1.6.

21. In the next chapter we will formally define A^T, the **transpose** of A, by the statement

"The entry in the (i, j) position of A^T is a_{ji},"

and think of A^T as the matrix obtained from A by interchanging rows and columns. For

$$A = \begin{bmatrix} 1 & 2 \\ 2 & 0 \\ -1 & 1 \end{bmatrix} \quad \text{and} \quad B = \begin{bmatrix} 1 & -1 & 2 \\ 2 & 1 & 0 \end{bmatrix},$$

find A^T, B^T, A^TB^T, B^TA^T, $(AB)^T$, and A^TA.

22. Prove the following properties of the transpose (see Exercise 21).

(a) $(A^T)^T = A$ **(b)** $(kA)^T = k(A^T)$

(c) $(A + B)^T = A^T + B^T$ **(d)** $(AB)^T = B^TA^T$

(e) $(A^T)^{-1} = (A^{-1})^T$

23. Let A be a square matrix and let n be a positive integer. Show that $(A^n)^T = (A^T)^n$.

24. Let A and B be square matrices of the same size. Show that $(AB)^2 - A^2B^2 = A(AB - BA)B$. When is $(AB)^2 = A^2B^2$?

25. Find all 3×3 diagonal matrices D which have the property that $D^2 = I$.

26. Let X and Y be $n \times n$ nonsingular matrices. Show that $XY^{-1} = Y^{-1}X$ if, and only if, $XY = YX$.

27. Given the matrices,

$$A = \begin{bmatrix} 1 & 4 \\ -2 & 3 \\ 1 & -2 \end{bmatrix}, B = \begin{bmatrix} 2 & 0 & 1 \\ 0 & 1 & 1 \end{bmatrix}, \text{and } C = \begin{bmatrix} 8 & 5 & 9 \\ 6 & 1 & 4 \\ -4 & -1 & -3 \end{bmatrix},$$

find a matrix X such that $AXB = C$.

28. Let A and B be nonsingular $n \times n$ matrices. Assuming that all necessary inverses exist, show that

(a) $(A^{-1} + B^{-1})^{-1} = A(A + B)^{-1}B$ **(b)** $(I + AB)^{-1} = A(I + BA)^{-1}A^{-1}$.

1.7 ROW EQUIVALENCE AND MATRIX MULTIPLICATION

> How are matrix inverses to be computed?
> Is there a relationship between elementary row operations and matrix multiplication?

If the coefficient matrix of the linear system

$$AX = K$$

is invertible, then multiplying the equation on the left by A^{-1} yields

$$A^{-1}AX = A^{-1}K \text{ which is equivalent to } X = A^{-1}K.$$

This formula for the solution of the linear system is useful theoretically, but should not be seriously considered as a solution technique; the elimination techniques introduced earlier turn out to be a much more efficient way to solve the linear system $AX = K$.

Consider now the more general matrix equation

$$AX = B,$$

where B is not necessarily a column matrix. Let us suppose that B is $n \times r$ and let us denote the columns of B and X respectively by B_i and X_i so that, in partitioned form,

$$B = [B_1 \quad B_2 \quad \cdots \quad B_r] \quad \text{and} \quad X = [X_1 \quad X_2 \quad \cdots \quad X_r].$$

From Equation (1.9) of Section 1.5 we have

$$\text{Col}_j(AX) = A\text{Col}_j(X) = AX_j,$$

so the columns of AX are just A times the columns of X; that is,

$$AX = [AX_1 \quad AX_2 \quad \cdots \quad AX_r].$$

If we now look at the matrix equation $AX = B$, column by column, we see that it is equivalent to the set of linear systems

$$AX_i = B_i; \quad i = 1, 2, \ldots, r.$$

We saw in Section 1.4 that a set of systems like this could be solved efficiently with a single row reduction. In particular, if A^{-1} exists and

$$[A \mid B] \xrightarrow[\text{reduce}]{\text{row}} [I \mid X], \tag{1.12}$$

then $X = A^{-1}B$ is the unique solution of the equation $AX = B$. If A is the not row equivalent to the identity matrix, then the reduction stops with a set of systems, whose coefficient matrix is in reduced row-echelon form, each system of which is easy to solve. In this case the solution may not exist and, if it does, it may not be unique.

EXAMPLE 1 **Solution of a Matrix Equation**

Solve the equation $AX = B$ if

$$A = \begin{bmatrix} 1 & 2 & 3 \\ 3 & 1 & -2 \\ 4 & 5 & 6 \end{bmatrix} \quad \text{and} \quad B = \begin{bmatrix} 1 & 1 & 4 & 6 \\ 1 & 3 & 3 & 0 \\ 1 & -2 & 0 & 12 \end{bmatrix}.$$

The augmented matrix for this system is the 3×7 matrix

$$[A \mid B] = \left[\begin{array}{ccc|cccc} 1 & 2 & 3 & 1 & 1 & 4 & 6 \\ 3 & 1 & -2 & 1 & 3 & 3 & 0 \\ 4 & 5 & 6 & 1 & -2 & 0 & 12 \end{array}\right],$$

which the operations $(R_2 \leftarrow -3R_1 + R_2)$ and $(R_3 \leftarrow -4R_1 + R_3)$, followed by $(R_2 \leftarrow -2R_3 + R_2)$ and $(R_3 \leftarrow -R_3)$, reduce to

$$\left[\begin{array}{ccc|cccc} 1 & 2 & 3 & 1 & 1 & 4 & 6 \\ 0 & -5 & -11 & -2 & 0 & -9 & -18 \\ 0 & -3 & -6 & -3 & -6 & -16 & -12 \end{array}\right] \quad \text{and then to}$$

$$\left[\begin{array}{ccc|cccc} 1 & 2 & 3 & 1 & 1 & 4 & 6 \\ 0 & 1 & 1 & 4 & 12 & 23 & 6 \\ 0 & 3 & 6 & 3 & 6 & 16 & 12 \end{array}\right].$$

The operations $(R_3 \leftarrow -3R_2 + R_3)$ and $(R_1 \leftarrow -2R_2 + R_1)$ now yield

$$\left[\begin{array}{ccc|cccc} 1 & 0 & 1 & -7 & -23 & -42 & -6 \\ 0 & 1 & 1 & 4 & 12 & 23 & 6 \\ 0 & 0 & 3 & -9 & -30 & -53 & -6 \end{array}\right].$$

Finally, $(R_3 \leftarrow \frac{1}{3}R_3)$, $(R_2 \leftarrow -R_3 + R_2)$ and $(R_1 \leftarrow -R_3 + R_1)$ lead to

$$\left[\begin{array}{ccc|cccc} 1 & 0 & 0 & -4 & -13 & -\frac{73}{3} & -4 \\ 0 & 1 & 0 & 7 & 22 & \frac{122}{3} & 8 \\ 0 & 0 & 1 & -3 & -10 & -\frac{53}{3} & -2 \end{array}\right] = [I \mid X] = [I \mid A^{-1}B].$$

Since the coefficient matrix is row equivalent to I, the system has the unique solution

$$X = A^{-1}B = \begin{bmatrix} -4 & -13 & -\frac{73}{3} & -4 \\ 7 & 22 & \frac{122}{3} & 8 \\ -3 & -10 & -\frac{53}{3} & -2 \end{bmatrix}.$$

The particular case $B = I$ of the above problem is worthy of special attention. If A is invertible, then the unique solution of the equation $AX = I$ is $X = A^{-1}$ and it can be computed by Equation (1.12),

$$[A \mid I] \xrightarrow[\text{reduce}]{\text{row}} [I \mid A^{-1}]. \tag{1.13}$$

Equation (1.13) describes the most efficient general purpose method for computing the inverse of a nonsingular matrix. The method is nothing more than the Gauss-Jordan elimination method for solving the matrix equation $AX = I$. Note that this method is the brute-force method described in Example 3 of Section 1.6.

EXAMPLE 2 Inverse Calculation

Find the inverse of $C = \begin{bmatrix} 1 & -3 & 0 & -2 \\ 3 & -12 & -2 & -6 \\ -2 & 10 & 2 & 5 \\ -1 & 6 & 1 & 3 \end{bmatrix}$.

The augmented matrix of the system $CX = I$ is

$$[C|I] = \left[\begin{array}{rrrr|rrrr} 1 & -3 & 0 & -2 & 1 & 0 & 0 & 0 \\ 3 & -12 & -2 & -6 & 0 & 1 & 0 & 0 \\ -2 & 10 & 2 & 5 & 0 & 0 & 1 & 0 \\ -1 & 6 & 1 & 3 & 0 & 0 & 0 & 1 \end{array}\right] \rightarrow \left[\begin{array}{rrrr|rrrr} 1 & -3 & 0 & -2 & 1 & 0 & 0 & 0 \\ 0 & -3 & -2 & 0 & -3 & 1 & 0 & 0 \\ 0 & 4 & 2 & 1 & 2 & 0 & 1 & 0 \\ 0 & 3 & 1 & 1 & 1 & 0 & 0 & 1 \end{array}\right],$$

where the first column has been reduced by the three row operations $(R_2 \leftarrow -3R_1 + R_2)$, $(R_3 \leftarrow 2R_1 + R_3)$, and $(R_4 \leftarrow R_1 + R_4)$. The operation $(R_2 \leftarrow R_3 + R_2)$ produces a 1 in the (2, 2) position without introducing fractions, and then the row operations $(R_1 \leftarrow 3R_2 + R_1)$, $(R_3 \leftarrow -4R_2 + R_3)$, and $(R_4 \leftarrow -3R_2 + R_4)$ lead to

$$\left[\begin{array}{rrrr|rrrr} 1 & 0 & 0 & 1 & -2 & 3 & 3 & 0 \\ 0 & 1 & 0 & 1 & -1 & 1 & 1 & 0 \\ 0 & 0 & 2 & -3 & 6 & -4 & -3 & 0 \\ 0 & 0 & 1 & -2 & 4 & -3 & -3 & 1 \end{array}\right] \rightarrow \left[\begin{array}{rrrr|rrrr} 1 & 0 & 0 & 1 & -2 & 3 & 3 & 0 \\ 0 & 1 & 0 & 1 & -1 & 1 & 1 & 0 \\ 0 & 0 & 1 & -2 & 4 & -3 & -3 & 1 \\ 0 & 0 & 0 & 1 & -2 & 2 & 3 & -2 \end{array}\right],$$

where the second matrix above comes from the first by the operations $(R_3 \leftarrow -2R_4 + R_3)$ and $(R_3 \leftrightarrow R_4)$.

Finally, operations $(R_3 \leftarrow 2R_4 + R_3)$, $(R_2 \leftarrow -R_4 + R_2)$, and $(R_1 \leftarrow -R_4 + R_1)$ yield

$$\left[\begin{array}{cccc|cccc} 1 & 0 & 0 & 0 & 0 & 1 & 0 & 2 \\ 0 & 1 & 0 & 0 & 1 & -1 & -2 & 2 \\ 0 & 0 & 1 & 0 & 0 & 1 & 3 & -3 \\ 0 & 0 & 0 & 1 & -2 & 2 & 3 & -2 \end{array}\right] = [I \mid C^{-1}]. \quad \blacksquare$$

It should be clear that the inversion algorithm of Equation (1.13) involves considerably more work than the solution of a single linear system with coefficient matrix A; thus one should not try to solve a single linear system $AX = K$ by finding A^{-1} and then computing $X = A^{-1}K$. If one needs to solve a set of linear systems $AX = B_i$; $i = 1, 2, \ldots s$, then the inverse may be useful, especially if not all the B_i are available initially. In Section 1.8 we will describe a still better way of dealing with the case of multiple right-hand sides.

In order to give more insight into why some matrices have inverses and others do not, we need to investigate the relationship between matrix multiplication and the elementary row operations used to solve linear systems. The key tool in the development to follow was given in Equation (1.10), which gives a relation between the rows of the product AB and the rows of B; that is,

$$\text{Row}_i(AB) = \text{Row}_i(A)B.$$

If we write this last equation in terms of the rows of B it becomes

$$\text{Row}_i(AB) = a_{i1}\,\text{Row}_1(B) + a_{i2}\,\text{Row}_2(B) + \cdots + a_{in}\,\text{Row}_n(B), \qquad \textbf{(1.14)}$$

which can be described as saying that the ith row of the product is a linear combination of the rows of the second factor with coefficients from the ith row of the first factor. This last observation can be used to obtain the following theorem which gives a connection between row equivalence and matrix multiplication. Recall Exercise 13 in Section 1.5.

THEOREM 1.7 An elementary row operation on a matrix A can be accomplished by multiplying on the left by an identity matrix on which the elementary row operation has been performed. Symbolically, if E is an elementary row operation and $E(A)$ is the result of performing E on A, then

$$E(A) = [E(I)]A.$$

A formal proof of this theorem divides naturally into three parts, one for each type of elementary row operation. We give the details for the Type III operation $(R_j \leftarrow kR_i + R_j)$. In this case the matrix $E = E(I)$ differs from I only in the (j, i) position, where there is a k. If $i < j$,

$$EA = \begin{bmatrix} 1 & \cdot & 0 & \cdot & \cdot & 0 & \cdot & 0 \\ \cdot & \cdot & \cdot & \cdot & \cdot & \cdot & \cdot & \cdot \\ 0 & \cdot & 1 & \cdot & \cdot & 0 & \cdot & 0 \\ \cdot & \cdot & \cdot & \cdot & \cdot & \cdot & \cdot & \cdot \\ 0 & \cdot & k & \cdot & \cdot & 1 & \cdot & 0 \\ \cdot & \cdot & \cdot & \cdot & \cdot & \cdot & \cdot & \cdot \\ 0 & \cdot & 0 & \cdot & \cdot & 0 & \cdot & 1 \end{bmatrix} \begin{bmatrix} \text{Row}_1(A) \\ \cdot \\ \text{Row}_i(A) \\ \cdot \\ \text{Row}_j(A) \\ \cdot \\ \text{Row}_n(\text{A}) \end{bmatrix}.$$

(Row $i \rightarrow$ marks the third displayed row; Row $j \rightarrow$ marks the fifth displayed row.)

From Equation (1.14) it follows that

$$\text{Row}_j(EA) = \text{Row}_j(E)A = k\text{Row}_i(A) + \text{Row}_j(A)$$

and that all other rows of EA and A agree. This establishes the assertion of the theorem for Type III operations. The other proofs are similar and will be left to the exercises. □

DEFINITION 1.12 An **elementary matrix** is a matrix that has been obtained from the identity matrix by applying one elementary row operation. □

Theorem 1.7 says that elementary row operations can be accomplished by multiplying on the left (premultiplying) by elementary matrices.

We list below examples of 4 by 4 elementary matrices corresponding respectively to the operations $(R_3 \leftarrow 5R_3)$, $(R_2 \leftrightarrow R_3)$, and $(R_4 \leftarrow 2R_3 + R_4)$.

$$\underset{R_3 \leftarrow 5R_3}{\begin{bmatrix} 1 & 0 & 0 & 0 \\ 0 & 1 & 0 & 0 \\ 0 & 0 & 5 & 0 \\ 0 & 0 & 0 & 1 \end{bmatrix}} \quad \underset{R_2 \leftrightarrow R_3}{\begin{bmatrix} 1 & 0 & 0 & 0 \\ 0 & 0 & 1 & 0 \\ 0 & 1 & 0 & 0 \\ 0 & 0 & 0 & 1 \end{bmatrix}} \quad \underset{R_4 \leftarrow 2R_3 + R_4}{\begin{bmatrix} 1 & 0 & 0 & 0 \\ 0 & 1 & 0 & 0 \\ 0 & 0 & 1 & 0 \\ 0 & 0 & 2 & 1 \end{bmatrix}}$$

For our purposes, the most important fact about elementary matrices is that they are always invertible and that their inverses are also elementary matrices. This follows easily from the observation in Section 1.3 that every elementary row operation can be reversed by an elementary row operation of the same type.

The inverse of the elementary matrix for $(R_i \leftarrow kR_i)$ is the elementary matrix for $(R_i \leftarrow \frac{1}{k}R_i)$, for example,

$$\begin{bmatrix} 1 & 0 & 0 \\ 0 & k & 0 \\ 0 & 0 & 1 \end{bmatrix}^{-1} = \begin{bmatrix} 1 & 0 & 0 \\ 0 & \frac{1}{k} & 0 \\ 0 & 0 & 1 \end{bmatrix}.$$

The inverse of the elementary matrix for $(R_i \leftrightarrow R_j)$ is itself; for example, if $i = 2$ and $j = 3$, then

$$\begin{bmatrix} 1 & 0 & 0 \\ 0 & 0 & 1 \\ 0 & 1 & 0 \end{bmatrix}^{-1} = \begin{bmatrix} 1 & 0 & 0 \\ 0 & 0 & 1 \\ 0 & 1 & 0 \end{bmatrix}.$$

The inverse of the elementary matrix for $(R_j \leftarrow kR_i + R_j)$ is the elementary matrix for $(R_j \leftarrow -kR_i + R_j)$, for example, if $i = 1$ and $j = 2$, then

$$\begin{bmatrix} 1 & 0 & 0 \\ k & 1 & 0 \\ 0 & 0 & 1 \end{bmatrix}^{-1} = \begin{bmatrix} 1 & 0 & 0 \\ -k & 1 & 0 \\ 0 & 0 & 1 \end{bmatrix}.$$

Suppose now that a sequence of elementary row operations reduces the matrix A to a matrix B; that is, A is row equivalent to B. Let $E_1, E_2, \ldots, E_r$ be the elementary matrices corresponding to the elementary row operations used to reduce A to B. From Theorem 1.7 we have

$$E_r E_{r-1} \cdots E_2 E_1 A = B.$$

The matrix

$$P = E_r E_{r-1} \cdots E_2 E_1 = E_r E_{r-1} \cdots E_2 E_1 I$$

is nonsingular, because it is a product of elementary matrices that are each nonsingular, and satisfies the equation $PA = B$. If we write $P = PI$ we see that P can be obtained by performing on I exactly the same operations which changed A to B. Thus the matrix P constitutes a "record" of the elementary row operations used to reduce A to B.

THEOREM 1.8 If A and B are row equivalent, then there exists a nonsingular matrix P such that $PA = B$. The matrix P can be computed, as A is being reduced to B, via

$$[A \mid I] \xrightarrow[\text{reduce}]{\text{row}} [B \mid P]. \tag{1.15}$$

If $B = I$, then $PA = I$ and $P = A^{-1}$. □

■ EXAMPLE 3 Matrix Multiplication and Row Reduction

Consider the following row reduction:

$$\left[\begin{array}{ccc|ccc} 1 & 2 & 3 & 1 & 0 & 0 \\ 3 & 1 & -2 & 0 & 1 & 0 \\ 4 & 5 & 6 & 0 & 0 & 1 \end{array}\right] \xrightarrow[\substack{R_2 \leftarrow -3R_1 + R_2 \\ R_3 \leftarrow -4R_1 + R_3}]{} \left[\begin{array}{ccc|ccc} 1 & 2 & 3 & 1 & 0 & 0 \\ 0 & -5 & -11 & -3 & 1 & 0 \\ 0 & -3 & -6 & -4 & 0 & 1 \end{array}\right]$$

$$\xrightarrow[R_3 \leftarrow -\frac{3}{5}R_2 + R_3]{} \left[\begin{array}{ccc|ccc} 1 & 2 & 3 & 1 & 0 & 0 \\ 0 & -5 & -11 & -3 & 1 & 0 \\ 0 & 0 & \frac{3}{5} & -\frac{11}{5} & -\frac{3}{5} & 1 \end{array}\right].$$

The information about the operations used is recorded in the last three columns and, from Theorem 1.8, it follows that

$$\begin{bmatrix} 1 & 0 & 0 \\ -3 & 1 & 0 \\ -\frac{11}{5} & -\frac{3}{5} & 1 \end{bmatrix}\begin{bmatrix} 1 & 2 & 3 \\ 3 & 1 & -2 \\ 4 & 5 & 6 \end{bmatrix} = \begin{bmatrix} 1 & 2 & 3 \\ 0 & -5 & -11 \\ 0 & 0 & \frac{3}{5} \end{bmatrix},$$

which you should verify by direct calculation. ■

The final theorem of this section provides several conditions equivalent to the nonsingularity of an $n \times n$ matrix. Several other conditions will be discovered later and a more general theorem will then be stated (see Theorem 3.12 on page 181).

THEOREM 1.9 For an $n \times n$ matrix A the following statements are equivalent:

(1) A is nonsingular.
(2) A has a left inverse (there exists a matrix B such that $BA = I$).
(3) $AX = 0$ only if $X = 0$.
(4) A is row equivalent to I.
(5) A is a product of elementary matrices.

Proof We will prove the equivalence of the five conditions by establishing the chain of implications $1 \to 2 \to 3 \to 4 \to 5 \to 1$.

$(1 \to 2)$ Obvious.
$(2 \to 3)$ If there exists B such that $BA = I$ and $AX = 0$, then

$$X = B(AX) = B0 = 0.$$

$(3 \to 4)$ If A is not row equivalent to I, then the solution process of Section 1.3 will result in a zero row and consequently will leave one of the unknowns arbitrary; hence nonzero solutions will exist. It follows that A must be row equivalent to I.
$(4 \to 5)$ Since A is row equivalent to I, there exist elementary matrices $E_1 E_2 \cdots E_k$ such that $E_k \cdots E_2 E_1 A = I$. Then

$$A = (E_k \cdots E_2 E_1)^{-1} I = E_1^{-1} E_2^{-1} \cdots E_k^{-1}$$

and, since each elementary matrix has an elementary inverse, we have succeeded in expressing A as a product of elementary matrices.
$(5 \to 1)$ This follows from the fact that each elementary matrix is invertible and the product of invertible matrices is invertible. □

EXAMPLE 4 **Factor a Matrix into Elementary Factors**

Express $A = \begin{bmatrix} 1 & 3 \\ 2 & 8 \end{bmatrix}$ as a product of elementary matrices.

Consider the following reduction of A to I:

$$\begin{bmatrix} 1 & 3 \\ 2 & 8 \end{bmatrix} \to \begin{bmatrix} 1 & 3 \\ 0 & 2 \end{bmatrix} \to \begin{bmatrix} 1 & 3 \\ 0 & 1 \end{bmatrix} \to \begin{bmatrix} 1 & 0 \\ 0 & 1 \end{bmatrix}.$$

The elementary matrices

$$E_1 = \begin{bmatrix} 1 & 0 \\ -2 & 1 \end{bmatrix}, E_2 = \begin{bmatrix} 1 & 0 \\ 0 & \frac{1}{2} \end{bmatrix}, \text{ and } E_3 = \begin{bmatrix} 1 & -3 \\ 0 & 1 \end{bmatrix}$$

correspond respectively to the three elementary row operations $(R_2 \leftarrow -2R_1 + R_2)$, $(R_2 \leftarrow \frac{1}{2}R_2)$, and $(R_1 \leftarrow -3R_2 + R_1)$ used in the above reduction. It follows, as in the proof of Theorem 1.9, that

$$E_3E_2E_1A = \begin{bmatrix} 1 & -3 \\ 0 & 1 \end{bmatrix}\begin{bmatrix} 1 & 0 \\ 0 & \frac{1}{2} \end{bmatrix}\begin{bmatrix} 1 & 0 \\ -2 & 1 \end{bmatrix}\begin{bmatrix} 1 & 3 \\ 2 & 8 \end{bmatrix} = \begin{bmatrix} 1 & 0 \\ 0 & 1 \end{bmatrix}$$

and then that

$$\begin{aligned} \begin{bmatrix} 1 & 3 \\ 2 & 8 \end{bmatrix} &= \left(\begin{bmatrix} 1 & -3 \\ 0 & 1 \end{bmatrix}\begin{bmatrix} 1 & 0 \\ 0 & \frac{1}{2} \end{bmatrix}\begin{bmatrix} 1 & 0 \\ -2 & 1 \end{bmatrix}\right)^{-1} I \\ &= \begin{bmatrix} 1 & 0 \\ -2 & 1 \end{bmatrix}^{-1}\begin{bmatrix} 1 & 0 \\ 0 & \frac{1}{2} \end{bmatrix}^{-1}\begin{bmatrix} 1 & -3 \\ 0 & 1 \end{bmatrix}^{-1} \\ &= \begin{bmatrix} 1 & 0 \\ 2 & 1 \end{bmatrix}\begin{bmatrix} 1 & 0 \\ 0 & 2 \end{bmatrix}\begin{bmatrix} 1 & 3 \\ 0 & 1 \end{bmatrix} = A. \end{aligned}$$

This is the desired factorization of A as a product of elementary matrices; in general such factorizations will not be unique. ■

In the final example of this section we will show how matrices and their inverses can be used in constructing codes which are very difficult to break.

EXAMPLE 5 **Matrix Codes**

Consider a message coding process which associates with the letters A through Z the numbers 1 through 26 and associates with a blank space the number 0. After the initial coding the numbers are split into groups of three and then scrambled using multiplication by the matrix

$$A = \begin{bmatrix} -3 & 5 & 6 \\ -1 & 2 & 2 \\ 1 & -1 & -1 \end{bmatrix}, \text{ for which } A^{-1} = \begin{bmatrix} 0 & 1 & 2 \\ -1 & 3 & 0 \\ 1 & -2 & 1 \end{bmatrix}.$$

The message "SUPPLIES LOW" would initially be coded as the string of integers

19, 21, 16, 16, 12, 9, 5, 19, 0, 12, 15, 23,

arranged as a set of 3×1 column matrices, and then scrambled. These steps lead to

$$\begin{bmatrix} -3 & 5 & 6 \\ -1 & 2 & 2 \\ 1 & -1 & -1 \end{bmatrix}\begin{bmatrix} 19 \\ 21 \\ 16 \end{bmatrix} = \begin{bmatrix} 144 \\ 55 \\ -18 \end{bmatrix}, A\begin{bmatrix} 16 \\ 12 \\ 9 \end{bmatrix} = \begin{bmatrix} 66 \\ 26 \\ -5 \end{bmatrix},$$

$$A\begin{bmatrix} 5 \\ 19 \\ 0 \end{bmatrix} = \begin{bmatrix} 80 \\ 33 \\ -14 \end{bmatrix}, \text{ and } A\begin{bmatrix} 12 \\ 15 \\ 23 \end{bmatrix} = \begin{bmatrix} 177 \\ 64 \\ -26 \end{bmatrix}.$$

The message would then be transmitted as the string of integers

$$144, 55, -18, 66, 26, -5, 80, 33, -14, 177, 64, -26.$$

Note that this code would be very difficult to break because a letter is not always coded the same way.

The recipient of the message would unscramble the message using A^{-1} and then assign letters to the integers as above. In order to translate the coded message

$$217, 79, -39, 78, 27, -13, 125, 42, -21, 26, 10, -5, 1, 5, 9$$

we arrange the given message into columns and unscramble using A^{-1}. Thus the unscrambled message would be

$$\begin{bmatrix} 0 & 1 & 2 \\ -1 & 3 & 0 \\ 1 & -2 & 1 \end{bmatrix}\begin{bmatrix} 217 \\ 79 \\ -39 \end{bmatrix} = \begin{bmatrix} 1 \\ 20 \\ 20 \end{bmatrix},$$

$$A^{-1}\begin{bmatrix} 78 \\ 27 \\ -13 \end{bmatrix} = \begin{bmatrix} 1 \\ 3 \\ 11 \end{bmatrix}, A^{-1}\begin{bmatrix} 125 \\ 42 \\ -21 \end{bmatrix} = \begin{bmatrix} 0 \\ 1 \\ 20 \end{bmatrix},$$

$$A^{-1}\begin{bmatrix} 26 \\ 10 \\ -5 \end{bmatrix} = \begin{bmatrix} 0 \\ 4 \\ 1 \end{bmatrix}, \text{ and } A^{-1}\begin{bmatrix} 1 \\ 5 \\ 9 \end{bmatrix} = \begin{bmatrix} 23 \\ 14 \\ 0 \end{bmatrix}.$$

The string

$$1, 20, 20, 1, 3, 11, 0, 1, 20, 0, 4, 1, 23, 14, 0$$

then yields the message

A T T A C K A T D A W N.

The complexity of such a code would depend on the size of the matrix used. There would certainly be advantages to having a coding matrix with integer entries whose inverse also had integer entries. It would be a simple matter to design a computer code to unscramble an incoming message. It would also be easy to change the code by changing the coding matrix. ■

EXERCISES 1.7

Solve the following matrix equations for X.

1. $\begin{bmatrix} 1 & 2 \\ 3 & 4 \end{bmatrix} X = \begin{bmatrix} 5 & 4 & 3 \\ 1 & 2 & 0 \end{bmatrix}$

2. $\begin{bmatrix} 1 & 2 & 1 \\ -1 & -1 & 1 \\ 0 & 1 & 3 \end{bmatrix} X = \begin{bmatrix} 3 & 0 & 2 \\ 1 & 0 & 3 \\ -2 & 0 & 7 \end{bmatrix}$

3. $\begin{bmatrix} 5 & 5 & -3 \\ 2 & 2 & -1 \\ -1 & -1 & 1 \end{bmatrix} X = \begin{bmatrix} 10 & 2 & 7 \\ 4 & 1 & 3 \\ -2 & 0 & -1 \end{bmatrix}$

4. $\begin{bmatrix} 1 & 2 \\ 3 & 6 \end{bmatrix} X = \begin{bmatrix} 3 & 0 & 1 \\ 9 & 0 & 1 \end{bmatrix}$

Find the inverses of the following matrices, if possible.

5. $A = \begin{bmatrix} 1 & 1 & 1 \\ 0 & 2 & 1 \\ 1 & 0 & 1 \end{bmatrix}$

6. $A = \begin{bmatrix} 3 & 2 & 1 \\ 2 & -1 & -1 \\ 1 & 4 & 0 \end{bmatrix}$

7. $B = \begin{bmatrix} -3 & 5 & 6 \\ -1 & 2 & 2 \\ 1 & -1 & -1 \end{bmatrix}$

8. $J = \begin{bmatrix} 3 & 0 & 0 \\ 1 & 3 & 0 \\ 0 & 1 & 3 \end{bmatrix}$

9. $T = \begin{bmatrix} 1 & 2 & 6 \\ 0 & -1 & 4 \\ 0 & 0 & 3 \end{bmatrix}$

10. $A = \begin{bmatrix} 1 & 3 & 3 \\ 1 & 3 & 4 \\ 1 & 4 & 3 \end{bmatrix}$

11. $B = \begin{bmatrix} 2 & 1 & -1 & 2 \\ 1 & 3 & 2 & -3 \\ -1 & 2 & 1 & -1 \\ 2 & -3 & -1 & 4 \end{bmatrix}$

12. $C = \begin{bmatrix} 2 & 4 & 3 & 2 \\ 3 & 6 & 5 & 2 \\ 2 & 5 & 2 & -3 \\ 4 & 5 & 14 & 14 \end{bmatrix}$

13. $R = \begin{bmatrix} \cos\theta & -\sin\theta \\ \sin\theta & \cos\theta \end{bmatrix}$

14. $S = \frac{1}{6}\begin{bmatrix} 3 & 3 & 3 & 3 \\ 3 & -5 & 1 & 1 \\ 3 & 1 & -5 & 1 \\ 3 & 1 & 1 & -5 \end{bmatrix}$

15. $L = \begin{bmatrix} 1 & 0 & 0 & 0 \\ 2 & 1 & 0 & 0 \\ -3 & 5 & 1 & 0 \\ 4 & 2 & 3 & 1 \end{bmatrix}$

16. Express each matrix in Exercises 6–11 as a product of elementary matrices.

17. Find a nonsingular matrix P such that $PA = B$, where

$$A = \begin{bmatrix} 2 & 3 & 4 \\ 4 & 3 & 1 \\ 1 & 2 & 4 \end{bmatrix} \quad \text{and} \quad B = \begin{bmatrix} 1 & 2 & -1 \\ -1 & 1 & 2 \\ 2 & -1 & 1 \end{bmatrix}.$$

18. Let $A = \begin{bmatrix} 4 & 1 & 2 & 6 \\ 1 & 2 & 4 & 1 \\ 0 & -1 & -2 & 1 \\ 0 & 6 & 12 & 0 \end{bmatrix}$. Find a nonsingular matrix P such that PA is in row-echelon form. Is P unique?

19. Let P be an approximation to A^{-1} (PA differs from I in only a few places). Show how to find A^{-1} from P and PA. (*Hint:* Row reduce $[PA \mid P]$.)

†20. Show that the inverse of a lower triangular matrix is also a lower triangular matrix.

21. Use the matrix code of Example 5 to decode the following messages.

(a) 22, 9, 1, 24, 8, 4, 135, 47, −21, 113, 38, −19, 75, 25, −6, 36, 18, 1, 56, 21, −6, −60, −20, 20

(b) 169, 63, −22, 113, 46, −21, 89, 30, −11, 121, 42, −19

(c) −37, −12, 16, 57, 19, −4, 78, 26, −6, 1, 2, 3, 7, 4, 2, 121, 42, −19

(d) 48, 22, −3, −39, −13, 19, 45, 21, −3, 151, 57, −19, 97, 37, −16, −20, −4, 12, 153, 56, −21, 181, 65, −32, 64, 23, 0, −57, −19, 19

22. Define **elementary column operations** by replacing "row" by "column" everywhere it occurs in Definition 1.2. Let $C(A)$ be the result of performing the column operation C on the matrix A.

(a) Show that $C(A) = A[C(I)]$.

(b) Show that every elementary column matrix $C(I)$ is an elementary (row) matrix.

(c) Show that if $\begin{bmatrix} A \\ I \end{bmatrix} \xrightarrow[\text{reduce}]{\text{column}} \begin{bmatrix} B \\ Q \end{bmatrix}$, then $AQ = B$.

(d) Use the result of part (c) to compute the inverse of

$$A = \begin{bmatrix} 1 & 4 & 0 \\ 2 & -1 & -1 \\ 3 & 2 & 1 \end{bmatrix}.$$

23. Use matrix A of Exercise 10 to code the following messages.

(a) RETREAT SLOWLY **(b)** PARTY AT SUNSET

24. Write a computer program which will unscramble messages coded with a given $n \times n$ matrix A. Test your program on the messages in Exercise 23.

25. Find a matrix X that satisfies the following equations.

(a) $\begin{bmatrix} 1 & -2 \\ -1 & 3 \end{bmatrix} X - X \begin{bmatrix} 1 & -2 \\ -1 & 3 \end{bmatrix} = \begin{bmatrix} 8 & -14 \\ -1 & -8 \end{bmatrix}$

(b) $\begin{bmatrix} 3 & -1 \\ -1 & 2 \end{bmatrix} X - X \begin{bmatrix} 1 & 4 \\ 2 & 0 \end{bmatrix} = \begin{bmatrix} 4 & -4 \\ 10 & 8 \end{bmatrix}$

26. Let U be the $n \times n$ matrix each of whose entries is 1. Verify that $U^2 = nU$ and that $(I - U)^{-1} = I - \frac{1}{n-1}U$.

27. Let A be a nonsingular matrix for which $A^{-1} = \begin{bmatrix} -2 & 7 & -1 \\ -4 & 5 & -1 \\ 4 & -3 & -1 \end{bmatrix}$. Find A.

28. Find elementary matrices E_1, E_2, and E_3 so that $E_3E_2E_1A$ is in row-echelon form where $A = \begin{bmatrix} 1 & 3 & 3 & 8 \\ 1 & 4 & 10 & 16 \\ -2 & -5 & 1 & -8 \end{bmatrix}$.

1.8 THE LU FACTORIZATION*

Several of our earlier examples and exercises have been designed to suggest that it is frequently, but not always, possible to factor a nonsingular square matrix into the product of a lower triangular matrix and an upper triangular matrix (see Exercises 4 and 12 of Section 1.5 and Example 3 of Section 1.7). Before we investigate such factorizations let us see what use we might make of them. Thus consider the linear system $AX = K$ and let us assume that we have a factorization $A = LU$ where L is a lower triangular matrix and U is an upper triangular matrix. If we write

$$AX = L(UX) = K$$

and let $Y = UX$, then we must have $LY = K$. Thus the solution process has been split into three parts:

> **1.** Factor $A = LU$;
> **2.** Solve the lower triangular system $LY = K$ for Y; and
> **3.** Solve the upper triangular system $UX = Y$ for X.

Step (2) is called the **forward substitution** and step (3) is called the **back substitution**. We have already observed that the solution of a triangular system is relatively easy, requiring only about $\frac{n^2}{2}$ multiplications (see Exercise 10 of Section 1.3). It turns out that the method suggested above is exactly equivalent to Gaussian elimination for a single linear system since the forward substitution is equivalent to the reduction of the last column of the augmented matrix. The advertised gain in efficiency comes when we have a set of systems to solve, perhaps with not all the right-hand sides available at the same time. Once one has the factorization, then the solution of a new system with the same coefficient matrix requires only the solution of two triangular systems and this requires only about n^2 multiplications compared with $\frac{n^3}{3}$ multiplications required to solve the system from scratch by Gaussian elimination.

*This section contains optional material which may be omitted without loss of continuity.

EXAMPLE 1 **Using the LU Factorization**

Given the factorization

$$A = \begin{bmatrix} 2 & 6 & 2 \\ 6 & 14 & 8 \\ 0 & -2 & -1 \end{bmatrix} = LU = \begin{bmatrix} 1 & 0 & 0 \\ 3 & 1 & 0 \\ 0 & \frac{1}{2} & 1 \end{bmatrix}\begin{bmatrix} 2 & 6 & 2 \\ 0 & -4 & 2 \\ 0 & 0 & -2 \end{bmatrix} \text{ and } K = \begin{bmatrix} 1 \\ 1 \\ 1 \end{bmatrix}.$$

Solve the linear systems $AX_1 = K$, and $AX_2 = X_1$.

The system $LY = K$ of step 2, for the first system, is

$$\begin{aligned} y_1 \qquad\qquad &= 1 \\ 3y_1 + y_2 \qquad &= 1, \\ \frac{1}{2}y_2 + y_3 &= 1 \end{aligned} \quad \text{for which} \quad \begin{aligned} y_1 &= 1 \\ y_2 &= -2. \\ y_3 &= 2 \end{aligned}$$

The system $UX = Y$ of step 3, for the first system, is

$$\begin{aligned} 2x_1 + 6x_2 + 2x_3 &= 1 \\ -4x_2 + 2x_3 &= -2, \\ -2x_3 &= 2 \end{aligned} \quad \text{for which} \quad \begin{aligned} x_1 &= \frac{3}{2} \\ x_2 &= 0; \\ x_3 &= -1 \end{aligned} \quad \text{therefore } X_1 = \begin{bmatrix} \frac{3}{2} \\ 0 \\ -1 \end{bmatrix}.$$

This is the solution of the first system and the input for the second system. We now repeat steps 2 and 3 for the second system. The forward substitution, the solution of the lower triangular system

$$\begin{bmatrix} 1 & 0 & 0 \\ 3 & 1 & 0 \\ 0 & \frac{1}{2} & 1 \end{bmatrix}\begin{bmatrix} y_1 \\ y_2 \\ y_3 \end{bmatrix} = \begin{bmatrix} \frac{3}{2} \\ 0 \\ -1 \end{bmatrix}, \text{ yields } Y = \begin{bmatrix} \frac{3}{2} \\ -\frac{9}{2} \\ \frac{5}{4} \end{bmatrix}$$

and the back substitution, the solution of the upper triangular system

$$\begin{bmatrix} 2 & 6 & 2 \\ 0 & -4 & 2 \\ 0 & 0 & -2 \end{bmatrix}\begin{bmatrix} x_1 \\ x_2 \\ x_3 \end{bmatrix} = \begin{bmatrix} \frac{3}{2} \\ -\frac{9}{2} \\ \frac{5}{4} \end{bmatrix}, \text{ yields } X_2 = \begin{bmatrix} -\frac{17}{16} \\ \frac{13}{16} \\ -\frac{5}{8} \end{bmatrix}.$$

This is the solution of the second system. ■

Example 1 shows how a factorization of the coefficient matrix into the product of a lower triangular matrix and an upper triangular matrix would be useful in dealing with the case of multiple right-hand sides. We will now discuss how to find this factorization and the conditions under which it exists. We will first show that the desired factorization exists if, and only if, no row interchanges are required in the Gaussian reduction of the coefficient matrix. If no zero diagonal elements occur at inopportune times, then only Type III elementary row operations are required and all of them correspond to elementary matrices which are lower triangular. Thus we have, from Theorem 1.7, the factorization

$$L_k \cdots L_2 L_1 A = U,$$

where U is an upper triangular matrix and the L_i are lower triangular elementary matrices corresponding to the elementary row operations used in the reduction. Now the product of lower triangular matrices is still lower triangular (see Exercise 13 of Section 1.6) and the inverse of a lower triangular matrix is also lower triangular (see Exercise 20 of Section 1.7). Thus,

$$A = LU \quad \text{where } L = (L_k \cdots L_2L_1)^{-1} = L_1^{-1}L_2^{-1} \cdots L_k^{-1} \tag{1.16}$$

is a lower triangular matrix.

From Equation (1.16) we see that we could compute L, as we are reducing A to U, by successively computing the products

$$IL_1^{-1} = L_1^{-1},\ L_1^{-1}L_2^{-1},\ (L_1^{-1}L_2^{-1})L_3^{-1},\ \ldots,\ (L_1^{-1}L_2^{-1} \cdots L_{k-1}^{-1})L_k^{-1} = L.$$

The right multiplications by the elementary matrices L_i^{-1} are greatly facilitated by the result of Exercise 22a in Section 1.7. This exercise, which is the column analog of Theorem 1.7, asserts that elementary column operations can be accomplished by multiplying on the right by an identity matrix on which the elementary column operation has been performed. That is, if C is an elementary column operation, then $C(A) = AC(I)$. Note that the elementary matrix which corresponds to the row operation $(R_i \leftarrow kR_j + R_i)$ also corresponds to the elementary column operation $(C_j \leftarrow kC_i + C_j)$. Thus if we do the row operation $(R_i \leftarrow kR_j + R_i)$ on the rows of A, we need to do the column operation $(C_j \leftarrow -kC_i + C_j)$ on the array where L is being constructed.

EXAMPLE 2 Finding the LU Factorization

Find the LU factorization of $A = \begin{bmatrix} 2 & -4 & 4 \\ -2 & 2 & 0 \\ 4 & 1 & -8 \end{bmatrix}$.

The first row operation needed is $(R_2 \leftarrow 1R_1 + R_2)$ which corresponds to the elementary matrix

$$L_1 = \begin{bmatrix} 1 & 0 & 0 \\ 1 & 1 & 0 \\ 0 & 0 & 1 \end{bmatrix} \text{ whose inverse is } L_1^{-1} = \begin{bmatrix} 1 & 0 & 0 \\ -1 & 1 & 0 \\ 0 & 0 & 1 \end{bmatrix},$$

which corresponds to the column operation $(C_1 \leftarrow -1C_2 + C_1)$.

The second row operation is $(R_3 \leftarrow -2R_1 + R_3)$ and we have

$$L_2(L_1A) = \begin{bmatrix} 1 & 0 & 0 \\ 0 & 1 & 0 \\ -2 & 0 & 1 \end{bmatrix}(L_1A) = \begin{bmatrix} 2 & -4 & 4 \\ 0 & -2 & 4 \\ 0 & 9 & -16 \end{bmatrix} \quad \text{and}$$

$$L_1^{-1}L_2^{-1} = \begin{bmatrix} 1 & 0 & 0 \\ -1 & 1 & 0 \\ 0 & 0 & 1 \end{bmatrix}\begin{bmatrix} 1 & 0 & 0 \\ 0 & 1 & 0 \\ 2 & 0 & 1 \end{bmatrix} = \begin{bmatrix} 1 & 0 & 0 \\ -1 & 1 & 0 \\ 2 & 0 & 1 \end{bmatrix}.$$

Note that forming the product $L_1^{-1}L_2^{-1}$ amounts to performing the column operation $(C_1 \leftarrow 2C_3 + C_1)$ on the matrix L_1^{-1}.

The final row operation required is $(R_3 \leftarrow \frac{9}{2}R_2 + R_3)$ and we then have

$$L_3(L_2L_1A) = \begin{bmatrix} 1 & 0 & 0 \\ 0 & 1 & 0 \\ 0 & \frac{9}{2} & 1 \end{bmatrix}(L_2L_1A) = \begin{bmatrix} 2 & -4 & 4 \\ 0 & -2 & 4 \\ 0 & 0 & 2 \end{bmatrix} = U.$$

To complete the calculation of L we compute

$$(L_1^{-1}L_2^{-1})L_3^{-1} = (L_1^{-1}L_2^{-1})\begin{bmatrix} 1 & 0 & 0 \\ 0 & 1 & 0 \\ 0 & -\frac{9}{2} & 1 \end{bmatrix} = \begin{bmatrix} 1 & 0 & 0 \\ -1 & 1 & 0 \\ 2 & -\frac{9}{2} & 1 \end{bmatrix} = L.$$

Note that the final product in the calculation of L amounts to performing the column operation $(C_2 \leftarrow -\frac{9}{2}C_3 + C_2)$. We now have the factorization

$$A = \begin{bmatrix} 2 & -4 & 4 \\ -2 & 2 & 0 \\ 4 & 1 & -8 \end{bmatrix} = LU = \begin{bmatrix} 1 & 0 & 0 \\ -1 & 1 & 0 \\ 2 & -\frac{9}{2} & 1 \end{bmatrix}\begin{bmatrix} 2 & -4 & 4 \\ 0 & -2 & 4 \\ 0 & 0 & 2 \end{bmatrix}. \blacksquare$$

In order to see what the matrix L is like we consider the factorization of a general nonsingular 3×3 matrix A.

$$\begin{bmatrix} a_{11} & a_{12} & a_{13} \\ a_{21} & a_{22} & a_{23} \\ a_{31} & a_{32} & a_{33} \end{bmatrix} \to \begin{bmatrix} a_{11} & a_{12} & a_{13} \\ 0 & b_{22} & b_{23} \\ 0 & b_{32} & b_{33} \end{bmatrix} \to \begin{bmatrix} a_{11} & a_{12} & a_{13} \\ 0 & b_{22} & b_{23} \\ 0 & 0 & c_{33} \end{bmatrix}$$

In this reduction we have not changed the row leaders to 1, so only three elementary row operations are needed to reduce A to upper triangular form; they are of the form $(R_2 \leftarrow aR_1 + R_2)$, $(R_3 \leftarrow bR_1 + R_3)$, and $(R_3 \leftarrow cR_2 + R_3)$, where $a = -\frac{a_{21}}{a_{11}}$, $b = -\frac{a_{31}}{a_{11}}$, and $c = -\frac{b_{32}}{b_{22}}$. From Equation (1.16) we have

$$\begin{aligned} L &= \left(\begin{bmatrix} 1 & 0 & 0 \\ 0 & 1 & 0 \\ 0 & c & 1 \end{bmatrix}\begin{bmatrix} 1 & 0 & 0 \\ 0 & 1 & 0 \\ b & 0 & 1 \end{bmatrix}\begin{bmatrix} 1 & 0 & 0 \\ a & 1 & 0 \\ 0 & 0 & 1 \end{bmatrix}\right)^{-1} \\ &= \begin{bmatrix} 1 & 0 & 0 \\ a & 1 & 0 \\ 0 & 0 & 1 \end{bmatrix}^{-1}\begin{bmatrix} 1 & 0 & 0 \\ 0 & 1 & 0 \\ b & 0 & 1 \end{bmatrix}^{-1}\begin{bmatrix} 1 & 0 & 0 \\ 0 & 1 & 0 \\ 0 & c & 1 \end{bmatrix}^{-1} \\ &= \begin{bmatrix} 1 & 0 & 0 \\ -a & 1 & 0 \\ 0 & 0 & 1 \end{bmatrix}\begin{bmatrix} 1 & 0 & 0 \\ 0 & 1 & 0 \\ -b & 0 & 1 \end{bmatrix}\begin{bmatrix} 1 & 0 & 0 \\ 0 & 1 & 0 \\ 0 & -c & 1 \end{bmatrix} \\ &= \begin{bmatrix} 1 & 0 & 0 \\ -a & 1 & 0 \\ -b & -c & 1 \end{bmatrix}. \end{aligned}$$

Note that the first operation in the reduction was ($R_2 \leftarrow aR_1 + R_2$), which was designed to produce a zero in the (2, 1) position, and the scalar $-a$ appears in the (2, 1) position of L. Moreover, the second operation used was ($R_3 \leftarrow bR_1 + R_3$) and the scalar $-b$ appears in the (3, 1) position of L while the third operation used was ($R_3 \leftarrow cR_2 + R_3$) and the scalar $-c$ appears in the (3, 2) position of L.

The situation in general is just like in the above 3×3 example: L has ones on the diagonal and the elements below the diagonal are the negatives of the multipliers used in the reduction. In most computer implementations of the Gaussian elimination algorithm, the original matrix A is overwritten at each step by the reduced matrix (see Exercise 8 in Section 1.3). In the case of the above 3×3 nonsingular matrix the algorithm changes A to the upper triangular matrix of the form

$$U = \begin{bmatrix} a_{11} & a_{12} & a_{13} \\ 0 & b_{22} & b_{23} \\ 0 & 0 & c_{33} \end{bmatrix}.$$

An efficient way to find the LU factorization of A is to use the above observation about the elements of L to construct L during the reduction, that is, store the negatives of the multipliers in the zeroed positions. For the reduction described above we would have

$$A \rightarrow \begin{bmatrix} a_{11} & a_{12} & a_{13} \\ -a & b_{22} & b_{23} \\ -b & b_{32} & b_{33} \end{bmatrix} \rightarrow \begin{bmatrix} a_{11} & a_{12} & a_{13} \\ -a & b_{22} & b_{23} \\ -b & -c & c_{33} \end{bmatrix},$$

from which we can recover both L and U. Note that it is not necessary to store the diagonal elements of L since they are always 1; such a matrix will be called **unit lower triangular.** We illustrate this compact storage strategy in our next example. Note that the matrix to be factored is the same matrix that we factored in Example 2.

EXAMPLE 3 Compact Storage Strategy

Use the compact storage strategy to find the LU factorization of

$$A = \begin{bmatrix} 2 & -4 & 4 \\ -2 & 2 & 0 \\ 4 & 1 & -8 \end{bmatrix}.$$

According to the procedure suggested above we make the following reduction:

$$\begin{bmatrix} 2 & -4 & 4 \\ -2 & 2 & 0 \\ 4 & 1 & -8 \end{bmatrix} \xrightarrow[R_3 \leftarrow -2R_1 + R_3]{R_2 \leftarrow R_1 + R_2} \begin{bmatrix} 2 & -4 & 4 \\ -1 & -2 & 4 \\ 2 & 9 & -16 \end{bmatrix}$$

$$\xrightarrow{R_3 \leftarrow (\frac{9}{2})R_2 + R_3} \begin{bmatrix} 2 & -4 & 4 \\ -1 & -2 & 4 \\ 2 & -\frac{9}{2} & 2 \end{bmatrix}.$$

From this reduction we recover the factorization

$$A = \begin{bmatrix} 2 & -4 & 4 \\ -2 & 2 & 0 \\ 4 & 1 & -8 \end{bmatrix} = LU = \begin{bmatrix} 1 & 0 & 0 \\ -1 & 1 & 0 \\ 2 & -\frac{9}{2} & 1 \end{bmatrix} \begin{bmatrix} 2 & -4 & 4 \\ 0 & -2 & 4 \\ 0 & 0 & 2 \end{bmatrix}.$$

Note that at the second step of the above reduction we reduced only the 2×2 submatrix in the lower right-hand corner; we did not change the multipliers in the first column at the second step. Clearly this procedure is more efficient than the method used in Example 2. ■

The above procedure requires some modifications if an untimely zero appears on the diagonal, necessitating a row interchange. The key to understanding how to revise the above calculation, in order to handle the case where row interchanges are necessary to avoid zero diagonal elements, lies in the following simple calculation:

$$PLP^{-1} = \begin{bmatrix} 1 & 0 & 0 & 0 \\ 0 & 0 & 1 & 0 \\ 0 & 1 & 0 & 0 \\ 0 & 0 & 0 & 1 \end{bmatrix} \begin{bmatrix} 1 & 0 & 0 & 0 \\ a & 1 & 0 & 0 \\ b & 0 & 1 & 0 \\ c & 0 & 0 & 1 \end{bmatrix} \begin{bmatrix} 1 & 0 & 0 & 0 \\ 0 & 0 & 1 & 0 \\ 0 & 1 & 0 & 0 \\ 0 & 0 & 0 & 1 \end{bmatrix}$$

$$= \begin{bmatrix} 1 & 0 & 0 & 0 \\ b & 1 & 0 & 0 \\ a & 0 & 1 & 0 \\ c & 0 & 0 & 1 \end{bmatrix} = L'.$$

Here P is the elementary matrix associated with the row interchange $(R_2 \leftrightarrow R_3)$, and L is the product of the elementary matrices associated with the elementary row operations $(R_2 \leftarrow aR_1 + R_2)$, $(R_3 \leftarrow bR_1 + R_3)$, and $(R_3 \leftarrow cR_1 + R_4)$. Note that the only difference between L and L' is that the second and third entries of the first column have been interchanged.

Suppose that after the first column of a 4×4 matrix A has been reduced we encounter a zero in the (2, 2) position; for example,

$$L_1 A = U_1 = \begin{bmatrix} x & x & x & x \\ 0 & 0 & x & x \\ 0 & x & x & x \\ 0 & x & x & x \end{bmatrix}.$$

Interchanging rows 2 and 3 leads to

$$PL_1A = PU_1,$$

which can be rewritten as

$$PL_1A = PL_1(P^{-1}P)A = (PL_1P^{-1})(PA) = PU_1$$

or, using the calculation above,

$$L_1'(PA) = PU_1 = \begin{bmatrix} x & x & x & x \\ 0 & x & x & x \\ 0 & 0 & x & x \\ 0 & x & x & x \end{bmatrix}.$$

The second column can now be reduced to yield

$$L_2L_1'(PA) = \begin{bmatrix} x & x & x & x \\ 0 & x & x & x \\ 0 & 0 & x & x \\ 0 & 0 & x & x \end{bmatrix} = U_2.$$

If the (3, 3) entry is zero and the (3, 4) entry is not zero, then we would need to interchange rows 3 and 4 which would lead to a matrix of the form

$$P'L_2L_1'(PA) = P'U_2 = \begin{bmatrix} x & x & x & x \\ 0 & x & x & x \\ 0 & 0 & x & x \\ 0 & 0 & 0 & x \end{bmatrix} = U.$$

Proceeding as above we write

$$P'L_2L_1'(PA) = [P'(L_2L_1')P'^{-1}]P'(PA) = L''(P'PA)$$

and, since $P'^{-1} = P'$, because P' is an elementary matrix of Type II, L'' is still a lower triangular matrix. In fact L'' differs from L_2L' only because the lower triangular elements in the last two rows have been interchanged. For our example,

$$L_2L_1' = \begin{bmatrix} 1 & 0 & 0 & 0 \\ b & 1 & 0 & 0 \\ a & 0 & 1 & 0 \\ c & d & 0 & 1 \end{bmatrix} \quad \text{and} \quad L'' = \begin{bmatrix} 1 & 0 & 0 & 0 \\ b & 1 & 0 & 0 \\ c & d & 1 & 0 \\ a & 0 & 0 & 1 \end{bmatrix}.$$

We now define a **permutation matrix** to be any product of elementary matrices of Type II; such a matrix is simply the identity matrix I with its rows rearranged.

By generalizing the above calculations in a straightforward way we have the following theorem.

THEOREM 1.10 If A is an $n \times n$ nonsingular matrix, then there exists a permutation matrix P such that PA can be factored as $PA = LU$ where L is a unit lower triangular matrix and U is upper triangular. □

The modified form of the factorization works just as well for solving the linear system $AX = K$ as the original form, since the system $AX = K$ is equivalent to the system $PAX = PK$, which is just the original system with the equations (rows of the augmented matrix) reordered. The solution of the second system involves the following three steps:

1. Factor $PA = LU$;
2. Solve the lower triangular system $LY = PK$ for Y; and
3. Solve the upper triangular system $UX = Y$ for X.

Note that one needs to know the permutation matrix P in order to rearrange the right-hand side in the forward substitution step of the solution process.

The compact form of the reduction needs only to be slightly modified to accommodate row interchanges; one must interchange rows in the entire matrix while remembering to use Type III operations only on the unreduced part of the matrix.

EXAMPLE 4 *LU* Factorization with Row Swaps

Find the LU factorization for either A or PA for a suitable permutation matrix P.

$$A = \begin{bmatrix} 2 & 3 & -1 & 2 \\ -4 & -6 & 2 & 1 \\ 2 & 4 & 4 & -1 \\ 4 & 8 & 2 & 7 \end{bmatrix} \qquad P = \begin{bmatrix} 1 & 0 & 0 & 0 \\ 0 & 1 & 0 & 0 \\ 0 & 0 & 1 & 0 \\ 0 & 0 & 0 & 1 \end{bmatrix}$$

$$\downarrow \begin{cases} R_2 \leftarrow 2R_1 + R_2 \\ R_3 \leftarrow -1R_1 + R_3 \\ R_4 \leftarrow -2R_1 + R_4 \end{cases}$$

$$\begin{bmatrix} 2 & 3 & -1 & 2 \\ -2 & 0 & 0 & 5 \\ 1 & 1 & 5 & -3 \\ 2 & 2 & 4 & 3 \end{bmatrix}$$

$$\downarrow \{R_2 \leftrightarrow R_3$$

$$\begin{bmatrix} 2 & 3 & -1 & 2 \\ 1 & 1 & 5 & -3 \\ -2 & 0 & 0 & 5 \\ 2 & 2 & 4 & 3 \end{bmatrix} \qquad P = \begin{bmatrix} 1 & 0 & 0 & 0 \\ 0 & 0 & 1 & 0 \\ 0 & 1 & 0 & 0 \\ 0 & 0 & 0 & 1 \end{bmatrix}$$

$$\downarrow \{R_4 \leftarrow -2R_2 + R_4$$

$$\begin{bmatrix} 2 & 3 & -1 & 2 \\ 1 & 1 & 5 & -3 \\ -2 & 0 & 0 & 5 \\ 2 & 2 & -6 & 9 \end{bmatrix}$$

$$\downarrow \{R_3 \leftrightarrow R_4$$

$$\begin{bmatrix} 2 & 3 & -1 & 2 \\ 1 & 1 & 5 & -3 \\ 2 & 2 & -6 & 9 \\ -2 & 0 & 0 & 5 \end{bmatrix} \qquad P = \begin{bmatrix} 1 & 0 & 0 & 0 \\ 0 & 0 & 1 & 0 \\ 0 & 0 & 0 & 1 \\ 0 & 1 & 0 & 0 \end{bmatrix}$$

$$\text{Thus } PA = \begin{bmatrix} 2 & 3 & -1 & 2 \\ 2 & 4 & 4 & -1 \\ 4 & 8 & 2 & 7 \\ -4 & -6 & 2 & 1 \end{bmatrix} = LU$$

$$= \begin{bmatrix} 1 & 0 & 0 & 0 \\ 1 & 1 & 0 & 0 \\ 2 & 2 & 1 & 0 \\ -2 & 0 & 0 & 1 \end{bmatrix} \begin{bmatrix} 2 & 3 & -1 & 2 \\ 0 & 1 & 5 & -3 \\ 0 & 0 & -6 & 9 \\ 0 & 0 & 0 & 5 \end{bmatrix},$$

as the reader should verify by direct multiplication. ■

EXAMPLE 5 **Using the General Factorization**

Use the factorization in Example 4 to solve the system

$$AX = K = \begin{bmatrix} 4 \\ 6 \\ 9 \\ 0 \end{bmatrix}.$$

The system $LY = PK$ is

$$\begin{bmatrix} 1 & 0 & 0 & 0 \\ 1 & 1 & 0 & 0 \\ 2 & 2 & 1 & 0 \\ -2 & 0 & 0 & 1 \end{bmatrix} Y = \begin{bmatrix} 4 \\ 9 \\ 0 \\ 6 \end{bmatrix}, \quad \text{for which} \quad Y = \begin{bmatrix} 4 \\ 5 \\ -18 \\ 14 \end{bmatrix}$$

and the system $UX = Y$ is

$$\begin{bmatrix} 2 & 3 & -1 & 2 \\ 0 & 1 & 5 & -3 \\ 0 & 0 & -6 & 9 \\ 0 & 0 & 0 & 5 \end{bmatrix} X = \begin{bmatrix} 4 \\ 5 \\ -18 \\ 14 \end{bmatrix}, \quad \text{for which} \quad X = \begin{bmatrix} 36.7 \\ -22.6 \\ 7.2 \\ 2.8 \end{bmatrix}. \quad ■$$

Note that the product LU can be written as $L(DU')$ where D is a diagonal matrix and U' is a unit upper triangular matrix. If the parentheses are moved, we obtain the factorization

$$(LD)U' = L'U'$$

in which L' is lower triangular and U' is unit upper triangular. This factorization is called the **Crout decomposition** or simply the LU factorization. This factorization will work just as well as the factorization described in the text for solving a set of systems of linear equations with the same coefficient matrix.

The TI-85 calculator has a built in LU factorization command "LU(A,L,U,P)." This command returns the Crout factors rather than the ones we have described in this section. For the matrix of Example 5, the TI-85 yields the factorization

$$\begin{bmatrix} 0 & 1 & 0 & 0 \\ 0 & 0 & 0 & 1 \\ 0 & 0 & 1 & 0 \\ 1 & 0 & 0 & 0 \end{bmatrix} A = \begin{bmatrix} -4 & 0 & 0 & 0 \\ 4 & 2 & 0 & 0 \\ 2 & 1 & 3 & 0 \\ 2 & 0 & 0 & 2.5 \end{bmatrix} \begin{bmatrix} 1 & 1.5 & -.5 & -.25 \\ 0 & 1 & 2 & 4 \\ 0 & 0 & 1 & -1.5 \\ 0 & 0 & 0 & 1 \end{bmatrix}$$

Both MATALG ("M, LU,A") and MATLAB ("[L,U,P] = lu(A)") compute the factorization we have described in the text, that is, L is unit lower triangular. All of these automatic decomposition routines will do more row interchanges than we have described above. The reason for these extra row interchanges will be explained in the next section.

EXERCISES 1.8

Find the LU factorization of the following matrices.

1. $A = \begin{bmatrix} 1 & 2 & 3 \\ 4 & 5 & 6 \\ 7 & 8 & 0 \end{bmatrix}$

2. $B = \begin{bmatrix} 1 & 1 & 1 \\ 3 & -2 & -17 \\ 3 & 2 & -1 \end{bmatrix}$

3. $C = \begin{bmatrix} 1 & 2 & 4 \\ 2 & 3 & 7 \\ 1 & 4 & 7 \end{bmatrix}$

4. $D = \begin{bmatrix} 1 & -1 & 2 & 3 \\ 2 & -1 & 0 & 2 \\ 4 & 1 & -11 & -1 \\ 1 & 2 & 33 & 83 \end{bmatrix}$

5. $E = \begin{bmatrix} 4 & 1 & 1 & 0 & 0 \\ 1 & 4 & 1 & 1 & 0 \\ 1 & 1 & 4 & 1 & 1 \\ 0 & 1 & 1 & 4 & 1 \\ 0 & 0 & 1 & 1 & 4 \end{bmatrix}$

6. Show that the inverse of a unit lower triangular matrix is also unit lower triangular.

7. Show that matrix $A = \begin{bmatrix} 2 & 4 & 3 & 2 \\ 3 & 6 & 5 & 2 \\ 2 & 5 & 2 & -3 \\ 4 & 5 & 14 & 14 \end{bmatrix}$ does not have an LU factorization, but that the matrix B, obtained from A by interchanging rows 2 and 3 does have an LU factorization.

8. Given $A = \begin{bmatrix} 1 & 1 & 3 \\ 1 & -2 & 1 \\ 3 & 1 & 3 \end{bmatrix}$ and $B = \begin{bmatrix} 1 \\ 2 \\ 3 \end{bmatrix}$, find the LU factorization of A and use it to find X_1 such that $AX_1 = B$, X_2 such that $AX_2 = X_1$, and X_3 such that $AX_3 = X_2$.

9. Solve the system $DX = [1 \quad 2 \quad 3 \quad 4]^T$ using the factorization from Exercise 4.

For the following matrices, find a permutation matrix P and triangular matrices L and U so that PA = LU.

10. $A = \begin{bmatrix} 1 & 2 & 3 \\ 4 & 5 & 6 \\ 7 & 8 & 0 \end{bmatrix}$

11. $A = \begin{bmatrix} 1 & 2 & 3 \\ 2 & 4 & 5 \\ 3 & 5 & 6 \end{bmatrix}$

12. $A = \begin{bmatrix} 1 & -1 & 2 & 3 \\ 2 & -2 & 0 & 2 \\ 4 & 1 & -11 & -1 \\ 1 & 2 & 33 & 83 \end{bmatrix}$

13. $A = \begin{bmatrix} 2 & 4 & 3 & 2 \\ 3 & 6 & 5 & 2 \\ 2 & 5 & 2 & -3 \\ 4 & 5 & 14 & 15 \end{bmatrix}$

14. Solve the system $AX = [1 \quad 1 \quad 1]^T$ using the factorization from Exercise 1.

15. Given that A has an LU decomposition, show that it has a decomposition of the form $A = LDR$ where L is unit lower triangular, R is unit upper triangular, and D is diagonal. Show that if A is nonsingular, then all three factors are unique.

1.9 PARTIAL PIVOTING*

In this section we will look a little more closely at the Gaussian elimination algorithm and will describe an important refinement which should be included in a good general purpose algorithm for solving the linear system $AX = B$ on a modern digital computer.

In order to understand the reasons for the refinements to be made we need to take a brief look at how a typical computer represents numbers and does arithmetic. Numbers in a computer are represented in normalized floating-point form. Thus a typical machine number is in the form

$$x = \pm 0.d_1 d_2 \cdots d_k \times 10^n$$

where the digits $d_1, d_2, \ldots, d_k$ are integers between 0 and 9, $d_1 \neq 0$, and n is an integer. The number k and the allowable values of n are characteristic of a particular machine and will vary from machine to machine. Many machines use number systems with bases other than 10; 2, 8, and 16 are common. We will

*This section contains optional material which may be omitted without loss of continuity.

restrict our attention to the familiar case of base 10. Some examples of numbers and their machine representations, for a machine with $k = 5$, follow.

Number	*Machine Representation*
312.52	$.31252 \times 10^3$
$-.002371$	$-.23710 \times 10^{-2}$
$\pi = 3.141592654\ldots$	$.31416 \times 10^1$
$\frac{40}{3} = 13.33333\ldots$	$.13333 \times 10^2$
$-.126786$	$-.12679 \times 10^0$

Note that some real numbers cannot be represented exactly in the machine; they can only be approximated correct to k digits. The difference between a number and its machine representation is an example of **round-off error;** it may occur whenever data is entered into the machine. Round-off error also occurs when two machine numbers are added or multiplied. Since the true sum or product will usually have more than k digits, some rounding is necessary to change the sum or product into a machine representable number. For example, on a machine with $k = 5$, the sum

$$(.31252 \times 10^3) + (.23710 \times 10^{-3}) = (.3125202371 \times 10^3)$$

would need to be rounded to $(.31252 \times 10^3)$ before it could be stored. Similarly, the product

$$(.31252 \times 10^3) \times (.23711 \times 10^{-3}) = (7.4101617 \times 10^{-2})$$

would be rounded to $.74102 \times 10^{-1}$. The difference

$$(.31252 \times 10^3) - (.31248 \times 10^3) = (.00004 \times 10^3)$$

would be stored as $(.40000 \times 10^{-1})$. Note that this difference has only one significant figure. This kind of loss of accuracy is a source of many inaccuracies in machine calculations; it is called **catastrophic cancellation** and has nothing to do with round-off error.

Because of round-off error the machine will have a difficult time distinguishing between two very small numbers. Thus it is best to avoid testing to see if two machine numbers are equal or testing to see if a machine number is zero. The first example of this section illustrates what can happen to the Gaussian elimination algorithm when it is implemented on a computer.

EXAMPLE 1 A Worst Case Scenario

Solve the system

$$\begin{aligned} 1.00 \times 10^{-4}x + 1.00y &= 1.00 \\ 1.00x + 1.00y &= 2.00. \end{aligned}$$

The true solution of this system, correct to the number of digits shown, is $x = 1.00010001$, $y = .99989999$. Suppose that we attempt to solve this system on a machine that carries only 3 significant figures; that is, the results of all calculations are rounded to 3 significant figures.

The operation $(R_2 \leftarrow -10^4 R_1 + R_2)$ reduces the augmented matrix

$$\begin{bmatrix} 1.00 \times 10^{-4} & 1.00 & 1.00 \\ 1.00 & 1.00 & 2.00 \end{bmatrix} \text{ to}$$

$$\begin{bmatrix} 1.00 \times 10^{-4} & 1.00 & 1.00 \\ 0 & -1.00 \times 10^4 & -1.00 \times 10^4 \end{bmatrix},$$

from which we obtain, by back substitution, $y = 1.00$ and $x = 0.00$. This is a very bad result by any reasonable standard.

If we interchange rows 1 and 2 and then use the operation $(R_2 \leftarrow -1.00 \times 10^{-4} R_1 + R_2)$, we obtain the reduction

$$\begin{bmatrix} 1.00 \times 10^{-4} & 1.00 & 1.00 \\ 1.00 & 1.00 & 2.00 \end{bmatrix} \rightarrow \begin{bmatrix} 1.00 & 1.00 & 2.00 \\ 1.00 \times 10^{-4} & 1.00 & 1.00 \end{bmatrix}$$

$$\rightarrow \begin{bmatrix} 1.00 & 1.00 & 2.00 \\ 0 & 1.00 & 1.00 \end{bmatrix}.$$

From this reduction, we obtain $x = 1.00$ and $y = 1.00$. This last result is an acceptable approximation to the true solution. ■

We have previously observed that a **zero pivot,** that is, a zero appearing on the diagonal during the Gaussian elimination algorithm, forces us to modify the algorithm to obtain a nonzero element in the pivot position. This modification takes the form of an elementary row operation of the form $(R_i \leftrightarrow R_j)$. Example 1 illustrates that small pivots are as troublesome as zero pivots since they lead to multiplications by very large numbers and loss of accuracy due to round-off error. For this reason it is important to incorporate into the Gaussian elimination algorithm for solving linear systems a process called **partial pivoting,** which protects against small elements appearing on the diagonal. In partial pivoting the element with the largest absolute value in the column to be reduced is moved to the diagonal position *before* the column is reduced. The new diagonal element is called the **pivot** for that stage of the reduction. It follows that all the row multipliers will be in absolute value less than or equal to 1. The net effect of this partial-pivoting procedure is that the numbers will stay small and the growth of round-off error will be controlled. This is clearly a machine procedure, not a hand-calculation procedure.

■ EXAMPLE 2 **Partial Pivoting**

Solve the following system by Gaussian elimination with partial pivoting.

$$\begin{aligned} 2x_1 + 3x_2 - x_3 + 2x_4 &= 3 \\ -4x_1 - 6x_2 + 2x_3 + x_4 &= 4 \\ 2x_1 + 4x_2 + 4x_3 - x_4 &= 0 \\ 4x_1 + 8x_2 + 2x_3 + 7x_4 &= 12 \end{aligned}$$

The augmented matrix for this system is

$$\begin{bmatrix} 2 & 3 & -1 & 2 & 3 \\ -4 & -6 & 2 & 1 & 4 \\ 2 & 4 & 4 & -1 & 0 \\ 4 & 8 & 2 & 7 & 12 \end{bmatrix}.$$

The element with the largest absolute value in the first column is the -4 in the second row. We select this element as the first pivot, move it to the $(1, 1)$ position using the elementary row operation $(R_1 \leftrightarrow R_2)$, and then use it to reduce the first column. This series of steps yields

$$\xrightarrow[\substack{R_1 \leftrightarrow R_2 \\ R_2 \leftarrow .5R_1 + R_2 \\ R_3 \leftarrow .5R_1 + R_3 \\ R_4 \leftarrow R_1 + R_4}]{} \begin{bmatrix} -4 & -6 & 2 & 1 & 4 \\ 0 & 0 & 0 & 2.5 & 5 \\ 0 & 1 & 5 & -.5 & 2 \\ 0 & 2 & 4 & 8 & 16 \end{bmatrix}.$$

The largest element in the second column, excluding the first row, is the 2 in the (4, 2) position. We select this element as the next pivot and move it to the diagonal position with the operation $(R_2 \leftrightarrow R_4)$. One Type III operation then completes the reduction:

$$\begin{bmatrix} -4 & -6 & 2 & 1 & 4 \\ 0 & 2 & 4 & 8 & 16 \\ 0 & 1 & 5 & -.5 & 2 \\ 0 & 0 & 0 & 2.5 & 5 \end{bmatrix} \xrightarrow[R_3 \leftarrow -.5R_2 + R_3]{} \begin{bmatrix} -4 & -6 & 2 & 1 & 4 \\ 0 & 2 & 4 & 8 & 16 \\ 0 & 0 & 3 & -4.5 & -6 \\ 0 & 0 & 0 & 2.5 & 5 \end{bmatrix}.$$

The routine back substitution now yields

$$x_4 = 2, \quad x_3 = 1, \quad x_2 = -2, \quad \text{and} \quad x_1 = 3. \quad ■$$

For a more complete discussion of partial pivoting and other methods for the control of round-off error you are urged to consult a book on numerical analysis.*

In order to control round-off error in machine calculations, the partial-pivoting procedure, described at the beginning of this section, should be included in any general purpose algorithm for computing the LU factorization or the inverse of a matrix. The row interchanges are handled as we have described in Section 1.8. We illustrate the calculations in our next example.

---

* Golub, G. H. and VanLoan, C. F. *Matrix Calculations* 2d ed. (Baltimore: The Johns Hopkins University Press, 1989).

EXAMPLE 3 **LU Factorization with Partial Pivoting**

Use partial pivoting to find a permutation matrix P so that $PA = LU$ for the matrix A of Example 2.

$$A = \begin{bmatrix} 2 & 3 & -1 & 2 \\ -4 & -6 & 2 & 1 \\ 2 & 4 & 4 & -1 \\ 4 & 8 & 2 & 7 \end{bmatrix} \xrightarrow[R_1 \leftrightarrow R_2]{} \begin{bmatrix} -4 & -6 & 2 & 1 \\ 2 & 3 & -1 & 2 \\ 2 & 4 & 4 & -1 \\ 4 & 8 & 2 & 7 \end{bmatrix}$$

$$\xrightarrow[\substack{R_2 \leftarrow \frac{1}{2}R_1 + R_2 \\ R_3 \leftarrow \frac{1}{2}R_1 + R_3 \\ R_4 \leftarrow R_1 + R_4}]{} \begin{bmatrix} -4 & -6 & 2 & 1 \\ -.5 & 0 & 0 & 2.5 \\ -.5 & 1 & 5 & -.5 \\ -1 & 2 & 4 & 8 \end{bmatrix} \xrightarrow[R_2 \leftrightarrow R_4]{} \begin{bmatrix} -4 & -6 & 2 & 1 \\ -1 & 2 & 4 & 8 \\ -.5 & 1 & 5 & -.5 \\ -5 & 0 & 0 & 2.5 \end{bmatrix}$$

$$\xrightarrow[R_3 \leftarrow -\frac{1}{2}R_2 + R_3]{} \begin{bmatrix} -4 & -6 & 2 & 1 \\ -1 & 2 & 4 & 8 \\ -.5 & .5 & 3 & -4.5 \\ -.5 & 0 & 0 & 2.5 \end{bmatrix}$$

The permutation matrix P is obtained by performing the two row interchanges used above on the identity matrix I.

$$\begin{bmatrix} 1 & 0 & 0 & 0 \\ 0 & 1 & 0 & 0 \\ 0 & 0 & 1 & 0 \\ 0 & 0 & 0 & 1 \end{bmatrix} \xrightarrow[R_1 \leftrightarrow R_2]{} \begin{bmatrix} 0 & 1 & 0 & 0 \\ 1 & 0 & 0 & 0 \\ 0 & 0 & 1 & 0 \\ 0 & 0 & 0 & 1 \end{bmatrix} \xrightarrow[R_2 \leftrightarrow R_4]{} \begin{bmatrix} 0 & 1 & 0 & 0 \\ 0 & 0 & 0 & 1 \\ 0 & 0 & 1 & 0 \\ 1 & 0 & 0 & 0 \end{bmatrix} = P.$$

Thus

$$PA = \begin{bmatrix} 0 & 1 & 0 & 0 \\ 0 & 0 & 0 & 1 \\ 0 & 0 & 1 & 0 \\ 1 & 0 & 0 & 0 \end{bmatrix} \begin{bmatrix} 2 & 3 & -1 & 2 \\ -4 & -6 & 2 & 1 \\ 2 & 4 & 4 & -1 \\ 4 & 8 & 2 & 7 \end{bmatrix} = \begin{bmatrix} -4 & -6 & 2 & 1 \\ 4 & 8 & 2 & 7 \\ 2 & 4 & 4 & -1 \\ 2 & 3 & -1 & 2 \end{bmatrix}$$

$$= LU = \begin{bmatrix} 1 & 0 & 0 & 0 \\ -1 & 1 & 0 & 0 \\ -.5 & .5 & 1 & 0 \\ -.5 & 0 & 0 & 1 \end{bmatrix} \begin{bmatrix} -4 & -6 & 2 & 1 \\ 0 & 2 & 4 & 8 \\ 0 & 0 & 3 & -4.5 \\ 0 & 0 & 0 & 2.5 \end{bmatrix}$$

as you may verify by direct calculation. ■

We conclude this section by collecting, for easy reference, the operation counts for the matrix operations discussed in this chapter.

Operation	*Number of Multiplications*
AB where A is $m \times n$ and B is $n \times r$	mnr
Reduction of $m \times n$ matrix to row-echelon form $(m < n)$	$\frac{m-1}{6}(3mn - m^2 + 6n - 4m)$
Reduction of $m \times n$ matrix to reduced row-echelon form	$\frac{m-1}{6}(6mn - 4m^2 + 6n - 4m)$
LU factorization of $n \times n$ nonsingular matrix	$\frac{n^3}{3} - \frac{n}{3}$
Solution of a system with unit lower triangular coefficient matrix	$\frac{n^2}{2} - \frac{n}{2}$
Solution of a system with upper triangular coefficient matrix	$\frac{n^2}{2} + \frac{n}{2}$
A^{-1} for $n \times n$ matrix A by Gauss-Jordan elimination	n^3

The software uses partial pivoting when it computes the LU factorization of a matrix A. The MATALG command "M,LU,A" will generate the matrices P, L, and U. You will be given the opportunity to name and save the factors. The MATLAB command "[L,U,P] = lu(A)" will generate the factors and save them. The LU command on the TI-85 generates a factorization in which U is unit upper triangular and L is lower triangular. The automatic row reduction in MATMAN (the "z" command) uses partial pivoting.

EXERCISES 1.9

1. Use Gaussian elimination with partial pivoting to solve for X.

(a) $\begin{bmatrix} 2 & 3 & -6 \\ 3 & -2 & 1 \\ 1 & -6 & 8 \end{bmatrix} X = \begin{bmatrix} -1 \\ 2 \\ 2 \end{bmatrix}$ **(b)** $\begin{bmatrix} 1 & 5 & 0 \\ 5 & 1 & 5 \\ 0 & 5 & 1 \end{bmatrix} X = \begin{bmatrix} 3 \\ 1 \\ 7 \end{bmatrix}$

(c) $\begin{bmatrix} 1 & 3 & 2 & 1 \\ 2 & 1 & 2 & 3 \\ -4 & -2 & 1 & 2 \\ 1 & 2 & 4 & 0 \end{bmatrix} X = \begin{bmatrix} 5 \\ 7 \\ 0 \\ 11 \end{bmatrix}$ **(d)** $\begin{bmatrix} 2 & 3 & -1 & 2 \\ -4 & -6 & 7 & -3 \\ 6 & 9 & 22 & 14 \\ 2 & 3 & 9 & -2 \end{bmatrix} X = \begin{bmatrix} 5 \\ 9 \\ 0 \\ 1 \end{bmatrix}$

(e) $\begin{bmatrix} .4095 & .0124 & .3678 & .2945 \\ .3654 & .1902 & .3718 & .0634 \\ .1748 & .4000 & .2708 & .3925 \\ .2251 & .3842 & .4015 & .1129 \end{bmatrix} X = \begin{bmatrix} 3.142 \\ 5.312 \\ 1.000 \\ -4.312 \end{bmatrix}$

(f) $\begin{bmatrix} 1 & \frac{1}{2} & \frac{1}{3} & \frac{1}{4} \\ \frac{1}{2} & \frac{1}{3} & \frac{1}{4} & \frac{1}{5} \\ \frac{1}{3} & \frac{1}{4} & \frac{1}{5} & \frac{1}{6} \\ \frac{1}{4} & \frac{1}{5} & \frac{1}{6} & \frac{1}{7} \end{bmatrix} X = \begin{bmatrix} 2 \\ 5 \\ 6 \\ 1.2 \end{bmatrix}$

For the following matrices, use partial pivoting to find a permutation matrix P and triangular matrices L and U so that PA = LU.

2. $A = \begin{bmatrix} 1 & 2 & 3 \\ 4 & 5 & 6 \\ 7 & 8 & 0 \end{bmatrix}$ **3.** $A = \begin{bmatrix} 1 & 2 & 3 \\ 2 & 4 & 5 \\ 3 & 5 & 6 \end{bmatrix}$

4. $A = \begin{bmatrix} 1 & -1 & 2 & 3 \\ 2 & -1 & 0 & 2 \\ 4 & 1 & -11 & -1 \\ 1 & 2 & 33 & 83 \end{bmatrix}$ **5.** $A = \begin{bmatrix} 2 & 4 & 3 & 2 \\ 3 & 6 & 5 & 2 \\ 2 & 5 & 2 & -3 \\ 4 & 5 & 14 & 14 \end{bmatrix}$

6. Solve the system $AX = [1 \quad 1 \quad 1]^T$ using the factorization from Exercise 2.

7. Solve the system $AX = [1 \quad 2 \quad 3 \quad 4]^T$ using the factorization from Exercise 5.

8. Consider the system $\begin{bmatrix} -3 & 2 & -100{,}000 \\ 6 & 1 & 1 \\ 3 & -.5 & -9.2 \end{bmatrix}\begin{bmatrix} x_1 \\ x_2 \\ x_3 \end{bmatrix} = \begin{bmatrix} 100{,}000 \\ 14.0 \\ 13.7 \end{bmatrix}$. Assume that this system is solved on a digital computer which rounds each calculation to three significant digits.

(a) Find the solution if no pivoting is used.
(b) Find the solution using partial pivoting.
(c) Compare the solutions with the true solution $X = [2 \quad 3 \quad -1]^T$.
(d) Multiply the first equation by 10^{-5} and then repeat steps (a) and (b). Are the results any better?

9. Find the inverses of the following matrices using Gauss-Jordan elimination with partial pivoting. Round C^{-1} to three significant figures and compute CC^{-1} and $C^{-1}C$.

$$A = \begin{bmatrix} 1 & 0 & 2 \\ -3 & 4 & 6 \\ -1 & -2 & 3 \end{bmatrix} \qquad B = \begin{bmatrix} -3 & -2 & 0 & 2 \\ 2 & 1 & 0 & -1 \\ 1 & 0 & 1 & 2 \\ 2 & 1 & -3 & 1 \end{bmatrix}$$

$$C = \begin{bmatrix} 2.35 & 4.12 & 3.17 & 4.31 \\ -4.27 & 8.36 & 7.15 & 9.87 \\ 5.87 & 0.00 & -5.89 & 18.0 \\ 11.3 & 14.7 & 11.2 & 5.75 \end{bmatrix} \qquad D = \begin{bmatrix} 1 & \frac{1}{2} & \frac{1}{3} & \frac{1}{4} \\ \frac{1}{2} & \frac{1}{3} & \frac{1}{4} & \frac{1}{5} \\ \frac{1}{3} & \frac{1}{4} & \frac{1}{5} & \frac{1}{6} \\ \frac{1}{4} & \frac{1}{5} & \frac{1}{6} & \frac{1}{7} \end{bmatrix}$$

10. **(a)** Show that there is a one-to-one correspondence between $n \times n$ permutation matrices P and the row matrices $[1 \quad 2 \quad 3 \quad 4 \cdots n]P$.
(b) Show how the observation in part (a) could be used to save storage space in the LU factorization algorithm with partial pivoting.

11. **(a)** Modify the computer program from Exercise 8 of Section 1.3 so that it will compute the LU factorization of A assuming that no pivoting is required.
(b) Modify the program in part (a) to include partial pivoting.

CHAPTER 1 SUMMARY

Chapter 1 introduces the Gaussian and Gauss-Jordan elimination methods for solving systems of linear equations $AX = K$. The basic definitions of matrix algebra are given and the manipulative properties of this algebra are compared carefully with ordinary algebra and related to the elimination methods for solving linear systems.

KEY DEFINITIONS

Coefficient matrix and augmented matrix of a linear system. *9, 51*

Elementary row operations, row equivalence. *16, 31*

Row echelon matrix, row reduced echelon matrix. *24*

Matrix addition, scalar multiplication, matrix multiplication, linear combination of matrices. *47, 49, 51*

Associativity, distributivity, commutativity, identity elements, inverses. *58*

Nonsingular matrix, elementary matrices. *60, 70*

KEY FACTS

The number of solutions of $AX = K$ is always 0, 1, or ∞.

There is a one-to-one correspondence between linear systems $AX = K$ and their augmented matrices $[A \mid K]$.

Elementary row operations, on augmented matrices, do not change the solutions of the associated linear systems; $[A \mid K] \underset{R}{\sim} [B \mid H] \Rightarrow AX = K$ and $BX = H$ have the same solutions.

Matrix multiplication involves taking linear combinations of rows or of columns. $\text{Row}_i(AB) = \text{Row}_i(A)B$ and $\text{Col}_j(AB) = A\text{Col}_j(B)$.

Elementary row operations amount to multiplication of the left by elementary matrices: $[A \mid I] \underset{R}{\sim} [B \mid P] \Rightarrow PA = B$.

The number of arbitrary constants in the general solution of $AX = K$ is the number of unknowns less the number of nonzero rows in the row reduced matrix for A.

The matrix A is nonsingular if, and only if, $AX = 0 \Rightarrow X = 0$. See Theorem 1.9 for other equivalent conditions.

The two major differences between the algebra of matrices and ordinary algebra is that matrix multiplication is not commutative and that not all nonzero square matrices have inverses.

A good linear equation algorithm should include the LU factorization (for efficient handling of multiple right-hand sides) and partial pivoting (to help control the growth of round-off error).

COMPUTATIONAL PROCEDURES

Gaussian elimination: $[A \mid K] \underset{R}{\sim} [R \mid H]$, where R is in row-echelon form, and then solve $RX = H$ by back substitution.

Gauss-Jordan elimination $[A \mid K] \underset{R}{\sim} [I \mid X] \Rightarrow AX = K$ or $X = A^{-1}K$.

Compute A^{-1} by solving $AX = I$ by Gauss-Jordan elimination: $[A \mid I] \underset{R}{\sim} [I \mid P] \Rightarrow A^{-1} = P$.

Decompose a matrix A as a product of elementary matrices.

LU factorization of A using partial pivoting.

APPLICATIONS

Balancing chemical equations.

Curve fitting.

Matrix codes.

Mixture problems.

Population growth.

Production planning and scheduling.

Temperature distribution.

Transportation problems.

CHAPTER 1 REVIEW EXERCISES

1. Complete the following statements.

(a) A linear equation in n variables is an equation of the form ______________.

(b) The matrices A and B are row equivalent if ______________.

(c) An elementary matrix is a matrix which has been obtained from the identity matrix by ______________.

(d) The matrix B is a multiplicative inverse of A if ______________.

(e) A^{-1} exists if, and only if, A is row equivalent to ______________.

(f) A^{-1} exists if, and only if, A is a product of ______________.

(g) A^{-1} exists if, and only if, $AX = 0$ implies ______________.

(h) A^{-1} exists if, and only if, $AX = K$ has a ______________ solution.

2. Compute $A(B + C)$ for the matrices $A = \begin{bmatrix} 1 & 5 & -2 \\ 0 & 3 & 4 \\ 1 & 5 & 3 \\ 7 & -2 & 2 \end{bmatrix}$, $B = \begin{bmatrix} 1 & 5 \\ 2 & -3 \\ 0 & 4 \end{bmatrix}$, and

$C = \begin{bmatrix} -2 & 2 \\ 5 & 0 \\ 6 & -1 \end{bmatrix}$.

3. Given $A = \begin{bmatrix} 2 & 1 & 1 \\ 2 & 3 & 2 \\ 1 & 1 & 2 \end{bmatrix}$ compute $A^3 - 7A^2 + 11A - 4I$.

4. List two major differences between matrix algebra and ordinary algebra.

5. If a matrix B satisfies $B^3 - 7B^2 + 7B + 3I = 0$, express B^{-1} as a polynomial in B.

6. Write a formula for the entry in the $(3, 5)$ position of $A(B + C)$, given that A, B, and C are $n \times n$ matrices.

7. Use Gauss-Jordan elimination to find the inverse of $M = \begin{bmatrix} 1 & 2 & 3 \\ 2 & 4 & 5 \\ 3 & 5 & 6 \end{bmatrix}$.

8. Find all solutions of the system

$$\begin{aligned} x_1 - x_2 + 2x_3 - x_4 &= -1 \\ 2x_1 + x_2 - 2x_3 - 2x_4 &= -2 \\ -x_1 + 2x_2 - 4x_3 + x_4 &= 1 \\ 2x_1 + 2x_2 - 4x_3 - 2x_4 &= -2. \end{aligned}$$

9. Find all solutions of the homogeneous system $\begin{aligned} x + 2y + 3z &= 0 \\ 4x + 5y + 6z &= 0. \end{aligned}$

10. Find the LU factorization of the matrix $A = \begin{bmatrix} 1 & 2 & 4 \\ 2 & 3 & 7 \\ 1 & 4 & 7 \end{bmatrix}$.

11. Find all solutions of the matrix equation $\begin{bmatrix} 1 & 2 \\ 3 & 5 \end{bmatrix} X = \begin{bmatrix} 7 & 3 & 4 \\ 2 & -1 & 0 \end{bmatrix}$.

12. **(a)** Give an example of a 3×5 matrix A that is in row-echelon form but not in reduced row-echelon form.

(b) Give an example of a 4×4 upper triangular matrix that is not a diagonal matrix.

(c) Given $A = \begin{bmatrix} 2 & 1 \\ 0 & 2 \end{bmatrix}$, find all matrices $B = \begin{bmatrix} a & b \\ c & d \end{bmatrix}$ such that $AB = BA$.

13. **(a)** For $A = \begin{bmatrix} 1 & 2 \\ 3 & -4 \end{bmatrix}$, compute $A^2 - 5A + I$.

(b) Write a formula for the element in the $(5, 7)$ position of $D = (2A + 3B)C$, if A, B, and C are 10 by 10 matrices.

(c) Determine x so that

$$\begin{bmatrix} -3 & -2 & 0 & 2 \\ 2 & 1 & 0 & -1 \\ 1 & 0 & 1 & 2 \\ 2 & 1 & -3 & 1 \end{bmatrix}^{-1} = \frac{1}{8}\begin{bmatrix} 8 & 16 & 0 & 0 \\ -19 & -31 & 3 & 1 \\ x & -2 & 2 & -2 \\ -3 & -7 & 3 & 1 \end{bmatrix}.$$

14. Find the inverse of $B = \begin{bmatrix} 1 & 2 & 3 \\ 2 & 4 & 5 \\ 3 & 5 & 6 \end{bmatrix}$.

15. Given $A = \begin{bmatrix} 1 & -1 & 2 & -1 \\ 2 & 1 & -2 & -2 \\ -1 & 2 & -4 & 1 \\ 2 & 2 & -4 & -2 \end{bmatrix}$, $B = \begin{bmatrix} -1 \\ -2 \\ 1 \\ -2 \end{bmatrix}$, and $K = \begin{bmatrix} k_1 \\ k_2 \\ k_3 \\ k_4 \end{bmatrix}$,

(a) reduce $[A \,|\, B \,|\, K]$ to row-echelon form;

(b) from your work in (a), find all X such that $AX = B$;

(c) from your work in (a), find all K such that $AX = K$ is solvable; and

(d) from your work in (a), find all Y such that $AY = 0$.

16. Find the coefficient matrix and the augmented matrix for the system

$$\begin{aligned} 2x_1 + x_2 + 5x_3 + x_4 &= 5 \\ x_1 + x_2 - 3x_3 - 4x_4 &= -1 \\ 3x_1 + 6x_2 - 2x_3 + x_4 &= 8 \\ 2x_1 + 2x_2 + 2x_3 - 3x_4 &= 2 \end{aligned}$$

and then find all solutions of the system.

17. Use Gaussian elimination to find all solutions of the system

$$\begin{aligned} 3x_2 + x_3 &= 2 \\ x_1 + x_2 + 2x_3 &= -1 \\ 2x_1 - x_2 + 3x_3 &= -4. \end{aligned}$$

18. Let A be the augmented matrix for the system in Exercise 17. Reduce A to reduced row-echelon form.

19. Find all values of λ such that the following system has a nontrivial solution:

$$\begin{aligned} 5x + 4y &= \lambda x \\ x + 8y &= \lambda y \end{aligned}$$

20. Find the inverse of the matrix $A = \begin{bmatrix} 3 & 1 & 3 \\ 4 & 1 & 3 \\ 3 & 1 & 4 \end{bmatrix}$.

21. For $J = \begin{bmatrix} 2 & 0 & 0 \\ 1 & 2 & 0 \\ 0 & 0 & 3 \end{bmatrix}$ it is true that $J^3 - 7J^2 + 16J - 12I = 0$. Use this fact to find J^{-1}.

22. Balance the following chemical reactions:

(a) $N_2H_4 + N_2O_4 \rightarrow N_2 + H_2O$.

(b) $C_6H_6 + O_2 \rightarrow CO_2 + H_2O$.

(c) $SO_2 + NO_3 + H_2O \rightarrow H + SO_4 + NO$.

(d) $H_2SO_4 + MnS + As_2Cr_{10}O_3 \rightarrow HMnO_4 + AsH_3 + CrS_3O_2 + H_2O$.

2 Determinants

In several exercises of Chapter 1 we encountered the number $ad - bc$ associated with the 2×2 matrix $A = \begin{bmatrix} a & b \\ c & d \end{bmatrix}$. In particular (see Exercise 7 in Section 1.4), we found that the system $AX = 0$ has a nonzero solution if, and only if, $ad - bc = 0$, and (see Exercise 11 in Section 1.5) that if $ad - bc \neq 0$, then

$$A^{-1} = \frac{1}{ad - bc}\begin{bmatrix} d & -b \\ -c & a \end{bmatrix}.$$

In Chapter 2 we will generalize these results by defining a scalar-valued function on square matrices, called the **determinant,** which is a rule that associates a number with a square matrix. The definition we will use is recursive in nature; that is, the value of the function for an $n \times n$ matrix will depend on the value of the function for several $(n - 1) \times (n - 1)$ matrices. Recursive functions are particularly useful when one is evaluating a function by machine.

2.1 DEFINITIONS AND EXAMPLES

DEFINITION 2.1 Let A be an $n \times n$ matrix. The $(n - 1) \times (n - 1)$ submatrix of A obtained by deleting the ith row of A and the jth column of A is called **minor of the (i, j) position of A** and will be denoted by either M_{ij} or $M_{ij}(A)$. □

EXAMPLE 1 **Minors**

$$\text{For } A = \begin{bmatrix} a_{11} & a_{12} & a_{13} & a_{14} \\ a_{21} & a_{22} & a_{23} & a_{24} \\ a_{31} & a_{32} & a_{33} & a_{34} \\ a_{41} & a_{42} & a_{43} & a_{44} \end{bmatrix},$$

$$M_{11} = \begin{bmatrix} a_{11} & a_{12} & a_{13} & a_{14} \\ a_{21} & a_{22} & a_{23} & a_{24} \\ a_{31} & a_{32} & a_{33} & a_{34} \\ a_{41} & a_{42} & a_{43} & a_{44} \end{bmatrix} = \begin{bmatrix} a_{22} & a_{23} & a_{24} \\ a_{32} & a_{33} & a_{34} \\ a_{42} & a_{43} & a_{44} \end{bmatrix},$$

$$M_{34} = \begin{bmatrix} a_{11} & a_{12} & a_{13} & a_{14} \\ a_{21} & a_{22} & a_{23} & a_{24} \\ a_{31} & a_{32} & a_{33} & a_{34} \\ a_{41} & a_{42} & a_{43} & a_{44} \end{bmatrix} = \begin{bmatrix} a_{11} & a_{12} & a_{13} \\ a_{21} & a_{22} & a_{23} \\ a_{41} & a_{42} & a_{43} \end{bmatrix}, \text{ and}$$

$$M_{22}(M_{11}(A)) = \begin{bmatrix} a_{22} & a_{24} \\ a_{42} & a_{44} \end{bmatrix}. \quad \blacksquare$$

DEFINITION 2.2 For an $n \times n$ matrix A we define the **determinant** of A by

$$\begin{aligned} \det(A) &= \sum_{j=1}^{n} a_{1j}(-1)^{1+j} \det(M_{1j}) \\ &= a_{11}(-1)^{1+1} \det(M_{11}) + a_{12}(-1)^{1+2} \det(M_{12}) \\ &\quad + \cdots + a_{1n}(-1)^{1+n} \det(M_{1n}) \end{aligned} \tag{2.1}$$

and for a 2×2 matrix we define

$$\det \begin{bmatrix} a_{11} & a_{12} \\ a_{21} & a_{22} \end{bmatrix} = a_{11}a_{22} - a_{12}a_{21}. \quad \square \tag{2.2}$$

If we apply Definition 2.2 to a general 3×3 matrix A we obtain, using Equation (2.1),

$$\begin{aligned} \det(A) &= \det \begin{bmatrix} a_{11} & a_{12} & a_{13} \\ a_{21} & a_{22} & a_{23} \\ a_{31} & a_{32} & a_{33} \end{bmatrix} \\ &= a_{11}(-1)^{1+1} \det M_{11} + a_{12}(-1)^{1+2} \det M_{12} + a_{13}(-1)^{1+3} \det M_{13} \\ &= a_{11} \det \begin{bmatrix} a_{22} & a_{23} \\ a_{32} & a_{33} \end{bmatrix} - a_{12} \det \begin{bmatrix} a_{21} & a_{23} \\ a_{31} & a_{33} \end{bmatrix} + a_{13} \det \begin{bmatrix} a_{21} & a_{22} \\ a_{31} & a_{32} \end{bmatrix} \end{aligned}$$

and from Equation (2.2) we have the formula

$$\begin{aligned}\det(A) &= a_{11}(a_{22}a_{33} - a_{23}a_{32}) - a_{12}(a_{21}a_{33} - a_{23}a_{31}) + a_{13}(a_{21}a_{32} - a_{22}a_{31}) \\ &= a_{11}a_{22}a_{33} + a_{12}a_{23}a_{31} + a_{13}a_{21}a_{32} - a_{11}a_{23}a_{32} \\ &\quad - a_{12}a_{21}a_{33} - a_{13}a_{22}a_{31}, \end{aligned} \tag{2.3}$$

which consists of six terms, each the product of three numbers. Equation (2.3) is certainly not the sort of thing one should try to memorize. Fortunately there is a simple way of remembering Equation (2.3) that is illustrated in Figure 2.1. One proceeds by rewriting the first two columns of A to the right of A and then looking at products of elements on the same "slant." The algebraic sign of each product is as indicated in the diagram. Note that only complete slants are used; that is, each product contains three factors.

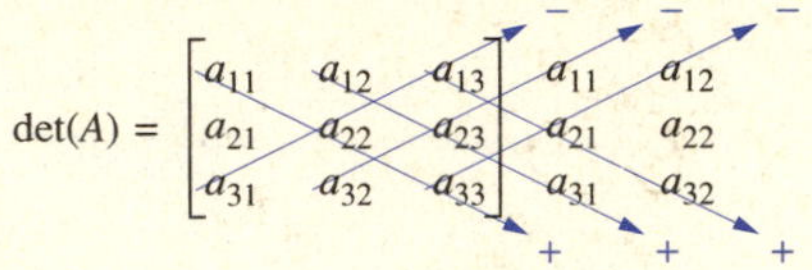

$$\det(A) = \begin{bmatrix} a_{11} & a_{12} & a_{13} \\ a_{21} & a_{22} & a_{23} \\ a_{31} & a_{32} & a_{33} \end{bmatrix} \begin{matrix} a_{11} & a_{12} \\ a_{21} & a_{22} \\ a_{31} & a_{32} \end{matrix}$$

FIGURE 2.1

We will refer to this "basket-weaving" procedure as the classical 3×3 trick. It is like the trick suggested in Equation (2.2) for remembering the determinant of a 2×2 matrix. It is very important to resist the temptation to overgeneralize; despite the efforts of many generations of students, **there is no 4×4 analog of the above trick!**

EXAMPLE 2 Basket Weaving

Use the classical 3×3 trick to evaluate

$$\det(A) = \det\begin{bmatrix} 1 & 0 & 2 \\ -3 & 4 & 6 \\ -1 & -2 & 3 \end{bmatrix}.$$

We rewrite the first two columns of A to obtain the array shown in Figure 2.2. The positive products are $1 \times 4 \times 3 = 12$, $0 \times 6 \times (-1) = 0$, and $2 \times (-3) \times (-2) = 12$ while the negative products are $(-1) \times 4 \times 2 = -8$, $(-2) \times 6 \times 1 = -12$, and $3 \times (-3) \times 0 = 0$. Thus $\det(A) = 12 + 0 + 12 - (-8 - 12 + 0) = 44$. ■

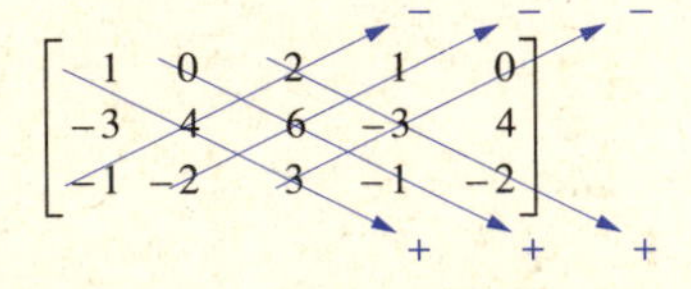

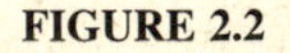

FIGURE 2.2

For a 4×4 matrix A, Definition 2.2 would express $\det(A)$ as a linear combination of the determinants of four 3×3 matrices, each of which involves 6 terms. Altogether there would be 24 terms, each the product of 4 numbers. Thus the evaluation of a 4×4 determinant directly from the definition requires $72 = 3 \times (4!)$ multiplications.*

EXAMPLE 3 **Determinant of a 4 × 4 Matrix**

Evaluate $\det B = \det \begin{bmatrix} -3 & -2 & 0 & 2 \\ 2 & 1 & 0 & -1 \\ 1 & 0 & 1 & 2 \\ 2 & 1 & -3 & 1 \end{bmatrix}$.

Directly from Equation (2.1) we have

$$\det B = -3 \det \begin{bmatrix} 1 & 0 & -1 \\ 0 & 1 & 2 \\ 1 & -3 & 1 \end{bmatrix} + 2 \det \begin{bmatrix} 2 & 0 & -1 \\ 1 & 1 & 2 \\ 2 & -3 & 1 \end{bmatrix}$$
$$+ 0 \det \begin{bmatrix} 2 & 1 & -1 \\ 1 & 0 & 2 \\ 2 & 1 & 1 \end{bmatrix} - 2 \det \begin{bmatrix} 2 & 1 & 0 \\ 1 & 0 & 1 \\ 2 & 1 & -3 \end{bmatrix},$$

and evaluating each of the 3×3 determinants directly from Equation (2.1) yields

$$\det B = -3\left(1 \det \begin{bmatrix} 1 & 2 \\ -3 & 1 \end{bmatrix} - 0 \det \begin{bmatrix} 0 & 2 \\ 1 & 1 \end{bmatrix} - 1 \det \begin{bmatrix} 0 & 1 \\ 1 & -3 \end{bmatrix}\right)$$
$$+ 2\left(2 \det \begin{bmatrix} 1 & 2 \\ -3 & 1 \end{bmatrix} - 0 \det \begin{bmatrix} 1 & 2 \\ 2 & 1 \end{bmatrix} - \det \begin{bmatrix} 1 & 1 \\ 2 & -3 \end{bmatrix}\right)$$
$$+ 0 - 2\left(2 \det \begin{bmatrix} 0 & 1 \\ 1 & -3 \end{bmatrix} - \det \begin{bmatrix} 1 & 1 \\ 2 & -3 \end{bmatrix} + 0 \det \begin{bmatrix} 1 & 0 \\ 2 & 1 \end{bmatrix}\right)$$
$$= -3[(1 + 6) - 0 - (0 - 1)] + 2[2(1 + 6) - 0 - (-3-2)]$$
$$+ 0 - 2[2(0 - 1) - (-3 - 2) + 0] = -24 + 38 - 6 = 8.$$

For an $n \times n$ matrix the evaluation procedure illustrated in Example 3 would involve the evaluation of a linear combination of

$$n,\ (n - 1) \times (n - 1) \text{ determinants,}$$

which would lead to

$$n(n - 1),\ (n - 2) \times (n - 2) \text{ determinants,}$$

*$n! = n(n - 1)(n - 2)(n - 3) \cdots (3)(2)(1)$ and is called ***n*-factorial.**

and then to

$$n(n-1)(n-2),\ (n-3) \times (n-3) \text{ determinants,}$$

and eventually to

$$n(n-1)(n-2) \cdots 4 \times 3,\ 2 \times 2 \text{ determinants,}$$

each of which involves two terms. Thus there would be $n!$ (n-factorial) terms, each of which would involve the product of n numbers. It follows that the evaluation of det(A), directly from Definition 2.2, would require $n! \times (n-1)$ multiplications. This is a prohibitive amount of arithmetic for even moderately large n and very fast computers (see Exercise 13 in Section 2.1).

The recursive use of Definition 2.2, for a general $n \times n$ matrix A leads, as above, to an expression of the form

$$\det(A) = \sum \pm a_{1i_1} a_{2i_2} \cdots a_{ni_n} \tag{2.4}$$

where the sum is taken over all $n!$ possible arrangements (called **permutations**) of the second subscripts. Note that this sum consists of all possible products obtained by taking one factor from each row and each column of A.

The classical way of deciding on the algebraic sign of each term in Equation (2.4) is to count the number of **transpositions** (interchange of two subscripts) required to restore the second subscripts to natural order; if the number is even, we assign a plus sign and if the number is odd, we assign a minus sign. For example, the term

$$a_{13}a_{24}a_{31}a_{42}$$

from the expansion of a 4×4 determinant would be positive because the sequence of second subscripts (3, 4, 1, 2) can be restored to natural order by two transpositions:

$$(3, 4, 1, 2) \longrightarrow (1, 4, 3, 2) \longrightarrow (1, 2, 3, 4).$$

Equation (2.4), with the above rule for choosing the algebraic sign of each term, constitutes the classical definition of the determinant function. We mention it here for historical interest and possibly to connect this discussion with your previous experience with determinants. We will have no further use for Equation (2.4).

If the determinant function is to be of any practical use, then it is clear that the first order of business is to find a more economical way of evaluating det(A). In the following sections we will develop such a method; it is like our methods for systems of equations in that it reduces the given problem to an equivalent problem for which the evaluation is easy.

$$A \xrightarrow{\textbf{simplify}} B \xrightarrow{\textbf{easy}} \det(B) \longrightarrow \det(A)$$

In order to implement such a procedure we need to answer two questions.

1. When is the evaluation of $\det(B)$ easy?
2. How can we simplify a matrix to take advantage of the answer to question 1?

Many of the exercises of this section are designed to prepare you to understand the answers to these questions, which will be presented in the next section.

We conclude this section by repeating the definition of the transpose of a matrix that was given in Exercise 21 of Section 1.6.

DEFINITION 2.3 For an $n \times m$ matrix A, the **transpose** of A, denoted as A^T, is the $m \times n$ matrix obtained from A by interchanging rows and columns. More precisely, if $B = A^T$, then

$$b_{ij} = a_{ji}; \qquad i = 1, 2, \ldots, m; \qquad \text{and} \qquad j = 1, 2, \ldots, n. \quad \square$$

For example, $\begin{bmatrix} 1 & 3 & -7 & 4 \\ 2 & 0 & 9 & 11 \end{bmatrix}^T = \begin{bmatrix} 1 & 2 \\ 3 & 0 \\ -7 & 9 \\ 4 & 11 \end{bmatrix}$ and $\begin{bmatrix} 3 \\ 6 \\ 7 \\ 8 \end{bmatrix} = [3 \quad 6 \quad 7 \quad 8]^T$.

The elementary properties of the transpose are given in the next theorem.

THEOREM 2.1 The matrix transpose has the following properties for all matrices A and B for which the indicated operations are defined.

(1) $(A^T)^T = A$
(2) $(A + B)^T = A^T + B^T$
(3) $(kA)^T = kA^T$
(4) $(AB)^T = B^T A^T$
(5) $(A^{-1})^T = (A^T)^{-1}$

The first three of these properties are quite transparent. A formal proof of (4) involves just a straightforward calculation using the definitions of matrix multiplication and the transpose. The details are given below where we assume A is $(m \times n)$, B is $(n \times s)$, and use $(C)_{ij}$ for c_{ij} when the matrix C has a name that is more complicated than a single letter.

$$(AB)^T_{ij} = (AB)_{ji} = \sum_{k=1}^{n} a_{jk} b_{ki} = \sum_{k=1}^{n} (A^T)_{kj} (B^T)_{ik}$$

$$= \sum_{k=1}^{n} (B^T)_{ik} (A^T)_{kj} = (B^T A^T)_{ij}$$

To prove (5), we use (4) to calculate

$$A^T (A^{-1})^T = (A^{-1} A)^T = I^T = I$$

Thus $(A^{-1})^T = (A^T)^{-1}$ as claimed. $\square$

The transpose has many uses, as we will see in the following sections and chapters. A theoretically unimportant use, that nonetheless has practical value, is the convention of writing a column matrix as the transpose of a row matrix; that is,

$$\begin{bmatrix} x_1 \\ x_2 \\ \vdots \\ x_n \end{bmatrix} = [x_1 \quad x_2 \quad \cdots \quad x_n]^T.$$

EXAMPLE 4 **$\det(A^T) = \det(A)$**

For $A = \begin{bmatrix} 1 & 0 & 2 \\ 3 & 4 & -7 \\ 1 & 5 & 6 \end{bmatrix}$ compute $\det(A)$ and $\det(A^T)$.

$$\det(A) = \det\begin{bmatrix} 4 & -7 \\ 5 & 6 \end{bmatrix} + 2\det\begin{bmatrix} 3 & 4 \\ 1 & 5 \end{bmatrix} = (24 + 35) + 2(15 - 4) = 81$$

$$\det(A^T) = \det\begin{bmatrix} 1 & 3 & 1 \\ 0 & 4 & 5 \\ 2 & -7 & 6 \end{bmatrix} = \det\begin{bmatrix} 4 & 5 \\ -7 & 6 \end{bmatrix} - 3\det\begin{bmatrix} 0 & 5 \\ 2 & 6 \end{bmatrix} + 4\det\begin{bmatrix} 0 & 4 \\ 2 & -7 \end{bmatrix}$$

$$= 59 + 30 - 8 = 81 = \det(A).$$ ■

For our current purposes, the most important property of the transpose is its relationship to the determinant. The result illustrated in Example 4 is true in general.

THEOREM 2.2 For any $n \times n$ matrix A, $\det(A^T) = \det(A)$. □

The proof, which involves calculations like those in Exercises 6 and 8, will be omitted. Many of the properties of the determinant function will be given in terms of the rows of the matrix. Theorem 2.2 is important because it will allow us to give a column analog of each row result that we obtain.

The software and the TI-85 calculator all have built in routines for computing the determinant. The MATALG command is "=det(A)" and the MATLAB command is "det(A)." On the TI-85 the det command is on the MATH submenu of the MATRIX menu. In the next section we will develop a more efficient way to evaluate the determinant function.

EXERCISES 2.1

1. For $A = \begin{bmatrix} 1 & 2 \\ 3 & 4 \end{bmatrix}$ and $B = \begin{bmatrix} 3 & 4 \\ -2 & 3 \end{bmatrix}$ compute $\det(A)$, $\det(B)$, $\det(AB)$, $\det(BA)$, $\det(A)\det(B)$, $\det(A^{-1})$, $\det(A^T)$.

2. For $A = \begin{bmatrix} 2 & 3 & 4 \\ 4 & 3 & 1 \\ 1 & 2 & 4 \end{bmatrix}$ and $B = \begin{bmatrix} 2 & -1 & 5 \\ 3 & 2 & 4 \\ 8 & 0 & -2 \end{bmatrix}$ compute $\det(A)$, $\det(B)$, $\det(AB)$, $\det(A)\det(B)$, and $\det(A^T)$.

3. Use the classical 3×3 trick to evaluate these determinants.

   (a) $\det \begin{bmatrix} 1 & 2 & 3 \\ 4 & 5 & 6 \\ 7 & 8 & 9 \end{bmatrix}$ **(b)** $\det \begin{bmatrix} x-2 & -3 & -4 \\ -4 & x-3 & -1 \\ -1 & -2 & x-4 \end{bmatrix}$

   (c) $\det \begin{bmatrix} x & -1 & 0 \\ 0 & x & -1 \\ -2 & -1 & x+2 \end{bmatrix}$

4. Show that one obtains a definition equivalent to Definition 2.2 if Equation (2.2) is replaced by $\det[a] = a$ for a 1×1 matrix $[a]$.

5. Evaluate the following, directly from Definition 2.2.

   (a) $\det \begin{bmatrix} 1 & 2 & 3 & 4 \\ -1 & 1 & 2 & 3 \\ 1 & -1 & 1 & 2 \\ -1 & 1 & -1 & 1 \end{bmatrix}$ **(b)** $\det \begin{bmatrix} 1 & -1 & 2 & 3 \\ 2 & 2 & 0 & 2 \\ 4 & 1 & -1 & 1 \\ 1 & 2 & 3 & 0 \end{bmatrix}$

   (c) $\det \begin{bmatrix} 1 & 2 & 3 & 4 \\ 1 & -1 & 1 & 2 \\ -1 & 1 & 2 & 3 \\ -1 & 1 & -1 & 11 \end{bmatrix}$

6. For a general 3×3 matrix A show that

   (a) $\det(A)$ can be computed from the elements in the first column of A and their minors via

   $$\det(A) = a_{11} \det M_{11} - a_{21} \det M_{21} + a_{31} \det M_{31}, \text{ and}$$

   (b) $\det(A)$ can be computed from the elements in the second row of A and their minors via

   $$\det(A) = -a_{21} \det M_{21} + a_{22} \det M_{22} - a_{23} \det M_{23}.$$

†7. Compute the determinants of the following matrices and express their value in terms of $\det(A)$.

   (a) $A = \begin{bmatrix} a & b \\ c & d \end{bmatrix}$ **(b)** $\begin{bmatrix} c & d \\ a & b \end{bmatrix} = A_{(R_1 \leftrightarrow R_2)}$

   (c) $\begin{bmatrix} a & b \\ kc & kd \end{bmatrix} = A_{(R_2 \leftrightarrow kR_2)}$ **(d)** $\begin{bmatrix} a & b \\ c+ka & d+kb \end{bmatrix} = A_{(R_2 \leftarrow kR_1 + R_2)}$

†**8.** For a general 4×4 matrix A,

(a) express $\det(A)$ as a linear combination of 2×2 determinants by using Definition 2.2 and then using the first formula in Exercise 6 for each of the remaining 3×3 determinants;

(b) express $a_{11} \det M_{11} - a_{21} \det M_{21} + a_{31} \det M_{31} - a_{41} \det M_{41}$ as a linear combination of 2×2 determinants and compare with part (a) to conclude that $\det(A)$ can be computed from the elements in the first column and their minors;

(c) show, by induction, that for any $n \times n$ matrix

$$\det(A) = \sum_{i=1}^{n} a_{i1}(-1)^{1+i} \det(M_{i1}); \text{ and}$$

(d) use part (c) to prove, by induction, that $\det(A^T) = \det(A)$.

†**9.** Let A be the general 3×3 matrix whose determinant is given in Equation (2.3). Evaluate the following determinants and express the answers in terms of $\det(A)$.

(a) $A_{(R_1 \leftrightarrow R_3)}$ **(b)** $A_{(R_2 \leftarrow kR_2)}$ **(c)** $A_{(R_3 \leftarrow kR_2 + R_3)}$

10. Show that a general 3×3 matrix $A = \begin{bmatrix} a & b & c \\ d & e & f \\ g & h & i \end{bmatrix}$ is invertible if $a \neq 0$, $ae - db \neq 0$, and $\det(A) \neq 0$. (*Hint:* Try to row reduce A to I without using row interchanges.) Which of the three given conditions can be eliminated if row interchanges are allowed?

†**11.** Consider the **trace** function on square matrices that can be defined recursively by the following.

(1) $\operatorname{tr}(A) = a_{11} + \operatorname{tr}(M_{11})$

(2) $\operatorname{tr}\begin{bmatrix} a & b \\ c & d \end{bmatrix} = a + d.$

(a) For the matrices A and B of Exercise 2 compute $\operatorname{tr}(A)$, $\operatorname{tr}(B)$, $\operatorname{tr}(AB)$, and $\operatorname{tr}(BA)$.

(b) Find a formula for $\operatorname{tr}(A)$ in terms of the elements of A.

(c) Prove in general that $\operatorname{tr}(AB) = \operatorname{tr}(BA)$.

(d) Show that $\operatorname{tr}(P^{-1}AP) = \operatorname{tr}(A)$ for any nonsingular matrix P.

12. Find a formula for the determinant of a general 4×4 matrix.

†**13.** If a large digital computer, like the DEC PDP10, can do a multiplication in 2.9×10^{-6} seconds, how long would it take to do the $49(50!)$ multiplications required to evaluate the determinant of a 50×50 matrix?

†**14.** Evaluate the determinants of the following triangular matrices.

(a) $\begin{bmatrix} a & 0 \\ c & d \end{bmatrix}$ **(b)** $\begin{bmatrix} a & 0 & 0 \\ b & d & 0 \\ c & e & f \end{bmatrix}$ **(c)** $\begin{bmatrix} a & b & c \\ 0 & d & e \\ 0 & 0 & f \end{bmatrix}$

15. Find the most general 3×3 matrix A such that $A^T = A$.

16. For $X = \begin{bmatrix} x_1 \\ x_2 \\ x_3 \end{bmatrix}$ and $A = \begin{bmatrix} 1 & -2 & 4 \\ -2 & 5 & 3 \\ 4 & 3 & 0 \end{bmatrix}$ compute X^TAX.

17. **(a)** For $X = [-2 \quad 3 \quad 4]^T$, compute X^TX and XX^T.
(b) For any $n \times 1$ matrix X show that $X^TX \geq 0$.

†**18.** Show that if either A or B is singular, then so is AB. (*Hint:* Use part 3 of Theorem 1.9 to show that if B is singular, then so is AB. Then use the transpose to do the other case.)

19. Find examples of 2×2 matrices A and B for which $\det(A + B) \neq \det(A) + \det(B)$.

20. Consider a parallelogram with vertices at $(0, 0)$, (a, b), (c, d), and $(a + c, b + d)$.

(a) Show that the area of the parallelogram is the absolute value of $\det\begin{bmatrix} a & b \\ c & d \end{bmatrix}$.

(b) Interpret the determinant of a 3×3 matrix as the volume of a three-dimensional solid.

21. Evaluate the determinants of

$$S = \begin{bmatrix} 0 & 0 & 0 & 0 & 9 \\ 0 & 0 & 0 & 1 & 0 \\ 0 & 0 & 2 & 0 & 0 \\ 0 & -3 & 0 & 0 & 0 \\ 4 & 0 & 0 & 0 & 0 \end{bmatrix} \quad \text{and} \quad T = \begin{bmatrix} 2 & 0 & 0 & 0 & 0 \\ 0 & 0 & 0 & 0 & -1 \\ 0 & 0 & 2 & 0 & 0 \\ 0 & 0 & 0 & 1 & 0 \\ 0 & \pi & 0 & 0 & 0 \end{bmatrix}.$$

22. Show that $\det\begin{bmatrix} 1 & a & a^2 \\ 1 & b & b^2 \\ 1 & c & c^2 \end{bmatrix} = (c - b)(c - a)(b - a)$.

23. Show that the following matrices all satisfy $A^T = A^{-1}$.

(a) $\begin{bmatrix} \cos\theta & \sin\theta \\ -\sin\theta & \cos\theta \end{bmatrix}$ **(b)** $\begin{bmatrix} \cos\theta & 0 & -\sin\theta \\ 0 & 1 & 0 \\ \sin\theta & 0 & \cos\theta \end{bmatrix}$

2.2 EVALUATION OF det(A): A BETTER WAY

> 1. When is the evaluation of det(B) easy?
> 2. How can we simplify a matrix to take advantage of the answer to question 1?

In this section we will use the Gaussian elimination techniques of the last chapter to describe an efficient algorithm for the evaluation of the determinant function. In the process we will discover several interesting and useful properties of the determinant function.

We begin by answering the first of the two questions raised above. "For what kind of matrices is the evaluation of det(B) easy?" As suggested in Exercise 14 of Section 2.1, the answer to this question is that it is easy to evaluate the determinant of a triangular matrix because the minor of the $(1, 1)$ position of a triangular matrix is another triangular matrix. For a lower triangular matrix there is only one element in the first row that can be nonzero, so Definition 2.1 yields

$$\begin{aligned}\det(T) &= t_{11} \det M_{11}(T) = t_{11}t_{22} \det M_{11}(M_{11}(T)) = \cdots \\ &= t_{11}t_{22} \cdots t_{n-2,\,n-2} \det\begin{bmatrix} t_{n-1,\,n-1} & 0 \\ t_{nn-1} & t_{nn} \end{bmatrix} \\ &= t_{11}t_{22} \cdots t_{n-1,\,n-1}t_{nn}.\end{aligned}$$

For example, $\det\begin{bmatrix} 5 & 0 & 0 \\ 11 & 3 & 0 \\ 2 & 0 & 2 \end{bmatrix} = 5 \times 3 \times 2 = 30$. For an upper triangular matrix U, the transpose U^T is lower triangular, so from Theorem 2.1 and the above calculation we have

$$\det(U) = \det(U^T) = u_{11}u_{22} \cdots u_{nn}.$$

PROPERTY 2.1 The determinant of a triangular matrix is the product of its diagonal elements. □

Because we have an orderly way of reducing A to an upper triangular matrix (Gaussian elimination), and since the evaluation of the determinant of a triangular matrix is easy, we will have found the desired "better way" if we can determine the effect of elementary row operations on the determinant of a matrix. Exercises 7 and 9 of Section 2.1 provide some experimental evidence to support the general results which follow.

PROPERTY 2.2 A Type II elementary row operation changes the sign of the determinant; that is, if B is obtained from A by interchanging two rows, then $\det(B) = -\det(A)$.

A careful proof of this property is quite involved and will be omitted.* We will illustrate it here for a general 3×3 matrix and the elementary row operation $(R_1 \leftrightarrow R_3)$. We use Equation (2.3) to evaluate the determinants:

$$\begin{aligned}\det(A) &= \det\begin{bmatrix} a_{11} & a_{12} & a_{13} \\ a_{21} & a_{22} & a_{23} \\ a_{31} & a_{32} & a_{33} \end{bmatrix} \\ &= a_{11}a_{22}a_{33} + a_{12}a_{23}a_{31} + a_{13}a_{21}a_{32} - a_{11}a_{23}a_{32} - a_{12}a_{21}a_{33} - a_{13}a_{22}a_{31}\end{aligned}$$

$$\begin{aligned}\det(B) &= \det\begin{bmatrix} a_{31} & a_{32} & a_{33} \\ a_{21} & a_{22} & a_{23} \\ a_{11} & a_{12} & a_{13} \end{bmatrix} \\ &= a_{31}a_{22}a_{13} + a_{32}a_{23}a_{11} + a_{33}a_{21}a_{12} - a_{11}a_{22}a_{33} - a_{12}a_{23}a_{31} - a_{13}a_{21}a_{32} \\ &= -\det(A).\end{aligned}$$ □

If the first row of A is zero, then it follows immediately from Definition 2.2 that $\det(A) = 0$. It also follows from Property 2.2 that if any row of A is zero,

*It is worth noting that this property follows easily from the classical definition of the determinant mentioned on page 102.

then $\det(A) = 0$; because the zero row can be moved to the first row by only changing the sign of the determinant. If a matrix A has two equal rows, then $\det(A) = -\det(A)$, since interchanging the two equal rows changes the sign of $\det(A)$ but does not change A. It follows that $\det(A) = 0$ if A has two equal rows. This observation establishes the following property.

PROPERTY 2.3 If the matrix A has a zero row or two equal rows, then $\det(A) = 0$. □

Our next definition is aimed primarily at simplifying the notation in the definition of the determinant function.

DEFINITION 2.4 The **cofactor** of the (i,j) position of A is

$$\text{cof}_{ij} = \text{cof}_{ij}(A) = (-1)^{i+j}\det(M_{ij}(A)). \quad \square$$

Note that $\text{cof}_{ij}(A)$ does not depend on the elements from the ith row of A or the jth column of A.

The definition of the determinant can now be rephrased as

$$\det(A) = a_{11}\,\text{cof}_{11}(A) + a_{12}\,\text{cof}_{12}(A) + \cdots + a_{1n}\,\text{cof}_{1n}(A); \tag{2.5}$$

that is, $\det(A)$ is the sum of the products of the elements of the first row of A times their cofactors.

The following checkerboard diagram is helpful in determining the signs associated with the cofactors. We show it for a 5×5 matrix; the generalization to any other size matrix is clear.

$$\begin{bmatrix} + & - & + & - & + \\ - & + & - & + & - \\ + & - & + & - & + \\ - & + & - & + & - \\ + & - & + & - & + \end{bmatrix}$$

Our next result shows that in Equation (2.5) the first row of A may be replaced by any other row of A.

PROPERTY 2.4 ***Laplace Expansion***

For any $n \times n$ matrix A and any integer $i = 1, 2, \ldots, n$, we can compute $\det(A)$ by taking the sum of the products of the elements in the ith row by their cofactors; that is,

$$\det(A) = \sum_{k=1}^{n} a_{ik}\,\text{cof}_{ik}(A) = a_{i1}\,\text{cof}_{i1}(A) + a_{i2}\,\text{cof}_{i2}(A) + \cdots + a_{in}\,\text{cof}_{in}(A).$$

To verify this property we move $\text{Row}_i(A)$ to the first position, without destroying the order of the other rows, by repeatedly interchanging it with the row directly above it until the original ith row reaches the first row. This requires $(i - 1)$ Type II operations and leads to the matrix

$$B = \begin{bmatrix} \text{Row}_i(A) \\ A \text{ with its} \\ i\text{th row} \\ \text{deleted} \end{bmatrix}$$

for which, by Property 2.2, $\det(A) = (-1)^{i-1} \det(B)$. Now, by Definition 2.2,

$$\det(B) = \sum_{k=1}^{n} b_{1k} \operatorname{cof}_{1k}(B) = \sum_{k=1}^{n} a_{ik}(-1)^{1+k} \det(M_{ik}(A)),$$

since $M_{ik}(A) = M_{1k}(B)$ by the way we constructed B. Finally,

$$\begin{aligned} \det(A) &= (-1)^{i-1} \det(B) = (-1)^{i-1} \sum_{k=1}^{n} a_{ik}(-1)^{1+k} \det(M_{ik}) \\ &= \sum_{k=1}^{n} a_{ik}(-1)^{i+k} \det(M_{ik}) = \sum_{k=1}^{n} a_{ik} \operatorname{cof}_{ik}(A). \quad \square \end{aligned}$$

Example 1 shows how Property 2.4 is useful when a row of the matrix has many zeros.

EXAMPLE 1 Expand by the Simplest Row

Evaluate $\det(A)$ for $A = \begin{bmatrix} 3 & 2 & 4 \\ 0 & 5 & 0 \\ 6 & 0 & 7 \end{bmatrix}$.

Since the second row of A is the simplest (because it has the most zeros) we use the Laplace expansion by the second row to obtain

$$\begin{aligned} \det(A) &= 0\operatorname{cof}_{21} + 5\operatorname{cof}_{22} + 0\operatorname{cof}_{23} = 5 \det\begin{bmatrix} 3 & 4 \\ 6 & 7 \end{bmatrix} \\ &= 5(-3) = -15. \quad ■ \end{aligned}$$

We can now show how the other two types of elementary row operations affect the determinant. If we use a Type I operation to multiply each element of the ith row by k, thus obtaining a matrix B, then an expansion by the ith row of B yields at once that $\det(B) = k \det(A)$. This result is most often used in the following form.

PROPERTY 2.5 If B is obtained from A by dividing each element of the ith row by k (factoring k from the ith row), then $\det(A) = k \det(B)$. $\square$

EXAMPLE 2 **Factoring a Constant from a Row or Column**

$$\det\begin{bmatrix} 5 & 7 & 9 \\ 3 & 6 & -9 \\ 4 & 2 & 6 \end{bmatrix} = 3\det\begin{bmatrix} 5 & 7 & 9 \\ 1 & 2 & -3 \\ 4 & 2 & 6 \end{bmatrix} = 6\det\begin{bmatrix} 5 & 7 & 9 \\ 1 & 2 & -3 \\ 2 & 1 & 3 \end{bmatrix},$$

where we have first factored a 3 from the second row of the matrix and then a 2 from the third row. ■

Before we can investigate the effect of a Type III operation on the determinant we need to establish the following property.

PROPERTY 2.6 If $i \neq j$, then the sum of the product of the elements of row i by the cofactors of row j is zero; that is,

$$\sum_{k=1}^{n} a_{ik}\operatorname{cof}_{jk}(A) = 0 \text{ if } i \neq j.$$

To see that this is so we let B be the matrix obtained from A by replacing $\text{Row}_j(A)$ by $\text{Row}_i(A)$. Thus $b_{jk} = a_{ik}$ and

$$B = \begin{bmatrix} \text{Row}_1(A) \\ \vdots \\ \text{Row}_i(A) \\ \vdots \\ \text{Row}_i(A) \\ \vdots \\ \text{Row}_n(A) \end{bmatrix} \begin{matrix} \\ \\ \leftarrow i\text{th row of } B \\ \\ \leftarrow j\text{th row of } B \\ \\ \\ \end{matrix}$$

Note that $\det(B) = 0$, since B has two equal rows, and that $\operatorname{cof}_{jk}(B) = \operatorname{cof}_{jk}(A)$. Now, using the Laplace expansion by the jth row, we have

$$0 = \det(B) = \sum_{k=1}^{n} b_{jk}\operatorname{cof}_{jk}(B) = \sum_{k=1}^{n} a_{ik}\operatorname{cof}_{jk}(A). \quad \square$$

We can now establish a property of determinants that is extremely useful for purposes of evaluation.

PROPERTY 2.7 If B is obtained from A by the Type III operation $(R_j \leftarrow cR_i + R_j)$, then $\det(B) = \det(A)$.

This is a very convenient property since most of the elementary row operations used in the reduction of a matrix to triangular form are of Type III. To verify this property we compute $\det(B)$ using the Laplace expansion by the jth row. Since the element in the (j, k) position of B is $ca_{ik} + a_{jk}$ we have

$$\det(B) = \sum_{k=1}^{n} b_{jk} \operatorname{cof}_{jk}(B) = \sum_{k=1}^{n} (ca_{ik} + a_{jk}) \operatorname{cof}_{jk}(A)$$
$$= c\sum_{k=1}^{n} a_{ik} \operatorname{cof}_{jk}(A) + \sum_{k=1}^{n} a_{jk} \operatorname{cof}_{jk}(A) = 0 + \det(A),$$

using Property 2.6 for the first sum and Property 2.4 for the second. □

We now have all the information necessary for an efficient algorithm for evaluating $\det(A)$: we reduce A to upper triangular form using elementary row operations and then evaluate the determinant of the resulting triangular matrix. Most of the operations used will be of Type III which do not change the determinant.

EXAMPLE 3 **Elimination Method for det(A)**

Evaluate $\det(A)$ where $A = \begin{bmatrix} 1 & -1 & 2 & 3 \\ 2 & 2 & 0 & 2 \\ 4 & 1 & 3 & -1 \\ 1 & 2 & 3 & 0 \end{bmatrix}$.

The first column of A is reduced by the three elementary row operations $(R_2 \leftarrow -2R_1 + R_2)$, $(R_3 \leftarrow -4R_1 + R_3)$, and $(R_4 \leftarrow -R_1 + R_4)$, none of which change the determinant, so that we have

$$\det(A) = \det \begin{bmatrix} 1 & -1 & 2 & 3 \\ 0 & 4 & -4 & -4 \\ 0 & 5 & -5 & -13 \\ 0 & 3 & 1 & -3 \end{bmatrix} = 4 \det \begin{bmatrix} 1 & -1 & 2 & 3 \\ 0 & 1 & -1 & -1 \\ 0 & 5 & -5 & -13 \\ 0 & 3 & 1 & -3 \end{bmatrix},$$

where the operation $(R_2 \leftarrow \frac{1}{4}R_2)$ was used to produce a 1 in the (2, 2) position. The operations $(R_3 \leftarrow -5R_2 + R_3)$ and $(R_4 \leftarrow -3R_2 + R_4)$ will now simplify the second column without changing the determinant. We then have

$$\det(A) = 4 \det \begin{bmatrix} 1 & -1 & 2 & 3 \\ 0 & 1 & -1 & -1 \\ 0 & 0 & 0 & -8 \\ 0 & 0 & 4 & 0 \end{bmatrix} = -4 \det \begin{bmatrix} 1 & -1 & 2 & 3 \\ 0 & 1 & -1 & -1 \\ 0 & 0 & 4 & 0 \\ 0 & 0 & 0 & -8 \end{bmatrix}.$$

The operation $(R_3 \leftrightarrow R_4)$ was used at the last step to yield the final upper triangular matrix. It now follows from Property 2.1 that

$$\det(A) = -4(-32) = 128. \blacksquare$$

The elimination method for evaluating det(A), illustrated in Example 3, is clearly more efficient than the direct use of the definition. The reduction process, which is equivalent to the reduction of A to row-echelon form, takes about $\frac{n^3}{3}$ multiplications compared with the $(n - 1)n!$ multiplications required using the definition. On a machine, like the DEC PDP10 that takes 2.9×10^{-6} seconds per multiplication, evaluating a 50×50 determinant requires less than 2 seconds, while performing $49 \times 50!$ multiplications requires 1.37×10^{53} years. This is a most dramatic saving.

Each row property stated above has a column analog that follows from the corresponding row property and Theorem 2.1. We leave it to you to formulate these column properties. You are strongly advised to use only row operations for evaluation purposes, at least until you have become proficient with the row operations. Column operations may be useful for hand calculations, but they are not needed for machine calculations. Remember, column operations should not be used in solving linear systems by Gaussian elimination.

EXERCISES 2.2

Use Gaussian reduction to evaluate the determinants of the following matrices.

1. $A = \begin{bmatrix} -1 & 1 & 3 \\ 2 & 1 & -1 \\ 4 & 2 & 2 \end{bmatrix}$

2. $B = \begin{bmatrix} 1 & 7 & 3 \\ 3 & -1 & 2 \\ 9 & -3 & 6 \end{bmatrix}$

3. $C = \begin{bmatrix} 1 & -1 & 2 \\ -3 & -3 & 3 \\ 4 & 0 & 5 \end{bmatrix}$

4. $D = \begin{bmatrix} 1 & 2 & 3 & 4 \\ 1 & -1 & 1 & 2 \\ -1 & 1 & -1 & 11 \\ -1 & 1 & 2 & 3 \end{bmatrix}$

5. $E = \begin{bmatrix} 1 & 0 & 2 & -1 \\ 3 & -2 & 6 & 4 \\ 2 & 2 & 5 & 6 \\ 5 & 4 & 3 & 0 \end{bmatrix}$

6. $F = \begin{bmatrix} 3 & -2 & 1 & 6 & 0 \\ 5 & -4 & 2 & 11 & 1 \\ 1 & -6 & 3 & 10 & 4 \\ -2 & 2 & -1 & -5 & 3 \\ 7 & -7 & 3 & 16 & -2 \end{bmatrix}$

7. $G = \begin{bmatrix} 1 & 1 & 2 & 1 \\ -1 & 1 & 0 & 1 \\ 2 & 1 & 1 & 0 \\ 1 & 3 & 1 & 0 \end{bmatrix}$

8. $M = \begin{bmatrix} 1 & 1 & 3 & 2 & 0 \\ 3 & 1 & 0 & 2 & 1 \\ 0 & 1 & 3 & 2 & 0 \\ 4 & -2 & 3 & 0 & 1 \\ 5 & 1 & 0 & 6 & 0 \end{bmatrix}$

9. Find the determinants of

$$A = \begin{bmatrix} 2 & 0 & 1 & 1 \\ 1 & -1 & 3 & 0 \\ 1 & 2 & 1 & 3 \\ 0 & 1 & -1 & 2 \end{bmatrix} \text{ and } B = \begin{bmatrix} 3 & 2 & 1 & 0 \\ 1 & 1 & 0 & -1 \\ 2 & 1 & -1 & 1 \\ 1 & 1 & 1 & 1 \end{bmatrix}.$$

10. For the matrix $A = \begin{bmatrix} 2 & 0 & 3 \\ 10 & 1 & 17 \\ 7 & 12 & -4 \end{bmatrix}$ verify that the Laplace expansion by the third row and the Laplace expansion by the second column yield the same value.

11. Evaluate the following determinants.

(a) $\det\begin{bmatrix} 0 & 0 & 0 & 3 \\ 0 & 0 & -1 & 5 \\ 0 & 4 & 6 & 11 \\ 2 & 8 & 9 & 0 \end{bmatrix}$ **(b)** $\det\begin{bmatrix} 1 & 0 & 3 & 5 & 9 \\ -4 & 2 & 3 & 5 & -12 \\ 0 & 0 & 2 & 0 & 0 \\ 0 & 0 & 3 & 2 & 0 \\ 0 & 0 & 1 & 2 & 5 \end{bmatrix}$

12. Evaluate the determinants of the following matrices.

(a) $M = \begin{bmatrix} 2 & 2 & 0 & -2 \\ 4 & 1 & -3 & 0 \\ 5 & 13 & 3 & 7 \\ 7 & 4 & -3 & 5 \end{bmatrix}$ **(b)** $N = \begin{bmatrix} 1 & 2 & 4 & 2 & 3 \\ 4 & 0 & 0 & 2 & 0 \\ 5 & 5 & 7 & 1 & 3 \\ 3 & 3 & 3 & -1 & 0 \\ 9 & 4 & 6 & 2 & 3 \end{bmatrix}$

13. Evaluate the following determinants without calculation.

(a) $\det\begin{bmatrix} 3 & 2 & 6 \\ 1.5 & 1 & 3 \\ 4 & 1 & 7 \end{bmatrix}$ **(b)** $\det\begin{bmatrix} 1 & 0 & 0 \\ 3 & 1 & 0 \\ 2 & 0 & 1 \end{bmatrix}$ **(c)** $\det\begin{bmatrix} 0 & 1 & 0 \\ 0 & 0 & 1 \\ 1 & 0 & 0 \end{bmatrix}$

14. Given that $\det\begin{bmatrix} a & b & c \\ d & e & f \\ g & h & i \end{bmatrix} = 5$, find:

(a) $\det\begin{bmatrix} 2a & 2b & 2c \\ 3d & 3e & 3f \\ -g & -h & -i \end{bmatrix}$ **(b)** $\det\begin{bmatrix} g & h & i \\ a & b & c \\ d & e & f \end{bmatrix}$

(c) $\det\begin{bmatrix} a+3d & b+3e & c+3f \\ d+3g & e+3h & f+3i \\ g & h & i \end{bmatrix}$

†**15.** For the matrices A and B of Exercises 1 and 2 show that **(a)** $\det(AB) = \det(A)\det(B)$; and **(b)** $\det(A+B) \neq \det(A) + \det(B)$.

16. State and prove column properties analogous to Properties 2.2, 2.3, 2.4, 2.5, 2.6, and 2.7.

17. Show that if any row of a square matrix A is a linear combination of the other rows of A, then $\det(A) = 0$.

18. Prove Property 2.2 assuming the Laplace expansion. (*Hint:* Let the elementary row operation be $(R_i \leftrightarrow R_j)$. Expand by any row p for which $p \neq i$ and $p \neq j$, and then use mathematical induction.)

†**19.** Show that if A is $n \times n$, then $\det(kA) = k^n \det(A)$.

20. What is the net effect of the following three elementary row operations?

$$R_1 \leftrightarrow R_2, \qquad R_2 \leftrightarrow R_j, \qquad R_2 \leftrightarrow R_1$$

21. What is the net effect of the following four elementary row operations?

$$R_j \leftarrow R_i + R_j, \qquad R_i \leftarrow -R_j + R_i, \qquad R_j \leftarrow R_i + R_j, \qquad R_i \leftarrow -R_i$$

22. How can one find $\det(A)$ from the LU factorization of A discussed in Section 1.8?

23. **(a)** Show that the equation $\det\begin{bmatrix} x & y & 1 \\ a & b & 1 \\ c & d & 1 \end{bmatrix} = 0$ represents the straight line through the points (a, b) and (c, d).

(b) What does the equation $\det\begin{bmatrix} x & y & z & 1 \\ 1 & 2 & 3 & 1 \\ 4 & 5 & 6 & 1 \\ 7 & 8 & 0 & 1 \end{bmatrix} = 0$ represent?

24. Modify the computer program of Exercise 8 in Section 1.3 (on page 35) to compute $\det(A)$.

25. Compare the area of the triangle with vertices (2, 2), (5, 2), and (4, 5) with the determinant of the matrix $A = \begin{bmatrix} 1 & 2 & 2 \\ 1 & 4 & 5 \\ 1 & 5 & 2 \end{bmatrix}$.

26. A matrix S is said to be **skew symmetric** if $S^T = -S$. Show that the determinant of a 3×3 skew-symmetric matrix must be 0.

27. Determine a scalar x so that $\det(A) = 1$.

(a) $A = \begin{bmatrix} 3 & 5 \\ -8 & x \end{bmatrix}$ **(b)** $A = \begin{bmatrix} 1 & 2 & 3 \\ -1 & 4 & 5 \\ 2 & 3 & x \end{bmatrix}$ **(c)** $A = \begin{bmatrix} 1 & x & 2 \\ 1 & 0 & x \\ 2 & 5 & 1 \end{bmatrix}$

28. Use a pocket calculator or MATMAN to evaluate the following determinants.

(a) $\det\begin{bmatrix} 1.21 & 3.52 & 4.81 \\ 3.56 & .00 & 7.10 \\ 4.36 & -2.07 & 5.23 \end{bmatrix}$ **(b)** $\det\begin{bmatrix} 1 & -.897 & .805 & -.721 \\ 1 & 1.23 & 1.52 & 2.30 \\ 1 & 2.45 & 6.01 & 14.7 \\ 1 & 8.12 & 65.9 & 535 \end{bmatrix}$

29. Use the fact that each of the integers 39950, 32215, 30413, 88638, and 11679 is a multiple of 17 to show that $\det\begin{bmatrix} 3 & 9 & 9 & 5 & 0 \\ 3 & 2 & 2 & 1 & 5 \\ 3 & 0 & 4 & 1 & 3 \\ 8 & 8 & 6 & 3 & 8 \\ 1 & 1 & 6 & 7 & 9 \end{bmatrix}$ is a multiple of 17.

2.3 ADDITIONAL PROPERTIES OF DETERMINANTS

1. Can the determinant be used to find matrix inverses and solutions of linear systems?
2. Are determinant methods more or less efficient than elimination methods?

In this section we will develop and illustrate some of the classical properties of the determinant function which are not directly related to the efficient evaluation of the function.

If the first row of the matrix can be written

$$\text{Row}_1(A) = hR + kS = h[r_1 \quad r_2 \quad \cdots \quad r_n] + k[s_1 \quad s_2 \quad \cdots \quad s_n],$$

then we have, directly from the definition of the determinant, that

$$\begin{aligned}\det(A) &= \det\begin{bmatrix} hr_1 + ks_1 \cdots hr_n + ks_n \\ \hline A \text{ with its first} \\ \text{row deleted}\end{bmatrix} \\ &= \sum_{j=1}^{n} (hr_j + ks_j)\text{cof}_{1j} = \sum_{j=1}^{n} hr_j \,\text{cof}_{1j} + \sum_{j=1}^{n} ks_j \,\text{cof}_{1j} \\ &= h \det(A') + k \det(A''),\end{aligned}$$

where A' is the matrix obtained from A by replacing the first row by R and A'' is the matrix obtained from A by replacing the first row by S. The above argument can be applied to any row of the matrix A to yield the following important property.

PROPERTY 2.8 The determinant is a linear function of its rows; for example, if $\text{Row}_i(A) = hR + kS$, then $\det(A) = h\det(A') + k\det(A'')$ where A' and A'' are obtained from A by replacing the ith row of A by R and S, respectively. □

EXAMPLE 1 **Linearity**

Consider the matrix $A = \begin{bmatrix} 4 & 6 & 7 \\ 3 & -14 & -10 \\ 8 & 9 & 3 \end{bmatrix}$.

We can write the second row of A as

$$[3 \quad -14 \quad -10] = 3[1 \quad 2 \quad 0] - 5[0 \quad 4 \quad 2]$$

so, from Property 2.8, we have

$$\det(A) = 3\det\begin{bmatrix} 4 & 6 & 7 \\ 1 & 2 & 0 \\ 8 & 9 & 3 \end{bmatrix} - 5\det\begin{bmatrix} 4 & 6 & 7 \\ 0 & 4 & 2 \\ 8 & 9 & 3 \end{bmatrix}.$$

We could also write the third row of A as $8[1 \quad 0 \quad 0] + 9[0 \quad 1 \quad 0] + 3[0 \quad 0 \quad 1]$. Using Property 2.8 we would then have

$$\det(A) = 8\det\begin{bmatrix} 4 & 6 & 7 \\ 3 & -14 & -10 \\ 1 & 0 & 0 \end{bmatrix} + 9\det\begin{bmatrix} 4 & 6 & 7 \\ 3 & -14 & -10 \\ 0 & 1 & 0 \end{bmatrix} + 3\det\begin{bmatrix} 4 & 6 & 7 \\ 3 & -14 & -10 \\ 0 & 0 & 1 \end{bmatrix}.$$

You should check the validity of these expansions by directly evaluating the determinants involved. ■

We next investigate the relationship between the determinant and the invertibility of a square matrix A. If A is invertible, then we know, from Theorem 1.9, that A is row equivalent to the identity matrix I. Since no elementary row operation can change the determinant from 0 to 1, it follows that $\det(A) \neq 0$. Conversely, if A is not row equivalent to I, then its row-echelon form must have a zero row and hence $\det(A) = 0$. It follows that if $\det(A) \neq 0$, then A is row equivalent to I and hence invertible. We state this result, which has applications in many areas, as our next theorem.

THEOREM 2.3 An $n \times n$ matrix A is invertible if, and only if, $\det(A) \neq 0$. □

The result of Theorem 2.3 can now be added to the list of conditions equivalent to the existence of A^{-1} given in Theorem 1.9.

Another useful property of the determinant function is the following multiplicative property, for which some experimental evidence was given in Exercise 15 of Section 2.2.

PROPERTY 2.9 For any $n \times n$ matrices A and B, $\det(AB) = \det(A)\det(B)$.

Proof If E is an elementary matrix, then its determinant is evident from Properties 2.2, 2.5, and 2.7. In particular

$$\det(I_{R_i \leftarrow kR_i}) = k, \qquad \det(I_{R_i \leftrightarrow R_j}) = -1, \qquad \det(I_{R_i \leftarrow cR_j + R_i}) = 1.$$

By Theorem 1.7, elementary row operations on B can be accomplished by multiplying on the left by the appropriate elementary matrix. It follows that $\det(EB) = \det(E)\det(B)$ if E is an elementary matrix. This establishes that Property 2.9 holds if A is an elementary matrix. More generally, if A is any invertible matrix, then by Theorem 1.9 it can be expressed as a product of elementary matrices $A = E_1E_2 \cdots E_k$. Now, by repeatedly using the fact that the theorem holds for elementary matrices, we have

$$\begin{aligned}\det(AB) &= \det(E_1(E_2 \cdots E_k B)) \\ &= \det(E_1)\det(E_2(\cdots E_k B)) \\ &= \cdots \\ &= (\det(E_1)\det(E_2)\cdots\det(E_k))\det(B) \\ &= \det(E_1 E_2 \cdots E_k)\det(B) \\ &= \det(A)\det(B).\end{aligned}$$

If A is singular, then so is AB (refer to Exercise 18 of Section 2.1) and in this case we have, from Theorem 2.3, $\det(AB) = 0 = 0\det(B) = \det(A)\det(B)$. □

The proof of the last property shows how elementary matrices can be used as the building blocks for all matrices. It is often possible to establish a general result by reducing it to the case of elementary matrices.

In the last section we established that

$$\det(A) = \sum_{j=1}^{n} a_{ij}\operatorname{cof}_{ij}(A) \quad \text{for } i = 1, 2, \ldots, n$$

and that

$$0 = \sum_{j=1}^{n} a_{ij}\operatorname{cof}_{kj}(A) \quad \text{for } i \neq k.$$

These equations look very much like the equations defining matrix multiplication. In fact, if we define a matrix B by

$$b_{ji} = \operatorname{cof}_{ij}(A),$$

then the first equation above says that each of the diagonal elements of AB is equal to $\det(A)$ while the second equation says that each nondiagonal entry of AB is zero. Together these two equations say that $AB = \det(A)I$. It follows, from the column versions of our row results, that we also have $BA = \det(A)I$. If $\det(A) \neq 0$, then we can divide both sides of the last equation by $\det(A)$ to obtain

$$\frac{1}{\det(A)}AB = A\left(\frac{1}{\det(A)}B\right) = I.$$

It follows from Theorem 1.9 that

$$A^{-1} = \frac{1}{\det(A)}B.$$

The matrix B is usually called the adjoint of A.

DEFINITION 2.5 For an $n \times n$ matrix A, the **adjoint of A** is the $n \times n$ matrix whose (i, j) entry is $\operatorname{cof}_{ji}(A)$. We will denote the adjoint of A by $\operatorname{Adj}(A)$. □

It is useful to think of the adjoint matrix as being constructed from A by first replacing each element of A by its cofactor and then transposing the resulting matrix. We summarize the discussion preceding Definition 2.5 with our next theorem.

THEOREM 2.4 For an $n \times n$ matrix A we have

$$A\ \mathrm{Adj}(A) = \mathrm{Adj}(A)A = \det(A)I,$$

and if $\det(A) \neq 0$,

$$A^{-1} = \frac{1}{\det(A)} \mathrm{Adj}(A). \quad \square$$

EXAMPLE 2 **The Adjoint Formula**

Use the adjoint formula to invert $A = \begin{bmatrix} 3 & 1 & -4 \\ 6 & 9 & -2 \\ 1 & 2 & 1 \end{bmatrix}$. The adjoint matrix of A is

$$\mathrm{Adj}(A) = \begin{bmatrix} +\det\begin{bmatrix} 9 & -2 \\ 2 & 1 \end{bmatrix} & -\det\begin{bmatrix} 6 & -2 \\ 1 & 1 \end{bmatrix} & +\det\begin{bmatrix} 6 & 9 \\ 1 & 2 \end{bmatrix} \\ -\det\begin{bmatrix} 1 & -4 \\ 2 & 1 \end{bmatrix} & +\det\begin{bmatrix} 3 & -4 \\ 1 & 1 \end{bmatrix} & -\det\begin{bmatrix} 3 & 1 \\ 1 & 2 \end{bmatrix} \\ +\det\begin{bmatrix} 1 & -4 \\ 9 & -2 \end{bmatrix} & -\det\begin{bmatrix} 3 & -4 \\ 6 & -2 \end{bmatrix} & +\det\begin{bmatrix} 3 & 1 \\ 6 & 9 \end{bmatrix} \end{bmatrix}^T$$

$$= \begin{bmatrix} 13 & -8 & 3 \\ -9 & 7 & -5 \\ 34 & -18 & 21 \end{bmatrix}^T = \begin{bmatrix} 13 & -9 & 34 \\ -8 & 7 & -18 \\ 3 & -5 & 21 \end{bmatrix}.$$

A direct multiplication now establishes that $A\ \mathrm{Adj}(A) = 19I$. It follows that $\det(A) = 19$ and that the inverse of A is

$$A^{-1} = \frac{1}{19}\begin{bmatrix} 13 & -9 & 34 \\ -8 & 7 & -18 \\ 3 & -5 & 21 \end{bmatrix}. \quad \blacksquare$$

The next theorem gives a classical formula for the solution of a system of linear equations in terms of determinants.

THEOREM 2.5 ***Cramer's Rule***

If A is a nonsingular $n \times n$ matrix, then the unique solution of the linear system $AX = K$ is given by

$$x_i = \frac{\det(A_i)}{\det(A)}; \quad i = 1, 2, \ldots, n,$$

where A_i is A with its ith column replaced by K. $\square$

The proof of this theorem amounts to writing the solution of the system as $X = A^{-1}K$ and using the adjoint formula for the inverse; the details are left to the exercises.

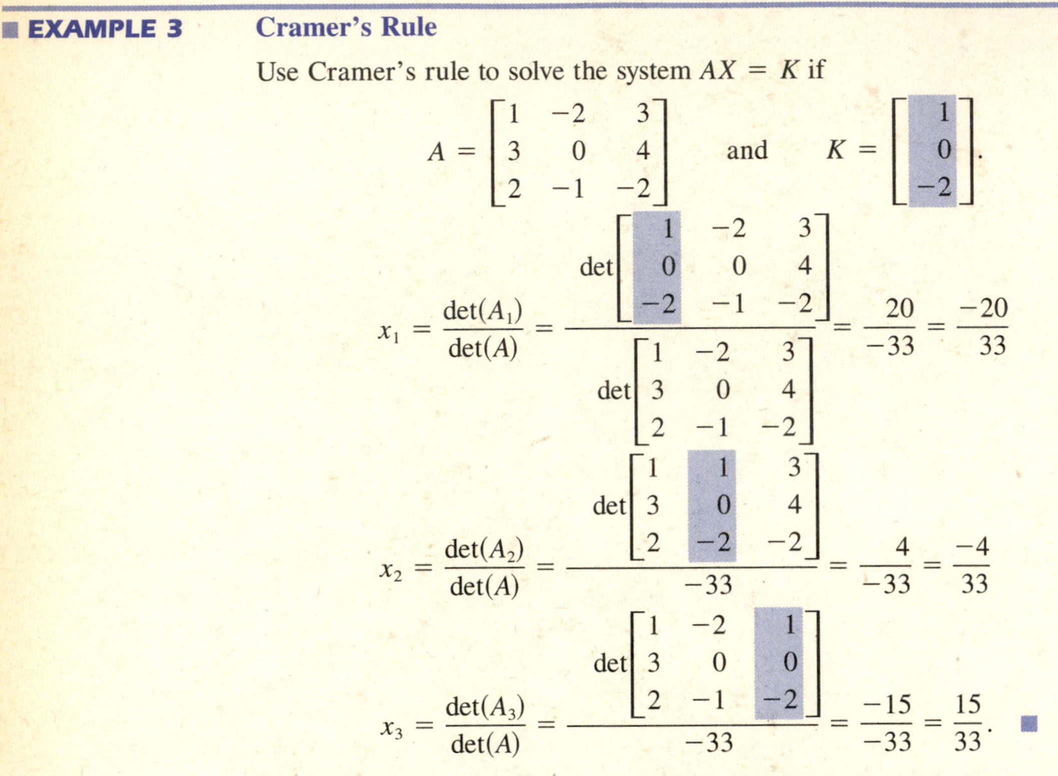

EXAMPLE 3 **Cramer's Rule**

Use Cramer's rule to solve the system $AX = K$ if

$$A = \begin{bmatrix} 1 & -2 & 3 \\ 3 & 0 & 4 \\ 2 & -1 & -2 \end{bmatrix} \quad \text{and} \quad K = \begin{bmatrix} 1 \\ 0 \\ -2 \end{bmatrix}.$$

$$x_1 = \frac{\det(A_1)}{\det(A)} = \frac{\det\begin{bmatrix} 1 & -2 & 3 \\ 0 & 0 & 4 \\ -2 & -1 & -2 \end{bmatrix}}{\det\begin{bmatrix} 1 & -2 & 3 \\ 3 & 0 & 4 \\ 2 & -1 & -2 \end{bmatrix}} = \frac{20}{-33} = \frac{-20}{33}$$

$$x_2 = \frac{\det(A_2)}{\det(A)} = \frac{\det\begin{bmatrix} 1 & 1 & 3 \\ 3 & 0 & 4 \\ 2 & -2 & -2 \end{bmatrix}}{-33} = \frac{4}{-33} = \frac{-4}{33}$$

$$x_3 = \frac{\det(A_3)}{\det(A)} = \frac{\det\begin{bmatrix} 1 & -2 & 1 \\ 3 & 0 & 0 \\ 2 & -1 & -2 \end{bmatrix}}{-33} = \frac{-15}{-33} = \frac{15}{33}.$$

It cannot be emphasized too strongly that the results of the last two theorems do not provide efficient computational tools. To solve an $n \times n$ system using Cramer's rule requires the evaluation of $(n + 1)$ $n \times n$ determinants. The evaluation of a single $n \times n$ determinant, by the most efficient method available, is roughly equivalent to the solution of the entire system by the elimination methods of Chapter 1. Thus Cramer's rule would require nearly $n + 1$ times as much work as Gaussian elimination. Similarly, the adjoint formula for the inverse is nearly n^2 times as much work as the Gauss-Jordan method of Chapter 1. Moreover, if A is singular, then Cramer's Rule provides no information about multiple solutions while the elimination methods of Chapter 1 find all solutions and flag the inconsistent cases. These results have important theoretical applications, but they should not be used as the basis for general purpose computational algorithms.

There are many applications of determinants to problems in geometry. In particular, note the results in Exercises 23 and 25 of Section 2.2 and Exercises 8, 24, 17, and 10 of Section 2.3. We conclude this section by showing how determinants can be used to compute areas. Additional applications to geometry will be considered in the next chapter.

EXAMPLE 4 **Area of a Triangle**

Show that the area of the triangle with vertices at the points (x_1, y_1), (x_2, y_2), and (x_3, y_3) is

$$\pm\frac{1}{2}\det\begin{bmatrix} 1 & 1 & 1 \\ x_1 & x_2 & x_3 \\ y_1 & y_2 & y_3 \end{bmatrix}.$$

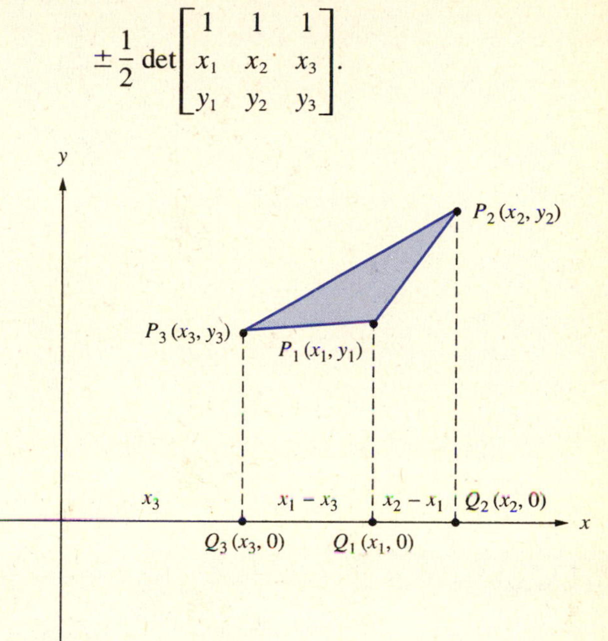

FIGURE 2.3

From Figure 2.3 we see that

$$\begin{aligned} \text{Area } (P_1 P_2 P_3) = \text{Area } (Q_3 Q_2 P_2 P_3) - \text{Area } (Q_3 Q_1 P_1 P_3) \\ - \text{Area } (Q_1 Q_2 P_2 P_1), \end{aligned}$$

where Area (P_1, P_2, P_3) is the area enclosed by the lines connecting points A, B, and C. Using the standard formula for the area of a trapezoid we have

$$\begin{aligned} \text{Area } (P_1 P_2 P_3) &= \frac{(x_2 - x_3)(y_2 + y_3)}{2} - \frac{(x_1 - x_3)(y_3 + y_1)}{2} \\ &\quad - \frac{(x_2 - x_1)(y_1 + y_2)}{2} \\ &= \frac{1}{2}\{x_2y_2 + x_2y_3 - x_3y_2 - x_3y_3 - x_1y_3 - x_1y_1 \\ &\quad + x_3y_3 + x_3y_1 - x_2y_1 - x_2y_2 + x_1y_1 + x_1y_2\} \\ &= \frac{1}{2}\{(x_2y_3 - x_3y_2) - (x_1y_3 - x_3y_1) + (x_1y_2 - x_2y_1)\} \\ &= \frac{1}{2}\det\begin{bmatrix} 1 & 1 & 1 \\ x_1 & x_2 & x_3 \\ y_1 & y_2 & y_3 \end{bmatrix}. \end{aligned}$$

Note that if P_1 is above the line P_2P_3, then the area is the negative of the determinant. ■

Determinants can be used to find the area of any plane figure that can be divided into triangles. For example the area of the six-sided polygon shown in Figure 2.4 could be computed by dividing the area into 4 triangles as shown.

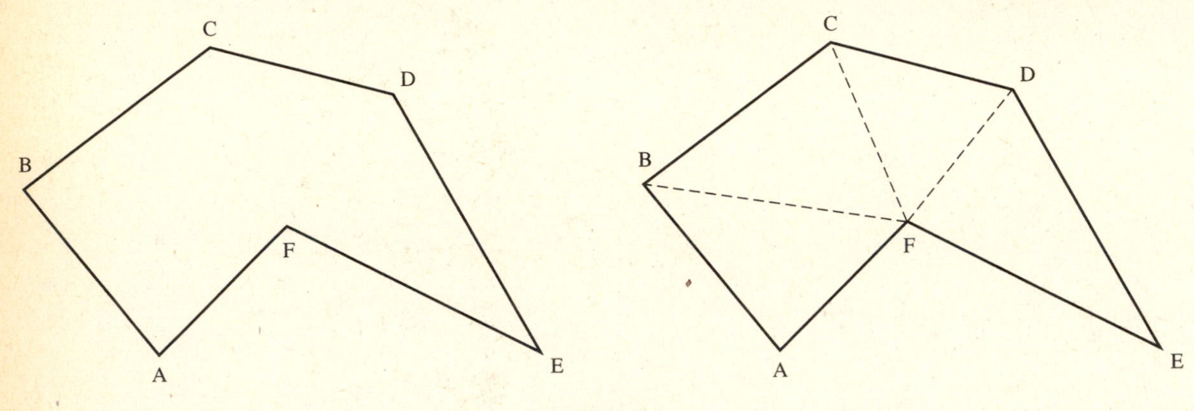

FIGURE 2.4

EXERCISES 2.3

1. Use the adjoint formula to find the inverses of the following matrices.

(a) $A = \begin{bmatrix} -1 & 1 & 3 \\ 2 & 1 & -1 \\ 4 & 2 & 2 \end{bmatrix}$ **(b)** $C = \begin{bmatrix} 1 & -1 & 2 \\ -3 & -3 & 3 \\ 4 & 0 & 5 \end{bmatrix}$

(c) $D = \begin{bmatrix} 1 & 2 & 3 & 4 \\ 1 & -1 & 1 & 2 \\ -1 & 1 & -1 & 11 \\ -1 & 1 & 2 & 3 \end{bmatrix}$ **(d)** $E = \begin{bmatrix} 1 & 0 & 2 & -1 \\ 3 & -2 & 6 & 4 \\ 2 & 2 & -5 & 6 \\ 5 & 4 & 3 & 0 \end{bmatrix}$

(e) $G = \begin{bmatrix} 1 & 1 & 2 & 1 \\ -1 & 1 & 0 & 1 \\ 2 & 1 & 1 & 0 \\ 1 & 3 & 1 & 0 \end{bmatrix}$

2. Solve the system $CX = \begin{bmatrix} 1 \\ 2 \\ 3 \end{bmatrix}$ by Cramer's rule, where $C = \begin{bmatrix} 1 & -1 & 2 \\ -3 & -3 & 3 \\ 4 & 0 & 5 \end{bmatrix}$.

3. Use Cramer's rule to find x_3 (the third component of the matrix X) for the system

$$GX = \begin{bmatrix} 1 \\ 0 \\ 1 \\ 0 \end{bmatrix}, \text{ where } G = \begin{bmatrix} 1 & 1 & 2 & 1 \\ -1 & 1 & 0 & 1 \\ 2 & 1 & 1 & 0 \\ 1 & 3 & 1 & 0 \end{bmatrix}.$$

4. **(a)** Prove that if $\det(A) = \pm 1$ and all the entries of A are integers, then all the entries of A^{-1} are integers.

(b) Let $AX = B$ be a system of n linear equations in n unknowns with integer coefficients and integer constants. Prove that if $\det(A) = \pm 1$, then the solution X has integer entries.

5. Use Cramer's rule to solve for z without solving for x, y, and w.

$$\begin{aligned} 4x + y + z + w &= 6 \\ 7x + 3y - 5z + 8w &= -3 \\ 3x + 7y - z + w &= 1 \\ x + y + z + 2w &= 3 \end{aligned}$$

6. **(a)** If A is partitioned as $A = \begin{bmatrix} A_{11} & 0 \\ 0 & A_{22} \end{bmatrix}$, where A_{11} and A_{22} are square, show that $\det(A) = \det(A_{11})\det(A_{22})$. (*Hint*: Verify the factorization $A = \begin{bmatrix} A_{11} & 0 \\ 0 & I \end{bmatrix}\begin{bmatrix} I & 0 \\ 0 & A_{22} \end{bmatrix}$, and then use Property 2.9.)

(b) Show that $\det\begin{bmatrix} A_{11} & A_{12} \\ 0 & A_{22} \end{bmatrix} = \det(A_{11})\det(A_{22})$.

7. If A and D are square and $\det(A) \neq 0$, verify that

$$\begin{bmatrix} I & 0 \\ -CA^{-1} & I \end{bmatrix}\begin{bmatrix} A & B \\ C & D \end{bmatrix} = \begin{bmatrix} A & B \\ 0 & D - CA^{-1}B \end{bmatrix},$$

and then show that

$$\det\begin{bmatrix} A & B \\ C & D \end{bmatrix} = \det(A)\det(D - CA^{-1}B).$$

8. **(a)** Prove that (x_1, y_1), (x_2, y_2), and (x_3, y_3) are collinear points if, and only if,

$$\det\begin{bmatrix} 1 & 1 & 1 \\ x_1 & x_2 & x_3 \\ y_1 & y_2 & y_3 \end{bmatrix} = 0.$$

(b) Show that the points (x_1, y_1, z_1), (x_2, y_2, z_2), (x_3, y_3, z_3), and (x_4, y_4, z_4) are coplanar (that is, lie on the same plane) if, and only if,

$$\det\begin{bmatrix} 1 & 1 & 1 & 1 \\ x_1 & x_2 & x_3 & x_4 \\ y_1 & y_2 & y_3 & y_4 \\ z_1 & z_2 & z_3 & z_4 \end{bmatrix} = 0.$$

9. Given that $\operatorname{Adj}(A) = \begin{bmatrix} 4 & 5 & 6 \\ 7 & 8 & -3 \\ 1 & 2 & 3 \end{bmatrix}$, find A.

10. For the triangle below, use trigonometry to show that

$$\begin{aligned} b\cos\psi + c\cos\phi &= a \\ c\cos\theta + a\cos\psi &= b \\ a\cos\phi + b\cos\theta &= c \end{aligned}$$

and then apply Cramer's rule to deduce the law of cosines:

$$\cos\theta = \frac{b^2 + c^2 - a^2}{2bc}.$$

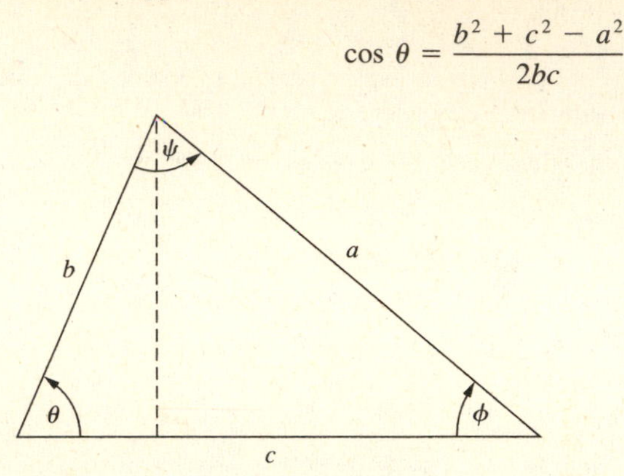

11. For the following matrices find the size of the largest submatrix with nonzero determinant.

$$A = \begin{bmatrix} 1 & -1 & 1 & 1 \\ 1 & 2 & -1 & -1 \\ 2 & -2 & 1 & -1 \\ 0 & -3 & -1 & -1 \end{bmatrix} \quad \text{and} \quad N = \begin{bmatrix} 0 & 0 & 0 & 0 \\ 1 & 0 & 0 & 0 \\ 0 & 1 & 0 & 0 \\ 0 & 0 & 1 & 0 \end{bmatrix}$$

12. **(a)** Show that the Gaussian elimination method for evaluating det(A) requires $\dfrac{n^3 + 2n - 3}{3}$ multiplications if A is $n \times n$.

(b) Find the number of multiplications required to solve an $n \times n$ system by Cramer's rule.

(c) Find the number of multiplications required to find the inverse of an $n \times n$ matrix using the adjoint formula.

13. Suppose that A is a 3×3 matrix for which $\det(A) = 5$. What are
(a) $\det(-A)$? **(b)** $\det(A^2)$? **(c)** $\det(2A)$? **(d)** $\det((\frac{1}{3}A)^3)$?

14. Use the result of Example 4 on page 121 to find the area of the triangle with the following vertices.
(a) (1, 1), (4, 5), and (3, 2) **(b)** (1, 1), (4, 5), and (3, 4)

15. Let A, B, and C be 4×4 matrices for which $C = AB$, $\det(A) = 2$, and $\det(C) = 3$. Find: **(a)** $\det(B)$ **(b)** $\det(ACB)$ **(c)** $\det((2A)^3(3C)^2)$ **(d)** $\det(BC^{-1})$.

16. If A and P are nonsingular $n \times n$ matrices, show that $\det(P^{-1}AP) = \det(A)$.

17. Find the area formed by joining the given points in order.
(a) (1, 1), (2, 3), (5, 4), (5, 0), (1, 1)
(b) (0, 0), (−2, 2), (2, 4), (4, 5), (5, 1), (3, 3), (0, 0)
(c) (1, −1), (2, 3), (1, 4), (3, 5), (4, 4), (6, 4), (5, 3), (6, −1), (3, 1), (1, −1)

18. An $n \times n$ square matrix A is called skew symmetric if $A = -A^T$. Show that if A is skew symmetric, then $\det(A) = (-1)^n \det(A)$ and hence that $\det(A) = 0$ if n is odd.

19. For an $n \times n$ nonsingular matrix A, show that $\det(\text{Adj}(A)) = \det(A)^{n-1}$ and hence that $\text{Adj}(\text{Adj}(A)) = \det(A)^{n-2}A$.

20. Show that the matrix $A = \begin{bmatrix} \sin^2\alpha & \cos^2\beta & \sin^2\gamma \\ \cos^2\alpha & \sin^2\beta & \cos^2\gamma \\ 2 & 2 & 2 \end{bmatrix}$ is singular.

21. Evaluate $\det(\lambda I - C)$, where $C = \begin{bmatrix} 0 & 0 & 0 & -a_0 \\ 1 & 0 & 0 & -a_1 \\ 0 & 1 & 0 & -a_2 \\ 0 & 0 & 1 & -a_3 \end{bmatrix}$.

(*Hint:* Use the elementary row operations $(R_1 \leftarrow \lambda R_2 + R_1)$, $(R_1 \leftarrow \lambda^2 R_3 + R_1)$, and $(R_1 \leftarrow \lambda^3 R_4 + R_1)$ followed by an expansion by the first row.)

22. Use Theorem 2.3 to find all values of λ for which $(\lambda I - A)$ is singular if A is the given matrix.

(a) $A = \begin{bmatrix} 5 & 1 \\ 4 & 8 \end{bmatrix}$ **(b)** $A = \begin{bmatrix} 2 & 1 & 1 \\ 2 & 3 & 2 \\ 1 & 1 & 2 \end{bmatrix}$

23. For what values of x do the following matrices fail to have inverses?

(a) $\begin{bmatrix} 2 & x \\ 1 & 4 \end{bmatrix}$ **(b)** $\begin{bmatrix} 1 & x \\ x & 16 \end{bmatrix}$ **(c)** $\begin{bmatrix} 2 & 1 & 3 \\ 1 & 0 & 2 \\ 0 & x & 1 \end{bmatrix}$ **(d)** $\begin{bmatrix} x & 1 & 1 \\ 1 & x & 1 \\ 1 & 1 & x \end{bmatrix}$

24. What do the following formulas represent?

(a) $\det\begin{bmatrix} 1 & x & y & x^2 + y^2 \\ 1 & x_1 & y_1 & x_1^2 + y_1^2 \\ 1 & x_2 & y_2 & x_2^2 + y_2^2 \\ 1 & x_3 & y_3 & x_3^2 + y_3^2 \end{bmatrix} = 0$

(b) $\det\begin{bmatrix} 1 & x & y & x^2 & y^2 \\ 1 & x_1 & y_1 & x_1^2 & y_1^2 \\ 1 & x_2 & y_2 & x_2^2 & y_2^2 \\ 1 & x_3 & y_3 & x_3^2 & y_3^2 \\ 1 & x_4 & y_4 & x_4^2 & y_4^2 \end{bmatrix} = 0$

(c) $\det\begin{bmatrix} 1 & x & y & x^2 & x^3 \\ 1 & x_1 & y_1 & x_1^2 & x_1^3 \\ 1 & x_2 & y_2 & x_2^2 & x_2^3 \\ 1 & x_3 & y_3 & x_3^2 & x_3^3 \\ 1 & x_4 & y_4 & x_4^2 & x_4^3 \end{bmatrix} = 0$

2.4 EIGENVALUES AND EIGENVECTORS*

Many applications (such as those in Exercises 20 and 21 of this section) require that one find a nonzero matrix X such that $AX = \lambda X$, where A is a given $n \times n$ matrix and λ is a scalar. This problem is called the algebraic eigenvalue problem

*This material can be delayed until the beginning of Chapter 5 if desired.

and is the second most frequently encountered matrix problem; the first is the solution of a system of linear equations. Additional motivation for this problem will be provided in Chapter 5, where we will use the techniques of this section to find a nonsingular matrix P such that $P^{-1}AP$ is a diagonal matrix. The computational tools are all available now and provide an important application which uses both determinants and the techniques of Chapter 1. Some of you may find immediate use for this material in other areas. You will no doubt note that many of these ideas have already been introduced in earlier exercises.

DEFINITION 2.6 Let A be any $n \times n$ matrix. A nonzero $n \times 1$ matrix X such that $AX = \lambda X$ is called an **eigenvector** of A and the scalar λ is called the **eigenvalue** of A, associated with X. □

EXAMPLE 1 **Eigenvalues and Eigenvectors**

For $A = \begin{bmatrix} 2 & 6 \\ 1 & 3 \end{bmatrix}$, $X = \begin{bmatrix} 4 \\ 2 \end{bmatrix}$, and $Z = \begin{bmatrix} 1 \\ 1 \end{bmatrix}$, we have

$$AX = \begin{bmatrix} 20 \\ 10 \end{bmatrix} = 5\begin{bmatrix} 4 \\ 2 \end{bmatrix} = 5X \text{ and } AZ = \begin{bmatrix} 8 \\ 4 \end{bmatrix} \neq \lambda\begin{bmatrix} 1 \\ 1 \end{bmatrix} \text{ for any } \lambda.$$

Thus, X is an eigenvector of A associated with the eigenvalue $\lambda = 5$ and Z is not an eigenvector of A. We leave it to you to verify that $Y = \begin{bmatrix} 3 \\ -1 \end{bmatrix}$ is an eigenvector of A associated with the eigenvalue $\lambda = 0$. ■

The defining equation, $AX = \lambda X$, is equivalent to $\lambda IX = AX$ or to $(\lambda I - A)X = 0$. This last equation sheds some light on the problem of how to find eigenvectors, since it shows that eigenvectors are solutions of certain homogeneous systems of linear equations. If we knew the eigenvalues, we could solve the homogeneous systems $(\lambda I - A)X = 0$ for the eigenvectors. If we knew the eigenvectors, then we could find the eigenvalues by comparing the entries of X and AX. If we know neither, and this is the usual case, then we can use Theorem 1.9 and Theorem 2.3 to observe the following.

> The homogeneous system $(\lambda I - A)X = 0$ has a nonzero solution if, and only if, $\det(\lambda I - A) = 0$, that is, the matrix $\lambda I - A$ is singular. The scalar λ is an eigenvalue of A if, and only if, $\det(\lambda I - A) = 0$.

For the matrix A of Example 1 we see that

$$\det(5I - A) = \det\begin{bmatrix} 3 & -6 \\ -1 & 2 \end{bmatrix} = 0 \text{ and}$$

$$\det(0I - A) = \det\begin{bmatrix} -2 & -6 \\ -1 & -3 \end{bmatrix} = 0.$$

These calculations show that $\lambda = 5$ and $\lambda = 0$ are eigenvalues of A. Note that $\lambda = 3$ is not an eigenvalue of A since

$$\det(3I - A) = \det\begin{bmatrix} 1 & -6 \\ -1 & 0 \end{bmatrix} = -6 \neq 0.$$

If we think of λ as a parameter, then $\det(\lambda I - A)$ is a polynomial in λ. This polynomial is called the **characteristic polynomial** of the matrix A (recall Exercises 21 and 22 of Section 2.3 and Exercise 3 of Section 2.1). For the matrix A of Example 1,

$$\begin{aligned}\det(\lambda I - A) &= \det\begin{bmatrix} \lambda - 2 & -6 \\ -1 & \lambda - 3 \end{bmatrix} \\ &= (\lambda - 2)(\lambda - 3) - 6 \\ &= \lambda^2 - 5\lambda = \lambda(\lambda - 5).\end{aligned}$$

The problem of finding the eigenvalues and eigenvectors of a matrix A can thus be reduced to the following three steps.

1. Find the characteristic polynomial $c(\lambda) = \det(\lambda I - A)$.
2. Find the eigenvalues of A by finding the roots λ_j of the characteristic equation $c(\lambda) = 0$.
3. For each eigenvalue λ_j, find an associated eigenvector of A by solving the homogeneous system $(\lambda_j I - A)X = 0$.

If the matrix A is large, then each of these three steps can involve a substantial amount of calculation. We now illustrate these calculations for a 3×3 matrix.

EXAMPLE 2 Eigenvalues and Eigenvectors

Find the eigenvalues and eigenvectors of

$$A = \begin{bmatrix} 0 & 1 & 0 \\ 0 & 0 & 1 \\ 2 & 1 & -2 \end{bmatrix}.$$

Step 1 The characteristic polynomial of A is

$$\begin{aligned}c(\lambda) = \det(\lambda I - A) &= \det\left(\lambda\begin{bmatrix} 1 & 0 & 0 \\ 0 & 1 & 0 \\ 0 & 0 & 1 \end{bmatrix} - \begin{bmatrix} 0 & 1 & 0 \\ 0 & 0 & 1 \\ 2 & 1 & -2 \end{bmatrix}\right) \\ &= \det\begin{bmatrix} \lambda & -1 & 0 \\ 0 & \lambda & -1 \\ -2 & -1 & \lambda + 2 \end{bmatrix}.\end{aligned}$$

Using the expansion by the first row we have

$$\det(\lambda I - A) = \lambda \det\begin{bmatrix} \lambda & -1 \\ -1 & \lambda + 2 \end{bmatrix} + \det\begin{bmatrix} 0 & -1 \\ -2 & \lambda + 2 \end{bmatrix}$$
$$+ 0 \det\begin{bmatrix} 0 & \lambda \\ -2 & -1 \end{bmatrix}$$
$$= \lambda(\lambda^2 + 2\lambda - 1) - 2 = \lambda^3 + 2\lambda^2 - \lambda - 2.$$

Step 2 From the easily verified factorization

$$\lambda^3 + 2\lambda^2 - \lambda - 2 = (\lambda - 1)(\lambda + 1)(\lambda + 2),$$

we see that the roots of the characteristic equation $c(\lambda) = 0$ are $\lambda_1 = 1$, $\lambda_2 = -1$, and $\lambda_3 = -2$. These roots are the eigenvalues of A.

Step 3 In order to find an eigenvector for $\lambda_1 = 1$ we solve the homogeneous system $(\lambda_1 I - A)X = (I - A)X = 0$, for which the augmented matrix and reduction to reduced row-echelon form follow:

$$\begin{bmatrix} 1 & -1 & 0 & 0 \\ 0 & 1 & -1 & 0 \\ -2 & -1 & 3 & 0 \end{bmatrix} \to \begin{bmatrix} 1 & -1 & 0 & 0 \\ 0 & 1 & -1 & 0 \\ 0 & -3 & 3 & 0 \end{bmatrix} \to \begin{bmatrix} 1 & 0 & -1 & 0 \\ 0 & 1 & -1 & 0 \\ 0 & 0 & 0 & 0 \end{bmatrix}.$$

The system equivalent to this final matrix is

$$x_1 - x_3 = 0$$
$$x_2 - x_3 = 0.$$

There are many solutions to this system; if we set $x_3 = 1$, we find $x_1 = x_2 = 1$ so that

$$X_1 = [1 \quad 1 \quad 1]^T \text{ is an eigenvector of } A \text{ for } \lambda_1 = 1.$$

Note that, for any $c \neq 0$, cX_1 is also an eigenvector for $\lambda = 1$. In order to find an eigenvector of A, for $\lambda_2 = -1$, we solve the system $(-I - A)X = 0$ via the row reduction

$$\begin{bmatrix} -1 & -1 & 0 & 0 \\ 0 & -1 & -1 & 0 \\ -2 & -1 & 1 & 0 \end{bmatrix} \to \begin{bmatrix} 1 & 1 & 0 & 0 \\ 0 & -1 & -1 & 0 \\ 0 & 1 & 1 & 0 \end{bmatrix}$$
$$\to \begin{bmatrix} 1 & 0 & -1 & 0 \\ 0 & 1 & 1 & 0 \\ 0 & 0 & 0 & 0 \end{bmatrix} \to \begin{cases} x_1 - x_3 = 0 \\ x_2 + x_3 = 0. \end{cases}$$

If we set $x_3 = 1$, we have $x_1 = 1$ and $x_2 = -1$ as one solution. Thus $X_2 = [1 \quad -1 \quad 1]^T$ is an eigenvector of A, for $\lambda_2 = -1$. Similar calculations, which you should carry out in detail, show that $X_3 = [1 \quad -2 \quad 4]^T$ is an eigenvector of A, for $\lambda_3 = -2$. ■

Several comments about the solution process illustrated in Example 2 are in order.

(1) For small matrices, it is best to evaluate the determinant, which defines the characteristic equation, by repeated use of the Laplace expansion (choose the row with the most zeros) rather than using elementary row operations, as recommended for matrices of constants. If A is $n \times n$, then the characteristic equation is a polynomial equation of degree n, of the form

$$\det(\lambda I - A) = \lambda^n - \operatorname{tr}(A)\lambda^{n-1} + \cdots + (-1)^n \det(A) = 0.$$

The observation that the coefficient of λ^{n-1} is $-\operatorname{tr}(A) = -(a_{11} + a_{22} + \cdots + a_{nn})$ is a useful check.* Note that the characteristic polynomial is always a **monic polynomial;** that is, the coefficient of λ^n is always 1.

(2) Finding all the roots of a polynomial equation is in general quite difficult since no general formulas exist for n greater than 4; for $n = 3$ and $n = 4$ these formulas are very complicated, difficult to use, and almost impossible to remember. If a monic polynomial equation with integer coefficients has any rational roots, then they must be integers that are factors of the constant term. This restriction provides a reasonably small set of candidates for roots; these candidates may then be checked individually. For hand calculations, the algebraic technique of synthetic division is an efficient way of doing the polynomial evaluations necessary to check the various candidates. If you are not familiar with synthetic division, please work Exercise 18 before you proceed further. If this "guess and check" method does not locate the roots, then one must resort to numerical approximation methods such as the bisection method or Newton's method (see Exercise 24 of this section).

(3) Eigenvectors are solutions of homogeneous systems and as such are never unique. The set of eigenvectors of A, associated with a given eigenvalue, is closed under addition and scalar multiplication. For small problems with integer data, it is frequently possible to choose the eigenvectors to have integer entries.

(4) Polynomial equations with real coefficients may very well have some complex roots. If there are complex roots, then they occur in conjugate pairs of the form $a \pm bi$ where the complex number $i = \sqrt{-1}$ satisfies $i^2 = -1$. If an eigenvalue is complex, then the corresponding eigenvector will also have complex entries. We will, for computational ease, restrict our examples and exercises largely to problems with real eigenvalues, but you should be alerted to the fact that many important applications, including most oscillatory phenomena, involve complex eigenvalues.

---

* The trace function was defined in Exercise 11 of Section 2.1.

EXAMPLE 3 **Synthetic Division and Eigenvalues**

Find the eigenvalues and all eigenvectors of

$$B = \begin{bmatrix} 5 & 4 & 2 \\ 4 & 5 & 2 \\ 2 & 2 & 2 \end{bmatrix}.$$

The characteristic polynomial of B is $c(\lambda) = \det(\lambda I - B)$

$$= \det \begin{bmatrix} \lambda - 5 & -4 & -2 \\ -4 & \lambda - 5 & -2 \\ -2 & -2 & \lambda - 2 \end{bmatrix}$$

$$= (\lambda - 5)[(\lambda - 5)(\lambda - 2) - 4] + 4[-4(\lambda - 2) - 4] - 2[8 + 2(\lambda - 5)]$$

$$= \lambda^3 - 12\lambda^2 + 21\lambda - 10 = c(\lambda).$$

The only rational possibilities for the eigenvalues are factors of the constant term: ± 1, ± 2, ± 5, and ± 10. From the synthetic divisions

$$10)\ \begin{array}{rrrr} 1 & -12 & 21 & -10 \\ & 10 & -20 & 10 \\ \hline 1 & -2 & 1 & 0 \end{array} \quad \text{and} \quad 2)\ \begin{array}{rrrr} 1 & -12 & 21 & -10 \\ & 2 & -20 & 2 \\ \hline 1 & -10 & 1 & -8 \end{array}$$

we see that $c(10) = 0$ and $c(2) = -8$. Thus 10 is an eigenvalue of B but 2 is not. Moreover, from the first division we can see that the characteristic equation of B factors as

$$\begin{aligned} \lambda^3 - 12\lambda^2 + 21\lambda - 10 &= (\lambda - 10)(\lambda^2 - 2\lambda + 1) \\ &= (\lambda - 10)(\lambda - 1)(\lambda - 1) = 0. \end{aligned}$$

The eigenvalues of B are thus $\lambda_1 = 10$ and $\lambda_2 = \lambda_3 = 1$.

The augmented matrix for the system $(10I - B)X = 0$ is

$$\begin{bmatrix} 5 & -4 & -2 & 0 \\ -4 & 5 & -2 & 0 \\ -2 & -2 & 8 & 0 \end{bmatrix} \to \begin{bmatrix} 1 & 1 & -4 & 0 \\ 0 & 9 & -18 & 0 \\ 0 & 0 & 0 & 0 \end{bmatrix}$$

$$\to \begin{bmatrix} 1 & 0 & -2 & 0 \\ 0 & 1 & -2 & 0 \\ 0 & 0 & 0 & 0 \end{bmatrix} \to \begin{cases} x_1 - 2x_3 = 0 \\ x_2 - 2x_3 = 0. \end{cases}$$

From this row reduction we see that the general solution of the system $(10I - B)X = 0$; that is, the most general eigenvector of B associated with the eigvenvalue 10 is

$$X = \begin{bmatrix} 2c \\ 2c \\ c \end{bmatrix} = c \begin{bmatrix} 2 \\ 2 \\ 1 \end{bmatrix}.$$

In order to find the eigenvectors associated with the repeated eigenvalue $\lambda = 1$ we consider the system $(I - B)X = 0$ for which the augmented matrix and row reduction are as follows:

$$\begin{bmatrix} -4 & -4 & -2 & 0 \\ -4 & -4 & -2 & 0 \\ -2 & -2 & -1 & 0 \end{bmatrix} \to \begin{bmatrix} 2 & 2 & 1 & 0 \\ 0 & 0 & 0 & 0 \\ 0 & 0 & 0 & 0 \end{bmatrix} \to 2x_1 + 2x_2 + x_3 = 0.$$

The general solution of this 1×3 system, which involves two arbitrary constants, is

$$X = \begin{bmatrix} c_1 \\ c_2 \\ -2c_1 - 2c_2 \end{bmatrix} = c_1 \begin{bmatrix} 1 \\ 0 \\ -2 \end{bmatrix} + c_2 \begin{bmatrix} 0 \\ 1 \\ -2 \end{bmatrix}$$

and these are all of the eigenvectors of B for $\lambda = 1$. ■

Note that in Example 3 we found a set of eigenvectors with two degrees of freedom, associated with the double eigenvalue $\lambda = 1$. *This frequently happens, but is not always the case.* Note that the matrix

$$J = \begin{bmatrix} 3 & 1 & 0 \\ 0 & 3 & 1 \\ 0 & 0 & 3 \end{bmatrix}$$

has $\lambda = 3$ as a triple root, but the set of eigenvectors has only one degree of freedom (see Exercise 22). We will return to the study of eigenvalues and eigenvectors in Chapter 5.

The software and the TI-85 have built in functions for doing the calculations of this section. The TI-85 cannot find the characteristic polynomial, but it can plot its graph. The eigenvalue commands on the TI-85 appear on the MATH submenu of the MATRIX menu.

	MATALG	MATLAB	TI-85
characteristic polynomial	p, A	poly(A)	use the graph menu on det(x * ident(n) − A)
eigenvalues	= eval(A)	eig(A)	eigVl(A)
eigenvectors	= evec(A)	[P, D] = eig(A)	eigVc(A)

EXERCISES 2.4

1. Determine if either X or Y is an eigenvector for A and if so, determine the eigenvalue.

(a) $A = \begin{bmatrix} 11 & -9 \\ 4 & -2 \end{bmatrix}$, $X = \begin{bmatrix} 1 \\ 1 \end{bmatrix}$, $Y = \begin{bmatrix} 1 \\ 2 \end{bmatrix}$

(b) $A = \begin{bmatrix} 2 & 1 & 1 \\ 2 & 3 & 2 \\ 1 & 1 & 2 \end{bmatrix}$, $X = \begin{bmatrix} -1 \\ 1 \\ 0 \end{bmatrix}$, $Y = \begin{bmatrix} -1 \\ 0 \\ 1 \end{bmatrix}$

For the matrices in Exercises 2–10 find the characteristic equation, the real eigenvalues, and the corresponding eigenvectors.

2. $\begin{bmatrix} 1 & 3 \\ -1 & 5 \end{bmatrix}$ **3.** $\begin{bmatrix} 16 & -14 \\ -14 & 9 \end{bmatrix}$ **4.** $\begin{bmatrix} 2 & 1 & 1 \\ 2 & 3 & 2 \\ 1 & 1 & 2 \end{bmatrix}$

5. $\begin{bmatrix} 1 & 0 & -4 \\ 0 & 5 & 4 \\ -4 & 4 & 3 \end{bmatrix}$ **6.** $\begin{bmatrix} 5 & -1 & 1 \\ -1 & 5 & -1 \\ 1 & -1 & 5 \end{bmatrix}$ **7.** $\begin{bmatrix} 2 & 1 & 0 \\ 0 & 2 & 1 \\ 2 & 1 & 0 \end{bmatrix}$

8. $\begin{bmatrix} -3 & 3 & -2 \\ 3 & 2 & -3 \\ -2 & -3 & -3 \end{bmatrix}$ **9.** $\begin{bmatrix} -1 & -1 & -3 \\ -1 & 1 & 1 \\ -3 & 1 & -1 \end{bmatrix}$ **10.** $\begin{bmatrix} 2 & 1 & 5 \\ 1 & -2 & -1 \\ 5 & -1 & 2 \end{bmatrix}$

11. Consider the matrices A of Exercises 2, 8, 9 and 10. Let P be a matrix whose columns are the eigenvectors of A. Compute $P^{-1}AP$.

12. Recall the definition of tr(A) from Exercise 11 of Section 2.1 and verify that $\det(\lambda I - A) = \lambda^n - \text{tr}(A)\lambda^{n-1} + \cdots + (-1)^n \det(A)$ for a general 3×3 matrix A.

13. What are the eigenvalues of
(a) a diagonal matrix?
(b) a triangular matrix?

14. Show that $\lambda = 0$ is an eigenvalue of a matrix A if, and only if, A is singular.

15. Show that A and $B = P^{-1}AP$ have the same characteristic polynomial.

16. The characteristic polynomial of the matrix $B = \begin{bmatrix} 0 & 1 & 0 \\ 0 & 0 & 1 \\ 3 & -5 & 2 \end{bmatrix}$ is $\lambda^3 - 2\lambda^2 + 5\lambda - 3$. Show that $B^3 - 2B^2 + 5B - 3I = 0$.

17. By multiplying the equation $AX = \lambda X$ on the left by A, show that if λ is an eigenvalue of A, then λ^2 is an eigenvalue of A^2. What is the corresponding eigenvector? What are the eigenvalues of A^3?

†18. Consider the polynomial function

$$p(\lambda) = a_k\lambda^k + \cdots + a_2\lambda^2 + a_1\lambda + a_0.$$

(a) Show that the value of $p(r)$ may be evaluated by the algorithm
(1) $b_{k-1} = a_k$
(2) For $i = k-2, k-3, \ldots, 3, 2, 1$ $\quad b_i = b_{i+1}r + a_{i+1}$
(3) $p(r) = b_0 r + a_0$.
(*Hint:* Rewrite $p(r)$ in the form
$p(r) = [(\cdots([(a_k r + a_{k-1})r + a_{k-2}]r + a_{k-3})r + \cdots + a_2)r + a_1]r + a_0$.)

(b) Explain why this is an efficient way to evaluate $p(r)$ for a given number r, especially on a pocket calculator with at least one storage location.

(c) Show that $p(\lambda) = (\lambda - r)(b_{k-1}\lambda^{k-1} + \cdots + b_1\lambda + b_0) + p(r)$.

(d) Show that the algorithm can be arranged in the following way for hand calculation.

$$\begin{array}{r|ccccccc} r) & a_k & a_{k-1} & a_{k-2} & \cdots & a_2 & a_1 & a_0 \\ & & rb_{k-1} & rb_{k-2} & \cdots & rb_2 & rb_1 & rb_0 \\ \hline & b_{k-1} & b_{k-2} & b_{k-3} & & b_1 & b_0 & rb_0 + a_0 \end{array}$$

(e) This process is called **synthetic division.** For $p(\lambda) = \lambda^3 - 6\lambda^2 + 9\lambda - 4$, evaluate $p(3)$, $p(5)$, and $p(1)$. Find the quotient when $p(\lambda)$ is divided by $\lambda - 1$.

(f) Delete the subscripts on the b_i in the algorithm of part (a). What does the resulting algorithm do?

19. Use Exercise 6 in Section 2.3 to find the characteristic polynomial of

$$A = \begin{bmatrix} 2 & 0 & 0 & 0 \\ 1 & 2 & 0 & 0 \\ 0 & 0 & 0 & 1 \\ 0 & 0 & -6 & 5 \end{bmatrix}.$$

†20. Let A be an $n \times n$ matrix and let $Y(t)$ be an $n \times 1$ matrix of unknown functions. Show that

$$Y(t) = Xe^{\lambda t} \text{ is a solution of } \frac{d}{dt} Y(t) = AY(t)$$

if, and only if, $AX = \lambda X$.

†21. As we saw in Example 5 of Section 1.5, population growth can be modeled by a matrix difference equation of the form

$$X^{k+1} = AX^k$$

where A is a fixed matrix, characteristic of the species and its environment, and X^k is the population distribution at the kth time step. A population distribution X^k is said to be **stable** if $X^{k+1} = \lambda X^k$ for some scalar λ. Show that the population distribution X is stable if, and only if, X is an eigenvector of A.

22. Show that the characteristic polynomial of $J = \begin{bmatrix} 3 & 0 & 0 \\ 0 & 3 & 1 \\ 0 & 0 & 3 \end{bmatrix}$ is $(\lambda - 3)^3$ and that the set of eigenvectors of J has only one degree of freedom.

23. Show that if $P^{-1}AP = D$ is a diagonal matrix, then the columns of P are eigenvectors of A. (*Hint:* Look at the equation $AP = PD$ column by column.)

24. **Newton's method** for finding a solution of the equation $f(x) = 0$ is

(1) x_0 arbitrary;

(2) for $i = 0, 1, 2, 3, 4, \ldots$

(3) $$x_{i+1} = x_i - \frac{f(x_i)}{f'(x_i)}.$$

The sequence $\{x_i\}$ will converge to a root of $f(x) = 0$ if x_0 is sufficiently close to a root. Normally the sequence will converge to the root which is nearest to x_0.

(a) Use Newton's method to find the roots of $x^2 + 5x - 11 = 0$.

(b) Use Newton's method to find the roots of $x^4 + 9x^3 + 11x^2 - 34x - 22 = 0$.

(c) Use Newton's method to find the eigenvalues of the matrices

$$A = \begin{bmatrix} 1 & 1 & 3 \\ 1 & -2 & 1 \\ 3 & 1 & 3 \end{bmatrix} \quad \text{and} \quad B = \begin{bmatrix} 2 & -1 & 1 \\ -1 & 2 & -1 \\ 1 & -1 & 1 \end{bmatrix}.$$

(d) Delete all the subscripts in the Newton algorithm. Show that the resulting algorithm is equivalent to the original. What does this implementation of the algorithm mean for storage in a computer program?

CHAPTER 2 SUMMARY

Chapter 2 introduces and studies the determinant function on square matrices. Applications to geometry are considered and eigenvalues and eigenvectors are introduced as an application of determinants and homogenous systems.

KEY DEFINITIONS

The determinant function. *99*
Minor and cofactor of (i, j) position of A. *98, 109*
Transpose of a matrix. *103*
Adjoint of a matrix. *118*
Eigenvalues, eigenvectors, and characteristic polynomial of A. *126*

KEY FACTS

The determinant of a triangular matrix is the product of the diagonal entries.
Transposition does not change the determinant.
Elementary row operations have simple and predictable effects on the determinant.
The Laplace expansion
The determinant is a linear function of the rows of the matrix.
A matrix A has an inverse if, and only if, $\det(A) \neq 0$.
$\det(AB) = \det(A)\det(B)$.
Adjoint formula: $A^{-1} = (\frac{1}{\det(A)})\text{Adj}(A)$ if $\det(A) \neq 0$.
Cramer's rule for solving $AX = K$.
λ is an eigenvalue of A if, and only if, $\det(\lambda I - A) = 0$.

COMPUTATIONAL PROCEDURES

Basket-weaving trick for $n = 2$ or 3 only.
Gaussian elimination method for evaluating determinants.
Synthetic division.
Find characteristic polynomial, the eigenvalue, and the eigenvectors of A.

APPLICATIONS

Area of a triangle and other polyhedra.
Eigenvalues and eigenvectors.

CHAPTER 2 REVIEW EXERCISES

1. Are the following statements *true* or *false*?

(a) $\det(AB) = \det(A)\det(B)$
(b) $\det(A + B) = \det(A) + \det(B)$
(c) $\det(kA) = k\det(A)$
(d) $\det(kA) = k^n \det(A)$
(e) $\det(A^{-1}) = \frac{1}{\det(A)}$
(f) $\det(A^T) = \det(A)$
(g) If A is row equivalent to I, then $\det(A) = 1$.
(h) If $\det(A) = 0$, then A is nonsingular.
(i) If $\det(A) = 0$, then $AX = 0$ has a nonzero solution.
(j) $\det(\text{Adj}(A)) = \det(A)^n$

2. (a) Evaluate the determinants of

$$A = \begin{bmatrix} 1 & 1 & 2 & 1 \\ 1 & 1 & 0 & -1 \\ 0 & 1 & 1 & 2 \\ 0 & 3 & 1 & 1 \end{bmatrix} \quad \text{and} \quad B = \begin{bmatrix} 1 & 1 & 1 & 1 \\ 1 & 1 & 2 & 1 \\ 1 & 3 & 1 & 1 \\ 4 & 1 & 1 & 1 \end{bmatrix}.$$

(b) Find the entry in the (3, 2) position of the adjoint of A.

3. Find the characteristic equation of $M = \begin{bmatrix} 1 & 1 & 1 \\ 0 & 2 & 1 \\ 1 & 0 & 1 \end{bmatrix}$.

4. Find the adjoint of $C = \begin{bmatrix} 3 & 2 & -1 \\ 1 & 6 & 3 \\ 2 & -4 & 0 \end{bmatrix}$ and use it to find C^{-1}.

5. Use Cramer's rule to solve the linear system

$$\begin{aligned} x - 2y + 3z &= 1 \\ 2x - y - 2z &= -2 \\ 3x \qquad + 4z &= 0. \end{aligned}$$

6. Show that the eigenvalues of $A = \begin{bmatrix} a & b \\ b & c \end{bmatrix}$ must be real numbers.

7. Given $A = \begin{bmatrix} 7 & 4 \\ -1 & 2 \end{bmatrix}$, find the characteristic polynomial, the eigenvalues, and the eigenvectors.

8. Let A be a matrix with the property that the sum of the elements in each column is 0. Show that $\det(A) = 0$.

9. Use Cramer's rule to solve for x_1 and y_1 in terms of x and y:

$$\begin{aligned} x &= x_1 \cos\theta - y_1 \sin\theta \\ y &= x_1 \sin\theta + y_1 \cos\theta. \end{aligned}$$

10. Use the adjoint formula to show that the inverse of an invertible upper triangular matrix is also upper triangular.

11. **(a)** Evaluate $\det(B)$ given that $B = \begin{bmatrix} 1 & 2 & 3 & 1 \\ 2 & 5 & 3 & 2 \\ -3 & 9 & 2 & 1 \\ 1 & 6 & 5 & 2 \end{bmatrix}$.

(b) Find the entry in the $(1, 4)$ position of the adjoint of B.

12. Show that the area of the parallelogram with vertices at $(0, 0)$, (a, b), (c, d), and $(a + c, b + d)$ is the absolute value of the determinant of the matrix $\begin{bmatrix} a & b \\ c & d \end{bmatrix}$.

13. **(a)** Find the characteristic equation of $A = \begin{bmatrix} 1 & 0 & 1 \\ 0 & 1 & 2 \\ 1 & 2 & 5 \end{bmatrix}$.

(b) Find an eigenvector of A associated with the eigenvalue $\lambda = 6$.

14. Show that $\det(V) = (x_2 - x_1)(x_3 - x_1)(x_4 - x_1)(x_3 - x_2)(x_4 - x_3)$ if

$$V = \begin{bmatrix} 1 & 1 & 1 & 1 \\ x_1 & x_2 & x_3 & x_4 \\ x_1^2 & x_2^2 & x_3^2 & x_4^2 \\ x_1^3 & x_2^3 & x_3^3 & x_4^3 \end{bmatrix}.$$

This matrix is called the Vandermonde matrix.

3 Vector Spaces

Many measurable quantities that occur in applications can be described by a single scalar; among these quantities are temperature, mass, height, and military time. Other quantities, like force, velocity, acceleration, wind, size of a matrix, dates, and civilian time cannot be described by a single number; force, velocity, acceleration, and wind require the specification of both a magnitude and a direction, while the size of a matrix requires two numbers, a date requires 3 numbers, and civilian time requires a number and either an A.M. or P.M. specification. Quantities that require more than one scalar for their definition are called **vector quantities.** As a mathematics student, you have most likely encountered the notion of a vector before, perhaps in an elementary physics course or in a beginning calculus course. In this chapter we treat vectors from an algebraic, rather than a geometric, point of view. The techniques and results of Chapters 1 and 2 will be used extensively in Chapter 3, but the emphasis shifts from performing calculations to interpreting the results of calculations. At this point you will find it helpful to use the software, or a good calculator, to automate routine matrix calculations such as inversion, row reduction, and determinant evaluation. We begin with a brief discussion of two- and three-dimensional vectors from the geometric point of view. This discussion will provide some geometric insight into, and a source of terminology for, the more general discussion that follows.

3.1 $\Re^2$ AND $\Re^3$: OLD FRIENDS

We can represent vectors as directed line segments, or arrows, in either 2-space (the plane) or 3-space; the direction of the arrow represents the direction of the vector and the length of the arrow represents the magnitude of the vector. In

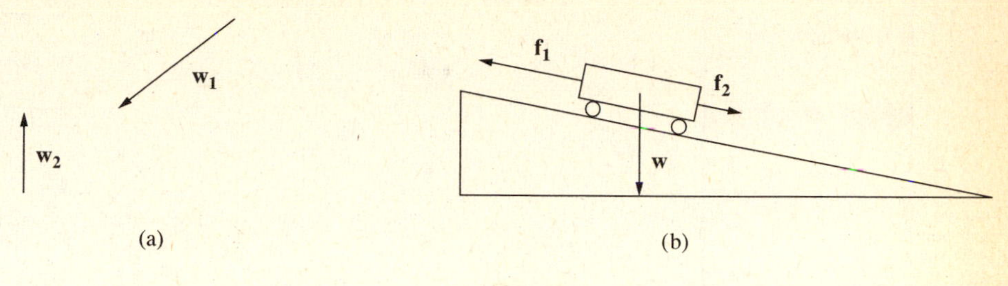

FIGURE 3.1

Figure 3.1a the vector $\mathbf{w}_1$ represents a 20 mile per hour wind from the northeast while the vector $\mathbf{w}_2$ represents a 10 mile per hour breeze from the south. Figure 3.1b shows a cart being moved up a hill (inclined plane). The vector $\mathbf{w}$ represents the weight of the cart while the vector $\mathbf{f}_2$ represents the force tending to move the cart down the hill. The vector $\mathbf{f}_1$ represents a force acting to move the cart up the hill. Since $\mathbf{f}_1$ is longer than $\mathbf{f}_2$ the cart will move up the hill.

We will denote vectors in two ways: either as boldface lowercase letters like $\mathbf{u}$, $\mathbf{v}$, $\mathbf{w}$, and $\mathbf{x}$, or as two boldface uppercase letters representing the initial and terminal points of the vector. Thus, $\mathbf{AB}$ is the vector extending from point A to point B while $\mathbf{BA}$ is the vector extending from B to A. Two vectors are defined to be **equal** if they have the same direction and the same magnitude, even if the arrows that represent them have different initial points (called *tails*) and terminal points (known as *heads*).* Hence, in Figure 3.2 the vectors $\mathbf{u} = \mathbf{AB}$ and $\mathbf{v} = \mathbf{CD}$ are equal, the vectors $\mathbf{x}_1$ and $\mathbf{x}_2$ are equal, but $\mathbf{v} \neq \mathbf{x}_1$ because, even though the two vectors have the same lengths, they have different directions.

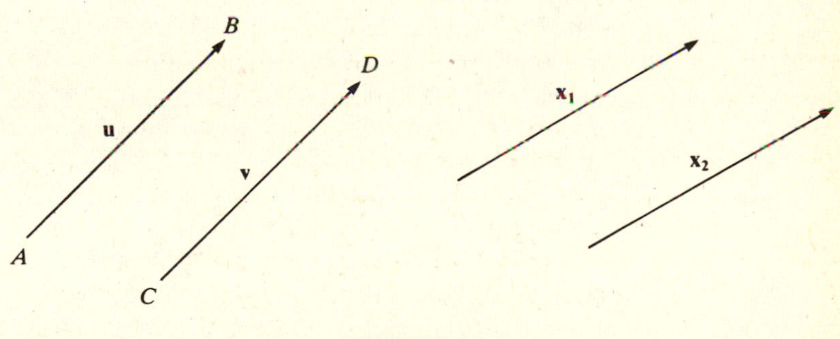

FIGURE 3.2

DEFINITION 3.1 If $\mathbf{u}$ and $\mathbf{v}$ are two vectors, then the **sum** of the two vectors is obtained by positioning $\mathbf{v}$ so that its tail coincides with the head of $\mathbf{u}$ and then defining $\mathbf{u} + \mathbf{v}$ to be the vector from the initial point of $\mathbf{u}$ to the terminal point of $\mathbf{v}$, as shown in Figure 3.3. □

*See Exercise 20 of this section.

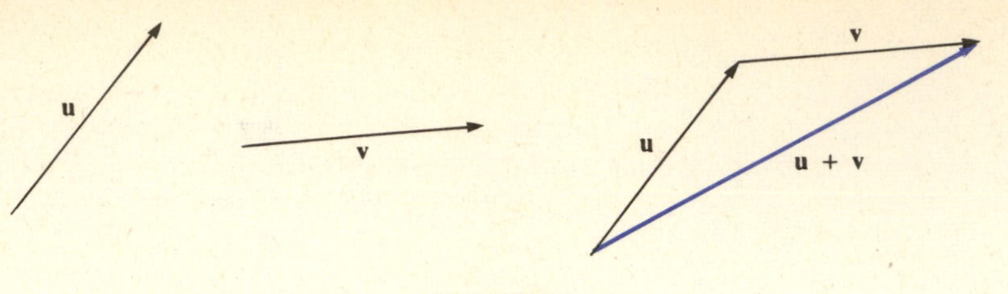

FIGURE 3.3

Definition 3.1 is frequently referred to as the **parallelogram law for vector addition** because, as shown in Figure 3.4, $\mathbf{u} + \mathbf{v}$ is the diagonal of the parallelogram with adjacent sides $\mathbf{u}$ and $\mathbf{v}$. This diagram also shows that vector addition is commutative; that is $\mathbf{u} + \mathbf{v} = \mathbf{v} + \mathbf{u}$.

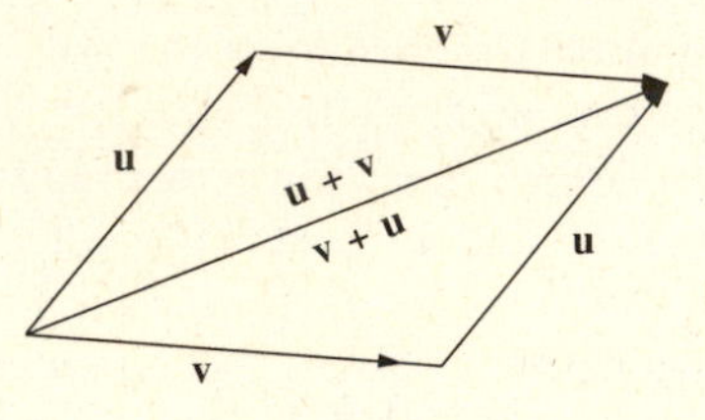

FIGURE 3.4

The geometric definition of vector addition given in Definition 3.1 can be motivated by the physical experiment pictured in Figure 3.5(a). In this experiment, force $\mathbf{F}_3$ balances the total effect of the forces $\mathbf{F}_1$ and $\mathbf{F}_2$. The magnitudes of the forces are read from the scales of the spring balances and the forces are plotted as in Figure 3.5(b), where the lengths of the arrows represent the magnitudes of the forces. If good equipment is used and the plotting is done carefully, you find that the force $\mathbf{F}_3$ is equal in magnitude and opposite in direction to the diagonal of the parallelogram whose sides are the vector representatives of the forces $\mathbf{F}_1$ and $\mathbf{F}_2$.

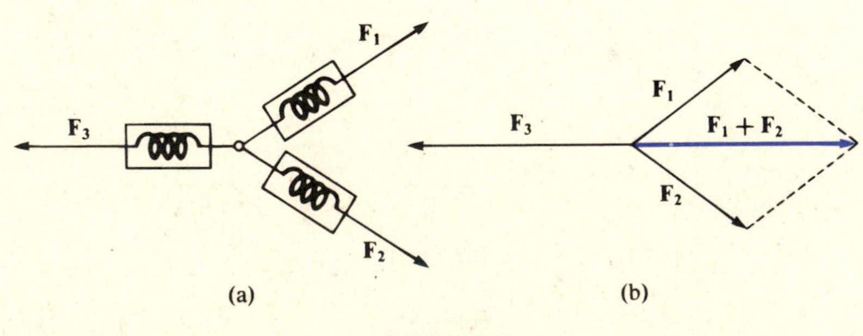

FIGURE 3.5

A second physical example of vector addition is described in Figure 3.6. If the vector $\mathbf{W_T}$ represents the true wind and $\mathbf{W_B}$ represents the wind generated by the motion of a sailboat ($\mathbf{W_B}$ is equal in magnitude and opposite in direction to the vector **v** which represents the velocity of the boat), then the apparent wind $\mathbf{W_A}$ (the wind the sailor actually feels) can be represented by the sum of the vectors $\mathbf{W_T}$ and $\mathbf{W_B}$. Note that $\mathbf{W_A}$ has greater magnitude than $\mathbf{W_T}$ and that the length of $\mathbf{W_A}$ becomes much greater than the length of $\mathbf{W_T}$ as $\mathbf{W_B}$ increases. This observation helps to explain why iceboats and some catamarans are able to travel much faster than the true wind. Essentially the same phenomenon is observed when riding a bicycle or driving a car in a strong crosswind. In these cases, the wind (air resistance) felt by the cyclist or the car comes from much further forward than the true wind and is of greater strength.

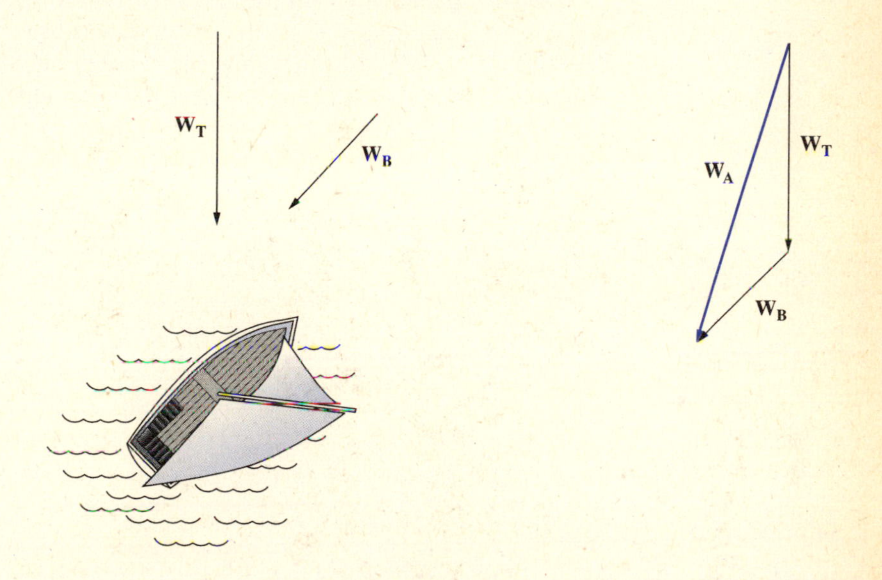

FIGURE 3.6

For any vector **v** we define its **negative,** $-\mathbf{v}$**,** to be the vector obtained from **v** by interchanging its head and tail. In the sailboat example, $\mathbf{W_B} = -\mathbf{v}$ where **v** is the velocity of the boat. Clearly,

$$\mathbf{v} + (-\mathbf{v}) = \mathbf{0},$$

where **0** is the **zero vector,** a vector with length 0.

For any vector **v** and any scalar k we define the vector $k\mathbf{v}$ to be the vector whose length is $|k|$ times the length of **v** and whose direction is either the same as that of **v** if $k > 0$ or the same as that of $-\mathbf{v}$ if $k < 0$. The vector $k\mathbf{v}$ is called

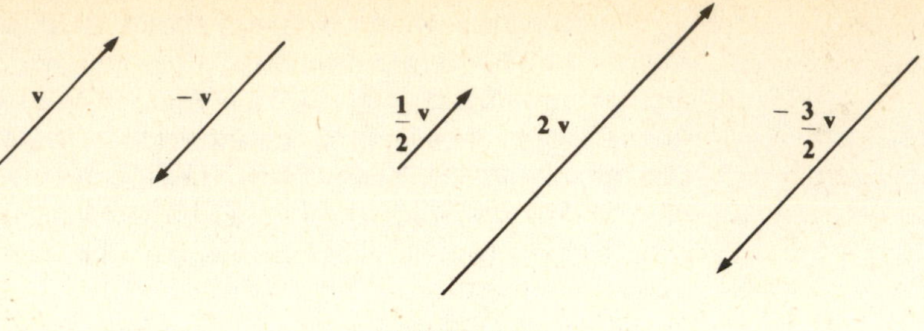

FIGURE 3.7

a **scalar multiple** of **v**. Figure 3.7 illustrates these ideas. Note that the vector $-\mathbf{v}$ is $(-1)\mathbf{v}$ and that two vectors are parallel if one is a scalar multiple of the other.

By the **difference** $\mathbf{v} - \mathbf{u}$ we mean the vector $\mathbf{v} + (-\mathbf{u})$. Figure 3.8 shows two ways of obtaining $\mathbf{v} - \mathbf{u}$: the first method uses the parallelogram law to compute $\mathbf{v} - \mathbf{u} = \mathbf{v} + (-\mathbf{u})$ while the second method uses head-to-tail positioning to find $\mathbf{v} - \mathbf{u}$ as the vector that must be added to **u** to produce **v**.

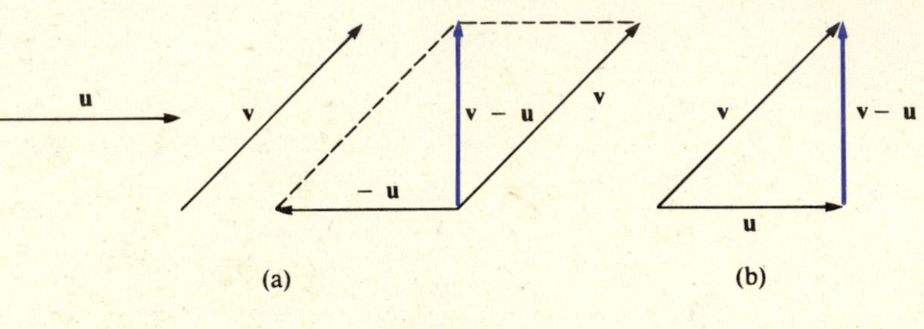

FIGURE 3.8

EXAMPLE 1 Linear Combination of Vectors

Given the vectors **u**, **v**, and **w** of Figure 3.9(a), show how to determine constants a and b so that $\mathbf{w} = a\mathbf{u} + b\mathbf{v}$.

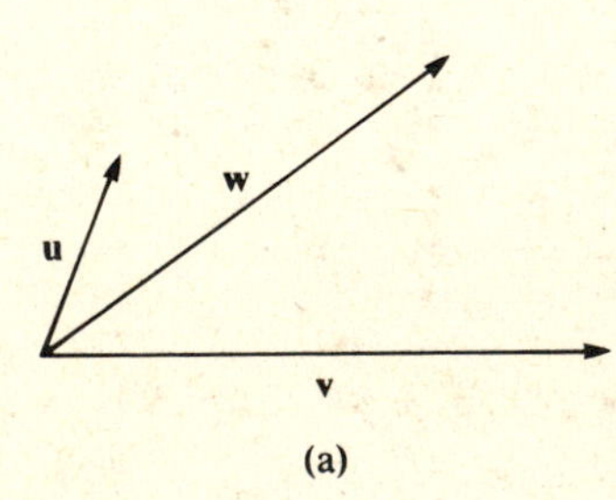

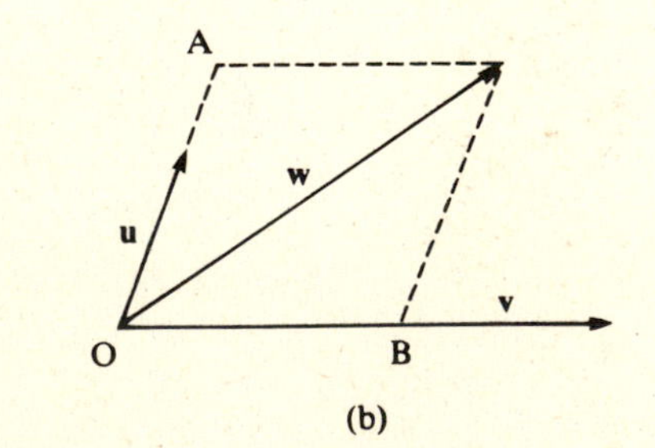

FIGURE 3.9

We first draw a line through the head of **w**, parallel to **v**, until it meets the extension of **u**. We next draw a line through the head of **w**, parallel to **u**, until it meets **v** extended, as shown in Figure 3.9(b). Since vectors **OA** + **OB** = **w** = a**u** + b**v**, **OA** and **OB** are the vectors a**u** and b**v** respectively. The scalar a can now be determined by comparing the lengths of **u** and a**u**; similarly, b can be determined by comparing the lengths of **v** and b**v**. Similar arguments would work for other orientations of the three vectors. ■

The discussion in Example 1 is a geometric proof of the following assertion.

> If the vectors **u** and **v** are not parallel, then any vector in their plane can be written as a linear combination of **u** and **v**.

Later in this section we will provide algebraic techniques for handling this problem and will also give generalizations of the preceding assertion.

The **length,** or the **norm,** of the vector **v** will be denoted by $\|\mathbf{v}\|$. A vector of length one is called a **unit vector.** Note that for any nonzero vector **u** the vector $(1/\|\mathbf{u}\|)\mathbf{u}$ is a unit vector.

We will now define a scalar-valued product of two vectors called the dot product. This product is useful in problems involving the angle between vectors, in determining projections of one vector onto another, and in computing lengths of vectors.

DEFINITION 3.2 For any pair of vectors **u** and **v** we define the **dot product**

$$\mathbf{u} \cdot \mathbf{v} = \|\mathbf{u}\|\,\|\mathbf{v}\| \cos \theta,$$

where θ is the smaller angle between **u** and **v**, as shown in Figure 3.10. □

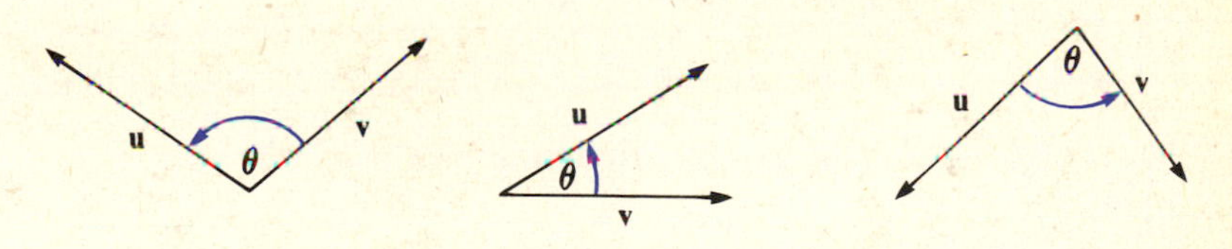

FIGURE 3.10

Since $\cos 90° = \cos\frac{\pi}{2} = 0$, it follows that the vectors **u** and **v** are perpendicular if, and only if, $\mathbf{u} \cdot \mathbf{v} = 0$. Moreover, since $\cos 0 = 1$ it follows that $\mathbf{u} \cdot \mathbf{u} = \|\mathbf{u}\|^2$ or that

$$\|\mathbf{u}\| = \sqrt{(\mathbf{u} \cdot \mathbf{u})}.$$

Let **i** and **j** be unit vectors in the coordinate directions of the plane and let **u** be a vector that makes an angle θ with **i** as shown in Figure 3.11. In this case

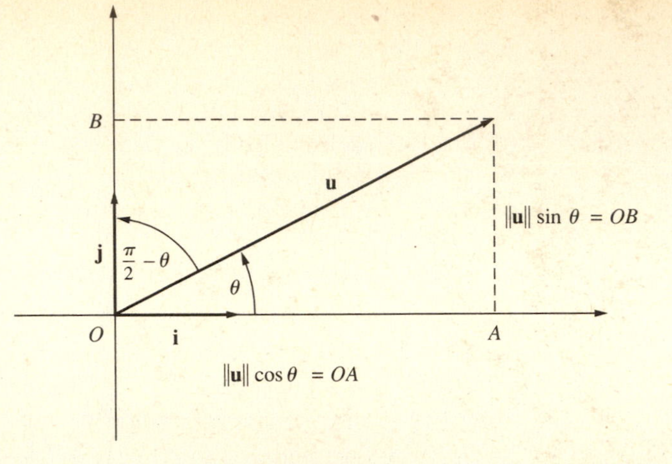

FIGURE 3.11

$\mathbf{u} \cdot \mathbf{i} = \|\mathbf{u}\| \cos\theta = OA$ which is the projection of **u** on **i**. Similarly, $\mathbf{u} \cdot \mathbf{j} = \|\mathbf{u}\| \cos(\frac{\pi}{2} - \theta) = \|\mathbf{u}\| \sin\theta = OB$ which is the projection of **u** on **j**. Note that we also have

$$\mathbf{u} = \|\mathbf{u}\| \cos\theta\, \mathbf{i} + \|\mathbf{u}\| \sin\theta\, \mathbf{j}.$$

The dot product is very useful in problems involving projections. Let **u** and **v** be positioned as shown in Figure 3.12 and let **w** be the orthogonal projection of **u** on **v**. Then $\cos\theta = \|\mathbf{w}\|/\|\mathbf{u}\|$, so

$$\|\mathbf{w}\| = \|\mathbf{u}\| \cos\theta = \frac{\mathbf{u} \cdot \mathbf{v}}{\|\mathbf{v}\|} \tag{3.1}$$

is the length of the orthogonal projection of the vector **u** on **v**. Moreover, the vector **w** is $\|\mathbf{w}\|$ times the unit vector in the direction of **v**; that is,

$$\mathbf{w} = \|\mathbf{w}\|\left(\frac{\mathbf{v}}{\|\mathbf{v}\|}\right) = \left(\frac{\mathbf{u} \cdot \mathbf{v}}{\|\mathbf{v}\|}\right)\left(\frac{\mathbf{v}}{\|\mathbf{v}\|}\right) = \left(\frac{\mathbf{u} \cdot \mathbf{v}}{\mathbf{v} \cdot \mathbf{v}}\right)\mathbf{v} \tag{3.2}$$

is the **vector projection** of **u** onto **v**.

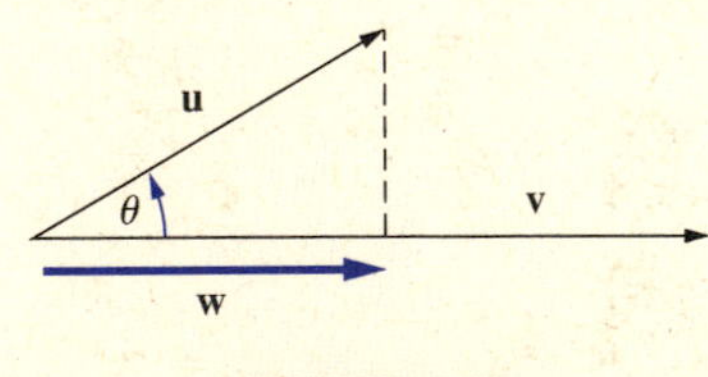

FIGURE 3.12

EXAMPLE 2 Projections

Let **u** be a vector of length 4 that makes an angle of 30° with the x-axis and let **v** be a vector of length 7 that makes an angle of 60° with the x-axis. Find the length of the projection of **v** on **u** and the vector projection of **u** on **v**.

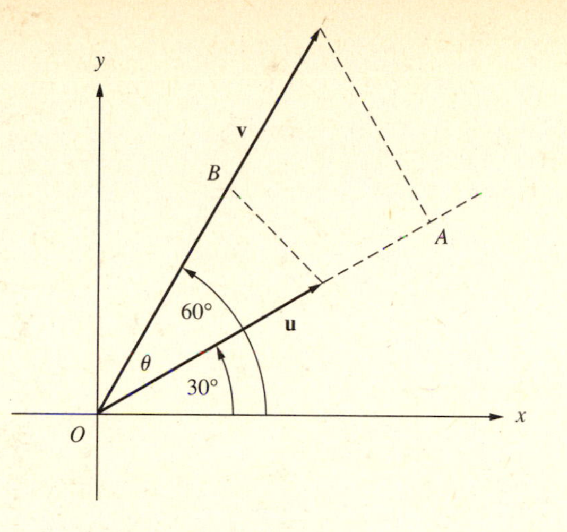

FIGURE 3.13

The vectors are shown in Figure 3.13. The angle θ between **u** and **v** is 30°, so the dot product is

$$\mathbf{u} \cdot \mathbf{v} = \|\mathbf{u}\| \|\mathbf{v}\| \cos\theta = 4 \cdot 7 \cdot \frac{\sqrt{3}}{2} = 14\sqrt{3}.$$

From Equation (3.1) it follows that the length of OA, the projection of **v** on **u**, is $\mathbf{u} \cdot \mathbf{v}/\|\mathbf{v}\| = \frac{14\sqrt{3}}{4} = \frac{7\sqrt{3}}{2}$.

The vector projection of **u** on **v**, as computed using Equation (3.2) is

$$\mathbf{OB} = \left(\frac{\mathbf{u} \cdot \mathbf{v}}{\mathbf{v} \cdot \mathbf{v}}\right)\mathbf{v} = \frac{14\sqrt{3}}{49}\mathbf{v} = \frac{2\sqrt{3}}{7}\mathbf{v}. \quad ■$$

All of the ideas discussed above for two dimensions (the plane) are equally applicable to vectors in three dimensions (space), the ideas are the same; only the pictures are harder to draw.

Since vectors are equal if they have the same length and direction, every vector is equal to a vector whose tail is at the origin, as shown in Figure 3.14. Thus there is a one-to-one correspondence between vectors, with their tails at the origin, and the coordinates of their heads. This correspondence is both the key to the algebraic treatment of vectors and the springboard to useful generalizations of the vector concept.

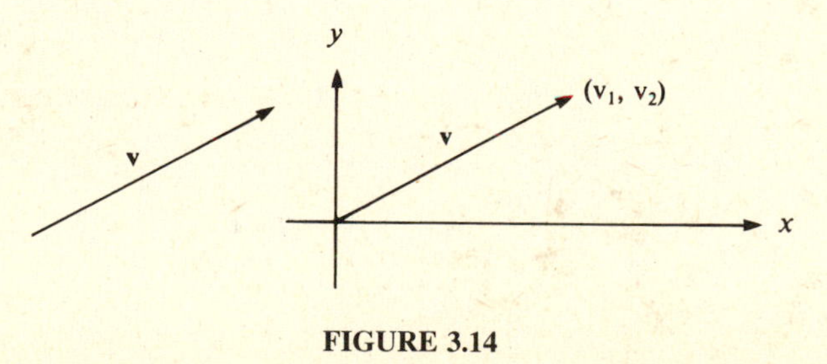

FIGURE 3.14

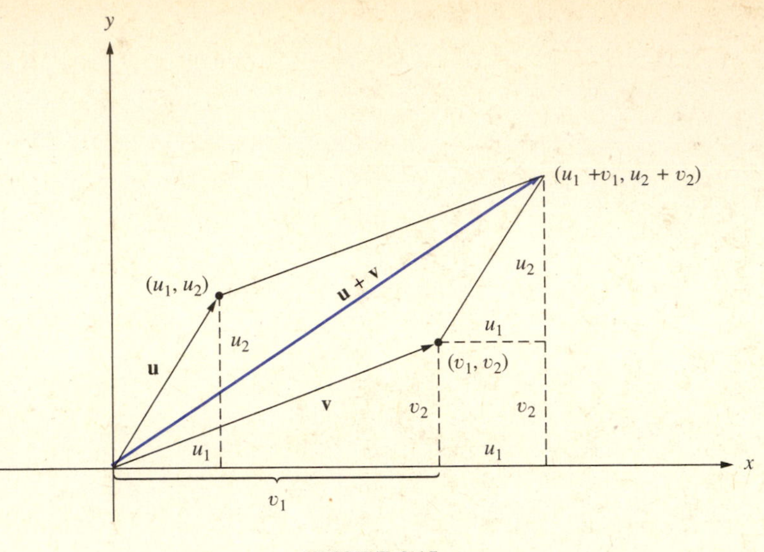

FIGURE 3.15

If **v** is the vector from the origin to the point (v_1, v_2) and **u** is the vector from the origin to the point (u_1, u_2) then $\mathbf{v} + \mathbf{u}$ is the vector from the origin to the point $(v_1 + u_1, v_2 + u_2)$, as is shown in Figure 3.15. Moreover, $k\mathbf{v}$ is the vector from the origin to the point (kv_1, kv_2).

Note we have associated vectors in the plane with 1×2 matrices and have observed that vector addition, by the parallelogram law, corresponds to matrix addition. Moreover, scalar multiplication of vectors corresponds to scalar multiplication of matrices. Again, these observations are as valid in three dimensions as they are in two. From now on we will write $\mathbf{v} = [a \quad b \quad c]$ to indicate that **v** is the vector from the origin to the point (a, b, c).

EXAMPLE 3 **Algebraic Addition of Vectors**

For the vectors $\mathbf{u} = [1 \quad 3 \quad -2]$ and $\mathbf{v} = [2 \quad 0 \quad 5]$:

$$\mathbf{u} + \mathbf{v} = [1 \quad 3 \quad -2] + [2 \quad 0 \quad 5] = [3 \quad 3 \quad 3];$$
$$2\mathbf{u} + 3\mathbf{v} = 2[1 \quad 3 \quad -2] + 3[2 \quad 0 \quad 5] = [8 \quad 6 \quad 11];$$
$$\mathbf{u} - \mathbf{v} = [1 \quad 3 \quad -2] - [2 \quad 0 \quad 5] = [-1 \quad 3 \quad -7]; \text{ and}$$
$$\mathbf{u} - \mathbf{u} = [1 \quad 3 \quad -2] - [1 \quad 3 \quad -2] = \mathbf{0}. \blacksquare$$

Although every vector is equal to a vector with its tail at the origin, it is often convenient to be able to handle vectors without first making the translation to the origin. Consider the vector **AB** where the coordinates of A are (a_1, a_2, a_3) and the coordinates of B are (b_1, b_2, b_3). As you can see from Figure 3.16,

$\mathbf{OA} + \mathbf{AB} = \mathbf{OB}$, so

$$\mathbf{AB} = \mathbf{OB} - \mathbf{OA} = [b_1 \quad b_2 \quad b_3] - [a_1 \quad a_2 \quad a_3]$$
$$= [b_1 - a_1 \quad b_2 - a_2 \quad b_3 - a_3].$$

Thus the algebraic representative of the vector **AB** is obtained by subtracting the coordinates of A (the tail) from the coordinates of B (the head).

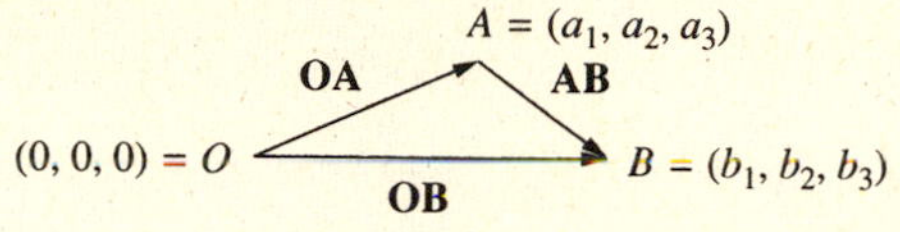

$A = (a_1, a_2, a_3)$
OA
AB
$(0, 0, 0) = O$
$B = (b_1, b_2, b_3)$
OB

FIGURE 3.16

The distance formula of analytic geometry, or the Pythagorean theorem, allows us to compute the length of a vector algebraically. In two dimensions we have

$$\|\mathbf{u}\| = \sqrt{u_1^2 + u_2^2} \qquad \text{if } \mathbf{u} = [u_1 \quad u_2],$$

and in three dimensions we have

$$\|\mathbf{v}\| = \sqrt{v_1^2 + v_2^2 + v_3^2} \qquad \text{if } \mathbf{v} = [v_1 \quad v_2 \quad v_3],$$

as shown in Figure 3.17.

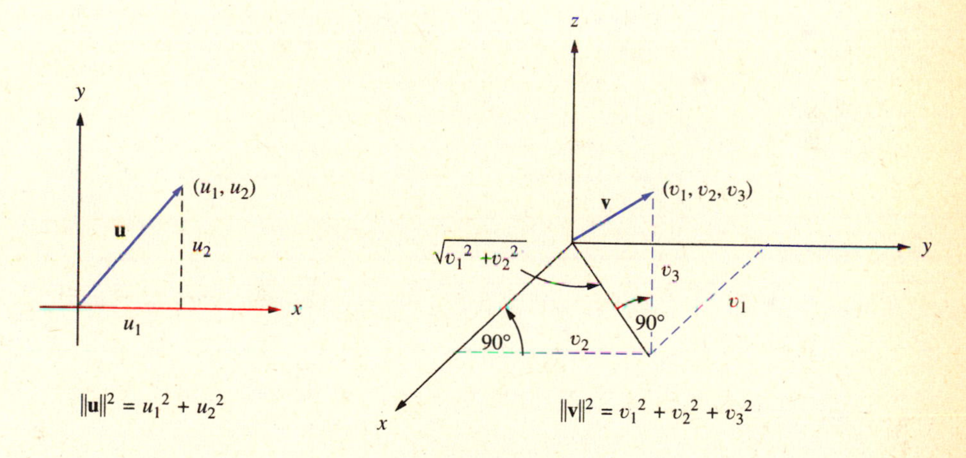

FIGURE 3.17

The dot product can be easily computed from the algebraic representation of vectors:

$$\text{if } \mathbf{v} = [v_1 \quad v_2 \quad v_3] \qquad \text{and} \qquad \mathbf{u} = [u_1 \quad u_2 \quad u_3], \text{ then}$$
$$\mathbf{v} \cdot \mathbf{u} = v_1 u_1 + v_2 u_2 + v_3 u_3. \tag{3.3}$$

The result in Equation (3.3) follows from the law of cosines (recall Exercise 10 of Section 2.3) in a straightforward way. If **u** and **v** are positioned as shown in Figure 3.18, then the law of cosines states that

$$\|\mathbf{v} - \mathbf{u}\|^2 = \|\mathbf{v}\|^2 + \|\mathbf{u}\|^2 - 2\|\mathbf{v}\|\,\|\mathbf{u}\| \cos\theta,$$

from which we obtain

$$(v_1 - u_1)^2 + (v_2 - u_2)^2 + (v_3 - u_3)^2$$
$$= (v_1^2 + v_2^2 + v_3^2) + (u_1^2 + u_2^2 + u_3^2) - 2(\mathbf{v}\cdot\mathbf{u}).$$

After some algebraic simplification this reduces to

$$-2v_1u_1 - 2v_2u_2 - 2v_3u_3 = -2(\mathbf{v}\cdot\mathbf{u}) \quad \text{or} \quad \mathbf{v}\cdot\mathbf{u} = v_1u_1 + v_2u_2 + v_3u_3.$$

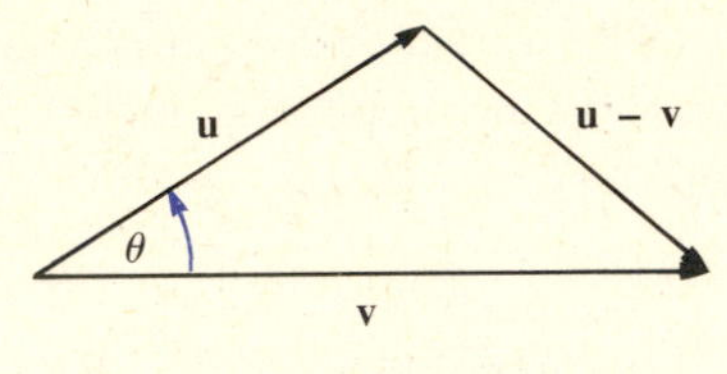

FIGURE 3.18

Equation (3.3) provides us with a much more convenient way to compute dot products than Definition 3.2. We will now be able to handle many vector problems involving angles and distances without the necessity of physically measuring these same distances and angles.

EXAMPLE 4 The Angle Between Two Vectors

Find the angle θ between the vectors $\mathbf{v} = [1 \quad 2]$ and $\mathbf{u} = [1 \quad 1]$.

We have $\|\mathbf{v}\| = \sqrt{5}$, $\|\mathbf{u}\| = \sqrt{2}$, and $\mathbf{v}\cdot\mathbf{u} = 1 \times 1 + 2 \times 1 = 3$, so it follows from Definition 3.2 that

$$\cos\theta = \frac{\mathbf{u}\cdot\mathbf{v}}{\|\mathbf{v}\|\,\|\mathbf{u}\|} = \frac{3}{\sqrt{10}} \quad \text{or} \quad \theta = 18.43°. \ ■$$

EXAMPLE 5 Projections of Vectors

Find the projection of the vector $\mathbf{u} = [5 \quad 2 \quad 1]$ on $\mathbf{v} = [3 \quad 4 \quad 12]$.

From Equation (3.2) we know that the projection of **u** on **v** is the vector

$$\mathbf{w} = \left(\frac{\mathbf{u}\cdot\mathbf{v}}{\mathbf{v}\cdot\mathbf{v}}\right)\mathbf{v}.$$

From Equation (3.3) we have

$$\mathbf{u}\cdot\mathbf{v} = 15 + 8 + 12 = 35 \quad \text{and} \quad \mathbf{v}\cdot\mathbf{v} = 9 + 16 + 144 = 169.$$

It follows that,

$$\mathbf{w} = \frac{35}{169}\mathbf{v} = \frac{35}{169}[3 \quad 4 \quad 12].$$

The length of **w** is the scalar projection of **u** on **v**. From Equation (3.2) it is

$$\|\mathbf{w}\| = \left(\frac{\mathbf{u}\cdot\mathbf{v}}{\mathbf{v}\cdot\mathbf{v}}\right)\|\mathbf{v}\| = \left(\frac{\mathbf{u}\cdot\mathbf{v}}{\mathbf{v}\cdot\mathbf{v}}\right)\sqrt{\mathbf{v}\cdot\mathbf{v}} = \frac{\mathbf{u}\cdot\mathbf{v}}{\sqrt{\mathbf{v}\cdot\mathbf{v}}} = \frac{35}{13}. \quad \blacksquare$$

In general we shall denote the set of all $m \times n$ real matrices by $\Re^{m\times n}$. We can then summarize the above discussion by saying that the set of vectors in the plane behaves algebraically like $\Re^{1\times 2}$ and that the set of vectors in 3-space behaves algebraically like $\Re^{1\times 3}$.

We could just as well have used column matrices as row matrices to represent vectors; in fact, in most of what follows it will be more convenient to use column matrices than row matrices. The only advantage of using row vectors rather than column vectors is that they take up less space on the printed page. We will sometimes write a column vector as the transpose of a row vector in order to save some space. For example

$$\begin{bmatrix} 1 \\ 3 \\ 5 \\ -7 \\ 9 \end{bmatrix} = [1 \quad 3 \quad 5 \quad -7 \quad 9]^T.$$

DEFINITION 3.3 We shall call the set of all $n \times 1$ real matrices **Euclidean *n*-space** and will denote it by $\Re^n$. □

We will study $\Re^n$ in the next section; it will be important to keep clearly in mind that we are generalizing the familiar spaces $\Re^2$ and $\Re^3$ for which we have geometric interpretations.

EXAMPLE 6 Linear Combination of Vectors

Find scalars a and b so that $\mathbf{w} = \begin{bmatrix} 1 \\ 6 \end{bmatrix} = a\mathbf{u} + b\mathbf{v}$, where $\mathbf{u} = \begin{bmatrix} 1 \\ 1 \end{bmatrix}$ and $\mathbf{v} = \begin{bmatrix} 2 \\ -3 \end{bmatrix}$.

This problem could be solved geometrically, as in Example 1. Here we give a more efficient algebraic solution. Since we want

$$a\mathbf{u} + b\mathbf{v} = \begin{bmatrix} a + 2b \\ a - 3b \end{bmatrix} = \begin{bmatrix} 1 \\ 6 \end{bmatrix},$$

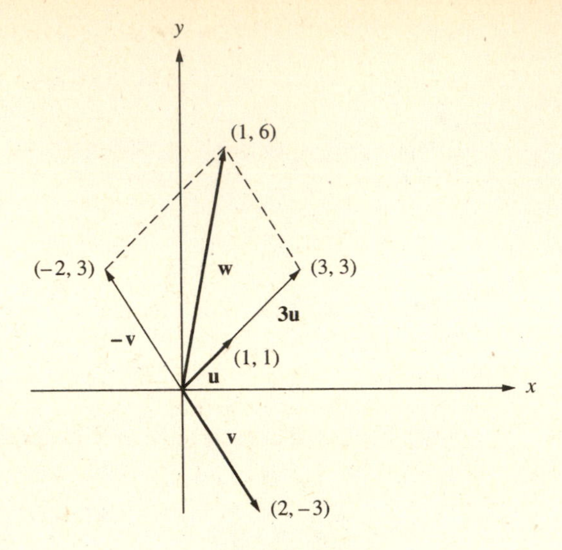

FIGURE 3.19

we must have

$$a + 2b = 1$$
$$a - 3b = 6,$$

from which it follows easily that $a = 3$ and $b = -1$. See Figure 3.19. ■

EXERCISES 3.1

1. Sketch the following vectors with initial point at the origin and then sketch them in red with initial point at $[3 \quad 4]$.
   (a) $\mathbf{u} = [2 \quad 5]$ **(b)** $\mathbf{v} = [-5 \quad -4]$ **(c)** $\mathbf{w} = [3 \quad -2]$
2. Find the vector **AB** if
   (a) A is (3, 5) and B is (2, 8);
   (b) A is (3, 1, 4) and B is (6, 4, 5); or
   (c) A is (2, −1, 7) and B is (−3, 4, 3).
3. Find a point Q so that the vector **PQ** is parallel to **w** if
   (a) $\mathbf{w} = [2 \quad 3]$ and P is (4, 2);
   (b) $\mathbf{w} = [3 \quad -2 \quad 0]$ and P is (1, 1, 2); or
   (c) $\mathbf{w} = [-2 \quad 3 \quad 5]$ and P is (2, −2, −3).
4. Show that the line segments AB and PQ are parallel if, and only if, the vectors **AB** and **PQ** are scalar multiples of each other.
5. **(a)** Show that the midpoint M of the line segment PQ is determined by $\mathbf{OM} = \mathbf{OP} + (\frac{1}{2})\mathbf{PQ}$.
   (b) If P is (x_1, x_2, x_3) and Q is (y_1, y_2, y_3), show that the midpoint of **PQ** is $\frac{1}{2}(x_1 + y_1, x_2 + y_2, x_3 + y_3)$.
   (c) Find the midpoint of PQ if P is (3, 2, 7) and Q is (5, −2, 3).
   (d) Find the midpoint of PQ if P is (7, −5) and Q is (4, −2).

6. Use the diagram below to show that vector addition is associative; that is, $\mathbf{u} + (\mathbf{v} + \mathbf{w}) = (\mathbf{u} + \mathbf{v}) + \mathbf{w}$.

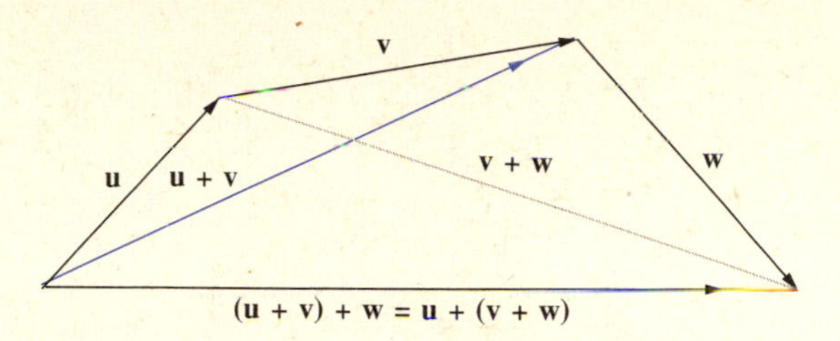

7. A cyclist with a long scarf rides due east at 15 mph. There is a cross wind from the north at 20 mph. In what direction does the scarf point? What is the magnitude of the wind felt by the cyclist?

8. For $\mathbf{u}$ and $\mathbf{v}$ in 3-space,
(a) show geometrically that $\|\mathbf{u} + \mathbf{v}\| \neq \|\mathbf{u}\| + \|\mathbf{v}\|$ and
(b) determine how $|\mathbf{u} \cdot \mathbf{v}|$ and $\|\mathbf{u}\| \, \|\mathbf{v}\|$ compare.

9. For $\mathbf{u}$ and $\mathbf{v}$ as given below, find $\|\mathbf{u}\|$, $\mathbf{u} \cdot \mathbf{v}$, a unit vector in the direction of $\mathbf{u}$, and $\cos \theta$, where θ is the angle between $\mathbf{u}$ and $\mathbf{v}$.
(a) $\mathbf{u} = [3 \quad 4]$, $\mathbf{v} = [4 \quad -3]$
(b) $\mathbf{u} = [3 \quad -4 \quad 12]$, $\mathbf{v} = [5 \quad 2 \quad 3]$.

10. Find the vector projection of $\mathbf{u}$ on $\mathbf{v}$ and the length of the projection if
(a) $\mathbf{u} = [5 \quad 2]$, $\mathbf{v} = [3 \quad 4]$;
(b) $\mathbf{u} = [1 \quad 7 \quad 4]$, $\mathbf{v} = [3 \quad 4 \quad -12]$; or
(c) $\mathbf{u} = [3 \quad 5 \quad 2]$, $\mathbf{v} = [1 \quad 1 \quad 1]$.

11. Express $\mathbf{w}$ as a linear combination of $\mathbf{u}$ and $\mathbf{v}$ if
(a) $\mathbf{w} = [-4 \quad 1]$, $\mathbf{u} = [1 \quad 5]$, $\mathbf{v} = [2 \quad 3]$; or
(b) $\mathbf{w} = [3 \quad 5]$, $\mathbf{u} = [1 \quad 0]$, $\mathbf{v} = [0 \quad 1]$.

12. Express $\mathbf{w}$ as a linear combination of $\mathbf{x}$, $\mathbf{y}$, and $\mathbf{z}$ if
(a) $\mathbf{w} = [2 \quad 1 \quad 7]$, $\mathbf{x} = [1 \quad 0 \quad 1]$, $\mathbf{y} = [0 \quad 1 \quad 2]$, $\mathbf{z} = [-1 \quad 3 \quad 0]$;
(b) $\mathbf{w} = [a \quad b \quad c]$, $\mathbf{x} = [1 \quad 0 \quad 0]$, $\mathbf{y} = [0 \quad 1 \quad 0]$, $\mathbf{z} = [0 \quad 0 \quad 1]$; or
(c) $\mathbf{w} = [5 \quad 7 \quad 8]$, $\mathbf{x} = [1 \quad 2 \quad 3]$, $\mathbf{y} = [4 \quad 5 \quad 6]$, $\mathbf{z} = [7 \quad 8 \quad 9]$.

13. For the vectors $\mathbf{x}$, $\mathbf{y}$, and $\mathbf{z}$ of Exercise 12(c) show that $\mathbf{z}$ is in the plane determined by $\mathbf{x}$ and $\mathbf{y}$ by finding scalars a and b so that $\mathbf{z} = a\mathbf{x} + b\mathbf{y}$.

14. Let $\mathbf{u}$ and $\mathbf{v}$ be nonzero vectors in 2- or 3-space. If $k = \|\mathbf{u}\|$ and $h = \|\mathbf{v}\|$, show geometrically that the vector $\mathbf{w} = (k\mathbf{v} + h\mathbf{v})$ bisects the angle between $\mathbf{u}$ and $\mathbf{v}$.

15. Show that any vector $\mathbf{u} = [u_1 \quad u_2 \quad u_3]$ in 3-space can be written as a linear combination of the unit vectors $\mathbf{i} = [1 \quad 0 \quad 0]$, $\mathbf{j} = [0 \quad 1 \quad 0]$, and $\mathbf{k} = [0 \quad 0 \quad 1]$.

16. For vectors $\mathbf{u} = [u_1 \quad u_2 \quad u_3]$ and $\mathbf{v} = [v_1 \quad v_2 \quad v_3]$ in 3-space, the **vector cross product** is defined by

$$\mathbf{u} \times \mathbf{v} = [u_2v_3 - u_3v_2, \quad u_3v_1 - u_1v_3, \quad u_1v_2 - u_2v_1].$$

(a) Show that

$$\mathbf{u} \times \mathbf{v} = \left[\det\begin{bmatrix} u_2 & u_3 \\ v_2 & v_3 \end{bmatrix} \quad -\det\begin{bmatrix} u_1 & u_3 \\ v_1 & v_3 \end{bmatrix} \quad \det\begin{bmatrix} u_1 & u_2 \\ v_1 & v_2 \end{bmatrix}\right]$$

$$= \det\begin{bmatrix} \mathbf{i} & \mathbf{j} & \mathbf{k} \\ u_1 & u_2 & u_3 \\ v_1 & v_2 & v_3 \end{bmatrix},$$

where $\mathbf{i} = [1 \quad 0 \quad 0]$, $\mathbf{j} = [0 \quad 1 \quad 0]$, and $\mathbf{k} = [0 \quad 0 \quad 1]$ are unit vectors in the coordinate (x, y, and z) directions. Note that the final form for the cross product is not a proper determinant since the elements of the first row are vectors, not scalars. This compact form for the cross product is a useful mnemonic device.

(b) For $\mathbf{u} = [1 \quad 2 \quad 3]$ and $\mathbf{v} = [3 \quad -1 \quad 4]$, compute $\mathbf{u} \times \mathbf{v}$ and $\mathbf{v} \times \mathbf{u}$.

(c) Show that $\mathbf{u} \cdot (\mathbf{u} \times \mathbf{v}) = \mathbf{v} \cdot (\mathbf{u} \times \mathbf{v}) = 0$ and interpret this result geometrically.

(d) Show that $\|\mathbf{u} \times \mathbf{v}\|^2 = \|\mathbf{u}\|^2 \|\mathbf{v}\|^2 - (\mathbf{u} \cdot \mathbf{v})^2$ and hence that $\|\mathbf{u} \times \mathbf{v}\| = \|\mathbf{u}\| \|\mathbf{v}\| \sin \theta$.

17. Show, from the result of Exercise 16(d), that $\|\mathbf{u} \times \mathbf{v}\|$ is the area of the parallelogram with $\mathbf{u}$ and $\mathbf{v}$ as adjacent sides.

18. Show that $|\mathbf{u} \cdot (\mathbf{v} \times \mathbf{w})|$ is the volume of the parallelepiped with sides $\mathbf{u}$, $\mathbf{v}$, and $\mathbf{w}$.

19. Given the vectors in the diagram below, find each of the following resultant vectors.

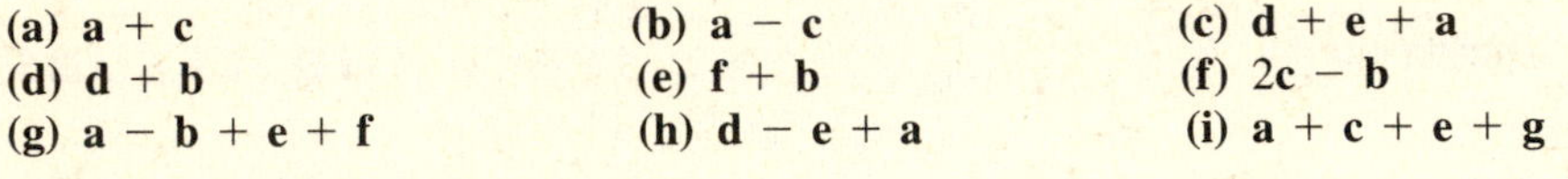

(a) $\mathbf{a} + \mathbf{c}$ **(b)** $\mathbf{a} - \mathbf{c}$ **(c)** $\mathbf{d} + \mathbf{e} + \mathbf{a}$
(d) $\mathbf{d} + \mathbf{b}$ **(e)** $\mathbf{f} + \mathbf{b}$ **(f)** $2\mathbf{c} - \mathbf{b}$
(g) $\mathbf{a} - \mathbf{b} + \mathbf{e} + \mathbf{f}$ **(h)** $\mathbf{d} - \mathbf{e} + \mathbf{a}$ **(i)** $\mathbf{a} + \mathbf{c} + \mathbf{e} + \mathbf{g}$

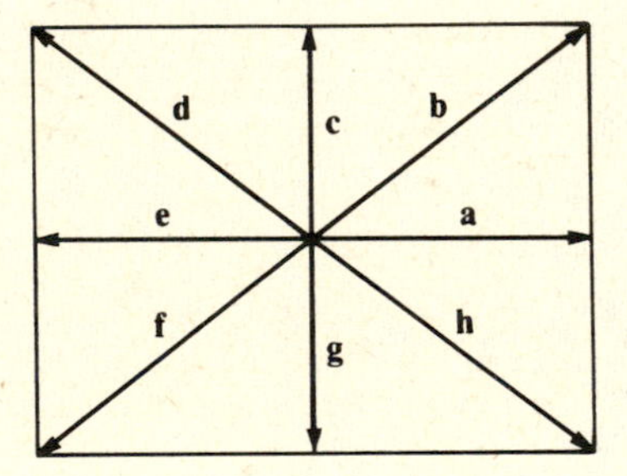

20. Define the following relation on arrows in the plane; $\mathbf{u} \sim \mathbf{v}$ if, and only if, $\mathbf{u}$ and $\mathbf{v}$ have the same length and direction.

(a) Show that $\sim$ is an equivalence relation; that is, show that $\sim$ is reflexive, symmetric, and transitive.

(b) Show that every equivalence class $\{\mathbf{v} \mid \mathbf{v} \sim \mathbf{u}\}$ contains precisely one arrow with tail at the origin.

(c) Compare equality of these equivalence classes with equality of vectors as defined in the text.

21. Which of the following statements are correct? Explain your reasoning.

(a) $\Re^2$ is a subset of $\Re^3$.

(b) $\Re^3$ has a subset that is "like" $\Re^2$.

(c) $\Re^3$ has many subsets that are "like" $\Re^2$.

22. Show that any plane in $\mathfrak{R}^3$ can be described as

$$\mathcal{W} = \{\mathbf{u} + \mathbf{v_p} \mid \mathbf{u} \text{ in } \mathcal{H}\},$$

where $\mathcal{H}$ is a plane through the origin and $\mathbf{v_p} = \mathbf{OP}$, where P is a fixed point on the plane $\mathcal{W}$.

23. **(a)** Show that the vector $\mathbf{v} = \begin{bmatrix} a \\ b \end{bmatrix}$ is perpendicular to the line $ax + by + d = 0$.

(*Hint*: Let A and B be any two points on the line. Then show that $\mathbf{v} \cdot \mathbf{AB} = 0$.)

(b) Show that the distance from the point $B = (x_1, y_1)$ to the line $ax + by + d = 0$ is

$$\frac{|ax_1 + by_1 + d|}{\sqrt{a^2 + b^2}}$$

by finding the length of the projection of $\mathbf{AB}$ on $\mathbf{v}$ where $A = (x_0, y_0)$ is any point on the line

24. Show that the vector $\mathbf{v} = \begin{bmatrix} a \\ b \\ c \end{bmatrix}$ is perpendicular to the plane $ax + by + cz = d$.

(*Hint*: Let A and B be any two points on the plane. Then show that $\mathbf{v} \cdot \mathbf{AB} = 0$.)

3.2 EUCLIDEAN n-SPACE AND ITS SUBSPACES

In the last section we defined $\mathfrak{R}^n$ to be the set of all $n \times 1$ matrices and we took a brief look at $\mathfrak{R}^2$ and $\mathfrak{R}^3$ from a geometric viewpoint. In this section we will study $\mathfrak{R}^n$, the natural generalization of $\mathfrak{R}^2$ and $\mathfrak{R}^3$. Since it is not possible to deal with $\mathfrak{R}^n$ geometrically, we have no choice but to proceed algebraically, although we will use geometric terminology and insight from the cases $n = 2$ and $n = 3$. We will refer to the elements of $\mathfrak{R}^n$ as either points or vectors. We will denote the elements of $\mathfrak{R}^n$ by boldface letters like **u**, **v**, and **w**. The entries, or components, of the vector **u** will be denoted $u_1, u_2, \ldots, u_n$. In order to save space we will sometimes write a vector as the transpose of a row matrix, that is

$$\mathbf{u} = \begin{bmatrix} u_1 \\ u_2 \\ \vdots \\ u_n \end{bmatrix} = [u_1 \quad u_2 \quad \cdots \quad u_n]^T.$$

Since the vector addition and scalar multiplication in $\mathfrak{R}^n$ are matrix addition and matrix scalar multiplication, we already know quite a bit about the algebraic structure of $\mathfrak{R}^n$. The pertinent results from Chapter 1 are collected in the following theorem.

THEOREM 3.1 If **u**, **v**, and **w** are any elements of $\mathfrak{R}^n$ and a and b are any scalars, then $\mathbf{u} + \mathbf{v}$ and $a\mathbf{u}$ are in $\mathfrak{R}^n$ and the following properties hold:

(1) For any vectors **u**, **v**, and **w** in $\mathfrak{R}^n$, $\mathbf{u} + (\mathbf{v} + \mathbf{w}) = (\mathbf{u} + \mathbf{v}) + \mathbf{w}$. (Vector addition is *associative.*)
(2) For any vectors **u** and **v** in $\mathfrak{R}^n$, $\mathbf{u} + \mathbf{v} = \mathbf{v} + \mathbf{u}$. (Vector addition is *commutative.*)
(3) There is a unique vector $\mathbf{0}$ in $\mathfrak{R}^n$ such that $\mathbf{u} + \mathbf{0} = \mathbf{u}$ for every **u** in $\mathfrak{R}^n$. (The vector $\mathbf{0}$ is called the *additive identity element* of $\mathfrak{R}^n$.)
(4) For each **u** in $\mathfrak{R}^n$, there is a unique vector $-\mathbf{u}$ in $\mathfrak{R}^n$ such that $\mathbf{u} + (-\mathbf{u}) = \mathbf{0}$. (The vector $-\mathbf{u}$ is called the *additive inverse* of **u**.)
(5) For any vector **u** and any scalars a and b, $(a + b)\mathbf{u} = a\mathbf{u} + b\mathbf{u}$. (Scalar multiplication *distributes* over scalar addition.)
(6) For any scalar a and any vectors **u** and **v**, $a(\mathbf{u} + \mathbf{v}) = a\mathbf{u} + a\mathbf{v}$. (Scalar multiplication *distributes* over vector addition.)
(7) For any scalars a and b and any vector **u**, $a(b\mathbf{u}) = (ab)\mathbf{u}$. (Scalar multiplication is *associative.*)
(8) $1\mathbf{u} = \mathbf{u}$ for every **u** in $\mathfrak{R}^n$. (The real number 1 is an *identity element* for scalar multiplication.) □

The properties of $\mathfrak{R}^n$ collected in Theorem 3.1 are all clear for $\mathfrak{R}^n$ but should be carefully examined since they will be the basis for further generalizations of the vector concept. The primary reason for our interest in $\mathfrak{R}^n$ is that the solutions of the linear system $AX = K$ come from $\mathfrak{R}^n$ if A is an $m \times n$ matrix. In that case A can be profitably viewed as defining a function (multiplication by A) from $\mathfrak{R}^n$ to $\mathfrak{R}^m$; we will pursue this functional point of view in Chapter 4.

The geometric definition of the dot product (Definition 3.2) for $\mathfrak{R}^2$ and $\mathfrak{R}^3$ does not generalize directly to $\mathfrak{R}^n$ since we do not yet have any notions of length or angle in $\mathfrak{R}^n$. Equation (3.3) gives a purely algebraic way to compute the dot product in $\mathfrak{R}^3$. This equation provides us with a natural way of extending the definition of the dot product to $\mathfrak{R}^n$.

DEFINITION 3.4 The **dot product, or Euclidean inner product,** of

$$\mathbf{u} = \begin{bmatrix} u_1 \\ u_2 \\ \vdots \\ u_n \end{bmatrix} \quad \text{and} \quad \mathbf{v} = \begin{bmatrix} v_1 \\ v_2 \\ \vdots \\ v_n \end{bmatrix}$$

in $\mathfrak{R}^n$ is defined by

$$\mathbf{u} \cdot \mathbf{v} = u_1v_1 + u_2v_2 + \cdots + u_nv_n. \quad \square$$

Some texts use the notation $(\mathbf{u}, \mathbf{v})$ or $\langle \mathbf{u}, \mathbf{v} \rangle$ instead of $\mathbf{u} \cdot \mathbf{v}$ for the Euclidean inner product.

EXAMPLE 1 Dot Product Calculations

For the vectors $\mathbf{u} = \begin{bmatrix} 1 \\ 2 \\ 3 \\ 4 \end{bmatrix}$, $\mathbf{v} = \begin{bmatrix} 4 \\ -2 \\ 0 \\ 5 \end{bmatrix}$, and $\mathbf{w} = \begin{bmatrix} 4 \\ 0 \\ 0 \\ -1 \end{bmatrix}$ from $\Re^4$ we have

$$\mathbf{u} \cdot \mathbf{v} = 1 \times 4 + 2 \times (-2) + 3 \times 0 + 4 \times 5 = 20;$$
$$\mathbf{u} \cdot \mathbf{u} = 1^2 + 2^2 + 3^2 + 4^2 = 30; \text{ and}$$
$$\mathbf{u} \cdot \mathbf{w} = 1 \times 4 + 2 \times 0 + 3 \times 0 + 4 \times (-1) = 0. \quad ■$$

Note that we can write the dot product $\mathbf{u} \cdot \mathbf{v}$ as the matrix product $\mathbf{u}^T\mathbf{v}$. Thus, we can use the properties of matrix multiplication and the transpose, that were developed earlier, to establish some elementary properties of the dot product in $\Re^n$.

THEOREM 3.2 For any vectors $\mathbf{u}$, $\mathbf{v}$, and $\mathbf{w}$ in $\Re^n$ and for any scalars a and b, the following statements are true:

(1) $\mathbf{u} \cdot \mathbf{v} = \mathbf{v} \cdot \mathbf{u}$;
(2) $\mathbf{u} \cdot (\mathbf{v} + \mathbf{w}) = \mathbf{u} \cdot \mathbf{v} + \mathbf{u} \cdot \mathbf{w}$;
(3) $\mathbf{u} \cdot (a\mathbf{v}) = (a\mathbf{u}) \cdot \mathbf{v} = a(\mathbf{u} \cdot \mathbf{v})$;
(4) $\mathbf{u} \cdot (a\mathbf{v} + b\mathbf{w}) = a(\mathbf{u} \cdot \mathbf{v}) + b(\mathbf{u} \cdot \mathbf{w})$;
(5) $\mathbf{u} \cdot \mathbf{0} = 0$; and
(6) $\mathbf{u} \cdot \mathbf{u} \geq 0$ with $\mathbf{u} \cdot \mathbf{u} = 0$ if, and only if, $\mathbf{u} = \mathbf{0}$. □

For a vector $\mathbf{u}$ in $\Re^3$ we know, from Section 3.1, that the length of $\mathbf{u}$ is given by

$$\|\mathbf{u}\| = \sqrt{(\mathbf{u} \cdot \mathbf{u})} = \sqrt{u_1^2 + u_2^2 + u_3^2}.$$

For a vector $\mathbf{u}$ in $\Re^n$ we can define the *length*, or *norm*, of $\mathbf{u}$ by generalizing the preceding formula.

DEFINITION 3.5 For the vector $\mathbf{u}$ in $\Re^n$ the **length,** or **norm,** of $\mathbf{u}$ is defined by

$$\|\mathbf{u}\| = \sqrt{(\mathbf{u} \cdot \mathbf{u})} = \sqrt{u_1^2 + u_2^2 + \cdots + u_n^2} = \sqrt{\textstyle\sum_{i=1}^{n} u_i^2}. \quad □$$

We will call the vector $\mathbf{u}$ a **unit vector** if $\|\mathbf{u}\| = 1$. Since $\|a\mathbf{u}\| = \sqrt{(a\mathbf{u}) \cdot (a\mathbf{u})} = |a|\sqrt{(\mathbf{u} \cdot \mathbf{u})} = |a|\,\|\mathbf{u}\|$, it follows that

for any vector $\mathbf{v} \neq \mathbf{0}$, the vector $\dfrac{1}{\|\mathbf{v}\|}\mathbf{v}$ is a unit vector.

We can also generalize directly from $\Re^3$ to define what it means for two vectors in $\Re^n$ to be perpendicular or orthogonal.

DEFINITION 3.6 The vectors $\mathbf{u}$ and $\mathbf{v}$ from $\Re^n$ are **orthogonal** if $\mathbf{u} \cdot \mathbf{v} = 0$. □

EXAMPLE 2 **Orthogonal Vectors**

Consider the vectors $\mathbf{u} = \begin{bmatrix} 2 \\ -1 \\ 0 \\ 3 \end{bmatrix}$ and $\mathbf{v} = \begin{bmatrix} 3 \\ 0 \\ 5 \\ -2 \end{bmatrix}$ from $\Re^4$.

We have

$$\mathbf{u} \cdot \mathbf{u} = 4 + 1 + 0 + 9 = 14,$$

so that $\|\mathbf{u}\| = \sqrt{14}$ and

$$\left(\frac{1}{\sqrt{14}}\right)\mathbf{u} = \left(\frac{1}{\sqrt{14}}\right)\begin{bmatrix} 2 \\ -1 \\ 0 \\ 3 \end{bmatrix}$$

is a unit vector. Moreover, $\mathbf{u} \cdot \mathbf{v} = 6 + 0 + 0 - 6 = 0$, so the vectors $\mathbf{u}$ and $\mathbf{v}$ are orthogonal. ■

We could carry the generalizations from $\Re^3$ to $\Re^n$ further by defining the angle between two vectors in $\Re^n$ to be the angle θ between 0 and π for which $\cos\theta = (\mathbf{u} \cdot \mathbf{v})/\|\mathbf{u}\|\,\|\mathbf{v}\|$* and by defining the distance between two points, $\mathbf{P}$ and $\mathbf{Q}$, of $\Re^n$ as $\|\mathbf{Q} - \mathbf{P}\|$.

We will have many occasions to consider certain subsets of $\Re^n$. Of particular interest to us are subsets associated with a linear system $AX = K$. If A is an $m \times n$ matrix, then the following subsets will be of particular interest:

(1) the *solution set* $\{X \mid AX = K\} \subset \Re^n$;
(2) the *null space* $\{X \mid AX = 0\} \subset \Re^n$;
(3) the *column space* $\{K \mid AX = K \text{ is solvable}\} \subset \Re^m$; and
(4) the *eigenspace* $\{Y \mid AY = \lambda Y \text{ for a fixed real } \lambda\} \subset \Re^n$.

As an example of a subset of $\Re^3$ consider

$$\mathcal{W} = \left\{\begin{bmatrix} x \\ y \\ 0 \end{bmatrix}\right\} = \left\{\begin{bmatrix} x \\ y \\ z \end{bmatrix} \middle|\, z = 0\right\}.$$

Although it is not strictly correct to say that $\mathcal{W} = \Re^2$, it should be clear geometrically that $\mathcal{W}$ and $\Re^2$ are very much alike. In fact, $\mathcal{W}$ is a copy of $\Re^2$ embedded in $\Re^3$. In this case, $\mathcal{W}$ is said to be a subspace of $\Re^3$.

*Provided that we could show that $|(\mathbf{u} \cdot \mathbf{v})/(\|\mathbf{u}\|\,\|\mathbf{v}\|)| \leq 1$.

DEFINITION 3.7 A nonempty subset $\mathcal{W}$ of $\Re^n$ is called a **subspace** of $\Re^n$ if $\mathcal{W}$ is closed under vector addition and scalar multiplication; that is,

$$\text{if } \mathbf{u} \text{ and } \mathbf{v} \text{ are in } \mathcal{W}, \text{ then so are } \mathbf{u} + \mathbf{v} \text{ and } k\mathbf{u},$$

where k is any scalar. □

EXAMPLE 3 Check a Subset for Closure

Is $\mathcal{W} = \left\{ X \text{ in } \Re^3 \,\middle|\, \begin{bmatrix} 1 & 0 & 2 \\ 0 & 1 & -3 \\ 0 & 0 & 0 \end{bmatrix} X = \begin{bmatrix} 3 \\ -1 \\ 0 \end{bmatrix} \right\}$ a subspace of $\Re^3$?

Since the coefficient matrix of the system is already in reduced row-echelon form we see at once that the general solution of the system, and hence a typical element of the set $\mathcal{W}$, is given by

$$X = \begin{bmatrix} 3 - 2c \\ -1 + 3c \\ c \end{bmatrix} = \begin{bmatrix} 3 \\ -1 \\ 0 \end{bmatrix} + c \begin{bmatrix} -2 \\ 3 \\ 1 \end{bmatrix}.$$

If we take $c = 1$ and then $c = 2$ we find that the two solutions

$$X_1 = \begin{bmatrix} 1 \\ 2 \\ 1 \end{bmatrix} \quad \text{and} \quad X_2 = \begin{bmatrix} -1 \\ 5 \\ 2 \end{bmatrix}$$

are elements of the set $\mathcal{W}$. In order to determine whether or not $X_1 + X_2$ is in $\mathcal{W}$, we compute

$$X_1 + X_2 = \begin{bmatrix} 0 \\ 7 \\ 3 \end{bmatrix} \quad \text{and}$$

$$A(X_1 + X_2) = \begin{bmatrix} 1 & 0 & 2 \\ 0 & 1 & -3 \\ 0 & 0 & 0 \end{bmatrix} \begin{bmatrix} 0 \\ 7 \\ 3 \end{bmatrix} = \begin{bmatrix} 6 \\ -2 \\ 0 \end{bmatrix} \neq \begin{bmatrix} 3 \\ -1 \\ 0 \end{bmatrix}.$$

This calculation shows that $X_1 + X_2$ is not a solution of the given system and hence not an element of the set $\mathcal{W}$. Thus, the set $\mathcal{W}$ is not closed under vector addition and therefore $\mathcal{W}$ is not a subspace of $\Re^3$. You should check that $2X_1$ is not in $\mathcal{W}$ and hence that $\mathcal{W}$ also fails to be closed under scalar multiplication. ■

The two closure conditions of Definition 3.7 can be replaced by the single condition that $\mathcal{W}$ be nonempty and closed under arbitrary linear combinations; that is, for any $\mathbf{u}$ and $\mathbf{v}$ in $\mathcal{W}$ and any a and b in $\Re$, $a\mathbf{u} + b\mathbf{v}$ is in $\mathcal{W}$. Note that if $\mathbf{u}$ is in the subspace $\mathcal{W}$, then $\mathcal{W}$ must also contain the vectors $-\mathbf{u} = (-1)\mathbf{u}$ and $\mathbf{0} = \mathbf{u} - \mathbf{u} = 0\mathbf{u}$.

If A is any $m \times n$ matrix, then the set of all solutions of the homogeneous system $AX = 0$ is a very important example of a subspace. This subspace is called the **null space** of A and will be denoted by $\mathcal{NS}(A)$:

$$\mathcal{NS}(A) = \{\mathbf{u} \text{ in } \Re^n \mid A\mathbf{u} = \mathbf{0}\}.$$

THEOREM 3.3 For any $m \times n$ matrix A, the null space of A is a subspace of $\Re^n$.

Proof To prove this theorem we need to show that $\mathcal{NS}(A)$ is closed under addition and scalar multiplication. Thus, suppose that $\mathbf{u}$ and $\mathbf{v}$ are in $\mathcal{NS}(A)$, which means that $A\mathbf{u} = \mathbf{0}$ and $A\mathbf{v} = \mathbf{0}$. To decide whether $\mathbf{u} + \mathbf{v}$ is in $\mathcal{NS}(A)$ we compute

$$A(\mathbf{u} + \mathbf{v}) = A\mathbf{u} + A\mathbf{v} = \mathbf{0} + \mathbf{0} = \mathbf{0},$$

where the first equality is a consequence of the fact that matrix multiplication distributes over matrix addition. This calculation shows that $\mathbf{u} + \mathbf{v}$ is also in $\mathcal{NS}(A)$ and, hence, that $\mathcal{NS}(A)$ is closed under vector addition. Similarly, to show that $\mathcal{NS}(A)$ is closed under scalar multiplication we let k be any scalar and compute

$$A(k\mathbf{u}) = k(A\mathbf{u}) = k\mathbf{0} = \mathbf{0}.$$

Thus, $k\mathbf{u}$ is in $\mathcal{NS}(A)$ and we have shown that $\mathcal{NS}(A)$ is closed under scalar multiplication. This completes the demonstration that $\mathcal{NS}(A)$ is a subspace of $\Re^n$. □

The key to success in deciding whether a subset $\mathcal{W}$ of $\Re^n$ is also a subspace is to clearly understand the criteria for membership in the set $\mathcal{W}$; in Theorem 3.3 the criterion for membership in $\mathcal{NS}(A)$ was that $\mathbf{u}$ was in $\mathcal{NS}(A)$ if, and only if, $A\mathbf{u} = \mathbf{0}$.

Note that one example of vectors $\mathbf{u}$ and $\mathbf{v}$ in $\mathcal{W}$ for which $\mathbf{u} + \mathbf{v}$ is not in $\mathcal{W}$ was enough to show, in Example 3, that $\mathcal{W}$ fails to be a subspace. To show that a set $\mathcal{W}$ is a subspace requires that we show that $\mathbf{u} + \mathbf{v}$ is in $\mathcal{W}$ *for every choice* of $\mathbf{u}$ and $\mathbf{v}$ in $\mathcal{W}$ and that $k\mathbf{u}$ is in $\mathcal{W}$ for every choice of $\mathbf{u}$ in $\mathcal{W}$ and every choice of k in $\Re$.

EXAMPLE 4 This Subset Is not a Subspace

Determine whether $\mathcal{W} = \{\mathbf{u} \text{ in } \Re^3 \mid \|\mathbf{u}\| = 1\}$ is a subspace of $\Re^3$.

The vectors $\mathbf{u} = \begin{bmatrix} 1 \\ 0 \\ 0 \end{bmatrix}$ and $\mathbf{v} = \begin{bmatrix} 0 \\ 1 \\ 0 \end{bmatrix}$ are each members of $\mathcal{W}$, but $\mathbf{u} + \mathbf{v} = \begin{bmatrix} 1 \\ 1 \\ 0 \end{bmatrix}$ is not in $\mathcal{W}$ because

$$\|\mathbf{u} + \mathbf{v}\| = \sqrt{(1^2 + 1^2)} = \sqrt{2} \neq 1.$$

Thus, $\mathcal{W}$ is not closed under vector addition and hence is not a subspace of $\Re^3$. Alternately, one could observe that if $\mathbf{u}$ is in $\mathcal{W}$ then so is $-\mathbf{u}$, but $\mathbf{u} + (-\mathbf{u}) = \mathbf{0}$ is not in $\mathcal{W}$ since $\|\mathbf{0}\| = 0 \neq 1$.

A third alternative for this example is to show that if $\mathbf{u}$ is in $\mathcal{W}$ then $k\mathbf{u}$ is in $\mathcal{W}$ only if $k = \pm 1$ and, hence, that $\mathcal{W}$ is not closed under all scalar multiplications. ■

EXAMPLE 5 This Subset Is a Subspace

Is $\mathcal{W} = \{\mathbf{u} = [0 \quad u_2 \quad u_3 \quad u_2 + u_3]^T \,|\, u_i \text{ in } \Re\}$ a subspace of $\Re^4$?

The membership criteria for the set $\mathcal{W}$ are not as clearly indicated as in Example 4. For this set $\mathcal{W}$, $\mathbf{u} = [u_1 \quad u_2 \quad u_3 \quad u_4]^T$ is in $\mathcal{W}$ if, and only if, $u_1 = 0$ and $u_4 = u_2 + u_3$. Thus, $\mathbf{v}_1 = [0 \quad 2 \quad 3 \quad 5]^T$ and $\mathbf{v}_2 = [0 \quad 0 \quad 3 \quad 3]^T$ are in $\mathcal{W}$ but $\mathbf{v}_3 = [1 \quad 2 \quad 3 \quad 5]^T$ and $\mathbf{v}_4 = [0 \quad 4 \quad 2 \quad 7]^T$ are not in $\mathcal{W}$. Observe that $\mathbf{v}_1 + \mathbf{v}_2 = [0 \quad 2 \quad 6 \quad 8]^T$ is also in $\mathcal{W}$, which suggests, *but does not prove,* that $\mathcal{W}$ is closed under vector addition. To prove that $\mathcal{W}$ is a subspace we suppose that $\mathbf{u} = [u_1 \quad u_2 \quad u_3 \quad u_4]^T$ and $\mathbf{v} = [v_1 \quad v_2 \quad v_3 \quad v_4]^T$ are any two vectors in $\mathcal{W}$. Therefore, by the membership criteria for the set $\mathcal{W}$,

$$u_1 = v_1 = 0, \quad u_4 = u_2 + u_3, \quad \text{and} \quad v_4 = v_2 + v_3.$$

Now the vector $\mathbf{x} = \mathbf{u} + \mathbf{v} = [u_1 + v_1 \quad u_2 + v_2 \quad u_3 + v_3 \quad u_4 + v_4]^T$. To see if $\mathbf{x}$ is in $\mathcal{W}$ we check the first and fourth components:

$$\begin{aligned} x_1 &= u_1 + v_1 = 0 \quad \text{since} \quad u_1 = 0 \quad \text{and} \quad v_1 = 0 \\ x_4 &= u_4 + v_4 = (u_2 + u_3) + (v_2 + v_3) = (u_2 + v_2) + (u_3 + v_3) \\ &= x_2 + x_3, \quad \text{since} \quad u_2 + u_3 = u_4 \quad \text{and} \quad v_2 + v_3 = v_4. \end{aligned}$$

Therefore, $\mathbf{x}$ is in $\mathcal{W}$ and we have shown that $\mathcal{W}$ is closed under vector addition. Similarly, $\mathbf{y} = k\mathbf{u} = [ku_1 \quad ku_2 \quad ku_3 \quad ku_4]^T$ is in $\mathcal{W}$ because

$$y_1 = ku_1 = k0 = 0 \quad \text{and} \quad y_4 = k(u_2 + u_3) = ku_2 + ku_3 = y_2 + y_3.$$

Thus, $\mathcal{W}$ is also closed under scalar multiplication, and it follows from Definition 3.7 that $\mathcal{W}$ is a subspace of $\Re^4$. ■

EXERCISES 3.2

1. For the vectors $\mathbf{u} = \begin{bmatrix} 3 \\ 0 \\ 1 \\ 2 \end{bmatrix}$ and $\mathbf{v} = \begin{bmatrix} 2 \\ 0 \\ 1 \\ 1 \end{bmatrix}$ in $\Re^4$:

(a) compute $\|\mathbf{u}\|$, $\mathbf{u} \cdot \mathbf{v}$, $\|\mathbf{v}\|$, and $\|\mathbf{u} + \mathbf{v}\|$; and
(b) find a vector orthogonal to $\mathbf{u}$.

2. For the vectors $\mathbf{u}$ and $\mathbf{v}$ in $\mathfrak{R}^4$ show that

(a) $(\mathbf{u} + \mathbf{v}) \cdot (\mathbf{u} + \mathbf{v}) = \mathbf{u} \cdot \mathbf{u} + 2(\mathbf{u} \cdot \mathbf{v}) + \mathbf{v} \cdot \mathbf{v}$;

(b) $(\mathbf{u} + \mathbf{v}) \cdot (\mathbf{u} - \mathbf{v}) = \|\mathbf{u}\|^2 - \|\mathbf{v}\|^2$; and

(c) $\|\mathbf{u}\|^2 + \|\mathbf{v}\|^2 = \|\mathbf{u} - \mathbf{v}\|^2$ if, and only if, $\mathbf{u} \perp \mathbf{v}$.

†**3.** **(a)** Prove that $a\mathbf{0} = \mathbf{0}$ and $0\mathbf{u} = \mathbf{0}$ for every scalar a and every vector $\mathbf{u}$ in $\mathfrak{R}^n$.

(b) Prove that $a\mathbf{u} = \mathbf{0}$ implies that either the vector $\mathbf{u} = \mathbf{0}$ or the scalar $a = 0$.

(c) Prove that $a\mathbf{u} = b\mathbf{u}$ if, and only if, either the vector $\mathbf{u} = \mathbf{0}$ or the scalar $a = b$.

4. Show that neither of the following subsets of $\mathfrak{R}^4$ is a subspace.

(a) $\left\{ \begin{bmatrix} x_1 \\ 1 + x_1 \\ x_3 \\ 5 \end{bmatrix} \middle| \, x_1 \text{ and } x_3 \text{ are real} \right\}$ **(b)** $\left\{ \mathbf{x} = \begin{bmatrix} x_1 \\ x_2 \\ x_3 \\ x_4 \end{bmatrix} \middle| \, x_1 + x_3 \leq x_4 \right\}$

5. Show that each of the following subsets of $\mathfrak{R}^4$ is a subspace.

(a) $\mathcal{W} = \{\mathbf{u} \text{ in } \mathfrak{R}^4 \mid A\mathbf{u} = 2\mathbf{u}\}$ (A is a fixed 4×4 matrix)

(b) $\mathcal{L} = \left\{ \mathbf{x} = \begin{bmatrix} x_1 \\ x_2 \\ x_3 \\ x_4 \end{bmatrix} \middle| \, x_1 + x_2 + x_4 = 0 \right\}$

†**6.** Let $\mathbf{u}$, $\mathbf{v}$, and $\mathbf{w}$ be any three vectors from $\mathfrak{R}^n$. Show that $\mathcal{W} = \{a\mathbf{u} + b\mathbf{v} + c\mathbf{w} \mid a, b, \text{ and } c \text{ real numbers}\}$ is a subspace of $\mathfrak{R}^n$.

7. Which of the following subsets of $\mathfrak{R}^4$ are subspaces?

(a) $\left\{ \begin{bmatrix} x_1 \\ x_2 \\ x_1 + 3x_2 \\ 5x_1 - 7x_2 \end{bmatrix} \middle| \, x_1, x_2 \text{ real} \right\}$

(b) $\{\mathbf{u} \text{ in } \mathfrak{R}^4 \mid \mathbf{u} \cdot \mathbf{v} = 0\}$ ($\mathbf{v}$ fixed in $\mathfrak{R}^4$)

(c) $\left\{ \mathbf{x} = \begin{bmatrix} x_1 \\ x_2 \\ x_3 \\ x_4 \end{bmatrix} \middle| \, x_1^2 + x_2^2 = 25 \right\}$

(d) $\left\{ \mathbf{x} = \begin{bmatrix} x_1 \\ x_2 \\ x_3 \\ x_4 \end{bmatrix} \middle| \, x_1 x_2 x_3 x_4 = 0 \right\}$

(e) $\left\{ \mathbf{x} = \begin{bmatrix} x_1 \\ x_2 \\ x_3 \\ x_4 \end{bmatrix} \middle| \, \frac{x_3}{x_4} = \frac{5x_2}{x_1} \right\}$

(f) $\left\{ \begin{bmatrix} a \\ a^{1/2} \\ \sin a \\ 2^a \end{bmatrix} \middle| \, 0 < a \right\}$

8. Is the dot product in $\mathfrak{R}^n$ associative?

9. For vectors $\mathbf{u}$ and $\mathbf{v}$ in $\mathfrak{R}^n$ we can generalize the results in $\mathfrak{R}^2$ to define the **orthogonal projection** of $\mathbf{u}$ on $\mathbf{v}$ to be the vector

$$\mathbf{w} = \left(\frac{\mathbf{u} \cdot \mathbf{v}}{\mathbf{v} \cdot \mathbf{v}}\right)\mathbf{v}.$$

(a) Find the projection of $\mathbf{u} = \begin{bmatrix} 1 \\ 2 \\ 4 \\ -2 \end{bmatrix}$ on $\mathbf{v} = \begin{bmatrix} 5 \\ 2 \\ -1 \\ 0 \end{bmatrix}$.

(b) Find the length of the projection of $\mathbf{x} = \begin{bmatrix} 1 \\ 1 \\ 1 \\ 1 \\ 1 \end{bmatrix}$ on $\mathbf{y} = \begin{bmatrix} 1 \\ 0 \\ 2 \\ 0 \\ 2 \end{bmatrix}$.

†**10.** Let $\mathcal{W}$ be a subspace of $\mathfrak{R}^n$. Show that

$$\mathcal{W}^\perp = \{\mathbf{u} \text{ in } \mathfrak{R}^n \,|\, \mathbf{u} \cdot \mathbf{v} = 0 \text{ for every } \mathbf{v} \text{ in } \mathcal{W}\}$$

is a subspace of $\mathfrak{R}^n$.

11. Prove Theorem 3.2.

12. Show that every proper subspace of $\mathfrak{R}^3$ is either a line or a plane through the origin.

13. Which of the following subsets of $\mathfrak{R}^4$ are subspaces of $\mathfrak{R}^4$?

(a) $\mathcal{W}_1 = \left\{ \begin{bmatrix} a + b + c \\ 2 + a \\ 3a - 4b \\ 5 \end{bmatrix} \middle| \, a, b, c \text{ are in } \mathfrak{R} \right\}$

(b) $\mathcal{W}_2 = \left\{ \begin{bmatrix} \sqrt{a} + 2b \\ 3a^2 + 5c \\ 4 \\ 5 \end{bmatrix} \middle| \, a, b, c \text{ are in } \mathfrak{R} \right\}$

(c) $\mathcal{W}_3 = \left\{ \begin{bmatrix} a + 2b - c \\ 3a + b + 5c \\ -5a + 2b + 3c \\ 0 \end{bmatrix} \middle| \, a, b, c \text{ are in } \mathfrak{R} \right\}$

3.3 SUBSPACES OF $\mathfrak{R}^n$—CONTINUED

Since a subspace $\mathcal{W}$ of $\mathfrak{R}^n$ must be closed under scalar multiplication, it follows that a nonzero subspace of $\mathfrak{R}^n$ must contain an infinite number of vectors. Thus a finite set $\mathcal{S} = \{\mathbf{u}_1, \mathbf{u}_2, \ldots, \mathbf{u}_s\} \subset \mathfrak{R}^n$ cannot be a subspace.

> What must be added to the set $\mathcal{S} = \{\mathbf{u}_1, \mathbf{u}_2, \ldots, \mathbf{u}_s\} \subset \mathfrak{R}^n$ to make it a subspace?

Since a subspace must be closed under linear combinations, any subspace containing $\mathcal{S}$ must include all vectors in the set $\mathcal{W} = \{\Sigma_{k=1}^{s} a_k\mathbf{u}_\mathbf{k} \,|\, a_k \text{ real}\}$. Since

$$\sum_{k=1}^{s} a_k\mathbf{u}_\mathbf{k} + \sum_{k=1}^{s} b_k\mathbf{u}_\mathbf{k} = \sum_{k=1}^{s} (a_k + b_k)\mathbf{u}_\mathbf{k} \qquad \text{and} \qquad c\sum_{k=1}^{s} b_k\mathbf{u}_\mathbf{k} = \sum_{k=1}^{s} cb_k\mathbf{u}_\mathbf{k}$$

it follows that $\mathcal{W}$ is closed under vector addition and scalar multiplication. It then follows from Definition 3.7 that $\mathcal{W}$ is a subspace of $\mathfrak{R}^n$. Moreover, $\mathcal{W}$ is the smallest subspace of $\mathfrak{R}^n$ containing $\mathcal{S}$. We now make the following definition:

DEFINITION 3.8 If $\mathcal{S} = \{\mathbf{u}_1, \mathbf{u}_2, \ldots, \mathbf{u}_s\}$ is any set of vectors from $\mathfrak{R}^n$, then the **subspace spanned** by the set $\mathcal{S}$ is defined by

$$\text{SPAN}(\mathcal{S}) = \text{SPAN}\{\mathbf{u}_1, \mathbf{u}_2, \ldots, \mathbf{u}_s\} = \left\{\sum_{k=1}^{s} a_k\mathbf{u}_\mathbf{k} \,\middle|\, a_k \text{ real}\right\};$$

that is, the subspace consists of all possible linear combinations of the vectors of $\mathcal{S}$. □

We noted, following Example 1 in Section 3.1, that for $\mathbf{u}$ and $\mathbf{v}$ in $\mathfrak{R}^3$ the plane containing the vectors $\mathbf{u}$ and $\mathbf{v}$ consists of all linear combinations of $\mathbf{u}$ and $\mathbf{v}$; that is, it is precisely the subspace SPAN$\{\mathbf{u}, \mathbf{v}\}$. Thus, we can think of the "subspace spanned by a set of vectors in $\mathfrak{R}^n$" as a generalization of the "plane generated by a pair of vectors in $\mathfrak{R}^3$." Another useful point of view is to think of SPAN$\{\mathbf{u}_1, \mathbf{u}_2, \ldots, \mathbf{u}_n\}$ as the "smallest" subspace of $\mathfrak{R}^n$ containing the vectors $\mathbf{u}_1, \mathbf{u}_2, \ldots, \mathbf{u}_\mathbf{n}$.

To determine if a given vector $\mathbf{w}$ is an element of the subspace SPAN$\{\mathbf{u}_1, \mathbf{u}_2, \ldots, \mathbf{u}_n\}$, we need to decide whether the vector $\mathbf{w}$ can be written as a linear combination of the vectors of the spanning set. This investigation normally leads to a question about the solvability of a system of linear equations; such questions can be answered efficiently using the results of Chapter 1.

EXAMPLE 1 **Test a Vector for Membership in SPAN($\mathcal{S}$)**

Is the vector $\mathbf{v}_1 = \begin{bmatrix} 4 \\ 7 \\ 6 \end{bmatrix}$ in $\mathcal{W} = \text{SPAN}\left\{\mathbf{u}_1 = \begin{bmatrix} 1 \\ 2 \\ 1 \end{bmatrix}, \mathbf{u}_2 = \begin{bmatrix} 2 \\ 3 \\ 4 \end{bmatrix}\right\}$?

To determine whether $\mathbf{v}_1$ is in $\mathcal{W}$ we need to determine if there are scalars x and y such that $x\mathbf{u}_1 + y\mathbf{u}_2 = \mathbf{v}_1$; that is, we want

$$x\begin{bmatrix}1\\2\\1\end{bmatrix} + y\begin{bmatrix}2\\3\\4\end{bmatrix} = \begin{bmatrix}4\\7\\6\end{bmatrix}, \text{ which is equivalent to } \begin{bmatrix}x + 2y\\2x + 3y\\x + 4y\end{bmatrix} = \begin{bmatrix}4\\7\\6\end{bmatrix}.$$

Thus, to determine whether the vector $\mathbf{v}_1$ is in the subspace SPAN$\{\mathbf{u}_1, \mathbf{u}_2\}$, we need to decide whether or not the linear system

$$AX = \begin{bmatrix}1 & 2\\2 & 3\\1 & 4\end{bmatrix}\begin{bmatrix}x\\y\end{bmatrix} = \begin{bmatrix}4\\7\\6\end{bmatrix} = \mathbf{v}_1$$

has a solution. Row reducing the augmented matrix of this system, we obtain

$$\begin{bmatrix}1 & 2 & 4\\2 & 3 & 7\\1 & 4 & 6\end{bmatrix} \to \begin{bmatrix}1 & 2 & 4\\0 & -1 & -1\\0 & 2 & 2\end{bmatrix} \to \begin{bmatrix}1 & 2 & 4\\0 & 1 & 1\\0 & 0 & 0\end{bmatrix} \to \begin{bmatrix}1 & 0 & 2\\0 & 1 & 1\\0 & 0 & 0\end{bmatrix}$$

and hence determine that $x = 2$ and $y = 1$ is the unique solution of the system. Our calculations show that $\mathbf{v}_1 = 2\mathbf{u}_1 + \mathbf{u}_2$, so that $\mathbf{v}_1$ is an element of the subspace $\mathcal{W} = $ SPAN$\{\mathbf{u}_1, \mathbf{u}_2\}$. ■

If we need to test several vectors for membership in the subspace SPAN$\{\mathbf{u}_1, \mathbf{u}_2, \ldots, \mathbf{u}_n\}$, then we can efficiently use the methods of Section 1.4 for solving sets of linear systems with the same coefficient matrix.

■ EXAMPLE 2 **Testing Several Vectors at Once**

Using the vectors $\mathbf{u}_1$ and $\mathbf{u}_2$ given in Example 1, which of

$$\mathbf{w}_1 = \begin{bmatrix}3\\5\\5\end{bmatrix}, \mathbf{w}_2 = \begin{bmatrix}1\\1\\3\end{bmatrix}, \mathbf{w}_3 = \begin{bmatrix}2\\9\\5\end{bmatrix}, \text{ and } \mathbf{w}_4 = \begin{bmatrix}0\\1\\-2\end{bmatrix} \text{ are in SPAN}\{\mathbf{u}_1, \mathbf{u}_2\}?$$

To decide if the $\mathbf{w}_i$ are in SPAN$\{\mathbf{u}_1, \mathbf{u}_2\}$ we need to investigate the solvability of the systems

$$AX = \begin{bmatrix}1 & 2\\2 & 3\\1 & 4\end{bmatrix}\begin{bmatrix}x\\y\end{bmatrix} = \mathbf{w}_i; \quad i = 1, 2, 3, 4.$$

We have here a set of four systems of linear equations, all with the same coefficient matrix. Using the techniques of Section 1.4, we row reduce the fourfold augmented matrix:

$$\left[\begin{array}{cc:cccc} 1 & 2 & 3 & 1 & 2 & 0 \\ 2 & 3 & 5 & 1 & 9 & 1 \\ 1 & 4 & 5 & 3 & 5 & -2 \end{array}\right] \rightarrow \left[\begin{array}{cc:cccc} 1 & 2 & 3 & 1 & 2 & 0 \\ 0 & -1 & -1 & -1 & 5 & 1 \\ 0 & 2 & 2 & 2 & 3 & -2 \end{array}\right]$$

$$\rightarrow \left[\begin{array}{cc:cccc} 1 & 2 & 3 & 1 & 2 & 0 \\ 0 & 1 & 1 & 1 & -5 & -1 \\ 0 & 0 & 0 & 0 & \boxed{13} & 0 \end{array}\right].$$

The nonzero entry (13) in the (3, 5) position of the last matrix indicates that the system $AX = \mathbf{w}_3$ is not solvable and hence that $\mathbf{w}_3$ is not an element of SPAN$\{\mathbf{u}_1, \mathbf{u}_2\}$. Continuing the above row reduction we obtain the reduced row-echelon matrix

$$\left[\begin{array}{cc:cccc} 1 & 0 & 1 & -1 & 0 & 2 \\ 0 & 1 & 1 & 1 & 0 & -1 \\ 0 & 0 & 0 & 0 & 1 & 0 \end{array}\right].$$

From the third column of the last matrix we conclude that

$$\mathbf{w}_1 = \mathbf{u}_1 + \mathbf{u}_2;$$

from the fourth column we obtain

$$\mathbf{w}_2 = -\mathbf{u}_1 + \mathbf{u}_2;$$

and from the last column we conclude that

$$\mathbf{w}_4 = 2\mathbf{u}_1 - \mathbf{u}_2. \quad \blacksquare$$

Our next example illustrates that finding a spanning set for a subspace often involves finding the general solution of some system of linear equations.

EXAMPLE 3 **Spanning Set for the Null Space**

Find a spanning set for the null space of $A = \begin{bmatrix} 1 & -2 & 0 & 0 & 3 \\ 0 & 0 & 1 & 0 & 7 \\ 0 & 0 & 0 & 1 & -2 \\ 0 & 0 & 0 & 0 & 0 \\ 0 & 0 & 0 & 0 & 0 \end{bmatrix}$.

Recall that $\mathcal{NS}(A) = \{\mathbf{X} \text{ in } \Re^5 \mid A\mathbf{X} = \mathbf{0}\}$. Since the matrix A is already in echelon form, the system is easy to solve. The variables x_2 and x_5 are free and can be assigned arbitrary values; the other variables are then determined in terms of

$x_2 = c_1$ and $x_5 = c_2$. The general solution is then

$$\mathbf{X} = \begin{bmatrix} 2c_1 - 3c_2 \\ c_1 \\ -7c_2 \\ 2c_2 \\ c_2 \end{bmatrix} = c_1 \begin{bmatrix} 2 \\ 1 \\ 0 \\ 0 \\ 0 \end{bmatrix} + c_2 \begin{bmatrix} -3 \\ 0 \\ -7 \\ 2 \\ 1 \end{bmatrix} = c_1\mathbf{X}_1 + c_2\mathbf{X}_2.$$

Thus we see that $\mathcal{NS}(A) = \text{SPAN}\{\mathbf{X}_1, \mathbf{X}_2\}$. ■

There are several subspaces that are useful in the study of a given $m \times n$ matrix A and the linear systems with coefficient matrix A. We have already mentioned the null space of A, and we now define another subspace called the column space of A.

DEFINITION 3.9 If A is an $m \times n$ matrix, the **column space** of A is the subspace of $\Re^m$ spanned by the columns of A. We will denote the column space of A by $\mathcal{CS}(A)$, so that

$$\mathcal{CS}(A) = \text{SPAN}\{\text{Col}_1(A), \text{Col}_2(A), \ldots, \text{Col}_n(A)\}. \quad \square$$

The subspace $\mathcal{CS}(A)$ is sometimes called the **range** of A, because it is the range of the function $\mathbf{T}_\mathbf{A}$ from $\Re^n$ to $\Re^m$ defined by $\mathbf{T}_\mathbf{A}(\mathbf{u}) = A\mathbf{u}$. In order to see why this is so, recall from Chapter 1 that the columns of the matrix product AB are linear combinations of the columns of A. A typical element of the column space of A is

$$x_1\text{Col}_1(A) + x_2\text{Col}_2(A) + \cdots + x_n\text{Col}_n(A),$$

which can be written in the form $A\mathbf{X}$, where $\mathbf{X}$ is the element of $\Re^n$ made up of the coefficients in the linear combination. Thus, for an $m \times n$ matrix A,

$$\mathcal{CS}(A) = \{A\mathbf{X} \mid \mathbf{X} \text{ in } \Re^n\} \tag{3.4}$$

which is the range of the function $\mathbf{T}_\mathbf{A}$.

Observe that the linear system $A\mathbf{X} = \mathbf{K}$ has a solution if, and only if, $\mathbf{K}$ is an element of the column space of A. We state this important result as our next theorem.

THEOREM 3.4 The column space of A consists of precisely those vectors $\mathbf{K}$ for which the linear system $A\mathbf{X} = \mathbf{K}$ has a solution; that is,

$$\mathcal{CS}(A) = \{\mathbf{K} \text{ in } \Re^m \mid A\mathbf{X} = \mathbf{K} \text{ is solvable}\}. \quad \square$$

Knowing a spanning set for a given subspace is often very useful, but, as illustrated in Examples 1 and 2, it requires some calculations to determine if a particular vector is in the subspace or not. In the next example we illustrate how we can find an alternate characterization of a subspace that is determined by a spanning set. This alternate characterization involves determining specific membership criteria for the subspace.

EXAMPLE 4 Find Membership Criteria for a Subspace

Find an alternate description of the column space of

$$A = \begin{bmatrix} 1 & -2 & 1 & -4 \\ 1 & 3 & 7 & 2 \\ 1 & -12 & -11 & -16 \\ 2 & -14 & -10 & -20 \end{bmatrix}.$$

Using Theorem 3.4 we can find the desired description for the column space of A by determining those vectors $\mathbf{K}$ in $\Re^4$ for which the linear system $A\mathbf{X} = \mathbf{K}$ is solvable. The augmented matrix of this system is

$$[A|\mathbf{K}] = \left[\begin{array}{cccc|c} 1 & -2 & 1 & -4 & k_1 \\ 1 & 3 & 7 & 2 & k_2 \\ 1 & -12 & -11 & -16 & k_3 \\ 2 & -14 & -10 & -20 & k_4 \end{array}\right].$$

From the straightforward row reduction

$$\left[\begin{array}{cccc|c} 1 & -2 & 1 & -4 & k_1 \\ 1 & 3 & 7 & 2 & k_2 \\ 1 & -12 & -11 & -16 & k_3 \\ 2 & -14 & -10 & -20 & k_4 \end{array}\right] \to \left[\begin{array}{cccc|c} 1 & -2 & 1 & 4 & k_1 \\ 0 & 5 & 6 & 6 & k_2 - k_1 \\ 0 & -10 & -12 & -12 & k_3 - k_1 \\ 0 & -10 & -12 & -12 & k_4 - 2k_1 \end{array}\right]$$

$$\to \left[\begin{array}{cccc|c} 1 & 0 & \frac{17}{5} & -\frac{8}{5} & k_1 + \frac{2}{5}(k_2 - k_1) \\ 0 & 1 & \frac{6}{5} & \frac{6}{5} & \frac{1}{5}(k_2 - k_1) \\ 0 & 0 & 0 & 0 & k_3 - k_1 + 2(k_2 - k_1) \\ 0 & 0 & 0 & 0 & k_4 - 2k_1 + 2(k_2 - k_1) \end{array}\right],$$

we see that the system $AX = K$ has a solution if, and only if, the elements in the (3, 5) and (4, 5) positions of the last matrix are zero, that is, if

$$-3k_1 + 2k_2 + k_3 = 0 \qquad \text{and} \qquad -4k_1 + 2k_2 + k_4 = 0.$$

We have thus shown that

$$\mathscr{CS}(A) = \{\mathbf{K} \text{ in } \Re^4 | k_3 = 3k_1 - 2k_2 \text{ and } k_4 = 4k_1 - 2k_2\}.$$

The alternate description of $\mathscr{CS}(A)$ just obtained makes much clearer the membership criteria for the subspace. Using these criteria it would be quite easy to determine whether or not a particular vector $\mathbf{K}$ was in $\mathscr{CS}(A)$.

It is also worth noting that $\mathscr{CS}(A) = \mathscr{NS}(B)$, where

$$B = \begin{bmatrix} -3 & 2 & 1 & 0 \\ -4 & 2 & 0 & 1 \end{bmatrix}.$$ ■

It is a fact, although we will not prove it now, that every subspace of $\Re^n$ is the null space of some matrix. This result would follow, as in Example 4, if we could show that any subspace of $\Re^n$ had a spanning set; this fact will be demonstrated in Section 3.5.

Consider now the subspace of $\Re^n$ spanned by $\mathcal{S} = \{\mathbf{u}_1, \mathbf{u}_2, \mathbf{u}_3, \mathbf{u}_4\}$. In general we cannot expect that the spanning set will be unique. Suppose that we knew that the vectors in the spanning set satisfied the linear relation

$$2\mathbf{u}_1 - 5\mathbf{u}_2 + \mathbf{u}_3 - 3\mathbf{u}_4 = \mathbf{0}. \tag{3.5}$$

In this case we could solve for any one of the $\mathbf{u}_i$ in terms of the others. For example

$$\mathbf{u}_3 = -2\mathbf{u}_1 + 5\mathbf{u}_2 + 3\mathbf{u}_4. \tag{3.6}$$

A typical vector in SPAN($\mathcal{S}$) is

$$\mathbf{v} = a_1\mathbf{u}_1 + a_2\mathbf{u}_2 + a_3\mathbf{u}_3 + a_4\mathbf{u}_4$$

which, using Equation (3.6), we can rewrite as

$$\begin{aligned}\mathbf{v} &= a_1\mathbf{u}_1 + a_2\mathbf{u}_2 + a_3(-2\mathbf{u}_1 + 5\mathbf{u}_2 + 3\mathbf{u}_4) + a_4\mathbf{u}_4\\ &= (a_1 - 2a_3)\mathbf{u}_1 + (a_2 + 5a_3)\mathbf{u}_2 + (a_4 + 3a_3)\mathbf{u}_4.\end{aligned}$$

We have shown that the vectors $\{\mathbf{u}_1, \mathbf{u}_2, \mathbf{u}_4\}$ are sufficient to generate every vector in SPAN($\mathcal{S}$). It follows that

$$\text{SPAN}\{\mathbf{u}_1, \mathbf{u}_2, \mathbf{u}_3, \mathbf{u}_4\} = \text{SPAN}\{\mathbf{u}_1, \mathbf{u}_2, \mathbf{u}_4\}.$$

Since we could have solved Equation (3.5) for any one of the $\mathbf{u}_i$ in terms of the others it follows that we could have deleted any one of the four vectors from the spanning set without making the space any smaller.

In Example 4 we see, from the first three columns of the matrices, that

$$\text{Col}_3(A) = \frac{17}{5}\,\text{Col}_1(A) + \frac{6}{5}\,\text{Col}_2(A).$$

Thus $\text{Col}_3(A)$ could be deleted from the spanning set. Similarly, from columns 1, 2, and 4 of the matrices we see that

$$\text{Col}_4(A) = -\frac{8}{5}\,\text{Col}_1(A) + \frac{6}{5}\,\text{Col}_2(A).$$

Thus $\text{Col}_4(A)$ can also be deleted from the spanning set. It follows, for the matrix of Example 4, that

$$\mathcal{CS}(A) = \text{SPAN}\{\text{Col}_1(A), \text{Col}_2(A)\}.$$

Since $\text{Col}_2(A)$ is not a scalar multiple of $\text{Col}_1(A)$, we have found a minimal spanning set $\mathcal{CS}(A)$.

We summarize the above discussion with the following theorem.

THEOREM 3.5 If $\mathcal{W} = \text{SPAN}\{\mathbf{u}_1, \mathbf{u}_2, \ldots, \mathbf{u}_{k-1}, \mathbf{u}_k\}$ and $\mathbf{u}_k$ is in $\text{SPAN}\{\mathbf{u}_1, \mathbf{u}_2, \ldots, \mathbf{u}_{k-1}\}$, then $\mathbf{u}_k$ can be deleted from the spanning set and $\mathcal{W} = \text{SPAN}\{\mathbf{u}_1, \mathbf{u}_2, \ldots, \mathbf{u}_{k-1}\}$. □

EXERCISES 3.3

1. Show that $\mathbf{w} = \begin{bmatrix} 0 \\ 1 \\ -1 \end{bmatrix}$ is in the subspace of $\mathfrak{R}^3$ spanned by $\mathbf{u} = \begin{bmatrix} 1 \\ 2 \\ 0 \end{bmatrix}$ and $\mathbf{v} = \begin{bmatrix} 3 \\ 5 \\ 1 \end{bmatrix}$.

2. Show that every vector in $\mathfrak{R}^2$ is a linear combination of the vectors $\begin{bmatrix} 1 \\ 2 \end{bmatrix}$ and $\begin{bmatrix} 5 \\ 7 \end{bmatrix}$.

3. Show that every vector in $\mathfrak{R}^3$ is a linear combination of the vectors

$$\mathbf{u} = \begin{bmatrix} 1 \\ 0 \\ 1 \end{bmatrix}, \quad \mathbf{v} = \begin{bmatrix} 1 \\ 1 \\ 1 \end{bmatrix}, \text{ and } \mathbf{w} = \begin{bmatrix} 0 \\ 1 \\ 1 \end{bmatrix}.$$

4. Find a spanning set for the null spaces of the following matrices.

$$\textbf{(a)} \begin{bmatrix} 3 & 2 & 16 & 5 \\ 0 & 2 & 10 & 8 \\ 1 & 1 & 7 & 3 \end{bmatrix} \qquad \textbf{(b)} \begin{bmatrix} 2 & 1 & 3 \\ 1 & 2 & 0 \\ 0 & 1 & 1 \end{bmatrix} \qquad \textbf{(c)} \begin{bmatrix} 1 & 1 & -3 & 1 \\ 2 & 0 & 1 & -1 \\ 1 & 3 & -10 & 4 \end{bmatrix}$$

5. Determine which of the following vectors belong to $\text{SPAN}\{\mathbf{u}_1, \mathbf{u}_2\}$:

$$\mathbf{v}_1 = \begin{bmatrix} 1 \\ 3 \\ 4 \end{bmatrix}, \quad \mathbf{v}_2 = \begin{bmatrix} 1 \\ 4 \\ 7 \end{bmatrix}, \quad \mathbf{v}_3 = \begin{bmatrix} 3 \\ 5 \\ 7 \end{bmatrix}, \text{ where } \mathbf{u}_1 = \begin{bmatrix} 1 \\ 2 \\ 3 \end{bmatrix}, \text{ and } \mathbf{u}_2 = \begin{bmatrix} 0 \\ 1 \\ 2 \end{bmatrix}.$$

6. Determine which of the following vectors belong to $\text{SPAN}\{\mathbf{x}, \mathbf{y}, \mathbf{z}\}$:

$$\mathbf{v}_1 = \begin{bmatrix} 0 \\ 1 \\ 4 \\ 3 \end{bmatrix}, \quad \mathbf{v}_2 = \begin{bmatrix} 1 \\ 0 \\ 3 \\ 2 \end{bmatrix}, \quad \text{and} \quad \mathbf{v}_3 = \begin{bmatrix} 1 \\ -1 \\ 3 \\ 1 \end{bmatrix}, \text{ where}$$

$$\mathbf{x} = \begin{bmatrix} 0 \\ 1 \\ 2 \\ 3 \end{bmatrix}, \quad \mathbf{y} = \begin{bmatrix} 1 \\ -1 \\ 2 \\ -1 \end{bmatrix}, \quad \text{and} \quad \mathbf{z} = \begin{bmatrix} -1 \\ 1 \\ 0 \\ 1 \end{bmatrix}.$$

7. Prove that the following two sets span the same subspace of $\mathfrak{R}^3$:
(a) $\mathbf{u}_1 = [1 \quad -1 \quad 2]^T$ and $\mathbf{u}_2 = [3 \quad 0 \quad 1]^T$
(b) $\mathbf{v}_1 = [-1 \quad -2 \quad 3]^T$ and $\mathbf{v}_2 = [3 \quad 3 \quad -4]^T$.
(*Hint:* Show that $\mathbf{u}_1$ and $\mathbf{u}_2$ are in $\text{SPAN}\{\mathbf{v}_1, \mathbf{v}_2\}$ and that $\mathbf{v}_1$ and $\mathbf{v}_2$ are in $\text{SPAN}\{\mathbf{u}_1, \mathbf{u}_2\}$.)

8. Prove that the only nontrivial ($\neq \{\mathbf{0}\}$ or $\Re^3$) subspaces of $\Re^3$ are lines and planes through the origin.

†9. Find a matrix B such that $\mathcal{NS}(B) = \mathcal{CS}(A)$ if A if the matrix given.

(a) $A = \begin{bmatrix} 1 & 2 & 3 \\ 4 & 5 & 6 \\ 7 & 8 & 9 \end{bmatrix}$ **(b)** $A = \begin{bmatrix} 1 & 1 & 4 \\ -1 & 0 & 2 \\ 3 & 1 & 3 \\ 0 & 0 & 6 \end{bmatrix}$

(c) $A = \begin{bmatrix} 1 & 0 & 0 & 1 \\ 1 & 1 & -1 & 0 \\ 1 & 2 & -2 & -1 \\ 1 & 3 & -3 & -2 \end{bmatrix}$

†10. Show that if $\mathbf{u}_1 + 2\mathbf{u}_2 - 3\mathbf{u}_3 + 4\mathbf{u}_4 = \mathbf{0}$, then

$$\text{SPAN}\{\mathbf{u}_1, \mathbf{u}_2, \mathbf{u}_3, \mathbf{u}_4\} = \text{SPAN}\{\mathbf{u}_2, \mathbf{u}_3, \mathbf{u}_4\}.$$

11. Determine whether the given set of vectors spans $\Re^3$:

(a) $\begin{bmatrix} 1 \\ 1 \\ 3 \end{bmatrix}, \begin{bmatrix} 1 \\ 1 \\ 0 \end{bmatrix}, \begin{bmatrix} 2 \\ 0 \\ 0 \end{bmatrix}.$

(b) $\begin{bmatrix} 2 \\ 2 \\ 2 \end{bmatrix}, \begin{bmatrix} 0 \\ 1 \\ 1 \end{bmatrix}, \begin{bmatrix} 0 \\ 0 \\ 4 \end{bmatrix}.$

(c) $\begin{bmatrix} 1 \\ 2 \\ 6 \end{bmatrix}, \begin{bmatrix} 4 \\ 3 \\ 1 \end{bmatrix}, \begin{bmatrix} 3 \\ 3 \\ 1 \end{bmatrix}, \begin{bmatrix} 3 \\ 4 \\ 1 \end{bmatrix}.$

12. (a) Show that elementary column operations on a matrix A do not change the column space of A. (*Hint:* Recalling Exercise 22 of Section 1.7, show that $\mathcal{CS}(AQ) \subseteq \mathcal{CS}(A)$ and, if Q^{-1} exists, that $\mathcal{CS}(A) \subseteq \mathcal{CS}(AQ)$.)

(b) How do elementary row operations affect the column space of A? (*Hint:* Consider $A = \begin{bmatrix} 1 & 3 \\ 2 & 6 \end{bmatrix}$ and the matrix B obtained from A by the operation $R_2 \leftarrow -2R_1 + R_2$. What do $\mathcal{CS}(A)$ and $\mathcal{CS}(B)$ represent geometrically in $\Re^2$?)

13. Find an equation for the plane spanned by $\mathbf{u} = \begin{bmatrix} 1 \\ 2 \\ 3 \end{bmatrix}$, and $\mathbf{v} = \begin{bmatrix} 3 \\ -1 \\ 2 \end{bmatrix}$.

In Exercises 14–17, let $\mathcal{W}$ be the set of all vectors of the type shown where a, b, and c are arbitrary real numbers. In each case either find a spanning set or give an example to show that $\mathcal{W}$ is not a subspace.

14. $\begin{bmatrix} 3a - 5b \\ 4a + 6b \\ 5a - 3c \end{bmatrix}$ **15.** $\begin{bmatrix} a + b + c \\ 3 \\ 2b + c \end{bmatrix}$

16. $\begin{bmatrix} 4a + 2b - 3c \\ 0 \\ 7a - c \\ a + b + 4 \end{bmatrix}$

17. $\begin{bmatrix} a + b - c \\ 2a - 3b + 4c \\ b - 3c \\ 4c \end{bmatrix}$

18. Show that

$$\text{SPAN}\left\{\mathbf{u}_1 = \begin{bmatrix} 1 \\ 6 \\ 4 \end{bmatrix}, \mathbf{u}_2 = \begin{bmatrix} 2 \\ 4 \\ -1 \end{bmatrix}, \mathbf{u}_3 = \begin{bmatrix} -1 \\ 2 \\ 5 \end{bmatrix}\right\}$$

$$= \text{SPAN}\left\{\mathbf{v}_1 = \begin{bmatrix} -1 \\ 2 \\ 5 \end{bmatrix}, \mathbf{v}_2 = \begin{bmatrix} 0 \\ 8 \\ 9 \end{bmatrix}\right\}.$$

19. Are the following vectors coplanar in $\mathfrak{R}^3$?

(a) $\begin{bmatrix} 1 \\ 1 \\ 1 \end{bmatrix}, \begin{bmatrix} 1 \\ 2 \\ 3 \end{bmatrix}, \begin{bmatrix} 1 \\ 4 \\ 6 \end{bmatrix}$ **(b)** $\begin{bmatrix} -1 \\ 2 \\ -3 \end{bmatrix}, \begin{bmatrix} -4 \\ 5 \\ -6 \end{bmatrix}, \begin{bmatrix} -7 \\ 8 \\ -9 \end{bmatrix}$ **(c)** $\begin{bmatrix} 1 \\ 2 \\ 1 \end{bmatrix}, \begin{bmatrix} -4 \\ 2 \\ 3 \end{bmatrix}, \begin{bmatrix} -2 \\ 6 \\ 5 \end{bmatrix}$

3.4 LINEAR INDEPENDENCE AND DEPENDENCE IN $\mathfrak{R}^n$

At the end of the last section we discussed one situation in which we could reduce the number of vectors in the spanning set for a subspace $\mathcal{W}$ of $\mathfrak{R}^n$. In general, if the set of vectors $\{\mathbf{u}_1, \mathbf{u}_2, \ldots, \mathbf{u}_k\}$ spans the subspace $\mathcal{W}$, then we know, from Definition 3.8, that every vector $\mathbf{v}$ in $\mathcal{W}$ can be expressed as a linear combination of the $\mathbf{u}_i$:

$$\mathbf{v} = a_1\mathbf{u}_1 + a_2\mathbf{u}_2 + \cdots + a_k\mathbf{u}_k. \tag{3.7}$$

Under what conditions are the coefficients in Equation (3.7) unique?

Suppose that in addition to the relation in Equation (3.7) we also had

$$\mathbf{v} = b_1\mathbf{u}_1 + b_2\mathbf{u}_2 + \cdots + b_k\mathbf{u}_k,$$

where $a_i \neq b_i$ for at least one i. Simple algebraic manipulation, using the properties of vector algebra listed in Theorem 3.1, leads to

$$\mathbf{0} = \mathbf{v} - \mathbf{v} = (a_1 - b_1)\mathbf{u}_1 + (a_2 - b_2)\mathbf{u}_2 + \cdots + (a_k - b_k)\mathbf{u}_k.$$

Since some of the coefficients in the last expression are different from zero, the $\mathbf{u}_i$ are dependent in the sense that at least one of them can be expressed as a linear combination of the others. We will use the above discussion as motivation for the following important definition.

DEFINITION 3.10 Let $\mathcal{S} = \{\mathbf{u}_1, \mathbf{u}_2, \ldots, \mathbf{u}_k\}$ be a set of vectors from the subspace $\mathcal{W}$. The set $\mathcal{S}$ is said to be **linearly dependent** if there exist scalars a_i, not all zero, such that

$$a_1\mathbf{u}_1 + a_2\mathbf{u}_2 + \cdots + a_k\mathbf{u}_k = \mathbf{0}. \tag{3.8}$$

The set $\mathcal{S}$ is **linearly independent** if it is not linearly dependent. □

Since $a\mathbf{u} + b\mathbf{v} = \mathbf{0}$, with $a \neq 0$, implies that $\mathbf{u} = -(\frac{b}{a})\mathbf{v}$, it follows that two vectors are linearly dependent if, and only if, one of them is a scalar multiple of the other; this case is always easy to determine. Note that if $\mathbf{x}$, $\mathbf{y}$, and $\mathbf{z}$ are three vectors in $\Re^3$, then a nontrivial relation of the form $a\mathbf{x} + b\mathbf{y} + c\mathbf{z} = \mathbf{0}$ means that one of the vectors is in the plane determined by the other two. In particular, if $a \neq 0$ then $\mathbf{x} = (-\frac{b}{a})\mathbf{y} + (-\frac{c}{a})\mathbf{z}$ is in the plane determined by $\mathbf{y}$ and $\mathbf{z}$. Thus we can view linear dependence as being a generalization of the geometric notion of three vectors being coplanar in $\Re^3$, or two vectors being collinear in $\Re^3$ or $\Re^2$.

Normally we think of linear independence as being a desirable condition and linear dependence as being undesirable. In order to investigate the linear independence or dependence of a given set of vectors we try to deduce from Equation (3.8) whether the coefficients must all be zero. This process always leads to a homogeneous system of linear equations, for which we need to determine whether a nonzero solution exists. Recall from our earlier work that the homogeneous system $AX = O$ has a nonzero solution if any one of the following conditions hold:

(1) A has fewer rows than columns;
(2) A is row equivalent to a matrix with fewer nonzero rows than columns;
(3) A is square and singular; or
(4) A is square and $\det(A) = 0$.

EXAMPLE 1 Test a Dependent Set for Independence

Determine if the following vectors from $\Re^4$ are linearly independent or linearly dependent.

$$\mathbf{u}_1 = \begin{bmatrix} 1 \\ 2 \\ -1 \\ 2 \end{bmatrix}, \quad \mathbf{u}_2 = \begin{bmatrix} -2 \\ -5 \\ 3 \\ 0 \end{bmatrix}, \quad \mathbf{u}_3 = \begin{bmatrix} 1 \\ 0 \\ 1 \\ 10 \end{bmatrix}.$$

The vector equation $a_1\mathbf{u}_1 + a_2\mathbf{u}_2 + a_3\mathbf{u}_3 = \mathbf{0}$ is equivalent to

$$a_1\begin{bmatrix} 1 \\ 2 \\ -1 \\ 2 \end{bmatrix} + a_2\begin{bmatrix} -2 \\ -5 \\ 3 \\ 0 \end{bmatrix} + a_3\begin{bmatrix} 1 \\ 0 \\ 1 \\ 10 \end{bmatrix} = \begin{bmatrix} 0 \\ 0 \\ 0 \\ 0 \end{bmatrix}.$$

By comparing the components of the vectors in the last equation we obtain the linear system

$$\begin{aligned} a_1 - 2a_2 + a_3 &= 0 \\ 2a_1 - 5a_2 &= 0 \\ -a_1 + 3a_2 + a_3 &= 0 \\ 2a_1 + 10a_3 &= 0 \end{aligned} \quad \text{or} \quad \begin{bmatrix} 1 & -2 & 1 \\ 2 & -5 & 0 \\ -1 & 3 & 1 \\ 2 & 0 & 10 \end{bmatrix} \begin{bmatrix} a_1 \\ a_2 \\ a_3 \end{bmatrix} = \begin{bmatrix} 0 \\ 0 \\ 0 \\ 0 \end{bmatrix}.$$

The question of the independence or dependence of the $\mathbf{u}_i$ has thus been reduced to the question of whether this system of linear equations has a nonzero solution; if it does, the vectors are dependent, and if it does not, the vectors are independent. To decide if this system has any nonzero solutions we row reduce the augmented matrix as follows:

$$\left[\begin{array}{rrr|r} 1 & -2 & 1 & 0 \\ 2 & -5 & 0 & 0 \\ -1 & 3 & 1 & 0 \\ 2 & 0 & 10 & 0 \end{array}\right] \to \left[\begin{array}{rrr|r} 1 & -2 & 1 & 0 \\ 0 & -1 & -2 & 0 \\ 0 & 1 & 2 & 0 \\ 0 & 4 & 8 & 0 \end{array}\right] \to \left[\begin{array}{rrr|r} 1 & 0 & 5 & 0 \\ 0 & 1 & 2 & 0 \\ 0 & 0 & 0 & 0 \\ 0 & 0 & 0 & 0 \end{array}\right].$$

The homogeneous system associated with the final matrix has only two equations in three unknowns and is therefore certain to have a nonzero solution; $a_3 = -1$, $a_2 = 2$, and $a_1 = 5$ is one such solution. Thus the vectors $\mathbf{u}_1$, $\mathbf{u}_2$, and $\mathbf{u}_3$ are linearly dependent. You should verify directly that $5\mathbf{u}_1 + 2\mathbf{u}_2 - \mathbf{u}_3 = \mathbf{0}$. ■

EXAMPLE 2 **Test an Independent Set for Dependence**

Determine if the following vectors from $\Re^3$ are linearly independent or linearly dependent.

$$\mathbf{v}_1 = \begin{bmatrix} -1 \\ 0 \\ 1 \end{bmatrix}, \quad \mathbf{v}_2 = \begin{bmatrix} -2 \\ 1 \\ 1 \end{bmatrix}, \quad \mathbf{v}_3 = \begin{bmatrix} 2 \\ 3 \\ 1 \end{bmatrix}.$$

The vector equation

$$a\mathbf{v}_1 + b\mathbf{v}_2 + c\mathbf{v}_3 = \mathbf{0} \quad \text{or} \quad a\begin{bmatrix} -1 \\ 0 \\ 1 \end{bmatrix} + b\begin{bmatrix} -2 \\ 1 \\ 1 \end{bmatrix} + c\begin{bmatrix} 2 \\ 3 \\ 1 \end{bmatrix} = \begin{bmatrix} 0 \\ 0 \\ 0 \end{bmatrix}$$

is equivalent to the homogeneous system

$$\begin{aligned} -a - 2b + 2c &= 0 \\ b + 3c &= 0 \\ a + b + c &= 0 \end{aligned} \quad \text{or} \quad \begin{bmatrix} -1 & -2 & 2 \\ 0 & 1 & 3 \\ 1 & 1 & 1 \end{bmatrix} \begin{bmatrix} a \\ b \\ c \end{bmatrix} = \begin{bmatrix} 0 \\ 0 \\ 0 \end{bmatrix}.$$

We have again reduced the vector question to an equivalent question about a system of linear equations. To decide if this system has nonzero solutions we row reduce the augmented matrix:

$$\begin{bmatrix} -1 & -2 & 2 & 0 \\ 0 & 1 & 3 & 0 \\ 1 & 1 & 1 & 0 \end{bmatrix} \to \begin{bmatrix} 1 & 2 & -2 & 0 \\ 0 & 1 & 3 & 0 \\ 0 & -1 & 3 & 0 \end{bmatrix}$$

$$\to \begin{bmatrix} 1 & 2 & -2 & 0 \\ 0 & 1 & 3 & 0 \\ 0 & 0 & 6 & 0 \end{bmatrix} \to \begin{bmatrix} 1 & 0 & 0 & 0 \\ 0 & 1 & 0 & 0 \\ 0 & 0 & 1 & 0 \end{bmatrix}.$$

From either of the last two matrices we observe that the system has only the zero solution. Hence the vectors $\mathbf{v}_1$, $\mathbf{v}_2$, and $\mathbf{v}_3$ are linearly independent elements of $\Re^3$. Geometrically, our calculations show that the three vectors are not coplanar in $\Re^3$. ■

EXAMPLE 3 Set Is too Big to be Independent

Determine if the following vectors from $\Re^4$ are linearly independent or dependent.

$$Y_1 = \begin{bmatrix} 1 \\ 2 \\ 3 \\ 4 \end{bmatrix}, \quad Y_2 = \begin{bmatrix} 1 \\ -1 \\ -1 \\ 1 \end{bmatrix}, \quad Y_3 = \begin{bmatrix} 1 \\ 3 \\ 0 \\ 4 \end{bmatrix}, \quad Y_4 = \begin{bmatrix} 0 \\ 2 \\ 2 \\ 0 \end{bmatrix}, \quad Y_5 = \begin{bmatrix} 1 \\ 1 \\ 2 \\ 3 \end{bmatrix}.$$

The vector equation

$$x_1 Y_1 + x_2 Y_2 + x_3 Y_3 + x_4 Y_4 + x_5 Y_5 = 0$$

is equivalent to

$$x_1 \begin{bmatrix} 1 \\ 2 \\ 3 \\ 4 \end{bmatrix} + x_2 \begin{bmatrix} 1 \\ -1 \\ -1 \\ 1 \end{bmatrix} + x_3 \begin{bmatrix} 1 \\ 3 \\ 0 \\ 4 \end{bmatrix} + x_4 \begin{bmatrix} 0 \\ 2 \\ 2 \\ 0 \end{bmatrix} + x_5 \begin{bmatrix} 1 \\ 1 \\ 2 \\ 3 \end{bmatrix} = \begin{bmatrix} 0 \\ 0 \\ 0 \\ 0 \end{bmatrix}$$

or

$$\begin{bmatrix} x_1 + x_2 + x_3 + x_5 \\ 2x_1 - x_2 + 3x_3 + 2x_4 + x_5 \\ 3x_1 - x_2 + 2x_4 + 2x_5 \\ 4x_1 + x_2 + 4x_3 + 3x_5 \end{bmatrix} = \begin{bmatrix} 0 \\ 0 \\ 0 \\ 0 \end{bmatrix}.$$

This last matrix equation leads to the following system of four homogeneous equations in five unknowns:

$$\begin{aligned} x_1 + x_2 + x_3 \phantom{{}+ 2x_4} + x_5 &= 0 \\ 2x_1 - x_2 + 3x_3 + 2x_4 + x_5 &= 0 \\ 3x_1 - x_2 \phantom{{}+ 3x_3} + 2x_4 + 2x_5 &= 0 \\ 4x_1 + x_2 + 4x_3 \phantom{{}+ 2x_4} + 3x_5 &= 0. \end{aligned}$$

Since this system has fewer equations than unknowns, by Theorem 1.3 it is certain to have a nonzero solution. Thus the five given vectors are linearly dependent vectors in $\Re^4$. ■

As illustrated in Examples 1–3, the procedure for testing a set of vectors $\{\mathbf{u}_1, \ldots, \mathbf{u}_k\}$ for linear independence or dependence requires three main steps.

■ TEST FOR LINEAR INDEPENDENCE OR DEPENDENCE ■

1. Suppose that $x_1\mathbf{u}_1 + x_2\mathbf{u}_2 + \cdots + x_k\mathbf{u}_k = \mathbf{0}$.
2. Interpret the vector equation in step 1 as a homogeneous system of linear equations $AX = 0$.
3. Using the techniques of Chapter 1, determine if the system in step 2 has a nonzero solution. If the system has only the trivial solution, then the $\mathbf{u}_i$ are linearly independent; otherwise they are linearly dependent.

The next two theorems summarize various ways of testing the columns of a matrix for linear independence. These results follow directly from the above test procedure using the results of Chapters 1 and 2.

THEOREM 3.6 Let A be an $m \times n$ real matrix. If $m < n$, then the columns of A are linearly dependent. If $m \geq n$, then the following statements are equivalent:

(1) The columns of A are linearly independent.
(2) The homogeneous system $AX = 0$ has only the zero solution.
(3) The number of nonzero rows of the row-reduced echelon matrix for A is n.
(4) The columns of the row-reduced echelon matrix for A are linearly independent. □

In the special case in which the matrix A is square we have the following theorem.

THEOREM 3.7 Let A be an $n \times n$ real matrix. Then the following statements are equivalent:

(1) The columns of A are linearly independent.
(2) $\det(A) \neq 0$.
(3) A is row equivalent to I.
(4) A^{-1} exists. □

EXERCISES 3.4

Determine if the sets of vectors in Exercises 1–11 are linearly independent or linearly dependent. In each case first rephrase the vector question as an equivalent question about a system of linear equations.

1. $\mathbf{v}_1 = \begin{bmatrix} 1 \\ 2 \end{bmatrix}$, $\mathbf{v}_2 = \begin{bmatrix} 3 \\ 5 \end{bmatrix}$.

2. $\mathbf{v}_1 = \begin{bmatrix} 1 \\ 2 \end{bmatrix}$, $\mathbf{v}_2 = \begin{bmatrix} 3 \\ 5 \end{bmatrix}$, $\mathbf{v}_3 = \begin{bmatrix} 6 \\ 7 \end{bmatrix}$.

3. $\mathbf{u}_1 = \begin{bmatrix} 1 \\ 1 \\ 3 \end{bmatrix}$, $\mathbf{u}_2 = \begin{bmatrix} 5 \\ -1 \\ 2 \end{bmatrix}$, $\mathbf{u}_3 = \begin{bmatrix} -3 \\ 0 \\ 4 \end{bmatrix}$.

4. $\mathbf{u}_1 = \begin{bmatrix} 1 \\ 1 \\ 3 \end{bmatrix}$, $\mathbf{u}_2 = \begin{bmatrix} 5 \\ -1 \\ 2 \end{bmatrix}$, $\mathbf{u}_3 = \begin{bmatrix} -3 \\ 0 \\ 4 \end{bmatrix}$, $\mathbf{u}_4 = \begin{bmatrix} 1 \\ 1 \\ 1 \end{bmatrix}$.

5. $\mathbf{x}_1 = \begin{bmatrix} 1 \\ 0 \\ 2 \\ 5 \end{bmatrix}$, $\mathbf{x}_2 = \begin{bmatrix} 0 \\ 3 \\ 0 \\ 2 \end{bmatrix}$, $\mathbf{x}_3 = \begin{bmatrix} 0 \\ 0 \\ 6 \\ 0 \end{bmatrix}$, $\mathbf{x}_4 = \begin{bmatrix} 0 \\ 0 \\ 0 \\ 2 \end{bmatrix}$.

6. $\mathbf{x}_1 = \begin{bmatrix} 1 \\ 0 \\ 2 \\ 5 \end{bmatrix}$, $\mathbf{x}_2 = \begin{bmatrix} 0 \\ 3 \\ 0 \\ 2 \end{bmatrix}$, $\mathbf{x}_3 = \begin{bmatrix} 2 \\ 3 \\ 4 \\ 12 \end{bmatrix}$.

7. The columns of the matrix $B = \begin{bmatrix} 1 & 4 & 0 & 3 \\ 2 & -1 & 2 & 6 \\ 1 & 5 & 3 & -1 \\ 3 & 8 & 7 & 3 \end{bmatrix}$.

8. The columns of the matrix $T = \begin{bmatrix} 2 & 1 & 0 & 0 \\ 1 & 2 & 1 & 0 \\ 0 & 1 & 2 & 1 \\ 0 & 0 & 1 & 2 \end{bmatrix}$.

9. The columns of the matrix $C = \begin{bmatrix} 1 & 1 & -2 & 3 \\ 2 & 3 & 2 & 4 \\ 4 & 9 & -2 & 14 \\ 8 & 27 & 2 & 34 \end{bmatrix}$.

10. The columns of the matrix $V = \begin{bmatrix} 1 & 1 & 1 & 1 \\ 1 & 2 & -1 & -2 \\ 1 & 4 & 4 & 1 \\ 1 & 8 & -8 & -1 \end{bmatrix}$.

11. The columns of the matrix $H = \begin{bmatrix} 5 & 4 & 3 & 2 & 1 \\ 4 & 4 & 3 & 2 & 1 \\ 0 & 3 & 3 & 2 & 1 \\ 0 & 0 & 2 & 2 & 1 \\ 0 & 0 & 0 & 1 & 1 \end{bmatrix}$.

12. Prove that any set of $n + 1$ vectors from $\mathfrak{R}^n$ must be linearly dependent.

13. Are the following sets of vectors from $\mathfrak{R}^3$ coplanar?

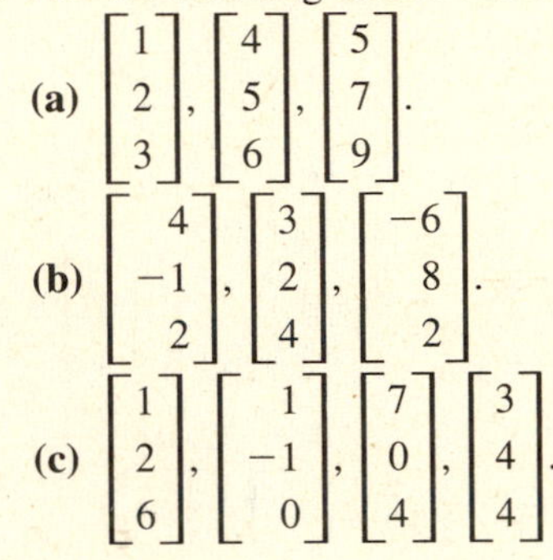

(a) $\begin{bmatrix} 1 \\ 2 \\ 3 \end{bmatrix}, \begin{bmatrix} 4 \\ 5 \\ 6 \end{bmatrix}, \begin{bmatrix} 5 \\ 7 \\ 9 \end{bmatrix}$.

(b) $\begin{bmatrix} 4 \\ -1 \\ 2 \end{bmatrix}, \begin{bmatrix} 3 \\ 2 \\ 4 \end{bmatrix}, \begin{bmatrix} -6 \\ 8 \\ 2 \end{bmatrix}$.

(c) $\begin{bmatrix} 1 \\ 2 \\ 6 \end{bmatrix}, \begin{bmatrix} 1 \\ -1 \\ 0 \end{bmatrix}, \begin{bmatrix} 7 \\ 0 \\ 4 \end{bmatrix}, \begin{bmatrix} 3 \\ 4 \\ 4 \end{bmatrix}$.

14. Which of the eight properties of vector algebra given in Theorem 3.1 are used in deducing from $\mathbf{v} = a_1\mathbf{u}_1 + a_2\mathbf{u}_2 + \cdots + a_k\mathbf{u}_k = b_1\mathbf{u}_1 + b_2\mathbf{u}_2 + \cdots + b_k\mathbf{u}_k$ that $\mathbf{0} = \mathbf{v} - \mathbf{v} = (a_1 - b_1)\mathbf{u}_1 + (a_2 - b_2)\mathbf{u}_2 + \cdots + (a_k - b_k)\mathbf{u}_k$?

15. Show that the vectors $\mathbf{u}_1 = \begin{bmatrix} 0 \\ 3 \\ -1 \\ 1 \end{bmatrix}, \mathbf{u}_2 = \begin{bmatrix} 4 \\ -7 \\ 3 \\ 1 \end{bmatrix}, \mathbf{u}_3 = \begin{bmatrix} 6 \\ 0 \\ 1 \\ 5 \end{bmatrix}$ are linearly dependent in $\mathfrak{R}^4$ and express each vector as a linear combination of the other two.

16. Let $\mathcal{S} = \{\mathbf{u}_1, \mathbf{u}_2, \ldots, \mathbf{u}_r\}$ be a subset of $\mathfrak{R}^n$ and let $\mathcal{S}_1$ be a subset of $\mathcal{S}$. Show that

(a) if $\mathcal{S}$ is linearly independent then so is $\mathcal{S}_1$.

(b) if $\mathcal{S}_1$ is linearly dependent then so is $\mathcal{S}$.

17. Show that if $\mathbf{v}_4$ is in SPAN$\{\mathbf{v}_1, \mathbf{v}_2, \mathbf{v}_3\}$, then SPAN$\{\mathbf{v}_1, \mathbf{v}_2, \mathbf{v}_3\}$ = SPAN$\{\mathbf{v}_1, \mathbf{v}_2, \mathbf{v}_3, \mathbf{v}_4\}$.

18. Show that if $\mathcal{S} = \{\mathbf{u}_1, \mathbf{u}_2, \mathbf{u}_3\}$ is linearly independent and $\mathbf{u}_4$ is not in SPAN$\{\mathcal{S}\}$, then $\{\mathbf{u}_1, \mathbf{u}_2, \mathbf{u}_3, \mathbf{u}_4\}$ is linearly independent.

3.5 BASIS AND DIMENSION

At the beginning of Section 3.4 we showed that if the vectors of a spanning set for the subspace $\mathcal{W}$ of $\mathfrak{R}^n$ are linearly independent, then every vector in the space is uniquely expressible as a linear combination of the spanning vectors. Many problems dealing with vector spaces are greatly simplified if we can find a linearly independent spanning set for the space. We shall show that finding such a set has the effect of introducing a coordinate system into the subspace $\mathcal{W}$.

DEFINITION 3.11 A **basis** for the subspace $\mathcal{W}$ of $\Re^n$ is a linearly independent set of vectors which span $\mathcal{W}$. If $\{\mathbf{v}_1, \mathbf{v}_2, \ldots, \mathbf{v}_r\}$ is a basis for $\mathcal{W}$, we will write

$$\mathcal{W} = \text{BASIS}\{\mathbf{v}_1, \mathbf{v}_2, \ldots, \mathbf{v}_r\}. \quad \square$$

We summarize the discussion at the beginning of Section 3.4 with the following theorem, the converse of which is left to the exercises.

THEOREM 3.8 If the vectors $\{\mathbf{v}_1, \mathbf{v}_2, \ldots, \mathbf{v}_r\}$ form a basis for the subspace $\mathcal{W}$ of $\Re^n$, then every vector in $\mathcal{W}$ can be expressed uniquely as a linear combination of the $\mathbf{v}_i$. Conversely, if every vector in $\mathcal{W}$ can be expressed uniquely as a linear combination of the $\mathbf{v}_i$ then the vectors $\{\mathbf{v}_1, \mathbf{v}_2, \ldots, \mathbf{v}_r\}$ form a basis for the subspace $\mathcal{W}$. $\square$

The way in which a basis for $\mathcal{W}$ leads to a coordinate system in the subspace $\mathcal{W}$ is described in the next definition.

DEFINITION 3.12 If the set $\mathcal{B} = \{\mathbf{v}_1, \mathbf{v}_2, \ldots, \mathbf{v}_r\}$ is a basis for the subspace $\mathcal{W}$ and

$$\mathbf{u} = a_1\mathbf{v}_1 + a_2\mathbf{v}_2 + \cdots + a_r\mathbf{v}_r$$

is any vector from $\mathcal{W}$, then the uniquely determined scalars a_i are called the **coordinates of u relative to the basis** $\mathcal{B}$. $\square$

In general, finding the coordinates of a vector with respect to a given basis involves solving a system of linear equations. By Theorem 3.8 this system must have a unique solution. These systems must therefore have square, nonsingular coefficient matrices.

In order to establish that a given set of vectors is a basis for the subspace $\mathcal{W}$ we will normally use the converse part of Theorem 3.8; this involves showing that a certain linear system has a unique solution for all possible right-hand sides. Recall that the system $AX = K$, where A is square, has a unique solution if, and only if, the coefficient matrix A is nonsingular. Among the conditions equivalent to the nonsingularity of the matrix A are $A \underset{R}{\sim} I$ and $\det(A) \neq 0$.

EXAMPLE 1 The Standard Basis for $\Re^n$

The vectors $\mathbf{e}_1 = \begin{bmatrix} 1 \\ 0 \\ 0 \\ 0 \end{bmatrix}$, $\mathbf{e}_2 = \begin{bmatrix} 0 \\ 1 \\ 0 \\ 0 \end{bmatrix}$, $\mathbf{e}_3 = \begin{bmatrix} 0 \\ 0 \\ 1 \\ 0 \end{bmatrix}$, and $\mathbf{e}_4 = \begin{bmatrix} 0 \\ 0 \\ 0 \\ 1 \end{bmatrix}$ form a basis for $\Re^4$ since any element $\mathbf{x}$ of $\Re^4$ can be written uniquely as

$$\mathbf{x} = \begin{bmatrix} x_1 \\ x_2 \\ x_3 \\ x_4 \end{bmatrix} = x_1\mathbf{e}_1 + x_2\mathbf{e}_2 + x_3\mathbf{e}_3 + x_4\mathbf{e}_4. \quad \blacksquare$$

The vectors $\mathbf{e}_i$ of Example 1 will be referred to as the *standard basis* for $\Re^4$. Note that each $\mathbf{e}_i$ is a unit vector. The obvious generalization to $\Re^n$ will be called the **standard basis** for $\Re^n$. We will reserve the notation $\mathbf{e}_i$ for these vectors. Note that if $\mathcal{S}$ is the standard basis for $\Re^n$, then its standard coordinates are the same as the components of the vector.

EXAMPLE 2 **A Nonstandard Basis for $\Re^3$**

Show that $\mathcal{B} = \left\{ \mathbf{u}_1 = \begin{bmatrix} 1 \\ 3 \\ -2 \end{bmatrix}, \mathbf{u}_2 = \begin{bmatrix} -3 \\ -12 \\ 10 \end{bmatrix}, \mathbf{u}_3 = \begin{bmatrix} -2 \\ -6 \\ 5 \end{bmatrix} \right\}$ is a basis for $\Re^3$ and find the $\mathcal{B}$-coordinates of $\mathbf{e}_1$.

In order to show that $\mathcal{B}$ is a basis for $\Re^3$ we will show that any vector in $\Re^3$ can be written in a unique way as a linear combination of the $\mathbf{u}_i$. Thus we let $\mathbf{w} = \begin{bmatrix} a \\ b \\ c \end{bmatrix}$ be an arbitrary vector in $\Re^3$ and try to find scalars x_i such that $x_1\mathbf{u}_1 + x_2\mathbf{u}_2 + x_3\mathbf{u}_3 = \mathbf{w}$, or equivalently

$$x_1 \begin{bmatrix} 1 \\ 3 \\ -2 \end{bmatrix} + x_2 \begin{bmatrix} -3 \\ -12 \\ 10 \end{bmatrix} + x_3 \begin{bmatrix} -2 \\ -6 \\ 5 \end{bmatrix} = \begin{bmatrix} a \\ b \\ c \end{bmatrix}.$$

This vector equation is equivalent to the system of linear equations

$$\begin{aligned} x_1 - 3x_2 - 2x_3 &= a \\ 3x_1 - 12x_2 - 6x_3 &= b \\ -2x_1 + 10x_2 + 5x_3 &= c. \end{aligned}$$

In matrix form this system of equations is

$$AX = \begin{bmatrix} 1 & -3 & -2 \\ 3 & -12 & -6 \\ -2 & 10 & 5 \end{bmatrix} \begin{bmatrix} x_1 \\ x_2 \\ x_3 \end{bmatrix} = \begin{bmatrix} a \\ b \\ c \end{bmatrix} = \mathbf{w}.$$

Note that the vectors $\mathbf{u}_i$ are the columns of the matrix A. The row reduction

$$\begin{bmatrix} 1 & -3 & -2 & 1 \\ 3 & -12 & -6 & 0 \\ -2 & 10 & 5 & 0 \end{bmatrix} \to \begin{bmatrix} 1 & -3 & -2 & 1 \\ 0 & -3 & 0 & -3 \\ 0 & 4 & 1 & 2 \end{bmatrix} \to \begin{bmatrix} 1 & -3 & -2 & 1 \\ 0 & 1 & 1 & -1 \\ 0 & 4 & 1 & 2 \end{bmatrix}$$

$$\to \begin{bmatrix} 1 & 0 & 1 & -2 \\ 0 & 1 & 1 & -1 \\ 0 & 0 & -3 & 6 \end{bmatrix} \to \begin{bmatrix} 1 & 0 & 0 & 0 \\ 0 & 1 & 0 & 1 \\ 0 & 0 & 1 & -2 \end{bmatrix},$$

shows that the coefficient matrix A is nonsingular, because A is row equivalent to I, and also solves the system in the special case $\mathbf{w} = \mathbf{e}_1$. Since A^{-1} exists, the system $AX = \mathbf{w}$ has a unique solution for any $\mathbf{w}$ in $\mathfrak{R}^3$ and, by Theorem 3.8, it follows that $\mathcal{B}$ is a basis for $\mathfrak{R}^3$. The above row reduction also shows that $\mathbf{e}_1 = \mathbf{u}_2 - 2\mathbf{u}_3$, so that the $\mathcal{B}$-coordinates of $\mathbf{e}_1$ are 0, 1, and -2. ■

In Example 2 we found a basis for $\mathfrak{R}^3$ that contains three vectors. Another basis for $\mathfrak{R}^3$ is the standard basis $\{\mathbf{e}_1, \mathbf{e}_2, \mathbf{e}_3\}$. Note that the number of vectors in the two bases is the same. That this is true in general is an immediate consequence of our next theorem.

THEOREM 3.9 If $\{\mathbf{v}_1, \mathbf{v}_2, \ldots, \mathbf{v}_k\}$ is a basis for the subspace $\mathcal{W}$, then any set of vectors from $\mathcal{W}$, having more than k elements, must be linearly dependent.

Proof Let $\{\mathbf{w}_1, \mathbf{w}_2, \ldots, \mathbf{w}_r\}$ be any set of $r > k$ vectors from $\mathcal{W}$ and suppose that

$$x_1\mathbf{w}_1 + x_2\mathbf{w}_2 + \cdots + x_r\mathbf{w}_r = \mathbf{0}. \tag{3.9}$$

In order to show that a nonzero solution exists for Equation (3.9) we first express each of the $\mathbf{w}_j$ as a linear combination of the basis vectors $\mathbf{v}_i$:

$$\mathbf{w}_j = a_{1j}\mathbf{v}_1 + a_{2j}\mathbf{v}_2 + \cdots + a_{kj}\mathbf{v}_k; \quad j = 1, 2, \ldots, r.$$

If we substitute these expressions for the $\mathbf{w}_j$ into Equation (3.9) we obtain

$$\begin{aligned} x_1(a_{11}\mathbf{v}_1 + \cdots + a_{k1}\mathbf{v}_k) + x_2(a_{12}\mathbf{v}_1 + \cdots + a_{k2}\mathbf{v}_k) + \cdots \\ + x_r(a_{1r}\mathbf{v}_1 + \cdots + a_{kr}\mathbf{v}_k) = \mathbf{0} \end{aligned}$$

which, after some rearrangement, reduces to

$$\begin{aligned} (a_{11}x_1 + \cdots + a_{1r}x_r)\mathbf{v}_1 + (a_{21}x_1 + \cdots + a_{2r}x_r)\mathbf{v}_2 + \cdots \\ + (a_{k1}x_1 + \cdots + a_{kr}x_r)\mathbf{v}_k = \mathbf{0}. \end{aligned}$$

Since the $\mathbf{v}_i$ are assumed to form a basis for $\mathcal{W}$, they must be linearly independent. Thus, each of the coefficients in the last equation must be zero. This observation leads us to conclude that the x_i must satisfy the homogeneous system $AX = 0$, where the coefficient matrix A is $k \times r$. Since $k < r$, this homogeneous system has fewer equations than unknowns and hence is certain to have a nonzero solution. It follows that the $\mathbf{w}_i$ are linearly dependent as claimed and the proof is complete. □

Theorem 3.9 shows that any four vectors from $\mathfrak{R}^3$ must be linearly dependent and that any $n + 1$ vectors from $\mathfrak{R}^n$ must be linearly dependent. Suppose now that we have two different bases for the subspace $\mathcal{W}$:

$$\mathcal{B} = \{\mathbf{u}_1, \mathbf{u}_2, \ldots, \mathbf{u}_n\} \qquad \text{and} \qquad \mathcal{B}' = \{\mathbf{v}_1, \mathbf{v}_2, \ldots, \mathbf{v}_k\}.$$

If $n < k$ then Theorem 3.9 says that $\mathcal{B}'$ is dependent; this is a contradiction, so we must have $n \geq k$. Similarly, assuming that $k < n$ leads to the conclusion that $\mathcal{B}$ is dependent; from this contradiction we conclude that $k \geq n$. It follows that

$k = n$; that is, the two bases must contain the same number of vectors. We state this important result as our next theorem.

THEOREM 3.10 Any two bases for a subspace $\mathcal{W}$ of $\mathfrak{R}^n$ must contain the same number of vectors; that is, if $\{\mathbf{v}_1, \mathbf{v}_2, \ldots, \mathbf{v}_r\}$ is a basis for the subspace $\mathcal{W}$, then any other basis of $\mathcal{W}$ must contain precisely r vectors. □

The number of vectors in a basis of $\mathcal{W}$ is an important property of the subspace $\mathcal{W}$. Theorem 3.10 shows that this number does not depend on how the basis is chosen.

DEFINITION 3.13 The number of vectors in a basis for the subspace $\mathcal{W}$ of $\mathfrak{R}^n$ is called the **dimension** of $\mathcal{W}$ and is denoted by $\dim(\mathcal{W})$. □

Example 1 shows that $\dim(\mathfrak{R}^4) = 4$, and the obvious generalization to $\mathfrak{R}^n$ shows that $\dim(\mathfrak{R}^n) = n$.

In order to find the dimension of the subspace $\mathcal{W}$ of $\mathfrak{R}^n$ one normally has to find a basis for that subspace.

EXAMPLE 3 Find a Basis from a Spanning Set

Find a basis for the subspace $\mathcal{W}$ of $\mathfrak{R}^4$ spanned by

$$\mathbf{u}_1 = \begin{bmatrix} 1 \\ 2 \\ 3 \\ 1 \end{bmatrix}, \quad \mathbf{u}_2 = \begin{bmatrix} 3 \\ 2 \\ 1 \\ 1 \end{bmatrix}, \quad \mathbf{u}_3 = \begin{bmatrix} 0 \\ 2 \\ 4 \\ 1 \end{bmatrix}, \quad \mathbf{u}_4 = \begin{bmatrix} 1 \\ 1 \\ 1 \\ 1 \end{bmatrix}.$$

The spanning set can fail to be a basis only if the spanning vectors are linearly dependent. Since $\mathbf{u}_1 \neq k\mathbf{u}_2$, $\{\mathbf{u}_1, \mathbf{u}_2\}$ is an independent set. We ask next if $\mathbf{u}_3$ is in SPAN$\{\mathbf{u}_1, \mathbf{u}_2\}$; that is, can we find scalars a and b so that

$$\mathbf{u}_3 = \begin{bmatrix} 0 \\ 2 \\ 4 \\ 1 \end{bmatrix} = a \begin{bmatrix} 1 \\ 2 \\ 3 \\ 1 \end{bmatrix} + b \begin{bmatrix} 3 \\ 2 \\ 1 \\ 1 \end{bmatrix} = a\mathbf{u}_1 + b\mathbf{u}_2?$$

This is equivalent to the system

$$\begin{aligned} a + 3b &= 0 \\ 2a + 2b &= 2 \\ 3a + b &= 4 \\ a + b &= 1 \end{aligned} \qquad \text{or} \qquad \begin{bmatrix} 1 & 3 \\ 2 & 2 \\ 3 & 1 \\ 1 & 1 \end{bmatrix} \begin{bmatrix} a \\ b \end{bmatrix} = \begin{bmatrix} 0 \\ 2 \\ 4 \\ 1 \end{bmatrix}.$$

From the row reduction

$$\begin{bmatrix} 1 & 3 & 0 \\ 2 & 2 & 2 \\ 3 & 1 & 4 \\ 1 & 1 & 1 \end{bmatrix} \rightarrow \begin{bmatrix} 1 & 3 & 0 \\ 0 & -4 & 2 \\ 0 & -8 & 4 \\ 0 & -2 & 1 \end{bmatrix} \rightarrow \begin{bmatrix} 1 & 0 & \frac{3}{2} \\ 0 & 1 & -\frac{1}{2} \\ 0 & 0 & 0 \\ 0 & 0 & 0 \end{bmatrix},$$

we see that $a = \frac{3}{2}$ and $b = -\frac{1}{2}$ is the solution of the system. Thus, $\mathbf{u}_3 = \frac{3}{2}\mathbf{u}_1 - \frac{1}{2}\mathbf{u}_2$ is in SPAN$\{\mathbf{u}_1, \mathbf{u}_2\}$. It follows that SPAN$\{\mathbf{u}_1, \mathbf{u}_2, \mathbf{u}_3\}$ = SPAN$\{\mathbf{u}_1, \mathbf{u}_2\}$, so we can discard $\mathbf{u}_3$ from the spanning set for $\mathcal{W}$. Finally, we ask if $\mathbf{u}_4$ is in SPAN$\{\mathbf{u}_1, \mathbf{u}_2\}$. If

$$\mathbf{u}_4 = \begin{bmatrix} 1 \\ 1 \\ 1 \\ 1 \end{bmatrix} = c\begin{bmatrix} 1 \\ 2 \\ 3 \\ 1 \end{bmatrix} + d\begin{bmatrix} 3 \\ 2 \\ 1 \\ 1 \end{bmatrix} = c\mathbf{u}_1 + d\mathbf{u}_2,$$

then

$$\begin{aligned} c + 3d &= 1 \\ 2c + 2d &= 1 \\ 3c + d &= 1 \\ c + d &= 1 \end{aligned} \qquad \text{or} \qquad \begin{bmatrix} 1 & 3 \\ 2 & 2 \\ 3 & 1 \\ 1 & 1 \end{bmatrix}\begin{bmatrix} c \\ d \end{bmatrix} = \begin{bmatrix} 1 \\ 1 \\ 1 \\ 1 \end{bmatrix}.$$

The row reduction

$$\begin{bmatrix} 1 & 3 & 1 \\ 2 & 2 & 1 \\ 3 & 1 & 1 \\ 1 & 1 & 1 \end{bmatrix} \rightarrow \begin{bmatrix} 1 & 3 & 1 \\ 0 & -4 & -1 \\ 0 & -8 & -2 \\ 0 & -2 & 0 \end{bmatrix} \rightarrow \begin{bmatrix} 1 & 0 & 1 \\ 0 & 1 & 0 \\ 0 & 0 & -1 \\ 0 & 0 & -2 \end{bmatrix} \rightarrow \begin{bmatrix} 1 & 0 & 0 \\ 0 & 1 & 0 \\ 0 & 0 & 1 \\ 0 & 0 & 0 \end{bmatrix},$$

shows that this system is inconsistent and hence that $\mathbf{u}_4$ is not in SPAN$\{\mathbf{u}_1, \mathbf{u}_2\}$. It follows that the set $\{\mathbf{u}_1, \mathbf{u}_2, \mathbf{u}_4\}$ is linearly independent and hence forms a basis for $\mathcal{W}$. The dimension of $\mathcal{W}$ is 3. ■

One of the most important examples of a subspace is the set of all solutions of a homogeneous system $AX = 0$. Recall that this is the null space of the coefficient matrix of the system. Our next example illustrates how to find a basis for the null space of a given matrix A.

EXAMPLE 4 Bases for the Null Space and the Column Space

Find bases for the null space and the column space of

$$A = \begin{bmatrix} 3 & -1 & -1 & -3 \\ -2 & 2 & -2 & 2 \\ -1 & -1 & 3 & 1 \end{bmatrix}.$$

Since $\mathcal{NS}(A)$ consists of all solutions of the homogeneous system $A\mathbf{X} = \mathbf{0}$ we need to find the general solution of that system. From the row reduction

$$[A|0] = \begin{bmatrix} 3 & -1 & -1 & -3 & 0 \\ -2 & 2 & -2 & 2 & 0 \\ -1 & -1 & 3 & 1 & 0 \end{bmatrix} \rightarrow \begin{bmatrix} 0 & -4 & 8 & 0 & 0 \\ 0 & 4 & -8 & 0 & 0 \\ -1 & -1 & 3 & 1 & 0 \end{bmatrix}$$
$$\rightarrow \begin{bmatrix} 1 & 0 & -1 & -1 & 0 \\ 0 & 1 & -2 & 0 & 0 \\ 0 & 0 & 0 & 0 & 0 \end{bmatrix} = [B|0],$$

we see that $\mathbf{X}$ is in $\mathcal{NS}(A)$ if, and only if, $B\mathbf{X} = \mathbf{0}$; that is,

$$\begin{aligned} x_1 - x_3 - x_4 &= 0 \\ x_2 - 2x_3 &= 0. \end{aligned}$$

The most general solution of this system is

$$\mathbf{X} = \begin{bmatrix} c_1 + c_2 \\ 2c_1 \\ c_1 \\ c_2 \end{bmatrix} = c_1 \begin{bmatrix} 1 \\ 2 \\ 1 \\ 0 \end{bmatrix} + c_2 \begin{bmatrix} 1 \\ 0 \\ 0 \\ 1 \end{bmatrix} = c_1\mathbf{X}_1 + c_2\mathbf{X}_2,$$

where c_1 and c_2 are arbitrary. From this last equation we see that

$$\mathcal{NS}(A) = \text{SPAN}\{\mathbf{X}_1, \mathbf{X}_2\}$$

and, since $\mathbf{X}_1$ and $\mathbf{X}_2$ are clearly independent, they form a basis for $\mathcal{NS}(A)$.

In order to find a basis for $\mathcal{CS}(A)$ we will remove the dependent vectors from the spanning set

$$\mathcal{CS}(A) = \text{SPAN}\{\text{Col}_1(A), \text{Col}_2(A), \text{Col}_3(A), \text{Col}_4(A)\}.$$

Since $\mathbf{X}_1$ is a solution of the homogeneous system $A\mathbf{X} = \mathbf{0}$ we have, from the fact that $A\mathbf{X}$ is a linear combination of the columns of A, $\text{Col}_1(A) + 2\text{Col}_2(A) + \text{Col}_3(A) = \mathbf{0}$. Thus $\text{Col}_3(A) = -\text{Col}_1(A) - 2\text{Col}_2(A)$ and $\text{Col}_3(A)$ can be removed from the spanning set for $\mathcal{CS}(A)$. We then have

$$\mathcal{CS}(A) = \text{SPAN}\{\text{Col}_1(A), \text{Col}_2(A), \text{Col}_4(A)\}.$$

Similarly, because $\mathbf{X}_2$ is a solution of the homogeneous system $A\mathbf{X} = \mathbf{0}$ we have $\text{Col}_1(A) + \text{Col}_4(A) = \mathbf{0}$. It follows that $\text{Col}_4(A) = -\text{Col}_1(A)$ and $\text{Col}_4(A)$ can also be removed from the spanning set for $\mathcal{CS}(A)$. Thus

$$\mathcal{CS}(A) = \text{SPAN}\{\text{Col}_1(A), \text{Col}_2(A)\}.$$

Since the vectors of this reduced spanning set are clearly independent they form a basis for $\mathcal{CS}(A)$. ■

In Chapter 1 we observed that the number of degrees of freedom in the general solution of the linear system $AX = K$ is equal to the number of unknowns less the number of nonzero rows in the echelon matrix. If we interpret this result in terms of the concepts introduced in this chapter, observing that the number of degrees of freedom in the solution is $\dim(\mathcal{NS}(A))$, we have the following theorem.

THEOREM 3.11 If A is an $m \times n$ matrix, then $\dim(\mathcal{NS}(A)) = n - r$ where r is the number of nonzero rows in the echelon matrix for A. □

The number of nonzero rows in the echelon matrix for A is frequently called the **rank** of A and denoted $\text{rank}(A)$. Theorem 3.11 can then be rephrased as

$$\dim(\mathcal{NS}(A)) + \text{rank}(A) = \text{number of columns of } A.$$

The normal way of computing the rank of a matrix is to row reduce the matrix to echelon form and then count the number of nonzero rows. For the matrix A of Example 4 we see that $\text{rank}(A) = 2$, since the echelon matrix B has two nonzero rows. The number of columns of A is four, so it follows that $\dim(\mathcal{NS}(A)) = 4 - 2 = 2$, which is the number of arbitrary constants in the general solution of the homogeneous system $AX = 0$.

For an $n \times n$ matrix A we know, from Theorem 1.9, that A^{-1} exists if, and only if, the homogeneous system $AX = 0$ has only the zero solution, that is only if $\mathcal{NS}(A) = \{\mathbf{0}\}$. It then follows from the last equation that A^{-1} exists if, and only if, $\text{rank}(A) = n$. We conclude this section by expanding Theorem 1.9 to include all the conditions so far discussed that are equivalent to the nonsingularity of the matrix A.

THEOREM 3.12 For an $n \times n$ matrix A the following statements are equivalent:

(1) A is nonsingular;
(2) A has either a right or a left inverse;
(3) $\mathcal{NS}(A) = \{\mathbf{0}\}$;
(4) A is row equivalent to I;
(5) A is a product of elementary matrices;
(6) $\det(A) \neq 0$;
(7) $\text{rank}(A) = n$;
(8) The columns of A are linearly independent; and
(9) $AX = K$ has a unique solution for every K. □

EXERCISES 3.5

1. Which of the following sets of vectors form a basis for $\mathfrak{R}^3$?

(a) $\begin{bmatrix} 1 \\ -3 \\ 2 \end{bmatrix}, \begin{bmatrix} 1 \\ 1 \\ 4 \end{bmatrix}, \begin{bmatrix} 1 \\ -7 \\ 0 \end{bmatrix}$

(b) $\begin{bmatrix} 2 \\ -3 \\ 1 \end{bmatrix}, \begin{bmatrix} 2 \\ 2 \\ 0 \end{bmatrix}$

(c) $\begin{bmatrix} 4 \\ 6 \\ 1 \end{bmatrix}, \begin{bmatrix} 5 \\ 2 \\ -1 \end{bmatrix}, \begin{bmatrix} 1 \\ 4 \\ 2 \end{bmatrix}$

(d) $\begin{bmatrix} 1 \\ 1 \\ 1 \end{bmatrix}, \begin{bmatrix} 0 \\ 1 \\ 1 \end{bmatrix}, \begin{bmatrix} 1 \\ 1 \\ 0 \end{bmatrix}, \begin{bmatrix} 1 \\ 0 \\ 1 \end{bmatrix}$

(e) $\begin{bmatrix} -1 \\ 2 \\ 5 \end{bmatrix}, \begin{bmatrix} 2 \\ 4 \\ -1 \end{bmatrix}, \begin{bmatrix} 1 \\ 6 \\ 4 \end{bmatrix}$ **(f)** $\begin{bmatrix} 1 \\ 2 \\ 3 \end{bmatrix}, \begin{bmatrix} 8 \\ 9 \\ 0 \end{bmatrix}, \begin{bmatrix} 5 \\ 3 \\ -9 \end{bmatrix}$

2. Find a basis for $\mathfrak{R}^2$ containing $\begin{bmatrix} 1 \\ 1 \end{bmatrix}$.

3. Find a basis for $\mathfrak{R}^3$ containing $\begin{bmatrix} 1 \\ 1 \\ 1 \end{bmatrix}$ and $\begin{bmatrix} 1 \\ 0 \\ 1 \end{bmatrix}$.

4. Find a basis for $\mathcal{NS}(A)$.

(a) $A = \begin{bmatrix} 1 & 1 & 1 \\ 4 & 3 & -1 \\ 3 & 2 & -2 \\ 6 & 5 & 1 \end{bmatrix}$ **(b)** $A = \begin{bmatrix} 1 & 4 & 3 & 6 \\ 1 & 3 & 2 & 5 \\ 1 & -1 & -2 & 1 \end{bmatrix}$

5. Show that the following set of vectors is a basis for $\mathfrak{R}^4$ and find the coordinates of each of the standard basis vectors with respect to this basis:

$$\mathbf{u}_1 = \begin{bmatrix} 1 \\ 2 \\ -3 \\ 4 \end{bmatrix}, \quad \mathbf{u}_2 = \begin{bmatrix} 0 \\ 1 \\ 2 \\ 0 \end{bmatrix}, \quad \mathbf{u}_3 = \begin{bmatrix} 0 \\ 0 \\ 1 \\ 3 \end{bmatrix}, \quad \mathbf{u}_4 = \begin{bmatrix} 0 \\ 0 \\ 0 \\ 1 \end{bmatrix}.$$

6. Explain why neither of the following sets is a basis for $\mathfrak{R}^4$.

(a) $\begin{bmatrix} 1 \\ 1 \\ 1 \\ -3 \end{bmatrix}, \begin{bmatrix} 3 \\ -2 \\ -17 \\ 16 \end{bmatrix}, \begin{bmatrix} 1 \\ 1 \\ 1 \\ 1 \end{bmatrix}$ **(b)** $\begin{bmatrix} 1 \\ 2 \\ 4 \\ 1 \end{bmatrix}, \begin{bmatrix} -1 \\ -1 \\ 1 \\ 1 \end{bmatrix}, \begin{bmatrix} 2 \\ 0 \\ -11 \\ 3 \end{bmatrix}, \begin{bmatrix} 3 \\ 2 \\ -1 \\ 8 \end{bmatrix}, \begin{bmatrix} 0 \\ 0 \\ 0 \\ 1 \end{bmatrix}$

7. Find the rank of the following matrices:

$$A = \begin{bmatrix} 1 & 3 \\ 4 & -1 \\ 3 & 7 \end{bmatrix}, \quad B = \begin{bmatrix} 1 & 1 & 1 & 3 \\ 3 & -2 & -17 & 16 \\ 3 & 2 & -1 & -4 \end{bmatrix}, \quad C = \begin{bmatrix} 1 & -1 & 2 & 3 \\ 2 & 2 & 0 & 2 \\ 4 & 1 & 3 & -1 \\ 1 & 2 & 3 & 0 \end{bmatrix}.$$

8. Find rank(A) and $\dim(\mathcal{NS}(A))$ if $A = \begin{bmatrix} 1 & 2 & -1 & 4 \\ -1 & -2 & 6 & -7 \\ 2 & 4 & 3 & 5 \end{bmatrix}$.

9. Find bases for $\mathcal{NS}(A)$ and $\mathcal{CS}(A)$ if $A = \begin{bmatrix} 1 & 0 & 4 & 2 \\ 0 & 1 & 5 & 3 \\ 0 & 0 & 0 & 0 \end{bmatrix}$.

10. Given that $A = \begin{bmatrix} 1 & 3 & 0 & 2 & -1 \\ 3 & 9 & 1 & 7 & 5 \\ 2 & 6 & 1 & 5 & 2 \\ 5 & 15 & 2 & 12 & 8 \end{bmatrix} \tilde{R} \begin{bmatrix} 1 & 3 & 0 & 2 & 0 \\ 0 & 0 & 1 & 1 & 0 \\ 0 & 0 & 0 & 0 & 1 \\ 0 & 0 & 0 & 0 & 0 \end{bmatrix} = R,$

find a basis for $\mathcal{NS}(A)$ and for $\mathcal{CS}(A)$.

11. Find bases for $\mathcal{NS}(A)$ and $\mathcal{CS}(A)$ if $A = \begin{bmatrix} 1 & 2 & -5 & 11 & -3 \\ 2 & 4 & -5 & 15 & 2 \\ 1 & 2 & 0 & 4 & 5 \\ 4 & 8 & -5 & 23 & 3 \end{bmatrix}$.

12. Using an argument like that in Example 4, show that the obvious spanning set for $\mathcal{CS}(A)$ can be reduced to a basis for $\mathcal{CS}(A)$ by deleting $\dim(\mathcal{NS}(A))$ vectors. Then show that $\dim(\mathcal{CS}(A)) = \text{rank}(A)$.

3.6 ORTHOGONALITY IN $\Re^n$

In this section we will present some special results about the inner product in $\Re^n$.

1. What is the relationship between orthogonality and linear independence?
2. Does every subspace of $\Re^n$ have a basis made up of mutually orthogonal unit vectors?

Recall, from Section 3.2, that we have defined the length, or norm, of a vector $\mathbf{u}$ in $\Re^n$ by

$$\|\mathbf{u}\| = \sqrt{\mathbf{u} \cdot \mathbf{u}} = \sqrt{\sum_{i=1}^{n} u_i^2}$$

and have defined two vectors, $\mathbf{u}$ and $\mathbf{v}$, to be orthogonal if

$$\mathbf{u} \cdot \mathbf{v} = \sum_{i=1}^{n} u_i v_i = 0.$$

As the first theorem of this section we list some elementary, but important, properties of the length, or norm, of a vector. The proofs will be left to the exercises.

THEOREM 3.13 For any vector $\mathbf{u}$ in $\Re^n$ and any scalar k,

(1) $\|\mathbf{u}\| \geq 0$;
(2) $\|\mathbf{u}\| = 0$ if, and only if, $\mathbf{u} = \mathbf{0}$; and
(3) $\|k\mathbf{u}\| = |k|\|\mathbf{u}\|$. □

For vectors $\mathbf{u}$ and $\mathbf{v}$ in $\Re^3$ we have the alternate geometric formula for the inner product

$$\mathbf{u} \cdot \mathbf{v} = \|\mathbf{u}\| \, \|\mathbf{v}\| \cos \theta.$$

Since $|\cos \theta| \leq 1$, it follows at once that

$$|\mathbf{u} \cdot \mathbf{v}| \leq \|\mathbf{u}\| \, \|\mathbf{v}\|.$$

This result is true for vectors in $\mathfrak{R}^n$, although the preceding argument is not valid in the general case since we have discussed no geometric way of defining angles and trigonometric functions in $\mathfrak{R}^n$. These ideas can be defined algebraically in $\mathfrak{R}^n$ using the above formulas; that is, we could define $\theta = \cos^{-1}(\mathbf{u} \cdot \mathbf{v}/\|\mathbf{u}\|\,\|\mathbf{v}\|)$, provided that we could show that $\mathbf{u} \cdot \mathbf{v}/\|\mathbf{u}\|\,\|\mathbf{v}\| \le 1$.

THEOREM 3.14 ***Cauchy-Schwarz Inequality***

For any $\mathbf{u}$ and $\mathbf{v}$ in $\mathfrak{R}^n$ we have $|\mathbf{u} \cdot \mathbf{v}| \le \|\mathbf{u}\|\,\|\mathbf{v}\|$.

Proof We first consider the case where $\mathbf{u} = \begin{bmatrix} u_1 \\ \vdots \\ u_n \end{bmatrix}$ and $\mathbf{v} = \begin{bmatrix} v_1 \\ \vdots \\ v_n \end{bmatrix}$ are unit vectors so that

$$\|\mathbf{u}\|^2 = \sum_{i=1}^{n} u_i^2 = 1 \qquad \text{and} \qquad \|\mathbf{v}\|^2 = \sum_{i=1}^{n} v_i^2 = 1.$$

In this case we have

$$0 \le \|\mathbf{u} - \mathbf{v}\|^2 = \sum_{i=1}^{n} u_i^2 - 2\sum_{i=1}^{n} u_i v_i + \sum_{i=1}^{n} v_i^2 = 2 - 2\mathbf{u} \cdot \mathbf{v}$$

and

$$0 \le \|\mathbf{u} + \mathbf{v}\|^2 = \sum_{i=1}^{n} u_i^2 + 2\sum_{i=1}^{n} u_i v_i + \sum_{i=1}^{n} v_i^2 = 2 + 2\mathbf{u} \cdot \mathbf{v}.$$

It follows from the first inequality that $\mathbf{u} \cdot \mathbf{v} \le 1$ and from the second that $-1 \le \mathbf{u} \cdot \mathbf{v}$; together these two inequalities yield $|\mathbf{u} \cdot \mathbf{v}| \le 1$. If $\mathbf{u}$ and $\mathbf{v}$ are both nonzero, then $\mathbf{u}/\|\mathbf{u}\|$ and $\mathbf{v}/\|\mathbf{v}\|$ are unit vectors, for which we have just shown that

$$\left| \frac{\mathbf{u} \cdot \mathbf{v}}{\|\mathbf{u}\|\,\|\mathbf{v}\|} \right| = \left| \left(\frac{\mathbf{u}}{\|\mathbf{u}\|} \right) \cdot \left(\frac{\mathbf{v}}{\|\mathbf{v}\|} \right) \right| \le 1.$$

It follows at once that

$$|\mathbf{u} \cdot \mathbf{v}| \le \|\mathbf{u}\|\,\|\mathbf{v}\|.$$

Since the theorem is trivially true for either $\mathbf{u} = \mathbf{0}$ or $\mathbf{v} = \mathbf{0}$, the proof is complete. □

An immediate consequence of the last theorem is a result that is usually referred to as the triangle inequality; in the case $n = 3$ it says that one side of a triangle is less than the sum of the other two sides; see Figure 3.20.

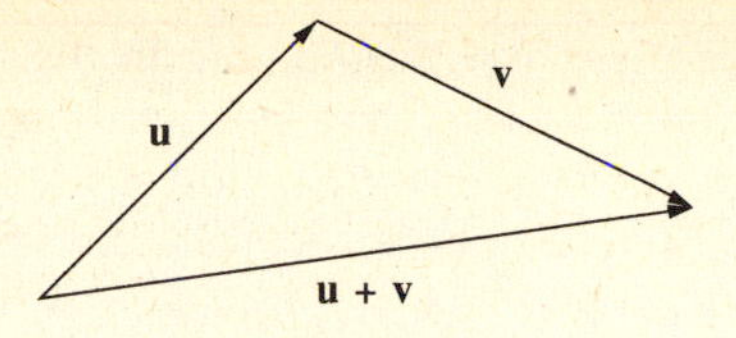

FIGURE 3.20

THEOREM 3.15 ***Triangle Inequality***

For any vectors **u** and **v** in $\Re^n$ we have

$$\|\mathbf{u} + \mathbf{v}\| \leq \|\mathbf{u}\| + \|\mathbf{v}\|.$$

Proof Since

$$\begin{aligned}\|\mathbf{u} + \mathbf{v}\|^2 &= (\mathbf{u} + \mathbf{v}) \cdot (\mathbf{u} + \mathbf{v}) = \mathbf{u} \cdot \mathbf{u} + \mathbf{u} \cdot \mathbf{v} + \mathbf{v} \cdot \mathbf{u} + \mathbf{v} \cdot \mathbf{v} \\ &= \|\mathbf{u}\|^2 + 2(\mathbf{u} \cdot \mathbf{v}) + \|\mathbf{v}\|^2,\end{aligned}$$

it follows from the Cauchy-Schwarz inequality that

$$\|\mathbf{u} + \mathbf{v}\|^2 \leq \|\mathbf{u}\|^2 + 2\|\mathbf{u}\|\,\|\mathbf{v}\| + \|\mathbf{v}\|^2 = (\|\mathbf{u}\| + \|\mathbf{v}\|)^2.$$

Taking square roots, we see that $\|\mathbf{u} + \mathbf{v}\| \leq \|\mathbf{u}\| + \|\mathbf{v}\|$ as claimed. □

DEFINITION 3.14 Two vectors, **u** and **v**, from $\Re^n$ are **orthogonal** if $\mathbf{u} \cdot \mathbf{v} = 0$. A set of vectors, $\mathcal{S} = \{\mathbf{u}_1, \mathbf{u}_2, \ldots, \mathbf{u}_k\}$ from $\Re^n$ is said to be **mutually orthogonal** if the vectors are pairwise orthogonal, that is, $\mathbf{u}_i \cdot \mathbf{u}_j = 0$ for $i \neq j$. If the vectors of the mutually orthogonal set $\mathcal{S}$ are all unit vectors, then we say that the set is **orthonormal;** that is, $\mathbf{u}_i \cdot \mathbf{u}_j = 0$ for $i \neq j$ and $\mathbf{u}_i \cdot \mathbf{u}_i = 1$ for $i = 1, 2, \ldots, k$. An orthonormal set of vectors that also forms a basis for $\Re^n$ is called an **orthonormal basis** for $\Re^n$. □

EXAMPLE 1 **The Standard Basis Is Orthonormal**

The standard basis for $\Re^n$, which we denote by

$$\mathbf{e}_1 = \begin{bmatrix} 1 \\ 0 \\ \cdot \\ \cdot \\ \cdot \\ 0 \end{bmatrix}, \mathbf{e}_2 = \begin{bmatrix} 0 \\ 1 \\ \cdot \\ \cdot \\ \cdot \\ 0 \end{bmatrix}, \ldots, \mathbf{e}_n = \begin{bmatrix} 0 \\ 0 \\ \cdot \\ \cdot \\ \cdot \\ 1 \end{bmatrix},$$

is easily seen to be an orthonormal basis. For $n = 3$ these vectors are frequently denoted **i**, **j**, and **k**. ■

EXAMPLE 2 **A Nonstandard Orthonormal Basis**

In $\Re^3$ the vectors $\mathbf{u}_1 = \begin{bmatrix} 1 \\ 1 \\ 1 \end{bmatrix}$, $\mathbf{u}_2 = \begin{bmatrix} 0 \\ 1 \\ -1 \end{bmatrix}$, and $\mathbf{u}_3 = \begin{bmatrix} -2 \\ 1 \\ 1 \end{bmatrix}$ are mutually orthogonal since

$$\mathbf{u}_1 \cdot \mathbf{u}_2 = (0 + 1 - 1) = 0, \quad \mathbf{u}_1 \cdot \mathbf{u}_3 = (-2 + 1 + 1) = 0, \quad \text{and}$$
$$\mathbf{u}_2 \cdot \mathbf{u}_3 = (0 + 1 - 1) = 0.$$

These vectors are not orthonormal since they are not unit vectors:

$$\mathbf{u}_1 \cdot \mathbf{u}_1 = 3 \neq 1, \quad \mathbf{u}_2 \cdot \mathbf{u}_2 = 2, \quad \text{and} \quad \mathbf{u}_3 \cdot \mathbf{u}_3 = 6.$$

We can easily modify the $\mathbf{u}_i$ to obtain an orthonormal set by dividing each of the vectors by its length. Thus, the vectors

$$\mathbf{v}_1 = \frac{\mathbf{u}_1}{\|\mathbf{u}_1\|} = \frac{1}{\sqrt{3}}\begin{bmatrix} 1 \\ 1 \\ 1 \end{bmatrix}, \quad \mathbf{v}_2 = \frac{1}{\sqrt{2}}\begin{bmatrix} 0 \\ 1 \\ -1 \end{bmatrix}, \quad \text{and} \quad \mathbf{v}_3 = \frac{1}{\sqrt{6}}\begin{bmatrix} -2 \\ 1 \\ 1 \end{bmatrix}$$

form an orthonormal set in $\Re^3$. We leave it to you to verify that these vectors are linearly independent and hence form an orthonormal basis for $\Re^3$. ■

Three mutually orthogonal vectors in $\Re^3$ are clearly not coplanar; thus they are linearly independent. In general, a mutually orthogonal set of nonzero vectors will always be linearly independent. This is the content of the next theorem.

THEOREM 3.16 Any set of mutually orthogonal nonzero vectors from $\Re^n$ is linearly independent.

Proof Let $\mathcal{S} = \{\mathbf{u}_1, \mathbf{u}_2, \ldots, \mathbf{u}_r\}$ be a mutually orthogonal set of nonzero vectors, thus $\mathbf{u}_i \cdot \mathbf{u}_j = 0$ for $i \neq j$. In order to test these vectors for linear independence we suppose that

$$a_1\mathbf{u}_1 + a_2\mathbf{u}_2 + \cdots + a_r\mathbf{u}_r = \mathbf{0}.$$

Taking the inner product of both sides of the last equation with $\mathbf{u}_i$ yields, using the properties of the dot product in Theorem 3.2,

$$a_1(\mathbf{u}_i \cdot \mathbf{u}_1) + a_2(\mathbf{u}_i \cdot \mathbf{u}_2) + \cdots + a_r(\mathbf{u}_i \cdot \mathbf{u}_r) = \mathbf{u}_i \cdot \mathbf{0} = 0.$$

This equation, because of the orthogonality of the $\mathbf{u}_i$, reduces to

$$a_i(\mathbf{u}_i \cdot \mathbf{u}_i) = a_i\|\mathbf{u}_i\|^2 = 0.$$

Since $\mathbf{u}_i \neq \mathbf{0}$ implies $\|\mathbf{u}_i\| \neq 0$, it follows that $a_i = 0$. This calculation can be repeated for each $i = 1, 2, \ldots, r$ to show that all the $a_i = 0$. Therefore the vectors $\mathbf{u}_i$ are linearly independent as claimed. □

In many applications it is convenient to work with an orthonormal basis for the subspace of $\Re^n$ under investigation. One of the advantages of an orthonormal basis is that it is very easy to compute coordinates. Thus let $\mathcal{B} = \{\mathbf{w_1}, \mathbf{w_2}, \ldots, \mathbf{w_k}\}$ be an orthonormal basis for a subspace $\mathcal{W}$ of $\Re^n$ and let $\mathbf{z}$ be any vector in $\mathcal{W}$. We know that there exist unique scalars a_i such that

$$\mathbf{z} = a_1\mathbf{w_1} + a_2\mathbf{w_2} + \cdots + a_k\mathbf{w_k}.$$

We could find these coefficients by solving a system of linear equations, as illustrated in Example 2 of Section 3.5, or by using the technique used in proving Theorem 3.16. We adopt the second point of view and compute the inner product of both sides of the preceding equation with $\mathbf{w_i}$ to obtain, after using the orthonormality of $\mathcal{B}$,

$$\mathbf{w_i} \cdot \mathbf{z} = \sum_{j=1}^{k} a_j(\mathbf{w_i} \cdot \mathbf{w_j}) = a_i(\mathbf{w_i} \cdot \mathbf{w_i}) = a_i; \quad i = 1, \ldots, k.$$

Thus we can find the coordinates of the vector $\mathbf{z}$ with respect to the orthonormal basis $\mathcal{B}$ by simply taking the inner product of $\mathbf{z}$ with the basis vectors. Thus we have

$$\mathbf{z} = \sum_{j=1}^{k} (\mathbf{z} \cdot \mathbf{w_j})\mathbf{w_j} = (\mathbf{z} \cdot \mathbf{w_1})\mathbf{w_1} + \cdots + (\mathbf{z} \cdot \mathbf{w_k})\mathbf{w_k}. \tag{3.10}$$

We have proven the following useful theorem.

THEOREM 3.17 If $\mathcal{B} = \{\mathbf{w_1}, \mathbf{w_2}, \ldots, \mathbf{w_k}\}$ is an orthonormal basis for a subspace $\mathcal{W}$ of $\Re^n$ and $\mathbf{z}$ is any vector in $\mathcal{W}$, then

$$\mathbf{z} = \sum_{j=1}^{k} (\mathbf{z} \cdot \mathbf{w_j})\mathbf{w_j} = (\mathbf{z} \cdot \mathbf{w_1})\mathbf{w_1} + \cdots + (\mathbf{z} \cdot \mathbf{w_k})\mathbf{w_k}. \quad \square$$

EXAMPLE 3 Coordinates with Respect to an Orthonormal Basis

Find the coordinates of the vector $\mathbf{z} = \begin{bmatrix} 1 \\ 2 \\ 3 \end{bmatrix}$ with respect to the orthonormal basis of $\Re^3$ obtained in Example 2.

Using the result in Equation (3.10) we see that the coordinates of $\mathbf{z}$ are

$$\mathbf{z} \cdot \mathbf{v_1} = \frac{1}{\sqrt{3}}(1 + 2 + 3) = \frac{6}{\sqrt{3}};$$

$$\mathbf{z} \cdot \mathbf{v_2} = \frac{1}{\sqrt{2}}(0 + 2 - 3) = \frac{-1}{\sqrt{2}}; \text{ and}$$

$$\mathbf{z} \cdot \mathbf{v_3} = \frac{1}{\sqrt{6}}(-2 + 2 + 3) = \frac{3}{\sqrt{6}}.$$

Thus $\mathbf{z} = \frac{6}{\sqrt{3}}\mathbf{v_1} + -\frac{1}{\sqrt{2}}\mathbf{v_2} + \frac{3}{\sqrt{6}}\mathbf{v_3}$ as you may check directly. ■

We will now show how to construct an orthonormal basis for any subspace of $\mathfrak{R}^n$. The procedure is essentially a replacement process in which the vectors of one basis are replaced, one at a time, by the vectors of an orthonormal basis. The basic step is described next. If $\{\mathbf{u}_1, \ldots, \mathbf{u}_k\}$ is a mutually orthogonal set and $\mathbf{v}_{k+1}$ is not in SPAN$\{\mathbf{u}_1, \ldots, \mathbf{u}_k\}$ then we try to find a vector $\mathbf{u}_{k+1}$ in SPAN$\{\mathbf{u}_1, \ldots, \mathbf{u}_k, \mathbf{v}_{k+1}\}$, which is orthogonal to each of the $\mathbf{u}_i$. Thus we need to choose scalars a_i so that

$$\mathbf{u}_{k+1} = a_1\mathbf{u}_1 + a_2\mathbf{u}_2 + \cdots + a_k\mathbf{u}_k + \mathbf{v}_{k+1} \tag{3.11}$$

is orthogonal to each of the $\mathbf{u}_i$, $i = 1, 2, \ldots, k$. There is no loss of generality in choosing the coefficient of $\mathbf{v}_{k+1}$ to be 1. We determine the coefficients in Equation (3.11) by taking the inner product of Equation (3.11) with each of the orthogonal vectors $\mathbf{u}_i$, $i = 1, 2, \ldots, k$. This step leads to

$$\begin{aligned} 0 = (\mathbf{u}_i \cdot \mathbf{u}_{k+1}) &= \sum_{j=1}^{k} a_j(\mathbf{u}_i \cdot \mathbf{u}_j) + (\mathbf{u}_i \cdot \mathbf{v}_{k+1}) \\ &= a_i(\mathbf{u}_i \cdot \mathbf{u}_i) + (\mathbf{u}_i \cdot \mathbf{v}_{k+1}) \end{aligned}$$

from which we obtain

$$a_i = -\frac{\mathbf{u}_i \cdot \mathbf{v}_{k+1}}{\mathbf{u}_i \cdot \mathbf{u}_i}; \quad i = 1, 2, \ldots, k. \tag{3.12}$$

Equation (3.12) determines the coefficients in Equation (3.11) and hence the vector $\mathbf{u}_{k+1}$.

The next theorem could be proven by direct verification, but the preceding discussion shows how the result is obtained.

THEOREM 3.18 If the vectors $\{\mathbf{u}_1, \ldots, \mathbf{u}_k\}$ are mutually orthogonal in $\mathfrak{R}^n$ and $\mathbf{v}_{k+1}$ is not in SPAN$\{\mathbf{u}_1, \ldots, \mathbf{u}_k\}$, then the vector

$$\mathbf{u}_{k+1} = \mathbf{v}_{k+1} - \sum_{i=1}^{k} \left(\frac{\mathbf{u}_i \cdot \mathbf{v}_{k+1}}{\mathbf{u}_i \cdot \mathbf{u}_i}\right)\mathbf{u}_i$$

is orthogonal to each of the $\mathbf{u}_i$; $i = 1, 2, \ldots, k$. □

It is worth noting the geometric interpretation of Equations (3.11) and (3.12) in the case $n = 3$ and $k = 2$. If $\mathbf{u}_1$ and $\mathbf{u}_2$ are orthogonal and $\mathbf{v}_3$ is not in the plane $\mathcal{P} = $ SPAN$\{\mathbf{u}_1, \mathbf{u}_2\}$, then a vector $\mathbf{u}_3$ orthogonal to both $\mathbf{u}_1$ and $\mathbf{u}_2$ is obtained by taking $\mathbf{u}_3$ to be the difference between $\mathbf{v}_3$ and the projection of $\mathbf{v}_3$ on the plane $\mathcal{P}$. Thus $\mathbf{u}_3 = \mathbf{v}_3 - (b_1\mathbf{u}_1 + b_2\mathbf{u}_2)$ where

$$b_1 = -a_1 = \frac{(\mathbf{u}_1 \cdot \mathbf{v}_3)}{(\mathbf{u}_1 \cdot \mathbf{u}_1)} \quad \text{and} \quad b_2 = -a_2 = \frac{(\mathbf{u}_2 \cdot \mathbf{v}_3)}{(\mathbf{u}_2 \cdot \mathbf{u}_2)}$$

are the scalar projections of $\mathbf{v}_3$ on $\mathbf{u}_1$ and $\mathbf{u}_2$, respectively. Equivalently, $\mathbf{u}_3$ is the projection of $\mathbf{v}_3$ on a line perpendicular to the plane determined by $\mathbf{u}_1$ and $\mathbf{u}_2$. See Figure 3.21.

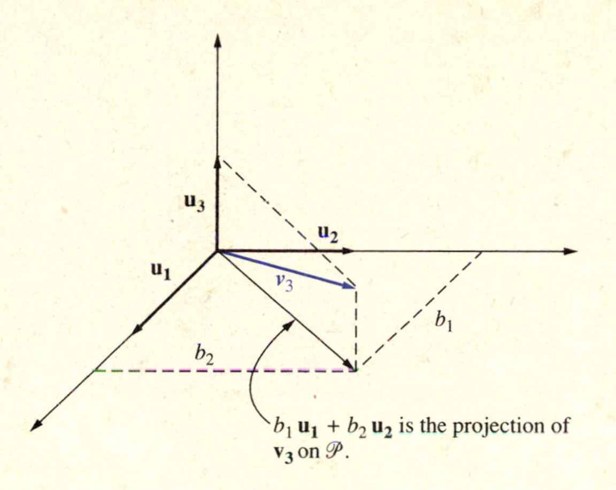

FIGURE 3.21

EXAMPLE 4 Find a Vector Orthogonal to a Plane

Given the orthogonal vectors $\mathbf{u}_1 = \begin{bmatrix} 1 \\ 1 \\ 2 \end{bmatrix}$ and $\mathbf{u}_2 = \begin{bmatrix} -2 \\ 0 \\ 1 \end{bmatrix}$ and the vector $\mathbf{v}_3 = \begin{bmatrix} -1 \\ 1 \\ 4 \end{bmatrix}$, which is not in SPAN$\{\mathbf{u}_1, \mathbf{u}_2\}$, find a vector $\mathbf{u}_3$ that is orthogonal to both $\mathbf{u}_1$ and $\mathbf{u}_2$.

Following the procedure described in the proof of Theorem 3.18, we write

$$\mathbf{u}_3 = a_1\mathbf{u}_1 + a_2\mathbf{u}_2 + \mathbf{v}_3$$

and take the dot product with $\mathbf{u}_1$ to obtain

$$0 = (\mathbf{u}_1 \cdot \mathbf{u}_3) = a_1(\mathbf{u}_1 \cdot \mathbf{u}_1) + a_2(\mathbf{u}_1 \cdot \mathbf{u}_2) + (\mathbf{u}_1 \cdot \mathbf{v}_3) = 6a_1 + 0 + 8,$$

from which it follows that $a_1 = -\frac{8}{6}$. Similarly, taking the dot product with $\mathbf{u}_2$ yields

$$0 = (\mathbf{u}_2 \cdot \mathbf{u}_3) = a_1(\mathbf{u}_2 \cdot \mathbf{u}_1) + a_2(\mathbf{u}_2 \cdot \mathbf{u}_2) + (\mathbf{u}_2 \cdot \mathbf{v}_3) = 0 + 5a_2 + 6,$$

from which we find $a_2 = -\frac{6}{5}$. Thus we conclude that

$$\mathbf{u}_3 = -\frac{4}{3}\mathbf{u}_1 - \frac{6}{5}\mathbf{u}_2 + \mathbf{v}_3$$
$$= -\frac{4}{3}\begin{bmatrix}1\\1\\2\end{bmatrix} - \frac{6}{5}\begin{bmatrix}-2\\0\\1\end{bmatrix} + \begin{bmatrix}-1\\1\\4\end{bmatrix}$$
$$= \frac{1}{15}\begin{bmatrix}1\\-5\\2\end{bmatrix}.$$

Note that $15\mathbf{u}_3 = \begin{bmatrix}1\\-5\\2\end{bmatrix}$ is also orthogonal to $\mathbf{u}_1$ and $\mathbf{u}_2$ and would be a bit more convenient to work with. ■

To find an orthonormal basis for a subspace $\mathcal{W}$ of $\mathfrak{R}^n$, one may proceed as follows:

■ THE GRAM-SCHMIDT ORTHOGONALIZATION PROCESS ■

1. Find any basis $\{\mathbf{v}_1, \mathbf{v}_2, \ldots, \mathbf{v}_k\}$ for $\mathcal{W}$.
2. Take $\mathbf{u}_1 = \mathbf{v}_1$.
3. Find $\mathbf{u}_2$ in SPAN$\{\mathbf{u}_1, \mathbf{v}_2\}$ orthogonal to $\mathbf{u}_1$.
4. Find $\mathbf{u}_3$ in SPAN$\{\mathbf{u}_1, \mathbf{u}_2, \mathbf{v}_3\}$ orthogonal to $\mathbf{u}_1$ and $\mathbf{u}_2$.
5. Find $\mathbf{u}_4$ in SPAN$\{\mathbf{u}_1, \mathbf{u}_2, \mathbf{u}_3, \mathbf{v}_4\}$ orthogonal to $\mathbf{u}_1$, $\mathbf{u}_2$, and $\mathbf{u}_3$.

⋮

Stop when $\mathbf{v}_k$ has been replaced by $\mathbf{u}_k$ and then divide each $\mathbf{u}_i$ by its length to obtain the desired orthonormal basis.

■ EXAMPLE 5 The Gram-Schmidt Process

Find an orthonormal basis for the column space of $A = \begin{bmatrix}1&1&1\\1&0&1\\1&0&0\\1&0&0\end{bmatrix}$.

Let the columns of A be denoted by $\mathbf{v}_1$, $\mathbf{v}_2$, and $\mathbf{v}_3$. Using the Gram-Schmidt process we set $\mathbf{u}_1 = \mathbf{v}_1$ and then use Theorem 3.18 to find a vector $\mathbf{u}_2$ in

SPAN$\{\mathbf{u}_1, \mathbf{v}_2\}$ orthogonal to $\mathbf{u}_1$. By Theorem 3.18,

$$\mathbf{u}_2 = \mathbf{v}_2 - \frac{(\mathbf{u}_1 \cdot \mathbf{v}_2)}{(\mathbf{u}_1 \cdot \mathbf{u}_1)}\mathbf{u}_1 = \mathbf{v}_2 - \frac{1}{4}\mathbf{u}_1$$

$$= \begin{bmatrix} 1 \\ 0 \\ 0 \\ 0 \end{bmatrix} - \frac{1}{4}\begin{bmatrix} 1 \\ 1 \\ 1 \\ 1 \end{bmatrix} = \frac{1}{4}\begin{bmatrix} 3 \\ -1 \\ -1 \\ -1 \end{bmatrix}$$

which is orthogonal to $\mathbf{u}_1$. We also have SPAN$\{\mathbf{u}_1, \mathbf{u}_2\}$ = SPAN$\{\mathbf{v}_1, \mathbf{v}_2\}$. Next we construct a vector $\mathbf{u}_3$ in SPAN$\{\mathbf{u}_1, \mathbf{u}_2, \mathbf{v}_3\}$ orthogonal to both $\mathbf{u}_1$ and $\mathbf{u}_2$. By Theorem 3.18,

$$\mathbf{u}_3 = \mathbf{v}_3 - \frac{(\mathbf{u}_1 \cdot \mathbf{v}_3)}{(\mathbf{u}_1 \cdot \mathbf{u}_1)}\mathbf{u}_1 - \frac{(\mathbf{u}_2 \cdot \mathbf{v}_3)}{(\mathbf{u}_2 \cdot \mathbf{u}_2)}\mathbf{u}_2$$

$$= \mathbf{v}_3 - \frac{2}{4}\mathbf{u}_1 - \frac{\frac{2}{4}}{\frac{12}{16}}\mathbf{u}_2$$

$$= \mathbf{v}_3 - \frac{1}{2}\mathbf{u}_1 - \frac{2}{3}\mathbf{u}_2 = -\frac{1}{6}\begin{bmatrix} 0 \\ -4 \\ 2 \\ 2 \end{bmatrix} = -\frac{1}{3}\begin{bmatrix} 0 \\ -2 \\ 1 \\ 1 \end{bmatrix}$$

which is orthogonal to both $\mathbf{u}_1$ and $\mathbf{u}_2$. Note that $-3\mathbf{u}_3 = [0 \quad -2 \quad 1 \quad 1]^T$ is also orthogonal to both $\mathbf{u}_1$ and $\mathbf{u}_2$. The desired orthonormal basis can now be obtained by dividing each of the $\mathbf{u}_i$ by its length to obtain a unit vector. Since $\|\mathbf{u}_1\| = 2$, $\|\mathbf{u}_2\| = \frac{\sqrt{6}}{3}$, and $\|\mathbf{u}_3\| = \frac{\sqrt{12}}{4}$ this final step leads to the vectors

$$\mathbf{w}_1 = \frac{\mathbf{u}_1}{\|\mathbf{u}_1\|} = \frac{1}{2}\begin{bmatrix} 1 \\ 1 \\ 1 \\ 1 \end{bmatrix}, \quad \mathbf{w}_2 = \frac{\mathbf{u}_2}{\|\mathbf{u}_2\|} = \frac{1}{\sqrt{12}}\begin{bmatrix} 3 \\ -1 \\ -1 \\ -1 \end{bmatrix},$$

$$\mathbf{w}_3 = \frac{\mathbf{u}_3}{\|\mathbf{u}_3\|} = \frac{-3\mathbf{u}_3}{\|-3\mathbf{u}_3\|} = \frac{1}{\sqrt{6}}\begin{bmatrix} 0 \\ -2 \\ 1 \\ 1 \end{bmatrix}$$

which form an orthonormal basis for $\mathscr{CS}(A)$. ■

There are many possible variations on the basic Gram-Schmidt process. The way that we have described the process works well for hand calculations. Since scaling (scalar multiplication) does not affect the orthogonality, one is free to scale the vectors in any way that is convenient; for example, in Example 5 it would have been helpful to replace $\mathbf{u}_2$ by $4\mathbf{u}_2$ in order to avoid fractions. One can change the vectors to unit vectors at any time during the process. If one is working by machine, or writing a program, then it is advisable to normalize the vectors at each step. In this case Equation (3.12) reduces to $a_i = -(\mathbf{u}_i \cdot \mathbf{v}_{k+1})$.

Let us look again at the calculation in Example 5. The equations relating the basis vectors $\{\mathbf{v}_1, \mathbf{v}_2, \mathbf{v}_3\}$ and $\{\mathbf{u}_1, \mathbf{u}_2, \mathbf{u}_3\}$ are

$$\begin{aligned} \mathbf{u}_1 &= \mathbf{v}_1 \\ \mathbf{u}_2 &= \mathbf{v}_2 - \frac{1}{4}\mathbf{u}_1 \\ \mathbf{u}_3 &= \mathbf{v}_3 - \frac{1}{2}\mathbf{u}_1 - \frac{2}{3}\mathbf{u}_2 \end{aligned} \qquad \text{or} \qquad \begin{aligned} \mathbf{v}_1 &= \mathbf{u}_1 \\ \mathbf{v}_2 &= \frac{1}{4}\mathbf{u}_1 + \mathbf{u}_2 \\ \mathbf{v}_3 &= \frac{1}{2}\mathbf{u}_1 + \frac{2}{3}\mathbf{u}_2 + \mathbf{u}_3. \end{aligned}$$

We can use this last set of equations to obtain a factorization of the matrix A:

$$A = [\mathbf{v}_1 \quad \mathbf{v}_2 \quad \mathbf{v}_3] = [\mathbf{u}_1 \quad \mathbf{u}_2 \quad \mathbf{u}_3]\begin{bmatrix} 1 & \frac{1}{4} & \frac{1}{2} \\ 0 & 1 & \frac{2}{3} \\ 0 & 0 & 1 \end{bmatrix}.$$

If we write the last set of equations in terms of the orthonormal basis $\{\mathbf{w}_1, \mathbf{w}_2, \mathbf{w}_3\}$ we obtain, since $\mathbf{u}_1 = 2\mathbf{w}_1$, $\mathbf{u}_2 = \frac{\sqrt{3}}{2}\mathbf{w}_2$, and $\mathbf{u}_3 = \frac{\sqrt{6}}{3}\mathbf{w}_3$:

$$\begin{aligned} \mathbf{v}_1 &= \mathbf{u}_1 = 2\mathbf{w}_1, \\ \mathbf{v}_2 &= \frac{1}{4}\mathbf{u}_1 + \mathbf{u}_2 = \frac{1}{2}\mathbf{w}_1 + \frac{\sqrt{3}}{2}\mathbf{w}_2, \\ \mathbf{v}_3 &= \frac{1}{2}\mathbf{u}_1 + \frac{2}{3}\mathbf{u}_2 + \mathbf{u}_3 = \mathbf{w}_1 + \frac{\sqrt{3}}{3}\mathbf{w}_2 + \frac{\sqrt{6}}{3}\mathbf{w}_3. \end{aligned}$$

The corresponding factorization of the matrix A is

$$\begin{aligned} A &= [\mathbf{w}_1 \quad \mathbf{w}_2 \quad \mathbf{w}_3]\begin{bmatrix} 2 & \frac{1}{2} & 1 \\ 0 & \frac{\sqrt{3}}{2} & \frac{\sqrt{3}}{3} \\ 0 & 0 & \frac{\sqrt{6}}{3} \end{bmatrix} \\ &= \begin{bmatrix} \frac{1}{2} & \frac{3}{\sqrt{12}} & 0 \\ \frac{1}{2} & -\frac{1}{\sqrt{12}} & -\frac{2}{\sqrt{6}} \\ \frac{1}{2} & -\frac{1}{\sqrt{12}} & \frac{1}{\sqrt{6}} \\ \frac{1}{2} & -\frac{1}{\sqrt{12}} & \frac{1}{\sqrt{6}} \end{bmatrix}\begin{bmatrix} 2 & \frac{1}{2} & 1 \\ 0 & \frac{\sqrt{3}}{2} & \frac{\sqrt{3}}{3} \\ 0 & 0 & \frac{\sqrt{6}}{3} \end{bmatrix} = QR \end{aligned}$$

where the matrix Q has orthonormal columns and the matrix R is upper triangular. Such a factorization is normally referred to as the "QR factorization of A." As the above calculation shows, the QR factorization is equivalent to the Gram-Schmidt process applied to the columns of the matrix A. Both MATALG and MATLAB have built in functions for computing this factorization. The MATALG command is "m QR A" and the MATLAB command is "[Q, R] = qr(A)."

A matrix whose columns form an orthonormal basis for $\Re^n$ is called an **orthogonal matrix.** Since the entry in the (i, j) position of the product A^TA is the dot product $(\text{Col}_i(A) \cdot \text{Col}_j(A))$, it follows that a matrix A is orthogonal if, and only if, $A^TA = I$. Thus for an orthogonal matrix A, $A^{-1} = A^T$.

A frequently occurring problem is to find an orthonormal basis for $\Re^n$ that has a prescribed first vector. This problem is equivalent to finding an orthogonal matrix whose first column is the prescribed unit vector. An attractive alternative to the Gram-Schmidt process in this case is described in the next theorem.

THEOREM 3.19 If $x_1 \neq 1$* and $\mathbf{u}_1 = \begin{bmatrix} x_1 \\ x_2 \\ \cdot \\ \cdot \\ \cdot \\ x_n \end{bmatrix} = \begin{bmatrix} x_1 \\ Y \end{bmatrix}$ is a unit vector in $\Re^n$, then the matrix

$$H = \left[\begin{array}{c|c} x_1 & Y^T \\ \hline Y & I - \frac{1}{1 - x_1} YY^T \end{array}\right]$$

is an orthogonal matrix and its columns form an orthonormal basis for $\Re^n$ containing $\mathbf{u}_1$.

Proof The proof amounts to showing, by a direct computation, that $H^TH = I$. Note that H is symmetric, $H^T = H$, and that, because $\mathbf{u}_1$ is a unit vector, $x_1^2 + Y^TY = 1$. Now

$$H^TH = H^2 = \left[\begin{array}{c|c} x_1 & Y^T \\ \hline Y & I - \frac{1}{1 - x_1} YY^T \end{array}\right]\left[\begin{array}{c|c} x_1 & Y^T \\ \hline Y & I - \frac{1}{1 - x_1} YY^T \end{array}\right]$$

$$= \left[\begin{array}{c|c} x_1^2 + Y^TY & x_1 Y^T + Y^T - \frac{1}{1 - x_1}(Y^TY)Y^T \\ \hline Yx_1 + Y - \frac{1}{1 - x_1} Y(Y^TY) & YY^T + I - \frac{2}{1 - x_1} YY^T + \frac{1}{(1 - x_1)^2} Y(Y^TY)Y^T \end{array}\right].$$

Using the fact that $YY^T = 1 - x_1^2$ this last equation reduces to

$$H^TH = \left[\begin{array}{c|c} x_1^2 + Y^TY & x_1 Y^T + Y^T - \frac{1}{1 - x_1}(1 - x_1^2)Y^T \\ \hline Yx_1 + Y - \frac{1}{1 - x_1} Y(1 - x_1^2) & YY^T + I - \frac{2}{1 - x_1} YY^T + \frac{1}{(1 - x_1)^2} Y(1 - x_1^2)Y^T \end{array}\right]$$

$$= \left[\begin{array}{c|c} 1 & x_1 Y^T + Y^T - (1 + x_1)Y^T \\ \hline x_1 Y + Y - (1 + x_1)Y & \frac{1 - x_1}{1 - x_1} YY^T + I - \frac{2}{1 - x_1} YY^T + \frac{1 + x_1}{(1 - x_1)} YY^T \end{array}\right] = \begin{bmatrix} 1 & 0 \\ 0 & I \end{bmatrix} = I.$$

Thus the matrix H is orthogonal and its columns form the desired orthonormal basis for $\Re^n$. □

---

*If $x_1 = 1$, then $\mathbf{u}_1 = \mathbf{e}_1$ and the identity matrix is an orthogonal matrix whose first column is $\mathbf{u}_1$.

EXAMPLE 6 **Illustrating Theorem 3.19**

Use Theorem 3.19 to find an orthonormal basis for $\Re^4$ containing

$$\mathbf{u}_1 = \frac{1}{6}\begin{bmatrix} 3 \\ 1 \\ 1 \\ -5 \end{bmatrix}.$$

The given vector $\mathbf{u}_1$ is of the form $\begin{bmatrix} x_1 \\ Y \end{bmatrix}$ with $Y = \begin{bmatrix} \frac{1}{6} \\ \frac{1}{6} \\ -\frac{5}{6} \end{bmatrix}$ and $x_1 = \frac{1}{2}$. Thus

$$YY^T = \frac{1}{36}\begin{bmatrix} 1 & 1 & -5 \\ 1 & 1 & -5 \\ -5 & -5 & 25 \end{bmatrix}, \quad \frac{1}{1 - x_1} = 2 \quad \text{and}$$

$$I - \frac{1}{1 - x_1}YY^T = \begin{bmatrix} 1 & 0 & 0 \\ 0 & 1 & 0 \\ 0 & 0 & 1 \end{bmatrix} - \frac{1}{18}\begin{bmatrix} 1 & 1 & -5 \\ 1 & 1 & -5 \\ -5 & -5 & 25 \end{bmatrix}$$

$$= \frac{1}{18}\begin{bmatrix} 17 & -1 & 5 \\ -1 & 17 & 5 \\ 5 & 5 & -7 \end{bmatrix}.$$

Thus, the matrix of Theorem 3.19 is

$$H = \frac{1}{18}\begin{bmatrix} 9 & 3 & 3 & -15 \\ 3 & 17 & -1 & 5 \\ 3 & -1 & 17 & 5 \\ -15 & 5 & 5 & -7 \end{bmatrix}.$$

You should verify directly that H is an orthogonal matrix; that is $H^TH = I$. The columns of H are an orthonormal set of vectors in $\Re^4$ and thus form an orthonormal basis for $\Re^4$. ■

EXERCISES 3.6

1. For $\mathbf{u} = [1 \quad 2 \quad 3 \quad -1 \quad 4]^T$ and $\mathbf{v} = [3 \quad 0 \quad -4 \quad 2 \quad 5]^T$, find $\mathbf{u} \cdot \mathbf{v}$, $\|\mathbf{u}\|$, $\|\mathbf{v}\|$, and $\|\mathbf{u} + \mathbf{v}\|$. Which is larger, $\|\mathbf{u}\| + \|\mathbf{v}\|$ or $\|\mathbf{u} + \mathbf{v}\|$? Find a scalar a in $\Re$ such that $\|a\mathbf{u}\| = 1$. Determine h so that $\mathbf{y} = [1 \quad 1 \quad 1 \quad 1 \quad h]^T$ is orthogonal to $\mathbf{u}$. Determine a and b so that $\mathbf{x} = [1 \quad 1 \quad 1 \quad a \quad b]^T$ is orthogonal to both $\mathbf{u}$ and $\mathbf{v}$. Determine c so that $\mathbf{z} = c\mathbf{u} + \mathbf{v}$ is orthogonal to $\mathbf{v}$.
2. Which of the following sets are orthonormal?

(a) $\frac{1}{\sqrt{2}}\begin{bmatrix} 1 \\ 0 \\ 1 \end{bmatrix}, \frac{1}{\sqrt{3}}\begin{bmatrix} 1 \\ 1 \\ -1 \end{bmatrix}, \frac{1}{\sqrt{6}}\begin{bmatrix} -1 \\ 2 \\ 1 \end{bmatrix}$ (b) $\begin{bmatrix} 1 \\ 0 \\ 0 \end{bmatrix}, \begin{bmatrix} 0 \\ 1 \\ 1 \end{bmatrix}, \begin{bmatrix} 0 \\ 0 \\ 1 \end{bmatrix}$

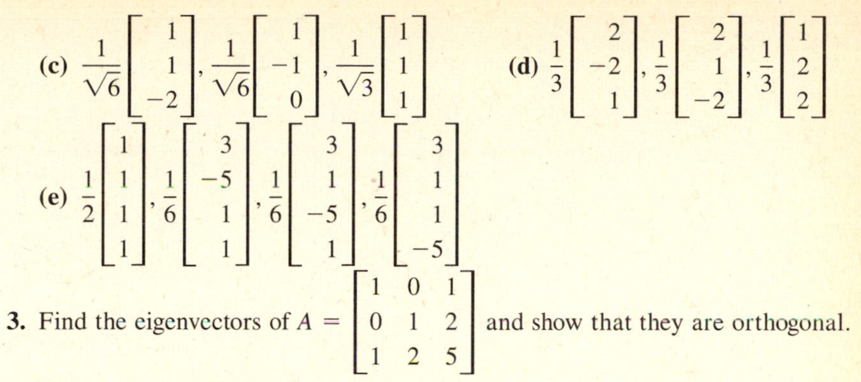

(c) $\frac{1}{\sqrt{6}}\begin{bmatrix}1\\1\\-2\end{bmatrix}, \frac{1}{\sqrt{6}}\begin{bmatrix}1\\-1\\0\end{bmatrix}, \frac{1}{\sqrt{3}}\begin{bmatrix}1\\1\\1\end{bmatrix}$ **(d)** $\frac{1}{3}\begin{bmatrix}2\\-2\\1\end{bmatrix}, \frac{1}{3}\begin{bmatrix}2\\1\\-2\end{bmatrix}, \frac{1}{3}\begin{bmatrix}1\\2\\2\end{bmatrix}$

(e) $\frac{1}{2}\begin{bmatrix}1\\1\\1\\1\end{bmatrix}, \frac{1}{6}\begin{bmatrix}3\\-5\\1\\1\end{bmatrix}, \frac{1}{6}\begin{bmatrix}3\\1\\-5\\1\end{bmatrix}, \frac{1}{6}\begin{bmatrix}3\\1\\1\\-5\end{bmatrix}$

3. Find the eigenvectors of $A = \begin{bmatrix}1 & 0 & 1\\0 & 1 & 2\\1 & 2 & 5\end{bmatrix}$ and show that they are orthogonal.

4. Prove Theorem 3.13.

5. Let $\mathcal{B}$ be the orthonormal basis for $\mathfrak{R}^4$ given in Exercise 2(e). Find the coordinates, with respect to the $\mathcal{B}$ basis, of each of the following vectors.

(a) $[1 \quad 0 \quad 0 \quad 0]^T$ **(b)** $[1 \quad 2 \quad 3 \quad 4]^T$ **(c)** $[-1 \quad 0 \quad 4 \quad 11]^T$

6. Show that if the matrix A is orthogonal, the $\det(A) = 1$ or -1.

7. Show that the vectors $\mathbf{u}$ and $\mathbf{v}$ in $\mathfrak{R}^n$ are orthogonal if, and only if, $\|\mathbf{u} - \mathbf{v}\| = \|\mathbf{u} + \mathbf{v}\|$. Interpret geometrically for $n = 3$.

†8. Show that if P_1 and P_2 are orthogonal matrices, then so is $\begin{bmatrix}P_1 & 0\\0 & P_2\end{bmatrix}$.

†9. Show that if A is an $m \times n$ matrix, then A and A^TA have the same rank. (*Hint:* Show that $\mathcal{NS}(A) = \mathcal{NS}(A^TA)$.)

10. Let $\{\mathbf{u}_1, \mathbf{u}_2, \ldots, \mathbf{u}_n\}$ be an orthonormal basis for $\mathcal{W}$. Show that for any $\mathbf{w}$ in $\mathcal{W}$

$$\|\mathbf{w}\|^2 = (\mathbf{w} \cdot \mathbf{u}_1)^2 + (\mathbf{w} \cdot \mathbf{u}_2)^2 + \cdots + (\mathbf{w} \cdot \mathbf{u}_n)^2.$$

11. Find an orthonormal basis for the subspace of $\mathfrak{R}^3$ spanned by

$$\mathbf{u}_1 = [1 \quad 2 \quad 2]^T \quad \text{and} \quad \mathbf{u}_2 = [7 \quad 1 \quad 0]^T.$$

12. Find an orthonormal basis for the subspace of $\mathfrak{R}^4$ spanned by

$$\mathbf{u}_1 = [1 \quad 2 \quad 1 \quad 1]^T, \quad \mathbf{u}_2 = [1 \quad -1 \quad 0 \quad 2]^T, \quad \text{and}$$
$$\mathbf{u}_3 = [2 \quad 0 \quad 1 \quad 1]^T.$$

13. Use the Gram-Schmidt process to find an orthonormal basis for $\mathcal{CS}(A)$ if A is the given matrix.

(a) $A = \begin{bmatrix}1 & 0 & 0\\1 & 1 & 0\\1 & 1 & 1\end{bmatrix}$ **(b)** $A = \begin{bmatrix}3 & 0 & 5\\0 & 1 & -2\\4 & 1 & 1\end{bmatrix}$

14. Find a unit vector in $\mathfrak{R}^3$ orthogonal to both $\mathbf{u}_1 = [1 \quad -1 \quad 3]^T$ and $\mathbf{u}_2 = [3 \quad 0 \quad -1]^T$.

15. Find an orthonormal basis for $\mathfrak{R}^4$ containing scalar multiples of $\mathbf{u}_1 = [1 \quad 2 \quad 1 \quad 1]^T$ and $\mathbf{u}_2 = [-1 \quad 0 \quad 2 \quad -1]^T$.

16. Find an orthonormal basis for the following subspaces and extend it to an orthonormal basis for the whole space.

(a) SPAN $\left\{\begin{bmatrix}1\\1\\0\end{bmatrix}, \begin{bmatrix}0\\1\\1\end{bmatrix}\right\}$ **(b)** SPAN $\left\{\begin{bmatrix}1\\0\\1\\0\end{bmatrix}, \begin{bmatrix}1\\-1\\-1\\1\end{bmatrix}, \begin{bmatrix}1\\0\\0\\1\end{bmatrix}\right\}$

(c) SPAN $\left\{\begin{bmatrix}1\\1\\0\end{bmatrix}, \begin{bmatrix}2\\-1\\-3\end{bmatrix}, \begin{bmatrix}1\\-1\\1\end{bmatrix}\right\}$ **(d)** SPAN $\left\{\begin{bmatrix}3\\0\\4\\12\end{bmatrix}, \begin{bmatrix}1\\0\\1\\0\end{bmatrix}, \begin{bmatrix}1\\1\\1\\1\end{bmatrix}\right\}$

17. Find an orthonormal basis for $\mathcal{NS}(A)$ if A is the matrix shown.

(a) $A = \begin{bmatrix}1 & 0 & 5 & -2 & -1\\0 & 1 & -2 & 4 & 2\end{bmatrix}$ **(b)** $A = \begin{bmatrix}1 & 0 & 0 & 3 & -4\\0 & 1 & 0 & 4 & 2\\0 & 0 & 1 & 5 & 0\end{bmatrix}$

18. Show that every 2×2 orthogonal matrix is of the form

$$\begin{bmatrix}\cos\theta & \sin\theta\\ \sin\theta & -\cos\theta\end{bmatrix} \quad \text{or} \quad \begin{bmatrix}\cos\theta & -\sin\theta\\ \sin\theta & \cos\theta\end{bmatrix}.$$

19. Use Theorem 3.19 to find an orthonormal basis for $\mathfrak{R}^4$ containing each of the following.
(a) $\mathbf{u}_1 = [3 \;\; 0 \;\; 4 \;\; 0]^T/5$
(b) $\mathbf{u}_1 = [2 \;\; -3 \;\; 1 \;\; 4]^T/\sqrt{30}$ (use a calculator or MATALG)
(c) $\mathbf{u}_1 = [1 \;\; -1 \;\; 1 \;\; -1]^T/2$

20. Definition: The **orthogonal complement** of $\mathcal{W} \subset \mathfrak{R}^n$ is the set

$$\mathcal{W}^{\perp} = \{\mathbf{u} \text{ in } \mathfrak{R}^n \mid \mathbf{u}\cdot\mathbf{x} = 0 \text{ for every } \mathbf{x} \text{ in } \mathcal{W}\}.$$

(a) Show that $\mathcal{W}^{\perp}$ is a subspace of $\mathfrak{R}^n$.
(b) Show that for any matrix A, $\mathcal{NS}(A^T) = \mathcal{CS}(A)^{\perp}$.
(c) Find an orthonormal basis for the orthogonal complement of each of the subspaces in Exercise 16.

21. Let $\mathbf{w}$ be a unit vector in $\mathfrak{R}^n$. Show that $H = I - 2\mathbf{w}\mathbf{w}^T$ is an orthogonal matrix. What is H if $\mathbf{w} = [\frac{3}{13} \;\; \frac{4}{13} \;\; \frac{12}{13}]^T$?

22. Let $[x_1 \;\; Y]^T$ be a unit vector in $\mathfrak{R}^n$. Show that

$$\mathbf{w} = (1/\sqrt{2(1 - x_1)})\,[x_1 - 1 \quad Y]^T$$

is also a unit vector in $\mathfrak{R}^n$. Find $H = I - 2\mathbf{w}\mathbf{w}^T$ and compare with Theorem 3.19.

23. Explain why there is no loss of generality in choosing the coefficient of $\mathbf{v}_{k+1}$ in Equation (3.11) to be 1.

24. Write a computer program that will produce the QR factorization, described following Example 5, of a given matrix A.

25. Find the QR factorization for each of the matrices of Exercise 13.

26. For $\mathbf{u} = [a_1 \quad a_2 \quad \cdots \quad a_n]^T$ in $\Re^n$ define

$$\|\mathbf{u}\|_1 = \sum_{i=1}^{n} |a_i| \quad \text{and} \quad \|\mathbf{u}\|_\infty = \max_i\{|a_i|\}.$$

Show that for each of these norms, the conclusions of Theorems 3.13 and 3.15 are satisfied.

†**27.** Prove the following generalization of the Pythagorean theorem: for X and Y in $\Re^n$, $\|X\|^2 + \|Y\|^2 = \|X - Y\|^2$ if, and only if, X and Y are orthogonal.

28. Use the software to find the QR factorization of the following matrices and then find an orthonormal basis for the column spaces of the matrices:

$$A = \begin{bmatrix} 8 & 2 & 8 \\ 10 & 10 & 7 \\ 4 & 7 & 1 \end{bmatrix}, \quad B = \begin{bmatrix} 2 & 9 & 0 & 0 \\ 0 & 4 & 1 & 4 \\ 7 & 5 & 5 & 1 \\ 7 & 8 & 7 & 4 \end{bmatrix},$$

$$C = \begin{bmatrix} 7 & 5 & 0 \\ 6 & 1 & 7 \\ 9 & 7 & 3 \\ 8 & 4 & 6 \end{bmatrix}, \quad D = \begin{bmatrix} 2 & 4 & 5 & 4 & 5 \\ 0 & 5 & 7 & 7 & 1 \\ 7 & 8 & 0 & 6 & 7 \\ 7 & 0 & 4 & 9 & 4 \\ 9 & 1 & 1 & 8 & 7 \end{bmatrix}.$$

3.7 VECTOR SPACES—THE GENERAL CONCEPT

In this section we will introduce and study a class of algebraic systems called vector spaces. The abstract vector concept, which we study here, is a fruitful generalization of the vectors in $\Re^2$, $\Re^3$, and $\Re^n$ that we studied in the last six sections. The generalization of the vector concept that we undertake here is much like the multistep generalization of the number concept that you have been exposed to throughout your mathematical education. For most readers, studying this concept will be a new mathematical experience because we will define our abstract vectors by how they behave, rather than by what they are; this is consistent with the way most people define political terms such as "liberal" or "conservative." Generalizations of this type are most useful if there are many interesting and important examples, making it worthwhile to study all the systems at the same time. It is also useful to have a familiar, motivating example to provide insight and a source of terminology.

The starting place for our generalized definition of a vector is the list of properties of $\Re^n$ contained in Theorem 3.1. Thus, $\Re^n$ will automatically satisfy our general definition and will have the distinction of being our "motivating example." The abstract systems we are going to study will be systems that "behave like $\Re^n$," even if their elements are very different from those of $\Re^n$.

In order for a system to "behave like $\Re^n$," there must be defined, within the system, a way of combining two elements to get another element of the system-like vector addition in $\Re^n$. There must also be a way of combining real numbers and elements of the system to get another element of the system-like scalar multiplication in $\Re^n$. The operations in the system must also be like those in $\Re^n$ in the sense that they satisfy the properties listed in Theorem 3.1. Thus, we are led to the following formal definition.

DEFINITION 3.15 A **real vector space** $(\mathcal{V}, +, \cdot)$ is a set $\mathcal{V}$ of objects that we will call vectors and denote by boldface letters like **u**, **v**, **w**, and **x**, along with two ways of combining them called vector addition $(+)$ and scalar multiplication $(\cdot)$, which satisfy the following ten axioms.

(1) For any $\mathbf{u}$ and $\mathbf{v}$ in $\mathcal{V}$, $\mathbf{u} + \mathbf{v}$ is also an element of $\mathcal{V}$. (The set $\mathcal{V}$ is *closed* under vector addition.)
(2) For any vectors $\mathbf{u}$, $\mathbf{v}$, and $\mathbf{w}$ in $\mathcal{V}$, $\mathbf{u} + (\mathbf{v} + \mathbf{w}) = (\mathbf{u} + \mathbf{v}) + \mathbf{w}$. (Vector addition is *associative*.)
(3) For any vectors $\mathbf{u}$ and $\mathbf{v}$ in $\mathcal{V}$, $\mathbf{u} + \mathbf{v} = \mathbf{v} + \mathbf{u}$. (Vector addition is *commutative*.)
(4) There is a unique vector $\mathbf{0}$ in $\mathcal{V}$ such that $\mathbf{u} + \mathbf{0} = \mathbf{u}$ for every $\mathbf{u}$ in $\mathcal{V}$. (The vector $\mathbf{0}$ is called the *additive identity element* of $\mathcal{V}$.)
(5) For each vector $\mathbf{u}$ in $\mathcal{V}$, there is a unique vector $(-\mathbf{u})$ in $\mathcal{V}$ such that $\mathbf{u} + (-\mathbf{u}) = \mathbf{0}$. (The vector $-\mathbf{u}$ is called the *additive inverse* of $\mathbf{u}$.)
(6) For any vector $\mathbf{u}$ and any real number k, $k \cdot \mathbf{u}$ is in $\mathcal{V}$. (The set $\mathcal{V}$ is *closed* under scalar multiplication.)
(7) For any vector $\mathbf{u}$ and any scalars h and k, $(h + k) \cdot \mathbf{u} = h \cdot \mathbf{u} + k \cdot \mathbf{u}$. (Scalar multiplication *distributes* over scalar addition.)
(8) For any scalar h and any vectors $\mathbf{u}$ and $\mathbf{v}$, $h \cdot (\mathbf{u} + \mathbf{v}) = h \cdot \mathbf{u} + h \cdot \mathbf{v}$. (Scalar multiplication *distributes* over vector addition.)
(9) For any scalars h and k and any vector $\mathbf{u}$, $h \cdot (k \cdot \mathbf{u}) = (hk) \cdot \mathbf{u}$. (Scalar multiplication is *associative*.)
(10) $1 \cdot \mathbf{u} = \mathbf{u}$ for every $\mathbf{u}$ in $\mathcal{V}$. (The real number 1 is an *identity element* for scalar multiplication.) □

We call $(\mathcal{V}, +, \cdot)$ a *real* vector space because the scalars we use are real numbers; there are further generalizations of Definition 3.15 to complex vector spaces and beyond, but we will restrict ourselves to real scalars in this text, except for a few exercises. We will adopt the common practice of referring to the vector space $(\mathcal{V}, +, \cdot)$ as the vector space $\mathcal{V}$, when there is no danger of confusion about the operations to be used. Also, we will normally use juxtaposition to indicate scalar multiplication; that is, we will write $k\mathbf{u}$ instead of $k \cdot \mathbf{u}$.

In order to test whether or not a given set is a real vector space, we must first identify a suitable vector addition and a suitable scalar multiplication. Then we must verify that each of the ten axioms of Definition 3.15 is satisfied. We will now give some examples of systems which have the desired structure and, hence, are real vector spaces. Additional examples are contained in the exercises.

EXAMPLE 1 **Our Motivating Example**

$\mathfrak{R}^n = \mathfrak{R}^{n\times 1}$ is our motivating example. Theorem 3.1 shows that $\mathfrak{R}^n$ is a real vector space for any positive integer n. For $n = 1$, $\mathfrak{R}^n = \mathfrak{R}$, so $\mathfrak{R}$ is itself a real vector space. ■

EXAMPLE 2 **Subspaces of $\mathfrak{R}^n$ Are Vector Spaces**

Any subspace of $\mathfrak{R}^n$ is a real vector space since the closure axioms (1 and 6) are satisfied as a consequence of Definition 3.7 and all of the remaining properties are inherited from $\mathfrak{R}^n$. ■

EXAMPLE 3 **Row Vectors**

$\mathfrak{R}^{m\times n}$, the set of all real $m \times n$ matrices, is a vector space by Theorem 1.4. An important special case is $\mathfrak{R}^{1\times n}$, a space that is "very much like" $\mathfrak{R}^n = \mathfrak{R}^{n\times 1}$. We will call the elements of $\mathfrak{R}^{1\times n}$ **row vectors.** ■

EXAMPLE 4 **Polynomials Are Vectors**

The set $\mathcal{P} = \{a_0 + a_1x + \cdots + a_nx^n \mid n \text{ is finite}\}$ of all polynomials with addition and scalar multiplication defined as in elementary algebra is a vector space. In other words, if

$$\begin{aligned} \mathbf{p} &= a_0 + a_1x + a_2x^2 + \cdots + a_nx^n \qquad \text{and} \\ \mathbf{q} &= b_0 + b_1x + b_2x^2 + \cdots + b_nx^n, \end{aligned}$$

then the sum of the polynomials $\mathbf{p}$ and $\mathbf{q}$ is defined to be

$$\mathbf{p} + \mathbf{q} = (a_0 + b_0) + (a_1 + b_1)x + (a_2 + b_2)x^2 + \cdots + (a_n + b_n)x^n,$$

and scalar multiplication is defined by

$$k\mathbf{p} = ka_0 + ka_1x + ka_2x^2 + \cdots + ka_nx^n.$$

Note that the definition of addition of polynomials does not require that $\mathbf{p}$ and $\mathbf{q}$ have the same degree (the degree of $\mathbf{p}$ is n if $a_n \neq 0$) since we can always add terms with zero coefficients. The operations in this set are familiar to you and are known to satisfy all of the axioms of Definition 3.15. Thus $\mathcal{P}$ is an example of a real vector space. ■

Our next example is also a familiar one because the elements of the set are functions, the objects studied in trigonometry, analytic geometry, and calculus.

EXAMPLE 5 **Functions Are Vectors**

Let $\mathcal{F} = \mathcal{F}(a, b)$ be the set of all real-valued functions defined on the interval $[a, b] = \{x \mid a \leq x \leq b\}$.

In dealing with the vector space $\mathcal{F}$ it is important to distinguish carefully between a function

$$\mathbf{f} = \{(x, f(x)) \mid a \leq x \leq b\}$$

and the value, $f(x)$, of the function at an element x of the domain of the function. Equality in $\mathcal{F}$ is defined by

$$\mathbf{f} = \mathbf{g} \text{ if, and only if, } f(x) = g(x) \text{ for all } x \text{ in the domain } [a, b].$$

Thus two functions are equal if, and only if, they have the same value for every element of their domain. Addition in $\mathcal{F}$ is defined by

$$(\mathbf{f} + \mathbf{g})(x) = f(x) + g(x),$$

and scalar multiplication by

$$(a\mathbf{f})(x) = a(f(x)).$$

The zero vector in $\mathcal{F}$ is the function $\mathbf{0}$ whose value is 0 for every x and the additive inverse of $\mathbf{f}$ is the function $-\mathbf{f}$ which satisfies

$$(-\mathbf{f})(x) = -(f(x)).$$

It is routine to verify any of the remaining vector space axioms for the set $\mathcal{F}$. For example, to show that addition in $\mathcal{F}$ is associative, we calculate

$$((\mathbf{f} + \mathbf{g}) + \mathbf{h})(x) = (\mathbf{f} + \mathbf{g})(x) + h(x) = [f(x) + g(x)] + h(x)$$

and

$$(\mathbf{f} + (\mathbf{g} + \mathbf{h}))(x) = f(x) + (\mathbf{g} + \mathbf{h})(x) = f(x) + [g(x) + h(x)],$$

and then observe that the last two expressions are equal by the associativity of addition in $\Re$. Thus, $\mathbf{f} + (\mathbf{g} + \mathbf{h}) = (\mathbf{f} + \mathbf{g}) + \mathbf{h}$. ■

Note that the elements of the vector space $\mathcal{P}$ of Example 4 can be viewed as polynomial functions and, hence, as elements of the vector space $\mathcal{F}$ of Example 5. Thus, $\mathcal{F}$ contains $\mathcal{P}$ as a subset.

EXAMPLE 6 **$\Re^{1\times 2}$ with Different Operations**

Consider $\mathcal{V} = \{[a \quad b] \mid a \text{ and } b \text{ in } \Re\}$, the set of all 1×2 real matrices, with addition defined by $[a \quad b] \oplus [c \quad d] = [a + c + 2 \quad b + d]$ and with scalar multiplication defined by $k : [a \quad b] = [ka \quad b]$. Is $(\mathcal{V}, \oplus, :)$ a real vector space?

To answer this question we need to either establish that the ten axioms of Definition 3.15 are satisfied (in which case $\mathcal{V}$ is a vector space) or find one of the ten axioms which is not satisfied (thus showing that $\mathcal{V}$ is not a vector space). The closure properties (1 and 6) are clearly satisfied. To check the associativity of $\oplus$ we compute

$$\begin{aligned}[a \quad b] \oplus ([c \quad d] \oplus [e \quad f]) &= [a \quad b] \oplus [c + e + 2 \quad d + f] \\ &= [a + (c + e + 2) + 2 \quad b + (d + f)]\end{aligned}$$

and

$$\begin{aligned}([a \quad b] \oplus [c \quad d]) \oplus [e \quad f] &= [a + c + 2 \quad b + d] \oplus [e \quad f] \\ &= [(a + c + 2) + e + 2 \quad (b + d) + f].\end{aligned}$$

These two vectors are the same because addition in $\mathfrak{R}$ is associative, so we have established that $\oplus$ is associative in $\mathcal{V}$ (Axiom 2 is satisfied). Since

$$\begin{aligned}[a \quad b] \oplus [c \quad d] = [a + c + 2 \quad b + d] &= [c + a + 2 \quad d + b] \\ &= [c \quad d] \oplus [a \quad b],\end{aligned}$$

we see that $\oplus$ is also commutative in $\mathcal{V}$ (Axiom 3 is satisfied).

In order to find an identity element for $\oplus$ we let $\mathbf{z} = [x \quad y]$ and try to determine x and y so that $[a \quad b] \oplus [x \quad y] = [a \quad b]$. This statement is equivalent to

$$[a + x + 2 \quad b + y] = [a \quad b], \qquad \text{or to}$$
$$a + x + 2 = a \qquad \text{and} \qquad b + y = b.$$

It follows that $x = -2$ and $y = 0$. Thus, $\mathbf{z} = [-2 \quad 0]$ is the identity element for $\oplus$ and Axiom 4 is satisfied. To verify Axiom 5 we need to find scalars x and y so that

$$[a \quad b] \oplus [x \quad y] = \mathbf{z} = [-2 \quad 0].$$

This expression is the same as

$$[a + x + 2 \quad b + y] = [-2 \quad 0],$$

so we must have $x = -4 - a$ and $y = -b$. Thus, $[-4 - a \quad -b]$ is the additive inverse of $[a \quad b]$. So far we have shown that $(\mathcal{V}, \oplus, :)$ satisfies all the additive properties of a vector space. In order to check Axiom 7 we compute $(h + k) : [a \quad b] = [(h + k)a \quad b]$ and $h : [a \quad b] \oplus k : [a \quad b] = [ha \quad b] \oplus [ka \quad b] = [ha + ka + 2 \quad b + b]$. Since $(h + k) : [a \quad b] \neq h : [a \quad b] \oplus k : [a \quad b]$, we see that Axiom 7 fails to hold and, hence, that $(\mathcal{V}, \oplus, :)$ is not a real vector space. You should check that Axiom 8 also fails to hold and that Axioms 9 and 10 are both satisfied for this system. Thus, $(\mathcal{V}, \oplus, :)$ fails to be a vector space because scalar multiplication does not distribute over scalar addition and scalar multiplication does not distribute over vector addition. ■

In our next theorem we show that some familiar algebraic properties hold for arbitrary vector spaces. These results are certainly obvious for all of the examples already encountered; the question before us now is whether or not they can be deduced from the axioms of Definition 3.15. We illustrate the strength of the abstract approach by showing that the results hold in general. The proofs are not difficult, but they require careful attention if you have not dealt previously with arguments of this kind.

THEOREM 3.20 If $\mathbf{u}$ is any element of the vector space $\mathcal{V}$ and k is any real number, then the following assertions are true:

(1) $0\mathbf{u} = \mathbf{0}$.
(2) $k\mathbf{0} = \mathbf{0}$.
(3) $(-1)\mathbf{u} = -\mathbf{u}$.
(4) If $k\mathbf{u} = \mathbf{0}$, then either $\mathbf{u} = \mathbf{0}$ or $k = 0$.

Proof We will prove the first assertion carefully and outline the proofs of the other parts. For parts 2, 3, and 4 you should supply a reason for each step in the argument.

(1) Since $0 + 0 = 0$, we can write $0\mathbf{u} = (0 + 0)\mathbf{u}$ and then use Axiom 7 to write

$$0\mathbf{u} = 0\mathbf{u} + 0\mathbf{u}.$$

Now, by Axiom 5, there is a vector $(-0\mathbf{u})$ in $\mathcal{V}$ such that $0\mathbf{u} + (-0\mathbf{u}) = \mathbf{0}$. Adding this vector to both sides of the last equation yields

$$\mathbf{0} = 0\mathbf{u} + (-0\mathbf{u}) = (0\mathbf{u} + 0\mathbf{u}) + (-0\mathbf{u}),$$

which, by the associativity of addition in $\mathcal{V}$, can be rewritten as

$$\mathbf{0} = 0\mathbf{u} + (0\mathbf{u} + (-0\mathbf{u})) = 0\mathbf{u} + \mathbf{0} = 0\mathbf{u}.$$

Thus, $0\mathbf{u} = \mathbf{0}$ as desired.

(2)

$$\begin{aligned} k\mathbf{0} &= k(\mathbf{0} + \mathbf{0}) = k\mathbf{0} + k\mathbf{0} \\ \mathbf{0} &= k\mathbf{0} + (-k\mathbf{0}) = (k\mathbf{0} + k\mathbf{0}) + (-k\mathbf{0}) \\ &= k\mathbf{0} + (k\mathbf{0} + (-k\mathbf{0})) = k\mathbf{0} + \mathbf{0} = k\mathbf{0}. \end{aligned}$$

(3) $\mathbf{u} + (-1)\mathbf{u} = 1\mathbf{u} + (-1)\mathbf{u} = (1 - 1)\mathbf{u} = 0\mathbf{u} = \mathbf{0}$. Therefore, $(-1)\mathbf{u} = -\mathbf{u}$.

(4) If $k\mathbf{u} = \mathbf{0}$ and $k \neq 0$, then

$$\mathbf{0} = \left(\frac{1}{k}\right)\mathbf{0} = \left(\frac{1}{k}\right)k\mathbf{u} = \left(\left(\frac{1}{k}\right)k\right)\mathbf{u} = 1\mathbf{u} = \mathbf{u}. \quad \square$$

EXAMPLE 7 **Algebraic Manipulation with Vectors**

Let **u**, **v**, and **w** be elements of the real vector space $\mathcal{V}$. Use the axioms of Definition 3.15 and the results of Theorem 3.20 to simplify

$$3\mathbf{u} + 2\mathbf{v} - 7\mathbf{w} = 5\mathbf{v} + 3\mathbf{u} - 8\mathbf{w}.$$

Adding $-(5\mathbf{v} + 3\mathbf{u} - 8\mathbf{w}) = -5\mathbf{v} - 3\mathbf{u} + 8\mathbf{w}$ to both sides of the given equation yields

$$3\mathbf{u} + 2\mathbf{v} - 7\mathbf{w} - 5\mathbf{v} - 3\mathbf{u} + 8\mathbf{w} = \mathbf{0}$$

or, using the commutativity and distributivity properties,

$$(3 - 3)\mathbf{u} + (2 - 5)\mathbf{v} + (-7 + 8)\mathbf{w} = \mathbf{0},$$

which is the same as

$$0\mathbf{u} + (-3)\mathbf{v} + (1)\mathbf{w} = \mathbf{0}.$$

Thus, by Theorem 3.20 and Axiom 10 we have

$$-3\mathbf{v} + \mathbf{w} = \mathbf{0} \quad \text{or} \quad \mathbf{w} = 3\mathbf{v}.$$

EXERCISES 3.7

In Exercises 1–10 a set of objects $\mathcal{V}$, an addition, and a scalar multiplication are specified. Determine whether $\mathcal{V}$ is a real vector space with the operations given. Note that in some cases familiar symbols have nonstandard meanings.

1. $\mathcal{V}_1 = \{(a, b)\}$, with $(a, b) + (c, d) = (a + c, b + d)$ and $k(a, b) = (ka, 0)$

2. $\mathcal{V}_2 = \{(a, b)\}$, with $(a, b) + (c, d) = (a + c, b + d)$ and $k(a, b) = (0, 0)$

3. $\mathcal{V}_3 = \{(a, b)\}$, with $(a, b) + (c, d) = (a + c, b + d)$ and $k(a, b) = (2ka, 2kb)$

4. $\mathcal{V}_4 = \{n \times n$ diagonal matrices$\}$, with $A + B = AB$, and ordinary scalar multiplication

5. $\mathcal{V}_5 = \{A \text{ in } \mathfrak{R}^{n\times n} \mid A^T = A\}$, with the usual matrix addition and scalar multiplication

6. $\mathcal{V}_6 = \{(a, b)\}$, with $(a, b) + (c, d) = (a + c + 1, b + d - 1)$ and $k(a, b) = (ka, kb)$

7. $\mathcal{V}_7 = \{a \mid a > 0\}$, with $a + b = ab$ and $ka = a^k$

8. $\mathcal{V}_8 = \{(a, b, c) \mid a \geq 0, b \geq 0\}$, with usual matrix operations

9. $\mathcal{V}_9 = \{p(x) \text{ in } \mathcal{P} \mid \text{degree of } p(x) \text{ is greater than } 5\}$

10. $\mathcal{V}_{10} = \{\text{subsets of } \mathfrak{R}^3\}$ with $S_1 + S_2 = S_1 \cup S_2$ and $kS = S$

In Exercises 11–18 a subset $\mathcal{W}$ of a vector space $\mathcal{V}$ is given. Determine whether $\mathcal{W}$ is closed under addition and scalar multiplication.

11. $\mathcal{W}_1 = \{a_1\mathbf{u}_1 + a_2\mathbf{u}_2 \mid \mathbf{u}_1 \text{ and } \mathbf{u}_2 \text{ fixed in } \mathcal{V};\ a_1 \text{ and } a_2 \text{ real}\}$

12. $\mathcal{W}_2 = \{\mathbf{f} \text{ in } \mathcal{F}[a, b] \mid f(a) = 0\}$

13. $\mathcal{W}_3 = \left\{ \begin{bmatrix} x_1 \\ x_2 \end{bmatrix} \text{ in } \Re^2 \,\middle|\, x_1 \geq 0 \right\}$

14. $\mathcal{W}_4 = \{p(x) \text{ in } \mathcal{P} \mid \text{degree of } p(x) \text{ is even}\}$

15. $\mathcal{W}_5 = \left\{ \begin{bmatrix} x_1 \\ x_2 \\ x_3 \end{bmatrix} \text{ in } \Re^3 \,\middle|\, x_1^2 + x_2^2 + x_3^2 = 1 \right\}$

16. $\mathcal{W}_6 = \{A \text{ in } \Re^{n \times n} \mid AB = BA\}$ (B fixed in $\Re^{n \times n}$)

17. $\mathcal{W}_7 = \{\mathbf{f} \text{ in } \mathcal{F} \mid \mathbf{f} \text{ has a second derivative}\}$

18. $\mathcal{W}_8 = \{\text{eigenvectors of } A\}$ where A is a fixed $n \times n$ matrix

19. Verify that vector addition in $\mathcal{F} = \mathcal{F}[a, b]$ is commutative and that scalar multiplication distributes over vector addition.

20. Let $\mathcal{S}$ be any set. Show that the set of all real valued functions with domain $\mathcal{S}$ is a real vector space.

21. Show how to define vector addition and scalar multiplication so that any set with a single element becomes a real vector space.

22. Let **u** and **v** be elements of an arbitrary real vector space. Prove that if $4\mathbf{u} + 3\mathbf{v} = 3\mathbf{v} + 2\mathbf{u}$, then $\mathbf{u} = \mathbf{0}$.

23. Let **u**, **v**, and **w** be elements of a real vector space $\mathcal{V}$. Show that if $\mathbf{u} + \mathbf{v} = \mathbf{w} + \mathbf{u}$, then $\mathbf{v} = \mathbf{w}$.

24. Show that Definition 3.15 may be weakened by deleting the word "unique" in Axioms 4 and 5.

25. Prove that $-(-\mathbf{u}) = \mathbf{u}$ for any vector **u**.

†26. Let $\mathcal{V}$ be any real vector space and let $\mathcal{V}^{m \times n}$ be the set of all $m \times n$ matrices with entries from $\mathcal{V}$. Show that $\mathcal{V}^{m \times n}$ is a real vector space.

27. Is $\left\{ \begin{bmatrix} a & a + 2b \\ a - 3b & b \end{bmatrix} \,\middle|\, a, b \text{ real} \right\}$ a vector space with the usual matrix operations?

28. Is $\{[x \quad y \quad z] \mid xyz > 0\}$ a vector space with the ordinary operations?

3.8 SUBSPACES

Many important examples of vector spaces are obtained by considering subspaces of the vector spaces encountered in the last section. We have already considered the notion of a subspace of $\Re^n$ in Sections 3.2 and 3.3; the general definition follows. Note that this definition is not just a rewrite of the one given in Section 3.2. The equivalence of the two definitions is established in Theorem 3.21.

DEFINITION 3.16 A subset $\mathcal{W}$ of the vector space $(\mathcal{V}, +, \cdot)$ is called a **subspace** of $\mathcal{V}$ if $(\mathcal{W}, +, \cdot)$ is itself a vector space, with the addition and scalar multiplication inherited from $\mathcal{V}$. □

To check whether a given subset $\mathcal{W}$ of the vector space $\mathcal{V}$ is a subspace, it is not necessary to check all ten axioms. Since the operations in $\mathcal{W}$ are inherited from $\mathcal{V}$, all the manipulative properties hold in the larger set $\mathcal{V}$ and hence also in the smaller set $\mathcal{W}$. It is therefore only necessary to check the closure properties (1 and 6) and the existence properties (4 and 5). Thus, $\mathcal{W}$ is a subspace of $\mathcal{V}$ if:

(1) for every $\mathbf{u}$ and $\mathbf{v}$ in $\mathcal{W}$, $\mathbf{u} + \mathbf{v}$ is in $\mathcal{W}$;
(2) for every $\mathbf{u}$ in $\mathcal{W}$ and every a in $\mathfrak{R}$, $a\mathbf{u}$ is in $\mathcal{W}$;
(3) for any $\mathbf{u}$ in $\mathcal{W}$, $-\mathbf{u}$ is in $\mathcal{W}$; and
(4) the zero vector $\mathbf{0}$ is in $\mathcal{W}$.

If any one of these four properties fails to hold, then the subset $\mathcal{W}$ is not a subspace of $\mathcal{V}$. Note that property 4 implies that a subspace $\mathcal{W}$ must be nonempty. Many interesting subsets of the vector space $\mathcal{V}$ fail to be subspaces because property 4 fails; it is always worth checking this property first. For example, this strategy shows that a plane in $\mathfrak{R}^3$ is not a subspace unless it contains the origin, and that the set of all solutions of the nonhomogeneous system $AX = K$, $K \neq 0$ is not a subspace. Since $-\mathbf{u} = (-1)\mathbf{u}$ and $\mathbf{0} = 0\mathbf{u}$, we see that the existence properties (3 and 4) follow from the closure properties, provided that $\mathcal{W}$ is nonempty. If we combine the two closure properties into a single condition, then we have the following theorem which provides an efficient criterion for determining if a given subset of a vector space is a subspace. Note that this theorem also establishes the equivalence of the definition of a subspace given here and the one given in Section 3.2 for the special case $\mathcal{V} = \mathfrak{R}^n$. Definition 3.16 would not have made sense in Section 3.2 because we had not yet introduced the general notion of a vector space.

THEOREM 3.21 Let $\mathcal{V}$ be a real vector space. The subspaces of $\mathcal{V}$ are precisely those nonempty subsets of $\mathcal{V}$ that are closed under arbitrary linear combinations; that is, a nonempty subset $\mathcal{W}$ is a subspace of $\mathcal{V}$ if, and only if, $a\mathbf{u} + b\mathbf{v}$ is in $\mathcal{W}$ for all vectors $\mathbf{u}$ and $\mathbf{v}$ in $\mathcal{W}$ and all scalars a and b in $\mathfrak{R}$. □

EXAMPLE 1 The Set of all Singular Matrices

Is $\mathcal{W} = \{A \text{ in } \mathfrak{R}^{3\times 3} \mid \det(A) = 0\}$ a subspace of $\mathfrak{R}^{3\times 3}$?

In general, $\det(A + B) \neq \det(A) + \det(B)$, so we do not expect this set to be closed under addition. In fact, the matrices

$$A = \begin{bmatrix} 1 & 0 & 0 \\ 0 & 1 & 0 \\ 0 & 0 & 0 \end{bmatrix} \quad \text{and} \quad B = \begin{bmatrix} 0 & 0 & 0 \\ 0 & 1 & 0 \\ 0 & 0 & 1 \end{bmatrix} \text{ are in } \mathcal{W}$$

since $\det(A) = \det(B) = 0$, but $\det(A + B) = \det\begin{bmatrix} 1 & 0 & 0 \\ 0 & 2 & 0 \\ 0 & 0 & 1 \end{bmatrix} = 2 \neq 0.$

This shows that $A + B$ is not in $\mathcal{W}$ and hence that $\mathcal{W}$ is not a subspace of the vector space $\Re^{3\times3}$. This example generalizes easily to show that the set of all $n \times n$ singular matrices is not a subspace of the vector space $\Re^{n\times n}$. ■

The next example requires a little knowledge of calculus. If you have never studied calculus, please go on to Example 3.

■ EXAMPLE 2 **Differential Equations often Lead to Subspaces**

Is $\mathcal{W} = \{\mathbf{f} \text{ in } \mathcal{F}(a, b) \mid f''(x) + 25f(x) = 0\}$ a subspace of the function space $\mathcal{F}(a, b)$?

Clearly $\mathbf{0}$ is in $\mathcal{W}$ so $\mathcal{W}$ is not empty. If $\mathbf{f}$ and $\mathbf{g}$ are elements of $\mathcal{W}$ then $f''(x) + 25f(x) = 0$ and $g''(x) + 25g(x) = 0$. We now let h and k be any scalars and try to determine if $h\mathbf{f} + k\mathbf{g}$ is in $\mathcal{W}$. To do this we use the familiar linearity properties of the derivative to compute

$$\begin{aligned}(hf(x) + kg(x))'' &+ 25(hf(x) + kg(x)) \\ &= (hf''(x) + kg''(x)) + 25hf(x) + 25kg(x) \\ &= h(f''(x) + 25f(x)) + k(g''(x) + 25g(x)) \\ &= h0 + k0 = 0.\end{aligned}$$

Thus $h\mathbf{f} + k\mathbf{g}$ is in $\mathcal{W}$ and, from Theorem 3.21, we conclude that $\mathcal{W}$ is a subspace of the function space $\mathcal{F}$. It is easy to verify that $\cos 5x$ and $\sin 5x$ are elements of $\mathcal{W}$; in fact, $\mathcal{W}$ can be shown to consist of all functions $\mathbf{h}$ such that $h(x) = a \cos 5x + b \sin 5x$. ■

This last example, along with Theorem 3.3, points out one reason for the interest in subspaces: frequently the set of all solutions of an equation or system of equations can be shown to be a subspace of some vector space. It is precisely the linear homogeneous equations that have this property. This fact motivates the general definition of linearity to be given in the next chapter.

■ EXAMPLE 3 **Polynomials of Degree 5 or Less**

Is $\mathcal{P}_5 = \{p(x) = a_0 + a_1x + \cdots + a_nx^n \mid n \leq 5\}$ a subspace of $\mathcal{P}$?

The elements of $\mathcal{P}_5$ are the polynomials of degree 5 or less. Since multiplication by a scalar cannot increase the degree of a polynomial and since the sum of two polynomials of degree 5 or less cannot have degree greater than 5, it follows from Theorem 3.21 that $\mathcal{P}_5$ is a subspace of $\mathcal{P}$. ■

The notion of the subspace spanned by a given set of vectors, which we have already considered for the special case $\Re^n$ (see Definition 3.8), generalizes naturally to an arbitrary vector space $\mathcal{V}$.

DEFINITION 3.17 Let $\mathscr{V}$ be an arbitrary real vector space and let $\mathscr{S} = \{\mathbf{u_1}, \mathbf{u_2}, \ldots, \mathbf{u_s}\}$ be an arbitrary set of vectors from $\mathscr{V}$. The subspace spanned by the set $\mathscr{S}$ is defined by

$$\text{SPAN}\{\mathbf{u_1}, \mathbf{u_2}, \ldots, \mathbf{u_s}\} = \left\{\sum_{k=1}^{s} a_k\mathbf{u_k} \,\middle|\, a_k \text{ is real}\right\};$$

that is, the subspace consists of all possible linear combinations of the vectors of the set $\mathscr{S}$. A set $\mathscr{S}$ of vectors is said to span the subspace $\mathscr{W}$ if every element of $\mathscr{W}$ can be written as a linear combination of the vectors of the set $\mathscr{S}$. □

The calculation

$$r\left(\sum_{k=1}^{s} a_k\mathbf{u_k}\right) + h\left(\sum_{k=1}^{s} b_k\mathbf{u_k}\right) = \sum_{k=1}^{s} (ra_k + hb_k)\mathbf{u_k}$$

shows that a linear combination of two linear combinations of the $\mathbf{u_i}$ is another linear combination of the $\mathbf{u_i}$. It then follows, from Theorem 3.21, that SPAN$\{\mathbf{u_1}, \ldots, \mathbf{u_s}\}$ is indeed a subspace of $\mathscr{V}$.

It is useful to think of SPAN$\{\mathbf{u_1}, \ldots, \mathbf{u_s}\}$ as the "smallest" subspace of $\mathscr{V}$ containing the spanning set; that is, if $\mathscr{W}$ is any other subspace of $\mathscr{V}$ containing all the $\mathbf{u_i}$, then SPAN$\{\mathbf{u_1}, \ldots, \mathbf{u_s}\}$ is a subspace of $\mathscr{W}$ (see Exercise 11 of this section). Recall that for $\mathscr{V} = \Re^3$, SPAN$\{\mathbf{u}, \mathbf{v}\}$ can be thought of as the plane through the origin determined by the vectors $\mathbf{u}$ and $\mathbf{v}$.

EXAMPLE 4 **Spanning Set for Polynomials of Degree 5 or Less**

For the subspace $\mathscr{P}_5$ of $\mathscr{P}$, discussed in Example 3, it should be apparent that the set of polynomials (vectors in $\mathscr{P}$, if you please)

$$\mathscr{S} = \{1, x, x^2, x^3, x^4, x^5\}$$

spans $\mathscr{P}_5$. To see this we observe that a typical element of $\mathscr{P}_5$ is

$$a_0 + a_1x + a_2x^2 + a_3x^3 + a_4x^4 + a_5x^5,$$

which is clearly a linear combination of the elements of the set $\mathscr{S}$. ■

It is generally helpful to be able to find a spanning set for a given subspace; this is particularly true if the subspace consists of all solutions of a given problem. The subspace of Example 2 can be shown to be SPAN$\{\cos 5x, \sin 5x\}$ and this description of that subspace describes it clearly, even for those of you who have not studied calculus. *Finding a spanning set, for a subspace consisting of all solutions of a given problem, amounts to solving the problem. Be careful not to overgeneralize; not all problems have solution sets that are subspaces.*

EXAMPLE 5 A Spanning Set for the Null Space of A

Find a spanning set for $\mathcal{NS}(A) = \{X \mid AX = 0\}$ if

$$A = \begin{bmatrix} 1 & 0 & 2 & -4 \\ 0 & 1 & 5 & 3 \\ 0 & 0 & 0 & 0 \\ 0 & 0 & 0 & 0 \end{bmatrix}.$$

Finding the elements of the null space of A amounts to finding all solutions of the homogeneous linear system $AX = 0$. Since A is already in echelon form, the solution of the system $AX = 0$ is easy. The nonzero equations of the system are

$$\begin{aligned} x_1 + 2x_3 - 4x_4 &= 0 \\ x_2 + 5x_3 + 3x_4 &= 0 \end{aligned}$$

and, if we set $x_3 = c_1$ and $x_4 = c_2$, we obtain as the general solution,

$$X = \begin{bmatrix} -2c_1 + 4c_2 \\ -5c_1 + 3c_2 \\ c_1 \\ c_2 \end{bmatrix} = c_1 \begin{bmatrix} -2 \\ -5 \\ 1 \\ 0 \end{bmatrix} + c_2 \begin{bmatrix} 4 \\ -3 \\ 0 \\ 1 \end{bmatrix}.$$

This expression for the solution shows that $\mathcal{NS}(A) = \text{SPAN}\{X_1, X_2\}$

$$\text{where } X_1 = \begin{bmatrix} -2 \\ -5 \\ 1 \\ 0 \end{bmatrix} \quad \text{and} \quad X_2 = \begin{bmatrix} 4 \\ -3 \\ 0 \\ 1 \end{bmatrix}. \quad \blacksquare$$

EXAMPLE 6 Trigonometric Functions

Which of the functions $\sin^2 x$, $\sin x$, and $\cos^2 x$ are in the subspace of $\mathcal{F}$ spanned by $\{1, \cos 2x, \sin 2x\}$?

From the familiar trigonometric identities

$$\sin^2 x = \frac{(1 - \cos 2x)}{2} \quad \text{and} \quad \cos^2 x = \frac{(1 + \cos 2x)}{2},$$

it follows at once that both $\sin^2 x$ and $\cos^2 x$ are in the subspace SPAN $\{1, \cos 2x, \sin 2x\}$. In order to determine whether $\sin x$ is in this subspace we will try to find scalars a, b, and c so that

$$\sin x = a + b \cos 2x + c \sin 2x. \tag{3.13}$$

Equation (3.13) is a functional identity and must hold for all values of x. For each value of x the functional equation reduces to a linear equation; in particular for

$$x = 0, \qquad 0 = a + b;$$

$$x = \frac{\pi}{2}, \qquad 1 = a - b; \quad \text{and}$$

$$x = \frac{\pi}{6}, \qquad \frac{1}{2} = a + \frac{1}{2}b + \frac{\sqrt{3}}{2}c.$$

These three equations are easily solved to yield $a = \frac{1}{2}$, $b = -\frac{1}{2}$, and $c = \frac{1}{2\sqrt{3}}$. Thus, the only possible solution of Equation (3.13) is

$$\sin x = \frac{1}{2} - \frac{\cos 2x}{2} + \frac{\sin 2x}{2\sqrt{3}}.$$

Now we need to ask if this last equation is valid for all values of x as required. For $x = \frac{\pi}{4}$ this equation leads to

$$.707\cdots = \frac{\sqrt{2}}{2} \stackrel{?}{=} \frac{1}{2} - 0 + \frac{1}{2\sqrt{3}} = .788\cdots,$$

which is clearly not valid. Thus it is not possible to find scalars a, b, and c so that Equation (3.13) is satisfied for all x. Therefore, $\sin x$ is *not* an element of the subspace SPAN$\{1, \cos 2x, \sin 2x\}$. ■

EXAMPLE 7 Another Spanning Set for a Polynomial Space

Show that the polynomials

$$h_1(x) = x^2 - 3x + 2, \quad h_2(x) = x^2 - 2x, \quad \text{and} \quad h_3(x) = x^2 - x$$

span the subspace

$$\mathcal{P}_2 = \{a_0 + a_1x + a_2x^2 \mid a_i \text{ real}\}.$$

We need to demonstrate that any element of $\mathcal{P}_2$ can be expressed as a linear combination of $h_1(x)$, $h_2(x)$, and $h_3(x)$. Thus for $a_0 + a_1x + a_2x^2$, an arbitrary element of $\mathcal{P}_2$, we seek scalars c_1, c_2, and c_3 such that

$$c_1h_1(x) + c_2h_2(x) + c_3h_3(x) = a_0 + a_1x + a_2x^2.$$

This equation is equivalent to

$$c_1(x^2 - 3x + 2) + c_2(x^2 - 2x) + c_3(x^2 - x) = a_0 + a_1x + a_2x^2$$

or, if we collect the coefficients of like power of x,

$$(c_1 + c_2 + c_3)x^2 + (-3c_1 - 2c_2 - c_3)x + 2c_1 = a_0 + a_1x + a_2x^2.$$

From the last equation we see that c_1, c_2, and c_3 must satisfy the system

$$\begin{aligned} 2c_1 \qquad\qquad &= a_0 \\ -3c_1 - 2c_2 - c_3 &= a_1 \\ c_1 + c_2 + c_3 &= a_2 \end{aligned} \quad \text{or} \quad \begin{bmatrix} 2 & 0 & 0 \\ -3 & -2 & -1 \\ 1 & 1 & 1 \end{bmatrix} \begin{bmatrix} c_1 \\ c_2 \\ c_3 \end{bmatrix} = \begin{bmatrix} a_0 \\ a_1 \\ a_2 \end{bmatrix}.$$

Since the coefficient matrix of this system is nonsingular (you can check that its determinant is $-2 \neq 0$) it follows that this system has a *unique* solution for every choice of a_0, a_1, and a_2. Thus

$$\mathcal{P}_2 = \text{SPAN}\{h_1(x), h_2(x), h_3(x)\}. \quad \blacksquare$$

In Sections 3.2 and 3.3 we introduced two subspaces of $\mathfrak{R}^n$: $\mathcal{NS}(A)$ and $\mathcal{CS}(A)$, which are related to a given matrix A. A third subspace associated with the matrix A is the row space of A.

DEFINITION 3.18 If A is an $m \times n$ matrix, then the **row space** of A is the subspace of $\mathfrak{R}^{1\times n}$ spanned by the rows of A; that is,

$$\mathcal{RS}(A) = \text{SPAN}\{\text{Row}_1(A), \text{Row}_2(A), \ldots, \text{Row}_m(A)\}. \quad \square$$

EXAMPLE 8 Testing for Membership in $\mathcal{RS}(A)$

Which of the row vectors $\mathbf{r}_1 = [1 \ \ 1 \ \ 1 \ \ 1]$, $\mathbf{r}_2 = [1 \ \ 2 \ \ 1 \ \ 1]$, and $\mathbf{r}_3 = [2 \ \ 4 \ \ 2 \ \ -2]$ are in the row space of

$$A = \begin{bmatrix} 1 & 2 & 3 & -3 \\ 2 & 4 & 3 & 0 \\ 4 & 8 & 2 & 8 \end{bmatrix}?$$

To determine if $\mathbf{r}_1$ is in $\mathcal{RS}(A)$ we try to find scalars x, y, and z such that

$$\mathbf{r}_1 = x\text{Row}_1(A) + y\text{Row}_2(A) + z\text{Row}_3(A).$$

This statement is equivalent to

$$\begin{aligned} \mathbf{r}_1 = [1 \ \ 1 \ \ 1 \ \ 1] &= x[1 \ \ 2 \ \ 3 \ \ -3] + y[2 \ \ 4 \ \ 3 \ \ 0] + z[4 \ \ 8 \ \ 2 \ \ 8] \\ &= [x + 2y + 4z \quad 2x + 4y + 8z \quad 3x + 3y + 2z \quad -3x + 8z]. \end{aligned}$$

Thus x, y, and z must satisfy the 4×3 linear system

$$\begin{aligned} x + 2y + 4z &= 1 \\ 2x + 4y + 8z &= 1 \\ 3x + 3y + 2x &= 1 \\ -3x \qquad + 8z &= 1 \end{aligned} \quad \text{or} \quad \begin{bmatrix} 1 & 2 & 4 \\ 2 & 4 & 8 \\ 3 & 3 & 2 \\ -3 & 0 & 8 \end{bmatrix} \begin{bmatrix} x \\ y \\ z \end{bmatrix} = \begin{bmatrix} 1 \\ 1 \\ 1 \\ 1 \end{bmatrix}$$

which, we observe, can be written as $A^T X = \mathbf{r}_1^T$. Similarly, to determine if $\mathbf{r}_2$ and $\mathbf{r}_3$ are in $\mathcal{RS}(A)$, we must try to solve the linear systems

$$A^T X = \mathbf{r}_i^T; \quad i = 2, 3.$$

We can solve these three systems with the single row reduction

$$\left[\begin{array}{rrr|r|r|r} 1 & 2 & 4 & 1 & 1 & 2 \\ 2 & 4 & 8 & 1 & 2 & 4 \\ 3 & 3 & 2 & 1 & 1 & 2 \\ -3 & 0 & 8 & 1 & 1 & -2 \end{array}\right] \rightarrow \left[\begin{array}{rrr|r|r|r} 1 & 2 & 4 & 1 & 1 & 2 \\ 0 & 0 & 0 & -1 & 0 & 0 \\ 0 & -3 & -10 & -2 & -2 & -4 \\ 0 & 6 & 20 & 4 & 4 & 4 \end{array}\right]$$

$$\rightarrow \left[\begin{array}{rrr|r|r|r} 1 & 2 & 4 & 1 & 1 & -2 \\ 0 & 3 & 10 & 2 & 2 & 4 \\ 0 & 0 & 0 & \boxed{-1} & 0 & 0 \\ 0 & 0 & 0 & 0 & 0 & \boxed{-4} \end{array}\right] \rightarrow \left[\begin{array}{rrr|r|r|r} 1 & 0 & -\frac{8}{3} & 0 & -\frac{1}{3} & 0 \\ 0 & 1 & \frac{10}{3} & 0 & \frac{2}{3} & 0 \\ 0 & 0 & 0 & \boxed{1} & 0 & 0 \\ 0 & 0 & 0 & 0 & 0 & \boxed{1} \end{array}\right].$$

From this reduction we see that only the second system is consistent. Thus $\mathbf{r}_2$ is an element of $\mathcal{RS}(A)$ while $\mathbf{r}_1$ and $\mathbf{r}_3$ are not in $\mathcal{RS}(A)$. There are many solutions of the second system; among them are $z = 2$, $y = -6$, and $x = 5$. Thus,

$$\mathbf{r}_2 = [1 \quad 2 \quad 1 \quad 1] = 5\text{Row}_1(A) - 6\text{Row}_2(A) + 2\text{Row}_3(A),$$

as you may verify directly. ■

The preceding examples have illustrated the following observations:

1. Finding spanning sets usually requires solving a linear system.
2. Determining if a given vector is in a subspace, which is described by a spanning set, usually requires investigating whether a certain linear system has a solution.

We conclude this section with a theorem about row equivalence that will be useful in later sections.

THEOREM 3.22 If A and B are row equivalent matrices, then $\mathcal{RS}(A) = \mathcal{RS}(B)$.

Proof If $A \underset{R}{\sim} B$, then from Theorem 1.8 we know that there exists a nonsingular matrix P such that $PA = B$. Since the rows of B are linear combinations of the rows of A (see Equation (1.10)), it follows at once that each row of B is in $\mathcal{RS}(A)$, and hence that $\mathcal{RS}(B) \subseteq \mathcal{RS}(A)$. Since P is nonsingular we can write $A = P^{-1}B$ and observe that the rows of A are linear combinations of the rows of B. Thus we also have $\mathcal{RS}(A) \subseteq \mathcal{RS}(B)$; it follows that $\mathcal{RS}(A) = \mathcal{RS}(B)$. □

If a subspace $\mathcal{W}$ of $\mathfrak{R}^{1\times n}$ can be described as the row space of some matrix A, then Theorem 3.22 and the row reduction techniques of Chapter 1 can be used to find the simplest matrix R such that $\mathcal{W} = \mathcal{RS}(A) = \mathcal{RS}(R)$. The nonzero rows of the echelon matrix form a minimal spanning set for $\mathcal{W} = \mathcal{RS}(A) = \mathcal{RS}(R)$.

EXAMPLE 9 **An Alternate Description of $\mathcal{RS}(A)$**

Use Theorem 3.22 to obtain an alternate description of the row space of the matrix A of Example 8.

From the row reduction

$$A = \begin{bmatrix} 1 & 2 & 3 & -3 \\ 2 & 4 & 3 & 0 \\ 4 & 8 & 2 & 8 \end{bmatrix} \to \begin{bmatrix} 1 & 2 & 3 & -3 \\ 0 & 0 & -3 & 6 \\ 0 & 0 & -10 & 20 \end{bmatrix}$$
$$\to \begin{bmatrix} 1 & 2 & 0 & 3 \\ 0 & 0 & 1 & -2 \\ 0 & 0 & 0 & 0 \end{bmatrix} = R$$

it follows, using Theorem 3.22, that

$$\begin{aligned} \mathcal{RS}(A) &= \mathcal{RS}(R) = \text{SPAN}\{[1 \quad 2 \quad 0 \quad 3], [0 \quad 0 \quad 1 \quad -2]\} \\ &= \{[a \quad 2a \quad b \quad 3a - 2b] \mid a, b \text{ in } \Re\} \\ &= \{[x_1 \quad x_2 \quad x_3 \quad x_4] \mid x_2 = 2x_1 \text{ and } x_4 = 3x_1 - 2x_2\}. \end{aligned}$$

From the last characterization of $\mathcal{RS}(A)$ it is easy to see that $\mathbf{r}_1$ and $\mathbf{r}_3$ are not in $\mathcal{RS}(A)$, while $\mathbf{r}_2$ is an element of $\mathcal{RS}(A)$. ■

Theorem 3.22 says that elementary row operations leave $\mathcal{RS}(A)$ invariant. A similar argument would show that elementary column operations leave $\mathcal{CS}(A)$ invariant. Note however that the matrices

$$\begin{bmatrix} 1 & 2 \\ 2 & 4 \end{bmatrix} \quad \text{and} \quad \begin{bmatrix} 1 & 2 \\ 0 & 0 \end{bmatrix}$$

are row equivalent but have different column spaces. Thus we see that row operations change the column space. Similarly, column operations would change the row space.

EXERCISES 3.8

1. Which of the following subsets of $\Re^{1\times 3}$ are subspaces? Interpret each of these subsets geometrically.
   (a) $\{[x_1 \quad x_2 \quad x_3] \mid 3x_1 - 5x_2 + 6x_3 = 0\}$
   (b) $\{[x_1 \quad x_2 \quad x_3] \mid x_1^2 + x_2^2 + x_3^2 = 1\}$
   (c) $\{[x_1 \quad x_2 \quad 3x_1 + 5x_2] \mid x_1, x_2 \text{ real}\}$
   (d) $\{[x_1 \quad x_2 \quad x_3] \mid x_1 + x_2 + x_3 = 5\}$
   (e) $\{[x_1 \quad x_2 \quad x_3] \mid x_1 + x_2 > x_3\}$
   (f) $\{[x_1 \quad x_2 \quad x_3] \mid x_1^2 + x_2^2 + x_3^2 = 0\}$

2. Which of the following subsets of $\mathfrak{R}^{3\times 3}$ are subspaces?

(a) $\{A \mid A^T = A\}$

(b) $\{A \mid \text{tr}(A) = 0\}$ (Recall Exercise 11 in Section 2.1.)

(c) $\{A \mid \det(A) = 1\}$

(d) $\{A \mid AB = BA\}$ (B is a fixed 3×3 matrix.)

(e) $\{A \mid a_{31} + a_{22} + a_{13} = 0\}$

(f) $\{A \mid a_{21} + a_{22} + a_{23} = 5\}$

3. Which of the following subsets of $\mathcal{P}$ are subspaces?

(a) $\{p(x) \mid \text{constant term in } 0\}$

(b) $\{p(x) \mid \text{all coefficients are positive}\}$

(c) $\{p(x) \mid \text{the sum of the coefficients is zero}\}$

(d) $\{p(x) \mid p(x) \text{ is a monic polynomial}\}$ (A polynomial is **monic** if the coefficient of the highest power of x is $+1$.)

4. Which of the following subsets of $\mathcal{F}$ are subspaces?

(a) $\{\mathbf{f} \mid f(2) = 0\}$

(b) $\{\mathbf{f} \mid f(x) \text{ is a polynomial in } x\}$

(c) $\{\mathbf{f} \mid \mathbf{f} \text{ is a nondecreasing function}\}$

(d) $\{\mathbf{f} \mid \mathbf{f} \text{ is continuous}\} = \mathcal{C}^0[a, b]$

(e) $\{\mathbf{f} \mid \int_a^b f(x)\,dx = 0\}$

(f) $\{\mathbf{f} \mid f''(x) - 7f'(x) = \sin 3x\}$

5. Determine which of the following vectors from $\mathfrak{R}^{1\times 4}$ are elements of the subspace $\mathcal{W}$ spanned by the set

$$\{\mathbf{u}_1 = [1 \quad 1 \quad 1 \quad 1], \mathbf{u}_2 = [3 \quad -2 \quad 0 \quad 0], \mathbf{u}_3 = [2 \quad -5 \quad -4 \quad 11]\}.$$

You should be able to do all six parts with a single row reduction.

(a) $\mathbf{a} = [2 \quad 3 \quad 0 \quad 13]$
(b) $\mathbf{b} = [2 \quad 0 \quad 0 \quad 3]$

(c) $\mathbf{c} = [0 \quad 0 \quad 0 \quad 4]$
(d) $\mathbf{d} = [4 \quad -1 \quad 1 \quad 1]$

(e) $\mathbf{e} = [1 \quad 1 \quad 1 \quad -1]$
(f) $\mathbf{f} = [0 \quad 5 \quad 4 \quad -1]$

6. **(a)** Show that the function $\mathbf{f}$, for which $f(x) = x^2 + 2$, is not in SPAN$\{\cos x, \sin x\}$.

(b) Show that $\sin(x + 5)$ is in SPAN$\{\cos x, \sin x\}$.

(c) Is $\sin 2x$ in SPAN$\{\cos x, \sin x\}$?

7. Determine which of the following polynomials are elements of the subspace of $\mathcal{P}$ spanned by the set

$$\{p_1(x) = x^3 + 2x^2 + 1, p_2(x) = x^2 - 2, p_3(x) = x^3 + x\}.$$

(a) $x^2 - x + 3$
(b) $x^2 - 2x + 1$
(c) $4x^3 - 3x + 5$

(d) $x^4 + 1$
(e) $-x^3 + 5x^2 - 2x - 2$
(f) $x - 5$

8. Let A be any $n \times n$ matrix, let P be any $m \times n$ matrix, and let Q be any $n \times m$ matrix.

(a) Show that $\mathcal{NS}(A) \subseteq (PA)$ with equality if P is invertible.

(b) Use Equation (1.10) on page 52 to show that $\mathcal{RS}(PA) \subseteq \mathcal{RS}(A)$ with equality if P is invertible.

(c) Use Equation (1.9) on page 52 to show that $\mathcal{CS}(AQ) \subseteq \mathcal{CS}(A)$ with equality if Q is invertible.

9. Let $\mathcal{W}_1$ and $\mathcal{W}_2$ be subspaces of the vector space $\mathcal{V}$. Which of the following subsets of $\mathcal{V}$ are subspaces of $\mathcal{V}$?
(a) $\mathcal{W}_1 \cap \mathcal{W}_2 = \{\mathbf{u} \mid \mathbf{u} \text{ in } \mathcal{W}_1 \text{ and } \mathbf{u} \text{ in } \mathcal{W}_2\}$
(b) $\mathcal{W}_1 \cup \mathcal{W}_2 = \{\mathbf{u} \mid \mathbf{u} \text{ in } \mathcal{W}_1 \text{ or } \mathbf{u} \text{ in } \mathcal{W}_2\}$
(c) $\mathcal{W}_1 + \mathcal{W}_2 = \{\mathbf{u}_1 + \mathbf{u}_2 \mid \mathbf{u}_1 \text{ in } \mathcal{W}_1 \text{ and } \mathbf{u}_2 \text{ in } \mathcal{W}_2\}$

10. Find a matrix in reduced row-echelon form whose row space is the same as $\mathcal{RS}(A)$ if A is one of the following matrices. In each case provide several alternate descriptions of $\mathcal{RS}(A)$.

(a) $A = \begin{bmatrix} 1 & 2 & 3 \\ 4 & 5 & 6 \\ 7 & 8 & 9 \end{bmatrix}$ **(b)** $A = \begin{bmatrix} 1 & 1 & 2 & 6 \\ 1 & -1 & -4 & -8 \\ 3 & -2 & 5 & 11 \\ 2 & 5 & -2 & 3 \end{bmatrix}$

(c) $A = \begin{bmatrix} 3 & -1 & 2 & 1 \\ 2 & 1 & 1 & 1 \\ 1 & -3 & 0 & 1 \end{bmatrix}$

11. Find spanning sets for the null spaces of the following matrices.

$A = \begin{bmatrix} 1 & -2 & 0 & 4 \\ 0 & 0 & 1 & -3 \\ 0 & 0 & 0 & 0 \end{bmatrix}$ $B = [2 \ \ 5 \ \ 6 \ \ 7]$ $C = \begin{bmatrix} 1 & 1 & 1 & 1 \\ 2 & 2 & 2 & 3 \\ 3 & 3 & 3 & 4 \end{bmatrix}$

†**12.** Let $\mathcal{W}$ be any subspace of $\mathcal{V}$ containing the vectors $\{\mathbf{u}_1, \ldots \mathbf{u}_k\}$. Show that SPAN$\{\mathbf{u}_1, \ldots, \mathbf{u}_k\}$ is a subspace of $\mathcal{W}$ and hence that it is the smallest subspace of $\mathcal{V}$ containing $\{\mathbf{u}_1, \ldots, \mathbf{u}_k\}$.

13. Define an inner product in $\mathfrak{R}^{1\times n}$ in the obvious way and then find an orthonormal basis for the row space of $A = \begin{bmatrix} 1 & 0 & 1 & 0 \\ 1 & -1 & -1 & 1 \\ 1 & 0 & 0 & 0 \end{bmatrix}$.

3.9 LINEAR INDEPENDENCE, BASIS, AND DIMENSION

How do the notions of linear independence, basis, and dimension, which were introduced earlier for $\mathcal{V} = \mathfrak{R}^n$, extend to arbitrary vector spaces?

The concepts of linear independence, basis, and dimension, which were introduced in Sections 3.4 and 3.5 for the special case $\mathcal{V} = \mathfrak{R}^n$, generalize directly to the case of an arbitrary vector space. The definitions and theorems given earlier were phrased so that they would apply to the general case as well.

DEFINITION 3.19 Let $\mathcal{S} = \{\mathbf{u}_1, \mathbf{u}_2, \ldots, \mathbf{u}_k\}$ be a set of vectors from the vector space $\mathcal{V}$. The set $\mathcal{S}$ is said to be **linearly dependent** if there exist scalars a_i, not all zero, such that

$$a_1\mathbf{u}_1 + a_2\mathbf{u}_2 + \cdots + a_k\mathbf{u}_k = \mathbf{0}. \tag{3.14}$$

The set $\mathcal{S}$ is **linearly independent** if it is not linearly dependent. □

Since $a\mathbf{u} + b\mathbf{v} = \mathbf{0}$ $(a \neq 0)$ implies that $\mathbf{u} = -(\frac{b}{a})\mathbf{v}$, it follows that two vectors are linearly dependent if, and only if, one of them is a scalar multiple of the other; this case is always easy to determine. Recall that if $\mathbf{x}$, $\mathbf{y}$, and $\mathbf{z}$ are three vectors in $\Re^3$, then a nontrivial relation of the form $a\mathbf{x} + b\mathbf{y} + c\mathbf{z} = \mathbf{0}$ means that one of the vectors is in the plane determined by the other two. In particular, if $a \neq 0$ then $\mathbf{x} = -\frac{b}{a}\mathbf{y} - \frac{c}{a}\mathbf{z}$ is in the plane determined by $\mathbf{y}$ and $\mathbf{z}$. Thus we can view linear dependence as being a generalization of the notion of three vectors being coplanar in $\Re^3$, or two vectors being collinear in $\Re^2$ or $\Re^3$.

In order to investigate the linear independence or dependence of a given set of vectors we try to deduce from Equation (3.14) whether the coefficients must all be zero. This process normally leads to a homogeneous system of linear equations, for which we need to determine whether a nonzero solution exists. Recall from our earlier work that the homogeneous system $AX = 0$ has a nonzero solution if any one of the following conditions hold.

(1) A has fewer rows than columns;
(2) A is row equivalent to a matrix with fewer nonzero rows than columns;
(3) A is square and singular; or
(4) A is square and $\det(A) = 0$.

EXAMPLE 1 Test Matrices for Independence

Determine if the following matrices (vectors from $\Re^{2\times 2}$) are linearly independent or dependent.

$$M_1 = \begin{bmatrix} 1 & 2 \\ 3 & 4 \end{bmatrix}, \quad M_2 = \begin{bmatrix} 1 & -1 \\ -1 & 1 \end{bmatrix}, \quad M_3 = \begin{bmatrix} 1 & 3 \\ 0 & 4 \end{bmatrix}, \quad M_4 = \begin{bmatrix} 0 & 2 \\ 2 & 0 \end{bmatrix}, \quad M_5 = \begin{bmatrix} 1 & 1 \\ 2 & 3 \end{bmatrix}$$

The vector equation

$$x_1M_1 + x_2M_2 + x_3M_3 + x_4M_4 + x_5M_5 = 0$$

is equivalent to

$$x_1\begin{bmatrix} 1 & 2 \\ 3 & 4 \end{bmatrix} + x_2\begin{bmatrix} 1 & -1 \\ -1 & 1 \end{bmatrix} + x_3\begin{bmatrix} 1 & 3 \\ 0 & 4 \end{bmatrix} + x_4\begin{bmatrix} 0 & 2 \\ 2 & 0 \end{bmatrix} + x_5\begin{bmatrix} 1 & 1 \\ 2 & 3 \end{bmatrix} = \begin{bmatrix} 0 & 0 \\ 0 & 0 \end{bmatrix}$$

or

$$\begin{bmatrix} x_1 + x_2 + x_3 + x_5 & 2x_1 - x_2 + 3x_3 + 2x_4 + x_5 \\ 3x_1 - x_2 + 2x_4 + 2x_5 & 4x_1 + x_2 + 4x_3 + 3x_5 \end{bmatrix} = \begin{bmatrix} 0 & 0 \\ 0 & 0 \end{bmatrix}.$$

This last matrix equation leads to a system of four homogeneous equations in five unknowns:

$$\begin{aligned} x_1 + x_2 + x_3 + x_5 &= 0 \\ 2x_1 - x_2 + 3x_3 + 2x_4 + x_5 &= 0 \\ 3x_1 - x_2 + 2x_4 + 2x_5 &= 0 \\ 4x_1 + x_2 + 4x_3 + 3x_5 &= 0. \end{aligned}$$

Since this system has fewer equations than unknowns, by Theorem 1.3 it is certain to have a nonzero solution. Thus the five given matrices are linearly dependent vectors in $\mathfrak{R}^{2\times 2}$. ■

■ EXAMPLE 2 Test Polynomials for Independence

Determine if the following polynomials (vectors from $\mathcal{P}$) are linearly independent or linearly dependent:

$$p(x) = x^2 - 1, \quad q(x) = x^2 + x - 2, \qquad \text{and} \qquad r(x) = x^2 + 3x + 2.$$

The vector equation $ap(x) + bq(x) + cr(x) = 0$ is equivalent to

$$a(x^2 - 1) + b(x^2 + x - 2) + c(x^2 + 3x + 2) = 0,$$

which, after some rearranging, becomes

$$(-a - 2b + 2c) + (b + 3c)x + (a + b + c)x^2 = 0 = 0 + 0x + 0x^2.$$

By comparing coefficients on the two sides of the above equation we see that this polynomial equation is equivalent to the homogeneous system

$$\begin{aligned} -a - 2b + 2c &= 0 \\ b + 3c &= 0 \\ a + b + c &= 0 \end{aligned} \qquad \text{or} \qquad \begin{bmatrix} -1 & -2 & 2 \\ 0 & 1 & 3 \\ 1 & 1 & 1 \end{bmatrix} \begin{bmatrix} a \\ b \\ c \end{bmatrix} = \begin{bmatrix} 0 \\ 0 \\ 0 \end{bmatrix}.$$

We have again reduced the vector question to an equivalent question about a system of linear equations. To decide if this system has nonzero solutions we row reduce the augmented matrix:

$$\begin{bmatrix} -1 & -2 & 2 & 0 \\ 0 & 1 & 3 & 0 \\ 1 & 1 & 1 & 0 \end{bmatrix} \to \begin{bmatrix} 1 & 2 & -2 & 0 \\ 0 & 1 & 3 & 0 \\ 0 & -1 & 3 & 0 \end{bmatrix}$$

$$\to \begin{bmatrix} 1 & 2 & -2 & 0 \\ 0 & 1 & 3 & 0 \\ 0 & 0 & 6 & 0 \end{bmatrix} \to \begin{bmatrix} 1 & 0 & 0 & 0 \\ 0 & 1 & 0 & 0 \\ 0 & 0 & 1 & 0 \end{bmatrix}.$$

From the last matrix we observe that the system has only the zero solution. Hence the polynomials $p(x)$, $q(x)$, and $r(x)$ are linearly independent polynomials (vectors in $\mathcal{P}$.) ■

It is easy to see that the nonzero rows of either a row-echelon matrix or a reduced row-echelon matrix are always linearly independent. For example, for the row-echelon matrix

$$B = \begin{bmatrix} 1 & 3 & 5 & 0 & 9 \\ 0 & 0 & 1 & 6 & 8 \\ 0 & 0 & 0 & 1 & 6 \\ 0 & 0 & 0 & 0 & 0 \end{bmatrix},$$

the equation

$$x_1\text{Row}_1(B) + x_2\text{Row}_2(B) + x_3\text{Row}_3(B) = [0 \quad 0 \quad 0 \quad 0 \quad 0]$$

or

$$\begin{aligned} x_1[1 \quad 3 \quad 5 \quad 0 \quad 9] + x_2[0 \quad 0 \quad 1 \quad 6 \quad 8] + x_3[0 \quad 0 \quad 0 \quad 1 \quad 6] \\ = [0 \quad 0 \quad 0 \quad 0 \quad 0] \end{aligned}$$

yields at once that $x_1 = x_2 = x_3 = 0$. Therefore the nonzero rows of B are linearly independent.

EXAMPLE 3 **Test Functions for Independence**

Determine if the following functions (vectors from $\mathcal{F}(1, \infty)$) are linearly independent or dependent.

$$f(x) = \sqrt{4x}, \quad g(x) = \frac{12}{x}, \quad h(x) = 2^x$$

The vector equation

$$a\mathbf{f} + b\mathbf{g} + c\mathbf{h} = \mathbf{0}$$

amounts to

$$\begin{aligned} &af(x) + bg(x) + ch(x) \\ &= a\sqrt{4x} + b\frac{12}{x} + c2^x \\ &= 0 \text{ for all } x \text{ such that } 1 \le x < \infty. \end{aligned}$$

We obtain linear equations for a, b, and c by choosing particular values for x in the last equation:

$$\begin{aligned} x = 1: &\quad 2a + 12b + 2c = 0 \\ x = 2: &\quad \sqrt{8}a + 6b + 4c = 0 \\ x = 4: &\quad 4a + 3b + 16c = 0 \end{aligned} \qquad \text{or} \qquad \begin{bmatrix} 2 & 12 & 2 \\ \sqrt{8} & 6 & 4 \\ 4 & 3 & 16 \end{bmatrix}\begin{bmatrix} a \\ b \\ c \end{bmatrix} = \begin{bmatrix} 0 \\ 0 \\ 0 \end{bmatrix}.$$

This homogeneous system has no nontrivial solution since

$$\det\begin{bmatrix} 2 & 12 & 2 \\ \sqrt{8} & 6 & 4 \\ 4 & 3 & 16 \end{bmatrix} = 312 - 186\sqrt{8} \neq 0.$$

Therefore, the functions **f**, **g**, and **h** are linearly independent vectors from $\mathcal{F}$. ■

As illustrated in the preceding examples, the procedure for testing a set of vectors $\{\mathbf{u}_1, \ldots, \mathbf{u}_k\}$ for linear independence or dependence requires three main steps.

■ TEST FOR LINEAR INDEPENDENCE OR DEPENDENCE ■

1. Suppose that $x_1\mathbf{u}_1 + x_2\mathbf{u}_2 + \cdots + x_k\mathbf{u}_k = \mathbf{0}$.
2. Interpret the vector equation in step 1 as a homogeneous system of linear equations $AX = 0$. (The technique depends on the nature of the vector space.)
3. Using the techniques of Chapter 1, determine if the system in step 2 has a nonzero solution. If the system has only the zero solution, then the $\mathbf{u}_i$ are linearly independent; otherwise they are linearly dependent.

The method just described cannot be used to establish the linear dependence of a set of functions (vectors from $\mathcal{F}$), since in this case it must be shown that the equation in step 1 holds for an infinite number of values of x. As illustrated in Example 3, the method works well for establishing the linear independence of a set of functions. You may consult a differential equations text for the definition of the Wronskian and a more complete discussion of linear dependence of functions.

The next theorem lists some useful facts about linear independence and linear dependence.

THEOREM 3.23 The following properties hold in any vector space $\mathcal{V}$:

(1) Any nonempty subset of an independent set is independent.

(2) Any set with a nonempty dependent subset is dependent.

(3) If the set $\{\mathbf{u}_1, \ldots, \mathbf{u}_k\}$ is independent and $\mathbf{u}_{k+1}$ is not in SPAN $\{\mathbf{u}_1, \ldots, \mathbf{u}_k\}$, then the set $\{\mathbf{u}_1, \ldots, \mathbf{u}_k, \mathbf{u}_{k+1}\}$ is linearly independent.

(4) If $\{\mathbf{u}_1, \mathbf{u}_2, \ldots, \mathbf{u}_n\}$ is a set of vectors such that $\mathbf{u}_1 \neq \mathbf{0}$ and $\mathbf{u}_{j+1}$ is not in SPAN$\{\mathbf{u}_1, \ldots, \mathbf{u}_j\}$ for $j = 1, 2, 3, \ldots, n - 1$, then $\{\mathbf{u}_1, \mathbf{u}_2, \ldots, \mathbf{u}_n\}$ is an independent set of vectors.

We will prove the third property; proofs of the other properties are left to the exercises.

Proof The set $\{\mathbf{u_1}, \ldots, \mathbf{u_k}, \mathbf{u_{k+1}}\}$ is either linearly independent or linearly dependent. If it is dependent, then there exist scalars, not all zero, such that

$$a_1\mathbf{u_1} + \cdots + a_k\mathbf{u_k} + a_{k+1}\mathbf{u_{k+1}} = \mathbf{0}.$$

If $a_{k+1}\mathbf{u_{k+1}} = \mathbf{0}$, then this equation says that $\{\mathbf{u_1}, \ldots, \mathbf{u_k}\}$ is a dependent set; this is a contradiction. Thus, $a_{k+1} \neq 0$ and it follows that

$$\mathbf{u_{k+1}} = \left(\frac{-1}{a_{k+1}}\right)(a_1\mathbf{u_1} + \cdots + a_k\mathbf{u_k}) \text{ is in SPAN}\{\mathbf{u_1}, \ldots, \mathbf{u_k}\}.$$

This is again a contradiction, so the assumption that the set $\{\mathbf{u_1}, \ldots, \mathbf{u_k}, \mathbf{u_{k+1}}\}$ is linearly dependent must be wrong. The only remaining possibility is that the set is linearly independent. This completes the proof of the third assertion of the theorem. The fourth assertion in the theorem is an immediate corollary of the third conclusion. □

From the second conclusion of Theorem 3.23 it follows at once that any set of vectors containing the zero vector must be linearly dependent.

EXAMPLE 4 **Reduce a Spanning Set to a Basis**

As an application of the fourth conclusion of Theorem 3.23, show how to identify an independent subset $\mathcal{S}_1$ of a set $\mathcal{S} = \{\mathbf{u_1}, \mathbf{u_2}, \ldots, \mathbf{u_m}\}$ such that SPAN$\{\mathcal{S}\}$ = SPAN$\{\mathcal{S}_1\}$.

We will construct $\mathcal{S}_1$ from $\mathcal{S}$ by working through the list of elements of $\mathcal{S}$, omitting any vector which can be expressed as a linear combination of those that precede it in the list. Thus we can construct $\mathcal{S}_1$ using the following algorithm:

(1) If $\mathbf{u_1} \neq \mathbf{0}$, include $\mathbf{u_1}$ in $\mathcal{S}_1$.
(2) For $j = 1, 2, \ldots, m - 1$
(3) If $\mathbf{u_{j+1}}$ is not in SPAN$\{\mathcal{S}_1\}$ = SPAN$\{\mathbf{u_1}, \ldots, \mathbf{u_j}\}$, then add $\mathbf{u_{j+1}}$ to $\mathcal{S}_1$.
(4) end.

By Theorem 3.23, the set $\mathcal{S}_1$ will be linearly independent. Since deleting dependent vectors from the spanning set will not make the subspace any smaller, we also have SPAN$\{\mathcal{S}_1\}$ = SPAN$\{\mathcal{S}\}$. ■

Suppose now that the vectors $\{\mathbf{u_1}, \mathbf{u_2}, \ldots, \mathbf{u_k}\}$ span a subspace $\mathcal{W}$ of the vector space $\mathcal{V}$. We know that every vector $\mathbf{v}$ in $\mathcal{W}$ = SPAN$\{\mathbf{u_1}, \mathbf{u_2}, \ldots, \mathbf{u_k}\}$ is expressible as a linear combination of the $\mathbf{u_i}$:

$$\mathbf{v} = a_1\mathbf{u_1} + a_2\mathbf{u_2} + \cdots + a_k\mathbf{u_k}.$$

Just as in the case $\mathcal{V} = \Re^n$ (Theorem 3.8), the coefficients a_i are unique if, and only if, the set $\{\mathbf{u_1}, \mathbf{u_2}, \ldots, \mathbf{u_k}\}$ is linearly independent.

DEFINITION 3.20 A **basis** for the subspace $\mathcal{W}$ of $\mathcal{V}$ is a linearly independent set of vectors from $\mathcal{W}$ which span $\mathcal{W}$. If $\{\mathbf{u}_1, \mathbf{u}_2, \ldots, \mathbf{u}_k\}$ is a basis for $\mathcal{W}$, we will write

$$\mathcal{W} = \text{BASIS}\{\mathbf{u}_1, \mathbf{u}_2, \ldots, \mathbf{u}_k\}. \quad \square$$

The next theorem summarizes the discussion preceding Definition 3.20. Theorem 3.8 is a special case of this theorem.

THEOREM 3.24 If the vectors $\{\mathbf{u}_1, \mathbf{u}_2, \ldots, \mathbf{u}_k\}$ form a basis for the subspace $\mathcal{W}$ of $\mathcal{V}$, then every vector in $\mathcal{W}$ can be expressed uniquely as a linear combination of the $\mathbf{u}_i$. Conversely, if every vector in $\mathcal{W}$ can be expressed uniquely as a linear combination of the $\mathbf{u}_i$ then the vectors $\{\mathbf{u}_1, \mathbf{u}_2, \ldots, \mathbf{u}_k\}$ form a basis for the subspace $\mathcal{W}$. $\square$

A basis for a vector space $\mathcal{V}$ leads to a coordinate system in the vector space $\mathcal{V}$ as described in the next definition.

DEFINITION 3.21 If the set $\mathcal{B} = \{\mathbf{u}_1, \mathbf{u}_2, \ldots, \mathbf{u}_r\}$ is a basis for the subspace $\mathcal{W}$ and

$$\mathbf{v} = a_1\mathbf{u}_1 + a_2\mathbf{u}_2 + \cdots + a_r\mathbf{u}_r$$

is any vector from $\mathcal{W}$, then the uniquely determined scalars a_i are called the **coordinates of v relative to the basis** $\mathcal{B}$. The column matrix $[a_1 \quad a_2 \quad \cdots \quad a_r]^T$ is called the **coordinate matrix of v relative to the basis** $\mathcal{B}$, or the $\mathcal{B}$-**coordinate matrix of v**, and will be denoted by $[\mathbf{v}]_{\mathcal{B}}$. $\square$

Note that the $\mathcal{B}$-coordinate matrix of a vector would be changed if we rearranged the elements of the basis $\mathcal{B}$, (see Exercise 39 of this section). Whenever we refer to a basis $\mathcal{B} = \{\mathbf{u}_1, \mathbf{u}_2, \ldots, \mathbf{u}_r\}$ we will mean that the vectors are to be taken in the indicated order; thus every basis which we consider is assumed to be an ordered basis.

In general, finding the coordinates of a vector with respect to a given basis involves solving a system of linear equations. By Theorem 3.24 this system must have a unique solution. These systems must therefore have square, nonsingular coefficient matrices.

In order to establish that a given set of vectors is a basis for the subspace $\mathcal{W}$ we will normally use the converse part of Theorem 3.24; this involves showing that a certain linear system has a unique solution for all possible right-hand sides. Recall that the system $AX = K$, where A is square, has a unique solution if, and only if, the coefficient matrix A is nonsingular. Among the conditions equivalent to the nonsingularity of the matrix A are $A \underset{R}{\sim} I$ and $\det(A) \neq 0$.

EXAMPLE 5 Coordinates with Respect to the Standard Basis of $\mathfrak{R}^{2\times 2}$

The matrices

$$\mathcal{U} = \left\{E_{11} = \begin{bmatrix} 1 & 0 \\ 0 & 0 \end{bmatrix},\ E_{12} = \begin{bmatrix} 0 & 1 \\ 0 & 0 \end{bmatrix},\ E_{21} = \begin{bmatrix} 0 & 0 \\ 1 & 0 \end{bmatrix},\ E_{22} = \begin{bmatrix} 0 & 0 \\ 0 & 1 \end{bmatrix}\right\}$$

form a basis for the vector space $\mathfrak{R}^{2\times 2}$ since a typical vector in $\mathfrak{R}^{2\times 2}$ is

$$\begin{bmatrix} a_{11} & a_{12} \\ a_{21} & a_{22} \end{bmatrix} = a_{11}E_{11} + a_{12}E_{12} + a_{21}E_{21} + a_{22}E_{22}.$$

The $\mathcal{U}$-coordinate matrix of $\begin{bmatrix} a_{11} & a_{12} \\ a_{22} & a_{22} \end{bmatrix}$ is $\begin{bmatrix} a_{11} \\ a_{12} \\ a_{21} \\ a_{22} \end{bmatrix}$. ■

EXAMPLE 6 **Coordinates in a Vector Space of Polynomials**

Show that $\mathcal{H} = \{x^2 - 3x + 2, x^2 - 2x, x^2 - x\}$ is a basis for $\mathcal{P}_2$ and find the $\mathcal{H}$-coordinate matrix of $x + 2$ with respect to this basis.

We need to determine scalars a, b, and c so that

$$a(x^2 - 3x + 2) + b(x^2 - 2x) + c(x^2 - x) = x + 2.$$

This last equation is equivalent to

$$(a + b + c)x^2 + (-3a - 2b - c)x + 2a = x + 2.$$

Comparing coefficients, we see that the scalars a, b, and c must satisfy

$$\begin{bmatrix} 2 & 0 & 0 \\ -3 & -2 & -1 \\ 1 & 1 & 1 \end{bmatrix} \begin{bmatrix} a \\ b \\ c \end{bmatrix} = \begin{bmatrix} 2 \\ 1 \\ 0 \end{bmatrix}.$$

From the row reduction

$$\left[\begin{array}{ccc|c} 2 & 0 & 0 & 2 \\ -3 & -2 & -1 & 1 \\ 1 & 1 & 1 & 0 \end{array}\right] \rightarrow \left[\begin{array}{ccc|c} 1 & 0 & 0 & 1 \\ 2 & 1 & 0 & -1 \\ 1 & 1 & 1 & 0 \end{array}\right]$$

$$\rightarrow \left[\begin{array}{ccc|c} 1 & 0 & 0 & 1 \\ 2 & 1 & 0 & -1 \\ -1 & 0 & 1 & 1 \end{array}\right] \rightarrow \left[\begin{array}{ccc|c} 1 & 0 & 0 & 1 \\ 0 & 1 & 0 & -3 \\ 0 & 0 & 1 & 2 \end{array}\right]$$

we see that the coefficient matrix of our linear system is invertible, and hence has a unique solution for any choice of the right-hand side. The solution of the above system is $a = 1$, $b = -3$, and $c = 2$. These are the desired coordinates and $[x + 2]_{\mathcal{H}} = [1 \quad -3 \quad 2]^T$ is the $\mathcal{H}$-coordinate matrix of $x + 2$. ■

The next two theorems are generalizations of Theorems 3.9 and 3.10 to the case of a general vector space. The proofs for the earlier theorems are adequate to prove these theorems and will not be repeated here.

THEOREM 3.25 If $\{\mathbf{v}_1, \mathbf{v}_2, \ldots, \mathbf{v}_k\}$ is a basis for the vector space $\mathcal{W}$, then any set of vectors from $\mathcal{W}$, having more than k elements, must be linearly dependent. □

THEOREM 3.26 If a vector space $\mathcal{V}$ has a finite basis, then any two bases for $\mathcal{V}$ must contain the same number of vectors; that is, if $\{\mathbf{v}_1, \mathbf{v}_2, \ldots, \mathbf{v}_n\}$ is a basis for $\mathcal{V}$, then any other basis of $\mathcal{V}$ must contain precisely n vectors. □

THEOREM 3.27 Any linearly independent set of vectors, in a finite dimensional vector space $\mathcal{V}$, can be completed to a basis for $\mathcal{V}$. □

The number of vectors in a basis of $\mathcal{V}$ is an important property of the space $\mathcal{V}$. Theorem 3.26 shows that this number does not depend on how the basis is chosen. Thus we can extend Definition 3.12 to the general case.

DEFINITION 3.22 If the vector space $\mathcal{V}$ has a finite basis, then the number of vectors in any basis for $\mathcal{V}$ is called the **dimension** of $\mathcal{V}$ and is denoted by $\dim(\mathcal{V})$. If $\mathcal{V}$ does not have a finite basis, then we will say that $\mathcal{V}$ is **infinite dimensional** and write $\dim(\mathcal{V}) = \infty$. □

Example 5 shows that $\dim(\mathfrak{R}^{2\times 2}) = 4$, and the obvious generalization to $\mathfrak{R}^{n\times n}$ shows that $\dim(\mathfrak{R}^{n\times n}) = n^2$. Example 6 shows that $\dim(\mathcal{P}_2) = 3$. Since the rows of an echelon matrix are independent they are a basis for the row space of the matrix. For any matrix A, it follows, from Theorem 3.22, that $\dim(\mathcal{RS}(A)) = \text{rank}(A)$ which is the number of nonzero rows in an echelon matrix row equivalent to A. Theorem 3.11 can be restated as

$$\dim(\mathcal{NS}(A)) + \dim(\mathcal{RS}(A)) = \text{number of columns of } A.$$

■ EXAMPLE 7 Using Row Reduction to Find a Basis

Find a basis for the subspace $\mathcal{W}$ of $\mathfrak{R}^{1\times 4}$ spanned by

$$\{\mathbf{u}_1 = [1 \quad 2 \quad 3 \quad 1],\ \mathbf{u}_2 = [3 \quad 2 \quad 1 \quad 1],\ \mathbf{u}_3 = [0 \quad 2 \quad 4 \quad 1],\ \mathbf{u}_4 = [1 \quad 1 \quad 1 \quad 1]\}.$$

We first observe that $\mathcal{W} = \mathcal{RS}(A)$, where

$$A = \begin{bmatrix} 1 & 2 & 3 & 1 \\ 3 & 2 & 1 & 1 \\ 0 & 2 & 4 & 1 \\ 1 & 1 & 1 & 1 \end{bmatrix}.$$

Since elementary row operations do not change the row space, we see from the row reduction

$$A = \begin{bmatrix} 1 & 2 & 3 & 1 \\ 3 & 2 & 1 & 1 \\ 0 & 2 & 4 & 1 \\ 1 & 1 & 1 & 1 \end{bmatrix} \to \begin{bmatrix} 1 & 2 & 3 & 1 \\ 0 & -4 & -8 & -2 \\ 0 & -1 & -2 & 0 \\ 0 & -1 & -2 & 0 \end{bmatrix}$$

$$\to \begin{bmatrix} 1 & 2 & 3 & 1 \\ 0 & 1 & 2 & 0 \\ 0 & 0 & 0 & 1 \\ 0 & 0 & 0 & 0 \end{bmatrix} \to \begin{bmatrix} 1 & 0 & -1 & 0 \\ 0 & 1 & 2 & 0 \\ 0 & 0 & 0 & 1 \\ 0 & 0 & 0 & 0 \end{bmatrix} = B$$

that a basis for the row space of A can be obtained from the nonzero rows of B. Thus dim $(\mathcal{RS}(A)) = 3$ and

$$\{\mathbf{v}_1 = [1 \quad 0 \quad -1 \quad 0], \quad \mathbf{v}_2 = [0 \quad 1 \quad 2 \quad 0], \quad \mathbf{v}_3 = [0 \quad 0 \quad 0 \quad 1]\}$$

is a basis for $\mathcal{W}$. The technique just illustrated produces a basis for $\mathcal{RS}(A)$ which is "simplest" in the sense that its vectors have many zero components. It is particularly easy to find coordinates with respect to such an "echelon" basis since the system to be solved will be triangular. ■

The simplest example of an infinite dimensional vector space is the vector space $\mathcal{P}$ of all polynomials of finite degree. To see that this space does not have a finite basis we suppose that

$$\{p_1(x), p_2(x), \ldots, p_k(x)\}$$

is a finite basis for $\mathcal{P}$. Note that each $p_i(x) \neq 0$, let d_i be the degree of $p_i(x)$, and let m be the maximum of the d_i, $i = 1, 2, \ldots, k$. Since no element of SPAN$\{p_1(x), p_2(x), \ldots, p_k(x)\}$ can have degree greater than m, it is clear that

$$x^{m+1} \text{ is in } \mathcal{P} \text{ but not in SPAN}\{p_1(x), p_2(x), \ldots, p_k(x)\}.$$

Thus SPAN$\{p_1(x), p_2(x), \ldots, p_k(x)\} \neq \mathcal{P}$ and it follows that $\mathcal{P}$ does not have a finite spanning set and hence is not finite dimensional.

EXERCISES 3.9

Determine if the sets of vectors in Exercises 1–14 are linearly independent or linearly dependent. In each case first rephrase the vector question as an equivalent question about a system of linear equations.

1. $\mathbf{x}_1 = [1 \quad -1 \quad 2 \quad 0]$, $\mathbf{x}_2 = [1 \quad 1 \quad 2 \quad 0]$, and $\mathbf{x}_3 = [4 \quad 1 \quad 2 \quad 1]$ in $\mathfrak{R}^{1\times 4}$

2. $\mathbf{y}_1 = [0 \quad 1 \quad 2 \quad 1]^T$, $\mathbf{y}_2 = [1 \quad 2 \quad 3 \quad 1]^T$, $\mathbf{y}_3 = [1 \quad 6 \quad 11 \quad 5]^T$, and $\mathbf{y}_4 = [-1 \quad 0 \quad 1 \quad 1]^T$ in $\mathfrak{R}^4$

3. $p_1(x) = x^3 + 2x + 5$, $p_2(x) = 3x^2 + 2$, $p_3(x) = 6x$, and $p_4(x) = 2$ in $\mathcal{P}$

4. The rows of the matrix $A = \begin{bmatrix} 1 & 1 & 3 \\ 5 & -1 & 2 \\ -3 & 0 & 4 \end{bmatrix}$

5. The columns of the matrix $B = \begin{bmatrix} 1 & 2 & 1 & 4 \\ 4 & -1 & 5 & 8 \\ 0 & 2 & 3 & 5 \\ 3 & 6 & -1 & 8 \end{bmatrix}$

6. $\sin x$, $\cos x$, and $e^{(x/\pi)}$ in $\mathcal{F}$

7. $\sqrt{x}$, x^2, and $|x|$ in $\mathcal{F}(0, \infty)$

8. $\sin(3x + 7)$, $\cos 3x$, and $\sin 3x$ in $\mathcal{F}$

9. x^2, $x^2 - 4x$, $x^2 + 3x$, and $x^2 - x - 12$ in $\mathcal{P}$

10. $q_1(x) = 3 + x + x^2$, $q_2(x) = 2 - x + 5x^2$, and $q_3(x) = 4 - 3x^2$ in $\mathcal{P}$

11. $\mathbf{z}_1 = [1 \quad 2 \quad 1 \quad -2]$, $\mathbf{z}_2 = [0 \quad 1 \quad 1 \quad 0]$, $\mathbf{z}_3 = [0 \quad 2 \quad 3 \quad 1]$, and $\mathbf{z}_4 = [1 \quad 0 \quad -1 \quad 2]$ in $\mathfrak{R}^{1\times4}$

12. $\mathbf{z}_1 = [1 \quad 3 \quad 3]^T$, $\mathbf{z}_2 = [0 \quad 1 \quad 4]^T$, $\mathbf{z}_3 = [5 \quad 6 \quad 3]^T$, and $\mathbf{z}_4 = [7 \quad 2 \quad 3]^T$ in $\mathfrak{R}^3$

13. $\begin{bmatrix} 1 & 0 \\ 1 & 0 \end{bmatrix}$, $\begin{bmatrix} 1 & 2 \\ 0 & 0 \end{bmatrix}$, $\begin{bmatrix} 5 & -1 \\ 4 & 0 \end{bmatrix}$, and $\begin{bmatrix} -1 & 7 \\ 6 & 0 \end{bmatrix}$ in $\mathfrak{R}^{2\times2}$

14. $\begin{bmatrix} 1 & 0 & 0 \\ 0 & 0 & 0 \end{bmatrix}$, $\begin{bmatrix} 1 & 1 & 0 \\ 0 & 0 & 0 \end{bmatrix}$, $\begin{bmatrix} 1 & 1 & 1 \\ 0 & 0 & 0 \end{bmatrix}$, $\begin{bmatrix} 0 & 0 & 0 \\ 1 & 0 & 0 \end{bmatrix}$, $\begin{bmatrix} 0 & 0 & 0 \\ 1 & 1 & 0 \end{bmatrix}$, and $\begin{bmatrix} 1 & 1 & 1 \\ 1 & 1 & 1 \end{bmatrix}$ in $\mathfrak{R}^{2\times3}$

15. Show that if $\mathbf{u}$, $\mathbf{v}$, and $\mathbf{w}$ are linearly independent, then so are the vectors $\mathbf{x}_1 = \mathbf{u} + \mathbf{v}$, $\mathbf{x}_2 = \mathbf{v} + \mathbf{w}$, $\mathbf{x}_3 = \mathbf{w} + \mathbf{u}$.

16. Which of the ten defining properties of a vector space are used in deducing from $\mathbf{v} = a_1\mathbf{u}_1 + a_2\mathbf{u}_2 + \cdots + a_n\mathbf{u}_n = b_1\mathbf{u}_1 + b_2\mathbf{u}_2 + \cdots + b_n\mathbf{u}_n$ that $\mathbf{0} = \mathbf{v} - \mathbf{v} = (a_1 - b_1)\mathbf{u}_1 + (a_2 - b_2)\mathbf{u}_2 + \cdots + (a_n - b_n)\mathbf{u}_n$?

17. Consider the three functions $\mathbf{f}$, $\mathbf{g}$, and $\mathbf{h}$ in $\mathcal{F}$:

$$f(t) = \cos^2 t, \quad g(t) = \sin^2 t, \qquad \text{and} \qquad h(t) = \sin 2t.$$

Assume that $x_1\mathbf{f} + x_2\mathbf{g} + x_3\mathbf{h} = \mathbf{0}$; that is,

$$x_1 \cos^2 t + x_2 \sin^2 t + x_3 \sin 2t = 0 \text{ for all } t.$$

Differentiate this relation twice to obtain two more linear relations. Write these three linear equations in matrix form and show that the determinant of the coefficient matrix is not zero for any t. What can we now conclude about the independence or dependence of the functions $\mathbf{f}$, $\mathbf{g}$, and $\mathbf{h}$?

18. Complete the proof of Theorem 3.23.

19. Show that any four polynomials from $\mathcal{P}_2$ are linearly dependent.

†20. Find a set of linearly independent 3×3 matrices that is as large as possible.

21. Find an independent subset $\mathcal{S}_1$ of $\mathcal{S} = \{[1 \quad 2 \quad 4 \quad 1], [2 \quad 3 \quad 7 \quad 1], [1 \quad 4 \quad 6 \quad 3]\}$ such that $\text{SPAN}\{\mathcal{S}_1\} = \text{SPAN}\{\mathcal{S}\}$.

†**22.** Show that the rows of a matrix A are linearly independent if, and only if, the columns of A^T are linearly independent. Then show that the rows of a square matrix A are linearly independent if, and only if, A is nonsingular.

23. Find an independent subset $\mathcal{S}_1$ of $\mathcal{S} = \{\sin^2 x, \cos^2 x, \cos 2x, e^x\}$ such that $\text{SPAN}\{\mathcal{S}_1\} = \text{SPAN}\{\mathcal{S}\}$.

24. Which of the following sets is a basis for $\Re^n$?

(a) $[2 \ \ 2 \ \ 2]^T, [1 \ \ 1 \ \ 0]^T, [3 \ \ 0 \ \ 0]^T; \ n = 3$

(b) $[1 \ \ 2 \ \ 3]^T, [4 \ \ 5 \ \ 6]^T, [7 \ \ 8 \ \ 9]^T; \ n = 3$

(c) $[1 \ \ 6 \ \ 4]^T, [1 \ \ -2 \ \ -5]^T, [2 \ \ 4 \ \ -1]^T; \ n = 3$

(d) $[3 \ \ -3 \ \ -1]^T, [-3 \ \ -3 \ \ 3]^T, [-1 \ \ 3 \ \ 3]^T; \ n = 3$

(e) $[1 \ \ 1 \ \ 1 \ \ -3]^T, [3 \ \ -2 \ \ -17 \ \ 16]^T, [3 \ \ 2 \ \ -1 \ \ -4]^T,$
$[1 \ \ 1 \ \ 1 \ \ 1]^T; \ n = 4$

(f) $[1 \ \ 2 \ \ 4 \ \ 1 \ \ 0]^T, [-1 \ \ -1 \ \ 1 \ \ 2 \ \ 0]^T, [2 \ \ 0 \ \ -11 \ \ 3 \ \ 0]^T,$
$[3 \ \ 2 \ \ -1 \ \ 8 \ \ 0]^T, [0 \ \ 0 \ \ 0 \ \ 1 \ \ 1]^T; \ n = 5$

25. Which of the following sets is a basis for $\mathcal{P}_2$?

(a) $x^2, \quad x^2 + x, \quad x^2 + x + 1$

(b) $2x^2 - 3x - 1, \quad 4x^2 + x + 1$

(c) $x^2 + 6x + 4, \quad 2x^2 + 4x - 1, \quad x^2 - 2x - 5, \quad 3x - 5$

(d) $x^2 - 3x + 2, \quad x^2 - 4x + 3, \quad x^2 - 5x + 6$

26. Show that the set

$$\mathcal{B} = \{\mathbf{u}_1 = [1 \ \ -1 \ \ 1 \ \ -1], \mathbf{u}_2 = [1 \ \ -1 \ \ 1 \ \ 2], \\ \mathbf{u}_3 = [0 \ \ 2 \ \ -1 \ \ 3], \mathbf{u}_4 = [0 \ \ 0 \ \ 1 \ \ 3]\}$$

forms a basis for $\Re^{1\times 4}$ and find $[\mathbf{v}_1]_{\mathcal{B}}$ and $[\mathbf{v}_2]_{\mathcal{B}}$ where $\mathbf{v}_1 = [2 \ \ 0 \ \ 2 \ \ 7]$ and $\mathbf{v}_2 = [2 \ \ -2 \ \ 4 \ \ 7]$.

27. Show that the set $\mathcal{B} = \{x^2 - 5x + 6, x^2 - 6x + 8, x^2 - 7x + 12\}$ is a basis for $\mathcal{P}_2$ and find $[x^2 + x + 1]_{\mathcal{B}}$, $[x]_{\mathcal{B}}$, and $[1]_{\mathcal{B}}$.

28. Show that the following matrices form a basis for $\Re^{2\times 2}$:

$$\mathcal{B} = \left\{A_1 = \begin{bmatrix} 1 & 2 \\ 1 & 0 \end{bmatrix} \ A_2 = \begin{bmatrix} 0 & 1 \\ 1 & 0 \end{bmatrix} \ A_3 = \begin{bmatrix} 0 & 2 \\ 3 & 1 \end{bmatrix} \ A_4 = \begin{bmatrix} -1 & 0 \\ 1 & -2 \end{bmatrix}\right\}.$$

Find the $\mathcal{B}$-coordinates of I and $T = \begin{bmatrix} 0 & 5 \\ 7 & -3 \end{bmatrix}$.

29. Find the echelon basis $\mathcal{E}$ for the row space of $A = \begin{bmatrix} 1 & 2 & 4 & 1 \\ 2 & 3 & 7 & 1 \\ 1 & 4 & 6 & 3 \end{bmatrix}$ and for $\mathbf{u}_1 = [3 \ \ -2 \ \ 4 \ \ -5]$ and $\mathbf{u}_2 = [1 \ \ -1 \ \ 1 \ \ -2]$. Find $[\mathbf{u}_1]_{\mathcal{E}}$ and $[\mathbf{u}_2]_{\mathcal{E}}$.

30. Find $\dim(\mathcal{RS}(A))$, $\dim(\mathcal{CS}(A))$, and $\dim(\mathcal{NS}(A))$ if $A = \begin{bmatrix} 1 & 2 & -1 & 4 \\ -1 & -2 & 6 & -7 \\ 2 & 4 & 3 & 5 \end{bmatrix}$.

31. Consider $\mathbf{w}_1 = [1 \ \ 1 \ \ 2 \ \ 4]$, $\mathbf{w}_2 = [1 \ \ -1 \ \ -4 \ \ 0]$, $\mathbf{w}_3 = [2 \ \ 1 \ \ 1 \ \ 6]$, and $\mathbf{w}_4 = [2 \ \ -1 \ \ -5 \ \ 2]$ from $\Re^{1\times 4}$. Let $\mathcal{V}$ denote the subspace spanned by $\{\mathbf{w}_1, \mathbf{w}_2, \mathbf{w}_3, \mathbf{w}_4\}$. Find a matrix A such that $\mathcal{V} = \mathcal{RS}(A)$ and then, by reducing A to echelon form, find a basis for $\mathcal{V}$. Also, find a basis for $\mathcal{V}$ which is a subset of $\{\mathbf{w}_1, \mathbf{w}_2, \mathbf{w}_3, \mathbf{w}_4\}$.

32. Let $\mathcal{W}$ be the subspace of $\mathfrak{R}^4$ spanned by $\mathbf{c}_1 = [0 \quad 1 \quad 2 \quad 1]^T$, $\mathbf{c}_2 = [1 \quad 2 \quad 3 \quad 1]^T$, $\mathbf{c}_3 = [3 \quad 6 \quad 9 \quad 3]^T$, and $\mathbf{c}_4 = [-2 \quad 0 \quad 2 \quad 2]^T$. Construct a matrix A such that $\mathcal{W} = \mathcal{CS}(A)$. By column reducing A, find a basis for $\mathcal{W}$. Explain why one cannot find a basis for $\mathcal{W}$ by row reducing the matrix A. Find a basis for $\mathcal{W}$ that is a subset of $\{\mathbf{c}_1, \mathbf{c}_2, \mathbf{c}_3, \mathbf{c}_4\}$.

33. **(a)** Show that $\dim(\mathfrak{R}^{m\times n}) = mn$. **(b)** Show that $\dim(\mathcal{F}) = \infty$.

34. **(a)** Show that any independent subset of a vector space $\mathcal{V}$ can be extended to a basis for $\mathcal{V}$. (*Hint:* Let $\{\mathbf{u}_1, \ldots, \mathbf{u}_k\}$ be independent and $\{\mathbf{v}_1, \ldots, \mathbf{v}_n\}$ be a basis. Show that $\{\mathbf{u}_1, \ldots, \mathbf{u}_k, \mathbf{v}_1, \ldots, \mathbf{v}_n\}$ spans $\mathcal{V}$. Then use the third property of Theorem 3.9.)

(b) Find a basis for $\mathfrak{R}^3$ containing $\mathbf{u}_1 = [1 \quad 2 \quad 3]^T$.

35. Find the rank of the following matrices.

$$A = \begin{bmatrix} 1 & 3 \\ 4 & -1 \\ 3 & -4 \end{bmatrix} \quad B = \begin{bmatrix} 1 & 1 & 1 & 3 \\ 3 & -2 & -17 & 16 \\ -1 & 4 & 19 & -10 \end{bmatrix} \quad C = \begin{bmatrix} 1 & -1 & 0 & 1 \\ 2 & 2 & 4 & 6 \\ 4 & 1 & 5 & 9 \\ 1 & 2 & 3 & 4 \end{bmatrix}$$

36. **(a)** Show that a subspace $\mathcal{W}$ of a finite dimensional space $\mathcal{V}$ is finite dimensional with $\dim(\mathcal{W}) \leq \dim(\mathcal{V})$.

(b) Show that if a subspace $\mathcal{W}$ of a finite dimensional space $\mathcal{V}$ satisfies $\dim(\mathcal{W}) = \dim(\mathcal{V})$, then $\mathcal{W} = \mathcal{V}$.

(c) Show that if a subspace $\mathcal{W}$ of $\mathcal{V}$ is infinite dimensional, then so is $\mathcal{V}$.

37. Find a basis for the null space of the following matrices. In each case find $\dim(\mathcal{NS}(A))$ and $\dim(\mathcal{RS}(A))$.

(a) $A = \begin{bmatrix} 1 & -3 & 4 & 3 & -2 \\ 7 & 4 & -2 & 6 & 1 \end{bmatrix}$ **(b)** $A = \begin{bmatrix} 1 & 1 & 1 & 3 \\ 3 & -2 & -17 & 16 \\ 3 & 2 & -1 & -4 \end{bmatrix}$

38. **(a)** Show that if $\dim(\mathcal{V}) = n$, and $\mathcal{S} = \{\mathbf{u}_1, \ldots, \mathbf{u}_n\}$ spans $\mathcal{V}$, then $\mathcal{S}$ is a basis for $\mathcal{V}$.

(b) Show that if $\dim(\mathcal{V}) = n$, and $\mathcal{S} = \{\mathbf{v}_1, \ldots, \mathbf{v}_n\}$ is independent, then $\mathcal{S}$ is a basis for $\mathcal{V}$.

†**39.** Let $\mathcal{B} = \{\mathbf{u}_1, \mathbf{u}_2, \mathbf{u}_3, \mathbf{u}_4\}$ be a basis for a four-dimensional subspace $\mathcal{W}$ and suppose that the $\mathcal{B}$-coordinate matrix of $\mathbf{v}$ is $[1 \quad 2 \quad -3 \quad 6]^T$. Find the $\mathcal{B}'$-coordinate of $\mathbf{v}$ with respect to the basis $\mathcal{B}' = \{\mathbf{u}_3, \mathbf{u}_1, \mathbf{u}_4, \mathbf{u}_2\}$.

40. Prove the converse part of Theorem 3.24.

41. Find bases for $\mathcal{NS}(A)$, $\mathcal{RS}(A)$, and $\mathcal{CS}(A)$ if $A = \begin{bmatrix} 1 & 0 & 4 & 2 \\ 0 & 1 & 5 & 3 \\ 0 & 0 & 0 & 0 \end{bmatrix}$.

42. Find bases for $\mathcal{NS}(A)$, $\mathcal{RS}(A)$, and $\mathcal{CS}(A)$ if

$$A = \begin{bmatrix} 1 & 1 & -3 & 7 & 9 & -9 \\ 1 & 2 & -4 & 10 & 13 & -12 \\ 1 & -1 & -1 & 1 & 1 & -3 \\ 1 & -3 & 1 & -5 & -7 & 3 \end{bmatrix} \tilde{R} \begin{bmatrix} 1 & 0 & -2 & 4 & 5 & -6 \\ 0 & 1 & -1 & 3 & 4 & -3 \\ 0 & 0 & 0 & 0 & 0 & 0 \\ 0 & 0 & 0 & 0 & 0 & 0 \end{bmatrix}.$$

43. Find bases for $\mathcal{NS}(A)$, $\mathcal{RS}(A)$, and $\mathcal{CS}(A)$ if $A = \begin{bmatrix} 1 & 3 & -5 & 1 & 5 \\ -2 & -5 & 8 & 0 & -17 \\ -1 & -2 & 3 & 1 & -12 \\ 1 & 7 & 5 & 4 & -2 \end{bmatrix}$.

CHAPTER 3 SUMMARY

Chapter 3 first treats vectors in $\mathfrak{R}^2$ and $\mathfrak{R}^3$ from a geometric point of view and then treats Euclidean n-space, $\mathfrak{R}^n$, from an algebraic point of view. Finally the abstract notion of a real vector space is introduced, illustrated, and studied.

KEY DEFINITIONS

Vector addition, scalar multiplication, dot product in $\mathfrak{R}^n$. *138, 140, 141*
Euclidean n-space. *147*
Subspace, subspace spanned by a set of vectors. *155, 160, 204, 207*
Linearly independent (dependent) set of vectors. *169, 214*
Basis for a subspace. Dimension of a subspace. *175, 220*
Coordinates of a vector with respect to a basis. *175, 220*
Rank of a matrix. *181*
Orthogonal vectors, orthonormal vectors. *185*
Orthogonal matrix. *193*
Real vector space. *198*

KEY FACTS

A nonempty subset is a subspace if, and only if, it is closed under vector addition and scalar multiplication.
SPAN$\{\mathbf{u}_1, \mathbf{u}_2, \ldots, \mathbf{u}_k\}$ consists of all possible linear combinations of the $\mathbf{u}_i$.
For a square matrix A, the columns of A are independent $\Leftrightarrow$ $\det A \neq 0 \Leftrightarrow A$ is row equivalent to I.
The coordinates of a vector with respect to a basis are unique.
Any two bases for a vector space have the same number of vectors.
$\dim(\mathcal{NS}(A)) + \text{rank}(A) =$ number of columns of A.
Cauchy-Schwartz inequality and the triangle inequality.
Every subspace of $\mathfrak{R}^n$ has an orthonormal basis which can be found using the Gram-Schmidt process.
Row equivalent matrices have the same row space and the same null space, but not the same column space.
The nonzero rows of an echelon matrix are always independent.
$\text{rank}(A) = \dim(\mathcal{RS}(A)) = \dim(\mathcal{CS}(A))$.

COMPUTATIONAL PROCEDURES

Find basis for $\mathcal{CS}(A)$, $\mathcal{NS}(A)$, $\mathcal{RS}(A)$.
Test a subset to see if it is a subspace.
Test a set of vectors for linear independence.
Gram-Schmidt orthogonalization process and the QR factorization.

APPLICATIONS

Systems of linear algebraic equations.
Differential equations.

CHAPTER 3 REVIEW EXERCISES

1. Are statements (a) through (i) *true* or *false?*
   (a) $A \tilde{_R} B \Rightarrow \mathcal{NS}(A) = \mathcal{NS}(B)$
   (b) $A \tilde{_R} B \Rightarrow \mathcal{CS}(B) = \mathcal{CS}(A)$
   (c) $A \tilde{_R} B \Rightarrow A$ and B have the same eigenvalues
   (d) $A \tilde{_R} I \Rightarrow \det(A) = 1$
   (e) $A \tilde{_R} B \Rightarrow \mathcal{RS}(A) = \mathcal{RS}(B)$
   (f) If A is an orthogonal matrix, then $\det(A) = \pm 1$.
   (g) Any independent subset of $\mathcal{V}$ is a basis for $\mathcal{V}$.
   (h) Any independent subset of $\mathcal{V}$ is a basis for a subspace of $\mathcal{V}$.
   (i) Any spanning set contains a basis.

2. Find a basis for the null space of $A = \begin{bmatrix} 1 & 2 & -3 & 10 \\ 2 & 3 & 8 & 17 \\ 1 & 3 & -5 & 13 \end{bmatrix}$.

3. Let A be an $m \times n$ matrix and let $K \neq 0$ be $m \times 3$. Let $\mathcal{W} = \{X \mid AX = K\}$. Explain why $\mathcal{W}$ is not a subspace of $\mathfrak{R}^{n \times 3}$.

4. For $\mathbf{u} = \begin{bmatrix} 3 \\ 4 \\ 12 \end{bmatrix}$ and $\mathbf{v} = \begin{bmatrix} 1 \\ 1 \\ 1 \end{bmatrix}$ in $\mathfrak{R}^3$, find the following:
   (a) $\mathbf{u} \cdot \mathbf{v}$; **(b)** $\|\mathbf{u}\|$; **(c)** a unit vector parallel to $\mathbf{u}$; and **(d)** the length of the projection of $\mathbf{u}$ on $\mathbf{v}$.

5. Given: $\mathbf{u}_1 = \begin{bmatrix} 1 \\ 2 \\ 1 \\ 1 \end{bmatrix}$, $\mathbf{u}_2 = \begin{bmatrix} 1 \\ 0 \\ -2 \\ 1 \end{bmatrix}$ and $\mathbf{u}_3 = \begin{bmatrix} 0 \\ 0 \\ 1 \\ 0 \end{bmatrix}$.
   (a) Which of these vectors are orthogonal?
   (b) Find an orthonormal basis for SPAN$\{\mathbf{u}_1, \mathbf{u}_2, \mathbf{u}_3\}$.

6. For questions (a)–(f) write an equivalent question of the form,

   "Does the system { } have a {nonzero, unique} solution?"

   Give the system explicitly and show where it came from. Do not try to answer the question!
   (a) Is $x^2 + 5x - 11$ in SPAN$\{x^2 + 5x, 3x^2 - 11, 5x + 7\}$?
   (b) Is the set $\{[1 \;\; 5 \;\; 3], [2 \;\; -1 \;\; 4], [3 \;\; 1 \;\; 7]\}$ linearly independent?
   (c) Is the set $\left\{ \begin{bmatrix} 1 & 2 \\ 3 & 4 \end{bmatrix}, \begin{bmatrix} 5 & 6 \\ 7 & 8 \end{bmatrix}, \begin{bmatrix} 9 & 0 \\ 1 & 2 \end{bmatrix}, \begin{bmatrix} 3 & 4 \\ 5 & 6 \end{bmatrix} \right\}$ a basis for $\mathfrak{R}^{2 \times 2}$?
   (d) Is $[1 \;\; 3 \;\; 7 \;\; 9]$ in the row space of $A = \begin{bmatrix} 1 & 3 & 5 & 9 \\ 2 & 8 & 6 & 4 \\ 3 & 0 & 5 & 1 \end{bmatrix}$?
   (e) Is the set $\{x^2 + 5x, 3x^2 - 11, 5x + 7\}$ a basis for $\mathcal{P}_2$?
   (f) Is $[7 \;\; 8 \;\; 10]$ in SPAN$\{[1 \;\; 5 \;\; 3], [2 \;\; -1 \;\; 4], [3 \;\; 1 \;\; 7]\}$?

7. **(a)** Find a basis for the null space of $A = \begin{bmatrix} 2 & 4 & 0 & -2 \\ 1 & 2 & -1 & -2 \\ 1 & 2 & 3 & 2 \end{bmatrix}$.

(b) What is the rank of A?

8. Let A be a 4×4 matrix whose characteristic polynomial is $(\lambda - 2)^2(\lambda + 7)^2$.
(a) Show that $\mathscr{E} = \{X \text{ in } \Re^4 \mid AX = 2X\}$ is a subspace of $\Re^4$.
(b) What are the possible values of $\dim(\mathscr{E})$?

9. For $\mathbf{u} = \begin{bmatrix} 1 \\ 3 \\ 5 \end{bmatrix}$ and $\mathbf{v} = \begin{bmatrix} 2 \\ -3 \\ 4 \end{bmatrix}$ in $\Re^3$ find

(a) $\mathbf{u} \cdot \mathbf{v}$; **(b)** $\|\mathbf{u}\|$; **(c)** a unit vector in the direction of $\mathbf{u} + \mathbf{v}$; and **(d)** the length of the projection of $\mathbf{u}$ on $\mathbf{v}$.

10. Let B be a fixed 3×3 matrix. Show that $\mathscr{W} = \{A \text{ in } \Re^{3\times3} \mid AB = 3A\}$ is a subspace of $\Re^{3\times3}$.

11. Given $\mathbf{u}_1 = \begin{bmatrix} 1 \\ 1 \\ 1 \\ 1 \end{bmatrix}$, $\mathbf{u}_2 = \begin{bmatrix} 3 \\ 1 \\ 1 \\ -5 \end{bmatrix}$ and $\mathbf{u}_3 = \begin{bmatrix} 12 \\ 0 \\ 0 \\ 0 \end{bmatrix}$ in $\Re^4$,

(a) show that $\mathbf{u}_1$ and $\mathbf{u}_2$ are orthogonal.
(b) use the Gram-Schmidt process to find an orthonormal basis for SPAN$\{\mathbf{u}_1, \mathbf{u}_2, \mathbf{u}_3\}$.

12. Find an orthonormal basis for the column space of $A = \begin{bmatrix} 2 & 4 \\ 0 & 1 \\ 1 & 0 \end{bmatrix}$.

13. Determine if the columns of the following matrices form a linearly dependent set. In each case find the rank of the matrix.

$$A = \begin{bmatrix} 0 & 2 & 6 & 8 \\ 1 & -3 & -4 & 1 \\ -1 & 4 & 7 & 3 \end{bmatrix}, \quad B = \begin{bmatrix} 1 & -1 & -3 & 0 \\ 0 & 1 & 5 & 4 \\ -1 & 2 & 8 & 5 \\ 3 & -1 & 1 & 3 \end{bmatrix}, \quad C = \begin{bmatrix} 1 & 1 & -3 & 7 & 9 & -9 \\ 0 & 1 & -1 & 3 & 4 & -3 \\ 0 & 0 & 0 & 1 & 1 & -2 \\ 0 & 0 & 0 & 0 & 0 & 0 \\ 0 & 0 & 0 & 0 & 0 & 0 \end{bmatrix}$$

14. Show that if the set $\{\mathbf{u}_1, \mathbf{u}_2, \mathbf{u}_3\}$ is linearly independent then so is the set $\{\mathbf{u}_1, \mathbf{u}_3\}$

15. Given that $A = \begin{bmatrix} 1 & 3 & 0 & 2 & -1 \\ 3 & 9 & 1 & 7 & 5 \\ 2 & 6 & 1 & 5 & 2 \\ 5 & 15 & 2 & 12 & 8 \end{bmatrix} \underset{R}{\sim} \begin{bmatrix} 1 & 3 & 0 & 2 & 0 \\ 0 & 0 & 1 & 1 & 0 \\ 0 & 0 & 0 & 0 & 1 \\ 0 & 0 & 0 & 0 & 0 \end{bmatrix} = R$, find a basis for $\mathscr{RS}(A)$, a basis for $\mathscr{NS}(A)$, and a basis for $\mathscr{CS}(A)$.

16. If $A = \begin{bmatrix} 2 & 1 & 1 \\ 1 & 2 & 1 \\ 1 & 1 & 2 \end{bmatrix}$, for what values of x are the rows of $xI - A$ linearly dependent?

17. Given: $A = \begin{bmatrix} 1 & 2 & -1 & -2 & 0 \\ 3 & 6 & 4 & -3 & 2 \\ 5 & 10 & 9 & -3 & 4 \\ 2 & 4 & 5 & -1 & 2 \end{bmatrix} \tilde{R} \begin{bmatrix} 1 & 2 & 0 & 0 & \frac{2}{7} \\ 0 & 0 & 1 & 0 & \frac{2}{7} \\ 0 & 0 & 0 & 1 & 0 \\ 0 & 0 & 0 & 0 & 0 \end{bmatrix}$.

(a) Find the dimension of the column space of A.
(b) Find the dimension of the null space of A.
(c) Find a basis for the null space of A.
(d) Find a basis for the row space of A.
(e) Find a subset of the columns of A which form a basis for $\mathscr{CS}(A)$.
(f) Express column 5 of A as a linear combination of the other columns of A.
(g) Find a vector from $\mathfrak{R}^{1\times 5}$ not in the row space of A.
(h) Find the dimension of the null space of A^T.

4 Linear Transformations

Let $\mathcal{V}$ and $\mathcal{W}$ be arbitrary, real vector spaces and suppose that for every vector $\mathbf{v}$ in $\mathcal{V}$ there is defined a unique vector $\mathbf{T}(\mathbf{v})$ in $\mathcal{W}$. We describe this situation by saying that $\mathbf{T}$ is a vector-valued function of the vector variable $\mathbf{v}$. We will denote such vector-valued functions by boldface capital letters. For example, if $\mathcal{V} = \mathcal{W} = \Re^3$, then

$$\mathbf{T}\begin{bmatrix} x \\ y \\ z \end{bmatrix} = \begin{bmatrix} 2x + 3y - z \\ 2 + \sin(x + y) \\ 3\,|x - 3z| \end{bmatrix}$$

is a vector-valued function; we will give other examples later. In this chapter we will study the most important kind of vector-valued functions, those that are linear.

4.1 DEFINITIONS AND EXAMPLES

We will use the following standard terminology, from algebra and calculus, in discussing functions from $\mathcal{V}$ to $\mathcal{W}$:

(1) The vector space $\mathcal{V}$ is called the **domain** of $\mathbf{T}$.
(2) The vector $\mathbf{T}(\mathbf{v})$ in $\mathcal{W}$ is the **value** of $\mathbf{T}$ at $\mathbf{v}$ or the **image** of $\mathbf{v}$ under $\mathbf{T}$.
(3) The **range** of $\mathbf{T}$ is the set $\mathcal{R}(\mathbf{T}) = \mathbf{T}(\mathcal{V}) = \{\mathbf{T}(\mathbf{v})\,|\,\mathbf{v} \text{ in } \mathcal{V}\} \subseteq \mathcal{W}$.
(4) The function $\mathbf{T}$ is **one-to-one** if $\mathbf{T}(\mathbf{u}) = \mathbf{T}(\mathbf{v})$ implies $\mathbf{u} = \mathbf{v}$, or, equivalently, if $\mathbf{u} \neq \mathbf{v}$ implies $\mathbf{T}(\mathbf{u}) \neq \mathbf{T}(\mathbf{v})$ (see Figure 4.1).
(5) The function $\mathbf{T}$ is **onto** $\mathcal{W}$ if $\mathcal{R}(\mathbf{T}) = \mathcal{W}$.

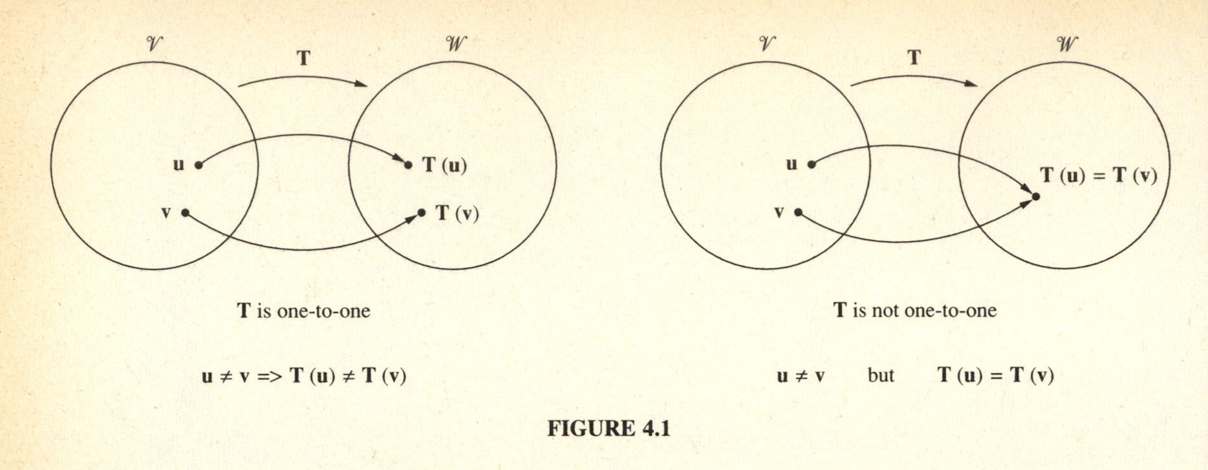

FIGURE 4.1

EXAMPLE 1 This Function Is not One-to-One

Let $\mathbf{T}$ be the function from $\mathfrak{R}$ to $\mathfrak{R}$ defined by $\mathbf{T}(x) = x^2 - 9$. The graph of $\mathbf{T}$ is the parabola shown in Figure 4.2.

Since $\mathbf{T}(3) = \mathbf{T}(-3) = 0$, this function is not one-to-one. Note that many horizontal lines cut the graph in two points. The range of $\mathbf{T}$ is the set $\mathbf{T}(\mathfrak{R}) = \{y \mid -9 \leq y < \infty\}$. Clearly $\mathbf{T}(\mathfrak{R}) \neq \mathfrak{R}$, so the function $\mathbf{T}$ is not onto $\mathfrak{R}$. ■

We will not attempt to study all vector-valued functions from $\mathcal{V}$ to $\mathcal{W}$; that would be an overly ambitious and difficult undertaking. We will restrict our attention to the linear functions, an important and useful subclass of all such functions.

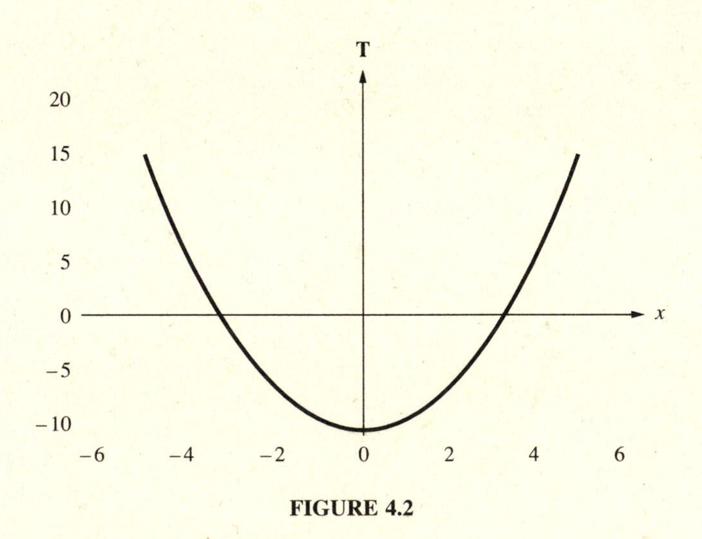

FIGURE 4.2

DEFINITION 4.1 Let $\mathcal{V}$ and $\mathcal{W}$ be vector spaces. A **linear transformation** from $\mathcal{V}$ to $\mathcal{W}$ is a function, $\mathbf{T}$, from $\mathcal{V}$ to $\mathcal{W}$ that satisfies

$$\mathbf{T}(k\mathbf{u} + h\mathbf{v}) = k\mathbf{T}(\mathbf{u}) + h\mathbf{T}(\mathbf{v}) \tag{4.1}$$

for all vectors $\mathbf{u}$ and $\mathbf{v}$ in $\mathcal{V}$ and all real scalars h and k. □

Note that Equation (4.1) is equivalent to the two conditions,

$$\mathbf{T}(\mathbf{u} + \mathbf{v}) = \mathbf{T}(\mathbf{u}) + \mathbf{T}(\mathbf{v}) \qquad \text{and} \qquad \mathbf{T}(k\mathbf{u}) = k\mathbf{T}(\mathbf{u})$$

for all $\mathbf{u}$ and $\mathbf{v}$ in $\mathcal{V}$ and all k in $\Re$.

Linear transformations are also called **linear functions** or **linear mappings.** In the special case in which $\mathcal{V} = \mathcal{W}$, a linear transformation is frequently called a **linear operator;** in the special case where $\mathcal{W} = \Re$, a linear transformation is often called a **linear functional.** In all of these situations "linear" means that Equation (4.1) is satisfied; that is, the functional value of a linear combination of vectors is the same linear combination of the functional values of those vectors.

You have already had a good bit of experience with linear transformations, as the following familiar examples will point out.

■ EXAMPLE 2 Matrix Multiplication Viewed as a Linear Transformation

Let $\mathcal{V} = \mathcal{W} = \Re^n$ and let the function $\mathbf{T}$ be defined as multiplication by the matrix A; that is, $\mathbf{T}(\mathbf{u}) = A\mathbf{u}$. In this case the previously established fact that matrix multiplication distributes over matrix addition (see Theorem 1.5) yields

$$\mathbf{T}(k\mathbf{u} + h\mathbf{v}) = A(k\mathbf{u} + h\mathbf{v}) = k(A\mathbf{u}) + h(A\mathbf{v}) = k\mathbf{T}(\mathbf{u}) + h\mathbf{T}(\mathbf{v}).$$

It follows at once that matrix multiplication is a linear function. To check whether or not $\mathbf{T}$ is one-to-one we let $\mathbf{u}$ and $\mathbf{v}$ be vectors in $\mathcal{V}$ such that $\mathbf{T}(\mathbf{u}) = \mathbf{T}(\mathbf{v})$. This is equivalent to $A\mathbf{u} = A\mathbf{v}$ or $A(\mathbf{u} - \mathbf{v}) = \mathbf{0}$. If A is invertible, it follows, from Theorem 3.12, that $\mathbf{u} - \mathbf{v} = \mathbf{0}$. Thus, $\mathbf{u} = \mathbf{v}$ and $\mathbf{T}$ is one-to-one. If A is singular, then the homogeneous system $A\mathbf{w} = \mathbf{0}$ has nonzero solutions and $\mathbf{T}$ is not one-to-one. Thus, $\mathbf{T}$ is one-to-one if, and only if, the matrix A is nonsingular. Since the $n \times n$ linear system $A\mathbf{u} = \mathbf{w}$ has a solution for every $\mathbf{w}$ in $\mathcal{W}$ if, and only if, A is nonsingular, it follows that $\mathbf{T}$ is onto $\mathcal{W}$ if, and only if, A is nonsingular. ■

The argument used in Example 2 carries over to the case where the defining matrix A is not square. We state the general result as our next theorem.

THEOREM 4.1 Let A be a fixed $m \times n$ matrix. The function $\mathbf{T}$, from $\Re^n$ to $\Re^m$ defined by $\mathbf{T}(\mathbf{u}) = A\mathbf{u}$ is linear. The transformation $\mathbf{T}$ is one-to-one if, and only if, $\text{rank}(A) = n$. The transformation $\mathbf{T}$ is onto $\Re^m$ if, and only if, $\text{rank}(A) = m$. If $m = n$, then $\mathbf{T}$ is both one-to-one and onto if, and only if, A is invertible.

Proof As in Example 2, the linearity is a consequence of the distributive properties of matrix multiplication. As in Example 2, **T** is one-to-one if, and only if, the homogeneous linear system $AX = 0$ has no nonzero solutions. This is equivalent to $\dim(\mathcal{NS}(A)) = 0$ which, by Theorem 3.11, is equivalent to $\text{rank}(A) = n$. The transformation **T** is onto $\Re^m$ if, and only if, the linear system $AX = Y$ has a solution for all Y in $\Re^m$, and this is equivalent to saying that $\mathcal{CS}(A) = \Re^m$. Since $\dim(\mathcal{CS}(A)) = \text{rank}(A)$ it follows that **T** is onto $\Re^m$ if, and only if, $\text{rank}(A) = m$. □

EXAMPLE 3 Testing for Linearity

Let $\mathcal{V} = \mathcal{W} = \Re^{1\times 3}$ and let **T** be the function defined by

$$\mathbf{T}[x \quad y \quad z] = [2x + 3y - z \quad 2 + x + y \quad 3|x - 3z|].$$

To get some feel for this function we will do some experimentation. If we let $\mathbf{u} = [1 \quad 2 \quad 3]$ and $\mathbf{v} = [-2 \quad 1 \quad 0]$ be a specific pair of vectors from $\mathcal{V}$, then by direct computation we have

$$\begin{aligned}
\mathbf{T}(\mathbf{u}) &= \mathbf{T}[1 \quad 2 \quad 3] = [2 + 6 - 3 \quad 2 + 1 + 2 \quad 3|1 - 9|] \\
&= [5 \quad 5 \quad 24]; \\
\mathbf{T}(\mathbf{v}) &= \mathbf{T}[-2 \quad 1 \quad 0] = [-1 \quad 1 \quad 6]; \\
\mathbf{T}(2\mathbf{v}) &= \mathbf{T}[-4 \quad 2 \quad 0] = [-2 \quad 0 \quad 12]; \\
\mathbf{T}(-2\mathbf{u}) &= \mathbf{T}[-2 \quad -4 \quad -6] = [-10 \quad -4 \quad 48]; \text{ and} \\
\mathbf{T}(\mathbf{u} + \mathbf{v}) &= \mathbf{T}[-1 \quad 3 \quad 3] = [4 \quad 4 \quad 30].
\end{aligned}$$

From these calculations we can see that

$$\begin{aligned}
\mathbf{T}(\mathbf{u} + \mathbf{v}) &= [4 \quad 4 \quad 30] \neq \mathbf{T}(\mathbf{u}) + \mathbf{T}(\mathbf{v}) \\
&= [5 \quad 5 \quad 24] + [-1 \quad 1 \quad 6] = [4 \quad 6 \quad 30]; \\
\mathbf{T}(2\mathbf{v}) &= [-2 \quad 0 \quad 12] \neq 2\mathbf{T}(\mathbf{v}) = 2[-1 \quad 1 \quad 6] \\
&= [-2 \quad 2 \quad 12]; \text{ and} \\
\mathbf{T}(-2\mathbf{u}) &= [-10 \quad -4 \quad 48] \neq -2\mathbf{T}(\mathbf{u}) = -2[5 \quad 5 \quad 24] \\
&= [-10 \quad -10 \quad -48].
\end{aligned}$$

Any one of the last three equations is sufficient to show that **T** is not a linear transformation. Note that **T** is not onto $\mathcal{W}$ since no vector with a negative third component is in the range of **T**. ■

Example 3 illustrates an important point: to show that a function is not linear, it is sufficient to find one special case in which Equation (4.1) fails to hold.

EXAMPLE 4 **Testing for Linearity**

Is the function **T**, from $\Re^2$ to $\Re^{1\times 2}$, defined by $\mathbf{T}\begin{bmatrix} x \\ y \end{bmatrix} = [x - 2y \quad 4x + 3y]$, a linear transformation?

Proceeding as in the last example we let $\mathbf{u} = \begin{bmatrix} 1 \\ 3 \end{bmatrix}$ and $\mathbf{v} = \begin{bmatrix} 2 \\ -1 \end{bmatrix}$ be a particular pair of vectors from $\Re^2$ and compute

$$\mathbf{T}(\mathbf{u}) = \mathbf{T}\begin{bmatrix} 1 \\ 3 \end{bmatrix} = [-5 \quad 13], \mathbf{T}(\mathbf{v}) = \mathbf{T}\begin{bmatrix} 2 \\ -1 \end{bmatrix} = [4 \quad 5];$$

$$\mathbf{T}(2\mathbf{u}) = \mathbf{T}\begin{bmatrix} 2 \\ 6 \end{bmatrix} = [-10 \quad 26] = 2\mathbf{T}(\mathbf{u}); \text{ and}$$

$$\mathbf{T}(\mathbf{u} + \mathbf{v}) = \mathbf{T}\begin{bmatrix} 3 \\ 2 \end{bmatrix} = [-1 \quad 18] = \mathbf{T}(\mathbf{u}) + \mathbf{T}(\mathbf{v}).$$

These calculations suggest, *but do not prove,* that **T** is a linear transformation. To establish this in general we let $\mathbf{u} = \begin{bmatrix} u_1 \\ u_2 \end{bmatrix}$ and $\mathbf{v} = \begin{bmatrix} v_1 \\ v_2 \end{bmatrix}$ be arbitrary vectors in $\mathcal{V}$ and compute

$$\mathbf{T}(k\mathbf{u} + h\mathbf{v}) = \mathbf{T}\begin{bmatrix} ku_1 + hv_1 \\ ku_2 + hv_2 \end{bmatrix}$$
$$= [(ku_1 + hv_1) - 2(ku_2 + hv_2) \quad 4(ku_1 + hv_1) + 3(ku_2 + hv_2)]$$

and $k\mathbf{T}(\mathbf{u}) + h\mathbf{T}(\mathbf{v}) = k[u_1 - 2u_2 \quad 4u_1 + 3u_2] + h[v_1 - 2v_2 \quad 4v_1 + 3v_2]$. Since these two vectors are equal, we have shown that Equation (4.1) holds for all vectors **u** and **v** in $\Re^2$. Thus, **T** is a linear transformation. ■

EXAMPLE 5 **Projections**

Consider the transformation **P**, from $\Re^3$ to $\Re^3$, defined by

$$\mathbf{P}\begin{bmatrix} x \\ y \\ z \end{bmatrix} = \begin{bmatrix} 1 & 0 & 0 \\ 0 & 1 & 0 \\ 0 & 0 & 0 \end{bmatrix}\begin{bmatrix} x \\ y \\ z \end{bmatrix} = \begin{bmatrix} x \\ y \\ 0 \end{bmatrix}.$$

Because it is defined by matrix multiplication, **P** is linear by Example 1. The range of this transformation is the xy-plane of $\Re^3$ and hence the transformation is not onto $\Re^3$; $\mathbf{e}_3 = [0 \quad 0 \quad 1]^T$ is not in the range of **P**. Since $\mathbf{P}(\mathbf{e}_3) = \mathbf{P}(\mathbf{0}) = \mathbf{0}$, **P** is not one-to-one. This transformation can be described geometrically as a **projection** of $\Re^3$ onto the xy-plane. ■

If you have studied calculus, you have already spent a good deal of time studying linear operators on vector spaces of functions. If you have not studied calculus, then you should disregard the next example.

EXAMPLE 6 **Linearity of the Derivative**

Let $\mathcal{V} = \mathcal{C}^1[a, b]$ be the subspace of $\mathcal{F}[a, b]$ which consists of all those functions from $\mathcal{F}[a, b]$ that have continuous derivatives on the interval $[a, b]$. Let $\mathbf{D}$ be the function from $\mathcal{V}$ to $\mathcal{F}[a, b]$ defined by

$$\mathbf{D}(f)(x) = f'(x) = \frac{d}{dx}(f(x)).$$

It is shown in every beginning calculus course that $\mathbf{D}$ is linear; that is,

$$\mathbf{D}(af(x) + bg(x)) = af'(x) + bg'(x) = a\mathbf{D}(f(x)) + b\mathbf{D}(g(x)).$$ ■

Many additional examples of linear transformations are contained in the exercises. We now list some elementary, but important, properties of any linear transformation.

THEOREM 4.2 If $\mathbf{T}$ is a linear transformation from $\mathcal{V}$ to $\mathcal{W}$, then for all vectors $\mathbf{v}$ and $\mathbf{w}$ in $\mathcal{V}$ and all scalars a_i we have:

(1) $\mathbf{T}(\mathbf{0}) = \mathbf{0}$;
(2) $\mathbf{T}(-\mathbf{v}) = -\mathbf{T}(\mathbf{v})$;
(3) $\mathbf{T}(\mathbf{v} - \mathbf{w}) = \mathbf{T}(\mathbf{v}) - \mathbf{T}(\mathbf{w})$;
(4) $\mathbf{T}(a_1\mathbf{v_1} + a_2\mathbf{v_2} + \cdots + a_k\mathbf{v_k}) = a_1\mathbf{T}(\mathbf{v_1}) + a_2\mathbf{T}(\mathbf{v_2}) + \cdots + a_k\mathbf{T}(\mathbf{v_k})$; and
(5) $\mathbf{T}$ is one-to-one if, and only if, $\mathbf{T}(\mathbf{u}) = \mathbf{0}$ implies $\mathbf{u} = \mathbf{0}$.

Proof The first four assertions of the theorem follow directly from Definition 4.1. If $\mathbf{T}$ is one-to-one and $\mathbf{T}(\mathbf{u}) = \mathbf{0}$, then, since we also have $\mathbf{T}(\mathbf{0}) = \mathbf{0}$, it follows that $\mathbf{u} = \mathbf{0}$. Conversely, suppose that $\mathbf{x}$ and $\mathbf{y}$ are vectors such that $\mathbf{T}(\mathbf{x}) = \mathbf{T}(\mathbf{y})$. It follows that $\mathbf{T}(\mathbf{x}) - \mathbf{T}(\mathbf{y}) = \mathbf{0}$ and, by the linearity of $\mathbf{T}$, that $\mathbf{T}(\mathbf{x} - \mathbf{y}) = \mathbf{0}$. It follows from the hypothesis that $\mathbf{x} - \mathbf{y} = \mathbf{0}$ so that $\mathbf{x} = \mathbf{y}$ and $\mathbf{T}$ is one-to-one. □

If $\mathcal{B} = \{\mathbf{u_1}, \mathbf{u_2}, \ldots, \mathbf{u_n}\}$ is a basis for the vector space $\mathcal{V}$ and $\mathbf{w}$ is any vector in $\mathcal{V}$, we know that we can express $\mathbf{w}$ as a linear combination of the basis vectors by solving a linear system:

$$\mathbf{w} = \sum_{i=1}^{n} a_i\mathbf{u_i} = a_1\mathbf{u_1} + \cdots + a_n\mathbf{u_n}.$$

If $\mathbf{T}$ is a linear transformation on $\mathcal{V}$, we can compute $\mathbf{T}(\mathbf{w})$ using

$$\mathbf{T}(\mathbf{w}) = \sum_{i=1}^{n} a_i\mathbf{T}(\mathbf{u_i}) = a_1\mathbf{T}(\mathbf{u_1}) + \cdots + a_n\mathbf{T}(\mathbf{u_n}).$$

This last equation shows that once we know the $\mathbf{T}(\mathbf{u_i})$, which are the images of the basis vectors, then we can compute the image, under $\mathbf{T}$, of any other vector in the space. Since $\mathbf{T}(\mathbf{w})$ is a typical vector in the range of $\mathbf{T}$, this equation also shows that the vectors $\mathbf{T}(\mathbf{u}_i)$ span the range of $\mathbf{T}$. This argument also shows that the range of $\mathbf{T}$ is a subspace of $\mathcal{W}$.

We summarize this discussion in the following theorem.

THEOREM 4.3 A linear transformation on a finite dimesional vector space $\mathcal{V}$ is completely determined by its values for the vectors of any basis for $\mathcal{V}$. Moreover, if $\mathcal{B} = \{\mathbf{u}_1, \mathbf{u}_2, \ldots, \mathbf{u}_n\}$ is a basis for $\mathcal{V}$, then $\{\mathbf{T}(\mathbf{u}_1), \mathbf{T}(\mathbf{u}_2), \ldots, \mathbf{T}(\mathbf{u}_n)\}$ is a spanning set for the range of $\mathbf{T}$. □

EXAMPLE 7 Transformation Determined by its Action on a Basis

Let $\mathbf{T}$ be a linear transformation from $\Re^3$ to $\Re^3$ satisfying

$$\mathbf{T}\begin{bmatrix}1\\1\\1\end{bmatrix} = \begin{bmatrix}2\\3\\3\end{bmatrix}, \quad \mathbf{T}\begin{bmatrix}1\\1\\0\end{bmatrix} = \begin{bmatrix}-2\\4\\3\end{bmatrix}, \quad \text{and} \quad \mathbf{T}\begin{bmatrix}0\\1\\1\end{bmatrix} = \begin{bmatrix}0\\7\\5\end{bmatrix}.$$

Use Theorem 4.3 to find $\mathbf{T}\begin{bmatrix}-1\\3\\6\end{bmatrix}$.

It is easy to see that the vectors

$$\left\{\mathbf{u}_1 = \begin{bmatrix}1\\1\\1\end{bmatrix}, \mathbf{u}_2 = \begin{bmatrix}1\\1\\0\end{bmatrix}, \mathbf{u}_3 = \begin{bmatrix}0\\1\\1\end{bmatrix}\right\}$$

form a basis for $\Re^3$. Thus, there must exist unique scalars a, b, and c such that

$$\mathbf{w} = \begin{bmatrix}-1\\3\\6\end{bmatrix} = a\begin{bmatrix}1\\1\\1\end{bmatrix} + b\begin{bmatrix}1\\1\\0\end{bmatrix} + c\begin{bmatrix}0\\1\\1\end{bmatrix} = a\mathbf{u}_1 + b\mathbf{u}_2 + c\mathbf{u}_3.$$

The scalars a, b, and c must satisfy the linear system

$$\begin{bmatrix}1 & 1 & 0\\1 & 1 & 1\\1 & 0 & 1\end{bmatrix}\begin{bmatrix}a\\b\\c\end{bmatrix} = \begin{bmatrix}-1\\3\\6\end{bmatrix}.$$

The row reduction

$$\begin{bmatrix}1 & 1 & 0 & -1\\1 & 1 & 1 & 3\\1 & 0 & 1 & 6\end{bmatrix} \to \begin{bmatrix}1 & 1 & 0 & -1\\0 & 0 & 1 & 4\\0 & -1 & 1 & 7\end{bmatrix} \to \begin{bmatrix}1 & 0 & 0 & 2\\0 & 1 & 0 & -3\\0 & 0 & 1 & 4\end{bmatrix}$$

shows that $a = 2$, $b = -3$, and $c = 4$, so that $\mathbf{w} = 2\mathbf{u}_1 - 3\mathbf{u}_2 + 4\mathbf{u}_3$. We can now compute $\mathbf{T}(\mathbf{w})$ using the linearity of $\mathbf{T}$:

$$\begin{aligned}\mathbf{T}(\mathbf{w}) &= \mathbf{T}(2\mathbf{u}_1 - 3\mathbf{u}_2 + 4\mathbf{u}_3)\\ &= 2\mathbf{T}(\mathbf{u}_1) - 3\mathbf{T}(\mathbf{u}_2) + 4\mathbf{T}(\mathbf{u}_3)\\ &= 2\begin{bmatrix}2\\3\\5\end{bmatrix} - 3\begin{bmatrix}-2\\4\\3\end{bmatrix} + 4\begin{bmatrix}0\\7\\5\end{bmatrix} = \begin{bmatrix}10\\22\\21\end{bmatrix}.\end{aligned}$$ ■

EXERCISES 4.1

1. Which of the following transformations are linear? In each case identify the domain of the function and the space $\mathcal{W}$ which contains the range.

(a) $\mathbf{T}\begin{bmatrix}x_1\\x_2\\x_3\end{bmatrix} = \begin{bmatrix}x_1 + x_2\\3x_1 - 7x_2 + x_3\\x_1 - 3x_2 + 2x_3\end{bmatrix}$

(b) $\mathbf{T}\begin{bmatrix}x_1\\x_2\\x_3\end{bmatrix} = \begin{bmatrix}0\\1\\x_1 + x_2\end{bmatrix}$

(c) $\mathbf{T}\begin{bmatrix}x_1\\x_2\\x_3\end{bmatrix} = \begin{bmatrix}x_1^2\\|x_2|\\\sin x_3\end{bmatrix}$

(d) $\mathbf{T}(a + bx + cx^2 + dx^3) = b + cx + dx^2$

(e) $\mathbf{T}(y) = \frac{d^2y}{dx^2} + 3y^2 - 6\frac{dy}{dx}$

(f) $\mathbf{T}\begin{bmatrix}a & b\\c & d\end{bmatrix} = \begin{bmatrix}a + b & 2b\\c + d & 2d\end{bmatrix}$

2. Which of the following functions are linear? In each case identify the domain of the function and the vector space $\mathcal{W}$ which contains the range.

(a) $\mathbf{T}[x \quad y] + \begin{bmatrix}2x\\y\end{bmatrix}$

(b) $\mathbf{T}[x \quad y \quad z] = x^2 + 2y + z$

(c) $\mathbf{T}[x \quad y] = [x + y \quad 2x - y + 2]^T$

(d) $\mathbf{T}[x \quad y \quad z] = \begin{bmatrix}2x + y & 3y\\4x - z & 2z + x\end{bmatrix}$

(e) $\mathbf{T}(a_0 + a_1x + a_2x^2) = a_0 + x + a_2x^2$

(f) $\mathbf{T}(A) = \det(A)$

(g) $\mathbf{T}(A) = \operatorname{tr}(A)$ (Recall Exercise 11 of Section 2.1.)

(h) $\mathbf{T}\begin{bmatrix}a & b\\c & d\end{bmatrix} = [a \quad b \quad c \quad d]$

(i) $\mathbf{T}(A) = A^TA$

3. Let $\mathcal{P}_k$ be the vector space of polynomials of degree less than or equal to k. Let $p(x) = a_0 + a_1x + a_2x^2 + \cdots + a_kx^k$. Which of the following functions on $\mathcal{P}_k$ are linear? In each case identify the range of the function.

(a) $\mathbf{T}(p(x)) = xp(x)$

(b) $\mathbf{T}(p(x)) = p(x)^2$

(c) $\mathbf{T}(p(x)) = 3a_3x^2$

(d) $\mathbf{T}(p(x)) = a_k + a_{k-1}x + a_{k-2}x^2 + \cdots + a_0x^k$

4. Which of the following mappings on the vector space $\mathfrak{R}^{2\times 2}$ are linear?

(a) $\mathbf{T}(A) = 3A^T$

(b) $\mathbf{T}(A) = A + M$ where M is a fixed matrix from $\mathfrak{R}^{2\times 2}$

(c) $\mathbf{T}(A) = AM$

5. Is the function $\mathbf{T}(\mathbf{x}) = a\mathbf{x} + b$ a linear operator on $\mathfrak{R}$?

6. Let $\mathbf{T}$ be a linear transformation from $\mathfrak{R}^2$ to $\mathfrak{R}^3$ such that

$\mathbf{T}(\mathbf{e}_1) = \begin{bmatrix} 3 \\ 2 \\ 1 \end{bmatrix}$ and $\mathbf{T}(\mathbf{e}_2) = \begin{bmatrix} -2 \\ 4 \\ 3 \end{bmatrix}$. Find $\mathbf{T}(7\mathbf{e}_1 + 5\mathbf{e}_2)$, $\mathbf{T}\begin{bmatrix} 3 \\ 2 \end{bmatrix}$, and $\mathbf{T}\begin{bmatrix} 6 \\ 5 \end{bmatrix}$.

7. Let $\mathbf{T}$ be a linear transformation from $\mathfrak{R}^2$ to $\mathfrak{R}^3$ which maps $\mathbf{u}_1 = \begin{bmatrix} 0 \\ 1 \end{bmatrix}$ into $\begin{bmatrix} 1 \\ 0 \\ 2 \end{bmatrix}$

and $\mathbf{u}_2 = \begin{bmatrix} 1 \\ 1 \end{bmatrix}$ into $\begin{bmatrix} 1 \\ 2 \\ 3 \end{bmatrix}$. Find $\mathbf{T}(3\mathbf{u}_1)$, $\mathbf{T}(-2\mathbf{u}_2)$, $\mathbf{T}(2\mathbf{u}_1 - 7\mathbf{u}_2)$, and $\mathbf{T}(3\mathbf{u}_1 + 2\mathbf{u}_2)$.

8. Let $\mathcal{S} = \{\mathbf{u}_1, \mathbf{u}_2, \ldots, \mathbf{u}_n\}$ be a set of n vectors from the vector space $\mathcal{V}$. Consider the function $\mathbf{L}$ from $\mathfrak{R}^n$ to $\mathcal{V}$ defined by

$$\mathbf{L}\begin{bmatrix} a_1 \\ a_2 \\ \cdot \\ \cdot \\ \cdot \\ (a_n) \end{bmatrix} = a_1\mathbf{u}_1 + a_2\mathbf{u}_2 + \cdots + a_n\mathbf{u}_n.$$

(a) Show that $\mathbf{L}$ is a linear transformation.

(b) Show that $\mathbf{L}$ is one-to-one if, and only if, $\mathcal{S}$ is linearly independent.

(c) Show that $\mathbf{L}$ is onto $\mathcal{V}$ if, and ony if, $\mathcal{S}$ spans $\mathcal{V}$.

9. A linear transformation $\mathbf{T}$, from $\mathfrak{R}^3$ to $\mathfrak{R}^{1\times 3}$, satisfies

$\mathbf{T}\begin{bmatrix} 1 \\ 0 \\ 0 \end{bmatrix} = [1 \quad 2 \quad 3]$, $\mathbf{T}\begin{bmatrix} 1 \\ 1 \\ 0 \end{bmatrix} = [-1 \quad 4 \quad 2]$, and $\mathbf{T}\begin{bmatrix} 1 \\ 1 \\ 1 \end{bmatrix} = [0 \quad 6 \quad 4]$.

Find $\mathbf{T}\begin{bmatrix} 3 \\ 2 \\ 2 \end{bmatrix}$, $\mathbf{T}\begin{bmatrix} 0 \\ 0 \\ 1 \end{bmatrix}$, and $\mathbf{T}\begin{bmatrix} 3 \\ 5 \\ -2 \end{bmatrix}$.

10. Let $\mathbf{T}$ be a linear transformation on a finite dimensional vector space $\mathcal{V}$. Show that $\mathbf{T}$ is zero on all of $\mathcal{V}$ if it is zero on a basis for $\mathcal{V}$.

11. Consider the transformation $\mathbf{U}$ from $\mathfrak{R}^4$ to $\mathfrak{R}^3$ defined by $\mathbf{U}(\mathbf{v}) = M\mathbf{v}$, where $M = \begin{bmatrix} 1 & 2 & 3 & 1 \\ 4 & 5 & 6 & 1 \\ 7 & 8 & 9 & 1 \end{bmatrix}$. Is $\mathbf{U}$ linear? Is it one-to-one? Is it onto $\mathfrak{R}^3$?

12. Show that $\left\{M_1 = \begin{bmatrix} 1 & 0 \\ 0 & 1 \end{bmatrix}, M_2 = \begin{bmatrix} 0 & 1 \\ 1 & 0 \end{bmatrix}, M_3 = \begin{bmatrix} 1 & 0 \\ 0 & -1 \end{bmatrix}, M_4 = \begin{bmatrix} 1 & 1 \\ 0 & 0 \end{bmatrix}\right\}$ is a basis for $\mathfrak{R}^{2\times 2}$. If $\mathbf{T}$ is a linear functional satisfying $\mathbf{T}(M_1) = 4$, $\mathbf{T}(M_2) = 1$, $\mathbf{T}(M_3) = 0$, and $\mathbf{T}(M_4) = 2$, find $\mathbf{T}\left(\begin{bmatrix} a & b \\ c & d \end{bmatrix}\right)$ and determine if $\mathbf{T}$ is one-to-one.

13. Let $\mathbf{T}$ be the linear operator on $\mathfrak{R}^3$ defined by $\mathbf{T}(\mathbf{u}) = A\mathbf{u}$, where $A = \begin{bmatrix} 1 & -1 & 0 \\ -1 & 2 & 3 \\ 0 & 3 & 0 \end{bmatrix}$. For $\mathbf{u} = \begin{bmatrix} 1 \\ 2 \\ 3 \end{bmatrix}$, compute $\mathbf{T}(\mathbf{u})$, $\mathbf{T}(\mathbf{T}(\mathbf{u}))$, $\mathbf{T}(\mathbf{T}(\mathbf{u}) + 3\mathbf{u})$ and $(A^2 + 3A)\mathbf{u}$.

14. Consider the transformation $\mathbf{S}$ from $\mathfrak{R}^{m\times m}$ to $\mathfrak{R}^{m\times m}$ defined by $\mathbf{S}(A) = U^{-1}AU$, where U is a fixed nonsingular matrix from $\mathfrak{R}^{m\times m}$. Is $\mathbf{S}$ linear? Is it one-to-one? Is it onto $\mathfrak{R}^{m\times m}$?

†**15.** **(a)** Let $\mathcal{V}$ be any real vector space and define the operator $\mathbf{I}$ on $\mathcal{V}$ by $\mathbf{I}(\mathbf{v}) = \mathbf{v}$. The transformation $\mathbf{I}$ is called the **identity mapping** on $\mathcal{V}$. Show that $\mathbf{I}$ is linear. Describe $\mathbf{I}$ as a matrix multiplication in the case $\mathcal{V} = \mathfrak{R}^n$.
(b) Let $\mathcal{V}$ be any real vector space, let k be any scalar and define the scalar multiplication operator $\mathbf{k}$ on $\mathcal{V}$ by $\mathbf{k}(\mathbf{v}) = k\mathbf{v}$. Show that $\mathbf{k}$ is linear.

16. Show that a function $\mathbf{F}$, from $\mathfrak{R}$ to $\mathfrak{R}$, is linear if, and only if, $\mathbf{F}(x) = kx$ for $k = \mathbf{F}(1)$.

17. Let $\mathbf{T}$ be a linear transformation on $\mathfrak{R}^{2\times 2}$ such that

$$\mathbf{T}\begin{bmatrix} 1 & 0 \\ 0 & 0 \end{bmatrix} = \begin{bmatrix} 2 \\ 3 \end{bmatrix}, \quad \mathbf{T}\begin{bmatrix} 0 & 1 \\ 0 & 0 \end{bmatrix} = \begin{bmatrix} 5 \\ -7 \end{bmatrix},$$

$$\mathbf{T}\begin{bmatrix} 0 & 0 \\ 1 & 0 \end{bmatrix} = \begin{bmatrix} 1 \\ 1 \end{bmatrix}, \quad \mathbf{T}\begin{bmatrix} 0 & 0 \\ 0 & 1 \end{bmatrix} = \begin{bmatrix} 0 \\ 0 \end{bmatrix}.$$

Find $\mathbf{T}\begin{bmatrix} 3 & -11 \\ 4 & 2 \end{bmatrix}$ and $\mathbf{T}\begin{bmatrix} 5 & 0 \\ 0 & 7 \end{bmatrix}$.

18. Can you find a one-to-one linear transformation from $\mathfrak{R}^3$ onto $\mathfrak{R}^4$? Can you find a one-to-one linear transformation from $\mathfrak{R}^3$ onto $\mathfrak{R}^2$?

19. Show that if $\{\mathbf{v}_1, \cdots, \mathbf{v}_k\}$ is a linearly independent set in $\mathcal{V}$ and $\mathcal{B}$ is any basis for $\mathcal{V}$, then the coordinate matrices

$$\{[\mathbf{v}_1]_{\mathcal{B}}, [\mathbf{v}_2]_{\mathcal{B}}, \ldots, [\mathbf{v}_k]_{\mathcal{B}}\}$$

form a linearly independent set in $\mathfrak{R}^n$.

20. Prove parts 1–4 of Theorem 4.2.

21. Show geometrically that the function **R** on $\mathfrak{R}^2$, defined by rotation through an angle θ, is a linear transformation on $\mathfrak{R}^2$.

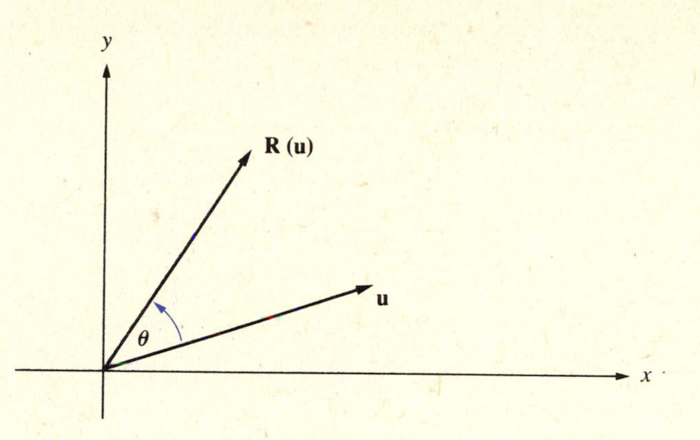

22. Let **T** be the linear transformation of Exercise 13. Find **x** in $\mathfrak{R}^3$ such that **(a)** $\mathbf{T}(\mathbf{x}) = [1 \quad 0 \quad 0]^T$ and **(b)** $\mathbf{T}(\mathbf{x}) = [2 \quad 3 \quad -7]^T$.

23. Let **D** be the derivative operator on $\mathcal{P}$, the vector space of all polynomials of finite degree. Show that **D** is onto $\mathcal{P}$ but not one-to-one. Find an example of a linear operator on $\mathcal{P}$ that is one-to-one but not onto $\mathcal{P}$.

†24. Let $\mathcal{V} = \mathcal{W} = \mathfrak{R}^n$ and let **T** be the linear transformation $\mathbf{T}(\mathbf{u}) = A\mathbf{u}$ where A is a fixed $n \times n$ matrix. Show that the following statements are equivalent:

(1) **T** is one-to-one;
(2) A is nonsingular; and
(3) **T** is onto $\mathcal{W}$.

25. Let **T** be a linear transformation from $\mathcal{V}$ to $\mathcal{W}$ and let

$$\mathcal{B} = \{\mathbf{u}_1, \mathbf{u}_2, \ldots, \mathbf{u}_n\}$$

be a basis for $\mathcal{V}$. Show that if

$$\mathbf{T}(\mathcal{B}) = \{\mathbf{T}(\mathbf{u}_1), \mathbf{T}(\mathbf{u}_2), \ldots, \mathbf{T}(\mathbf{u}_n)\}$$

is a basis for $\mathcal{W}$, then **T** is one-to-one and onto $\mathcal{W}$.

26. Let $\mathcal{V}$ be any n-dimensional real vector space and let $\mathcal{B} = \{\mathbf{v}_1, \mathbf{v}_2, \ldots, \mathbf{v}_n\}$ be a basis for $\mathcal{V}$. If **w** and **u** are any vectors in $\mathcal{V}$, show that $[\mathbf{w} + \mathbf{u}]_{\mathcal{B}} = [\mathbf{w}]_{\mathcal{B}} + [\mathbf{u}]_{\mathcal{B}}$ and that $[k\mathbf{w}]_{\mathcal{B}} = k[\mathbf{w}]_{\mathcal{B}}$, thus showing that the mapping **C** from $\mathcal{V}$ to $\mathfrak{R}^n$ defined by $\mathbf{C}(\mathbf{w}) = [\mathbf{w}]_{\mathcal{B}}$ is a linear transformation. Show further that the coordinate mapping $\mathbf{C} = [\ ]_{\mathcal{B}}$ is a one-to-one linear transformation from $\mathcal{V}$ onto $\mathfrak{R}^n$.

4.2 THE RANGE AND NULL SPACE OF A LINEAR TRANSFORMATION

1. What does it mean to say that a problem is "linear"?
2. Which subspaces of $\mathcal{V}$ and $\mathcal{W}$ are useful in studying a linear transformation $\mathbf{T}$ from $\mathcal{V}$ to $\mathcal{W}$?

Let us consider the linear transformation $\mathbf{T}$, from $\Re^n$ to $\Re^m$, that is defined by multiplication by the $m \times n$ matrix A, that is, $\mathbf{T}(\mathbf{u}) = A\mathbf{u}$. Transformations of this type are called **matrix transformations**; they serve as the motivating example for the study of linear transformations. Solving the linear system

$$AX = K \qquad \text{or} \qquad \mathbf{T}(\mathbf{X}) = \mathbf{K}$$

with A as its coefficient matrix, can be viewed as the problem of finding all vectors $\mathbf{X}$ in $\Re^n$, whose image under the transformation $\mathbf{T}$ is the vector $\mathbf{K}$ in $\Re^m$.

Now that we have introduced the notion of linearity, we can describe precisely what it means for a given problem to be linear. The **general linear problem** can be phrased as follows: given a linear transformation $\mathbf{T}$ from a vector space $\mathcal{V}$ to a vector space $\mathcal{W}$ and a fixed vector $\mathbf{w}$ in $\mathcal{W}$, find all vectors $\mathbf{x}$ in $\mathcal{V}$ such that $\mathbf{T}(\mathbf{x}) = \mathbf{w}$.

If $\mathbf{T}$ is matrix multiplication, then the associated linear problem is the linear equations problem of Chapter 1; if $\mathbf{T}$ is differentiation, then the linear problem is a simple differential equations problem that can be solved by integration. Other examples occur frequently in higher mathematics and in many areas of application.

The linear problem $\mathbf{T}(\mathbf{x}) = \mathbf{w}$ is said to be **homogeneous** if $\mathbf{w} = \mathbf{0}$ and **nonhomogeneous** otherwise. In Figure 4.3 the vectors $\mathbf{u}_1$, $\mathbf{u}_2$, and $\mathbf{u}_3$ are all solutions of the nonhomogeneous linear problem $\mathbf{T}(\mathbf{x}) = \mathbf{w}$, while the vectors $\mathbf{v}_1$, $\mathbf{v}_2$, and $\mathbf{0}$ are all solutions of the associated homogeneous problem $\mathbf{T}(\mathbf{x}) = \mathbf{0}$.

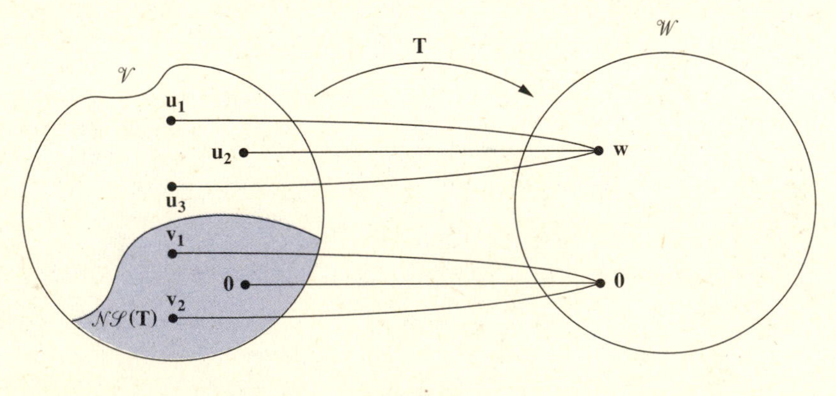

FIGURE 4.3

In the last chapter we introduced several subspaces associated with the matrix A. We can just as easily associate these subspaces with the linear transformation $\mathbf{T}$ defined by $\mathbf{T}(\mathbf{u}) = A\mathbf{u}$. In particular, recall that the null space of the $m \times n$ matrix A and the column space of A were defined by

$$\mathcal{NS}(A) = \{X \text{ in } \Re^n \mid AX = 0\} \qquad \text{and} \qquad \mathcal{CS}(A) = \{AX \mid X \text{ in } \Re^n\}.$$

In terms of the matrix transformation $\mathbf{T}$ these definitions become

$$\mathcal{NS}(\mathbf{T}) = \{\mathbf{u} \text{ in } \Re^n \mid \mathbf{T}(\mathbf{u}) = \mathbf{0}\} \qquad \text{and}$$
$$\mathcal{CS}(A) = \mathbf{T}(\mathcal{V}) = \{\mathbf{T}(\mathbf{u}) \mid \mathbf{u} \text{ in } \Re^n\}.$$

Note that in this case $\mathcal{CS}(A) = \mathcal{R}(\mathbf{T})$, which is the range of the transformation $\mathbf{T}$. When dealing with linear transformations the null space is sometimes called the **kernel** of $\mathbf{T}$; we will not use that terminology here.

These notions extend easily to the general case.

DEFINITION 4.2 If $\mathbf{T}$ is a linear transformation from $\mathcal{V}$ to $\mathcal{W}$, then the **null space** of $\mathbf{T}$ is defined by

$$\mathcal{NS}(\mathbf{T}) = \{\mathbf{u} \text{ in } \mathcal{V} \mid \mathbf{T}(\mathbf{u}) = \mathbf{0}\}.$$

The **range** of $\mathbf{T}$ is defined by

$$\mathcal{R}(\mathbf{T}) = \mathbf{T}(\mathcal{V}) = \{\mathbf{T}(\mathbf{u}) \mid \mathbf{u} \text{ in } \mathcal{V}\}. \quad \square$$

The diagram in Figure 4.4 should be helpful in visualizing these subsets of $\mathcal{V}$ and $\mathcal{W}$. Recall, from part 5 of Theorem 4.1, that $\mathbf{T}$ is one-to-one if, and only if, $\mathcal{NS}(\mathbf{T})$ contains only the vector $\mathbf{0}$.

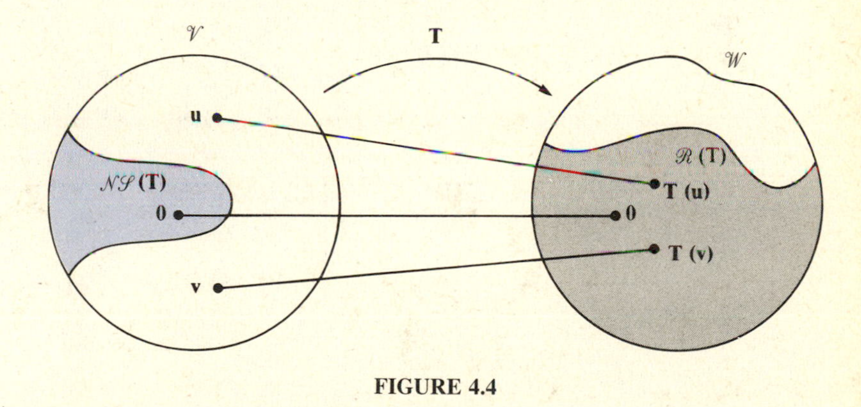

FIGURE 4.4

THEOREM 4.4 If $\mathbf{T}$ is a linear transformation from $\mathcal{V}$ to $\mathcal{W}$, then the null space of $\mathbf{T}$ is a subspace of the domain $\mathcal{V}$ and the range of $\mathbf{T}$ is a subspace of $\mathcal{W}$.

Proof If $\mathbf{u}$ and $\mathbf{v}$ are in $\mathcal{NS}(\mathbf{T})$, then $\mathbf{T}(\mathbf{u}) = \mathbf{T}(\mathbf{v}) = \mathbf{0}$. It follows from the linearity of $\mathbf{T}$ that $\mathbf{T}(a\mathbf{u} + b\mathbf{v}) = a\mathbf{T}(\mathbf{u}) + b\mathbf{T}(\mathbf{v}) = \mathbf{0}$, so $a\mathbf{u} + b\mathbf{v}$ is also in $\mathcal{NS}(\mathbf{T})$. Therefore, by Theorem 3.21, $\mathcal{NS}(\mathbf{T})$ is a subspace of $\mathcal{V}$. To see that

$\mathcal{R}(\mathbf{T})$ is a subspace of $\mathcal{W}$ we let $\mathbf{w}_1$ and $\mathbf{w}_2$ be any two vectors in $\mathcal{R}(\mathbf{T})$ and let a and b be any real numbers. We need to show that $a\mathbf{w}_1 + b\mathbf{w}_2$ is also in $\mathcal{R}(\mathbf{T})$. Since $\mathbf{w}_1$ and $\mathbf{w}_2$ are in $\mathcal{R}(\mathbf{T})$, there exist vectors $\mathbf{v}_1$ and $\mathbf{v}_2$ in $\mathcal{V}$ such that $\mathbf{T}(\mathbf{v}_1) = \mathbf{w}_1$ and $\mathbf{T}(\mathbf{v}_2) = \mathbf{w}_2$. In order to show that $a\mathbf{w}_1 + b\mathbf{w}_2$ is in $\mathcal{R}(\mathbf{T})$ we need to find a vector $\mathbf{x}$ in $\mathcal{V}$ such that $\mathbf{T}(\mathbf{x}) = a\mathbf{w}_1 + b\mathbf{w}_2$. The calculation

$$\mathbf{T}(a\mathbf{v}_1 + b\mathbf{v}_2) = a\mathbf{T}(\mathbf{v}_1) + b\mathbf{T}(\mathbf{v}_2) = a\mathbf{w}_1 + b\mathbf{w}_2$$

shows that we can choose $\mathbf{x} = a\mathbf{v}_1 + b\mathbf{v}_2$. Since $\mathbf{x}$ is in $\mathcal{V}$ the proof is completed. □

DEFINITION 4.3 For a linear transformation $\mathbf{T}$ we define the **rank** of $\mathbf{T}$ to be the dimension of the range of $\mathbf{T}$ and the **nullity** of $\mathbf{T}$ to be the dimension of the null space of $\mathbf{T}$:

$$\text{rank}(\mathbf{T}) = \dim(\mathcal{R}(\mathbf{T})) \qquad \text{and} \qquad \text{nullity}(\mathbf{T}) = \dim(\mathcal{NS}(\mathbf{T})). \quad \square$$

EXAMPLE 1 Bases for $\mathcal{NS}(\mathbf{T})$ and $\mathcal{R}(\mathbf{T})$

Find bases for $\mathcal{NS}(\mathbf{T})$ and $\mathcal{R}(\mathbf{T})$ if $\mathbf{T}$ is the linear transformation on $\mathfrak{R}^4$ defined by $\mathbf{T}(\mathbf{u}) = A\mathbf{u}$ with

$$A = \begin{bmatrix} 1 & -2 & 1 & -4 \\ 1 & 3 & 7 & 2 \\ 1 & -12 & -11 & -16 \\ 2 & -14 & -10 & -20 \end{bmatrix}.$$

In Example 4 of Section 3.3 we did the following row reduction:

$$[A|K] = \begin{bmatrix} 1 & -2 & 1 & -4 & k_1 \\ 1 & 3 & 7 & 2 & k_2 \\ 1 & -12 & -11 & -16 & k_3 \\ 2 & -14 & -10 & -20 & k_4 \end{bmatrix}$$

$$\rightarrow \begin{bmatrix} 1 & 0 & \frac{17}{5} & -\frac{8}{5} & k_1 + \frac{2}{5}(k_2 - k_1) \\ 0 & 1 & \frac{6}{5} & \frac{6}{5} & \frac{k_2 - k_1}{5} \\ 0 & 0 & 0 & 0 & k_3 - 3k_1 + 2k_2 \\ 0 & 0 & 0 & 0 & k_4 - 4k_1 + 2k_2 \end{bmatrix}.$$

From this row reduction, with $K = 0$, we see that the homogeneous system $AX = 0$ is equivalent to the system

$$x_1 + \frac{17}{5}x_3 - \frac{8}{5}x_4 = 0$$

$$x_2 + \frac{6}{5}x_3 + \frac{6}{5}x_4 = 0.$$

If we set $x_3 = 5c_1$ and $x_4 = 5c_2$, it follows that the general solution of $AX = 0$ is

$$X = c_1\begin{bmatrix} -17 \\ -6 \\ 5 \\ 0 \end{bmatrix} + c_2\begin{bmatrix} 8 \\ -6 \\ 0 \\ 5 \end{bmatrix} = c_1 X_1 + c_2 X_2.$$

It also follows that $\dim(\mathcal{NS}(\mathbf{T})) = 2$ and that a basis for $\mathcal{NS}(\mathbf{T})$ is

$$\mathcal{B}_1 = \left\{ X_1 = \begin{bmatrix} -17 \\ -6 \\ 5 \\ 0 \end{bmatrix}, X_2 = \begin{bmatrix} 8 \\ -6 \\ 0 \\ 5 \end{bmatrix} \right\}.$$

Since $\mathcal{R}(\mathbf{T}) = \mathcal{CS}(A)$ is the set of vectors K for which the system $AX = K$ has a solution, it follows from the above row reduction that K is in $\mathcal{R}(\mathbf{T})$ if, and only if,

$$k_3 - 3k_1 + 2k_2 = 0 \qquad \text{and} \qquad k_4 - 4k_1 + 2k_2 = 0.$$

The general solution of this 2×4 system is easily seen to be

$$K = k_1\begin{bmatrix} 1 \\ 0 \\ 3 \\ 4 \end{bmatrix} + k_2\begin{bmatrix} 0 \\ 1 \\ -2 \\ -2 \end{bmatrix}$$

so $\text{rank}(\mathbf{T}) = 2$ and a basis for $\mathcal{R}(\mathbf{T})$ is

$$\mathcal{B}_2 = \left\{ \begin{bmatrix} 1 \\ 0 \\ 3 \\ 4 \end{bmatrix}, \begin{bmatrix} 0 \\ 1 \\ -2 \\ -2 \end{bmatrix} \right\}.$$

A second way of finding a basis for $\mathcal{R}(\mathbf{T})$ is to observe, as in the proof of Theorem 4.3, that if $\{\mathbf{u}_1, \mathbf{u}_2, \ldots, \mathbf{u}_n\}$ is a basis for $\mathcal{V}$, then the set $\{\mathbf{T}(\mathbf{u}_1), \mathbf{T}(\mathbf{u}_2), \ldots, \mathbf{T}(\mathbf{u}_n)\}$ spans the range of $\mathbf{T}$. Thus

$$\left\{ \mathbf{T}(\mathbf{e}_1) = \begin{bmatrix} 1 \\ 1 \\ 1 \\ 2 \end{bmatrix}, \mathbf{T}(\mathbf{e}_2) = \begin{bmatrix} -2 \\ 3 \\ -12 \\ -14 \end{bmatrix}, \mathbf{T}(\mathbf{e}_3) = \begin{bmatrix} 1 \\ 7 \\ -11 \\ -10 \end{bmatrix}, \mathbf{T}(\mathbf{e}_4) = \begin{bmatrix} -4 \\ 2 \\ -16 \\ -20 \end{bmatrix} \right\}$$

spans $\mathcal{R}(\mathbf{T})$. A basis could be obtained from this spanning set as illustrated in Example 4 of Section 3.5. In this case, any two independent vectors from the spanning set will form a basis for $\mathcal{R}(\mathbf{T})$; since the first two vectors are independent they form a basis for $\mathcal{R}(\mathbf{T})$. ■

For the case of a matrix transformation $\mathbf{T}$, with $\mathbf{T}(\mathbf{u}) = A\mathbf{u}$, we saw in Theorem 3.11 that

$$\dim(\mathcal{NS}(A)) + \operatorname{rank}(A) = \text{number of columns of } A. \tag{4.2}$$

Note that we have defined the rank of a matrix A to be $\dim(\mathcal{RS}(A))$ while the rank of matrix transformation $\mathbf{T}$ is defined to be $\dim(\mathcal{R}(\mathbf{T})) = \dim(\mathcal{CS}(A))$. Our next theorem generalizes Equation (4.2) to any linear transformation on a finite dimensional vector space. A corollary (see Exercise 12 of this section) of this theorem is that $\dim(\mathcal{RS}(A)) = \dim(\mathcal{CS}(A))$ for any matrix A. The proof of this theorem is somewhat technical, but it is included because it is typical of many arguments about finite dimensional vector spaces.

THEOREM 4.5 If $\mathcal{V}$ is any finite dimensional vector space and $\mathbf{T}$ is a linear transformation from $\mathcal{V}$ to the vector space $\mathcal{W}$ then

$$\dim(\mathcal{NS}(\mathbf{T})) + \dim(\mathbf{T}(\mathcal{V})) = \dim(\mathcal{V})$$

or

$$\operatorname{nullity}(\mathbf{T}) + \operatorname{rank}(\mathbf{T}) = \dim(\mathcal{V}).$$

Proof Let $\mathcal{B}_1 = \{\mathbf{u_1}, \ldots, \mathbf{u_k}\}$ be a basis for $\mathcal{NS}(\mathbf{T})$. Since $\mathcal{NS}(\mathbf{T})$ is a subspace of $\mathcal{V}$, this basis for $\mathcal{NS}(\mathbf{T})$ can, by Theorem 3.27, be extended to a basis for all of $\mathcal{V}$. Suppose that $\dim(\mathcal{V}) = n$ and let

$$\mathcal{B} = \{\mathbf{u_1}, \cdots, \mathbf{u_k}, \mathbf{u_{k+1}}, \cdots, \mathbf{u_n}\}$$

be such a basis for $\mathcal{V}$. Now a typical vector in the range of $\mathbf{T}$ is $\mathbf{w} = \mathbf{T}(\mathbf{u})$, where $\mathbf{u}$ is arbitrary in $\mathcal{V}$. Since $\mathcal{B}$ is a basis for $\mathcal{V}$, $\mathbf{u}$ can be written as a linear combination of the vectors of $\mathcal{B}$; that is,

$$\mathbf{u} = a_1\mathbf{u_1} + \cdots + a_k\mathbf{u_k} + a_{k+1}\mathbf{u_{k+1}} + \cdots + a_n\mathbf{u_n}.$$

We then have, from the linearity of $\mathbf{T}$,

$$\mathbf{w} = \mathbf{T}(\mathbf{u}) = [a_1\mathbf{T}(\mathbf{u_1}) + \cdots + a_k\mathbf{T}(\mathbf{u_k})] + [a_{k+1}\mathbf{T}(\mathbf{u_{k+1}}) + \cdots + a_n\mathbf{T}(\mathbf{u_n})]$$

and, since the basis vectors $\mathbf{u_1}, \mathbf{u_2}, \ldots, \mathbf{u_k}$ are in $\mathcal{NS}(\mathbf{T})$,

$$\mathbf{w} = \mathbf{T}(\mathbf{u}) = a_{k+1}\mathbf{T}(\mathbf{u_{k+1}}) + \cdots + a_n\mathbf{T}(\mathbf{u_n}).$$

This last equation shows that $\mathcal{B}_2 = \{\mathbf{T}(\mathbf{u_{k+1}}), \ldots, \mathbf{T}(\mathbf{u_n})\}$ is a spanning set for the range of $\mathbf{T}$. The proof will be completed if we can show that $\mathcal{B}_2$ is independent and hence is a basis for $\mathcal{R}(\mathbf{T})$. If

$$b_{k+1}\mathbf{T}(\mathbf{u_{k+1}}) + \cdots + b_n\mathbf{T}(\mathbf{u_n}) = \mathbf{0}$$

then, by the linearity of $\mathbf{T}$,

$$\mathbf{T}(b_{k+1}\mathbf{u_{k+1}} + \cdots + b_n\mathbf{u_n}) = \mathbf{0}$$

which shows that $\mathbf{y} = (b_{k+1}\mathbf{u_{k+1}} + \cdots + b_n\mathbf{u_n})$ is in $\mathcal{NS}(\mathbf{T})$. Therefore, since $\mathcal{B}_1$ is a basis for $\mathcal{NS}(\mathbf{T})$, there exist scalars c_i such that

$$\mathbf{y} = b_{k+1}\mathbf{u_{k+1}} + \cdots + b_n\mathbf{u_n} = c_1\mathbf{u_1} + \cdots + c_k\mathbf{u_k}.$$

Since $\mathcal{B}$ is a basis for $\mathcal{V}$, its elements must be independent; hence the last equation can hold only if all the coefficients are zero; in particular $b_{k+1} = b_{k+2} = \cdots = b_n = 0$. Thus the spanning set $\mathcal{B}_2$ is independent and is therefore a basis for $\mathcal{R}(\mathbf{T})$. By counting the vectors in each of the bases we see that $\dim(\mathcal{NS}(\mathbf{T})) + \dim(\mathcal{R}(\mathbf{T})) = n = \dim(\mathcal{V})$ as claimed. □

If we think of the dimension of a subspace as measuring the size of the subspace, then Theorem 4.5 says that if $\mathcal{NS}(\mathbf{T})$ becomes larger, then the range of $\mathbf{T}$ must become smaller.

We now consider an example of a linear transformation on a vector space whose elements are not matrices.

■ EXAMPLE 2 Range and Null Space of D

Consider the linear transformation $\mathbf{D}$ on the polynomial subspace $\mathcal{P}_3 = \{a_0 + a_1x + a_2x^2 + a_3x^3\}$ of $\mathcal{P}$ defined by

$$\mathbf{D}(a_0 + a_1x + a_2x^2 + a_3x^3) = a_1 + 2a_2x + 3a_3x^2.^*$$

It follows from the definition of $\mathbf{D}$ that the polynomials $p(x)$ for which $\mathbf{D}(p(x)) = 0$ are the constants. Thus the nullity of $\mathbf{D}$ is 1 and

$$\mathcal{NS}(\mathbf{D}) = \{a_0\} = \text{BASIS}\{1\} \subset \mathcal{P}_3.$$

The range of $\mathbf{D}$ is

$$\begin{aligned}\mathcal{R}(\mathbf{D}) &= \{a_1 + 2a_2x + 3a_3x^2\}\\ &= \text{BASIS}\{1 = \mathbf{D}(x),\ 2x = \mathbf{D}(x^2),\ 3x^2 = \mathbf{D}(x^3)\}\\ &= \text{BASIS}\{1, x, x^2\},\end{aligned}$$

so $\text{rank}(\mathbf{D}) = 3$. Note that $\text{rank}(\mathbf{D}) + \text{nullity}(\mathbf{D}) = 3 + 1 = 4 = \dim(\mathcal{P}_3)$, as predicted by Theorem 4.5. ■

We have already seen that the set of all solutions of the homogeneous linear problem $\mathbf{T}(\mathbf{x}) = \mathbf{0}$ is a subspace of the domain of $\mathbf{T}$. Several exercises have indicated that the set of all solutions of the nonhomogeneous linear problem $\mathbf{T}(\mathbf{x}) = \mathbf{w}$ is not a subspace of $\mathcal{V}$. The final theorem of this section describes the relationship between the solutions sets of the homogeneous and nonhomogeneous problems. This theorem may be interpreted geometrically as saying that

* The calculus student should note that $\mathbf{D} = \frac{d}{dx}$.

the solution set of the nonhomogeneous problem is "parallel" to the subspace $\mathcal{NS}(\mathbf{T})$, or that the solution set of the nonhomogeneous problem is a translation of the subspace $\mathcal{NS}(\mathbf{T})$ (refer to Exercise 14 in Section 1.6 and Exercise 22 in Section 3.1). The diagram in Figure 4.5 illustrates the general situation.

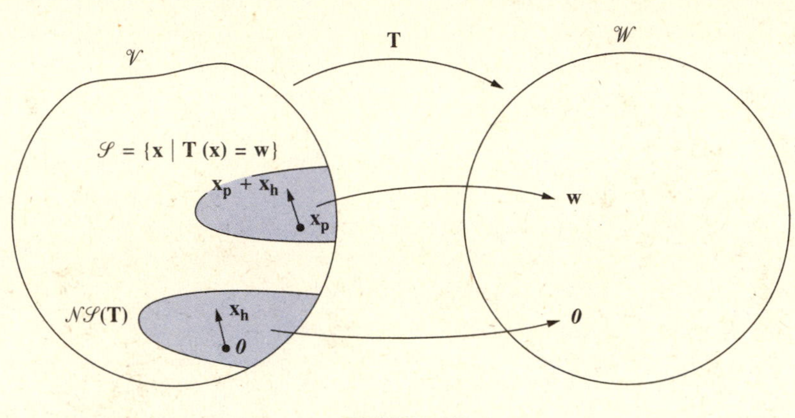

FIGURE 4.5

THEOREM 4.6 Let $\mathbf{T}$ be a linear transformation from $\mathcal{V}$ to $\mathcal{W}$ and let $\mathbf{w}$ be a fixed nonzero vector in $\mathcal{W}$. Then every solution of the linear problem $\mathbf{T}(\mathbf{x}) = \mathbf{w}$ is of the form $\mathbf{x} = \mathbf{x}_h + \mathbf{x}_p$, where $\mathbf{x}_h$ is in $\mathcal{NS}(\mathbf{T})$ and $\mathbf{x}_p$ is a particular solution of the nonhomogeneous problem $\mathbf{T}(\mathbf{x}) = \mathbf{w}$. Equivalently,

$$\{\mathbf{x} \text{ in } \mathcal{V} \,|\, \mathbf{T}(\mathbf{x}) = \mathbf{w}\} = \{\mathbf{x}_h + \mathbf{x}_p \,|\, \mathbf{x}_h \text{ in } \mathcal{NS}(\mathbf{T})\} = \mathcal{NS}(\mathbf{T}) + \mathbf{x}_p.$$

Proof If $\mathbf{x}_p$ is a particular solution of $\mathbf{T}(\mathbf{x}) = \mathbf{w}$, then $\mathbf{T}(\mathbf{x}_p) = \mathbf{w}$ and, from the linearity of $\mathbf{T}$, we have

$$\mathbf{T}(\mathbf{x}_h + \mathbf{x}_p) = \mathbf{T}(\mathbf{x}_h) + \mathbf{T}(\mathbf{x}_p) = \mathbf{0} + \mathbf{w} = \mathbf{w}.$$

Thus any vector of the form $\mathbf{x}_h + \mathbf{x}_p$ is a solution of $\mathbf{T}(\mathbf{x}) = \mathbf{w}$.

If $\mathbf{y}$ is any other solution of $\mathbf{T}(\mathbf{x}) = \mathbf{w}$ (that is, $\mathbf{T}(\mathbf{y}) = \mathbf{w}$), then we can write

$$\mathbf{y} = \mathbf{x}_p + (\mathbf{y} - \mathbf{x}_p)$$

and the calculation

$$\mathbf{T}(\mathbf{y} - \mathbf{x}_p) = \mathbf{T}(\mathbf{y}) - \mathbf{T}(\mathbf{x}_p) = \mathbf{w} - \mathbf{w} = \mathbf{0}$$

shows that $\mathbf{x}_h = \mathbf{y} - \mathbf{x}_p$ is in $\mathcal{NS}(\mathbf{T})$. Thus, $\mathbf{y} = \mathbf{x}_p + \mathbf{x}_h$ as claimed. □

The Gaussian elimination algorithm of Chapter 1 produces the general solution of the linear equations problem and it is easy to write the solution in the form $X = X_p + X_h$.

EXAMPLE 3 **Gaussian Elimination Revisited**

Express the general solution of the linear system

$$\begin{bmatrix} 1 & -2 & 1 & -4 \\ 1 & 3 & 7 & 2 \\ 1 & -12 & -11 & -16 \\ 2 & -14 & -10 & -20 \end{bmatrix} \begin{bmatrix} x_1 \\ x_2 \\ x_3 \\ x_4 \end{bmatrix} = \begin{bmatrix} 3 \\ 5 \\ -1 \\ 2 \end{bmatrix}$$

in the form $X = X_p + X_h$.

Since the coefficient matrix of this system is the matrix A of Example 1, we can use the reduction given there to see that the given system is equivalent to the triangular system

$$\begin{aligned} x_1 + \frac{12}{5}x_3 - \frac{8}{5}x_4 &= \frac{19}{5} \\ x_2 + \frac{6}{5}x_3 + \frac{6}{5}x_4 &= \frac{2}{5}. \end{aligned}$$

If we set $x_3 = 5c_1$ and $x_4 = 5c_2$, we obtain the general solution

$$X = \begin{bmatrix} \frac{19}{5} - 17c_1 + 8c_2 \\ \frac{2}{5} - 6c_1 - 6c_2 \\ 5c_1 \\ 5c_2 \end{bmatrix}$$

$$= \begin{bmatrix} \frac{19}{5} \\ \frac{2}{5} \\ 0 \\ 0 \end{bmatrix} + c_1 \begin{bmatrix} -17 \\ -6 \\ 5 \\ 0 \end{bmatrix} + c_2 \begin{bmatrix} 8 \\ -6 \\ 0 \\ 5 \end{bmatrix} = X_p + X_h.$$

If we compare this solution with the general solution of $AX = 0$ given in Example 1, we see that the particular solution is

$$X_p = \begin{bmatrix} \frac{19}{5} \\ \frac{2}{5} \\ 0 \\ 0 \end{bmatrix}$$

and that the general solution of the homogeneous system is

$$X_h = c_1X_1 + c_2X_2, \text{ where } X_1 = \begin{bmatrix} -17 \\ -6 \\ 5 \\ 0 \end{bmatrix} \quad \text{and} \quad X_2 = \begin{bmatrix} 8 \\ -6 \\ 0 \\ 5 \end{bmatrix}$$

are the basis vectors for $\mathcal{NS}(\mathbf{T})$ found in Example 1. ■

EXERCISES 4.2

1. For each of the following linear transformations find $\mathcal{NS}(\mathbf{T})$, $\mathcal{R}(\mathbf{T})$, rank($\mathbf{T}$), and nullity($\mathbf{T}$).
   (a) $\mathbf{T}\begin{bmatrix} x \\ y \end{bmatrix} = \begin{bmatrix} 1 & 2 \\ 2 & 4 \end{bmatrix}\begin{bmatrix} x \\ y \end{bmatrix}$
   (b) $\mathbf{D}^2(a_0 + a_1x + a_2x^2 + a_3x^3) = 2a_2 + 6a_3x$
   (c) $\mathbf{T}\begin{bmatrix} a & b \\ c & d \end{bmatrix} = \begin{bmatrix} a+b & 0 \\ 0 & c+d \end{bmatrix}$
   (d) $\mathbf{T}[x \quad y \quad z] = [x+y+z \quad y+z \quad x+z]$

2. For each of the following linear transformations find a basis for the null space of $\mathbf{T}$ and a basis for the range of $\mathbf{T}$.
   (a) $\mathbf{T}(\mathbf{u}) = \begin{bmatrix} 1 & 3 & 4 \\ 3 & 4 & 7 \\ -2 & 2 & 0 \end{bmatrix}\mathbf{u}$
   (b) $\mathbf{T}(\mathbf{u}) = \begin{bmatrix} 2 & 0 & -3 & 1 \\ 2 & 1 & 1 & -4 \\ 2 & -1 & -11 & 12 \end{bmatrix}\mathbf{u}$
   (c) $\mathbf{T}(\mathbf{u}) = \begin{bmatrix} 1 & 2 & 1 \\ 0 & 3 & 0 \\ 0 & 0 & 4 \end{bmatrix}\mathbf{u}$
   (d) $\mathbf{T}(\mathbf{u}) = \begin{bmatrix} 1 & 2 & 3 & 1 \\ 4 & 5 & 6 & 1 \\ 7 & 8 & 9 & 1 \end{bmatrix}\mathbf{u}$

3. (Calculus required) For each of the following linear transformations on $\mathcal{F}$ find a basis for $\mathcal{NS}(\mathbf{T})$.
   (a) $\mathbf{T}(y) = \frac{dy}{dt} - 2y$ **(b)** $\mathbf{T}(y) = \frac{d^2y}{dt^2}$ **(c)** $\mathbf{T}(y) = \frac{d^2y}{dt^2} + 4y$

4. For any linear transformation $\mathbf{T}$ from $\mathcal{V}$ to $\mathcal{W}$, show that if $\mathcal{S} = \{\mathbf{u}_1, \ldots, \mathbf{u}_m\}$ is a spanning set for $\mathcal{V}$, then $\{\mathbf{T}(\mathbf{u}_1), \mathbf{T}(\mathbf{u}_2), \ldots, \mathbf{T}(\mathbf{u}_m)\}$ spans $\mathcal{R}(\mathbf{T})$.

5. Let $\mathbf{T}$ be a one-to-one linear transformation of $\mathcal{V}$ onto $\mathcal{W}$. Show that $\dim(\mathcal{V}) = \dim(\mathcal{W})$. What can you say about $\dim(\mathcal{V})$ and $\dim(\mathcal{W})$ if $\mathbf{T}$ is one-to-one but not onto $\mathcal{W}$? What can you say if $\mathbf{T}$ is neither one-to-one nor onto $\mathcal{W}$?

6. Let $\mathbf{T}$ be the linear transformation from $\mathfrak{R}^{1\times3}$ to $\mathfrak{R}^{1\times3}$ defined by $\mathbf{T}[x_1 \quad x_2 \quad x_3] = [x_1 + x_2 + 2x_3 \quad 2x_1 + x_2 + x_3 \quad -x_1 - 2x_2 + 2x_3]$. Find a basis for the range of $\mathbf{T}$.

7. Let $\mathbf{T}$ be a linear transformation from $\mathcal{V}$ to $\mathcal{W}$.
   (a) Given that $\dim(\mathcal{V}) = \dim(\mathcal{R}(\mathbf{T})) = 2$, find $\dim(\mathcal{NS}(\mathbf{T}))$.
   (b) Given that $\dim(\mathcal{NS}(\mathbf{T})) = \dim(\mathcal{R}(\mathbf{T})) = 2$, find $\dim(\mathcal{V})$.
   (c) Given that $\dim(\mathcal{V}) = 7$, $\dim(\mathcal{W}) = 5$, and rank($\mathbf{T}$) $= 5$, find nullity($\mathbf{T}$) and $\mathcal{R}(\mathbf{T})$.
   (d) Given that $\dim(\mathcal{V}) = 5$, $\dim(\mathcal{W}) = 1$, and $\mathbf{T} \neq \mathbf{0}$, find rank($\mathbf{T}$) and nullity($\mathbf{T}$).

8. Let $\mathbf{T}$ be a matrix transformation from $\mathfrak{R}^n$ to $\mathfrak{R}^n$ with matrix A. What condition on A would ensure that for any $\mathbf{v}$ in $\mathfrak{R}^n$ there would exist a unique vector $\mathbf{u}$ in $\mathfrak{R}^n$ such that $\mathbf{T}(\mathbf{u}) = \mathbf{v}$?

9. Let $\mathbf{T}$ be the transformation $\mathbf{T}(\mathbf{u}) = A\mathbf{u}$. What is the relationship between $\mathcal{R}(A^T)$ and the row space of A?

10. Let $\mathbf{T}$ be a matrix transformation from $\mathfrak{R}^3$ to $\mathfrak{R}^3$ that has rank 2. Show that $\mathcal{NS}(\mathbf{T})$ is a line through the origin and that $\{\mathbf{u} \mid \mathbf{T}(\mathbf{u}) = \mathbf{w}\}$ is a line parallel to $\mathcal{NS}(\mathbf{T})$. What is the corresponding geometric interpretation if rank($\mathbf{T}$) $= 1$?

11. Express the general solutions of the following linear systems in the form $X = X_p + X_h$, as in Example 3.

(a) $\begin{bmatrix} 1 & 5 & 0 & 0 & 5 \\ 0 & 0 & 1 & 0 & 3 \\ 0 & 0 & 0 & 1 & 4 \\ 0 & 0 & 0 & 0 & 0 \end{bmatrix} X = \begin{bmatrix} -1 \\ 1 \\ 2 \\ 0 \end{bmatrix}$ **(b)** $\begin{bmatrix} 1 & 3 & -2 & 4 \\ 0 & 2 & 3 & 0 \\ 0 & 0 & 0 & 3 \\ 0 & 0 & 0 & 0 \end{bmatrix} X = \begin{bmatrix} 3 \\ 4 \\ 5 \\ 0 \end{bmatrix}$

(c) $\begin{bmatrix} 1 & -1 & 2 & 3 \\ 2 & -1 & 0 & 2 \\ 4 & 1 & -11 & -1 \\ 1 & 2 & 3 & 8 \end{bmatrix} X = \begin{bmatrix} 1 \\ 1 \\ 1 \\ 1 \end{bmatrix}$ **(d)** $\begin{aligned} x + 2y &= 5 \\ 2x + 3y &= 8 \\ x + 4y &= 9 \end{aligned}$

†**12.** Use Theorem 4.5 and Exercise 8 of Section 3.8 to show that $\dim(\mathcal{RS}(A)) = \dim(\mathcal{CS}(A))$ for any matrix A.

13. Given that a matrix transformation $\mathbf{T}$ maps $\Re^m$ onto $\Re^n$, what conclusion can you make about the size of m and n? What conclusions can you draw if it is given that $\mathbf{T}$ is one-to-one?

14. Let $\mathbf{T}$ be the matrix transformation on $\Re^3$ whose matrix is

$\mathbf{A} = \begin{bmatrix} 5 & 21 & 13 \\ 8 & 2 & -17 \\ 8 & 14 & 1 \end{bmatrix}$. Which of the vectors $\mathbf{u} = \begin{bmatrix} 2 \\ 4 \\ -3 \end{bmatrix}$, $\mathbf{v} = \begin{bmatrix} -5 \\ 3 \\ -2 \end{bmatrix}$,

$\mathbf{w} = \begin{bmatrix} 0 \\ 1 \\ -2 \end{bmatrix}$ are in $\mathcal{NS}(\mathbf{T})$?

15. **(a)** Is it possible to have a one-to-one linear transformation from $\Re^5$ to $\Re^4$?
(b) Is it possible to have a one-to-one linear transformation from $\Re^5$ to $\Re^{2\times 3}$?
(c) Is it possible to have a one-to-one linear transformation from $\Re^5$ to $\mathcal{P}_5$?
(d) Is it possible to have a one-to-one linear transformation from $\Re^{2\times 2}$ to $\Re^{1\times 4}$?

16. **(a)** Is it possible for a linear transformation from $\Re^4$ to $\Re^5$ to be onto?
(b) Is it possible for a linear transformation from $\Re^4$ to $\mathcal{P}_4$ to be onto?
(c) Is it possible for a linear transformation from $\Re^{2\times 2}$ to $\Re^4$ to be onto?
(d) Is it possible for a linear transformation from $\mathcal{P}_4$ to $\Re^{2\times 2}$ to be onto?

17. Consider the matrix transformation $\mathbf{T}$ defined by $\mathbf{T}(\mathbf{u}) = A\mathbf{u}$ where

$$A = \begin{bmatrix} 1 & 3 & 0 & 2 & -1 \\ 3 & 9 & 1 & 7 & 5 \\ 2 & 6 & 1 & 5 & 2 \\ 5 & 15 & 2 & 12 & 8 \end{bmatrix} \tilde{R} \begin{bmatrix} 1 & 3 & 0 & 2 & 0 \\ 0 & 0 & 1 & 1 & 0 \\ 0 & 0 & 0 & 0 & 1 \\ 0 & 0 & 0 & 0 & 0 \end{bmatrix} = R.$$

(a) Find a basis for $\mathcal{CS}(A)$. **(b)** Find a basis for $\mathcal{RS}(A)$.
(c) Find a basis for $\mathcal{NS}(\mathbf{T})$ **(d)** Find a basis for $\mathcal{R}(\mathbf{T})$.

18. Let A by any $m \times n$ real matrix.
(a) Show that $\dim(\mathcal{CS}(A)) + \dim(\mathcal{NS}(A^T)) = m$.
(b) Show that the equation $AX = B$ has a solution for all B in $\Re^m$ if, and only if, the equation $A^TX = 0$ has only the trivial solution.

19. Let $\mathbf{U}$ be the transformation of the vector space $\mathcal{P}$ of all polynomials defined by $\mathbf{U}(p(x)) = (a + bx)p(x)$. Show that $\mathbf{U}$ is linear. Find the image, under $\mathbf{U}$, of $\mathcal{P}_4$.

20. Let A, B, and C be $n \times n$ matrices such that $A = BC$.
   (a) Show that $\mathcal{NS}(C) \subset \mathcal{NS}(A)$.
   (b) Show that rank(A) is less than or equal to the minimum of rank(B) and rank(C).
   (c) Show that if A is nonsingular, then so are B and C.
   (d) Find 3×3 diagonal matrices which show that rank(A) can be less than both rank(B) and rank(C).

21. **(a)** Find the rank of a 6×4 matrix A for which $\mathcal{NS}(A) = \{0\}$.
   (b) Find the rank of a 5×6 matrix A for which $\dim(\mathcal{NS}(A)) = 3$.
   (c) Find the rank of a 7×5 matrix A for which $\dim(\mathcal{NS}(A^T)) = 3$.
   (d) Find the rank of a 3×5 matrix A for which $AX = B$ is solvable for all B in $\mathfrak{R}^3$.
   (e) Find the rank of an 8×3 matrix A for which A^TA is nonsingular.

22. Let **T** be the linear operator from $\mathfrak{R}^3$ to $\mathfrak{R}^3$ defined by $\mathbf{T}(\mathbf{u}) = \begin{bmatrix} 4 & 1 & -2 \\ 2 & 1 & 1 \\ 2 & 0 & -3 \end{bmatrix}\mathbf{u}$.
   (a) Show that the $\mathcal{NS}(\mathbf{T})$ is a line through the origin.
   (b) Show that the $\mathcal{R}(\mathbf{T})$ is a plane through the origin and find its equation.

23. Let **T** be the linear operator from $\mathfrak{R}^4$ to $\mathfrak{R}^3$ defined by

$$\mathbf{T}(\mathbf{u}) = \begin{bmatrix} 4 & 1 & -2 & -3 \\ 2 & 1 & 1 & -4 \\ 2 & 0 & -3 & 1 \end{bmatrix}\mathbf{u}.$$

   (a) Which of the vectors $\begin{bmatrix} 0 \\ 0 \\ 6 \end{bmatrix}, \begin{bmatrix} 1 \\ -1 \\ 2 \end{bmatrix}, \begin{bmatrix} 2 \\ 4 \\ 1 \end{bmatrix}$ are in $\mathcal{R}(\mathbf{T})$?
   (b) Which of the vectors $\begin{bmatrix} 3 \\ -8 \\ 2 \\ 0 \end{bmatrix}, \begin{bmatrix} 0 \\ 0 \\ 0 \\ 1 \end{bmatrix}, \begin{bmatrix} -1 \\ 10 \\ 0 \\ 2 \end{bmatrix}$ are in $\mathcal{NS}(\mathbf{T})$?

4.3 THE ALGEBRA OF LINEAR TRANSFORMATIONS

> How do the familiar matrix operations generalize to linear transformations?

Since the motivating example for the study of linear transformations is matrix multiplication and since there is a well-defined algebra of matrices, it is not surprising that there is also an algebra of linear transformations that generalizes matrix algebra. We begin by defining addition and scalar multiplication for linear transformations. Note that the definitions are like those given in Example 5 of Section 3.7 for the vector space of real-valued functions.

DEFINITION 4.4 Let $\mathbf{S}$ and $\mathbf{T}$ be linear transformations from $\mathcal{V}$ to $\mathcal{W}$. The **sum** $\mathbf{S} + \mathbf{T}$ is the function from $\mathcal{V}$ to $\mathcal{W}$ defined by

$$(\mathbf{S} + \mathbf{T})(\mathbf{u}) = \mathbf{S}(\mathbf{u}) + \mathbf{T}(\mathbf{u})$$

and the **scalar multiple** $k\mathbf{T}$ is the function defined by

$$(k\mathbf{T})(\mathbf{u}) = k(\mathbf{T}(\mathbf{u})). \quad \square$$

Note that for linear transformations from $\mathcal{V}$ to $\mathcal{W}$, the sum and the scalar multiple are defined using the vector operations in the space $\mathcal{W}$ and the ranges of the linear transformations are all subspaces of $\mathcal{W}$.

It is easy to see that both $\mathbf{S} + \mathbf{T}$ and $k\mathbf{T}$ are linear transformations from $\mathcal{V}$ to $\mathcal{W}$. For $\mathbf{S} + \mathbf{T}$ the calculation is as follows:

$$\begin{aligned}(\mathbf{S} + \mathbf{T})(a\mathbf{u} + b\mathbf{v}) &= \mathbf{S}(a\mathbf{u} + b\mathbf{v}) + \mathbf{T}(a\mathbf{u} + b\mathbf{v}) \\ &= \{a\mathbf{S}(\mathbf{u}) + b\mathbf{S}(\mathbf{v})\} + \{a\mathbf{T}(\mathbf{u}) + b\mathbf{T}(\mathbf{v})\} \\ &= a\{\mathbf{S}(\mathbf{u}) + \mathbf{T}(\mathbf{u})\} + b\{\mathbf{S}(\mathbf{v}) + \mathbf{T}(\mathbf{v})\} \\ &= a(\mathbf{S} + \mathbf{T})(\mathbf{u}) + b(\mathbf{S} + \mathbf{T})(\mathbf{v}).\end{aligned}$$

You should justify each step in the above calculation and then construct a similar argument to show that $k\mathbf{T}$ is linear.

DEFINITION 4.5 The set of all linear transformations from $\mathcal{V}$ to $\mathcal{W}$ will be denoted by $\mathcal{L}(\mathcal{V}, \mathcal{W})$. $\square$

If $\mathbf{S}$ and $\mathbf{T}$ are matrix transformations, with $\mathbf{S}(\mathbf{u}) = A\mathbf{u}$ and $\mathbf{T}(\mathbf{u}) = B\mathbf{u}$, then

$$(\mathbf{S} + \mathbf{T})(\mathbf{u}) = \mathbf{S}(\mathbf{u}) + \mathbf{T}(\mathbf{u}) = A\mathbf{u} + B\mathbf{u} = (A + B)\mathbf{u}$$

and

$$(k\mathbf{T})(\mathbf{u}) = k(\mathbf{T}(\mathbf{u})) = k(B\mathbf{u}) = (kB)\mathbf{u}.$$

These calculations show that for linear transformations, the sum and scalar multiple, as defined in Definition 4.4, are generalizations of familiar matrix operations. Thus, the following theorem is not at all surprising.

THEOREM 4.7 The set of all linear transformations from $\mathcal{V}$ to $\mathcal{W}$, $\mathcal{L}(\mathcal{V}, \mathcal{W})$, is a real vector space with the operations in Definition 4.4.

The proof of this theorem is very easy, involving only routine calculations like those in Example 5 of Section 3.7. Recall that functions are equal if and only if their values are equal for every element of the domain. We will prove that addition of transformations is associative; the remaining details will be left to the exercises. Thus we let **T**, **S**, and **U** be elements of $\mathscr{L}(\mathscr{V}, \mathscr{W})$ and compute, for a typical vector **v** in $\mathscr{V}$,

$$((\mathbf{S} + \mathbf{T}) + \mathbf{U})(\mathbf{v}) = (\mathbf{S} + \mathbf{T})(\mathbf{v}) + \mathbf{U}(\mathbf{v}) = (\mathbf{S}(\mathbf{v}) + \mathbf{T}(\mathbf{v})) + \mathbf{U}(\mathbf{v})$$

and

$$(\mathbf{S} + (\mathbf{T} + \mathbf{U}))(\mathbf{v}) = \mathbf{S}(\mathbf{v}) + (\mathbf{T} + \mathbf{U})(\mathbf{v}) = \mathbf{S}(\mathbf{v}) + (\mathbf{T}(\mathbf{v}) + \mathbf{U}(\mathbf{v})).$$

The associativity of vector addition in $\mathscr{W}$ shows that these last two expressions are equal for every vector **v** in $\mathscr{V}$, and hence that the functions $(\mathbf{S} + \mathbf{T}) + \mathbf{U}$ and $\mathbf{S} + (\mathbf{T} + \mathbf{U})$ are equal. Therefore addition of linear transformations is associative, as claimed. □

The next question we consider is whether there is a way to define a product for linear transformations that is a generalization of matrix multiplication. The definition of multiplication of functions does not involve the product of the functional values; rather, it involves the familiar function of a function notion (composition) encountered in elementary algebra and calculus. For example, the function $f(x) = \sqrt{\sin x}$ can be viewed as the product of the sine function and the square-root function and could be evaluated on many pocket calculators by entering x, pushing the sin key, and then pushing the square-root key.

DEFINITION 4.6 For **T** in $\mathscr{L}(\mathscr{V}, \mathscr{W})$ and **S** in $\mathscr{L}(\mathscr{W}, \mathscr{X})$ the **product ST** is the mapping from $\mathscr{V}$ to $\mathscr{X}$ defined by

$$\mathbf{ST}(\mathbf{u}) = \mathbf{S}(\mathbf{T}(\mathbf{u})). \quad \square$$

The simple diagrams in Figure 4.6 may be helpful in understanding Definition 4.6.

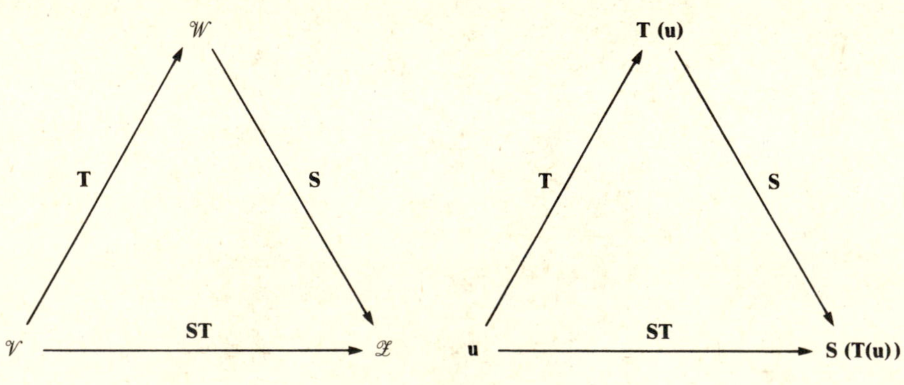

FIGURE 4.6

It is easy to see that the product of two linear transformations is another linear transformation; to verify this we compute, for arbitrary vectors **u** and **v** in $\mathcal{V}$ and arbitrary scalars a and b

$$\begin{aligned}\mathbf{ST}(a\mathbf{u} + b\mathbf{v}) &= \mathbf{S}(\mathbf{T}(a\mathbf{u} + b\mathbf{v}))\\ &= \mathbf{S}(a\mathbf{T}(\mathbf{u}) + b\mathbf{T}(\mathbf{v}))\\ &= a\mathbf{S}(\mathbf{T}(\mathbf{u})) + b\mathbf{S}(\mathbf{T}(\mathbf{v}))\\ &= a\mathbf{ST}(\mathbf{u}) + b\mathbf{ST}(\mathbf{v}).\end{aligned}$$

If **S** and **T** are matrix transformations with $\mathbf{S}(\mathbf{u}) = A\mathbf{u}$ and $\mathbf{T}(\mathbf{u}) = B\mathbf{u}$, then

$$\mathbf{ST}(\mathbf{u}) = \mathbf{S}(\mathbf{T}(\mathbf{u})) = A(B\mathbf{u}) = (AB)\mathbf{u},$$

so that **ST** is the matrix transformation with matrix AB. Thus, the product introduced in Definition 4.6 is a generalization of matrix multiplication. This observation means that *we cannot expect multiplication of transformations to satisfy properties not enjoyed by matrix multiplication. In particular, it must be the case that multiplication of transformations is not commutative, not always defined, and not always invertible.*

The obvious way to extend the definition of the matrix inverse to more general transformations is to call a transformation $\mathbf{S} = \mathbf{T}^{-1}$ if **S** is such that $\mathbf{ST} = \mathbf{TS} = \mathbf{I}$, where **I** is the **identity operator** on $\mathcal{V}$ defined by $\mathbf{I}(\mathbf{x}) = \mathbf{x}$ for every **x** in $\mathcal{V}$ (recall Exercise 15 in Section 4.1). It is clear, from Figure 4.6, that this can occur only if $\mathcal{V} = \mathcal{W} = \mathcal{X}$, that is, only linear operators (transformations from $\mathcal{V}$ to $\mathcal{V}$) can have inverses. We will extend the familiar terminology associated with matrix inverses to the case of transformations; we will say that **T** is **invertible** or **nonsingular** if **T** has a multiplicative inverse. A necessary and sufficient condition that **T** be nonsingular is that **T** be both one-to-one and onto (see Exercise 12 of this section).

The multiplicative properties of transformations are very much like the properties of matrix multiplication. The proofs are usually less involved in the more general case of linear transformations, since we are free from subscripts and double sums. For example, to show that multiplication of transformations is associative we compute, for any vector **v** in $\mathcal{V}$,

$$\mathbf{S}(\mathbf{TU})(\mathbf{v}) = \mathbf{S}(\mathbf{TU}(\mathbf{v})) = \mathbf{S}(\mathbf{T}(\mathbf{U}(\mathbf{v}))) \quad \text{and}$$
$$(\mathbf{ST})\mathbf{U}(\mathbf{v}) = \mathbf{ST}(\mathbf{U}(\mathbf{v})) = \mathbf{S}(\mathbf{T}(\mathbf{U}(\mathbf{v}))).$$

It follows that $\mathbf{S}(\mathbf{TU}) = (\mathbf{ST})\mathbf{U}$.

We have proved part 1 of the following theorem. Proofs of the other parts are left to the exercises.

THEOREM 4.8 If **S**, **T**, and **U** are linear operators on $\mathcal{V}$, then:

(1) $\mathbf{S}(\mathbf{TU}) = (\mathbf{ST})\mathbf{U}$;
(2) $\mathbf{S}(\mathbf{T} + \mathbf{U}) = \mathbf{ST} + \mathbf{SU}$;
(3) $(\mathbf{S} + \mathbf{T})\mathbf{U} = \mathbf{SU} + \mathbf{TU}$; and
(4) If $\dim(\mathcal{V}) < \infty$, then **T** is invertible if, and only if, **T** is one-to-one. □

EXAMPLE 1 **Algebraic Combinations of Matrix Transformations**

Let $\mathbf{T}$, $\mathbf{S}$, and $\mathbf{I}$ be linear operators on $\Re^n$ defined by $\mathbf{T}(\mathbf{u}) = A\mathbf{u}$, $\mathbf{S}(\mathbf{u}) = B\mathbf{u}$, and $\mathbf{I}(\mathbf{u}) = \mathbf{u}$. Show that $\mathbf{T}^2 + 5\mathbf{ST} - 6\mathbf{T} - 7\mathbf{S} - 12\mathbf{I}$ is a matrix transformation and find its standard matrix.

For any vector $\mathbf{u}$ in $\Re^n$ we have

$$
\begin{aligned}
&(\mathbf{T}^2 + 5\mathbf{ST} - 6\mathbf{T} - 7\mathbf{S} - 12\mathbf{I})(\mathbf{u}) \\
&\quad = \mathbf{T}^2(\mathbf{u}) + 5\mathbf{ST}(\mathbf{u}) + 6\mathbf{T}(\mathbf{u}) - 7\mathbf{S}(\mathbf{u}) - 12\mathbf{I}(\mathbf{u}) \\
&\quad = \mathbf{T}(\mathbf{T}(\mathbf{u})) - 5\mathbf{S}(\mathbf{T}(\mathbf{u})) + 6A\mathbf{u} - 7B\mathbf{u} - 12\mathbf{u} \\
&\quad = A(A\mathbf{u}) - 5B(A\mathbf{u}) + 6A\mathbf{u} - 7B\mathbf{u} - 12I\mathbf{u} \\
&\quad = (A^2 - 5BA + 6A - 7B - 12I)\mathbf{u}. \quad \blacksquare
\end{aligned}
$$

Our next example shows how these ideas can be applied to the linear operators studied in calculus.

EXAMPLE 2 **Algebraic Combinations of the Derivative Operator**

Let $\mathbf{D}$ be the differentiation operator $\mathbf{D}(y) = y' = \frac{dy}{dx}$. Compute $\mathbf{D}^2$, $(\mathbf{D} + 3)(\mathbf{D} - 5)$, and $\mathbf{DxD}$.

Since $\mathbf{D}^2 y = \mathbf{D}(\mathbf{D}y) = \mathbf{D}(y') = y''$, we see that $\mathbf{D}^2$ is the second derivative operator.

$$
\begin{aligned}
(\mathbf{D} + 3)(\mathbf{D} - 5)y &= (\mathbf{D} + 3)(y' - 5y) = \mathbf{D}(y' - 5y) + 3(y' - 5y) \\
&= y'' - 5y' + 3y' - 15y = \mathbf{D}^2 y - 2\mathbf{D}y - 15y \\
&= (\mathbf{D}^2 - 2\mathbf{D} - 15)y.
\end{aligned}
$$

Therefore $(\mathbf{D} + 3)(\mathbf{D} - 5) = \mathbf{D}^2 - 2\mathbf{D} - 15$.

The operator $\mathbf{DxD}$ is to be viewed as the product of three linear operators: first, differentiation; second, multiplication by x; and third, another differentiation. Thus,

$$\mathbf{DxD}(y) = \mathbf{D}(\mathbf{x}(y')) = \mathbf{D}(xy') = xy'' + y' = (\mathbf{xD}^2 + \mathbf{D})(y).$$

Note that we have used the product rule for the derivative in the above calculation. Note also that this calculation shows that $\mathbf{DxD} \neq \mathbf{xD}^2$. $\blacksquare$

EXERCISES 4.3

1. Consider $\mathbf{T}$ and $\mathbf{U}$ in $\mathcal{L}(\Re^2, \Re^{1\times 2})$ defined by

$$\mathbf{T}\begin{bmatrix} x \\ y \end{bmatrix} = [2x - 3y \quad 0] \quad \text{and} \quad \mathbf{U}\begin{bmatrix} x \\ y \end{bmatrix} = [3y \quad x + y].$$

Find:

(a) $(\mathbf{T} + \mathbf{U})(\mathbf{e}_1)$; **(b)** $(3\mathbf{T} - 5\mathbf{U})(\mathbf{e}_2)$; and **(c)** $(2\mathbf{T} + \mathbf{U})\begin{bmatrix} 3 \\ 2 \end{bmatrix}$.

2. Consider $\mathbf{T}$ and $\mathbf{U}$ in $\mathcal{L}(\mathfrak{R}^{2\times 2}, \mathfrak{R}^{2\times 2})$ defined by

$$\mathbf{T}(A) = A^T \quad \text{and} \quad \mathbf{U}\begin{bmatrix} a & b \\ c & d \end{bmatrix} = \begin{bmatrix} a+b & 0 \\ 0 & c+d \end{bmatrix}.$$

For $A = \begin{bmatrix} 1 & 2 \\ 3 & 4 \end{bmatrix}$, find **(a)** $(\mathbf{U} + 2\mathbf{T})(A)$; **(b)** $\mathbf{T}^2(A)$; **(c)** $(\mathbf{UT})(A)$; **(d)** $(\mathbf{TU})(A)$; and **(e)** $(\mathbf{U}^2 + \mathbf{T}^2)(A)$.

3. Consider $\mathbf{T}$ in $\mathcal{L}(\mathfrak{R}^3, \mathfrak{R}^3)$ defined by $\mathbf{T}(\mathbf{u}) = A\mathbf{u}$ where $A = \begin{bmatrix} 2 & 1 & 1 \\ 2 & 3 & 2 \\ 1 & 1 & 2 \end{bmatrix}$. Show that $(\mathbf{T}^3 + 7\mathbf{T}^2 + 11\mathbf{T})\mathbf{u} = 5\mathbf{u}$ for every $\mathbf{u}$ in $\mathfrak{R}^3$.

4. Consider $\mathbf{T}$ and $\mathbf{U}$ in $\mathcal{L}(\mathcal{P}, \mathcal{P})$ defined by

$$\mathbf{T}(a_0 + a_1x + a_2x^2 + \cdots + a_nx^n) = a_1 + 2a_2x + 3a_3x^2 + \cdots + na_nx^{n-1} \quad \text{and}$$

$$\mathbf{U}(a_0 + a_1x + a_2x^2 + \cdots + a_nx^n) = a_0x + \frac{a_1}{2}x^2 + \frac{a_2}{3}x^3 + \cdots + \frac{a_n}{(n+1)}x^{n+1}.$$

(a) Compute $\mathbf{TU}(x^2 + x + 5)$ and $\mathbf{UT}(x^2 + x + 5)$.
(b) Show that $\mathbf{TU} = \mathbf{I}$, but $\mathbf{UT} \neq \mathbf{I}$.

5. Let $\mathbf{D} = \frac{d}{dx}$ be the differentiation operator. Compute:
(a) $(\mathbf{D} - 1)(\mathbf{D} + 1)$
(b) $(\mathbf{D}^2 + 5\mathbf{D} - 11)(\mathbf{D} + 2)$
(c) $(x\mathbf{D} + 3)(\mathbf{D}^2 + 3\mathbf{D} + 1)$
(d) $(x\mathbf{D} + 3)(e^x\mathbf{D} + \sin 3x)$

6. Let $\mathbf{T}_1$ and $\mathbf{T}_2$ be in $\mathcal{L}(\mathcal{V}, \mathcal{V})$. How do $\mathcal{NS}(\mathbf{T}_1\mathbf{T}_2)$, $\mathcal{NS}(\mathbf{T}_1)$, and $\mathcal{NS}(\mathbf{T}_2)$ compare? What if $\mathbf{T}_1\mathbf{T}_2 = \mathbf{T}_2\mathbf{T}_1$?

7. Verify that if $\mathbf{T}$ is in $\mathcal{L}(\mathcal{V}, \mathcal{W})$, then so is $k\mathbf{T}$.

8. Verify that multiplication of transformations distributes over addition of transformations (parts 2 and 3 of Theorem 4.8).

9. Let $\mathbf{T}$ be a matrix transformation from $\mathfrak{R}^n$ to $\mathfrak{R}^n$ with matrix A. What condition on A would ensure that for any $\mathbf{v}$ in $\mathfrak{R}^n$ there would exist a unique vector $\mathbf{u}$ in $\mathfrak{R}^n$ such that $\mathbf{T}(\mathbf{u}) = \mathbf{v}$?

10. Let $\mathcal{V}$ and $\mathcal{W}$ be finite dimensional vector spaces. Show that $\dim(\mathcal{L}(\mathcal{V}, \mathcal{W})) = \dim(\mathcal{V}) \times \dim(\mathcal{W})$. (*Hint:* Let $\mathcal{B} = \{\mathbf{u}_1, \ldots, \mathbf{u}_n\}$ and $\mathcal{B}' = \{\mathbf{w}_1, \ldots, \mathbf{w}_m\}$ be bases for $\mathcal{V}$ and $\mathcal{W}$ respectively. Define $\mathbf{E}_{ij}$ in $\mathcal{L}(\mathcal{V}, \mathcal{W})$ by

$$\mathbf{E}_{ij}(\mathbf{u}_k) = \begin{cases} \mathbf{w}_i & \text{if } k = j \\ \mathbf{0} & \text{if } k \neq j. \end{cases}$$

Show that $\{\mathbf{E}_{ij} \mid i = 1, \ldots, n; j = 1, \ldots, m\}$ is a basis for $\mathcal{L}(\mathcal{V}, \mathcal{W})$.)

11. For $\mathbf{T}$ in $\mathcal{L}(\mathcal{V}, \mathcal{V})$ show that $\mathbf{T}^{-1}$ exists if, and only if, $\mathbf{T}$ is both one-to-one and onto. (*Hint:* If $\mathbf{T}(\mathbf{u}) = \mathbf{v}$, then define $\mathbf{T}^{-1}(\mathbf{v}) = \mathbf{u}$ and show that $\mathbf{T}^{-1}$ is linear.)

12. Let $\mathbf{T}$ be a linear operator on a finite dimensional space $\mathcal{V}$ for which

$$\{\mathbf{u}_1, \mathbf{u}_2, \ldots, \mathbf{u}_n\}$$

is a basis. Show that the following are equivalent:

(1) $\mathbf{T}^{-1}$ exists;
(2) if $\mathbf{T}(\mathbf{x}) = \mathbf{0}$, then $\mathbf{x} = \mathbf{0}$;

(3) $\mathbf{T}$ is one-to-one;
(4) $\{\mathbf{T}(\mathbf{u}_1), \mathbf{T}(\mathbf{u}_2), \ldots, \mathbf{T}(\mathbf{u}_n)\}$ is a basis for $\mathcal{V}$; and
(5) $\mathbf{T}$ is onto $\mathcal{V}$.

13. Let $\mathbf{T}_1$ and $\mathbf{T}_2$ be linear operators on $\mathfrak{R}^{2\times 2}$ defined by $\mathbf{T}_1(A) = A^T$ and $\mathbf{T}_2(A) = \text{tr}(A)I$. Compare the products $\mathbf{T}_1\mathbf{T}_2$ and $\mathbf{T}_2\mathbf{T}_1$.

14. Let $\mathbf{R}(\alpha)$ be the linear transformation which rotates $\mathfrak{R}^2$ through the angle α. Show geometrically that $\mathbf{R}(\alpha)\mathbf{R}(\beta) = \mathbf{R}(\alpha + \beta)$.

15. Let $\mathcal{V}$ and $\mathcal{W}$ be finite dimensional vector spaces that are unequal but which have the same dimension. Let $\mathbf{T}$ be in $\mathcal{L}(\mathcal{V}, \mathcal{W})$ and $\mathbf{U}$ in $\mathcal{L}(\mathcal{W}, \mathcal{V})$ be such that $\mathbf{UT} = \mathbf{I}_{\mathcal{V}}$. Show that $\mathbf{TU} = \mathbf{I}_{\mathcal{W}}$.

4.4 GEOMETRIC INTERPRETATION

1. Are all linear transformations on $\mathfrak{R}^n$ matrix transformations?
2. How can we interpret linear transformations on $\mathfrak{R}^3$ and $\mathfrak{R}^2$ geometrically?

The vector spaces $\mathfrak{R}^n$ hold a very prominent place in the development of linear algebra. Not only are they the motivating example for the study of more general vector spaces, but they are the range of an arbitrary finite dimensional vector space $\mathcal{V}$ under the linear transformation that associates with every vector in $\mathcal{V}$ its coordinate matrix with respect to a fixed basis (recall Definition 3.21 and Exercise 26 in Section 4.1). For this reason we call the spaces $\mathfrak{R}^n$ **coordinate spaces.** Since the coordinate mapping is both one-to-one and onto we call the spaces $\mathcal{V}$ and $\mathfrak{R}^n$ **isomorphic,** which means "same structure."

The matrix transformations on the coordinate spaces $\mathfrak{R}^n$ are more than just the motivating example for the study of linear transformations. We will now show that every linear transformation between coordinate spaces is a matrix transformation. To see why this is so we consider any linear transformation $\mathbf{T}$ from $\mathfrak{R}^n$ to $\mathfrak{R}^m$, and let

$$\mathcal{S} = \{\mathbf{e}_1, \mathbf{e}_2, \cdots, \mathbf{e}_n\}$$

be the standard basis for $\mathfrak{R}^n$. For any vector $\mathbf{u}$ in $\mathfrak{R}^n$ we can write

$$\mathbf{u} = \begin{bmatrix} u_1 \\ u_2 \\ \vdots \\ u_n \end{bmatrix} = u_1\mathbf{e}_1 + u_2\mathbf{e}_2 + \cdots + u_n\mathbf{e}_n$$

and then use the linearity of $\mathbf{T}$ to compute

$$\mathbf{T}(\mathbf{u}) = u_1\mathbf{T}(\mathbf{e}_1) + u_2\mathbf{T}(\mathbf{e}_2) + \cdots + u_n\mathbf{T}(\mathbf{e}_n).$$

Since each $\mathbf{T}(\mathbf{e}_i) = \begin{bmatrix} a_{1i} \\ a_{2i} \\ \vdots \\ a_{mi} \end{bmatrix}$ is an $m \times 1$ matrix, (an element of $\Re^m$) we can write

$$\mathbf{T}(\mathbf{u}) = \mathbf{T}\begin{bmatrix} u_1 \\ u_2 \\ \vdots \\ u_n \end{bmatrix} = [\mathbf{T}(\mathbf{e}_1) \quad \mathbf{T}(\mathbf{e}_2) \quad \cdots \quad \mathbf{T}(\mathbf{e}_n)]\begin{bmatrix} u_1 \\ u_2 \\ \vdots \\ u_n \end{bmatrix} = A\mathbf{u}.$$

Thus $\mathbf{T}$ is multiplication by the $m \times n$ matrix A and $\mathbf{T}$ is a matrix transformation as claimed. We will call the matrix

$$A = [\mathbf{T}(\mathbf{e}_1) \quad \mathbf{T}(\mathbf{e}_2) \quad \cdots \quad \mathbf{T}(\mathbf{e}_n)]$$

the **standard matrix of T.**

EXAMPLE 1 The Standard Matrix of a Linear Operator

For the linear operator on $\Re^3$ defined by

$$\mathbf{T}\begin{bmatrix} x_1 \\ x_2 \\ x_3 \end{bmatrix} = \begin{bmatrix} 3x_1 + 4x_3 \\ 2x_2 - 5x_3 \\ -x_1 + 2x_2 \end{bmatrix}$$

we have

$$\mathbf{T}(\mathbf{e}_1) = \mathbf{T}\begin{bmatrix} 1 \\ 0 \\ 0 \end{bmatrix} = \begin{bmatrix} 3 \\ 0 \\ -1 \end{bmatrix}, \quad \mathbf{T}(\mathbf{e}_2) = \begin{bmatrix} 0 \\ 2 \\ 2 \end{bmatrix}, \quad \text{and} \quad \mathbf{T}(\mathbf{e}_3) = \begin{bmatrix} 4 \\ -5 \\ 0 \end{bmatrix},$$

so that the standard matrix of $\mathbf{T}$ is

$$A = [\mathbf{T}(\mathbf{e}_1) \quad \mathbf{T}(\mathbf{e}_2) \quad \mathbf{T}(\mathbf{e}_3)] = \begin{bmatrix} 3 & 0 & 4 \\ 0 & 2 & -5 \\ -1 & 2 & 0 \end{bmatrix}. \quad ■$$

We summarize the discussion preceding Example 1 in the following theorem.

THEOREM 4.9 If $\mathbf{T}$ is a linear transformation from $\Re^n$ to $\Re^m$, then $\mathbf{T}$ is multiplication by the $m \times n$ matrix

$$A = [\mathbf{T}(\mathbf{e}_1) \quad \mathbf{T}(\mathbf{e}_2) \quad \cdots \quad \mathbf{T}(\mathbf{e}_n)]. \quad \square$$

EXAMPLE 2 **Standard Matrix of a Rotation**

Find the standard matrix of the transformation **R** that rotates $\Re^2$ through an angle θ (recall Exercise 21 in Section 4.1).

As we see from Figure 4.7,

$$\mathbf{R}(\mathbf{e}_1) = \mathbf{e}_1 \cos\theta + \mathbf{e}_2 \sin\theta = \begin{bmatrix} \cos\theta \\ \sin\theta \end{bmatrix}$$

and

$$\mathbf{R}(\mathbf{e}_2) = -\mathbf{e}_1 \sin\theta + \mathbf{e}_2 \cos\theta = \begin{bmatrix} -\sin\theta \\ \cos\theta \end{bmatrix},$$

so the standard matrix of **R** is $\begin{bmatrix} \cos\theta & -\sin\theta \\ \sin\theta & \cos\theta \end{bmatrix} = R$. Note that the standard matrix of a rotation is an orthogonal matrix; that is, $R^T R = I$. ■

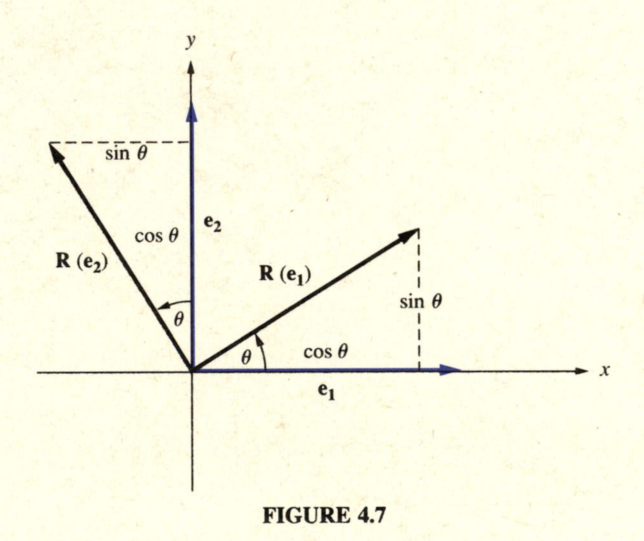

FIGURE 4.7

We will devote the rest of this section to a geometric interpretation of linear operators on $\Re^2$ and $\Re^3$. The main tools in this study will be Theorems 4.3, 4.9, and the fact, proven in Theorem 1.9, that every nonsingular matrix can be written as a product of elementary matrices.

Since a linear operator on $\Re^2$ maps points of the plane to points of the plane it will frequently be helpful to describe such a mapping by showing the image of the unit rectangle in the first quadrant (see Figure 4.8). Recall from Theorem 4.9 that **T** is completely determined by the images of the standard basis vectors; that is, by the vectors $\mathbf{T}(\mathbf{e}_1)$ and $\mathbf{T}(\mathbf{e}_2)$.

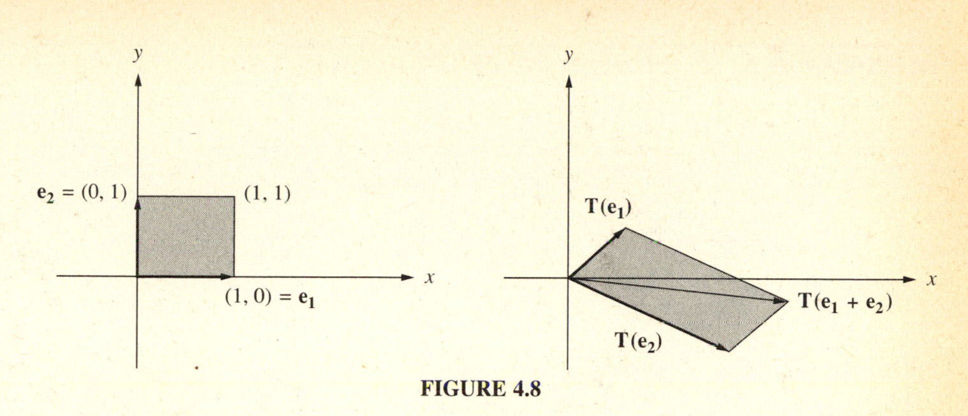

FIGURE 4.8

We begin by observing that a *linear transformation on $\mathfrak{R}^2$ or $\mathfrak{R}^3$ always sends straight lines to straight lines and straight lines through the origin to straight lines through the origin.* To see that this is the case we let P be any point on the line ℓ and let $\mathbf{u}$ be a vector parallel to the line. Then the line can be described as $\ell = \{\mathbf{OP} + k\mathbf{u} \mid k \text{ real}\}$ and the image of this line under $\mathbf{T}$ is the line $\mathbf{T}(\ell) = \{\mathbf{T}(\mathbf{OP}) + k\mathbf{T}(\mathbf{u}) \mid k \text{ real}\}$ which passes through the point $\mathbf{T}(\mathbf{OP})$ and is parallel to the vector $\mathbf{T}(\mathbf{u})$. See Figure 4.9.

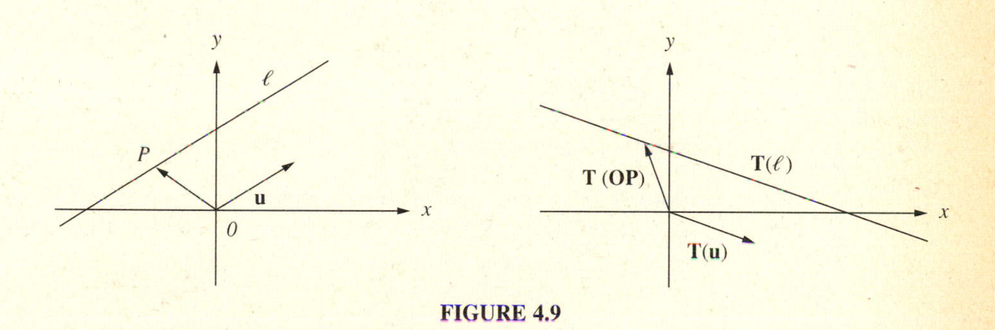

FIGURE 4.9

Since the origin is a fixed point under any linear operator (that is, $\mathbf{T}(\mathbf{0}) = \mathbf{0}$), a line through the origin always maps into another line through the origin or into the origin if the line is contained in the null space of the operator. Since two lines are parallel if, and only if, they are parallel to a common vector $\mathbf{u}$ (see Figure 4.10), it follows easily that parallel lines map into parallel lines and hence that parallelograms map into parallelograms. Thus, the image of the unit rectangle in Figure 4.8 must be a parallelogram.

From Theorem 1.9 we know that we can express any nonsingular matrix as a product of elementary matrices. Thus, from Section 4.3 it follows that any nonsingular linear operator on $\mathfrak{R}^2$ can be viewed as a product of linear operators associated with the 2×2 elementary matrices. Let us now look closely at the

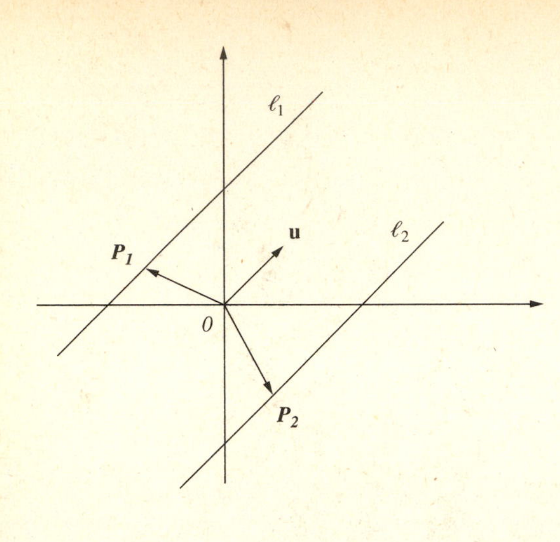

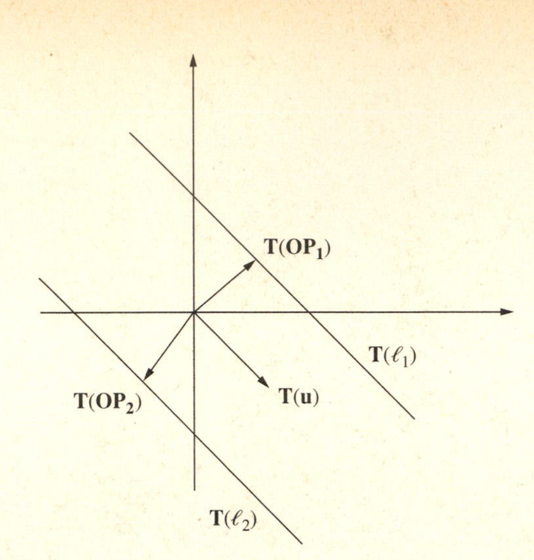

FIGURE 4.10

geometric interpretation of the matrix transformations determined by the elementary matrices

$$\begin{bmatrix} 0 & 1 \\ 1 & 0 \end{bmatrix}, \quad \begin{bmatrix} -1 & 0 \\ 0 & 1 \end{bmatrix}, \quad \begin{bmatrix} k & 0 \\ 0 & 1 \end{bmatrix}, \quad \begin{bmatrix} 1 & k \\ 0 & 1 \end{bmatrix}, \quad \text{and} \quad \begin{bmatrix} 1 & 0 \\ k & 1 \end{bmatrix}.$$

Reflections

If $\mathbf{T}\begin{bmatrix} x \\ y \end{bmatrix} = \begin{bmatrix} 0 & 1 \\ 1 & 0 \end{bmatrix}\begin{bmatrix} x \\ y \end{bmatrix} = \begin{bmatrix} y \\ x \end{bmatrix}$, then $\mathbf{T}(\mathbf{e}_1) = \mathbf{e}_2$ and $\mathbf{T}(\mathbf{e}_2) = \mathbf{e}_1$. Since the line connecting (a, b) and (b, a) has slope -1, it is perpendicular to the line $y = x$, and thus this transformation amounts to a **reflection in the line** $y = x$. Figure 4.11 shows the image of a 1×2 rectangle under this transformation.

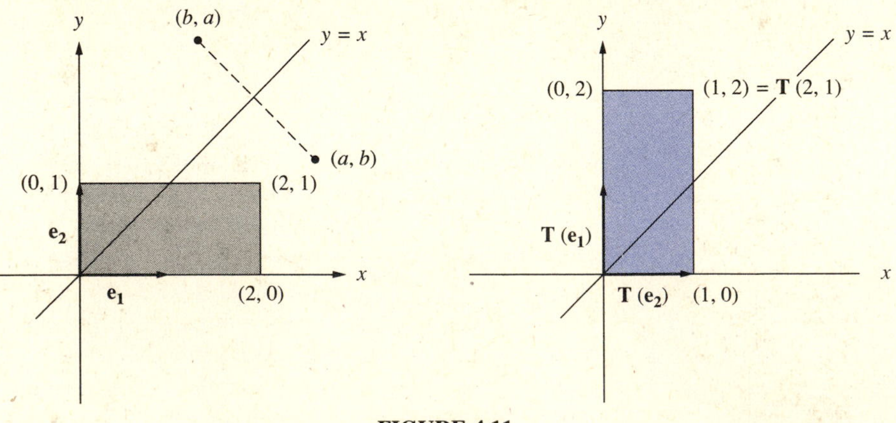

FIGURE 4.11

If $\mathbf{T}\begin{bmatrix} x \\ y \end{bmatrix} = \begin{bmatrix} -1 & 0 \\ 0 & 1 \end{bmatrix}\begin{bmatrix} x \\ y \end{bmatrix} = \begin{bmatrix} -x \\ y \end{bmatrix}$, then $\mathbf{T}(\mathbf{e}_1) = -\mathbf{e}_1$ and $\mathbf{T}(\mathbf{e}_2) = \mathbf{e}_2$. In this case $\mathbf{T}$ is a **reflection in the y-axis**, as shown in Figure 4.12. Similarly, the transformation with matrix $\begin{bmatrix} 1 & 0 \\ 0 & -1 \end{bmatrix}$ represents a **reflection in the x-axis.**

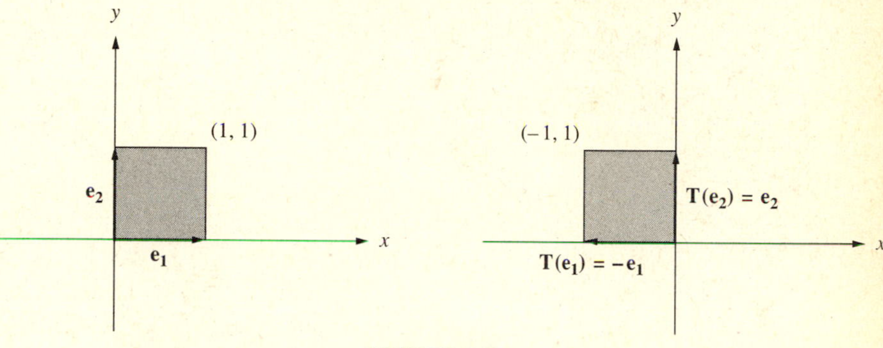

FIGURE 4.12

Expansion and Compressions

The matrix transformation with matrix $\begin{bmatrix} k & 0 \\ 0 & 1 \end{bmatrix}$, $k > 0$, sends $\mathbf{e}_1$ to $k\mathbf{e}_1$ and leaves $\mathbf{e}_2$ fixed. If $k > 1$, this transformation represents an **expansion in the x-direction by a factor of k**. If $k < 1$, this represents a **contraction in the x-direction by a factor of k**. Figure 4.13 illustrates these two cases. Similarly, the matrix $\begin{bmatrix} 1 & 0 \\ 0 & k \end{bmatrix}$ with $k > 0$, represents either an expansion or a contraction in the y-direction.

Shears

The matrix transformation with matrix $\begin{bmatrix} 1 & 0 \\ k & 1 \end{bmatrix}$ sends $\mathbf{e}_1$ to $\mathbf{e}_1 + k\mathbf{e}_2$ and $\mathbf{e}_2$ to $\mathbf{e}_2$. We will call this transformation a **shear in the y-direction with factor k**. The effect on the unit rectangle is shown in Figure 4.14.

Similarly, the matrix transformation with matrix $\begin{bmatrix} 1 & k \\ 0 & 1 \end{bmatrix}$ represents a **shear in the x-direction with factor k**.

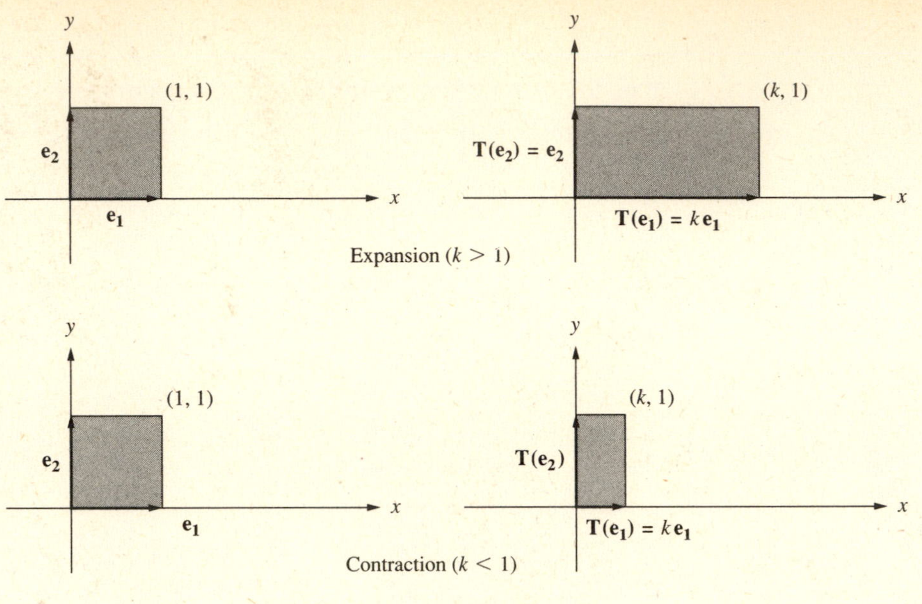

FIGURE 4.13

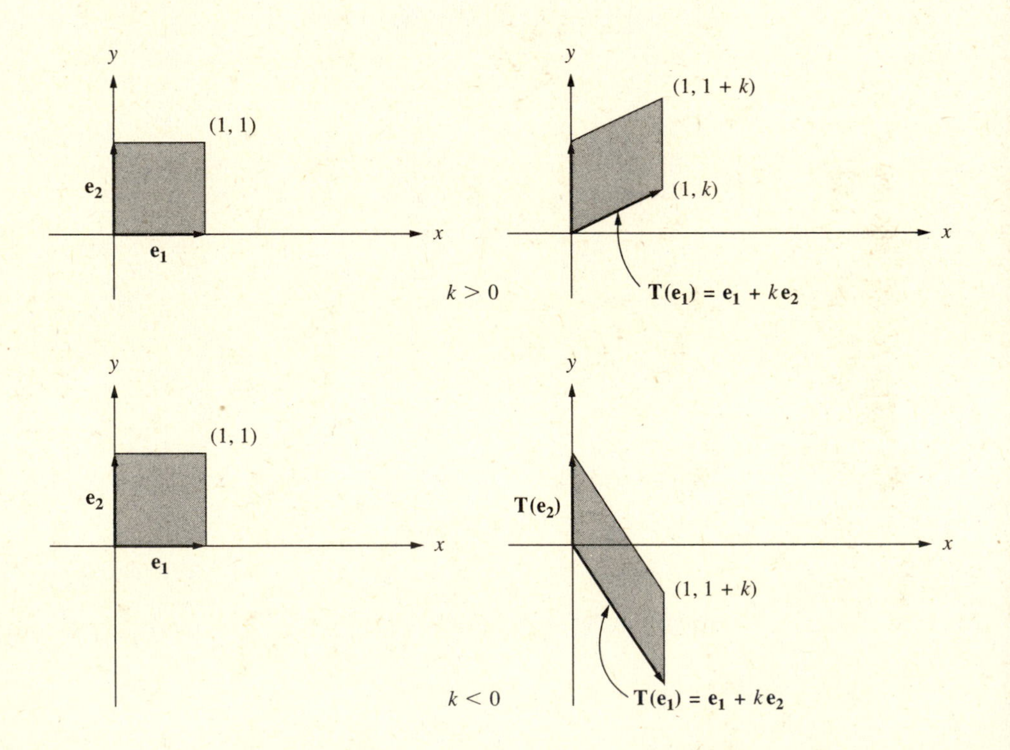

FIGURE 4.14

All of the 2×2 elementary matrices have either been included in the preceding discussion or are simple products of the preceding matrices. Thus we can describe the transformation associated with any nonsingular 2×2 matrix as a product of the simple transformations described above.

EXAMPLE 3 Decomposition into Elementary Transformations

Decompose the matrix transformation, with matrix $A = \begin{bmatrix} 1 & 2 \\ -1 & -4 \end{bmatrix}$ into a product of the elementary transformations (reflections, expansions, compressions, and shears) just described.

The row operations $(R_2 \leftarrow R_1 + R_2)$, $(R_2 \leftarrow -\frac{1}{2}R_2)$, and $(R_1 \leftarrow -2R_2 + R_1)$ reduce A to I, so we have, from Theorem 1.7,

$$\begin{bmatrix} 1 & -2 \\ 0 & 1 \end{bmatrix}\begin{bmatrix} 1 & 0 \\ 0 & -\frac{1}{2} \end{bmatrix}\begin{bmatrix} 1 & 0 \\ 1 & 1 \end{bmatrix}A = I.$$

It follows that
$$A = \begin{bmatrix} 1 & 0 \\ 1 & 1 \end{bmatrix}^{-1}\begin{bmatrix} 1 & 0 \\ 0 & -\frac{1}{2} \end{bmatrix}^{-1}\begin{bmatrix} 1 & -2 \\ 0 & 1 \end{bmatrix}^{-1}$$
$$= \begin{bmatrix} 1 & 0 \\ -1 & 1 \end{bmatrix}\begin{bmatrix} 1 & 0 \\ 0 & -2 \end{bmatrix}\begin{bmatrix} 1 & 2 \\ 0 & 1 \end{bmatrix}$$
$$= \begin{bmatrix} 1 & 0 \\ -1 & 1 \end{bmatrix}\begin{bmatrix} 1 & 0 \\ 0 & 2 \end{bmatrix}\begin{bmatrix} 1 & 0 \\ 0 & -1 \end{bmatrix}\begin{bmatrix} 1 & 2 \\ 0 & 1 \end{bmatrix} = E_1 E_2 E_3 E_4.$$

Note that we have written the Type I elementary matrix $\begin{bmatrix} 1 & 0 \\ 0 & -2 \end{bmatrix}$ as the product $\begin{bmatrix} 1 & 0 \\ 0 & 2 \end{bmatrix}\begin{bmatrix} 1 & 0 \\ 0 & -1 \end{bmatrix}$ which represents a reflection followed by an expansion. Thus, **T** can be viewed as the product of a shear in the x-direction (E_4), a reflection in the x-axis (E_3), an expansion in the y-direction (E_2) and a shear in the y-direction (E_1). The corners of the unit rectangle are successively transformed into

$$\begin{bmatrix} 1 \\ 0 \end{bmatrix} \to \begin{bmatrix} 1 \\ 0 \end{bmatrix} \to \begin{bmatrix} 1 \\ 0 \end{bmatrix} \to \begin{bmatrix} 1 \\ 0 \end{bmatrix} \to \begin{bmatrix} 1 \\ -1 \end{bmatrix} = \mathbf{T}\begin{bmatrix} 1 \\ 0 \end{bmatrix};$$
$$\begin{bmatrix} 0 \\ 1 \end{bmatrix} \to \begin{bmatrix} 2 \\ 1 \end{bmatrix} \to \begin{bmatrix} 2 \\ -1 \end{bmatrix} \to \begin{bmatrix} 2 \\ -2 \end{bmatrix} = \begin{bmatrix} 2 \\ -4 \end{bmatrix} = \mathbf{T}\begin{bmatrix} 0 \\ 1 \end{bmatrix}; \text{ and}$$
$$\begin{bmatrix} 1 \\ 1 \end{bmatrix} \to \begin{bmatrix} 3 \\ 1 \end{bmatrix} \to \begin{bmatrix} 3 \\ -1 \end{bmatrix} \to \begin{bmatrix} 3 \\ -2 \end{bmatrix} \to \begin{bmatrix} 3 \\ -5 \end{bmatrix} = \mathbf{T}\begin{bmatrix} 1 \\ 1 \end{bmatrix}.$$

The successive transformations of the unit rectangle are shown in Figure 4.15. ■

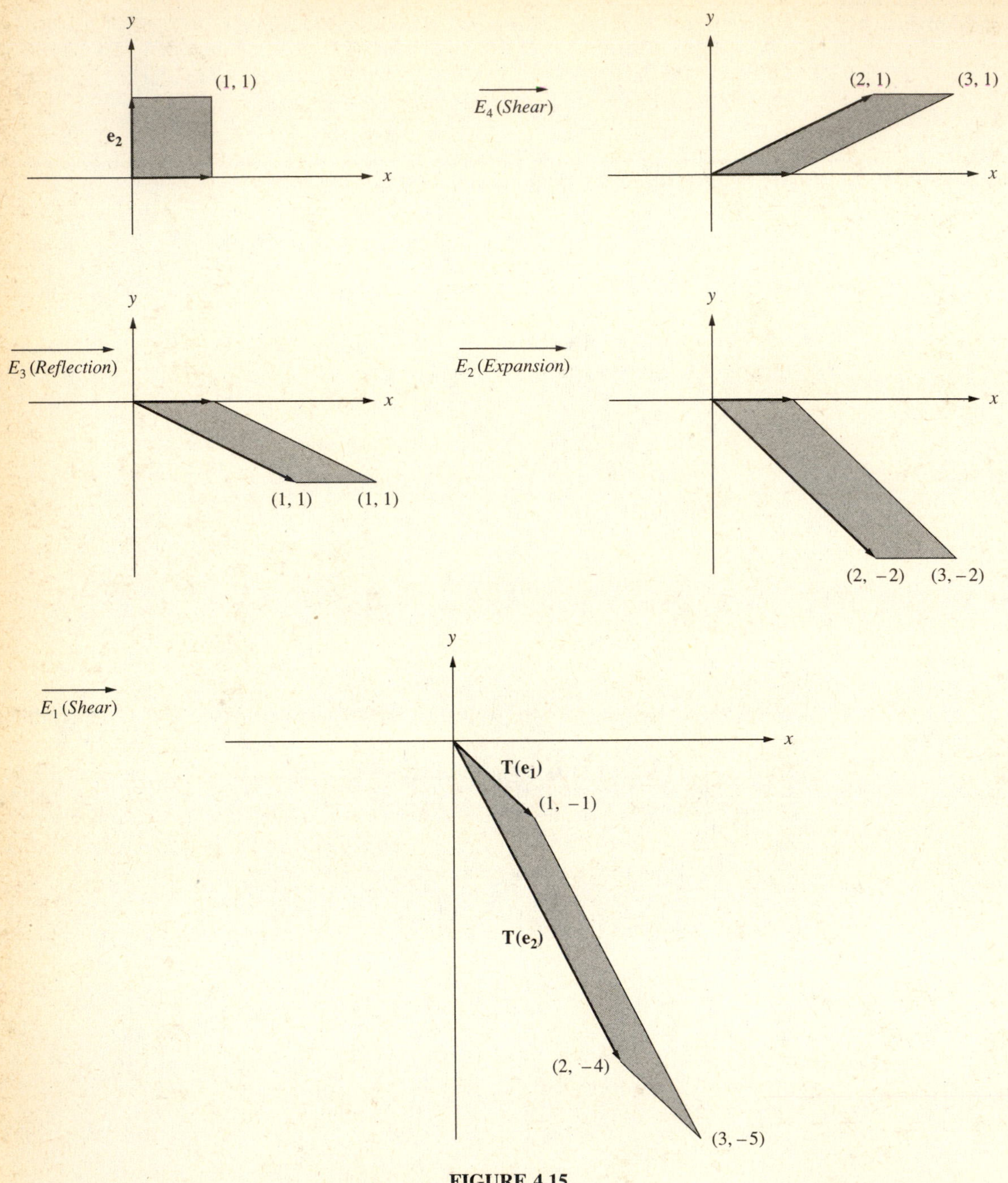
y
(1, 1)
e2
x
E4 (Shear)
y
(2, 1)
(3, 1)
x
y
E3 (Reflection)
x
(1, 1)
(1, 1)
y
E2 (Expansion)
x
(2, −2)
(3, −2)
y
E1 (Shear)
x
T(e1)
(1, −1)
T(e2)
(2, −4)
(3, −5)

FIGURE 4.15

EXAMPLE 4 **Construct a Transformation as a Product of Elementary Transformations**

Find the standard matrix of the transformation **T**, on $\mathfrak{R}^2$, that is obtained by the following sequence of elementary transformations:

(1) an expansion in the y-direction with factor 3;
(2) a shear in the x-direction with factor -3;
(3) a reflection in the y-axis;
(4) a compression in the x-direction with factor $\frac{1}{3}$;
(5) a shear in the y-direction with factor 2; and
(6) a reflection in the line $y = x$.

The matrices of the elementary transformations are

$$E_1 = \begin{bmatrix} 1 & 0 \\ 0 & 3 \end{bmatrix}, \quad E_2 = \begin{bmatrix} 1 & -3 \\ 0 & 1 \end{bmatrix}, \quad E_3 = \begin{bmatrix} -1 & 0 \\ 0 & 1 \end{bmatrix},$$

$$E_4 = \begin{bmatrix} \frac{1}{3} & 0 \\ 0 & 1 \end{bmatrix}, \quad E_5 = \begin{bmatrix} 1 & 0 \\ 2 & 1 \end{bmatrix}, \quad \text{and} \quad E_6 = \begin{bmatrix} 0 & 1 \\ 1 & 0 \end{bmatrix},$$

so $\mathbf{T}(\mathbf{u}) = E_6 E_5 E_4 E_3 E_2 E_1 \mathbf{u}$

$$= \begin{bmatrix} 0 & 1 \\ 1 & 0 \end{bmatrix}\begin{bmatrix} 1 & 0 \\ 2 & 1 \end{bmatrix}\begin{bmatrix} \frac{1}{3} & 0 \\ 0 & 1 \end{bmatrix}\begin{bmatrix} -1 & 0 \\ 0 & 1 \end{bmatrix}\begin{bmatrix} 1 & -3 \\ 0 & 1 \end{bmatrix}\begin{bmatrix} 1 & 0 \\ 0 & 3 \end{bmatrix}\begin{bmatrix} x \\ y \end{bmatrix}$$

$$= \begin{bmatrix} -\frac{2}{3} & 9 \\ -\frac{1}{3} & 3 \end{bmatrix}\begin{bmatrix} x \\ y \end{bmatrix}.$$

Note that the product of the E_i must be formed from right to left since the first transformation must go with the rightmost factor. ■

There is one other kind of transformation on $\mathfrak{R}^2$ that is worth mentioning: the projections onto the coordinate axes. The transformation

$$\mathbf{P_x}\begin{bmatrix} x \\ y \end{bmatrix} = \begin{bmatrix} 1 & 0 \\ 0 & 0 \end{bmatrix}\begin{bmatrix} x \\ y \end{bmatrix} = \begin{bmatrix} x \\ 0 \end{bmatrix}$$

is called a **projection onto the x-axis** while the transformation

$$\mathbf{P_y}\begin{bmatrix} x \\ y \end{bmatrix} = \begin{bmatrix} 0 & 0 \\ 0 & 1 \end{bmatrix}\begin{bmatrix} x \\ y \end{bmatrix} = \begin{bmatrix} 0 \\ y \end{bmatrix}$$

is called a **projection onto the y-axis**.

Note that these projections are not one-to-one mappings and hence are not invertible; their standard matrices are also singular. It is easy to see that $\mathcal{NS}(\mathbf{P_x})$ is the y-axis while the x-axis is the null space of $\mathbf{P_y}$.

There are more general types of projections than those mentioned above. Suppose that $\mathcal{W}_1$ is a subspace of $\mathfrak{R}^n$ with a basis $\{\mathbf{u_1}, \ldots, \mathbf{u_k}\}$, which has been completed to a basis $\{\mathbf{u_1}, \ldots, \mathbf{u_k}, \ldots, \mathbf{u_n}\}$ for $\mathfrak{R}^n$. Let $\mathcal{W}_2$ be the subspace of

$\mathfrak{R}^n$ spanned by $\{\mathbf{u}_{k+1}, \ldots, \mathbf{u}_n\}$ and let $\mathbf{v} = a_1\mathbf{u}_1 + \cdots + a_k\mathbf{u}_k + \cdots + a_n\mathbf{u}_n$ be a typical vector in $\mathfrak{R}^n$. The transformation $\mathbf{P}$ defined by

$$\mathbf{P}(\mathbf{v}) = a_1\mathbf{u}_1 + a_2\mathbf{u}_2 + \cdots + a_k\mathbf{u}_k$$

is called a **projection onto $\mathcal{W}_1$ parallel to $\mathcal{W}_2$**. Note that the range of $\mathbf{P}$ is $\mathcal{W}_1$, that the null space of $\mathbf{P}$ is $\mathcal{W}_2$, that $\mathbf{P}(\mathbf{w}) = \mathbf{w}$, for any $\mathbf{w}$ in $\mathcal{W}_1$, and that $\mathbf{P}^2(\mathbf{v}) = \mathbf{P}(\mathbf{v})$ for every $\mathbf{v}$ in $\mathfrak{R}^n$; that is, $\mathbf{P}^2 = \mathbf{P}$.

If $\{\mathbf{u}_1, \mathbf{u}_2, \ldots, \mathbf{u}_n\}$ is an orthonormal basis for $\mathfrak{R}^n$ then the subspace $\mathcal{W}_2$ is the orthogonal complement of $\mathcal{W}_1$ ($\mathcal{W}_2 = \mathcal{W}_1^\perp$) and $\mathbf{P}$ is called an **orthogonal projection onto $\mathcal{W}_1$**.

Each 3×3 elementary matrix can also be interpreted geometrically, so the above results generalize easily to $\mathfrak{R}^3$. For example, $\begin{bmatrix} 0 & 1 & 0 \\ 1 & 0 & 0 \\ 0 & 0 & 1 \end{bmatrix}$ is a reflection about the plane $y = x$; $\begin{bmatrix} 1 & 0 & 0 \\ 0 & 1 & 0 \\ a & 0 & 1 \end{bmatrix}$ is a shear in the z direction of the xz-plane with factor a; and $\begin{bmatrix} 1 & 0 & 0 \\ 0 & 3 & 0 \\ 0 & 0 & 1 \end{bmatrix}$ is an expansion in the y-direction with a factor of 3.

The rotation and projection matrices also generalize to $\mathfrak{R}^3$: $\begin{bmatrix} \cos\theta & 0 & -\sin\theta \\ 0 & 1 & 0 \\ \sin\theta & 0 & \cos\theta \end{bmatrix}$ is a rotation in the xz-plane and $\begin{bmatrix} 1 & 0 & 0 \\ 0 & 0 & 0 \\ 0 & 0 & 1 \end{bmatrix}$ is a projection onto the xz-plane. We leave the details of the three-dimensional case to the exercises. The ideas introduced in this section have many applications, e.g. to robotics, computer graphics, and metal deformation.

EXERCISES 4.4

1. Find the standard matrix of the linear operator $\mathbf{T}$ on $\mathfrak{R}^2$ given the following information.

   (a) $\mathbf{T}\begin{bmatrix} 1 \\ 3 \end{bmatrix} = \begin{bmatrix} 2 \\ 6 \end{bmatrix}$ and $\mathbf{T}\begin{bmatrix} 1 \\ -3 \end{bmatrix} = \begin{bmatrix} 3 \\ -9 \end{bmatrix}$

   $\left(\textit{Hint: } \mathbf{e}_1 = \frac{1}{2}\begin{bmatrix} 1 \\ 3 \end{bmatrix} + \frac{1}{2}\begin{bmatrix} 1 \\ -3 \end{bmatrix}\right)$

   (b) $\mathbf{T}\begin{bmatrix} 1 \\ 1 \end{bmatrix} = \begin{bmatrix} 1 \\ 1 \end{bmatrix}$ and $\mathbf{T}\begin{bmatrix} 2 \\ 3 \end{bmatrix} = \begin{bmatrix} 5 \\ 5 \end{bmatrix}$

   (c) $\mathbf{T}\begin{bmatrix} 1 \\ 2 \end{bmatrix} = \begin{bmatrix} 3 \\ 6 \end{bmatrix}$ and $\mathbf{T}\begin{bmatrix} 1 \\ 1 \end{bmatrix} = \begin{bmatrix} -2 \\ -2 \end{bmatrix}$

2. Find the standard matrix of the linear operator **T** on $\Re^3$ for which

$$\mathbf{T}\begin{bmatrix}1\\0\\0\end{bmatrix} = \begin{bmatrix}2\\3\\1\end{bmatrix}, \quad \mathbf{T}\begin{bmatrix}1\\1\\0\end{bmatrix} = \begin{bmatrix}0\\-3\\8\end{bmatrix}, \quad \text{and} \quad \mathbf{T}\begin{bmatrix}1\\1\\1\end{bmatrix} = \begin{bmatrix}7\\5\\3\end{bmatrix}.$$

3. Show that a product of two rotations in $\Re^2$ is another rotation in $\Re^2$.

4. Show that every linear transformation **T** from $\Re^{1\times n}$ to $\Re^{1\times m}$ is right multiplication by a suitable matrix A; that is,

$$\mathbf{T}(\mathbf{u}) = \mathbf{u}A \quad \text{for every } \mathbf{u} \text{ in } \Re^{1\times n}.$$

In each of the following problems a transformation **T** *on* $\Re^2$ *is described as a sequence of simple transformations. In each case find the standard matrix of* **T** *and the image of the unit rectangle under* **T**.

5. **T** is
- **(a)** a compression by a factor of $\frac{1}{4}$ in the x-direction;
- **(b)** a shear by a factor of 2 in the y-direction;
- **(c)** a shear by a factor of -3 in the x-direction; and
- **(d)** a reflection in the line $y = x$.

6. **T** is
- **(a)** a reflection in the y-axis;
- **(b)** an expansion by a factor of 3 in the y-direction;
- **(c)** a shear by a factor of 1 in the x-direction; and
- **(d)** a rotation through 45 degrees.

7. **T** is
- **(a)** a compression by a factor of $\frac{1}{3}$ in both directions;
- **(b)** a shear by a factor of -3 in the x-direction;
- **(c)** a shear by a factor of 2 in the y-direction; and
- **(d)** a reflection in the line $y = x$.

8. **T** is
- **(a)** an expansion by a factor of 2 in the x-direction;
- **(b)** a reflection in the x-axis;
- **(c)** a shear by a factor of $-\frac{1}{2}$ in the y-direction; and
- **(d)** a shear by a factor of 5 in the x-direction.

In each of the following problems describe the given matrix transformation of $\Re^2$ *as a product of expansions, contractions, reflections, and shears.*

9. $\mathbf{T}(\mathbf{u}) = \begin{bmatrix}4 & -3\\3 & -2\end{bmatrix}\mathbf{u}$

10. $\mathbf{T}(\mathbf{u}) = \begin{bmatrix}1 & 4\\2 & 9\end{bmatrix}\mathbf{u}$

11. $\mathbf{T}(\mathbf{u}) = \begin{bmatrix}\cos\theta & -\sin\theta\\\sin\theta & \cos\theta\end{bmatrix}\mathbf{u}$

12. $\mathbf{T}(\mathbf{u}) = \begin{bmatrix}3 & 1\\6 & 3\end{bmatrix}\mathbf{u}$

In each of the following problems a transformation **T** *on* $\Re^3$ *is described as a sequence of elementary transformations. Find the standard matrix of* **T**. *In each case find the image of the unit cube under* **T**.

13. **T** is
 (a) an expansion by a factor of 2 in the z-direction;
 (b) a shear by a factor of 3 in the z-direction of the xz-plane;
 (c) a reflection in the plane $x = y$; and
 (d) a projection onto the yz-plane.

14. **T** is
 (a) a rotation through 30 degrees in the xy-plane;
 (b) an expansion by the factor 2 in the y-direction;
 (c) a reflection in the xy-plane; and
 (d) a shear by a factor of 3 in the y-direction of the yz-plane.

In Exercises 15 and 16 describe the given matrix transformation on $\Re^3$ as a product of elementary transformations.

15. $\mathbf{T}(\mathbf{u}) = \begin{bmatrix} -1 & 1 & 3 \\ 2 & 1 & 1 \\ 4 & 2 & 3 \end{bmatrix}\mathbf{u}$ **16.** $\mathbf{T}(\mathbf{u}) = \begin{bmatrix} 1 & -1 & -1 \\ 5 & -9 & -13 \\ 3 & 1 & -3 \end{bmatrix}\mathbf{u}$

17. Determine a rotation matrix $R(\theta) = \begin{bmatrix} \cos\theta & -\sin\theta \\ \sin\theta & \cos\theta \end{bmatrix}$ so that

$$R(\theta)\begin{bmatrix} a & b \\ c & d \end{bmatrix} = \begin{bmatrix} a' & b' \\ 0 & d' \end{bmatrix}.$$

18. Determine a rotation matrix R in the yz-plane so that $RA = R\begin{bmatrix} a & b & c \\ d & e & f \\ g & h & i \end{bmatrix} =$

$\begin{bmatrix} a & b & c \\ d' & e' & f' \\ 0 & h' & i' \end{bmatrix}$. Show that RAR^T also has a zero in position 3, 1.

19. Show that every 2×2 orthogonal matrix is either a rotation or a reflection in a coordinate axis followed by a rotation.

20. **(a)** Determine a sequence of three plane rotations R_1, R_2, and R_3 so that

$$R_3R_2R_1A = R_3R_2R_1\begin{bmatrix} a & b & c \\ d & e & f \\ g & h & i \end{bmatrix} = \begin{bmatrix} * & * & * \\ 0 & * & * \\ 0 & 0 & * \end{bmatrix} = T$$

 (b) Show how this result can be used to find the QR factorization of a matrix A (see Section 3.6).
 (c) Show that if A is an orthogonal matrix, then T is a diagonal matrix.
 (d) Use part (b) to show that any 3×3 orthogonal matrix is a product of plane rotations and reflections.

21. **(a)** Show that each reflection matrix has eigenvalues 1 and -1.
 (b) Show that the matrices representing plane shears have two eigenvalues equal to 1, but only a one-dimensional subspace of eigenvectors.
 (c) Show that a plane rotation has no real eigenvalues.
 (d) Interpret each of statements (a) through (c) geometrically.

22. Let $\mathbf{w}$ be a unit vector in $\mathfrak{R}^n$ and let $H = I - 2\mathbf{w}\mathbf{w}^{\mathbf{T}}$.
(a) Show that $H\mathbf{w} = -\mathbf{w}$.
(b) Show that if $\mathbf{x}$ is orthogonal to $\mathbf{w}$, then $H\mathbf{x} = \mathbf{x}$.
(c) Show that H can be interpreted as a reflection in the subspace orthogonal to $\mathbf{w}$.

23. Find the orthogonal projection of $\mathbf{u} = [1 \quad 2 \quad 3]^T$ onto the subspace of $\mathfrak{R}^3$ spanned by $\{\mathbf{v}_1 = [1 \quad 5 \quad 2]^T, \mathbf{v}_2 = [7 \quad -1 \quad -1]^T\}$.

24. Consider the linear operator on $\mathfrak{R}^2$ defined by $\mathbf{T}\begin{bmatrix} x \\ y \end{bmatrix} = \begin{bmatrix} a & b \\ c & d \end{bmatrix}\begin{bmatrix} x \\ y \end{bmatrix}$.
(a) Show that the image of the unit square is a parallelogram.
(b) Show that the area of the parallelogram in part (a) is $|\det(A)|$. (*Hint:* Recall Example 4 in Section 2.3.)
(c) Let $\mathcal{S}$ be any plane area containing the origin and let $\mathbf{T}(\mathcal{S})$ be its image under $\mathbf{T}$. Show that the area of $\mathbf{T}(\mathcal{S})$ is equal to the area of $\mathcal{S}$ multiplied by $|\det(A)|$.
(d) If $\mathcal{S}$ is the unit circle and $A = \begin{bmatrix} a & 0 \\ 0 & d \end{bmatrix}$, find the area of $\mathbf{T}(\mathcal{S})$. Describe $\mathbf{T}(\mathcal{S})$ geometrically.
(e) What is the image of the unit circle under an arbitrary linear operator?

25. Let $\mathbf{u}$ and $\mathbf{v}$ be vectors in $\mathfrak{R}^3$ connecting the origin and the points A and B.
(a) Show that the vectors $t\mathbf{u} + (1 - t)\mathbf{v}$, with $0 \le t \le 1$, all connect the origin and some point on the line segment AB.
(b) Show that all the vectors $t\mathbf{u} + s\mathbf{v}$ with $0 \le t \le 1$ and $0 \le s \le 1$ connect the origin to points in the parallelogram determined by $\mathbf{u}$ and $\mathbf{v}$.
(c) Show that a linear operator on $\mathfrak{R}^3$ maps parallelograms into parallelograms.

26. Consider the linear transformation $\mathbf{U}$ from $\mathfrak{R}^3$ to $\mathfrak{R}^2$ which satisfies

$$\mathbf{U}(\mathbf{e}_1) = \begin{bmatrix} 1 \\ 4 \end{bmatrix}, \quad \mathbf{U}(\mathbf{e}_2) = \begin{bmatrix} 3 \\ 5 \end{bmatrix}, \quad \mathbf{U}(\mathbf{e}_3) = \begin{bmatrix} -2 \\ 7 \end{bmatrix}.$$

(a) Find the standard matrix of $\mathbf{U}$.
(b) Is $\mathbf{U}$ one-to-one?
(c) Is $\mathbf{U}$ onto $\mathfrak{R}^2$?

27. Consider the linear transformation $\mathbf{V}$ from $\mathfrak{R}^2$ to $\mathfrak{R}^3$ which satisfies

$$\mathbf{V}(\mathbf{e}_1) = \begin{bmatrix} 1 \\ 2 \\ 5 \end{bmatrix}, \quad \mathbf{V}(\mathbf{e}_2) = \begin{bmatrix} 3 \\ 5 \\ 1 \end{bmatrix}.$$

(a) Find the standard matrix of $\mathbf{V}$.
(b) Is $\mathbf{V}$ one-to-one?
(c) Is $\mathbf{V}$ onto $\mathfrak{R}^3$?

28. Verify the matrix identity $\begin{bmatrix} \sec\theta & -\tan\theta & 0 \\ 0 & 1 & 0 \\ 0 & 0 & 1 \end{bmatrix}\begin{bmatrix} 1 & 0 & 0 \\ \sin\theta & \cos\theta & 0 \\ 0 & 0 & 1 \end{bmatrix} =$ $\begin{bmatrix} \cos\theta & -\sin\theta & 0 \\ \sin\theta & \cos\theta & 0 \\ 0 & 0 & 1 \end{bmatrix}$. Show from this factorization that a rotation of $\mathfrak{R}^3$ can be accomplished as a product of shears and expansions.

29. Verify the matrix identity

$$\begin{bmatrix} 1 & -\tan(\frac{\alpha}{2}) & 0 \\ 0 & 1 & 0 \\ 0 & 0 & 1 \end{bmatrix} \begin{bmatrix} 1 & 0 & 0 \\ \sin\alpha & 1 & 0 \\ 0 & 0 & 1 \end{bmatrix} \begin{bmatrix} 1 & -\tan(\frac{\alpha}{2}) & 0 \\ 0 & 1 & 0 \\ 0 & 0 & 1 \end{bmatrix}$$
$$= \begin{bmatrix} \cos\alpha & -\sin\alpha & 0 \\ \sin\alpha & \cos\alpha & 0 \\ 0 & 2 & 1 \end{bmatrix}.$$

Show, using this factorization, that a plane rotation, on a vector in $\mathfrak{R}^3$, can be accomplished using only 3 multiplications rather than the 4 that would be required using the unfactored matrix. $\left(\textit{Hint: } \tan\frac{\alpha}{2} = \frac{1-\cos\alpha}{\sin\alpha} = \frac{\sin\alpha}{1+\cos\alpha}.\right)$

4.5 MATRICES AND LINEAR TRANSFORMATIONS

> In what sense can all linear transformations, on a finite dimensional vector space, be viewed as matrix multiplication?

In the last section we saw that every linear transformation between coordinate spaces can be described as multiplication by a suitable matrix. This is equivalent to saying that the vector space $\mathcal{L}(\mathfrak{R}^n, \mathfrak{R}^m)$, of all linear transformations from $\mathfrak{R}^n$ to $\mathfrak{R}^m$, is really just $\mathfrak{R}^{m\times n}$, the vector space of all $m \times n$ matrices; that is,

$$\mathcal{L}(\mathfrak{R}^n, \mathfrak{R}^m) = \mathfrak{R}^{m\times n}.$$

In this section we will see that there is also a close connection between matrix multiplication and linear transformations between arbitrary finite dimensional vector spaces.

Let $\mathbf{T}$ be any linear transformation from the vector space $\mathcal{V}$ to the vector space $\mathcal{W}$. If we choose a basis $\mathcal{B} = \{\mathbf{v}_1, \mathbf{v}_2, \ldots, \mathbf{v}_n\}$ for $\mathcal{V}$, then we have seen that the coordinate mapping $[\ ]_{\mathcal{B}}$, which sends the vector $\mathbf{u}$ in $\mathcal{V}$ to its $\mathcal{B}$-coordinate matrix $[\mathbf{u}]_{\mathcal{B}}$ in $\mathfrak{R}^n$, is a one-to-one linear mapping. Similarly, if we choose a basis $\mathcal{B}' = \{\mathbf{w}_1, \mathbf{w}_2, \ldots, \mathbf{w}_m\}$ for $\mathcal{W}$, then the coordinate mapping $[\ ]_{\mathcal{B}'}$, which sends $\mathbf{w}$ in $\mathcal{W}$ to $[\mathbf{w}]_{\mathcal{B}'}$ in $\mathfrak{R}^m$, is also one-to-one and linear. The diagrams in Figure 4.16 illustrate the various transformations.

What is the relationship between $[\mathbf{T}(\mathbf{u})]_{\mathcal{B}'}$, and $[\mathbf{u}]_{\mathcal{B}}$? It certainly seems reasonable to expect that $\mathbf{T}$ will "induce" a linear transformation $\mathbf{T}'$ from $\mathfrak{R}^n$ to $\mathfrak{R}^m$ such that $\mathbf{T}'([\mathbf{u}]_{\mathcal{B}}) = [\mathbf{T}(\mathbf{u})]_{\mathcal{B}'}$. According to Theorem 4.9, such a transformation must be a matrix transformation. Thus, we seek an $m \times n$ matrix A such that

$$A[\mathbf{u}]_{\mathcal{B}} = [\mathbf{T}(\mathbf{u})]_{\mathcal{B}'} \text{ for all } \mathbf{u} \text{ in } \mathcal{V}. \qquad \textbf{(4.3)}$$

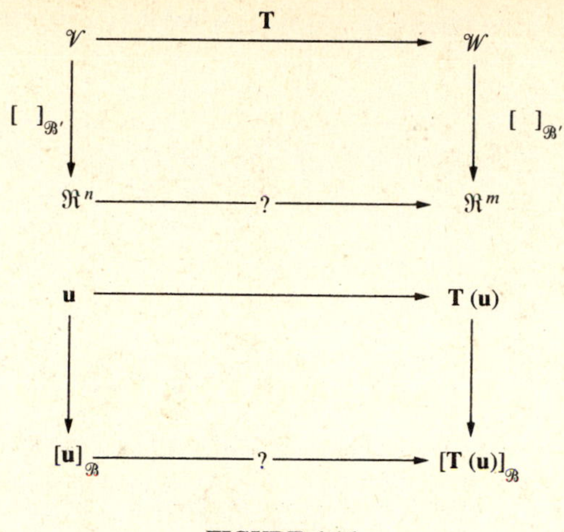

FIGURE 4.16

If we could find such a matrix, then the evaluation of $\mathbf{T}(\mathbf{u})$ could be accomplished by the following three steps, all of which could be easily handled on a digital computer.

(1) Find the coordinate matrix $[\mathbf{u}]_{\mathcal{B}}$; this normally involves solving a linear system.
(2) Multiply $[\mathbf{u}]_{\mathcal{B}}$ by A to find $[\mathbf{T}(\mathbf{u})]_{\mathcal{B}'} = A[\mathbf{u}]_{\mathcal{B}}$.
(3) Find $\mathbf{T}(\mathbf{u})$ from its coordinate matrix $[\mathbf{T}(\mathbf{u})]_{\mathcal{B}'}$.

We now attempt to find a matrix A which satisfies Equation (4.3) for all vectors $\mathbf{u}$ in $\mathcal{V}$. Since Equation (4.3) must hold for all vectors $\mathbf{u}$ in $\mathcal{V}$, it must hold for the vectors $\mathbf{v}_i$ of the basis $\mathcal{B}$. For $\mathbf{u} = \mathbf{v}_1$ we have

$$[\mathbf{v}_1]_{\mathcal{B}} = \mathbf{e}_1 = \begin{bmatrix} 1 \\ 0 \\ \vdots \\ 0 \end{bmatrix},$$

so that Equation (4.3) yields

$$[\mathbf{T}(\mathbf{v}_1)]_{\mathcal{B}'} = A[\mathbf{v}_1]_{\mathcal{B}} = A\mathbf{e}_1 = \begin{bmatrix} a_{11} \\ a_{21} \\ \vdots \\ a_{m1} \end{bmatrix} = \mathrm{Col}_1(A).$$

Thus, the first column of A is the $\mathcal{B}'$-coordinate matrix of the image of the first basis vector of $\mathcal{B}$.

Similarly, for $\mathbf{u} = \mathbf{v_i}$ we have $[\mathbf{v_i}]_{\mathcal{B}} = \mathbf{e_i}$, so from Equation (4.3)

$$[\mathbf{T}(\mathbf{v_i})]_{\mathcal{B}'} = A[\mathbf{v_i}]_{\mathcal{B}} = A\mathbf{e_i} = \text{Col}_i(A) = \begin{bmatrix} a_{1i} \\ a_{2i} \\ \vdots \\ a_{mi} \end{bmatrix}.$$

This calculation shows that the matrix A of Equation (4.3), if it exists, is unique and

$$A = \begin{bmatrix} [\mathbf{T}(\mathbf{v_1})]_{\mathcal{B}'} & [\mathbf{T}(\mathbf{v_2})]_{\mathcal{B}'} & \cdots & [\mathbf{T}(\mathbf{v_n})]_{\mathcal{B}'} \end{bmatrix}.$$

We now need to show that (4.3) holds for an arbitrary vector $\mathbf{x}$ in $\mathcal{V}$. Since the $\mathbf{v_i}$ form a basis for $\mathcal{V}$, there exists unique scalars b_i such that

$$\mathbf{x} = b_1\mathbf{v_1} + b_2\mathbf{v_2} + \cdots + b_n\mathbf{v_n}.$$

Since $\mathbf{T}$ is linear we have

$$\mathbf{T}(\mathbf{x}) = b_1\mathbf{T}(\mathbf{v_1}) + b_2\mathbf{T}(\mathbf{v_2}) + \cdots + b_n\mathbf{T}(\mathbf{v_n})$$

and, since the coordinate mapping $[\ ]_{\mathcal{B}'}$ is also linear (see Exercise 26 of Section 4.1), we see that

$$\begin{aligned}[\mathbf{T}(\mathbf{x})]_{\mathcal{B}'} &= b_1[\mathbf{T}(\mathbf{v_1})]_{\mathcal{B}'} + b_2[\mathbf{T}(\mathbf{v_2})]_{\mathcal{B}'} + \cdots + b_n[\mathbf{T}(\mathbf{v_n})]_{\mathcal{B}'} \\ &= \begin{bmatrix} [\mathbf{T}(\mathbf{v_1})]_{\mathcal{B}'} & [\mathbf{T}(\mathbf{v_2})]_{\mathcal{B}'} & \cdots & [\mathbf{T}(\mathbf{v_n})]_{\mathcal{B}'} \end{bmatrix} \begin{bmatrix} b_1 \\ b_2 \\ \vdots \\ b_n \end{bmatrix} = A[\mathbf{x}]_{\mathcal{B}}.\end{aligned}$$

We summarize this argument with our next theorem.

THEOREM 4.10 Let $\mathcal{V}$ and $\mathcal{W}$ be finite dimensional vector spaces and let

$$\mathcal{B} = \{\mathbf{v_1}, \cdots, \mathbf{v_n}\} \quad \text{and} \quad \mathcal{B}' = \{\mathbf{w_1}, \cdots, \mathbf{w_m}\}$$

be bases for $\mathcal{V}$ and $\mathcal{W}$ respectively. If $\mathbf{T}$ is any linear transformation from $\mathcal{V}$ to $\mathcal{W}$, then

$$A = \begin{bmatrix} [\mathbf{T}(\mathbf{v_1})]_{\mathcal{B}'} & [\mathbf{T}(\mathbf{v_2})]_{\mathcal{B}'} & \cdots & [\mathbf{T}(\mathbf{v_n})]_{\mathcal{B}'} \end{bmatrix}$$

is the unique $m \times n$ matrix satisfying

$$[\mathbf{T}(\mathbf{u})]_{\mathcal{B}'} = A[\mathbf{u}]_{\mathcal{B}} \quad \text{for every } \mathbf{u} \text{ in } \mathcal{V}. \quad \square$$

The matrix A of Theorem 4.10 is called the **matrix representative of T** with respect to the bases $\mathcal{B}$ and $\mathcal{B}'$. This matrix will be denoted by $[\mathbf{T}]_{\mathcal{B}'\mathcal{B}}$ when a compact notation is desired. In this notation, Equation (4.3) becomes

$$[\mathbf{T}(\mathbf{u})]_{\mathcal{B}'} = [\mathbf{T}]_{\mathcal{B}'\mathcal{B}}[\mathbf{u}]_{\mathcal{B}}.$$

In the important case where $\mathcal{V} = \mathcal{W}$, the matrix representative is a square matrix. In this case it is natural to take $\mathcal{B}' = \mathcal{B}$ and Equation (4.3) becomes

$$[\mathbf{T}(\mathbf{u})]_{\mathcal{B}} = [\mathbf{T}]_{\mathcal{B}\mathcal{B}}[\mathbf{u}]_{\mathcal{B}}.$$

If $\mathcal{V} = \mathcal{W} = \mathfrak{R}^n$ and the basis is the standard basis $\mathcal{S}$, then the matrix representative $[\mathbf{T}]_{\mathcal{S}\mathcal{S}}$ is the standard matrix of the transformation which was introduced in Section 4.4. In this case finding the coordinate vectors is easy because $[\mathbf{v}]_{\mathcal{S}} = \mathbf{v}$ for every $\mathbf{v}$ in $\mathcal{V}$.

In general, finding the matrix representative $[\mathbf{T}]_{\mathcal{B}'\mathcal{B}}$ involves the following steps.

TO FIND THE MATRIX REPRESENTATIVE $[\mathbf{T}]_{\mathcal{B}'\mathcal{B}}$

1. Find $\mathbf{T}(\mathbf{v}_i)$ for each $\mathbf{v}_i$ in the basis $\mathcal{B}$.
2. Find $[\mathbf{T}(\mathbf{v}_i)]_{\mathcal{B}'}$ for each $\mathbf{v}_i$ in $\mathcal{B}$ (this normally involves solving a set of linear systems, all with the same coefficient matrix).
3. Write down $[\mathbf{T}]_{\mathcal{B}'\mathcal{B}}$, using the fact that its columns are the coordinate matrices from step 2.

In general, the second step will require the bulk of the computational effort.

EXAMPLE 1 Matrix Representative of the Derivative Operator

Let $\mathbf{D}$ be the linear operator from $\mathcal{P}_3$ to $\mathcal{P}_2$ defined by

$$\mathbf{D}(a_0 + a_1x + a_2x^2 + a_3x^3) = a_1 + 2a_2x + 3a_3x^2.$$

Find the matrix representative of $\mathbf{D}$ with respect to the basis $\mathcal{B} = \{1, x, x^2, x^3\}$ of $\mathcal{P}_3$ and the basis $\mathcal{B}' = \{1, x, x^2\}$ of $\mathcal{P}_2$. Use this matrix to find $\mathbf{D}(5 - 7x + 11x^2 + 4x^3)$.

Direct calculation shows that $\mathbf{D}(1) = 0$, $\mathbf{D}(x) = 1$, $\mathbf{D}(x^2) = 2x$, and $\mathbf{D}(x^3) = 3x^2$, so that

$$[\mathbf{D}]_{\mathcal{B}'\mathcal{B}} = \Big[[0]_{\mathcal{B}'} \quad [1]_{\mathcal{B}'} \quad [2x]_{\mathcal{B}'} \quad [3x^2]_{\mathcal{B}'} \Big] = \begin{bmatrix} 0 & 1 & 0 & 0 \\ 0 & 0 & 2 & 0 \\ 0 & 0 & 0 & 3 \end{bmatrix}.$$

To find $\mathbf{D}(p(x))$ for $p(x) = 5 - 7x + 11x^2 + 4x^3$, we first note that $[p(x)]_{\mathcal{B}} = [5 \quad -7 \quad 11 \quad 4]^T$ and then compute

$$[\mathbf{D}]_{\mathcal{B}'\mathcal{B}}[p(x)]_{\mathcal{B}} = \begin{bmatrix} 0 & 1 & 0 & 0 \\ 0 & 0 & 2 & 0 \\ 0 & 0 & 0 & 3 \end{bmatrix} \begin{bmatrix} 5 \\ -7 \\ 11 \\ 4 \end{bmatrix} = \begin{bmatrix} -7 \\ 22 \\ 12 \end{bmatrix} = [\mathbf{D}(p(x))]_{\mathcal{B}'},$$

from which it follows that $\mathbf{D}(p(x)) = -7 + 22x + 12x^2$. ■

■ EXAMPLE 2 Matrix Representative with Respect to a Nonstandard Basis

Let $\mathbf{T}$ be the linear operator on $\mathfrak{R}^2$ defined by

$$\mathbf{T}(\mathbf{u}) = A\mathbf{u} = \begin{bmatrix} 2 & 1 \\ -2 & 5 \end{bmatrix} \begin{bmatrix} x \\ y \end{bmatrix}.$$

Find the matrix representative of $\mathbf{T}$ with respect to the basis

$$\mathcal{B} = \left\{ \mathbf{u}_1 = \begin{bmatrix} 1 \\ 2 \end{bmatrix}, \mathbf{u}_2 = \begin{bmatrix} 1 \\ 1 \end{bmatrix} \right\}.$$

The matrix A is both the standard matrix of $\mathbf{T}$ and the matrix representative relative to the standard basis for $\mathfrak{R}^2$. In order to find the matrix representative, relative to the $\mathcal{B}$ basis, we first need to compute

$$\mathbf{T}(\mathbf{u}_1) = A\mathbf{u}_1 = \begin{bmatrix} 2 & 1 \\ -2 & 5 \end{bmatrix} \begin{bmatrix} 1 \\ 2 \end{bmatrix} = \begin{bmatrix} 4 \\ 8 \end{bmatrix}$$

and

$$\mathbf{T}(\mathbf{u}_2) = A\mathbf{u}_2 = \begin{bmatrix} 2 & 1 \\ -2 & 5 \end{bmatrix} \begin{bmatrix} 1 \\ 2 \end{bmatrix} = \begin{bmatrix} 3 \\ 3 \end{bmatrix}.$$

The second step in finding $[\mathbf{T}]_{\mathcal{B}\mathcal{B}}$ is to find the $\mathcal{B}$-coordinate matrices of $\mathbf{T}(\mathbf{u}_1)$ and $\mathbf{T}(\mathbf{u}_2)$. In order to find the $\mathcal{B}$-coordinate matrix of $\mathbf{T}(\mathbf{u}_1)$ we need to find scalars a and b so that

$$\mathbf{T}(\mathbf{u}_1) = \begin{bmatrix} 4 \\ 8 \end{bmatrix} = a\mathbf{u}_1 + b\mathbf{u}_2 = a\begin{bmatrix} 1 \\ 2 \end{bmatrix} + b\begin{bmatrix} 1 \\ 1 \end{bmatrix}.$$

The solution of this system is obvious; it is $a = 4$ and $b = 0$, so that $\mathbf{T}(\mathbf{u}_1) = 4\mathbf{u}_1 + 0\mathbf{u}_2 = 4\mathbf{u}_1$. Similarly, we see that $\mathbf{T}(\mathbf{u}_2) = 0\mathbf{u}_1 + 3\mathbf{u}_2 = 3\mathbf{u}_2$. Thus, $[\mathbf{T}(\mathbf{u}_1)]_{\mathcal{B}} = \begin{bmatrix} 4 \\ 0 \end{bmatrix}$ and $[\mathbf{T}(\mathbf{u}_2)]_{\mathcal{B}} = \begin{bmatrix} 0 \\ 3 \end{bmatrix}$. These coordinate matrices are the columns of the matrix that represents $\mathbf{T}$ with respect to the $\mathcal{B}$ basis. That matrix is $[\mathbf{T}]_{\mathcal{B}\mathcal{B}} = \begin{bmatrix} 4 & 0 \\ 0 & 3 \end{bmatrix}$. Note that this is a much simpler matrix than the matrix A that was originally used to define the transformation.

We can use the factorization

$$\begin{bmatrix} 4 & 0 \\ 0 & 3 \end{bmatrix} = \begin{bmatrix} 4 & 0 \\ 0 & 1 \end{bmatrix}\begin{bmatrix} 1 & 0 \\ 0 & 3 \end{bmatrix}$$

to observe that $\mathbf{T}$ is the product of two expansions: by a factor of 3 in the $\mathbf{u}_2$ direction and by a factor of 4 in the $\mathbf{u}_1$ direction. Figure 4.17 shows the image of a 2×1 rectangle under this transformation. This simple geometric description of $\mathbf{T}$ was certainly not apparent from the original definition of $\mathbf{T}$. ■

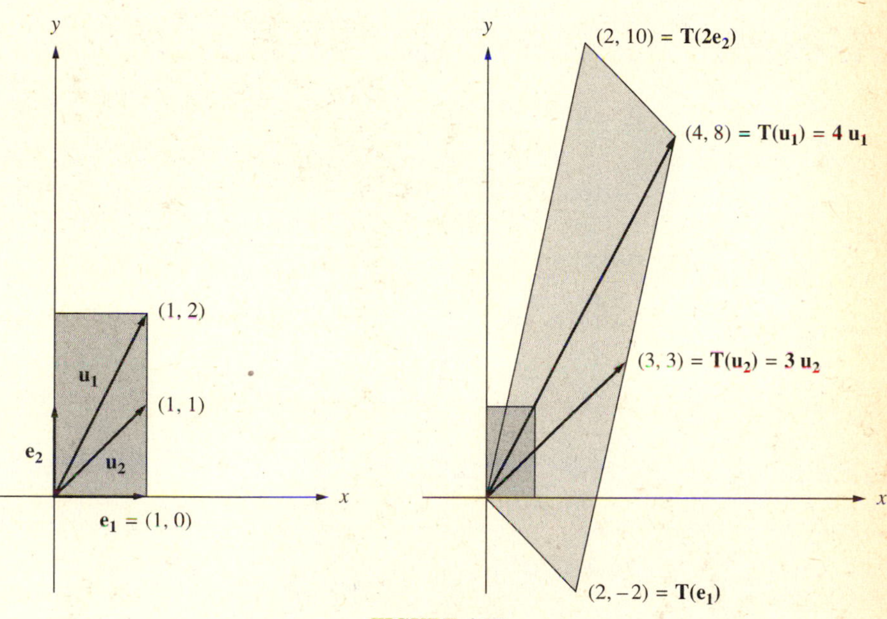

FIGURE 4.17

The hardest step in finding the matrix representative of a linear transformation is finding the coordinates of the images of the basis vectors. This has been easy in the first two examples, but in general it requires the solution of a set of systems of linear equations, all with the same coefficient matrix. We illustrate this process in our next example.

■ EXAMPLE 3 Matrix Representative for Spaces of Row Vectors

Let $\mathbf{T}$ be the linear transformation from $\mathcal{V} = \mathfrak{R}^{1\times 4}$ to $\mathcal{W} = \mathfrak{R}^{1\times 3}$ defined by

$$\mathbf{T}[x \quad y \quad z \quad w] = [3x + 4y - w \quad 2y + 5w \quad x + y + z + w].$$

Let $\mathcal{S}$ be the standard basis for $\mathcal{V}$ and let

$$\mathcal{B}' = \{\mathbf{u}_1 = [1 \quad 3 \quad 1], \mathbf{u}_2 = [1 \quad 1 \quad 0], \mathbf{u}_3 = [0 \quad 3 \quad 1]\}$$

be a basis for $\mathcal{W} = \mathfrak{R}^{1\times 3}$. Find the matrix representative of $\mathbf{T}$ with respect to the $\mathcal{S}$ and $\mathcal{B}'$ bases and use this matrix to compute $\mathbf{T}(\mathbf{v})$ if $\mathbf{v} = [1 \quad 2 \quad 3 \quad 4]$.

From Theorem 4.10 we know that

$$[\mathbf{T}]_{\mathcal{B}'\mathcal{S}} = [[\mathbf{T}(\mathbf{e}_1)]_{\mathcal{B}'} \quad [\mathbf{T}(\mathbf{e}_2)]_{\mathcal{B}'} \quad [\mathbf{T}(\mathbf{e}_3)]_{\mathcal{B}'} \quad [\mathbf{T}(\mathbf{e}_4)]_{\mathcal{B}'}].$$

Directly from the definition of $\mathbf{T}$ we find that $\mathbf{T}(\mathbf{e}_1) = [3 \quad 0 \quad 1]$, $\mathbf{T}(\mathbf{e}_2) = [4 \quad 2 \quad 1]$, $\mathbf{T}(\mathbf{e}_3) = [0 \quad 0 \quad 1]$, and $\mathbf{T}(\mathbf{e}_4) = [-1 \quad 5 \quad 1]$. In order to find $[\mathbf{T}(\mathbf{e}_1)]_{\mathcal{B}'}$, we need to find scalars x_i so that

$$\begin{aligned}[3 \quad 0 \quad 1] &= x_1\mathbf{u}_1 + x_2\mathbf{u}_2 + x_3\mathbf{u}_3 \\ &= x_1[1 \quad 3 \quad 1] + x_2[1 \quad 1 \quad 0] + x_3[0 \quad 3 \quad 1].\end{aligned}$$

This vector equation is equivalent to the system

$$AX = \begin{bmatrix} 1 & 1 & 0 \\ 3 & 1 & 3 \\ 1 & 0 & 1 \end{bmatrix}\begin{bmatrix} x_1 \\ x_2 \\ x_3 \end{bmatrix} = \begin{bmatrix} 3 \\ 0 \\ 1 \end{bmatrix} = \mathbf{T}(\mathbf{e}_1)^T.$$

To find the $\mathcal{B}'$-coordinates of the other $\mathbf{T}(\mathbf{e}_i)$ we need to solve the linear systems $AX = \mathbf{T}(\mathbf{e}_i)^T$; $i = 2, 3, 4$. These four systems can be solved by the single row reduction

$$\begin{bmatrix} 1 & 1 & 0 & 3 & 4 & 0 & -1 \\ 3 & 1 & 3 & 0 & 2 & 0 & 5 \\ 1 & 0 & 1 & 1 & 1 & 1 & 1 \end{bmatrix} \to \begin{bmatrix} 1 & 1 & 0 & 3 & 4 & 0 & -1 \\ 0 & -2 & 3 & -9 & -10 & 0 & 8 \\ 0 & -1 & 1 & -2 & -3 & 1 & 2 \end{bmatrix}$$

$$\to \begin{bmatrix} 1 & 0 & 0 & 0 & 0 & 0 & 0 \\ 0 & 0 & 1 & -5 & -4 & -2 & 4 \\ 0 & 1 & -1 & 2 & 3 & -1 & -2 \end{bmatrix}$$

$$\to \begin{bmatrix} 1 & 0 & 0 & 6 & 5 & 3 & -3 \\ 0 & 1 & 0 & -3 & -1 & -3 & 2 \\ 0 & 0 & 1 & -5 & -4 & -2 & 4 \end{bmatrix}.$$

From the above reduction it follows that

$$[\mathbf{T}(\mathbf{e}_1)]_{\mathcal{B}'} = \begin{bmatrix} 6 \\ -3 \\ -5 \end{bmatrix}, \quad [\mathbf{T}(\mathbf{e}_2)]_{\mathcal{B}'} = \begin{bmatrix} 5 \\ -1 \\ -4 \end{bmatrix}, \quad [\mathbf{T}(\mathbf{e}_3)]_{\mathcal{B}'} = \begin{bmatrix} 3 \\ -3 \\ -2 \end{bmatrix},$$

$$[\mathbf{T}(\mathbf{e}_4)]_{\mathcal{B}'} = \begin{bmatrix} -3 \\ 2 \\ 4 \end{bmatrix}$$

and that the matrix representative of $\mathbf{T}$ with respect to the given bases is

$$[\mathbf{T}]_{\mathcal{B}'\mathcal{S}} = \begin{bmatrix} 6 & 5 & 3 & -3 \\ -3 & -1 & -3 & 2 \\ -5 & -4 & -2 & 4 \end{bmatrix}.$$

In order to find $\mathbf{T}(\mathbf{v})$ we observe that $[\mathbf{v}]_{\mathscr{S}} = \mathbf{v}^T$ and compute

$$[\mathbf{T}(\mathbf{v})]_{\mathscr{S}} = [\mathbf{T}]_{\mathscr{B}'\mathscr{S}}[\mathbf{v}]_{\mathscr{S}} = \begin{bmatrix} 6 & 5 & 3 & -3 \\ -3 & -1 & -3 & 2 \\ -5 & -4 & -2 & 4 \end{bmatrix} \begin{bmatrix} 1 \\ 2 \\ 3 \\ 4 \end{bmatrix} = \begin{bmatrix} 13 \\ -6 \\ -3 \end{bmatrix}.$$

Thus,

$$\begin{aligned} \mathbf{T}(\mathbf{v}) &= 13\mathbf{u}_1 - 6\mathbf{u}_2 - 3\mathbf{u}_3 \\ &= 13[1 \quad 3 \quad 1] - 6[1 \quad 1 \quad 0] - 3[0 \quad 3 \quad 1] \\ &= [7 \quad 24 \quad 10], \end{aligned}$$

which agrees with the value of $\mathbf{T}(\mathbf{v})$ obtained by direct evaluation. ■

EXERCISES 4.5

1. Let $A = \begin{bmatrix} 3 & -2 & 1 & 0 \\ 1 & 6 & 2 & 1 \\ -3 & 0 & 7 & 1 \end{bmatrix}$ be the matrix representative of $\mathbf{T}$ in $\mathscr{L}(\Re^4, \Re^3)$ with respect to the bases $\mathscr{B} = \{\mathbf{v}_1, \mathbf{v}_2, \mathbf{v}_3, \mathbf{v}_4\}$ and $\mathscr{B}' = \{\mathbf{w}_1, \mathbf{w}_2, \mathbf{w}_3\}$ where

$$\mathbf{v}_1 = \begin{bmatrix} 0 \\ 1 \\ 1 \\ 1 \end{bmatrix}, \quad \mathbf{v}_2 = \begin{bmatrix} 2 \\ 1 \\ -1 \\ -1 \end{bmatrix}, \quad \mathbf{v}_3 = \begin{bmatrix} 1 \\ 4 \\ -1 \\ 2 \end{bmatrix}, \quad \mathbf{v}_4 = \begin{bmatrix} 6 \\ 9 \\ 4 \\ 2 \end{bmatrix},$$

$$\mathbf{w}_1 = \begin{bmatrix} 0 \\ 8 \\ 8 \end{bmatrix}, \quad \mathbf{w}_2 = \begin{bmatrix} -7 \\ 8 \\ 1 \end{bmatrix}, \quad \text{and} \quad \mathbf{w}_3 = \begin{bmatrix} -6 \\ 9 \\ 1 \end{bmatrix}.$$

(a) Find $[\mathbf{T}(\mathbf{v}_1)]_{\mathscr{B}'}$, $[\mathbf{T}(\mathbf{v}_2)]_{\mathscr{B}'}$, $[\mathbf{T}(\mathbf{v}_3)]_{\mathscr{B}'}$, and $[\mathbf{T}(\mathbf{v}_4)]_{\mathscr{B}'}$.
(b) Find $\mathbf{T}(\mathbf{v}_1)$, $\mathbf{T}(\mathbf{v}_2)$, $\mathbf{T}(\mathbf{v}_3)$, and $\mathbf{T}(\mathbf{v}_4)$.
(c) Find $\mathbf{T}\begin{bmatrix} 2 \\ 2 \\ 0 \\ 0 \end{bmatrix}$.

2. Let $\mathbf{T}$ be the linear transformation from $\Re^{1\times 4}$ to $\Re^{1\times 3}$ defined by $\mathbf{T}[x_1 \quad x_2 \quad x_3 \quad x_4] = [x_1 + 2x_2 - x_3 \quad 2x_1 - 5x_2 \quad 7x_1 - 3x_2]$. Find the matrix representative of $\mathbf{T}$ relative to the standard bases of $\Re^{1\times 4}$ and $\Re^{1\times 3}$.

3. Let $\mathbf{U}$ be the linear transformation from $\Re^3$ to $\Re^{1\times 5}$ defined by $\mathbf{U}[x_1 \quad x_2 \quad x_3]^T = [x_1 + x_2 \quad 0 \quad x_2 + x_3 \quad 0 \quad x_3 + x_1]$. Consider $\mathbf{u}_1 = [1 \quad 1 \quad 1]^T$, $\mathbf{u}_2 = [1 \quad 1 \quad 0]^T$, and $\mathbf{u}_3 = [0 \quad 1 \quad 1]^T$ as the basis for $\Re^3$ and the standard basis for $\Re^{1\times 5}$. Find the matrix representative of $\mathbf{U}$ with respect to these bases.

4. Define a linear operator $\mathbf{T}$ on $\mathfrak{R}^{1\times 4}$ by

$$\mathbf{T}[x_1 \quad x_2 \quad x_3 \quad x_4] = [x_1 + 2x_2 \quad x_1 - 3x_3 \quad x_1 + 4x_4 \quad x_2 + x_3].$$

Find the matrix representative of $\mathbf{T}$ with respect to the basis

$$\mathcal{B} = \{\mathbf{w}_1 = [1 \quad 2 \quad 0 \quad 3],\ \mathbf{w}_2 = [0 \quad 1 \quad -1 \quad 0],\ \mathbf{w}_3 = [0 \quad 0 \quad 1 \quad 2],\ \mathbf{w}_4 = [0 \quad 0 \quad 0 \quad 1]\}.$$

5. Let $\mathbf{T}$ be the linear operator on $\mathcal{P}_3$ considered in Example 1. Find the matrix representative of $\mathbf{T}$ with respect to the basis

$$\mathcal{B} = \{-3 + x^2, -3 + x, 7 - x - x^2, x^3\}.$$

6. Let $\mathbf{T}$ be a linear operator on $\mathfrak{R}^{1\times 3}$ which is represented, relative to the standard basis, by

$$A = \begin{bmatrix} 7 & -8 & -8 \\ 9 & -16 & -18 \\ -5 & 11 & 13 \end{bmatrix}.$$

The vectors $\mathbf{u}_1 = [0 \quad -1 \quad 1]$, $\mathbf{u}_2 = [1 \quad 3 \quad -2]$, and $\mathbf{u}_3 = [2 \quad 0 \quad 1]$ form a basis for $\mathfrak{R}^{1\times 3}$. Find the matrix representative of $\mathbf{T}$ with respect to this basis.

7. Let $\mathbf{I}$ in $\mathcal{L}(\mathcal{V}, \mathcal{V})$ be the identity operator on $\mathcal{V}$ defined by $\mathbf{I}(\mathbf{u}) = \mathbf{u}$, for all $\mathbf{u}$ in $\mathcal{V}$. Show that the matrix representative of $\mathbf{I}$, $[\mathbf{I}]_{\mathcal{B}\mathcal{B}'}$, is the identity matrix I, no matter which basis is chosen for $\mathcal{V}$.

8. Let $\mathbf{I}$ be the identity operator on $\mathcal{V}$ and let $\mathcal{B}$ and $\mathcal{B}'$ be two bases for $\mathcal{V}$. Describe the matrix $[\mathbf{I}]_{\mathcal{B}'\mathcal{B}}$.

9. Let $\mathbf{T}$ be linear operator on $\mathfrak{R}^{1\times 3}$ that is represented, relative to the standard basis, by $A = \begin{bmatrix} 2 & 1 & 1 \\ 2 & 3 & 2 \\ 1 & 1 & 2 \end{bmatrix}$. Find $[\mathbf{T}]_{\mathcal{B}\mathcal{B}}$ where

$$\mathcal{B} = \{\mathbf{u}_1 = [1 \quad -2 \quad 2], \mathbf{u}_2 = [-2 \quad 1 \quad 2], \mathbf{u}_3 = [2 \quad 2 \quad 1]\}.$$

10. Let $\mathbf{T}$ be a linear operator on $\mathcal{V}$. Call a subspace $\mathcal{W}$ of $\mathcal{V}$ **T-invariant** if $\mathbf{T}(\mathbf{u})$ is in $\mathcal{W}$ for every $\mathbf{u}$ in $\mathcal{W}$. Show that $\mathcal{NS}(\mathbf{T})$ and $\mathcal{E} = \{\mathbf{u} \mid \mathbf{T}(\mathbf{u}) = 2\mathbf{u}\}$ are $\mathbf{T}$-invariant subspaces of $\mathcal{V}$.

11. Find the matrix representative of the transposition operator $\mathbf{T}$ ($\mathbf{T}(A) = A^T$) on $\mathfrak{R}^{2\times 2}$ relative to the basis

$$\mathcal{S} = \left\{ \begin{bmatrix} 1 & 0 \\ 0 & 0 \end{bmatrix}, \begin{bmatrix} 0 & 1 \\ 0 & 0 \end{bmatrix}, \begin{bmatrix} 0 & 0 \\ 1 & 0 \end{bmatrix}, \begin{bmatrix} 0 & 0 \\ 0 & 1 \end{bmatrix} \right\} \quad \text{for } \mathfrak{R}^{2\times 2}.$$

12. Let $\mathcal{B}_1$ be a basis for the vector space $\mathcal{V}_1$, $\mathcal{B}_2$ a basis for the vector space $\mathcal{V}_2$, and $\mathcal{B}_3$ a basis for the vector space $\mathcal{V}_3$. Consider $\mathbf{T}_1$ in $\mathcal{L}(\mathcal{V}_1, \mathcal{V}_2)$ and $\mathbf{T}_2$ in $\mathcal{L}(\mathcal{V}_2, \mathcal{V}_3)$ so that $\mathbf{T}_2\mathbf{T}_1$ is in $\mathcal{L}(\mathcal{V}_1, \mathcal{V}_3)$. Show that

$$[\mathbf{T}_2\mathbf{T}_1]_{\mathcal{B}_3\mathcal{B}_1} = [\mathbf{T}_2]_{\mathcal{B}_3\mathcal{B}_2}[\mathbf{T}_1]_{\mathcal{B}_2\mathcal{B}_1}.$$

13. Let $\mathbf{T}$ be the matrix transformation on $\mathfrak{R}^3$ whose standard matrix is

$$C = \begin{bmatrix} 0 & 1 & 0 \\ 0 & 0 & 1 \\ 15 & -23 & 9 \end{bmatrix}.$$

Find the matrix representative of $\mathbf{T}$ relative to the basis

$$\mathcal{B} = \left\{ \mathbf{u}_1 = \begin{bmatrix} 1 \\ 5 \\ 25 \end{bmatrix}, \mathbf{u}_2 = \begin{bmatrix} 1 \\ 3 \\ 9 \end{bmatrix}, \mathbf{u}_3 = \begin{bmatrix} 1 \\ 1 \\ 1 \end{bmatrix} \right\}.$$

14. Explain why $[\mathbf{T}]_{\mathcal{B}'\mathcal{B}}$ is a better notation for the matrix representative than $[\mathbf{T}]_{\mathcal{B}\mathcal{B}'}$.

15. Let $\mathcal{V}$ and $\mathcal{W}$ be finite dimensional vector spaces with bases $\mathcal{B}$ and $\mathcal{B}'$ respectively; let $\dim(\mathcal{V}) = n$ and $\dim(\mathcal{W}) = m$. Consider the mapping which sends a linear transformation $\mathbf{T}$ to its matrix representative $[\mathbf{T}]_{\mathcal{B}'\mathcal{B}}$. Show that this mapping is a one-to-one linear transformation of $\mathcal{L}(\mathcal{V}, \mathcal{W})$ onto $\mathfrak{R}^{m\times n}$.

4.6 CHANGE OF BASIS*

> How are the coordinate matrices with respect to two different bases related?

Let $\mathcal{V}$ be an arbitrary n-dimensional vector space and let

$$\mathcal{B} = \{\mathbf{u}_1, \mathbf{u}_2, \cdots, \mathbf{u}_n\}$$

be a fixed basis for $\mathcal{V}$. For any vector $\mathbf{w}$ in $\mathcal{V}$ we can write

$$\mathbf{w} = a_1\mathbf{u}_1 + a_2\mathbf{u}_2 + \cdots + a_n\mathbf{u}_n$$

where the uniquely determined scalars a_i are called the $\mathcal{B}$-coordinates of the vector $\mathbf{w}$ and the column matrix $[\mathbf{w}]_{\mathcal{B}} = [a_1 \quad a_2 \quad \cdots \quad a_n]^T$ is called the $\mathcal{B}$-coordinate matrix of $\mathbf{w}$. Finding coordinates usually involves solving linear systems; the complexity of the solution process depends on the choice of the basis for $\mathcal{V}$. Finding coordinates is very easy for the standard basis for $\mathfrak{R}^n$, easy for any orthonormal basis for $\mathfrak{R}^n$ (Theorem 3.17), and easy for an echelon basis for the row space of a matrix A.

We list some important elementary properties of the coordinate matrix in our next theorem. These properties say that $[\ ]_{\mathcal{B}}$ is a one-to-one linear transformation from $\mathcal{V}$ onto $\mathfrak{R}^n$. The proof is left to the exercises (26 of Section 4.1 or 10 of this section).

THEOREM 4.11 If $\mathcal{B}$ is any basis for the finite dimensional vector space $\mathcal{V}$, then for any vectors $\mathbf{v}$ and $\mathbf{w}$ in $\mathcal{V}$ and any scalar k we have:

(1) $[\mathbf{w}]_{\mathcal{B}} = [\mathbf{v}]_{\mathcal{B}}$ if, and only if, $\mathbf{w} = \mathbf{v}$;
(2) $[\mathbf{w} + \mathbf{v}]_{\mathcal{B}} = [\mathbf{w}]_{\mathcal{B}} + [\mathbf{v}]_{\mathcal{B}}$;
(3) $[k\mathbf{w}]_{\mathcal{B}} = k[\mathbf{w}]_{\mathcal{B}}$; and
(4) for each X in $\mathfrak{R}^n$, there exists a unique vector $\mathbf{u}$ in $\mathcal{V}$ such that $[\mathbf{u}]_{\mathcal{B}} = X$. □

---

*This section is optional.

Suppose now that we have two bases for the vector space $\mathcal{V}$:

$$\mathcal{B} = \{\mathbf{u}_1, \mathbf{u}_2, \cdots, \mathbf{u}_n\} \qquad \text{and} \qquad \mathcal{B}' = \{\mathbf{v}_1, \mathbf{v}_2, \cdots, \mathbf{v}_n\}.$$

We will refer to $\mathcal{B}$ as the *old basis* and $\mathcal{B}'$ as the *new basis*. For a typical vector $\mathbf{w}$ in $\mathcal{V}$ we have the two representations

$$\mathbf{w} = a_1\mathbf{u}_1 + a_2\mathbf{u}_2 + \cdots + a_n\mathbf{u}_n \qquad \text{and}$$
$$\mathbf{w} = b_1\mathbf{v}_1 + b_2\mathbf{v}_2 + \cdots + b_n\mathbf{v}_n,$$

as well as the two coordinate matrices

$$[\mathbf{w}]_{\mathcal{B}} = \begin{bmatrix} a_1 \\ a_2 \\ \vdots \\ a_n \end{bmatrix} \qquad \text{and} \qquad [\mathbf{w}]_{\mathcal{B}'} = \begin{bmatrix} b_1 \\ b_2 \\ \vdots \\ b_n \end{bmatrix}.$$

It would be extremely naive to expect that these two coordinate matrices are equal. For example, consider the two bases for $\mathfrak{R}^2$:

$$\mathcal{B} = \left\{\mathbf{u}_1 = \begin{bmatrix} 1 \\ 1 \end{bmatrix}, \mathbf{u}_2 = \begin{bmatrix} 1 \\ 0 \end{bmatrix}\right\} \qquad \text{and} \qquad \mathcal{B}' = \left\{\mathbf{v}_1 = \begin{bmatrix} -1 \\ 1 \end{bmatrix}, \mathbf{v}_2 = \begin{bmatrix} 4 \\ -5 \end{bmatrix}\right\}.$$

For the vector $\mathbf{w} = \begin{bmatrix} 3 \\ 5 \end{bmatrix}$, it is easy to check that

$$\mathbf{w} = 5\mathbf{u}_1 - 2\mathbf{u}_2 = -35\mathbf{v}_1 - 8\mathbf{v}_2.$$

Thus we have $[\mathbf{w}]_{\mathcal{B}} = \begin{bmatrix} 5 \\ -2 \end{bmatrix}$ and $[\mathbf{w}]_{\mathcal{B}'} = \begin{bmatrix} -35 \\ -8 \end{bmatrix} \neq [\mathbf{w}]_{\mathcal{B}}$. Note that we also have $[\mathbf{u}_1]_{\mathcal{B}} = \begin{bmatrix} 1 \\ 0 \end{bmatrix}$ and $[\mathbf{u}_1]_{\mathcal{B}'} = \begin{bmatrix} -9 \\ -2 \end{bmatrix} \neq [\mathbf{u}_1]_{\mathcal{B}}$.

We will now investigate how the two coordinate matrices, $[\mathbf{w}]_{\mathcal{B}}$ and $[\mathbf{w}]_{\mathcal{B}'}$, are related. Since we can change one column matrix into another column matrix by multiplying by an $n \times n$ matrix, let us try to find a matrix P such that

$$P[\mathbf{w}]_{\mathcal{B}} = [\mathbf{w}]_{\mathcal{B}'}, \quad \text{for all vectors } \mathbf{w} \text{ in } \mathcal{V}. \tag{4.4}$$

If Equation (4.4) is to hold for all vectors $\mathbf{w}$ in $\mathcal{V}$, it must hold for the vectors $\mathbf{u}_i$ which make up the $\mathcal{B}$ basis (the old basis). For these vectors we have $[\mathbf{u}_i]_{\mathcal{B}} = \mathbf{e}_i$; so that, if Equation (4.4) is to hold for $\mathbf{w} = \mathbf{u}_i$, we must have

$$[\mathbf{u}_i]_{\mathcal{B}} = P[\mathbf{u}_i]_{\mathcal{B}} = P\mathbf{e}_i = \text{Col}_i(P); \quad i = 1, 2, \cdots, n.$$

Thus, we see that the only matrix that could satisfy Equation (4.4) *for all vectors* $\mathbf{w}$ in $\mathcal{V}$ is the matrix

$$P = \Big[[\mathbf{u}_1]_{\mathcal{B}'} \quad [\mathbf{u}_2]_{\mathcal{B}'} \quad \cdots \quad [\mathbf{u}_n]_{\mathcal{B}'}\Big]$$

whose columns are the coordinate matrices of the old basis vectors $\mathbf{u}_i$ with respect to the new basis $\mathcal{B}'$.

We now need to check that, with this choice of P, Equation (4.4) holds *for all* vectors $\mathbf{w}$ in $\mathcal{V}$. Thus if $\mathbf{w} = a_1\mathbf{u}_1 + a_2\mathbf{u}_2 + \cdots + a_n\mathbf{u}_n$ is any vector from $\mathcal{V}$, then we have

$$P[\mathbf{w}]_{\mathcal{B}} = P\begin{bmatrix} a_1 \\ a_2 \\ \cdot \\ \cdot \\ \cdot \\ a_n \end{bmatrix} = \Big[[\mathbf{u}_1]_{\mathcal{B}'} \quad [\mathbf{u}_2]_{\mathcal{B}'} \quad \cdots \quad [\mathbf{u}_n]_{\mathcal{B}'}\Big]\begin{bmatrix} a_1 \\ a_2 \\ \cdot \\ \cdot \\ \cdot \\ a_n \end{bmatrix}$$

$$= a_1[\mathbf{u}_1]_{\mathcal{B}'} + a_2[\mathbf{u}_2]_{\mathcal{B}'} + \cdots + a_n[\mathbf{u}_n]_{\mathcal{B}'}.$$

From parts 2 and 3 of Theorem 4.11 it follows that this last expression is

$$[a_1\mathbf{u}_1 + a_2\mathbf{u}_2 + \cdots + a_n\mathbf{u}_n]_{\mathcal{B}'} = [\mathbf{w}]_{\mathcal{B}'}.$$

This calculation establishes that Equation (4.4) holds for every $\mathbf{w}$ in $\mathcal{V}$.

If $P[\mathbf{w}]_{\mathcal{B}} = [\mathbf{w}]_{\mathcal{B}'} = 0$, then if follows that $\mathbf{w} = \mathbf{0}$ and $[\mathbf{w}]_{\mathcal{B}} = 0$. Thus, from Theorem 1.9, it follows that the change-of-basis matrix P must be nonsingular. This observation completes the proof of the following important theorem.

THEOREM 4.12 If $\mathcal{B} = \{\mathbf{u}_1, \mathbf{u}_2, \ldots, \mathbf{u}_n\}$ and $\mathcal{B}' = \{\mathbf{v}_1, \mathbf{v}_2, \ldots, \mathbf{v}_n\}$ are two bases for a finite dimensional vector space $\mathcal{V}$, then the matrix

$$P = \Big[[\mathbf{u}_1]_{\mathcal{B}'} \quad [\mathbf{u}_2]_{\mathcal{B}'} \quad \cdots \quad [\mathbf{u}_n]_{\mathcal{B}'}\Big],$$

whose columns are the new coordinates of the old basis vectors, is the unique nonsingular matrix P such that

$$P[\mathbf{w}]_{\mathcal{B}} = [\mathbf{w}]_{\mathcal{B}'}, \quad \text{for all vectors } \mathbf{w} \text{ in } \mathcal{V}. \quad \square$$

We will refer to the matrix P of Theorem 4.12 as the $\mathcal{B}$ to $\mathcal{B}'$ **change-of-basis matrix.** Note that P^{-1} must satisfy

$$[\mathbf{w}]_{\mathcal{B}} = P^{-1}[\mathbf{w}]_{\mathcal{B}'}, \quad \text{for all } \mathbf{w} \text{ in } \mathcal{V}$$

so that $P^{-1} = \Big[[\mathbf{v}_1]_{\mathcal{B}} \quad [\mathbf{v}_2]_{\mathcal{B}} \quad \cdots \quad [\mathbf{v}_n]_{\mathcal{B}}\Big]$ is the $\mathcal{B}'$ to $\mathcal{B}$ change-of-basis matrix.

EXAMPLE 1 **Change-of-Basis Matrix for a Space of Row Vectors**

Let $\mathcal{V} = \Re^{1\times 3}$. Find the $\mathcal{B}$ to $\mathcal{B}'$ change-of-basis matrix if

$$\mathcal{B} = \{\mathbf{u}_1 = [2 \quad 3 \quad 2], \mathbf{u}_2 = [7 \quad 10 \quad 6], \mathbf{u}_3 = [6 \quad 10 \quad 7]\}$$

and

$$\mathcal{B}' = \{\mathbf{v}_1 = [1 \quad 1 \quad 1], \mathbf{v}_2 = [0 \quad 1 \quad 1], \mathbf{v}_3 = [1 \quad 1 \quad 0]\}.$$

From Theorem 4.12 we know that the change-of-basis matrix we seek has as its columns the new coordinates of the old basis vectors; that is,

$$P = \begin{bmatrix} [\mathbf{u}_1]_{\mathcal{B}'} & [\mathbf{u}_2]_{\mathcal{B}'} & [\mathbf{u}_3]_{\mathcal{B}'} \end{bmatrix}.$$

In order to find P we need to express the $\mathbf{u}_i$ in terms of the $\mathbf{v}_i$. To find the coordinate matrix $[\mathbf{u}_1]_{\mathcal{B}'}$ we must solve the vector equation

$$x_1\mathbf{v}_1 + x_2\mathbf{v}_2 + x_3\mathbf{v}_3 = \mathbf{u}_1$$

or

$$x_1[1 \quad 1 \quad 1] + x_2[0 \quad 1 \quad 1] + x_3[1 \quad 1 \quad 0] = [2 \quad 3 \quad 2].$$

This vector equation is equivalent to the linear system

$$AX = \begin{bmatrix} 1 & 0 & 1 \\ 1 & 1 & 1 \\ 1 & 1 & 0 \end{bmatrix}\begin{bmatrix} x_1 \\ x_2 \\ x_3 \end{bmatrix} = \begin{bmatrix} 2 \\ 3 \\ 2 \end{bmatrix} = \mathbf{u}_1^T.$$

In order to find the coordinate matrices $[\mathbf{u}_2]_{\mathcal{B}'}$ and $[\mathbf{u}_3]_{\mathcal{B}'}$, we need to solve the systems

$$AX = \mathbf{u}_2^T \qquad \text{and} \qquad AX = \mathbf{u}_3^T.$$

Since these three systems have the same coefficient matrix, we can solve all three systems with the single row reduction

$$\begin{bmatrix} 1 & 0 & 1 & 2 & 7 & 6 \\ 1 & 1 & 1 & 3 & 10 & 10 \\ 1 & 1 & 0 & 2 & 6 & 7 \end{bmatrix} \to \begin{bmatrix} 1 & 0 & 1 & 2 & 7 & 6 \\ 0 & 1 & 0 & 1 & 3 & 4 \\ 0 & 1 & -1 & 0 & -1 & 1 \end{bmatrix}$$

$$\to \begin{bmatrix} 1 & 0 & 0 & 1 & 3 & 3 \\ 0 & 1 & 0 & 1 & 3 & 4 \\ 0 & 0 & 1 & 1 & 4 & 3 \end{bmatrix}.$$

From this row reduction it follows that

$$\mathbf{u}_1 = \mathbf{v}_1 + \mathbf{v}_2 + \mathbf{v}_3, \quad \mathbf{u}_2 = 3\mathbf{v}_1 + 3\mathbf{v}_2 + 4\mathbf{v}_3, \qquad \text{and}$$
$$\mathbf{u}_3 = 3\mathbf{v}_1 + 4\mathbf{v}_2 + 3\mathbf{v}_3.$$

Thus, $[\mathbf{u}_1]_{\mathcal{B}'} = \begin{bmatrix} 1 \\ 1 \\ 1 \end{bmatrix}$, $[\mathbf{u}_2]_{\mathcal{B}'} = \begin{bmatrix} 3 \\ 3 \\ 4 \end{bmatrix}$, and $[\mathbf{u}_3]_{\mathcal{B}'} = \begin{bmatrix} 3 \\ 4 \\ 3 \end{bmatrix}$, so that $P = \begin{bmatrix} 1 & 3 & 3 \\ 1 & 3 & 4 \\ 1 & 4 & 3 \end{bmatrix}$ is the $\mathcal{B}$ to $\mathcal{B}'$ change-of-basis matrix.

For the vector $\mathbf{w} = 3\mathbf{u}_1 + 5\mathbf{u}_2 - 6\mathbf{u}_3 = [5 \quad -1 \quad -6]$, we have $[\mathbf{w}]_{\mathcal{B}} = [3 \quad 5 \quad -6]^T$, so from Theorem 4.12 we must have

$$[\mathbf{w}]_{\mathcal{B}'} = P[\mathbf{w}]_{\mathcal{B}} = \begin{bmatrix} 1 & 3 & 3 \\ 1 & 3 & 4 \\ 1 & 4 & 3 \end{bmatrix} \begin{bmatrix} 3 \\ 5 \\ -6 \end{bmatrix} = \begin{bmatrix} 0 \\ -6 \\ 5 \end{bmatrix}.$$

Thus, we have $\mathbf{w} = -6\mathbf{v}_2 + 5\mathbf{v}_3 = [5 \quad -1 \quad 6]$, which checks with the above expression for $\mathbf{w}$. ■

EXAMPLE 2 **Change of Basis in a Vector Space of Polynomials**

Consider the subspace $\mathcal{P}_2$ of the vector space $\mathcal{P}$ of all polynomials and the two bases

$$\mathcal{S} = \{1, x, x^2\} \qquad \text{and}$$
$$\mathcal{H} = \{h_1(x) = x^2 - 3x + 2,\ h_2(x) = x^2 - 2x,\ h_3(x) = x^2 - x\}.$$

Find the $\mathcal{S}$ to $\mathcal{H}$ change-of-basis matrix Q and use it to express $p(x) = 12x^2 - 11x + 5$ as a linear combination of the $h_i(x)$.

The $\mathcal{S}$ to $\mathcal{H}$ change-of-basis matrix Q must satisfy $Q[p(x)]_{\mathcal{S}} = [p(x)]_{\mathcal{H}}$ for all polynomials $p(x)$. From Theorem 4.12 Q is the matrix

$$Q = \begin{bmatrix} [1]_{\mathcal{H}} & [x]_{\mathcal{H}} & [x^2]_{\mathcal{H}} \end{bmatrix}.$$

Thus we need to express 1, x, and x^2 as linear combinations of the $h_i(x)$. If

$$a_1h_1(x) + a_2h_2(x) + a_3h_3(x) = (1, x, \text{ or } x^2),$$

then

$$\begin{aligned} a_1(x^2 - 3x + 2) &+ a_2(x^2 - 2x) + a_3(x^2 - x) \\ &= (2a_1) + (-3a_1 - 2a_2 - a_3)x + (a_1 + a_2 + a_3)x^2 \\ &= (1, x, \text{ or } x^2), \end{aligned}$$

so that the a_i must satisfy the systems

$$\begin{bmatrix} 2 & 0 & 0 \\ -3 & -2 & -1 \\ 1 & 1 & 1 \end{bmatrix} \begin{bmatrix} a_1 \\ a_2 \\ a_3 \end{bmatrix} = \left(\begin{bmatrix} 1 \\ 0 \\ 0 \end{bmatrix}, \begin{bmatrix} 0 \\ 1 \\ 0 \end{bmatrix}, \text{ or } \begin{bmatrix} 0 \\ 0 \\ 1 \end{bmatrix} \right).$$

From the row reduction

$$\left[\begin{array}{rrr|rrr} 2 & 0 & 0 & 1 & 0 & 0 \\ -3 & -2 & -1 & 0 & 1 & 0 \\ 1 & 1 & 1 & 0 & 0 & 1 \end{array}\right] \to \left[\begin{array}{rrr|rrr} 1 & 0 & 0 & \frac{1}{2} & 0 & 0 \\ 0 & -2 & -1 & \frac{3}{2} & 1 & 0 \\ 0 & 1 & 1 & -\frac{1}{2} & 0 & 1 \end{array}\right]$$

$$\to \left[\begin{array}{rrr|rrr} 1 & 0 & 0 & \frac{1}{2} & 0 & 0 \\ 0 & 1 & 0 & -1 & -1 & -1 \\ 0 & 0 & 1 & \frac{1}{2} & 1 & 2 \end{array}\right],$$

we see that $1 = \frac{1}{2}h_1(x) - h_2(x) + \frac{1}{2}h_3(x)$, $x = -h_2(x) + h_3(x)$, and $x^2 = -h_2(x) + 2h_3(x)$. Thus,

$$[1]_{\mathcal{H}} = \begin{bmatrix} \frac{1}{2} \\ -1 \\ \frac{1}{2} \end{bmatrix}, \quad [x]_{\mathcal{H}} = \begin{bmatrix} 0 \\ -1 \\ 1 \end{bmatrix}, \quad \text{and} \quad [x^2]_{\mathcal{H}} = \begin{bmatrix} 0 \\ -1 \\ 2 \end{bmatrix},$$

so the $\mathcal{S}$ to $\mathcal{H}$ change-of-basis matrix is $Q = \begin{bmatrix} \frac{1}{2} & 0 & 0 \\ -1 & -1 & -1 \\ \frac{1}{2} & 1 & 2 \end{bmatrix}$.

In order to find $[12x^2 - 11x + 5]_{\mathcal{H}}$ we first observe that

$$[12x^2 - 11x + 5]_{\mathcal{S}} = \begin{bmatrix} 5 \\ -11 \\ 12 \end{bmatrix}$$

so

$$[12x^2 - 11x + 5]_{\mathcal{H}} = Q\begin{bmatrix} 5 \\ -11 \\ 12 \end{bmatrix} = \begin{bmatrix} \frac{5}{2} \\ -6 \\ \frac{31}{2} \end{bmatrix}.$$

Thus, $12x^2 - 11x + 5 = \frac{5}{2}h_1(x) - 6h_2(x) + \frac{31}{2}h_3(x)$, as you should check by direct multiplication. ■

We conclude this section with a theorem about the change-of-basis matrix between two orthonormal bases of $\mathfrak{R}^n$.

THEOREM 4.13 The change-of-basis matrix P, between two orthonormal bases of $\mathfrak{R}^n$, is an orthogonal matrix; that is, $P^TP = I$.

Proof Let $\mathcal{B} = \{\mathbf{u}_1, \mathbf{u}_2, \ldots, \mathbf{u}_n\}$ and $\mathcal{B}' = \{\mathbf{v}_1, \mathbf{v}_2, \ldots, \mathbf{v}_n\}$ be orthonormal bases for $\mathfrak{R}^n$ and let P be the $\mathcal{B}$ to $\mathcal{B}'$ change-of-basis matrix. Thus,

$$\mathbf{u}_i = p_{1i}\mathbf{v}_1 + p_{2i}\mathbf{v}_2 + \cdots + p_{ni}\mathbf{v}_n; \quad i = 1, 2, \ldots, n.$$

Since $\mathbf{u_i}$ is a unit vector we have, using the orthonormality of $\mathcal{B}'$,

$$\begin{aligned}1 = \mathbf{u_i} \cdot \mathbf{u_i} &= (p_{1i}\mathbf{v_1} + p_{2i}\mathbf{v_2} + \cdots + p_{ni}\mathbf{v_n})\cdot(p_{1i}\mathbf{v_1} + p_{2i}\mathbf{v_2} + \cdots + p_{ni}\mathbf{v_n}) \\ &= \sum_{s=1}^{n}\sum_{k=1}^{n} p_{ki}p_{si}(\mathbf{v_k} \cdot \mathbf{v_s}) = \sum_{k=1}^{n} p_{ki}p_{ki} \\ &= p_{1i}^2 + p_{2i}^2 + \cdots + p_{ni}^2.\end{aligned}$$

Thus $\text{Col}_i(P)$ is a unit vector in $\mathfrak{R}^n$ for each i.

Since $\mathbf{u_i}$ and $\mathbf{u_j}$ are orthogonal vectors, we have, from the orthonormality of the basis $\mathcal{B}'$

$$\begin{aligned}0 = \mathbf{u_i}\cdot\mathbf{u_j} &= (p_{1i}\mathbf{v_1} + p_{2i}\mathbf{v_2} + \cdots + p_{ni}\mathbf{v_n})\cdot(p_{1j}\mathbf{v_1} + p_{2j}\mathbf{v_2} + \cdots + p_{nj}\mathbf{v_n}) \\ &= \sum_{s=1}^{n}\sum_{k=1}^{n} p_{ki}p_{sj}(\mathbf{v_k} \cdot \mathbf{v_s}) = \sum_{k=1}^{n} p_{ki}p_{kj} \\ &= \text{Col}_i(P)\cdot\text{Col}_j(P).\end{aligned}$$

Thus the columns of P are mutually orthogonal unit vectors in $\mathfrak{R}^n$ and hence P is an orthogonal matrix ($P^TP = I$). □

EXERCISES 4.6

1. Find the $\mathcal{B}$ to $\mathcal{B}'$ change-of-basis matrix P in the following cases.

   (a) $\mathcal{B} = \left\{\mathbf{u_1} = \begin{bmatrix}1\\1\end{bmatrix}, \mathbf{u_2} = \begin{bmatrix}2\\1\end{bmatrix}\right\}$ and $\mathcal{B}' = \left\{\mathbf{v_1} = \begin{bmatrix}1\\3\end{bmatrix}, \mathbf{v_2} = \begin{bmatrix}1\\2\end{bmatrix}\right\}$

   (b) $\mathcal{B} = \{\mathbf{u_1} = [-1 \quad 0 \quad 1], \mathbf{u_2} = [3 \quad -2 \quad 1], \mathbf{u_3} = [1 \quad 6 \quad -1]\}$
   $\mathcal{B}' = \{\mathbf{v_1} = [3 \quad 1 \quad -5], \mathbf{v_2} = [1 \quad 1 \quad -3], \mathbf{v_3} = [1 \quad 0 \quad -2]\}$

   (c) $\mathcal{B} = \{p_1(x) = 2x^2 + x + 1, p_2(x) = 2x^2 - x + 1, p_3(x) = x^2 + 2x + 1\}$
   $\mathcal{B}' = \{q_1(x) = 2x^2 + 3x - 7, q_2(x) = x^2 + 3x - 2, q_3(x) = x^2 + x\}$

   (d) $\mathcal{B} = \left\{A_1 = \begin{bmatrix}1 & 0\\1 & 0\end{bmatrix}, A_2 = \begin{bmatrix}1 & 2\\0 & 0\end{bmatrix}, A_3 = \begin{bmatrix}5 & -1\\4 & 0\end{bmatrix}, A_4 = \begin{bmatrix}-1 & 7\\0 & 6\end{bmatrix}\right\}$

   $\mathcal{B}' = \left\{C_1 = \begin{bmatrix}1 & 0\\0 & 0\end{bmatrix}, C_2 = \begin{bmatrix}0 & 1\\0 & 0\end{bmatrix}, C_3 = \begin{bmatrix}0 & 0\\1 & 0\end{bmatrix}, C_4 = \begin{bmatrix}0 & 0\\0 & 1\end{bmatrix}\right\}$

   (e) $\mathcal{B} = \{\sin x, \cos x\}$, $\mathcal{B}' = \{\sin(x + 2), \cos(x + 3)\}$

2. Use the matrix from 1(a) to compute $[\mathbf{w}]_{\mathcal{B}'}$, where $\mathbf{w} = 3\mathbf{u_1} + 5\mathbf{u_2}$.
3. Use the matrix from 1(b) to compute $[\mathbf{w}]_{\mathcal{B}'}$, where $\mathbf{w} = [1 \quad 2 \quad 3]$.
4. Use the matrix from 1(c) to compute $[h(x)]_{\mathcal{B}'}$, where $h(x) = p_1(x) + 5p_2(x) - 7p_3(x)$.
5. Let $\mathcal{S}$ be the standard basis for $\mathfrak{R}^2$ and let $\mathcal{N}$ be the basis obtained by rotating the vectors of $\mathcal{S}$ through an angle θ. Find the $\mathcal{S}$ to $\mathcal{N}$ change-of-basis matrix and the $\mathcal{N}$ to $\mathcal{S}$ change-of-basis matrix.
6. Let $\mathcal{B}$, $\mathcal{B}'$, and $\mathcal{B}''$ be three bases for $\mathcal{V}$ and let P be the $\mathcal{B}$ to $\mathcal{B}'$ change-of-basis matrix and let Q be the $\mathcal{B}'$ to $\mathcal{B}''$ change-of-basis matrix. Show that QP is the $\mathcal{B}$ to $\mathcal{B}''$ change-of-basis matrix.

7. Let $\mathcal{B} = \{\mathbf{u}_1, \ldots, \mathbf{u}_n\}$ be an orthonormal basis for $\mathfrak{R}^n$ and let $\mathcal{S}$ be the standard basis for $\mathfrak{R}^n$. Find the $\mathcal{B}$ to $\mathcal{S}$ change-of-basis matrix and show that it is orthogonal. What is the $\mathcal{S}$ to $\mathcal{B}$ change-of-basis matrix?

8. Let $\mathcal{B}$ be an orthonormal basis for $\mathfrak{R}^n$.
  (a) Show that $\mathbf{u} \cdot \mathbf{v} = [\mathbf{u}]_{\mathcal{B}} \cdot [\mathbf{v}]_{\mathcal{B}}$ and, hence, that $\mathbf{u}$ and $\mathbf{v}$ are orthogonal if, and only if, $[\mathbf{u}]_{\mathcal{B}}$ and $[\mathbf{v}]_{\mathcal{B}}$ are orthogonal.
  (b) Use the result of part (a) to prove Theorem 4.13.

9. Let $\mathcal{B} = \{\mathbf{u}_1, \mathbf{u}_2, \mathbf{u}_3, \mathbf{u}_4\}$ and $\mathcal{B}' = \{\mathbf{u}_3, \mathbf{u}_1, \mathbf{u}_4, \mathbf{u}_2\}$ be two bases for a vector space $\mathcal{V}$. Find the $\mathcal{B}$ to $\mathcal{B}'$ change-of-basis matrix and the $\mathcal{B}'$ to $\mathcal{B}$ change-of-basis matrix.

10. Prove Theorem 4.11.

11. Let $\mathbf{u}_1 = \begin{bmatrix} x_1 \\ Y \end{bmatrix}$ be a unit vector in $\mathfrak{R}^n$ so that $x_1^2 + Y^TY = 1$. Show that the columns of the matrix $H = \left[\begin{array}{c|c} x_1 & Y^T \\ \hline Y & I - \frac{1}{1 - x_1} YY^T \end{array}\right]$ form an orthonormal basis for $\mathfrak{R}^n$ containing $\mathbf{u}_1$ by showing that $H^TH = I$.

4.7 THE CHANGE-OF-BASIS PROBLEM*

> How are the matrix representatives of a linear operator, with respect to two different bases, related?

In Section 4.5 you had several opportunities to observe that the matrix representative of a linear operator $\mathbf{T}$ in $\mathcal{L}(\mathcal{V}, \mathcal{V})$ depends on the choice of the basis for the vector space $\mathcal{V}$. In general, we have seen that changing the basis of the vector space changes the matrix representative; it would be naive to expect otherwise. In this section we will investigate the relationship between the various matrix representatives of the linear operator $\mathbf{T}$. This investigation will lead us to the study of similar matrices and will provide geometric motivation for a closer look at eigenvalues and eigenvectors, an important topic that was first introduced in Section 2.4.

Let $\mathbf{T}$ be a linear operator on the finite dimensional vector space $\mathcal{V}$ and let

$$\mathcal{B} = \{\mathbf{u}_1, \mathbf{u}_2, \cdots, \mathbf{u}_n\} \qquad \text{and} \qquad \mathcal{B}' = \{\mathbf{v}_1, \mathbf{v}_2, \cdots, \mathbf{v}_n\}$$

be two bases for $\mathcal{V}$; we will think of $\mathcal{B}$ as the *old basis* and $\mathcal{B}'$ as the *new basis*. Let $A = [\mathbf{T}]_{\mathcal{B}} = [\mathbf{T}]_{\text{old}}$ be the matrix that represents $\mathbf{T}$ with respect to the $\mathcal{B}$ basis, and let $A' = [\mathbf{T}]_{\mathcal{B}'} = [\mathbf{T}]_{\text{new}}$ be the matrix that represents $\mathbf{T}$ with respect to the $\mathcal{B}'$ basis. We will now investigate the relationship between the matrices

*This section is optional.

A and A'. Recall, from Theorem 4.12, that there is a unique $\mathcal{B}$ to $\mathcal{B}'$ (old to new) change-of-basis matrix P satisfying

$$P[\mathbf{w}]_{\mathcal{B}} = [\mathbf{w}]_{\mathcal{B}'} \quad \text{for any vector } \mathbf{w} \text{ in } \mathcal{V}, \tag{4.5}$$

and that P is given by

$$P = \begin{bmatrix} [\mathbf{u}_1]_{\mathcal{B}'} & [\mathbf{u}_2]_{\mathcal{B}'} & \cdots & [\mathbf{u}_n]_{\mathcal{B}'} \end{bmatrix}.$$

In other words, the columns of P are the new coordinates of the old basis vectors. We have also shown that the change-of-basis matrix P is nonsingular and that P^{-1} is the $\mathcal{B}'$ to $\mathcal{B}$ (new to old) change-of-basis matrix. Exercise 15 shows how to interpret P as a matrix representative of the identity operator on $\mathcal{V}$.

From Theorem 4.9 we know that the matrix $A' = [\mathbf{T}]_{\mathcal{B}'} = [\mathbf{T}]_{\text{new}}$ satisfies the relation

$$[\mathbf{T}(\mathbf{x})]_{\mathcal{B}'} = A'[\mathbf{x}]_{\mathcal{B}'} \quad \text{for all vectors } \mathbf{x} \text{ in } \mathcal{V}.$$

If we use Equation (4.5), first with $\mathbf{w} = \mathbf{x}$ and then with $\mathbf{w} = \mathbf{T}(\mathbf{x})$, to change to $\mathcal{B}$ (old) coordinates, then the last equation becomes

$$P[\mathbf{T}(\mathbf{x})]_{\mathcal{B}} = [\mathbf{T}(\mathbf{x})]_{\mathcal{B}'} = A'[\mathbf{x}]_{\mathcal{B}'} = A'(P[\mathbf{x}]_{\mathcal{B}}).$$

Since the change-of-basis matrix P is nonsingular, we can multiply on the left by P^{-1} to obtain

$$[\mathbf{T}(\mathbf{x})]_{\mathcal{B}} = (P^{-1}A'P)[\mathbf{x}]_{\mathcal{B}}.$$

This last relation holds for all vectors $\mathbf{x}$ in $\mathcal{V}$, so $P^{-1}A'P = A = [\mathbf{T}]_{\text{old}} = [\mathbf{T}]_{\mathcal{B}}$ is the matrix representative of $\mathbf{T}$ with respect to the $\mathcal{B}$ basis. This discussion is summarized in the following theorem.

THEOREM 4.14 Let $\mathbf{T}$ be a linear operator on the finite dimensional vector space $\mathcal{V}$, whose matrix representative with respect to the $\mathcal{B}$ basis is A and whose matrix representative with respect to the $\mathcal{B}'$ basis is A'. If P is the $\mathcal{B}$ to $\mathcal{B}'$ change-of-basis matrix, then

$$A = P^{-1}A'P \quad \text{and} \quad A' = PAP^{-1}. \quad \square \tag{4.6}$$

■ EXAMPLE 1 Change of Basis in $\Re^2$

Let $\mathbf{T}$ be the linear operator on $\Re^2$ defined by

$$\mathbf{T}(\mathbf{u}) = A\mathbf{u} = \begin{bmatrix} 1 & 2 \\ 3 & 4 \end{bmatrix}\begin{bmatrix} u_1 \\ u_2 \end{bmatrix}.$$

The matrix representative of $\mathbf{T}$ with respect to the standard basis of $\Re^2$ is $A = [\mathbf{T}]_{\mathcal{S}}$. Use Theorem 4.14 to find $A' = [\mathbf{T}]_{\mathcal{B}'}$, which is the matrix representative of $\mathbf{T}$ with respect to the basis

$$\mathcal{B}' = \left\{ \mathbf{v}_1 = \begin{bmatrix} 1 \\ 2 \end{bmatrix}, \mathbf{v}_2 = \begin{bmatrix} 2 \\ 3 \end{bmatrix} \right\}.$$

From Theorem 4.14 we know that A and A' satisfy

$$A' = [\mathbf{T}]_{\mathcal{B}'} = P^{-1}[\mathbf{T}]_{\mathcal{S}}P = P^{-1}AP,$$

where P must be the $\mathcal{B}'$ to $\mathcal{S}$ (new to old) change-of-basis matrix. Thus, P must satisfy

$$P[\mathbf{x}]_{\mathcal{B}'} = [\mathbf{x}]_{\mathcal{S}} \quad \text{for all } \mathbf{x} \text{ in } \Re^2.$$

From Equation (4.5) we have, since $[\mathbf{x}]_{\mathcal{S}} = \mathbf{x}$, for all vectors $\mathbf{x}$ in $\Re^2$,

$$P = \begin{bmatrix} [\mathbf{v}_1]_{\mathcal{S}} & [\mathbf{v}_2]_{\mathcal{S}} \end{bmatrix} = \begin{bmatrix} 1 & 2 \\ 2 & 3 \end{bmatrix}.$$

Thus,

$$A' = \begin{bmatrix} 1 & 2 \\ 2 & 3 \end{bmatrix}^{-1} \begin{bmatrix} 1 & 2 \\ 3 & 4 \end{bmatrix} \begin{bmatrix} 1 & 2 \\ 2 & 3 \end{bmatrix} = \begin{bmatrix} -3 & 2 \\ 2 & -1 \end{bmatrix} \begin{bmatrix} 5 & 8 \\ 11 & 18 \end{bmatrix} = \begin{bmatrix} 7 & 12 \\ -1 & -2 \end{bmatrix}$$

is the matrix representative of $\mathbf{T}$ with respect to the $\mathcal{B}'$ basis. From this last matrix we see that we must have

$$\mathbf{T}(\mathbf{v}_1) = 7\mathbf{v}_1 - \mathbf{v}_2 \qquad \text{and} \qquad \mathbf{T}(\mathbf{v}_2) = 12\mathbf{v}_1 - 2\mathbf{v}_2.$$

These last relations can be checked directly; for $\mathbf{v}_1$ we have

$$\mathbf{T}(\mathbf{v}_1) = \begin{bmatrix} 1 & 2 \\ 3 & 4 \end{bmatrix} \begin{bmatrix} 1 \\ 2 \end{bmatrix} = \begin{bmatrix} 5 \\ 11 \end{bmatrix} = 7\begin{bmatrix} 1 \\ 2 \end{bmatrix} - \begin{bmatrix} 2 \\ 3 \end{bmatrix} = 7\mathbf{v}_1 - \mathbf{v}_2.$$ ■

Theorem 4.14 provides motivation for the study of the important matrix relation introduced in the next definition.

DEFINITION 4.7 The $n \times n$ matrices A and B are said to be **similar** ($A \approx B$) if there exists a nonsingular matrix Q such that $Q^{-1}AQ = B$. □

We can now rephrase Theorem 4.14 to say that the different matrix representatives of a linear operator are similar matrices. The converse is also true, so we have the following result that provides geometric motivation for the study of similarity.

THEOREM 4.15 Two $n \times n$ matrices are similar if, and only if, they represent the same linear operator relative to different bases for some n-dimensional vector space $\mathcal{V}$. □

Note that Theorem 4.15 does not specify the space $\mathcal{V}$. If the matrices are $n \times n$, then any n-dimensional space will do; in particular we could take $\mathcal{V} = \Re^n$.

Similarity, although a less restrictive relation than equality, shares three important properties with the relations of equality and of row equivalence: similarity is *reflexive* ($A \approx A$), *symmetric* (if $A \approx B$, then $B \approx A$), and *transitive* (if $A \approx B$ and $B \approx C$, then $A \approx C$) (see Exercise 10 of this section).

It is easy to become confused when applying Theorems 4.14 or 4.15; is P the $\mathcal{B}$ to $\mathcal{B}'$ change-of-basis matrix or the $\mathcal{B}'$ to $\mathcal{B}$ change-of-basis matrix? The key to keeping it all straight is to remember Theorem 4.15, note that the *rightmost matrix is the first to have its effect felt,* and then decide what that effect must be. For example,

$$\begin{aligned}
[\mathbf{T}(\mathbf{x})]_{\mathcal{B}} &= P^{-1}A'P[\mathbf{x}]_{\mathcal{B}} && \text{where} \\
P[\mathbf{x}]_{\mathcal{B}} &= [\mathbf{x}]_{\mathcal{B}'} && (\mathcal{B} \text{ to } \mathcal{B}' \text{ coordinate change}); \\
A'[\mathbf{x}]_{\mathcal{B}'} &= [\mathbf{T}(\mathbf{x})]_{\mathcal{B}'} && (A' \text{ represents } \mathbf{T} \text{ in the } \mathcal{B}' \text{ basis}); \text{ and} \\
P^{-1}[\mathbf{T}(\mathbf{x})]_{\mathcal{B}'} &= [\mathbf{T}(\mathbf{x})]_{\mathcal{B}} && (\mathcal{B}' \text{ to } \mathcal{B} \text{ coordinate change}).
\end{aligned}$$

It may also be helpful to write the first equation of (4.6) as

A	$=$	P^{-1}	A'	P
[old matrix of **T**]	=	[new to old]	[new matrix of **T**]	[old to new].
[$\mathcal{B}$ basis]		[$\mathcal{B}'$ to $\mathcal{B}$]	[$\mathcal{B}'$ basis]	[$\mathcal{B}$ to $\mathcal{B}'$]

Remember to read the last line from right to left.

In general, finding the change-of-basis matrix of Theorem 4.14 involves solving a set of linear systems. The solution of these systems is easy if the old basis is the standard basis as in Example 1. Our next example illustrates the more general case.

EXAMPLE 2 Change of Basis in $\Re^3$

Let **T** be the linear operator on $\Re^3$ whose matrix representative with respect to the old basis,

$$\mathcal{B} = \left\{ \mathbf{u}_1 = \begin{bmatrix} 1 \\ 1 \\ 1 \end{bmatrix}, \mathbf{u}_2 = \begin{bmatrix} 1 \\ 0 \\ 1 \end{bmatrix}, \mathbf{u}_3 = \begin{bmatrix} 1 \\ 1 \\ 0 \end{bmatrix} \right\},$$

is $A = [\mathbf{T}]_{\mathcal{B}} = \begin{bmatrix} 11 & -21 & 6 \\ 3 & -5 & 2 \\ -2 & 5 & 1 \end{bmatrix}$.

Find the matrix representative of **T** with respect to the new basis

$$\mathcal{B}' = \left\{ \mathbf{v}_1 = \begin{bmatrix} -3 \\ -2 \\ -4 \end{bmatrix}, \mathbf{v}_2 = \begin{bmatrix} 6 \\ 4 \\ 7 \end{bmatrix}, \mathbf{v}_3 = \begin{bmatrix} 7 \\ 5 \\ 8 \end{bmatrix} \right\}.$$

From Theorem 4.15 we know that $[\mathbf{T}]_{\mathcal{B}'}$ is similar to A; thus there exists a nonsingular matrix Q such that

$$[\mathbf{T}]_{\mathcal{B}'} = Q^{-1}[\mathbf{T}]_{\mathcal{B}}Q = Q^{-1}AQ.$$

It follows that Q must be the $\mathcal{B}'$ to $\mathcal{B}$ (new to old) change-of-basis matrix

$$Q = \begin{bmatrix} [\mathbf{v}_1]_{\mathcal{B}} & [\mathbf{v}_2]_{\mathcal{B}} & [\mathbf{v}_3]_{\mathcal{B}} \end{bmatrix}.$$

In order to find the matrix Q we need to find the $\mathcal{B}$-coordinates of the $\mathbf{v}_i$. The three vector equations

$$x_1\mathbf{u}_1 + x_2\mathbf{u}_2 + x_3\mathbf{u}_3 = \mathbf{v}_i; \quad i = 1, 2, 3$$

lead easily to the three systems

$$\begin{bmatrix} 1 & 1 & 1 \\ 1 & 0 & 1 \\ 1 & 1 & 0 \end{bmatrix}\begin{bmatrix} x_1 \\ x_2 \\ x_3 \end{bmatrix} = \begin{bmatrix} -3 \\ -2 \\ -4 \end{bmatrix}, \begin{bmatrix} 6 \\ 4 \\ 7 \end{bmatrix}, \begin{bmatrix} 7 \\ 5 \\ 8 \end{bmatrix}.$$

We can solve these three systems by the single row reduction

$$\begin{bmatrix} 1 & 1 & 1 & -3 & 6 & 7 \\ 1 & 0 & 1 & -2 & 4 & 5 \\ 1 & 1 & 0 & -4 & 7 & 8 \end{bmatrix} \to \begin{bmatrix} 1 & 1 & 1 & -3 & 6 & 7 \\ 0 & -1 & 0 & 1 & -2 & -2 \\ 0 & 0 & -1 & -1 & 1 & 1 \end{bmatrix}$$

$$\to \begin{bmatrix} 1 & 0 & 0 & -3 & 5 & 6 \\ 0 & 1 & 0 & -1 & 2 & 2 \\ 0 & 0 & 1 & 1 & -1 & -1 \end{bmatrix}.$$

It follows that $\mathbf{v}_1 = -3\mathbf{u}_1 - \mathbf{u}_2 + \mathbf{u}_3$, $\mathbf{v}_2 = 5\mathbf{u}_1 + 2\mathbf{u}_2 - \mathbf{u}_3$, and $\mathbf{v}_3 = 6\mathbf{u}_1 + 2\mathbf{u}_2 - \mathbf{u}_3$. Thus, the $\mathcal{B}'$ to $\mathcal{B}$ change-of-basis matrix is $Q = \begin{bmatrix} -3 & 5 & 6 \\ -1 & 2 & 2 \\ 1 & -1 & -1 \end{bmatrix}$.

In Section 1.7 we saw that the most efficient way to compute $Q^{-1}AQ$ was to compute AQ and then use the row reduction

$$[Q \mid AQ] \to [I \mid Q^{-1}AQ].$$

In this case $AQ = \begin{bmatrix} -6 & 7 & 18 \\ -2 & 3 & 6 \\ 2 & -1 & -3 \end{bmatrix}$ and the row reduction

$$\begin{bmatrix} -3 & 5 & 6 & -6 & 7 & 18 \\ -1 & 2 & 2 & -2 & 3 & 6 \\ 1 & -1 & -1 & 2 & -1 & -3 \end{bmatrix} \to \begin{bmatrix} 0 & 2 & 3 & 0 & 4 & 9 \\ 0 & 1 & 1 & 0 & 2 & 3 \\ 1 & -1 & -1 & 2 & -1 & -3 \end{bmatrix}$$

$$\to \begin{bmatrix} 0 & 0 & 1 & 0 & 0 & 3 \\ 0 & 1 & 1 & 0 & 2 & 3 \\ 1 & 0 & 0 & 2 & 1 & 0 \end{bmatrix} \to \begin{bmatrix} 1 & 0 & 0 & 2 & 1 & 0 \\ 0 & 1 & 0 & 0 & 2 & 0 \\ 0 & 0 & 1 & 0 & 0 & 3 \end{bmatrix}$$

shows that $[\mathbf{T}]_{\mathcal{B}'} = Q^{-1}AQ = \begin{bmatrix} 2 & 1 & 0 \\ 0 & 2 & 0 \\ 0 & 0 & 3 \end{bmatrix} = [\mathbf{T}]_{\text{new}}$.

From the last matrix we see that

$$\mathbf{T}(\mathbf{v}_1) = 2\mathbf{v}_1, \quad \mathbf{T}(\mathbf{v}_2) = \mathbf{v}_1 + 2\mathbf{v}_2, \qquad \text{and} \qquad \mathbf{T}(\mathbf{v}_3) = 3\mathbf{v}_3.$$

It follows, as in Section 4.4, that **T** can be viewed geometrically as the product of expansions by a factor of 2 in the $\mathbf{v}_1$ and $\mathbf{v}_2$ directions, a shear in the $\mathbf{v}_1$ direction of the $\mathbf{v}_1\mathbf{v}_2$ plane with factor $\frac{1}{2}$, and an expansion by a factor of **3** in the $\mathbf{v}_3$ direction. ■

EXAMPLE 3 **A Change of Basis where the New Matrix Is Diagonal**

Let **T** be the linear operator on $\Re^{1\times 3}$ whose matrix representative, relative to the standard basis, is

$$A = [\mathbf{T}]_{\mathcal{S}} = \begin{bmatrix} 17 & 12 & 18 \\ -16 & -9 & -24 \\ -5 & -4 & -4 \end{bmatrix}.$$

Find the matrix representative of **T** with respect to the basis

$$\mathcal{E} = \{\mathbf{x}_1 = [10 \quad -8 \quad -3], \ \mathbf{x}_2 = [-3 \quad 3 \quad 1], \ \mathbf{x}_3 = [-3 \quad 2 \quad 1]\}.$$

From Equation (4.6) we have

$$C = [\mathbf{T}]_{\mathcal{E}} = P^{-1}[\mathbf{T}]_{\mathcal{S}}P = P^{-1}AP,$$

where P is the $\mathcal{E}$ to $\mathcal{S}$ change-of-basis matrix given by

$$P = \begin{bmatrix} [\mathbf{x}_1]_{\mathcal{S}} & [\mathbf{x}_2]_{\mathcal{S}} & [\mathbf{x}_3]_{\mathcal{S}} \end{bmatrix} = \begin{bmatrix} 10 & -3 & -3 \\ -8 & 3 & 2 \\ -3 & 1 & 1 \end{bmatrix}.$$

It follows that

$$C = \begin{bmatrix} 10 & -3 & -3 \\ -8 & 3 & 2 \\ -3 & 1 & 1 \end{bmatrix}^{-1} \begin{bmatrix} 17 & 12 & 18 \\ -16 & -9 & -24 \\ -5 & -4 & -4 \end{bmatrix} \begin{bmatrix} 10 & -3 & -3 \\ -8 & 3 & 2 \\ -3 & 1 & 1 \end{bmatrix}$$

$$= \begin{bmatrix} 10 & -3 & -3 \\ -8 & 3 & 2 \\ -3 & 1 & 1 \end{bmatrix}^{-1} \begin{bmatrix} 20 & 3 & -9 \\ -16 & -3 & 6 \\ -6 & -1 & 3 \end{bmatrix} = \begin{bmatrix} 2 & 0 & 0 \\ 0 & -1 & 0 \\ 0 & 0 & 3 \end{bmatrix}$$

and we have

$$\mathbf{T}(\mathbf{x}_1) = 2\mathbf{x}_1, \quad \mathbf{T}(\mathbf{x}_2) = -\mathbf{x}_2, \qquad \text{and} \qquad \mathbf{T}(\mathbf{x}_3) = 3\mathbf{x}_3. \ ■$$

It would certainly be easier to study the linear operator **T** of Example 3 relative to the $\mathcal{E}$ basis rather than relative to the standard basis. In particular, we could describe **T** as the product of two expansions (by a factor of 2 in the $\mathbf{x}_1$-direction and by a factor of 3 in the $\mathbf{x}_3$-direction) and a reflection in the $\mathbf{x}_1\mathbf{x}_3$-plane. The last example suggests some interesting questions about linear operators.

1. Is it always possible to find a basis for $\mathcal{V}$ relative to which a given linear operator is represented by a diagonal matrix? Equivalently, can we find, for any square matrix A, a nonsingular matrix P such that $P^{-1}AP$ is a diagonal matrix?
2. If so, how? If not, when?
3. If not, how close can we come to a diagonal representative?

These questions will be considered in Chapter 5.

EXERCISES 4.7

1. Let $\mathcal{B} = \{\mathbf{u}_1, \mathbf{u}_2, \mathbf{u}_3, \mathbf{u}_4\}$ and $\mathcal{B}' = \{\mathbf{v}_1, \mathbf{v}_2, \mathbf{v}_3, \mathbf{v}_4\}$ be two bases for a four-dimensional vector space $\mathcal{V}$. Define a linear operator $\mathbf{T}$ by

$$\mathbf{T}(\mathbf{u}_1) = 2\mathbf{u}_1 - 3\mathbf{u}_3 + \mathbf{u}_3 \qquad \mathbf{T}(\mathbf{u}_2) = 4\mathbf{u}_1 - 5\mathbf{u}_4$$
$$\mathbf{T}(\mathbf{u}_3) = \mathbf{u}_1 + 4\mathbf{u}_3 \qquad \mathbf{T}(\mathbf{u}_4) = 5\mathbf{u}_1 + \mathbf{u}_2 - \mathbf{u}_4.$$

(a) Find $[\mathbf{T}]_{\mathcal{B}} = [\mathbf{T}]_{\mathcal{B}\mathcal{B}}$.
(b) If $\mathbf{v}_1 = \mathbf{u}_1 - 2\mathbf{u}_2 + \mathbf{u}_3 + \mathbf{u}_4$, $\mathbf{v}_2 = \mathbf{u}_2 + \mathbf{u}_3 - 3\mathbf{u}_4$, $\mathbf{v}_3 = \mathbf{u}_3 + 5\mathbf{u}_4$, and $\mathbf{v}_4 = \mathbf{u}_4$, find P so that $P[\mathbf{w}]_{\mathcal{B}'} = [\mathbf{w}]_{\mathcal{B}}$ for all $\mathbf{w}$ in $\mathcal{V}$.
(c) Find $[\mathbf{T}]_{\mathcal{B}'}$.
(d) Express each $\mathbf{T}(\mathbf{v}_i)$ as a linear combination of the $\mathbf{v}_j$.

2. Let $\mathcal{V}$ be an arbitrary three-dimensional real vector space and let $\mathbf{T}$ be a linear operator on $\mathcal{V}$. You are given two bases for $\mathcal{V}$,

$$\mathcal{B} = \{\mathbf{v}_1, \mathbf{v}_2, \mathbf{v}_3\} \quad \text{and} \quad \mathcal{B}' = \{\mathbf{w}_1, \mathbf{w}_2, \mathbf{w}_3\},$$

and the following information.

$$\mathbf{T}(\mathbf{v}_1) = \mathbf{v}_1 - 3\mathbf{v}_3 \qquad \mathbf{w}_1 = 2\mathbf{v}_1 - \mathbf{v}_2 \qquad \mathbf{v}_1 = \mathbf{w}_1 + \mathbf{w}_2$$
$$\mathbf{T}(\mathbf{v}_2) = 2\mathbf{v}_2 + 5\mathbf{v}_3 \qquad \mathbf{w}_2 = -\mathbf{v}_1 + \mathbf{v}_2 \qquad \mathbf{v}_2 = \mathbf{w}_1 + 2\mathbf{w}_2$$
$$\mathbf{T}(\mathbf{v}_3) = 2\mathbf{v}_1 + 7\mathbf{v}_2 + \mathbf{v}_3 \qquad \mathbf{w}_3 = -\mathbf{v}_1 + \mathbf{v}_3 \qquad \mathbf{v}_3 = \mathbf{w}_1 + \mathbf{w}_2 + \mathbf{w}_3$$

(a) Find $[\mathbf{T}]_{\mathcal{B}}$, which is the matrix representative of $\mathbf{T}$ with respect to the $\mathcal{B}$ basis.
(b) Find a nonsingular matrix P such that $[\mathbf{T}]_{\mathcal{B}'} = P^{-1}[\mathbf{T}]_{\mathcal{B}}P$.
(c) Compute $[\mathbf{T}]_{\mathcal{B}'}$.

3. Let $\mathbf{T}$ be the linear operator on $\Re^2$ defined by

$$\mathbf{T}(\mathbf{v}) = \begin{bmatrix} 2 & -1 \\ 4 & 3 \end{bmatrix} \mathbf{v}.$$

(a) Use Theorem 4.14 to find the matrix representative of $\mathbf{T}$ with respect to the basis

$$\mathcal{B} = \{\mathbf{u}_1 = [4 \quad 3]^T, \mathbf{u}_2 = [3 \quad 2]^T\}.$$

(b) Find $[\mathbf{T}]_{\mathscr{B}}$ by finding $\mathbf{T}(\mathbf{u}_1)$ and $\mathbf{T}(\mathbf{u}_2)$ directly.

4. Let $\mathbf{T}$ be the linear operator on $\mathfrak{R}^2$ whose matrix representative relative to the standard basis is $[\mathbf{T}]_{\mathscr{S}} = \begin{bmatrix} 3 & 2 \\ -1 & 0 \end{bmatrix}$.

(a) Use Theorem 4.14 to find the matrix representative of $\mathbf{T}$ with respect to the basis

$$\mathscr{B} = \{\mathbf{u}_1 = [1 \quad -1]^T, \mathbf{u}_2 = [2 \quad -1]^T\}.$$

(b) Find $[\mathbf{T}]_{\mathscr{B}}$ by finding $\mathbf{T}(\mathbf{u}_1)$ and $\mathbf{T}(\mathbf{u}_2)$ directly.

5. Let $\mathbf{T}$ be the linear operator on $\mathfrak{R}^2$ whose matrix representative relative to the basis

$$\mathscr{B} = \left\{\mathbf{u}_1 = \begin{bmatrix} 1 \\ 2 \end{bmatrix}, \mathbf{u}_2 = \begin{bmatrix} 2 \\ 3 \end{bmatrix}\right\} \text{ is } [\mathbf{T}]_{\mathscr{B}} = \begin{bmatrix} 2 & 1 \\ 0 & 2 \end{bmatrix}.$$

Find the matrix representative of $\mathbf{T}$ with respect to the basis

$$\mathscr{B}' = \left\{\mathbf{w}_1 = \begin{bmatrix} 2 \\ 1 \end{bmatrix}, \mathbf{w}_2 = \begin{bmatrix} 1 \\ 0 \end{bmatrix}\right\}.$$

6. Consider the basis $\mathscr{B} = \{\mathbf{u}_1 = [1 \quad 2 \quad 3], \mathbf{u}_2 = [2 \quad 4 \quad 5], \mathbf{u}_3 = [3 \quad 5 \quad 6]\}$ and let $\mathbf{T}$ be the linear operator on $\mathfrak{R}^{1\times 3}$ for which $[\mathbf{T}]_{\mathscr{B}} = \begin{bmatrix} 1 & 3 & 3 \\ 1 & 4 & 3 \\ 1 & 3 & 4 \end{bmatrix}$. Find the matrix representative of $\mathbf{T}$ with respect to the standard basis.

7. Let $\mathbf{T}$ be a linear operator on $\mathfrak{R}^3$ whose standard matrix is $\begin{bmatrix} 2 & 1 & 1 \\ 2 & 3 & 2 \\ 1 & 1 & 2 \end{bmatrix}$. Find the matrix representative of $\mathbf{T}$ with respect to the basis

$$\mathscr{B} = \left\{\mathbf{u}_1 = \begin{bmatrix} 1 \\ 2 \\ 1 \end{bmatrix}, \mathbf{u}_2 = \begin{bmatrix} -1 \\ 1 \\ 0 \end{bmatrix}, \mathbf{u}_3 = \begin{bmatrix} 1 \\ 0 \\ -1 \end{bmatrix}\right\}$$

and describe $\mathbf{T}$ geometrically.

8. Let $\mathbf{T}$ be the linear operator on $\mathfrak{R}^3$ whose standard matrix is $\begin{bmatrix} 0 & 1 & 0 \\ 0 & 0 & 1 \\ 8 & -12 & 6 \end{bmatrix}$.

Find the matrix representative of $\mathbf{T}$ with respect to the basis

$$\mathscr{B} = \left\{\begin{bmatrix} 1 \\ 2 \\ 4 \end{bmatrix}, \begin{bmatrix} 0 \\ 1 \\ 4 \end{bmatrix}, \begin{bmatrix} 0 \\ 0 \\ 1 \end{bmatrix}\right\}.$$

9. Let **T** be a linear operator on $\mathfrak{R}^4$ whose standard matrix is

$$[\mathbf{T}]_{\mathcal{S}} = \begin{bmatrix} 6 & 0 & -3 & 0 \\ 4 & 2 & -1 & -1 \\ 3 & 1 & 1 & -1 \\ 6 & 2 & 0 & -1 \end{bmatrix}.$$

Find the matrix representative of **T** with respect to the basis

$$\mathcal{B} = \left\{ \begin{bmatrix} 1 \\ 1 \\ 1 \\ 2 \end{bmatrix}, \begin{bmatrix} 3 \\ 3 \\ 4 \\ 8 \end{bmatrix}, \begin{bmatrix} 3 \\ 4 \\ 3 \\ 6 \end{bmatrix}, \begin{bmatrix} 0 \\ 1 \\ 0 \\ 1 \end{bmatrix} \right\}.$$

Can you interpret **T** geometrically?

10. Verify that similarity is a symmetric, reflexive, and transitive relation on the set of all $n \times n$ matrices.

11. Show that similar matrices have the same characteristic polynomial; that is, show that if $B = P^{-1}AP$, then $\det(\lambda I - B) = \det(\lambda I - A)$.

†12. Show that similar matrices have the same rank, nullity, trace, and determinant.

13. **(a)** Show that if $B = P^{-1}AP$, then $B^k = P^{-1}A^kP$.

(b) Let $p(x) = a_0 + a_1x + a_2x^2 + \cdots + a_nx^n$ be any polynomial and define $p(A) = a_0I + a_1A + a_2A^2 + \cdots + a_nA^n$. Show that if A and B are similar, then so are $p(A)$ and $p(B)$.

14. Show that if A and B are similar, then A^T and B^T are also similar. Moreover, if A is nonsingular, then B is also nonsingular and A^{-1} and B^{-1} are similar.

15. **(a)** Let $\mathcal{B}$ and $\mathcal{B}'$ be two different bases for the vector space $\mathcal{V}$ and let $P = [\mathbf{I}]_{\mathcal{B}'\mathcal{B}}$ be the matrix that represents the identity operator on $\mathcal{V}$ with respect to the bases $\mathcal{B}$ and $\mathcal{B}'$(recall Section 4.5). Show that P is the $\mathcal{B}$ to $\mathcal{B}'$ change-of-basis matrix.

(b) Use Exercise 12 in Section 4.5 to show that

$$P^{-1}[\mathbf{T}]_{\mathcal{B}'}P = [\mathbf{I}]_{\mathcal{B}\mathcal{B}'}[\mathbf{T}]_{\mathcal{B}'}[\mathbf{I}]_{\mathcal{B}'\mathcal{B}} = [\mathbf{ITI}]_{\mathcal{B}\mathcal{B}} = [\mathbf{T}]_{\mathcal{B}}.$$

This point of view, and the diagram below, can be useful tools for remembering how to construct the $\mathcal{B}$ to $\mathcal{B}'$ change-of-basis matrix P.

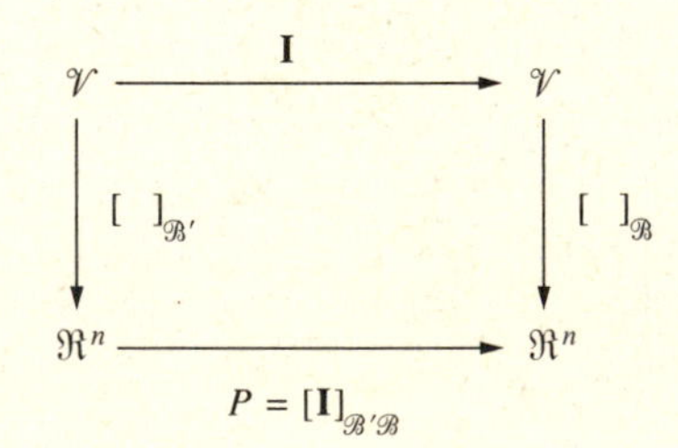

16. Use the result of Exercise 15 and the diagram below to describe the change-of-basis theorem.

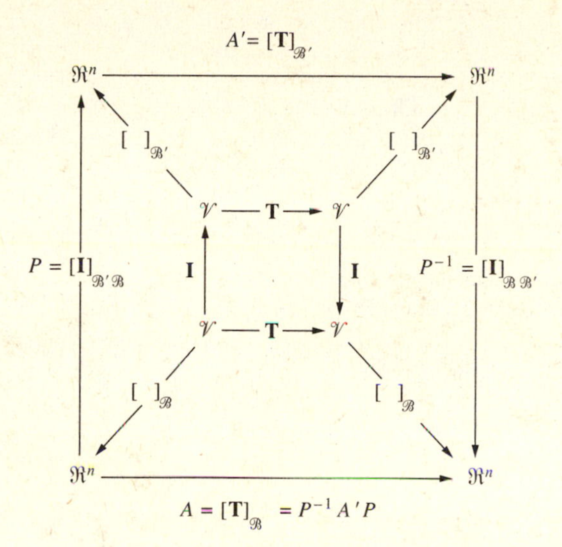

17. Let **T** be a linear transformation from $\mathcal{V}$ to $\mathcal{W} \neq \mathcal{V}$. Let $\mathcal{B}$ and $\mathcal{B}'$ be two bases for $\mathcal{V}$ and let $\mathcal{C}$ and $\mathcal{C}'$ be two bases for $\mathcal{W}$. Construct a diagram like that in Exercise 16 and then determine the relationship between the matrix representatives $[\mathbf{T}]_{\mathcal{C}\mathcal{B}}$ and $[\mathbf{T}]_{\mathcal{C}'\mathcal{B}'}$.

18. Consider the two bases for $\mathcal{P}_3$:

$$\mathcal{B} = \{1, 1 - x, x - x^2\} \qquad \text{and} \qquad \mathcal{B}' = \{1 + x^2, 1 - x^2, 2x\}.$$

Let **T** be a linear operator on $\mathcal{P}_3$ for which $[\mathbf{T}]_{\mathcal{B}} = \begin{bmatrix} 1 & -1 & 2 \\ 2 & 1 & 0 \\ -1 & 0 & 3 \end{bmatrix}$. Find $[\mathbf{T}]_{\mathcal{B}'}$.

CHAPTER 4 SUMMARY

In Chapter 4 we introduced and studied the linear transformations between a vector space $\mathcal{V}$ and a vector space $\mathcal{W}$. The motivating example here was matrix multiplication. We showed that every linear transformation between coordinate spaces was matrix multiplication and then that every linear transformation between finite dimensional vector spaces can be represented by matrix multiplication. The relationship between matrix representatives with respect to different bases was investigated.

KEY DEFINITIONS

One-to-one functions and functions which are onto $\mathcal{W}$. *231*
Linear transformation. *233*
The general linear problem. *242*
Null space and range of a linear transformation. *243*
nullity(**T**) and rank(**T**). *244*
$\mathcal{L}(\mathcal{V}, \mathcal{W})$. *253*
Algebraic operations on linear transformations: addition, scalar multiplication, multiplication. *253, 254*
Standard matrix of **T** in $\mathcal{L}(\mathfrak{R}^m, \mathfrak{R}^n)$. *259*
The elementary transformations of $\mathfrak{R}^2$; reflections, shears, expansions, contractions, projections, rotations. *262*
The matrix representative, $[\mathbf{T}]_{\mathcal{B}'\mathcal{B}}$. *274*
Similar matrices. *290*

KEY FACTS

Matrix multiplication is a linear function.
A linear transformation is one-to-one if, and only if, its null space is zero.
A linear transformation is completely determined by its action on a basis.
$\mathcal{L}(\mathcal{V}, \mathcal{W})$ is a real vector space and dim $(\mathcal{L}(\mathcal{V}, \mathcal{W}))$ = dim $(\mathcal{V})$ dim $(\mathcal{W})$.
T in $\mathcal{L}(\mathcal{V}, \mathcal{V})$ is invertible if, and only if, **T** is both one-to-one and onto $\mathcal{V}$.
If dim $(\mathcal{V})$ is finite, then **T** in $\mathcal{L}(\mathcal{V}, \mathcal{V})$ is invertible if, and only if, **T** is either one-to-one or onto $\mathcal{V}$.
Every **T** in $\mathcal{L}(\mathfrak{R}^m, \mathfrak{R}^n)$ is multiplication by some $m \times n$ matrix A.
Every linear operator on $\mathfrak{R}^2$ or $\mathfrak{R}^3$ is a product of elementary transformations.
$[\mathbf{T}(\mathbf{u})]_{\mathcal{B}'} = [\mathbf{T}]_{\mathcal{B}'\mathcal{B}}[\mathbf{u}]_{\mathcal{B}}$.
If $\mathcal{B} = \{\mathbf{u}_1, \mathbf{u}_2, \ldots, \mathbf{u}_n\}$ and $\mathcal{B}' = \{\mathbf{v}_1, \mathbf{v}_2, \ldots, \mathbf{v}_n\}$ are two bases for a finite dimensional vector space $\mathcal{V}$, then the matrix

$$P = \left[[\mathbf{u}_1]_{\mathcal{B}'} \quad [\mathbf{u}_2]_{\mathcal{B}'} \quad \cdots \quad [\mathbf{u}_n]_{\mathcal{B}'} \right],$$

whose columns are the new coordinates of the old basis vectors, is the unique nonsingular matrix P such that $P[\mathbf{w}]_{\mathcal{B}} = [\mathbf{w}]_{\mathcal{B}'}$, for all vectors **w** in $\mathcal{V}$.
The change-of-basis matrix P, between two orthonormal bases of $\mathfrak{R}^n$, is an orthogonal matrix; that is, $P^TP = I$.
Different matrix representatives of the same linear operator are similar matrices; $[\mathbf{T}]_{\mathcal{B}} = P^{-1}[\mathbf{T}]_{\mathcal{B}}P$ where P is the $\mathcal{B}$ to $\mathcal{B}'$ change-of-basis matrix.

COMPUTATIONAL PROCEDURES

Test a given function for linearity and find its domain and range.
Find basis for $\mathcal{R}(\mathbf{T})$, $\mathcal{NS}(\mathbf{T})$.
Decompose a linear operator on $\mathfrak{R}^2$ or $\mathfrak{R}^3$ into a product of elementary transformations.
Find the change-of-basis matrix.
Find $[\mathbf{T}]_{\mathcal{B}'}$, from $[\mathbf{T}]_{\mathcal{B}}$, $\mathcal{B}$, and $\mathcal{B}'$.

APPLICATIONS

Calculus, differential equations. Robotics, metal deformation. Computer graphics.

CHAPTER 4 REVIEW EXERCISES

1. Let **T** be a linear transformation from $\mathcal{V}$ to $\mathcal{W}$. Complete the following from the list on the right.

(a) Domain of **T** = _____
(b) $\mathcal{NS}(\mathbf{T})$ = _____
(c) Range of **T** = _____
(d) rank(**T**) = _____
(e) nullity(**T**) = _____

1. $\mathcal{W}$ **2.** $\mathcal{V}$ **3.** dim$(\mathcal{V})$ **4.** $\mathcal{RS}(A)$
5. $\mathcal{RS}(\mathbf{T})$ **6.** $\{\mathbf{x} \mid \mathbf{T}(\mathbf{x}) = 2\mathbf{x}\}$
7. $\{\mathbf{T}(\mathbf{u}) \mid \mathbf{u} \text{ in } \mathcal{V}\}$ **8.** $\{\mathbf{u} \mid \mathbf{T}(\mathbf{u}) = \mathbf{0}\}$
9. dim$(\mathcal{NS}(\mathbf{T}))$ **10.** dim$(\mathcal{R}(\mathbf{T}))$
11. dim $(\mathcal{V})$ − rank $(\mathbf{T})$ **12.** dim $(\mathcal{V})$ + rank $(\mathbf{T})$

2. Consider the function $\mathbf{T}$ from $\mathfrak{R}^3$ to $\mathfrak{R}^{2\times 2}$ defined by

$$\mathbf{T}\begin{bmatrix} x \\ y \\ z \end{bmatrix} = \begin{bmatrix} x+y & 3+z \\ 0 & 2x+3z \end{bmatrix}.$$

Determine if $\mathbf{T}$ is linear. Explain your reasoning.

3. Given $\mathbf{T}$, a function from $\mathfrak{R}^{2\times 2}$ to $\mathfrak{R}$ defined by $\mathbf{T}(A) = \det(A^T)$, determine whether $\mathbf{T}$ is a linear transformation. Explain carefully.

4. For $\mathbf{T}$ in $\mathcal{L}(\mathcal{V}, \mathcal{W})$, prove that $\mathcal{NS}(\mathbf{T})$ is a subspace of $\mathcal{V}$.

5. Is the transformation $\mathbf{T}\begin{bmatrix} x_1 \\ x_2 \\ x_3 \end{bmatrix} = \begin{bmatrix} 1+x_1 \\ x_1x_2 \\ x_1+x_2+x_3^3 \end{bmatrix}$ linear?

6. Let $\mathbf{T}$ be a linear transformation from $\mathfrak{R}^2$ to $\mathfrak{R}^3$ such that

$\mathbf{T}(\mathbf{e}_1) = \begin{bmatrix} 3 \\ 5 \\ 1 \end{bmatrix}$ and $\mathbf{T}(\mathbf{e}_2) = \begin{bmatrix} -1 \\ 4 \\ 3 \end{bmatrix}$. Find $\mathbf{T}(7\mathbf{e}_1 - 2\mathbf{e}_2)$, $\mathbf{T}\begin{bmatrix} 4 \\ 2 \end{bmatrix}$, and $\mathbf{T}\begin{bmatrix} 2 \\ -5 \end{bmatrix}$.

7. For the linear operator $\mathbf{Q}$ on $\mathcal{P}_3$ defined by

$$\mathbf{Q}(a + bx + cx^2 + dx^3) = b + cx^2,$$

find $\mathcal{NS}(\mathbf{Q})$, $\mathcal{R}(\mathbf{Q})$, rank($\mathbf{Q}$), and nullity($\mathbf{Q}$).

8. Let $\mathbf{T}$ in $\mathcal{L}(\mathfrak{R}^4, \mathfrak{R}^3)$ be defined by $\mathbf{T}(\mathbf{u}) = \begin{bmatrix} 1 & 2 & 3 & 1 \\ 4 & 5 & 6 & 1 \\ 7 & 8 & 9 & 1 \end{bmatrix}\mathbf{u}$.

(a) Find a basis for $\mathcal{NS}(\mathbf{T})$.

(b) Find the rank of $\mathbf{T}$.

(c) Find a basis for the range of $\mathbf{T}$.

9. Let $\mathbf{T}$ be a linear operator on $\mathfrak{R}^2$ which is described geometrically by the following sequence of elementary transformations. Find the standard matrix of $\mathbf{T}$.

(1) Compression by a factor of $\frac{1}{3}$ in the x-direction
(2) A shear, by a factor of -3, in the x-direction
(3) A shear, by a factor of 2, in the y-direction
(4) A reflection in the line $y = x$

10. Let $\mathbf{T}$ be a linear operator on $\mathfrak{R}^2$ which is described geometrically by the following sequence of elementary transformations. Find the standard matrix of $\mathbf{T}$.

(1) An expansion by a factor of 3 in the y-direction
(2) A shear by a factor of 2 in the y-direction
(3) A reflection in the line $y = x$
(4) A rotation through an angle of 45 degrees

11. Let $\mathbf{T}$ be a linear transformation from $\mathfrak{R}^3$ to $\mathcal{P}_4$ for which

$$\mathbf{T}\begin{bmatrix} 1 \\ 1 \\ 1 \end{bmatrix} = x^2 + x + 1, \quad \mathbf{T}\begin{bmatrix} 0 \\ 1 \\ 1 \end{bmatrix} = 2x^3 + x^2, \quad \mathbf{T}\begin{bmatrix} 0 \\ 1 \\ 2 \end{bmatrix} = x^3 + x + 1.$$

Find $\mathbf{T}\begin{bmatrix} 6 \\ 1 \\ 7 \end{bmatrix}$ and $\mathbf{T}\begin{bmatrix} a_1 \\ a_2 \\ a_3 \end{bmatrix}$.

12. A linear transformation $\mathbf{T}$ from $\mathfrak{R}^{2\times 2}$ to $\mathfrak{R}$ satisfies

$$\mathbf{T}\begin{bmatrix} 1 & 0 \\ 0 & 0 \end{bmatrix} = \mathbf{T}\begin{bmatrix} 0 & 0 \\ 0 & 1 \end{bmatrix} = 1 \quad \text{and} \quad \mathbf{T}\begin{bmatrix} 0 & 1 \\ 0 & 0 \end{bmatrix} = \mathbf{T}\begin{bmatrix} 0 & 0 \\ 1 & 0 \end{bmatrix} = 0.$$

Find $\mathbf{T}\begin{bmatrix} a & b \\ c & d \end{bmatrix}$ and show that $\mathbf{T}$ is not one-to-one.

13. Consider the two bases for $\mathcal{P}_3$:

$$\mathcal{S} = \{1, x, x^2\} \quad \text{and} \quad \mathcal{B} = \{1 - 2x + x^2, 3 - 5x + 4x^2, 2x + 3x^2\}.$$

Find the $\mathcal{S}$ to $\mathcal{B}$ change-of-basis matrix. What is the $\mathcal{B}$-coordinate matrix of $1 + x + 3x^2$?

14. Let $\mathcal{V}$ be an arbitrary three-dimensional real vector space and let $\mathbf{T}$ be a linear operator on $\mathcal{V}$. You are given

$$\mathcal{B} = \{\mathbf{v}_1, \mathbf{v}_2, \mathbf{v}_3\} \quad \text{and} \quad \mathcal{B}' = \{\mathbf{w}_1, \mathbf{w}_2, \mathbf{w}_3\}$$

as two bases for $\mathcal{V}$, and the following information.

$$\begin{array}{lll}
\mathbf{T}(\mathbf{v}_1) = \mathbf{v}_1 + 5\mathbf{v}_3 & \mathbf{v}_1 = -3\mathbf{w}_1 - 5\mathbf{w}_2 + 2\mathbf{w}_3 & \mathbf{w}_1 = \mathbf{v}_1 + \mathbf{v}_2 + \mathbf{v}_3 \\
\mathbf{T}(\mathbf{v}_2) = 2\mathbf{v}_2 - 5\mathbf{v}_3 & \mathbf{v}_2 = 2\mathbf{w}_1 + 2\mathbf{w}_2 - \mathbf{w}_3 & \mathbf{w}_2 = -\mathbf{v}_2 + \mathbf{v}_3 \\
\mathbf{T}(\mathbf{v}_3) = 2\mathbf{v}_1 - \mathbf{v}_2 + \mathbf{v}_3 & \mathbf{v}_3 = 2\mathbf{w}_1 + 3\mathbf{w}_2 - \mathbf{w}_3 & \mathbf{w}_3 = 2\mathbf{v}_1 - \mathbf{v}_2 + 4\mathbf{v}_3
\end{array}$$

(a) Find $[\mathbf{T}]_{\mathcal{B}}$, the matrix representative of $\mathbf{T}$ with respect to the $\mathcal{B}$ basis.
(b) Find a nonsingular matrix P such that $[\mathbf{T}]_{\mathcal{B}'} = P^{-1}[\mathbf{T}]_{\mathcal{B}}P$.

5 Similar Matrices and the Eigenvalue Problem

In this chapter we will introduce and study the relationship of similarity between matrices. In Section 4.7 we saw how this important relationship arises naturally in the study of the matrix representatives of a linear operator. That discussion provides geometric motivation for the further study of eigenvalues and eigenvectors, a topic which was discussed briefly in Section 2.4. Some of the special topics we will consider in this chapter include the eigenvalue problem for symmetric matrices, the special case of orthogonal similarity, and similarity to a triangular matrix. The chapter concludes with a brief discussion of the power method for computing eigenvalues and eigenvectors on a computer.

5.1 DIAGONALIZATION—EIGENVALUES AND EIGENVECTORS

What do eigenvalues and eigenvectors have to do with linear operators?

Eigenvalues and eigenvectors were first introduced in Section 2.4 as an application of determinants and of homogeneous systems. Section 2.4 was listed as optional in Chapter 2. If you omitted it then, it is now time to study it carefully. If you included Section 2.4 when you studied Chapter 2, you should review the basic definitions, theorems, and computational procedures of Section 2.4 at this time. Recall that a nonzero vector X from $\Re^n$ is an eigenvector of the $n \times n$ matrix A if $AX = \lambda X$ for some scalar λ; λ is called the eigenvalue associated with X. In this section we will provide some further motivation for the study of eigenvalues and eigenvectors as well as a good bit of additional information about them.

It is useful to give a slightly more general definition of eigenvalues and eigenvectors than that given in Chapter 2. Since we can now view square matrices as special kinds of linear operators, it is natural to extend these notions to linear operators on arbitrary vector spaces.

DEFINITION 5.1 Let **T** be any linear operator on the vector space $\mathcal{V}$ and let λ be a scalar. A nonzero vector **v** in $\mathcal{V}$ such that $\mathbf{T}(\mathbf{v}) = \lambda\mathbf{v}$ is called an **eigenvector (characteristic vector)** of **T**; the scalar λ is called the **eigenvalue (characteristic value)** of **T**, associated with the eigenvector **v**. □

■ EXAMPLE 1 Eigenvectors Exist

The vector $X = \begin{bmatrix} 1 \\ 1 \\ 1 \\ 1 \end{bmatrix}$ is an eigenvector of $A = \begin{bmatrix} 2 & 1 & 0 & 1 \\ 1 & 2 & 1 & 0 \\ 0 & 1 & 2 & 1 \\ 1 & 0 & 1 & 2 \end{bmatrix}$ since

$$AX = \begin{bmatrix} 4 \\ 4 \\ 4 \\ 4 \end{bmatrix} = 4X.$$

Students who have studied calculus will observe that the familiar differentiation formulas

$$\frac{d}{dt}(e^{at}) = ae^{at} \qquad \text{and} \qquad \frac{d^2}{dt^2}(\sin at) = -a^2 \sin at$$

say that the exponential function e^{at} is an eigenvector of the derivative operator with eigenvalue a and that the sine function, $\sin at$, is an eigenvector of the second derivative operator with eigenvalue $-a^2$. Note that the derivative operators have an infinite number of eigenvectors and eigenvalues. ■

Theorem 4.8 tells us that if $\mathcal{V} = \mathfrak{R}^n$, then **T** is a matrix transformation, $\mathbf{T}(\mathbf{v}) = A\mathbf{v}$, so an eigenvector of the transformation **T** must satisfy $\mathbf{T}(\mathbf{v}) = A\mathbf{v} = \lambda\mathbf{v}$. Thus, **v** is an eigenvector of the matrix A associated with the eigenvalue λ, and we see that Definition 5.1 is indeed a generalization of Definition 2.5. In our next example we will review the computational techniques for finding eigenvalues and eigenvectors of a square matrix. These techniques were introduced in Section 2.4.

■ EXAMPLE 2 Eigenvalues and Eigenvectors of a Matrix

Let **T** be the matrix transformation on $\mathfrak{R}^3$ whose standard matrix is

$$A = \begin{bmatrix} 2 & 1 & 1 \\ 2 & 3 & 2 \\ 1 & 1 & 2 \end{bmatrix}; \quad \text{that is, } \mathbf{T}(\mathbf{v}) = A\mathbf{v}.$$

Find the eigenvalues and eigenvectors of **T**.

Since λ is an eigenvalue of A if, and only if, $\det(\lambda I - A) = 0$, we first compute $\det(\lambda I - A)$, the characteristic polynomial of A.

$$\begin{aligned} c(\lambda) = \det(\lambda I - A) &= \det\begin{bmatrix} \lambda - 2 & -1 & -1 \\ -2 & \lambda - 3 & -2 \\ -1 & -1 & \lambda - 2 \end{bmatrix} \\ &= (\lambda - 2)\det\begin{bmatrix} \lambda - 3 & -2 \\ -1 & \lambda - 2 \end{bmatrix} + \det\begin{bmatrix} -2 & -2 \\ -1 & \lambda - 2 \end{bmatrix} \\ &\quad - \det\begin{bmatrix} -2 & \lambda - 3 \\ -1 & -1 \end{bmatrix} \\ &= (\lambda - 2)[(\lambda - 3)(\lambda - 2) - 2] - (-1)[-2(\lambda - 2) - 2] \\ &\quad + (-1)[2 + (\lambda - 3)] \\ &= \lambda^3 - 7\lambda^2 + 11\lambda - 5 \end{aligned}$$

To find the eigenvalues of A, we need to find the roots of the characteristic equation $c(\lambda) = 0$. The only rational possibilities are the factors of -5; that is, ± 1 and ± 5. Direct evaluation (synthetic division is an efficient way to do the evaluations) leads quickly to the factorization

$$c(\lambda) = \lambda^3 - 7\lambda^2 + 11\lambda - 5 = (\lambda - 5)(\lambda - 1)(\lambda - 1),$$

so the eigenvalues of **T** are $\lambda_1 = 5$ and $\lambda_2 = \lambda_3 = 1$.

In order to find an eigenvector for $\lambda_1 = 5$ we need to solve the homogeneous system $(\lambda_1 I - A)X = 0$. We see from the row reduction

$$\begin{aligned} [5I - A \mid 0] &= \left[\begin{array}{rrr|r} 3 & -1 & -1 & 0 \\ -2 & 2 & -2 & 0 \\ -1 & -1 & 3 & 0 \end{array}\right] \to \left[\begin{array}{rrr|r} 0 & 4 & -8 & 0 \\ 0 & -4 & 8 & 0 \\ 1 & 1 & -3 & 0 \end{array}\right] \\ &\to \left[\begin{array}{rrr|r} 1 & 1 & -3 & 0 \\ 0 & 1 & -2 & 0 \\ 0 & 0 & 0 & 0 \end{array}\right] \to \left[\begin{array}{rrr|r} 1 & 0 & -1 & 0 \\ 0 & 1 & -2 & 0 \\ 0 & 0 & 0 & 0 \end{array}\right] \\ &\Rightarrow \begin{cases} x_1 - x_3 = 0 \\ x_2 - 2x_3 = 0 \end{cases} \end{aligned}$$

that the general solution of $(5I - A)X = 0$ is $X = c_1\begin{bmatrix} 1 \\ 2 \\ 1 \end{bmatrix} = c_1\mathbf{v}_1$. Thus, an eigenvector of A associated with $\lambda_1 = 5$ is $\mathbf{v}_1 = \begin{bmatrix} 1 \\ 2 \\ 1 \end{bmatrix}$.

To find the eigenvectors for $\lambda = 1$ we solve the homogeneous system $(I - A)X = 0$ via the row reduction

$$[I - A|0] = \begin{bmatrix} -1 & -1 & -1 & 0 \\ -2 & -2 & -2 & 0 \\ -1 & -1 & -1 & 0 \end{bmatrix} \rightarrow \begin{bmatrix} 1 & 1 & 1 & 0 \\ 0 & 0 & 0 & 0 \\ 0 & 0 & 0 & 0 \end{bmatrix}$$

$$\Rightarrow x_1 + x_2 + x_3 = 0.$$

The general solution of $(I - A)X = 0$ is thus $X = c_2\begin{bmatrix} -1 \\ 1 \\ 0 \end{bmatrix} + c_3\begin{bmatrix} -1 \\ 0 \\ 1 \end{bmatrix}$. In this case we see that the set of all eigenvalues of the linear operator **T** (or of the matrix A), corresponding to the repeated eigenvalue $\lambda = 1$, is the two-dimensional subspace of $\Re^3$ spanned by the eigenvectors

$$\mathbf{v}_2 = \begin{bmatrix} -1 \\ 1 \\ 0 \end{bmatrix} \quad \text{and} \quad \mathbf{v}_3 = \begin{bmatrix} -1 \\ 0 \\ 1 \end{bmatrix}. \quad ■$$

The study of eigenvalues and eigenvectors is closely related to the study of a relationship on square matrices called similarity which was first introduced in Definition 4.7.

DEFINITION 5.2 The $n \times n$ matrices A and B are said to be **similar** ($A \approx B$) if there exists a nonsingular matrix Q such that $Q^{-1}AQ = B$. □

Similarity, although a less restrictive relation than equality, shares three important properties with the relations of equality and of row equivalence: similarity is *reflexive* ($A \approx A$), *symmetric* (if $A \approx B$, then $B \approx A$), and *transitive* (if $A \approx B$ and $B \approx C$, then $A \approx C$) (see Exercise 27 of this section).

In order to see how similarity is related to the study of eigenvalues and eigenvectors we suppose that $B = Q^{-1}AQ$ and that $AX = \lambda X$, that is X is an eigenvalue of A associated with the eigenvalue λ. If we make the substitution $X = QY$ and then multiply on the left by Q^{-1} we obtain

$$AX = \lambda X \Rightarrow AQY = \lambda QY \Rightarrow Q^{-1}AQY = \lambda Q^{-1}QY = \lambda Y \Rightarrow BY = \lambda Y.$$

This last calculation shows that λ is also an eigenvalue of B and that the associated eigenvector is $Y = QX$. Thus similar matrices have the same eigenvalues, but not the same eigenvectors. More is true as the next theorem asserts.

THEOREM 5.1 If A and $B = Q^{-1}AQ$ are similar matrices, then A and B have the same:

(1) characteristic polynomial,
(2) determinant,
(3) trace, and
(4) rank.

Moreover $B^k = Q^{-1}A^kQ$, so B^k and A^k are also similar for any positive integer k.

Proof Using Property 2.9 of the determinant function and the fact that $I = Q^{-1}IQ$ we see that the characteristic polynomial of B is

$$\begin{aligned}\det(\lambda I - B) &= \det(\lambda I - Q^{-1}AQ) = \det(\lambda Q^{-1}IQ - Q^{-1}AQ)\\ &= \det[Q^{-1}(\lambda I - A)Q] = \det(Q^{-1})\det(\lambda I - A)\det(Q)\\ &= \det(Q^{-1})\det(Q)\det(\lambda I - A) = \det(\lambda I - A).\end{aligned}$$

Thus A and B have the same characteristic polynomial and hence the same eigenvalues. Since $\det(A)$ and $\operatorname{tr}(A)$ are coefficients in the characteristic equation of A, the second and third assertions follow at once. Since similarity is a special case of equivalence and rank is invariant under equivalence, it follows that rank is also invariant under similarity.

Finally $B^2 = BB = Q^{-1}AQQ^{-1}AQ = Q^{-1}A^2Q$ and similar calculations show that $B^k = Q^{-1}A^kQ$ for any positive integer k. □

If you have read optional Section 4.7, you will note that Definition 5.2 is a repeat of Definition 4.7. In Section 4.7 we showed that if $\mathbf{T}$ is a linear operator on $\mathcal{V}$ and

$$\mathcal{B} = \{\mathbf{u}_1, \cdots, \mathbf{u}_n\} \qquad \text{and} \qquad \mathcal{B}' = \{\mathbf{v}_1, \cdots, \mathbf{v}_n\}$$

are two different bases for $\mathcal{V}$, then $[\mathbf{T}]_{\mathcal{B}'} = P^{-1}AP$ is the matrix representative of $\mathbf{T}$ with respect to the $\mathcal{B}'$ basis where $A = [\mathbf{T}]_{\mathcal{B}}$ and P is the $\mathcal{B}'$ to $\mathcal{B}$ change-of-basis matrix. Thus similar matrices represent the same linear operator relative to different bases of the underlying space. The discussion in Section 4.7, which we have just summarized, provides geometric motivation for the study of similarity. Additional motivation can be found in Example 5 of this section.

We will now investigate the possibility that a given matrix A be similar to a diagonal matrix.

THEOREM 5.2 The $n \times n$ matrix A is similar to a diagonal matrix

$$P^{-1}AP = \begin{bmatrix} \lambda_1 & & & & \\ & \lambda_2 & & & \\ & & \cdot & & \\ & & & \cdot & \\ & & & & \lambda_n \end{bmatrix} = \mathrm{Dg}\{\lambda_1, \lambda_2, \cdots, \lambda_n\},$$

if, and only if, the columns of P are a linearly independent set of eigenvectors for A; that is

$$A\,\mathrm{Col}_i(P) = \lambda_i\,\mathrm{Col}_i(P), \quad i = 1, 2, \cdots, n.$$

Proof Suppose that $P^{-1}AP = D = \mathrm{Dg}\{\lambda_1, \lambda_2, \ldots, \lambda_n\}$ so that $AP = PD$. If we partition P into its columns and look at this last equation column by column we see that

$$\mathrm{Col}_i(AP) = A\,\mathrm{Col}_i(P) = \mathrm{Col}_i(PD) = \lambda_i\,\mathrm{Col}_i(P).$$

Thus the columns of P are eigenvectors of A. Since P is nonsingular, its columns must be linearly independent and hence must form a basis for $\mathfrak{R}^n$. Conversely, let $X_1, X_2, \ldots, X_n$ be a linearly independent set of eigenvectors of A with $AX_i = \lambda_i X_i$. As in the first part of the proof, the matrix $P = [X_1 \quad X_2 \quad \cdots \quad X_n]$ satisfies $AP = PD$. Since the columns of P are linearly independent, the matrix P must be nonsingular and it follows that $P^{-1}AP = D$ is a diagonal matrix. □

The next theorem describes when a linear operator is represented by a diagonal matrix.

THEOREM 5.3 Let $\mathbf{T}$ be a linear operator on an n-dimensional vector space $\mathcal{V}$. If there exists a basis $\mathcal{B}' = \{\mathbf{v}_1, \mathbf{v}_2, \ldots, \mathbf{v}_n\}$ for $\mathcal{V}$, relative to which $\mathbf{T}$ is represented by the diagonal matrix $D = \mathrm{Dg}\{\lambda_1, \lambda_2, \ldots, \lambda_n\}$ then each of the basis vectors of $\mathcal{B}'$ is an eigenvector of $\mathbf{T}$; that is, $\mathbf{T}(\mathbf{v}_i) = \lambda_i \mathbf{v}_i$; $i = 1, 2, \ldots, n$.

Proof Suppose that $[\mathbf{T}]_{\mathcal{B}'}$, is a diagonal matrix:

$$[\mathbf{T}]_{\mathcal{B}'} = \begin{bmatrix} \lambda_1 & & & \\ & \lambda_2 & & \\ & & \ddots & \\ & & & \lambda_n \end{bmatrix} = \mathrm{Dg}\{\lambda_1, \lambda_2, \cdots, \lambda_n\}.$$

It follows, from the way that the matrix representative is constructed, that

$$\mathbf{T}(\mathbf{v}_1) = \lambda_1 \mathbf{v}_1, \quad \mathbf{T}(\mathbf{v}_2) = \lambda_2 \mathbf{v}_2, \cdots, \quad \mathbf{T}(\mathbf{v}_n) = \lambda_n \mathbf{v}_n;$$

that is, the basis $\mathcal{B}'$ consists of eigenvectors of the linear operator $\mathbf{T}$ and the diagonal elements of $[\mathbf{T}]_{\mathcal{B}'}$ are the eigenvalues of $\mathbf{T}$. □

Using the results of Section 4.7 it can be seen that Theorems 5.2 and 5.3 are corollaries one of the other. Exercise 38 discusses this equivalence.

EXAMPLE 3 Similarity to a Diagonal Matrix

Find a nonsingular matrix P such that $P^{-1}AP$ is a diagonal matrix, where $A = \begin{bmatrix} 2 & 1 & 1 \\ 2 & 3 & 2 \\ 1 & 1 & 2 \end{bmatrix}$ is the matrix of Example 1.

By Theorem 5.2 the columns of P must be eigenvectors of A. Using the eigenvectors found in Example 1 we see that a suitable matrix P is

$$P = \begin{bmatrix} 1 & -1 & -1 \\ 2 & 1 & 0 \\ 1 & 0 & 1 \end{bmatrix} = [\mathbf{v}_1 \quad \mathbf{v}_2 \quad \mathbf{v}_3].$$

For this matrix P we compute $\det(P) = 4$ so that P is nonsingular. As a consequence of Theorem 5.2, we then have

$$P^{-1}AP = \begin{bmatrix} \lambda_1 & 0 & 0 \\ 0 & \lambda_2 & 0 \\ 0 & 0 & \lambda_3 \end{bmatrix} = \begin{bmatrix} 5 & 0 & 0 \\ 0 & 1 & 0 \\ 0 & 0 & 1 \end{bmatrix}.$$

Since eigenvectors are not unique, the matrix P is not unique. Note in particular that rearranging the columns of P would rearrange the diagonal entries of $P^{-1}AP$. ■

Since the operator equation $\mathbf{T}(\mathbf{v}) = \lambda\mathbf{v}$ is equivalent to the equation $(\lambda\mathbf{I} - \mathbf{T})(\mathbf{v}) = \mathbf{0}$, we see that the set of all eigenvectors of $\mathbf{T}$, associated with the eigenvalue λ, is the subspace $\mathcal{NS}(\lambda\mathbf{I} - \mathbf{T})$ of $\mathcal{V}$. These subspaces will be called the **eigenspaces** of $\mathbf{T}$. Note that $\mathcal{NS}(\mathbf{T}) = \mathcal{NS}(-\mathbf{T})$ is the eigenspace associated with the eigenvalue $\lambda = 0$; thus the linear operator $\mathbf{T}$ is singular if, and only if, $\lambda = 0$ is an eigenvalue of $\mathbf{T}$.

A very important property of eigenvectors, which was not included in Section 2.4, is contained in the next theorem.

THEOREM 5.4 Let $\mathbf{T}$ be a linear operator on the vector space $\mathcal{V}$. Eigenvectors associated with distinct eigenvalues of $\mathbf{T}$ are linearly independent.

Proof Let $\lambda_1, \lambda_2, \ldots, \lambda_k$ be distinct eigenvalues of $\mathbf{T}$ and let $\mathbf{v}_1, \mathbf{v}_2, \ldots, \mathbf{v}_k$ be the associated eigenvectors; that is, $\mathbf{T}(\mathbf{v}_i) = \lambda_i\mathbf{v}_i$ and each $\mathbf{v}_i \neq \mathbf{0}$.

If $\{\mathbf{v}_1, \mathbf{v}_2, \ldots, \mathbf{v}_k\}$ is a linearly dependent set, then there must be a first vector $\mathbf{v}_s$ in the list such that $\{\mathbf{v}_1, \ldots, \mathbf{v}_{s-1}\}$ is independent and $\{\mathbf{v}_1, \ldots, \mathbf{v}_s\}$ is dependent. Thus, $\mathbf{v}_s$ is in $\text{SPAN}\{\mathbf{v}_1, \ldots, \mathbf{v}_{s-1}\}$ and there exist scalars a_i such that

$$\mathbf{v}_s = a_1\mathbf{v}_1 + a_2\mathbf{v}_2 + \cdots + a_{s-1}\mathbf{v}_{s-1}. \tag{5.1}$$

Taking the image of Equation (5.1) under the transformation $\mathbf{T}$ yields

$$\mathbf{T}(\mathbf{v}_s) = a_1\mathbf{T}(\mathbf{v}_1) + a_2\mathbf{T}(\mathbf{v}_2) + \cdots + a_{s-1}\mathbf{T}(\mathbf{v}_{s-1})$$

which, using the fact that the $\mathbf{v}_i$ are eigenvectors of $\mathbf{T}$, becomes

$$\lambda_s\mathbf{v}_s = a_1\lambda_1\mathbf{v}_1 + a_2\lambda_2\mathbf{v}_2 + \cdots + a_{s-1}\lambda_{s-1}\mathbf{v}_{s-1}. \tag{5.2}$$

Multiplying Equation (5.1) by λ_s yields the additional relation

$$\lambda_s\mathbf{v}_s = a_1\lambda_s\mathbf{v}_1 + a_2\lambda_s\mathbf{v}_2 + \cdots + a_{s-1}\lambda_s\mathbf{v}_{s-1}. \tag{5.3}$$

Now, subtracting Equation (5.3) from Equation (5.2) yields the equation

$$\mathbf{0} = a_1(\lambda_1 - \lambda_s)\mathbf{v}_1 + a_2(\lambda_2 - \lambda_s)\mathbf{v}_2 + \cdots + a_{s-1}(\lambda_{s-1} - \lambda_s)\mathbf{v}_{s-1}.$$

Since the set $\{\mathbf{v}_1, \mathbf{v}_2, \ldots, \mathbf{v}_{s-1}\}$ is linearly independent, each coefficient in the last equation must vanish. Thus, $a_i(\lambda_i - \lambda_s) = 0$ and, since $\lambda_i \neq \lambda_s$ for $i = 1, 2, \ldots, s - 1$, it follows that each $a_i = 0$ and hence, from Equation (5.1), that $\mathbf{v}_s = \mathbf{0}$. This is a contradiction that follows from our assumption that the set $\{\mathbf{v}_1, \mathbf{v}_2, \ldots, \mathbf{v}_k\}$ is dependent. That assumption must then be incorrect and so it follows that the set $\{\mathbf{v}_1, \mathbf{v}_2, \ldots, \mathbf{v}_k\}$ must be linearly independent. □

Theorem 5.4 allows us to conclude, in the case of distinct eigenvalues, that the matrix P, whose columns are eigenvectors of the matrix A, must be nonsingular and its columns form a basis for $\Re^n$. In this case we can be certain that A is similar to a diagonal matrix; $P^{-1}AP = D$.

When the matrix A has repeated eigenvalues it may or may not be possible to find a set of eigenvectors large enough to diagonalize A. Examples 2 and 3 illustrate the case where A is diagonalizable even though it has a repeated eigenvalue. If the $n \times n$ matrix A has repeated eigenvalues, then it may happen that A does not have enough linearly independent eigenvectors to make a basis for $\Re^n$. We illustrate this case in our next example.

EXAMPLE 4 A Nondiagonalizable Matrix

If possible, diagonalize $J = \begin{bmatrix} 2 & 1 & 0 \\ 0 & 2 & 1 \\ 0 & 0 & 2 \end{bmatrix}$.

If $P^{-1}JP$ is to be a diagonal matrix, then the columns of P must be three linearly independent eigenvectors of J. Since J and $\lambda I - J$ are upper triangular, it follows that the characteristic polynomial of J is:

$$\det(\lambda I - J) = \det\begin{bmatrix} \lambda - 2 & -1 & 0 \\ 0 & \lambda - 2 & -1 \\ 0 & 0 & \lambda - 2 \end{bmatrix} = (\lambda - 2)^3,$$

so $\lambda = 2$ is a repeated eigenvalue of J. The system $(2I - J)X = 0$ is $\begin{bmatrix} 0 & -1 & 0 \\ 0 & 0 & -1 \\ 0 & 0 & 0 \end{bmatrix}\begin{bmatrix} x_1 \\ x_2 \\ x_3 \end{bmatrix} = \begin{bmatrix} 0 \\ 0 \\ 0 \end{bmatrix}$, for which $X = c_1\begin{bmatrix} 1 \\ 0 \\ 0 \end{bmatrix}$ is the general solution. Thus, the eigenspace of J, for $\lambda = 2$, is only one-dimensional and we cannot find three linearly independent eigenvectors. Hence, J is not similar to a diagonal matrix. ■

We summarize this discussion in the final theorem of this section.

THEOREM 5.5 An $n \times n$ matrix A is similar to a diagonal matrix if, and only if, A has n linearly independent eigenvectors. If the eigenvalues of A are distinct, then A is similar to a diagonal matrix. □

In our next example we show how the results of this section can be used to uncouple the equations in a system of first-order linear differential equations. This example requires some knowledge of calculus.

EXAMPLE 5 **Coupled System of Linear Differential Equations**

Let $Y(t) = [y_1(t) \quad y_2(t) \quad \cdots \quad y_n(t)]^T$ be an $n \times 1$ matrix of unknown functions that satisfy the differential equations

$$\begin{aligned} y_1'(t) &= a_{11}y_1(t) + a_{12}y_2(t) + \cdots + a_{1n}y_n(t) \\ y_2'(t) &= a_{21}y_1(t) + a_{22}y_2(t) + \cdots + a_{2n}y_n(t) \\ &\vdots \\ y_n'(t) &= a_{n1}y_1(t) + a_{n2}y_2(t) + \cdots + a_{nn}y_n(t) \end{aligned}$$

as well as the initial conditions $y_i(0) = c_i$, $i = 1, 2, \ldots, n$.

We can write these equations as the single matrix equation

$$Y'(t) = AY(t) \quad \text{with initial condition} \quad Y(0) = C.$$

The solution of this system is easy if the coefficient matrix A is diagonal since it is known from calculus that the solution of $y'(t) = ay(t)$; $y(0) = c$ is the exponential function $y = ce^{at}$. If A is a diagonal matrix, then we say that the equations of the system are uncoupled. If the equations are coupled (that is, if A is not a diagonal matrix) we can try to uncouple them by making a linear substitution of the form $Y(t) = QZ(t)$, where Q is an $n \times n$ nonsingular matrix. It follows that $Y'(t) = QZ'(t)$, so the system $Y' = AY$ reduces to $QZ'(t) = AQZ(t)$ or equivalently, $Z'(t) = Q^{-1}AQZ(t)$ with $Z(0) = Q^{-1}C$. If Q is chosen so that $Q^{-1}AQ = \text{Dg}\{\lambda_1, \lambda_2, \ldots, \lambda_n\}$ is a diagonal matrix, then we have reduced the problem to that of solving the simple equations

$$z_i'(t) = \lambda_i z_i(t); \quad i = 1, 2, \ldots, n.$$

As noted earlier these solutions are $z_i(t) = z_i(0)e^{\lambda_i t}$, so that the solution of the original system is

$$Y(t) = QZ(t) = Q\begin{bmatrix} z_1(0)e^{\lambda_1 t} \\ z_2(0)e^{\lambda_2 t} \\ \vdots \\ z_n(0)e^{\lambda_n t} \end{bmatrix}.$$

If $\lambda = a + bi$ and $\bar{\lambda} = a - bi$ are a pair of complex eigenvalues of A, then you can use the Euler formula, $e^{i\theta} = \cos\theta + i\sin\theta$, to write $e^{\lambda t} = e^{at}(\cos bt + i\sin bt)$ and $e^{\bar{\lambda}t} = e^{at}(\cos bt - i\sin bt)$. The solution of the system will involve the two real functions $e^{at}\cos bt$ and $e^{at}\sin bt$.* ■

*For more details on applications to systems of linear differential equations you may consult C. G. Cullen, *Linear Algebra and Differential Equations, 2d ed.* (Boston: Prindle, Weber & Schmidt, 1992).

The MATALG command "P = evec(A)", the MATLAB command "[P, D] = eig(A)", or the TI-85 command "eigVc(A)", will produce a matrix P such that $AP = PD$. If P is nonsingular, then $P^{-1}AP = D$ where D is a diagonal matrix with the eigenvalues of A on the diagonal. The programs used by these machines will normally produce different, but equivalent, eigenvectors than those obtained by hand calculations.

EXERCISES 5.1

For each of the following matrices find the characteristic polynomial, the eigenvalues, and, if possible, a nonsingular diagonalizing matrix.

1. $\begin{bmatrix} 5 & 1 \\ 4 & 8 \end{bmatrix}$ **2.** $\begin{bmatrix} 0 & 1 \\ -4 & 4 \end{bmatrix}$ **3.** $\begin{bmatrix} 1 & 3 \\ -1 & 5 \end{bmatrix}$ **4.** $\begin{bmatrix} 1 & 7 \\ 3 & -4 \end{bmatrix}$

5. $\begin{bmatrix} 3 & -1 & 1 \\ -1 & 3 & -1 \\ 1 & -1 & 3 \end{bmatrix}$ **6.** $\begin{bmatrix} 5 & 0 & 1 \\ -2 & 2 & 0 \\ -2 & 0 & 2 \end{bmatrix}$ **7.** $\begin{bmatrix} 0 & 1 & 0 \\ 0 & 0 & 1 \\ 20 & -24 & 9 \end{bmatrix}$

8. $\begin{bmatrix} 5 & 6 & 2 \\ 0 & -1 & -8 \\ 1 & 0 & -2 \end{bmatrix}$ **9.** $\begin{bmatrix} 0 & 0 & 4 \\ 1 & 0 & 1 \\ 0 & 1 & -4 \end{bmatrix}$ **10.** $\begin{bmatrix} 2 & -1 & 0 \\ -1 & 2 & -1 \\ 0 & -1 & 2 \end{bmatrix}$

11. $\begin{bmatrix} 0 & 1 & 0 & 0 \\ 0 & 0 & 1 & 0 \\ 0 & 0 & 0 & 1 \\ -36 & 0 & 13 & 0 \end{bmatrix}$ **12.** $\begin{bmatrix} 1 & 1 & 1 & 1 \\ 1 & 1 & 1 & 1 \\ 1 & 1 & 1 & 1 \\ 1 & 1 & 1 & 1 \end{bmatrix}$ **13.** $\begin{bmatrix} 2 & -1 & 0 & 0 \\ -1 & 2 & -1 & 0 \\ 0 & -1 & 2 & -1 \\ 0 & 0 & -1 & 2 \end{bmatrix}$

The following matrices have some complex eigenvalues. Find them and then find a complex diagonalizing matrix.

14. $\begin{bmatrix} 2 & 3 \\ -3 & 2 \end{bmatrix}$ **15.** $\begin{bmatrix} 0 & 1 & 0 \\ 0 & 0 & 1 \\ 12 & -4 & 3 \end{bmatrix}$ **16.** $\begin{bmatrix} \cos\theta & -\sin\theta \\ \sin\theta & \cos\theta \end{bmatrix}$

17. **(a)** Determine how the eigenvalues and eigenvectors of A^k are related to those of A.

(b) Let $p(x)$ be any polynomial. Determine how the eigenvalues and eigenvectors of $p(A)$ are related to those of A.

18. Show that the eigenvectors of $A = \begin{bmatrix} 1 & 0 & 1 \\ 0 & 1 & 2 \\ 1 & 2 & 5 \end{bmatrix}$ are mutually orthogonal.

19. Show that if A is similar to a diagonal matrix, then A and A^T are similar.

20. How can one define the characteristic equation of a linear operator $\mathbf{T}$?

21. The **companion matrix** of the polynomial

$$p(x) = x^n + a_{n-1}x^{n-1} + \cdots + a_2x^2 + a_1x + a_0$$

is the $n \times n$ matrix $C = \begin{bmatrix} 0 & 1 & 0 & \cdots & 0 \\ 0 & 0 & 1 & \cdots & 0 \\ \vdots & \vdots & \vdots & & \vdots \\ 0 & 0 & 0 & \cdots & 1 \\ -a_0 & -a_1 & -a_2 & \cdots & -a_{n-1} \end{bmatrix}$.

Show that the characteristic polynomial of A is $p(\lambda)$. (*Hint:* Use the elementary column operations $C_1 \leftarrow \lambda C_2 + C_1$, $C_1 \leftarrow \lambda^2 C_3 + C_1$, $C_1 \leftarrow \lambda^3 C_4 + C_1, \ldots,$ $C_1 \leftarrow \lambda^{n-1}C_n + C_1$).

22. Let A be an $n \times n$ matrix, set $B_0 = I$, and for $k = 1, 2, 3, \ldots, n$, compute

$$A_k = AB_{k-1}$$

$$c_k = -\left(\frac{1}{k}\right)\text{tr}(A_k)$$

$$B_k = A_k + c_k I.$$

It can be shown that the characteristic polynomial of A is

$$c(\lambda) = \lambda^n + c_1\lambda^{n-1} + \cdots + c_{n-1}\lambda + c_n.$$

This is called the **Souriau-Frame** method and can be used to compute the characteristic equation on a digital computer. Show that this method requires about n^4 multiplications and use this method to compute the characteristic equations of the matrices in parts (a) and (b).

(a) $\begin{bmatrix} 2 & -2 & 3 \\ 1 & 1 & 1 \\ 1 & 3 & -1 \end{bmatrix}$ **(b)** $\begin{bmatrix} 2 & 1 & -1 & 2 \\ 1 & 3 & 2 & -3 \\ -1 & 2 & 1 & -1 \\ 2 & -3 & -1 & 4 \end{bmatrix}$

(c) Write a computer program to find the characteristic equation of an $n \times n$ matrix.

†**23.** Show that the eigenvalues of the real matrix $A = \begin{bmatrix} a & b \\ b & d \end{bmatrix}$ are real.

24. A matrix A is said to be **tridiagonal** if $a_{ij} = 0$ for $j \neq i - 1, i, i + 1$. The matrices of Exercises 10 and 13 are tridiagonal.

(a) Write down a general 5×5 tridiagonal matrix.

(b) Show that the characteristic polynomial of an $n \times n$ tridiagonal matrix is the polynomial $c_n(\lambda)$, where

$$c_0(\lambda) = 1, \qquad c_1(\lambda) = \lambda - a_{11}$$

and, for $i = 2, 3, \ldots, n$,

$$c_i(\lambda) = (\lambda - a_{ii})c_{i-1}(\lambda) - a_{i(i-1)}a_{(i-1)i}c_{i-2}(\lambda).$$

(*Hint:* Expand $\det(\lambda I - A)$ by the last column.)

(c) Use this recursion to find the characteristic polynomials of the matrices in Exercises 10 and 13.

(d) For the matrix of Exercise 13 use the above recursion to evaluate $c_4(3)$, $c_4(2)$, and $c_4(5)$.

25. Use the results of Exercise 3 to solve

$$Y'(t) = \begin{bmatrix} 1 & 3 \\ -1 & 5 \end{bmatrix} Y(t), \quad Y(0) = \begin{bmatrix} 1 \\ 4 \end{bmatrix}.$$

26. Use the results of Exercise 6 to solve

$$Y'(t) = \begin{bmatrix} 5 & 0 & 1 \\ -2 & 2 & 0 \\ -2 & 0 & 2 \end{bmatrix} Y(t), \quad Y(0) = \begin{bmatrix} 2 \\ -3 \\ 5 \end{bmatrix}.$$

27. Verify that similarity is a symmetric, reflexive, and transitive relation on the set of all $n \times n$ matrices.

28. Show that if A and B are similar, then A^T and B^T are also similar. Moreover, if A is nonsingular, then B is also nonsingular and A^{-1} and B^{-1} are similar.

29. For $A = \begin{bmatrix} 5 & 1 \\ 4 & 8 \end{bmatrix}$ and $B = \begin{bmatrix} 1 & 3 \\ -1 & 5 \end{bmatrix}$, find the eigenvalues of A, B, $A + B$, AB, and BA. What conclusions can you draw from this simple example?

30. **(a)** Given that the characteristic polynomial of A is $\lambda^4 + 5\lambda^3 + 3\lambda^2 - 7\lambda + 11$, find det($A$) and tr(A). (*Hint:* Recall Exercise 12 of Section 2.4.)

(b) Given that the characteristic polynomial of B is $\lambda^5 + 3\lambda^4 - 5\lambda^2 - 11\lambda + 5$, find det($B$) and tr($B$).

31. Show that A and A^T have the same characteristic polynomial.

32. Show that A and A^{-1} have the same eigenvectors. How are the eigenvalues related?

33. Find a 3×3 matrix whose eigenvalues are 1, 2, and 3 with corresponding eigenvectors

$$\begin{bmatrix} 1 \\ 1 \\ 1 \end{bmatrix}, \begin{bmatrix} 0 \\ 1 \\ -1 \end{bmatrix}, \begin{bmatrix} 0 \\ 1 \\ 1 \end{bmatrix}.$$

34. Find a 4×4 matrix whose eigenvalues are 2, -3, 0, and 3 with corresponding eigenvectors

$$\begin{bmatrix} 1 \\ 2 \\ 3 \\ -1 \end{bmatrix}, \begin{bmatrix} 4 \\ 2 \\ -1 \\ 0 \end{bmatrix}, \begin{bmatrix} 3 \\ 1 \\ 0 \\ 0 \end{bmatrix}, \begin{bmatrix} 1 \\ 0 \\ 0 \\ 0 \end{bmatrix}.$$

35. A nonnegative matrix is said to be a **stochastic matrix** if the sum of the entries in each column is 1. Find the eigenvalues and eigenvectors of the stochastic matrix $A = \begin{bmatrix} .5 & .2 & .3 \\ .3 & .8 & .3 \\ .2 & 0 & .4 \end{bmatrix}$.

36. Find the characteristic polynomial, the eigenvalues, and the eigenvectors of

$$A = \begin{bmatrix} 5 & 11 & 3 \\ 0 & 2 & 0 \\ 6 & 3 & -2 \end{bmatrix}.$$

37. How are the eigenvalues of A and B related to those of $\begin{bmatrix} A & 0 \\ 0 & B \end{bmatrix}$?

38. Show that Theorem 5.2 is a corollary of Theorem 5.3.

5.2 ORTHOGONAL SIMILARITY AND SYMMETRIC MATRICES

Do the eigenvalues and eigenvectors of symmetric matrices have any special properties?

The eigenvalue problem for an $n \times n$ real matrix A is complicated by the following facts:

(1) Some, or perhaps all, of the eigenvalues of A may be complex.
(2) Complex eigenvalues will lead to complex eigenvectors.* Although the theory is pretty much the same in the complex case, the computations are noticeably more difficult.
(3) Repeated eigenvalues may mean that the matrix is not diagonalizable; this is the case if there are not enough eigenvectors, as illustrated in Example 4 in Section 5.1.

Let $\lambda = a + bi$ be a complex eigenvalue of the real $n \times n$ matrix A and let X be the corresponding complex eigenvector of A. It is easy to see that we can write $X = X_1 + X_2 i$, where X_1 and X_2 are real $n \times 1$ matrices. We must then have

$$AX = A(X_1 + X_2 i) = (a + bi)(X_1 + X_2 i)$$

or

$$AX_1 + AX_2 i = (aX_1 - bX_2) + (bX_1 + aX_2)i.$$

By equating real and imaginary parts in the last equation we obtain the relations

$$AX_1 = aX_1 - bX_2 \qquad \text{and} \qquad AX_2 = bX_1 + aX_2.$$

It then follows that the complex vector $\overline{X} = X_1 - X_2 i$ satisfies

$$\begin{aligned} A\overline{X} = A(X_1 - X_2 i) &= AX_1 - AX_2 i = (aX_1 - bX_2) - (bX_1 + aX_2)i \\ &= (a - bi)(X_1 - X_2 i) = \overline{\lambda}\overline{X}, \end{aligned}$$

where $\overline{\lambda} = a - bi$ is the *complex conjugate* of $\lambda = a + bi$ and $\overline{X} = X_1 - X_2 i$ is the *complex conjugate* of the vector $X = X_1 + X_2 i$. Thus, complex eigenvalues of real matrices must occur in conjugate pairs and the eigenvectors also occur in conjugate pairs; that is, if $AX = \lambda X$, then we must have the conjugate relation $A\overline{X} = \overline{\lambda}\overline{X}$.

*Exercises 1–4 of this section provide a brief review of complex numbers.

EXAMPLE 1 **A Real Matrix with Complex Eigenvalues**

For the 3×3 matrix $A = \begin{bmatrix} 1 & 0 & 1 \\ -4 & 5 & -17 \\ -1 & 1 & -1 \end{bmatrix}$, the characteristic polynomial is

$$\det(\lambda I - A) = \lambda^3 - 5\lambda^2 + 17\lambda - 13 = (\lambda - 1)(\lambda^2 - 4\lambda + 13)$$
$$= (\lambda - 1)((\lambda - 2)^2 + 9).$$

It follows that the eigenvalues of A are $\lambda_1 = 1$, $\lambda_2 = 2 + 3i$, and $\lambda_3 = \overline{\lambda_2} = 2 - 3i$ and the matrix A is an example of a real matrix with some complex eigenvalues. An eigenvector for $\lambda_2 = 2 + 3i$ is $Y_2 = \begin{bmatrix} 1 \\ -5 + 12i \\ 1 + 3i \end{bmatrix} = \begin{bmatrix} 1 \\ -5 \\ 1 \end{bmatrix} + i \begin{bmatrix} 0 \\ 12 \\ 3 \end{bmatrix}$ as one can verify by computing

$$AY_2 = \begin{bmatrix} 2 + 3i \\ -46 + 9i \\ -7 + 9i \end{bmatrix} = (2 + 3i) \begin{bmatrix} 1 \\ -5 + 12i \\ 1 + 3i \end{bmatrix}.$$

Note that any complex scalar multiple of Y_2 is also an eigenvector of A for λ_2. The general discussion above asserts that $Y_3 = \overline{Y_2} = \begin{bmatrix} 1 \\ -5 - 12i \\ 1 - 3i \end{bmatrix} = \begin{bmatrix} 1 \\ -5 \\ 1 \end{bmatrix} - i \begin{bmatrix} 0 \\ 12 \\ 3 \end{bmatrix}$ is an eigenvector of A for $\lambda_3 = \overline{\lambda_2} = 2 - 3i$. This can be verified by the computation

$$AY_3 = \begin{bmatrix} 2 - 3i \\ -46 - 9i \\ -7 - 9i \end{bmatrix} = (2 - 3i) \begin{bmatrix} 1 \\ -5 - 12i \\ 1 - 3i \end{bmatrix}.$$ ■

The symmetric matrices, which arise frequently in applications, are a very important type of matrix with many interesting properties.

DEFINITION 5.3 An $n \times n$ real matrix A is said to be **symmetric** if $A^T = A$, that is, if $a_{ij} = a_{ji}$ for all i and j. □

The matrices

$$A = \begin{bmatrix} 1 & -2 & 3 \\ -2 & 0 & 5 \\ 3 & 5 & 6 \end{bmatrix}, \quad B = \begin{bmatrix} 2 & 3 \\ 3 & 0 \end{bmatrix}, \quad \text{and} \quad C = \begin{bmatrix} 2 & 1 & -1 & 4 \\ 1 & 3 & 2 & -3 \\ -1 & 2 & 1 & -1 \\ 4 & -3 & -1 & 4 \end{bmatrix}$$

are all examples of symmetric matrices. Note that elements that are symmetrically placed with respect to the main diagonal are equal; for example,

$$a_{23} = a_{32} = 5, \quad b_{12} = b_{21} = 3, \quad c_{14} = c_{41} = 4, \quad \text{and so forth.}$$

If λ is an eigenvalue of a symmetric matrix A and X is the associated eigenvector, then, if λ is complex, we have the two relations

$$AX = \lambda X \qquad \text{and} \qquad A\overline{X} = \overline{\lambda}\overline{X}$$

from which we obtain the two equations

$$\overline{X}^T AX = \overline{X}^T(AX) = \overline{X}^T(\lambda X) = \lambda\overline{X}^T X$$

and

$$\overline{X}^T AX = (\overline{X}^T A)X = (A\overline{X})^T X = (\overline{\lambda}\overline{X})^T X = \overline{\lambda}\overline{X}^T X.$$

It follows that

$$\lambda\overline{X}^T X = \overline{\lambda}\overline{X}^T X \qquad \text{or} \qquad (\lambda - \overline{\lambda})\overline{X}^T X = 0.$$

Since $\overline{X}^T X > 0$ by Exercise 4, we see that $\lambda = \overline{\lambda}$; that is, λ is a real number. We state this important property of symmetric matrices as our next theorem.

THEOREM 5.6 The eigenvalues of a symmetric matrix are all real. □

Another useful property of symmetric matrices will be demonstrated next. Some laboratory evidence in support of this theorem was provided in Exercise 18 of Section 5.1

THEOREM 5.7 Eigenvectors of a symmetric matrix A, associated with distinct eigenvalues, must be orthogonal.

Proof Suppose that $AX_1 = \lambda_1 X_1$ and $AX_2 = \lambda_2 X_2$, where $\lambda_1 \neq \lambda_2$. By direct calculation we have

$$X_2^T(AX_1) = X_2^T(\lambda_1 X_1) = \lambda_1(X_2^T X_1)$$

and

$$(X_2^T A)X_1 = (AX_2)^T X_1 = (\lambda_2 X_2)^T X_1 = \lambda_2(X_2^T X_1).$$

Since $\lambda_1 \neq \lambda_2$ it follows that $X_2^T X_1 = X_2 \cdot X_1 = 0$, and hence that the eigenvectors X_1 and X_2 are orthogonal as claimed. □

If the $n \times n$ matrix A is symmetric, with distinct eigenvalues, then Theorem 5.7 says that the eigenvectors of A form a basis for $\Re^n$, whose vectors are mutually orthogonal. If the eigenvectors are scaled so that they are unit vectors, then they will form an orthonormal basis for $\Re^n$. In this case the matrix P, whose columns are the orthonormal eigenvectors of A, will be an orthogonal matrix ($P^T P = I$), and we will have

$$P^T AP = P^{-1}AP = D,$$

where $D = \mathrm{Dg}\{\lambda_1, \ldots, \lambda_n\}$ is the $n \times n$ diagonal matrix whose diagonal entries are the eigenvalues of A.

DEFINITION 5.4 The matrices A and B are said to be **orthogonally similar** if there exists an orthogonal matrix P such that $P^TAP = B$. □

EXAMPLE 2 Orthogonal Similarity to a Diagonal Matrix

Find an orthogonal matrix P such that P^TAP is diagonal if

$$A = \begin{bmatrix} 5 & -2 \\ -2 & 8 \end{bmatrix}.$$

The characteristic polynomial of A is

$$\det(\lambda I - A) = \det\begin{bmatrix} \lambda - 5 & 2 \\ 2 & \lambda - 8 \end{bmatrix} = \lambda^2 - 13\lambda + 36 = (\lambda - 9)(\lambda - 4).$$

To find an eigenvector for $\lambda = 9$, we solve the homogeneous system $(9I - A)X = 0$ using the row reduction

$$\begin{bmatrix} 4 & 2 & 0 \\ 2 & 1 & 0 \end{bmatrix} \to \begin{bmatrix} 2 & 1 & 0 \\ 0 & 0 & 0 \end{bmatrix} \Rightarrow 2x_1 + x_2 = 0.$$

Thus $X_1 = \begin{bmatrix} 1 \\ -2 \end{bmatrix}$ is an eigenvector for $\lambda = 9$. Similarly, the reduction

$$\begin{bmatrix} -1 & 2 & 0 \\ 2 & -4 & 0 \end{bmatrix} \to \begin{bmatrix} 1 & -2 & 0 \\ 0 & 0 & 0 \end{bmatrix}$$ leads to the eigenvector $X_2 = \begin{bmatrix} 2 \\ 1 \end{bmatrix}$

for $\lambda_2 = 4$. Note that $X_1 \cdot X_2 = 0$, as guaranteed by Theorem 5.7. In order to change the basis $\{X_1, X_2\}$ into an orthonormal basis for $\mathfrak{R}^2$ we need only divide each of the basis vectors by its length to obtain the orthonormal basis

$$\left\{X_1' = \frac{1}{\sqrt{5}}\begin{bmatrix} 1 \\ -2 \end{bmatrix}, X_2' = \frac{1}{\sqrt{5}}\begin{bmatrix} 2 \\ 1 \end{bmatrix}\right\}.$$

It now follows that the orthogonal matrix $P = \dfrac{1}{\sqrt{5}}\begin{bmatrix} 1 & 2 \\ -2 & 1 \end{bmatrix}$ is such that

$$P^TAP = \begin{bmatrix} 9 & 0 \\ 0 & 4 \end{bmatrix}.$$ ■

In Chapter 4 we have observed that the change-of-basis matrix between two orthonormal bases of $\mathfrak{R}^n$ is an orthogonal matrix (see Theorem 4.13). If $\mathcal{B}$ and $\mathcal{B}'$ are two orthonormal bases for $\mathfrak{R}^n$ and $\mathbf{T}$ is a linear operator on $\mathfrak{R}^n$, then, from Theorem 4.13, we have

$$[\mathbf{T}]_{\mathcal{B}'} = P^{-1}[\mathbf{T}]_{\mathcal{B}}P = P^T[\mathbf{T}]_{\mathcal{B}}P.$$

This observation proves our next theorem.

THEOREM 5.8 Orthogonally similar matrices represent the same linear operator with respect to different orthonormal bases of $\mathfrak{R}^n$. □

If the matrix A is orthogonally similar to a diagonal matrix D, $P^TAP = D$, then

$$A = PDP^T \qquad \text{and} \qquad A^T = (P^T)^T D^T P^T = PDP^T = A,$$

so A is symmetric.

In Section 5.3 we will show that any symmetric matrix is orthogonally similar to a diagonal matrix; that is, there are always enough independent eigenvectors to form a basis. To find the orthogonal diagonalizing matrix, for a symmetric matrix A, one first finds bases for the eigenspaces of A and then uses the Gram-Schmidt process of Section 3.6 to find orthonormal bases for the eigenspaces of dimension greater than one. The one-dimensional eigenspaces are treated as in Example 2.

EXAMPLE 3 Orthogonal Diagonalization with Repeated Eigenvalues

Find an orthogonal matrix P such that P^TAP is a diagonal matrix if

$$A = \begin{bmatrix} 2 & -1 & 1 \\ -1 & 2 & -1 \\ 1 & -1 & 2 \end{bmatrix}.$$

The characteristic polynomial of A is

$$\begin{aligned} \det(\lambda I - A) &= \det\begin{bmatrix} \lambda - 2 & 1 & -1 \\ 1 & \lambda - 2 & 1 \\ -1 & 1 & \lambda - 2 \end{bmatrix} \\ &= \lambda^3 - 6\lambda^2 + 9\lambda - 4 = (\lambda - 1)^2(\lambda - 4), \end{aligned}$$

so the eigenvalues of A are $\lambda_1 = \lambda_2 = 1$, and $\lambda_3 = 4$. To find the eigenspace $\mathscr{E}_1 = \{X \mid AX = X\}$ we solve the homogeneous system $(I - A)X = 0$ using the row reduction

$$\left[\begin{array}{rrr|r} -1 & 1 & -1 & 0 \\ 1 & -1 & 1 & 0 \\ -1 & 1 & -1 & 0 \end{array}\right] \to \left[\begin{array}{rrr|r} 1 & -1 & 1 & 0 \\ 0 & 0 & 0 & 0 \\ 0 & 0 & 0 & 0 \end{array}\right] \Rightarrow x_1 - x_2 + x_3 = 0.$$

Thus, $\mathscr{E}_1 = \{X \mid x_1 = x_2 - x_3\} = \left\{ c_1\begin{bmatrix} 1 \\ 1 \\ 0 \end{bmatrix} + c_2\begin{bmatrix} -1 \\ 0 \\ 1 \end{bmatrix} \right\}$, so $X_1 = \begin{bmatrix} 1 \\ 1 \\ 0 \end{bmatrix}$ and $X_2 = \begin{bmatrix} -1 \\ 0 \\ 1 \end{bmatrix}$ form a basis for the eigenspace $\mathscr{E}_1$. Note that the vectors X_1 and

X_2 are not orthogonal. The eigenspace $\mathscr{E}_4 = \{X \mid AX = 4X\}$ is determined by solving the system $(4I - A)X = 0$. From the row reduction

$$\left[\begin{array}{ccc|c} 0 & 1 & 1 & 0 \\ 1 & 2 & 1 & 0 \\ 0 & 0 & 0 & 0 \end{array}\right] \to \left[\begin{array}{ccc|c} 1 & 0 & -1 & 0 \\ 0 & 1 & 1 & 0 \\ 0 & 0 & 0 & 0 \end{array}\right] \Rightarrow \begin{cases} x_1 - x_3 = 0 \\ x_2 + x_3 = 0 \end{cases}$$

we see that $\mathscr{E}_4$ is the one-dimensional subspace of $\mathfrak{R}^3$ spanned by $X_3 = [1 \quad -1 \quad 1]^T$. Note that X_3 is orthogonal to both X_1 and X_2, as guaranteed by Theorem 5.7, and hence to any vector in the subspace $\mathscr{E}_1$.

We now use the Gram-Schmidt process of Section 3.6 to find an orthonormal basis for $\mathscr{E}_1$. Thus we need to determine a scalar a so that the vector

$$X_2' = aX_1 + X_2,$$

which is in $\mathscr{E}_1$, is orthogonal to X_1. From the equation

$$0 = X_1 \cdot X_2' = aX_1 \cdot X_1 + X_1 \cdot X_2 = 2a - 1$$

we see that we must take $a = \frac{1}{2}$ and that

$$X_2' = \frac{1}{2}X_1 + X_2 = \frac{1}{2}\begin{bmatrix} -1 \\ 1 \\ 2 \end{bmatrix}$$

is orthogonal to X_1.

We now have the mutually orthogonal set of eigenvectors X_1, X_2', and X_3. Normalizing these vectors yields the following orthonormal set of eigenvectors of A, which form an orthonormal basis for $\mathfrak{R}^3$:

$$Y_1 = \frac{1}{\sqrt{2}}\begin{bmatrix} 1 \\ 1 \\ 0 \end{bmatrix}, \quad Y_2 = \frac{1}{\sqrt{6}}\begin{bmatrix} -1 \\ 1 \\ 2 \end{bmatrix}, \quad \text{and} \quad Y_3 = \frac{1}{\sqrt{3}}\begin{bmatrix} 1 \\ -1 \\ 1 \end{bmatrix}.$$

It follows that

$$P = [Y_1 \quad Y_2 \quad Y_3] = \frac{1}{\sqrt{6}}\begin{bmatrix} \sqrt{3} & -1 & \sqrt{2} \\ \sqrt{3} & 1 & -\sqrt{2} \\ 0 & 2 & \sqrt{2} \end{bmatrix}$$

is an orthogonal matrix ($P^TP = I$) such that

$$P^TAP = \begin{bmatrix} 1 & 0 & 0 \\ 0 & 1 & 0 \\ 0 & 0 & 4 \end{bmatrix}. \quad \blacksquare$$

For a symmetric matrix, the eigenvectors computed by MATLAB are always orthonormal. The eigenvectors computed by MATALG are of unit length but may not be orthogonal if the matrix has repeated eigenvalues. The eigenvectors computed by the TI-85 graphics calculator are not of unit length and may not be orthogonal if the matrix has repeated eigenvalues. MATLAB and the TI-85 can handle complex numbers. MATALG uses only real numbers but will find complex eigenvalues and eigenvectors for a real matrix.

EXERCISES 5.2

†**1.** For the complex numbers $z = 3 + 2i$ and $w = 4 - 5i$, compute the following values.

(a) $z + w$ **(b)** $2z - 3w$ **(c)** $z + \overline{w}$ **(d)** $\overline{z} + \overline{w}$

(e) $\overline{zw}$ **(f)** $\overline{z}\,\overline{w}$ **(g)** $\overline{z}z$ **(h)** $\begin{bmatrix} \overline{z} \\ \overline{w} \end{bmatrix}^T \begin{bmatrix} z \\ w \end{bmatrix}$

†**2.** If $z = a + bi$ and $w = c + di$ are complex numbers show that statements (a) through (e) are true.

(a) $\overline{\overline{z}} = z$ **(b)** $\overline{z + w} = \overline{z} + \overline{w}$ **(c)** $\overline{zw} = \overline{z}\,\overline{w}$ **(d)** $\overline{z}z = a^2 + b^2 \geq 0$

(e) $z = a + bi$ is real $(b = 0)$ if, and only if, $\overline{z} = z$, and z is pure imaginary $(a = 0)$ if, and only if, $\overline{z} = -z$.

(f) Show that $\mathscr{C}$, the set of all complex numbers, is a two-dimensional real vector space.

†**3.** Let A and B be matrices with complex entries.

(a) Show that there exist unique real matrices A_1 and A_2 such that $A = A_1 + A_2 i$.

(b) Let $\overline{A}$ be the matrix obtained from A by taking the complex conjugate of each entry of A and let $A^H = \overline{A}^T$ be the **Hermitian conjugate** of A. Show that $\overline{AB} = \overline{A}\,\overline{B}$ and that $(AB)^H = B^H A^H$.

(c) Show that conjugation and transposition are commutative, that is, $\overline{A}^T = \overline{A^T}$.

†**4.** Let X be an $n \times 1$ matrix with complex entries. Show that

$$\overline{X}^T X = X^H X = \overline{x}_1 x_1 + \overline{x}_2 x_2 + \cdots + \overline{x}_n x_n \geq 0$$

and hence that $X^H X = 0$ if, and only if, $X = 0$.

5. Find the eigenvalues and eigenvectors of the following matrices and verify directly that the eigenvectors are orthogonal.

(a) $\begin{bmatrix} 3 & 1 \\ 1 & 3 \end{bmatrix}$ **(b)** $\begin{bmatrix} 4 & 2 & 2 \\ 2 & 3 & 0 \\ 2 & 0 & 5 \end{bmatrix}$

For each of the following matrices, find an orthogonal diagonalizing matrix.

6. $\begin{bmatrix} 3 & 1 \\ 1 & 3 \end{bmatrix}$ **7.** $\begin{bmatrix} 16 & -12 \\ -12 & 9 \end{bmatrix}$ **8.** $\begin{bmatrix} 2 & -1 \\ -1 & 2 \end{bmatrix}$

9. $\begin{bmatrix} 4 & 2 & 2 \\ 2 & 3 & 0 \\ 2 & 0 & 5 \end{bmatrix}$ **10.** $\begin{bmatrix} 3 & 4 & 2 \\ 4 & 3 & 2 \\ 2 & 2 & 0 \end{bmatrix}$ **11.** $\begin{bmatrix} 2 & 0 & 0 \\ 0 & -2 & 2 \\ 0 & 2 & 1 \end{bmatrix}$

12. $\begin{bmatrix} 2 & 1 & 1 \\ 1 & 1 & 0 \\ 1 & 0 & 1 \end{bmatrix}$ **13.** $\begin{bmatrix} 3 & 1 & 0 & 0 \\ 1 & 3 & 0 & 0 \\ 0 & 0 & 16 & -12 \\ 0 & 0 & -12 & 9 \end{bmatrix}$ **14.** $\begin{bmatrix} 4 & 0 & -1 & 1 \\ 0 & 3 & 0 & 0 \\ -1 & 0 & 4 & -1 \\ 1 & 0 & -1 & 4 \end{bmatrix}$

15. Show that if A is symmetric, then so is P^TAP.

16. Let A be any $m \times n$ matrix.
(a) Show that A^TA is symmetric with nonnegative diagonal elements.
(b) Show that $\mathcal{NS}(A^TA) = \mathcal{NS}(A)$ and hence that $\text{rank}(A^TA) = \text{rank}(A)$.

17. Show that the eigenvalues of $A = \begin{bmatrix} 0 & b \\ -b & 0 \end{bmatrix}$ are pure imaginary.

18. A real matrix A is **skew symmetric** if $A^T = -A$.
(a) Show that the diagonal elements of a skew symmetric matrix are zero.
(b) Show that the eigenvalues of a skew symmetric matrix are pure imaginary.

19. Show that any $n \times n$ real matrix can be written, in a unique way, as the sum of a symmetric matrix and a skew symmetric matrix.

20. An $n \times n$ complex matrix A is called **Hermitian** if $A^H = A$.
(a) Show that the diagonal elements of a Hermitian matrix are real.
(b) Show that the eigenvalues of a Hermitian matrix are real.

21. Let A be an $n \times n$ symmetric matrix whose eigenvalues are all positive. Show that $X^TAX > 0$ for all $X \neq 0$ in $\Re^n$. (*Hint:* Express X as a linear combination of the eigenvectors of A.)

22. **(a)** Assume that A is similar to a diagonal matrix with $\lambda_1, \ldots, \lambda_k$ as distinct eigenvalues. Show that there exist matrices E_i such that

$$A = \lambda_1 E_1 + \lambda_2 E_2 + \cdots + \lambda_k E_k,$$

where $E_1 + E_2 + \cdots + E_k = I$, $E_iE_j = 0$ for $i \neq j$, and $E_i^2 = E_i$. (*Hint:* First consider the case where A is a diagonal matrix.)
(b) Use this decomposition to compute A^r and $\sqrt{A}$.
(c) How many different square roots of A are there?

23. Which of the following matrices are orthogonal?

$$A = \tfrac{1}{3}\begin{bmatrix} 1 & 2 & 2 \\ 2 & 1 & -2 \\ 2 & -2 & 1 \end{bmatrix}, \quad B = \frac{1}{8}\begin{bmatrix} 2 & 3 & -6 \\ 3 & -6 & -2 \\ 6 & 2 & 3 \end{bmatrix},$$

$$C = \begin{bmatrix} 1 & 1 & 1 & 1 & 1 \\ 1 & -1 & 1 & 1 & 1 \\ 1 & 0 & -2 & 1 & 1 \\ 1 & 0 & 0 & -3 & 1 \\ 1 & 0 & 0 & 0 & -4 \end{bmatrix}.$$

24. Show that an orthogonal triangular matrix with positive diagonal entries must be I. Use this fact to show that a factorization $A = QR$, where Q is orthogonal and R is upper triangular with positive diagonal entries, must be unique.

25. Show that the eigenvalues of $A = \begin{bmatrix} a & b \\ c & d \end{bmatrix}$ are

$$\lambda_1 = \frac{1}{2}[(a + d) + \sqrt{(a - d)^2 + 4bc}]$$

and $$\lambda_2 = \frac{1}{2}[(a + d) - \sqrt{(a - d)^2 + 4bc}].$$

Now show that A has two distinct real eigenvalues if $(a - d)^2 + 4bc > 0$, a repeated real eigenvalue if $(a - d)^2 + 4bc = 0$ and two complex eigenvalues if $(a - d)^2 + 4bc < 0$. Express the eigenvectors in terms of λ_1 and λ_2.

26. Show that if a matrix A is diagonalizable, then the rank of A is the number of nonzero eigenvalues. Construct a 2×2 example to show that the above assertion is false if A is not diagonalizable.

27. Find the eigenvalues and eigenvectors of the matrix $A = \begin{bmatrix} a & a & a & a \\ a & a & a & a \\ a & a & a & a \\ a & a & a & a \end{bmatrix}$.

28. Let A be an $n \times n$ matrix for which $AX = \lambda X$. If $m(x)$ is any polynomial, then show that $m(A)X = m(\lambda)X$. What does this result tell you about the eigenvalues of a matrix A such that $A^3 = A$?

29. Find A^{12} if $A = \begin{bmatrix} -18 & 30 \\ -10 & 17 \end{bmatrix}$.

30. Use the software or a graphics calculator to find the eigenvalues of the matrices

$$A = \begin{bmatrix} 10 & 7 & 8 & 7 \\ 7 & 5 & 6 & 5 \\ 8 & 6 & 10 & 9 \\ 7 & 5 & 9 & 10 \end{bmatrix}, \quad B = \begin{bmatrix} 2 & -1 & 0 & 0 \\ -1 & 2 & -1 & 0 \\ 0 & -1 & 2 & -1 \\ 0 & 0 & -1 & 2 \end{bmatrix},$$

$$C = \begin{bmatrix} 5 & 2 & 0 & 0 & 0 \\ 2 & -4 & 3 & 0 & 0 \\ 0 & 3 & 0 & 6 & 0 \\ 0 & 0 & 6 & 4 & 1 \\ 0 & 0 & 0 & 1 & 5 \end{bmatrix}.$$

In each case find the eigenvector matrix P and compute $P^{-1}AP$.

31. Find an orthogonal matrix P such that P^TAP is diagonal if $A = \begin{bmatrix} 1 & 1 & 1 & 1 \\ 1 & 1 & 1 & 1 \\ 1 & 1 & 1 & 1 \\ 1 & 1 & 1 & 1 \end{bmatrix}$.

32. Find the eigenvalues of the following linear operators on $\Re^2$:
(a) A shear in the x-direction with factor k.
(b) A reflection in the line $y = x$.
(c) A contraction in the y-direction by a factor k.
(d) An orthogonal projection onto the x-axis.
(e) A rotation through an angle θ.

5.3 SCHUR'S THEOREM*

What do we know for Schur?

In this section we will prove that every symmetric matrix is orthogonally similar to a diagonal matrix. If A is not symmetric, then, as shown in the last section, it cannot be orthogonally similar to a diagonal matrix. The next theorem describes the simplest matrix that is orthogonally similar to a general matrix A under the additional assumption that all the eigenvalues of A are real. The proof is a bit involved, but we include it because it can be used as the basis for a computer algorithm to do the reduction.

THEOREM 5.9 ***Schur's Theorem***

If A is an $n \times n$ matrix, all of whose eigenvalues are real, then A is orthogonally similar to an upper triangular matrix T; that is, there exists an orthogonal matrix P such that

$$P^T AP = T = \begin{bmatrix} \lambda_1 & t_{12} & t_{13} & \cdots & t_{1n} \\ 0 & \lambda_2 & t_{23} & \cdots & t_{2n} \\ 0 & 0 & \lambda_3 & \cdots & t_{3n} \\ \vdots & \vdots & \vdots & & \vdots \\ 0 & 0 & 0 & \cdots & \lambda_n \end{bmatrix}.$$

Proof The proof is by mathematical induction on n (a loop in the computer program). Since the assertion is obvious for $n = 1$, we need to show that if the theorem is true for $n = r - 1$, then it is also true for $n = r$. It will then follow that the theorem is true for all n. Let A be an $r \times r$ matrix with real eigenvalues and let λ_1 be an eigenvalue of A associated with the eigenvector X_1. We will assume that X_1 has been scaled so that it is a unit vector; thus we have

$$AX_1 = \lambda_1 X_1 \qquad \text{and} \qquad X_1^T X_1 = X_1 \cdot X_1 = 1.$$

Using either Theorem 3.19 or the Gram-Schmidt orthogonalization process, construct an orthogonal matrix P_1 whose first column is X_1. Let $P_1 = [X_1 \quad X_2 \quad \cdots \quad X_r]$ and compute

$$P_1^T AP_1 = \begin{bmatrix} X_1^T \\ X_2^T \\ \vdots \\ X_r^T \end{bmatrix} [AX_1 \quad AX_2 \quad \cdots \quad AX_r] = \begin{bmatrix} X_1^T \\ X_2^T \\ \vdots \\ X_r^T \end{bmatrix} [\lambda_1 X_1 \quad AX_2 \quad \cdots \quad AX_r]$$

*This section is optional.

$$= \begin{bmatrix} \lambda_1 X_1^T X_1 & X_1^T A X_2 & \cdots & X_1^T A X_r \\ \lambda_1 X_2^T X_1 & X_2^T A X_2 & \cdots & X_2^T A X_r \\ \vdots & \vdots & & \vdots \\ \lambda_1 X_r^T X_1 & X_r^T A X_2 & \cdots & X_r^T A X_r \end{bmatrix} = \begin{bmatrix} \lambda_1 & * \\ 0 & A_2 \end{bmatrix}.$$

Note that the subdiagonal elements in the first column are 0 because $X_j^T X_1 = X_j \cdot X_1 = 0$ for $j \neq 1$, by the orthogonality of P_1. Note also that the matrix A_2 is an $(r - 1) \times (r - 1)$ matrix whose eigenvalues are also eigenvalues of A and, hence, real. Since we have assumed that the theorem is true for $n = r - 1$, we can assert the existence of an $(r - 1) \times (r - 1)$ orthogonal matrix P_2 such that

$$P_2^T A_2 P_2 = \begin{bmatrix} \lambda_2 & & & * \\ 0 & \lambda_3 & & \\ \vdots & & \ddots & \\ 0 & 0 & \cdots & \lambda_r \end{bmatrix} = T_2$$

is an upper triangular matrix.

Now the $r \times r$ matrix $P_3 = \begin{bmatrix} 1 & 0 \\ 0 & P_2 \end{bmatrix}$ is also orthogonal and we have

$$P_3^T P_1^T A P_1 P_3 = (P_1 P_3)^T A (P_1 P_3) = \begin{bmatrix} 1 & 0 \\ 0 & P_2^T \end{bmatrix} \begin{bmatrix} \lambda_1 & * \\ 0 & A_2 \end{bmatrix} \begin{bmatrix} 1 & 0 \\ 0 & P_2 \end{bmatrix}$$

$$= \begin{bmatrix} \lambda_1 & * \\ 0 & P_2^T A_2 P_2 \end{bmatrix} = \begin{bmatrix} \lambda_1 & * \\ 0 & T_2 \end{bmatrix} = T.$$

Thus $P = P_1 P_3$ is an orthogonal matrix such that

$$P^T A P = T = \begin{bmatrix} \lambda_1 & t_{12} & t_{13} & \cdots & t_{1r} \\ 0 & \lambda_2 & t_{23} & \cdots & t_{2r} \\ 0 & 0 & \lambda_3 & \cdots & t_{3r} \\ \vdots & \vdots & \vdots & \ddots & \vdots \\ 0 & 0 & 0 & \cdots & \lambda_r \end{bmatrix}$$

is the desired triangular matrix. □

■ EXAMPLE 1 Orthogonal Triangularization

Determine if $A = \begin{bmatrix} -2 & 1 & 0 \\ -4 & 0 & 3 \\ 0 & 1 & -2 \end{bmatrix}$ is orthogonally similar to a triangular matrix, and if it is, find the transforming matrix.

The characteristic polynomial of A is

$$\det(\lambda I - A) = \lambda^3 + 4\lambda^2 + 5\lambda + 2 = (\lambda + 2)(\lambda + 1)^2,$$

so the eigenvalues of A are $\lambda_1 = -2$ and $\lambda_2 = \lambda_3 = -1$. Since the eigenvalues of A are all real, we can be Schur that A is orthogonally similar to a triangular matrix. A unit eigenvector for $\lambda_1 = -2$ is $X_1 = [\frac{3}{5} \ \ 0 \ \ \frac{4}{5}]^T$ and, using Theorem 3.19, we construct the orthogonal matrix

$$P_1 = \left[\begin{array}{c|c} \frac{3}{5} & 0 \quad \frac{4}{5} \\ \hline 0 & \\ \frac{4}{5} & I + \frac{-1}{1-\frac{3}{5}}\begin{bmatrix} 0 & 0 \\ 0 & \frac{16}{25} \end{bmatrix} \end{array}\right] = \frac{1}{5}\begin{bmatrix} 3 & 0 & 4 \\ 0 & 5 & 0 \\ 4 & 0 & -3 \end{bmatrix}.$$

whose first column is X_1. Direct multiplication yields

$$P_1^T A P_1 = \left[\begin{array}{c|cc} -2 & \frac{7}{5} & 0 \\ \hline 0 & 0 & -5 \\ 0 & \frac{1}{5} & -2 \end{array}\right] = \begin{bmatrix} \lambda_1 & * \\ 0 & A_2 \end{bmatrix}.$$

The eigenvalues of A_2 are $\lambda_2 = \lambda_3 = -1$ and a unit eigenvector of A_2 is $X_2 = \frac{1}{\sqrt{26}}\begin{bmatrix} 5 \\ 1 \end{bmatrix}$. The matrix $P_2 = \frac{1}{\sqrt{26}}\begin{bmatrix} 5 & 1 \\ 1 & -5 \end{bmatrix}$ is a 2×2 orthogonal matrix whose first column is X_2. Moreover,

$$P_2^T A_2 P_2 = \begin{bmatrix} -1 & \frac{338}{65} \\ 0 & -1 \end{bmatrix} \text{ is upper triangular.}$$

Thus,

$$P = P_1\begin{bmatrix} 1 & 0 \\ 0 & P_2 \end{bmatrix} = \frac{1}{5\sqrt{26}}\begin{bmatrix} 3\sqrt{26} & 4 & -20 \\ 0 & 25 & 5 \\ 4\sqrt{26} & -3 & 15 \end{bmatrix}$$

is such that P^TAP is upper triangular. Direct calculation yields

$$P^TAP = \begin{bmatrix} 1 & 0 \\ 0 & P_2^T \end{bmatrix}\begin{bmatrix} -2 & \frac{7}{5} & 0 \\ 0 & 0 & -5 \\ 0 & \frac{1}{5} & -2 \end{bmatrix}\begin{bmatrix} 1 & 0 \\ 0 & P_2 \end{bmatrix}$$

$$= \begin{bmatrix} -2 & \frac{7}{\sqrt{26}} & \frac{7}{5\sqrt{26}} \\ 0 & -1 & \frac{338}{65} \\ 0 & 0 & -1 \end{bmatrix}. \quad ■$$

The unpleasant arithmetic encountered in Example 1 is typical of the computations encountered in the orthogonal triangularization algorithm based on the proof of Schur's theorem. For this reason it is best to do the arithmetic by machine, using either a good pocket calculator or a computer program like MATALG or MATLAB.

If A is symmetric, then, by Theorem 5.6, the eigenvalues of A are real and from Theorem 5.9 we have an orthogonal matrix P such that $P^TAP = U$ is an upper triangular matrix. The calculation

$$U^T = (P^TAP)^T = P^TA^T(P^T)^T = P^TAP = U$$

shows that U is symmetric as well as upper triangular; these two properties are compatible only if U is a diagonal matrix. We have thus shown that every symmetric matrix is orthogonally similar to a diagonal matrix. This completes the proof of the following important theorem.

THEOREM 5.10 A real matrix A is orthogonally similar to a diagonal matrix if, and only if, A is symmetric. □

In order to find an orthogonal diagonalizing matrix, for a symmetric matrix A, it is computationally more convenient, at least when working by hand, to use the method described in Example 3 of the last section rather than the method used in Example 1 of this section.

Many applications involve polynomial expressions in a matrix A. If

$$q(x) = a_nx^n + a_{n-1}x^{n-1} + \cdots + a_2x^2 + a_1x + a_0$$

is any polynomial and A is any square matrix, then we define the value of the polynomial $q(x)$ at the matrix A to be the matrix

$$q(A) = a_nA^n + a_{n-1}A^{n-1} + \cdots + a_2A^2 + a_1A + a_0I.$$

We saw in Exercise 13 of Section 4.7 that if A and B are similar, then so are $q(A)$ and $q(B)$; that is, if $B = P^{-1}AP$, then $q(B) = P^{-1}q(A)P$. If A is a diagonal matrix, $A = \mathrm{Dg}\{\lambda_1, \lambda_2, \ldots, \lambda_n\}$ then it is easy to see that

$$q(A) = \mathrm{Dg}\{q(\lambda_1), q(\lambda_2), \cdots, q(\lambda_n)\}.$$

Similarly, if A is upper triangular,

$$A = \begin{bmatrix} \lambda_1 & & & \\ & \lambda_2 & * & \\ & 0 & \ddots & \\ & & & \lambda_n \end{bmatrix}, \text{ then } q(A) = \begin{bmatrix} q(\lambda_1) & & & \\ & q(\lambda_2) & * & \\ & 0 & \ddots & \\ & & & q(\lambda_n) \end{bmatrix}$$

is also an upper triangular matrix (see Exercise 12).

If $c(x) = \det(xI - A)$ is the characteristic polynomial of A, then $c(\lambda_i) = 0$ for each eigenvalue of A. If A is diagonalizable, with

$$P^{-1}AP = D = \mathrm{Dg}\{\lambda_1, \lambda_2, \cdots, \lambda_n\},$$

then

$$c(D) = \mathrm{Dg}\{c(\lambda_1), c(\lambda_2), \cdots, c(\lambda_n)\} = 0$$

and

$$c(A) = P^{-1}c(D)P = P^{-1}0P = 0.$$

We have proven, for the case of a diagonalizable matrix A, the following famous theorem. The proof for the nondiagonalizable case will be omitted.*

THEOREM 5.11 ***Cayley-Hamilton Theorem***

If A is any $n \times n$ matrix and

$$c(\lambda) = \lambda^n + a_{n-1}\lambda^{n-1} + \cdots + a_2\lambda^2 + a_1\lambda + a_0$$

is the characteristic polynomial of A, then $c(A) = 0$; that is,

$$A^n + a_{n-1}A^{n-1} + \cdots + a_2A^2 + a_1A + a_0I = 0. \quad \square$$

EXAMPLE 2 **Illustration of Theorem 5.11**

Verify the Cayley-Hamilton theorem for the matrix

X

$$A = \begin{bmatrix} 2 & 1 & 1 \\ -1 & 2 & -1 \\ -1 & 1 & 3 \end{bmatrix}.$$

The characteristic polynomial of A is

$$\det(\lambda I - A) = \lambda^3 - 7\lambda^2 + 19\lambda - 19,$$

which has two complex roots and one irrational real root. We compute

$$A^2 = \begin{bmatrix} 2 & 5 & 4 \\ -3 & 2 & -6 \\ -6 & 4 & 7 \end{bmatrix} \quad \text{and} \quad A^3 = \begin{bmatrix} -5 & 16 & 9 \\ -2 & -5 & -23 \\ -23 & 9 & 11 \end{bmatrix}$$

and then

$$A^3 - 7A^2 + 19A - 19I = \begin{bmatrix} 0 & 0 & 0 \\ 0 & 0 & 0 \\ 0 & 0 & 0 \end{bmatrix}.$$

This provides some experimental evidence in support of Theorem 5.11.

Note that we can write $A^3 = 7A^2 - 19A + 19I$ so that

$$\begin{aligned} A^4 &= AA^3 = 7A^3 - 19A^2 + 19A \\ &= 7(7A^2 - 19A + 19I) - 19A^2 + 19A \\ &= 49A^2 - 133A + 133I - 19A^2 + 19A = 30A^2 - 114A + 133I \end{aligned}$$

*For a complete proof you may consult C. G. Cullen, *Linear Algebra and Differential Equations, 2d ed.* (Boston: Prindle, Weber & Schmidt, 1992).

and

$$\begin{aligned} A^5 = AA^4 &= 30A^3 - 114A^2 + 133A \\ &= 30(7A^2 - 19A + 19I) - 114A^2 + 133A \\ &= 96A^2 - 437A + 570I. \end{aligned}$$

Clearly we would continue the above process, thus expressing any power of A, and hence any polynomial in A, as a polynomial of degree 2 in A.

We can also write $A^3 - 7A^2 - 19A = -19I$ or $-\frac{1}{19}(A^2 - 7A - 19I)A = I$ so that $A^{-1} = -\frac{1}{19}(A^2 - 7A - 19I)$ is also a second-degree polynomial in A. ■

The arguments used in Example 2 can be applied to any matrix to yield the following useful corollary of the Cayley-Hamilton theorem.

THEOREM 5.12 For any $n \times n$ matrix A the following statements are true:

(1) If A^{-1} exists, it is a polynomial in A of degree less than n.
(2) $A^k (k \geq n)$ is a polynomial in A of degree less than n.
(3) For any polynomial $p(x)$, there exists a polynomial $r(x)$ of degree less than n such that $p(A) = r(A)$. □

Both MATALG and MATLAB have easy ways to evaluate a polynomial $q(x)$ at a matrix A. If the coefficients of the polynomial $q(x)$ are stored in the vector $\mathbf{v}$, with the constant term last, then the MATLAB command "polyvalm(v, A)" will compute $q(A)$. The MATALG command for computing $q(A)$ is "= poly(v, A)."

EXERCISES 5.3

For each of the following matrices, determine if it is orthogonally similar to a triangular matrix, and if it is, find an orthogonal triangularizing matrix.

1. $\begin{bmatrix} 5 & 1 \\ 4 & 8 \end{bmatrix}$ **2.** $\begin{bmatrix} 3 & 2 \\ -2 & 4 \end{bmatrix}$ **3.** $\begin{bmatrix} -2 & 1 & 0 \\ -4 & 0 & 3 \\ 0 & 1 & 2 \end{bmatrix}$

4. $\begin{bmatrix} 0 & 1 & 0 \\ 0 & 0 & 1 \\ 26 & -21 & 6 \end{bmatrix}$ **5.** $\begin{bmatrix} 2 & 1 & 1 \\ 2 & 3 & 2 \\ 1 & 1 & 2 \end{bmatrix}$ **6.** $\begin{bmatrix} \cos\theta & 0 & -\sin\theta \\ 0 & 1 & 0 \\ \sin\theta & 0 & \cos\theta \end{bmatrix}$

7. Show that if A and B are orthogonal matrices, then so are AB and $\begin{bmatrix} A & 0 \\ 0 & B \end{bmatrix}$.

8. Write a computer subroutine that uses Theorem 3.19 to construct an orthogonal matrix whose first column is a scalar multiple of a given vector.

9. Let $\mathbf{w}$ be a unit vector in $\Re^n$ and define $H = I - 2\mathbf{w}\mathbf{w}^T$.
(a) Show that H is both symmetric and orthogonal.
(b) Compute $H\mathbf{w}$.
(c) Show that if $\mathbf{u}$ and $\mathbf{w}$ are orthogonal, then $H\mathbf{u} = \mathbf{u}$.
(d) Use parts (b) and (c) to show that, for $n = 3$, H is a reflection in the plane perpendicular to $\mathbf{w}$. For this reason these matrices are called **elementary reflectors.**
(e) Show that the orthogonal matrix of Theorem 3.19 is an elementary reflector with

$$\mathbf{w} = \frac{-1}{\sqrt{2(1 - x_1)}}\begin{bmatrix} x_1 - 1 \\ Y \end{bmatrix}.$$

10. Assuming the existence of a subroutine EIGEN(n, A, X, r) which will compute one eigenvalue $\lambda = r$ and the corresponding eigenvector X, write a computer program that will triangularize an $n \times n$ matrix.

11. A real matrix is called **normal** if $A^TA = AA^T$.
(a) Show that if A is normal, then so is P^TAP for any orthogonal matrix P.
(b) Show that a normal triangular matrix must be diagonal.
(c) Show, as a corollary of Schur's theorem, that every normal matrix with real eigenvalues is orthogonally similar to a diagonal matrix and hence must be symmetric.
(d) Show that any orthogonal matrix is also a normal matrix.

†12. Let T be an upper triangular matrix and let $p(x)$ be any polynomial.
(a) Show that the diagonal elements of T^k are $(t_{ii})^k$.
(b) Show that the diagonal elements of $p(T)$ are $p(t_{ii})$.

13. Verify the Cayley-Hamilton theorem for each of the matrices in Exercises 1–5.

14. For each of the matrices in Exercises 1–5, express A^{-1} as a polynomial in A.

15. If the polynomial $p(x)$ is divided by the polynomial $c(x)$, the result is a quotient $q(x)$ and a remainder $r(x)$ that is either zero or of degree lower than $c(x)$. These polynomials must satisfy

$$p(x) = q(x)c(x) + r(x).$$

Show that $p(A) = r(A)$ if $c(x)$ is the characteristic polynomial of A.

16. Let $p(x) = x^5 - 2x^4 + x^2 - 7x + 2$ and use the result of Exercise 15 to evaluate $p(A)$ for each of the matrices in Exercises 1–5.

17. Show that the inverse of a symmetric matrix must be symmetric.

†18. Show that if a matrix A is orthogonally similar to a diagonal matrix, then A must be symmetric.

19. Find J^k for $J = \begin{bmatrix} \lambda & 1 & 0 \\ 0 & \lambda & 1 \\ 0 & 0 & \lambda \end{bmatrix}$.

20. An $n \times n$ complex matrix U is called **unitary** if $U^HU = I$.
(a) Show that a real unitary matrix is orthogonal.
(b) Generalize Theorem 3.19 to the complex case. Two complex vectors X and Y are said to be orthogonal if $X^HY = 0$.
(c) Generalize Schur's theorem to the case where A might have some complex eigenvalues.

21. Find a diagonal matrix D such that PD is orthogonal if

$$P_1 = \begin{bmatrix} 2 & 3 & 6 \\ -3 & 6 & -2 \\ 6 & 2 & -3 \end{bmatrix}, \quad P_2 = \begin{bmatrix} 2 & 0 & -2 \\ 2 & 3 & 1 \\ 2 & -3 & 1 \end{bmatrix}, \quad P_3 = \begin{bmatrix} 1 & 3 & 3 & 3 \\ 1 & -5 & 1 & 1 \\ 1 & 1 & -5 & 1 \\ 1 & 1 & 1 & -5 \end{bmatrix}.$$

22. Find an orthogonal matrix Q such that $Q^{-1}CQ$ is diagonal if

$$C = \begin{bmatrix} 1 & 1 & 0 & 0 \\ 1 & 1 & 0 & 0 \\ 0 & 0 & 1 & 1 \\ 0 & 0 & 1 & 1 \end{bmatrix}.$$

23. Find a square root of $A = \begin{bmatrix} 2 & 1 & 1 \\ 2 & 3 & 2 \\ 1 & 1 & 2 \end{bmatrix}$.

24. Show, without finding the characteristic polynomial, that the eigenvalues of

$$\begin{bmatrix} x & y & z \\ x-a & y+a & z \\ x-b & y & z+b \end{bmatrix} \text{ are } a, b, \text{ and } x+y+z.$$

25. The matrix $A = \begin{bmatrix} 16 & 3 & 2 & 13 \\ 5 & 10 & 11 & 8 \\ 9 & 6 & 7 & 12 \\ 4 & 15 & 14 & 1 \end{bmatrix}$ is an example of a **magic square** (all row and column sums are the same). Find the eigenvalues of A and an invertible matrix Q such that $Q^{-1}AQ$ is a diagonal matrix.

5.4 THE POWER METHOD*

How can calculators and computers be used to compute eigenvalues and eigenvectors?

The methods described in the text for computing eigenvalues and eigenvectors, although important for understanding the concepts involved, do not provide good algorithms for use with a modern digital computer. In this section we will describe one method that works well on a computer. In most cases numerical algorithms are designed to produce a sequence of approximations to the solution of a problem rather than an exact solution. The generated sequence should converge to the exact solution of the problem, but the machine is incapable of

*This section is optional and requires some understanding of the limit process encountered in beginning calculus.

carrying the process to the limit; the best we can do is accept one of the terms of the approximating sequence as being "good enough." The error introduced by this approximation is called the **truncation error.** A simple example of such a sequence of approximations is

$$x_0 = 1$$
$$\text{for } n = 0, 1, 2, 3, \cdots$$
$$x_{n+1} = \frac{2}{3}x_n + \frac{a}{3x_n^2},$$

which comes from using Newton's method (recall Exercise 24 of Section 2.4), for the function $f(x) = x^3 - a$, to compute $\sqrt[3]{a}$. For $a = 10$ the terms of this sequence are as follows, correct to six significant figures:

$$x_1 = 4$$
$$x_2 = 2.875$$
$$x_3 = 2.31994$$
$$x_4 = 2.16596$$
$$x_5 = 2.15449$$
$$x_6 = 2.15443$$
$$x_7 = 2.15443.$$

If we accept 2.15443 as a "good enough" approximation, then we could check the accuracy of our approximation by computing the residual

$$10 - (2.15443)^3 = 10 - 9.999934693 = 6.5307 \times 10^{-5},$$

which indicates that our approximation is slightly in error. Now it can be shown that the sequence just listed converges to $\sqrt[3]{10}$, but we have stopped short of the limit and taken an approximation which is correct to six significant figures. If we use the very accurate approximation $\sqrt[3]{10} = 2.15443469003$, we see that the truncation error in our approximation is

$$|\sqrt[3]{10} - 2.15443|, \quad \text{which is about } .00000469003 = 4.69003 \times 10^{-6}.$$

This algorithm may well be the way your pocket calculator computes cube roots.

For more general algorithms the truncation error can only be approximated since the exact solution is unknown. A second kind of error that is always present in machine calculations is **round-off error.** This error comes from the fact that the machine represents every number as a decimal with a finite number of significant figures. Thus the machine can only approximate numbers like

$$\frac{1}{3} = 3.333333333\cdots, \quad \sqrt{2} = 1.414213562\cdots, \text{ and}$$
$$\pi = 3.141592654\cdots.$$

After each calculation the machine either rounds or chops the result to the number of places that is proper for that machine. A well-conceived algorithm should attempt to control both round-off error and truncation error. Detailed error analysis and the design of efficient algorithms is the subject matter of a branch of mathematics called numerical analysis. Readers who are interested in scientific computing should explore this area at the earliest possible opportunity.

If one has a matrix A and a nonzero vector X, then one can check to see if X is an eigenvector of A by comparing the entries of X and AX. If X is not an eigenvector, then it can be shown that AX is, in some sense, "closer" to an eigenvector than was X. This is the basis for the **power method** that we will now describe.

Suppose that the eigenvalues of A are $\{\lambda_1, \lambda_2, \ldots, \lambda_n\}$ and that λ_1 is largest in the sense that $|\lambda_1| > |\lambda_i|$, $i = 2, 3, \ldots, n$. In this case we say that λ_1 is the **dominant eigenvalue** of A.

The power method will normally produce approximations to the eigenvector associated with the dominant eigenvalue of A. The basic algorithm is described here.

POWER METHOD

Y_0 arbitrary nonzero vector
For $i = 0, 1, 2, 3, \cdots$

$$Y_{i+1} = AY_i \tag{5.4}$$

(Y_n approaches a dominant eigenvector of A)

EXAMPLE 1 **The Power Method**

First Solution: Use five iterations of the power method to estimate the dominant eigenvalue and eigenvector for the matrix $A = \begin{bmatrix} 1 & 2 \\ 2 & 3 \end{bmatrix}$.

If we take $Y_0 = \begin{bmatrix} 1 \\ 1 \end{bmatrix}$ as an initial approximation, then we have

$$Y_1 = AY_0 = \begin{bmatrix} 3 \\ 5 \end{bmatrix}, \quad Y_2 = AY_1 = \begin{bmatrix} 13 \\ 21 \end{bmatrix}, \quad Y_3 = AY_2 = \begin{bmatrix} 55 \\ 89 \end{bmatrix},$$

$$Y_4 = AY_3 = \begin{bmatrix} 233 \\ 377 \end{bmatrix}, \quad \text{and} \quad Y_5 = AY_4 = \begin{bmatrix} 987 \\ 1597 \end{bmatrix}.$$

The sequence of Y_i is clearly not converging. Observe, however, that $\frac{987}{233} = 4.23605$ and that $\frac{1597}{377} = 4.23607$. Thus, $Y_5 = AY_4$ is approximately $(4.2361)Y_4$. We conclude that $\lambda = 4.2361$ is an approximation to an eigenvalue of A and that Y_4 is an approximation to the associated eigenvector.

Since any scalar multiple of an eigenvector is also an eigenvector we can scale Y_4 to make the numbers smaller. Thus

$$X_1 = \frac{1}{233}Y_4 = \begin{bmatrix} 1.000 \\ 1.618 \end{bmatrix} \quad \text{is also an approximate eigenvector.}$$

For this simple problem it is easy to check the above calculations. The characteristic polynomial of A is $\lambda^2 - 4\lambda - 1$ and the roots, computed by the quadratic formula, are $\frac{4 \pm \sqrt{20}}{2}$. The largest root, correct to ten places, is $\lambda_1 = 4.236067977$, so we see that five iterations of the power method have produced an approximation that is correct to five significant figures. This degree of accuracy would be adequate for most applications. If more accurate results are needed, then one can compute more terms of the sequence. ■

As one can see, even for the simple matrix of Example 1, the elements of the Y_i may grow quickly. Since a scalar multiple of an eigenvector is still an eigenvector, it is advisable to scale the Y_i in order to keep the entries small. The scaling can be done in a number of ways; when working by hand it is best to do the scaling so that the largest entry of each Y_i is 1. When writing a program for a computer, it is best to scale by $\| Y_i \|$, thus turning each Y_i into a unit vector. In either case the sequence of scale factors approaches the absolute value of the dominant eigenvalue (refer to Exercise 6). If the dominant eigenvalue is negative, the signs of the eigenvector approximations will alternate.

■ POWER METHOD WITH SCALING ■

Y_0 arbitrary nonzero vector
For $i = 0, 1, 2, \cdots$

$$\begin{aligned} Y'_{i+1} &= AY_i \\ s_{i+1} &= \| Y'_{i+1} \| \quad \text{or} \quad s_{i+1} = \max_j \{ | [Y'_{i+1}]_j | \} \\ Y_{i+1} &= \frac{1}{s_{i+1}} Y'_{i+1} \end{aligned} \tag{5.5}$$

(Y_n approaches a dominant eigenvector of A and s_n approaches the absolute value of the dominant eigenvalue of A)

■ EXAMPLE 1 **The Power Method**

Second Solution: We will use the second scaling method of Algorithm (5.5) to obtain an alternate solution of Example 1.

As in the first solution we take $Y_0 = \begin{bmatrix} 1 \\ 1 \end{bmatrix}$ as our initial approximation. Then, using 5.5 we obtain

$$Y'_1 = AY_0 = \begin{bmatrix} 3 \\ 5 \end{bmatrix} \qquad s_1 = 5 \qquad Y_1 = \frac{1}{s_1} Y'_1 = \begin{bmatrix} 0.6 \\ 1.0 \end{bmatrix}$$

$$Y'_2 = AY_1 = \begin{bmatrix} 2.6 \\ 4.2 \end{bmatrix} \qquad s_2 = 4.2 \qquad Y_2 = \frac{1}{s_2} Y'_2 = \begin{bmatrix} 0.619 \\ 1.000 \end{bmatrix}$$

$$Y_3' = AY_2 = \begin{bmatrix} 2.619 \\ 4.238 \end{bmatrix} \qquad s_3 = 4.238 \qquad Y_3 = \begin{bmatrix} 0.61798 \\ 1.00000 \end{bmatrix}$$

$$Y_4' = AY_3 = \begin{bmatrix} 2.61798 \\ 4.23596 \end{bmatrix} \qquad s_4 = 4.23596 \qquad Y_4 = \begin{bmatrix} 0.618036 \\ 1.000000 \end{bmatrix}$$

$$Y_5' = AY_4 = \begin{bmatrix} 2.618036 \\ 4.236074 \end{bmatrix} \qquad s_5 = 4.236074 \qquad Y_5 = \begin{bmatrix} 0.6180335 \\ 1.0000000 \end{bmatrix}$$

Therefore, a dominant eigenvector of A is approximately $X_1 = Y_5$ and the dominant eigenvalue is about $\lambda_1 = 4.23607$. Note that the eigenvector obtained here is an approximate scalar multiple of the one obtained in the first solution. ■

■ EXAMPLE 2 Power Method with Scaling

Use the power method, with the first scaling method of (5.5), to find the dominant eigenvector and corresponding eigenvalue for $A = \begin{bmatrix} 1 & 1 & 3 \\ 1 & -2 & 1 \\ 3 & 1 & 3 \end{bmatrix}$.

We take $Y_0 = \begin{bmatrix} 1 \\ 1 \\ 1 \end{bmatrix}$ and, using either the software or a graphics calculator like the TI-85, compute the following sequence of unit vectors.

$$Y_1' = AY_0 = \begin{bmatrix} 5 \\ 0 \\ 7 \end{bmatrix}; \qquad s_1 = \sqrt{74} = 8.6023; \qquad Y_1 = \frac{1}{s_1} Y_1' = \begin{bmatrix} .5812 \\ 0 \\ .8137 \end{bmatrix}$$

$$Y_2' = \begin{bmatrix} 3.02244 \\ 1.39497 \\ 4.18491 \end{bmatrix}; \qquad s_2 = 5.34739; \qquad Y_2 = \begin{bmatrix} 5.65217 \\ .260870 \\ .782609 \end{bmatrix}$$

$$Y_3' = \begin{bmatrix} 3.17391 \\ .82609 \\ 4.30435 \end{bmatrix}; \qquad s_3 = 5.41143; \qquad Y_3 = \begin{bmatrix} .58652 \\ .15266 \\ .79542 \end{bmatrix}$$

$$Y_4' = \begin{bmatrix} 3.1254307 \\ 1.0766265 \\ 4.2984715 \end{bmatrix}; \qquad s_4 = 5.42257; \qquad Y_4 = \begin{bmatrix} .5763741 \\ .1985453 \\ .7926996 \end{bmatrix}$$

$$Y_5' = \begin{bmatrix} 3.1530182 \\ .9719831 \\ 4.3057664 \end{bmatrix}; \qquad s_5 = 5.42456; \qquad Y_5 = \begin{bmatrix} .58124818 \\ .17918177 \\ .793753390 \end{bmatrix}$$

$$\vdots \qquad\qquad \vdots \qquad\qquad \vdots$$

Continuing this iteration, we obtain, after thirty steps,

$$Y_{29} = Y_{30} = \begin{bmatrix} .5797702392 \\ .1849535588 \\ .7935103344 \end{bmatrix} \quad \text{and} \quad s_{29} = s_{30} = 5.425002162$$

as very accurate estimates of the dominant eigenvector and eigenvalue of the matrix A. ■

To understand how the power method works, we assume that A is a diagonalizable matrix and that $\{X_1, X_2, \ldots, X_n\}$ is a basis for $\mathfrak{R}^n$ consisting of eigenvectors of A, with $AX_i = \lambda_i X_i$, $i = 1, 2, \ldots, n$.

The starting vector Y_0 in Algorithm (5.4) can be written as a linear combination of the above basis vectors:

$$Y_0 = a_1X_1 + a_2X_2 + \cdots + a_nX_n.$$

If we repeatedly multiply this last equation by A we obtain

$$\begin{aligned}
Y_1 &= AY_0 = a_1\lambda_1X_1 + a_2\lambda_2X_2 + \cdots + a_n\lambda_nX_n \\
Y_2 &= AY_1 = a_1\lambda_1^2X_1 + a_2\lambda_2^2X_2 + \cdots + a_n\lambda_n^2X_n \\
&\vdots \\
Y_k &= AY_{k-1} = a_1\lambda_1^kX_1 + a_2\lambda_2^kX_2 + \cdots + a_n\lambda_n^kX_n \\
&= \lambda_1^k\left(a_1X_1 + a_2\left(\frac{\lambda_2}{\lambda_1}\right)^k X_2 + \cdots + a_n\left(\frac{\lambda_n}{\lambda_1}\right)^k X_n\right).
\end{aligned} \tag{5.6}$$

Since λ_1 is the dominant eigenvalue of A, we have, for $i \neq 1$, $|\frac{\lambda_i}{\lambda_1}| < 1$, so that $(\frac{\lambda_i}{\lambda_1})^k$ gets small as k gets large. Thus we see that Y_k approaches $\lambda_1^k a_1 X_1$, a scalar multiple of the dominant eigenvector X_1.

Equation (5.6) is the key to understanding what happens if A does not have a dominant eigenvalue ($\lambda_1 = \pm\lambda_2$) or if the dominant eigenvalue is complex. We leave the details, and some experimentation with the exceptional cases, to the exercises and to a course in numerical analysis.

Many applications require only the dominant eigenvalue. If one needs more eigenvalues, then one can use the technique introduced in the proof of Schur's theorem to transform the $n \times n$ matrix A into $\begin{bmatrix} \lambda_1 & * \\ 0 & A_2 \end{bmatrix}$ where A_2 is an $(n-1) \times (n-1)$ matrix whose eigenvalues are also eigenvalues of A. The power method can then be used to obtain the dominant eigenvalue of A_2 and hence a second eigenvalue of A. In theory, one could use this procedure to find all the eigenvalues of A. In practice, the truncation errors accumulate so that only the first few eigenvalues are found accurately.

There are available very powerful methods designed to accurately compute all of the eigenvalues and eigenvectors of an $n \times n$ matrix A. One such method is the QR algorithm. This algorithm has been implemented in a very fine software package called EISPACK. The details of this method are too advanced for this text, although EISPACK can be used as a "black box" without understanding the details of the method. Both MATALG and MATLAB use EISPACK subroutines to compute eigenvalues and eigenvectors.

EXERCISES 5.4

Use the power method, with the second scaling method of (5.5), to compute the dominant eigenvalue and dominant eigenvector of the matrices in Exercises 1–4.

1. $\begin{bmatrix} 0 & 1 \\ -7 & -6 \end{bmatrix}$ **2.** $\begin{bmatrix} 1 & 1 & 3 \\ 1 & -2 & 1 \\ 3 & 1 & 3 \end{bmatrix}$ **3.** $\begin{bmatrix} -3 & 5 & -2 \\ -6 & 8 & -2 \\ -4 & 4 & 1 \end{bmatrix}$ **4.** $\begin{bmatrix} 0 & 3 & 0 \\ 0 & 0 & 3 \\ 5 & 2 & 2 \end{bmatrix}$

5. Consider $A = \begin{bmatrix} 6 & -4 \\ 8 & -6 \end{bmatrix}$.

(a) Find the characteristic polynomial and show that the eigenvalues are $\lambda_1 = 2$ and $\lambda_2 = -2$. Find eigenvectors for each of these eigenvalues.

(b) Take $Y_0 = [2 \quad 1]^T$ and compute eight terms of the sequence Y_i, using the second scaling method of (5.5). Is this sequence converging?

(c) What is the subsequence $Y_0, Y_2, Y_4, Y_6, \ldots$ converging to? What is the subsequence $Y_1, Y_3, Y_5, Y_7, \ldots$ converging to?

(d) Show that in general the subsequence $Y_0, Y_2, Y_4, \ldots$ can be used to compute the dominant eigenvalue and dominant eigenvector of the $n \times n$ matrix A^2, provided that the dominant eigenvalue of A^2 is real.

(e) Take $A = \begin{bmatrix} -2 & -1 & 4 \\ 2 & 1 & -2 \\ -1 & -1 & 3 \end{bmatrix}$ and $Y_0 = \begin{bmatrix} 3 \\ 1 \\ 2 \end{bmatrix}$ and compute six terms of the sequence defined by (5.4). What do your calculations show?

†**6.** Show that if Y is an eigenvector of A and $\| Y \| = 1$, then $\| AY \|$ is the absolute value of the corresponding eigenvalue.

7. Consider $A = \begin{bmatrix} 2 & 4 \\ -4 & 3 \end{bmatrix}$.

(a) Find the characteristic polynomial and the eigenvalues of A.

(b) Take $Y_0 = [1 \quad 1]^T$ and compute eight terms of the sequence Y_i using the second scaling method of (5.5).

(c) Let $Z_7 = AY_6$ and $Z_8 = AZ_7$ and then compute $Z_8 - 5Z_7 + 22Y_6 = (A^2 - 5A + 22I)Y_6$. Does this procedure suggest a way of estimating the dominant complex eigenvalue of this matrix?

8. Suppose that the eigenvalues of A are $\{\lambda_1, \lambda_2, \ldots, \lambda_n\}$. Show that the eigenvalues of $A - kI$ are $\lambda_i - k$, $i = 1, 2, \ldots, n$. What is the dominant eigenvalue of $A - kI$ if k is an approximation for the dominant eigenvalue of A? Describe how this observation can be used to compute a second eigenvalue and eigenvector of A.

9. Let A be a symmetric matrix with eigenvalues $\lambda_1 > \lambda_2 > \cdots > \lambda_n$, and let X_1 be a unit eigenvector associated with the dominant eigenvalue λ_1. Let $X_1, X_2, \ldots, X_n$ be a full set of eigenvectors for A. Show that the matrix $A_1 = A - \lambda_1 X_1 X_1^T$ has eigenvalues $0, \lambda_2, \lambda_3, \ldots, \lambda_n$ and eigenvectors $X_1, X_2, \ldots, X_n$. Describe how this could be used, in conjunction with the power method, to find all the eigenvectors and all the eigenvalues of A.

10. Write a computer subroutine EIGEN(n, A, X, r) that will take a given matrix A and a given first approximation X and return an approximate eigenvector X and the corresponding eigenvalue r. Stop the iteration when $\| Y_{k+1} - Y_k \| \leq .0001$ or when $k = 50$, whichever comes first.

11. Use the power method to estimate the dominant eigenvalue of each of the following matrices:

$$A = \begin{bmatrix} 10 & 7 & 8 & 7 \\ 7 & 5 & 6 & 5 \\ 8 & 6 & 10 & 9 \\ 7 & 5 & 9 & 10 \end{bmatrix}, \quad B = \begin{bmatrix} 2 & -1 & 0 & 0 \\ -1 & 2 & -1 & 0 \\ 0 & -1 & 2 & -1 \\ 0 & 0 & -1 & 2 \end{bmatrix},$$

$$C = \begin{bmatrix} 5 & 2 & 0 & 0 & 0 \\ 2 & -4 & 3 & 0 & 0 \\ 0 & 3 & 0 & 6 & 0 \\ 0 & 0 & 6 & 4 & 1 \\ 0 & 0 & 0 & 1 & 5 \end{bmatrix}.$$

12. Combine the result of Exercise 10 and the program from Exercise 10 of Section 5.3 to obtain a program which will compute all the real eigenvalues of an $n \times n$ matrix A. Test your program on some of the matrices of this section.

13. What would the power method yield if applied to the matrix A^{-1}? What would the power method yield if applied to the matrix $(A - kI)^{-1}$?

14. For each of the following matrices use $X_0 = [1 \quad 1 \quad 1 \quad 1]^T$ and compute 20 terms of the sequence (5.5). Interpret the results and find as many eigenvalues and eigenvectors as possible from the iteration.

$$A = \begin{bmatrix} 7 & -2 & 2 & 2 \\ 5 & 0 & 2 & 2 \\ -4 & 2 & 2 & -3 \\ -4 & 2 & -1 & 0 \end{bmatrix}, \quad B = \begin{bmatrix} -3 & 6 & -12 & 0 \\ -5 & 8 & -12 & 0 \\ -5 & 4 & -3 & -2 \\ 1 & -2 & 6 & 1 \end{bmatrix},$$

$$C = \begin{bmatrix} -10 & 5 & 12 & -8 \\ -8 & 4 & 12 & -6 \\ -3 & 0 & 5 & -3 \\ 9 & -5 & -6 & 7 \end{bmatrix}$$

15. For $A = \begin{bmatrix} 30 & -4 & -2 & 34 \\ 34 & -8 & -2 & 34 \\ -20 & 4 & 10 & -32 \\ -20 & 4 & 1 & -23 \end{bmatrix}$ and $X_0 = \begin{bmatrix} 1 \\ 2 \\ 1 \\ 0 \end{bmatrix}$ compute 10 terms of the sequence (5.5), using the second scaling method, and then estimate the dominant eigenvalue. Compute another 400 terms of the sequence and estimate the dominant eigenvalue again. Now take $X_0 = [1 \quad 1 \quad 1 \quad 1]^T$, compute 25 terms of (5.5), and interpret the results. Explain what is going on here.

16. Let μ_1 be an approximation to the dominant eigenvalue of A. What is the dominant eigenvalue of $A - \mu_1 I$? Use this observation and the power method to find a second eigenvalue and eigenvector for the matrices A and C of Exercise 14. What happens if we apply the power method to $A + I$?

17. **(a)** Use the software to find the QR factorization of $A = \begin{bmatrix} 1 & 2 & 3 & 4 \\ 2 & 3 & 5 & 6 \\ 3 & 5 & 2 & 7 \\ 4 & 6 & 7 & 1 \end{bmatrix}$.

(b) Compute $A_1 = RQ$ and show that it is orthogonally similar to A.

(c) For $i = 1, 2, 3, 4, 5$

$$\text{Factor } A_i = Q_i R_i$$
$$\text{Compute } A_{i+1} = R_i Q_i$$

(d) Show that A_5 is similar to A.

(e) What seems to be happening to the sequence $\{A_i\}$?

CHAPTER 5 SUMMARY

In Chapter 5 we extend the study of the eigenvalue problem which we began in Section 2.4. Special features of the eigenvalue problem for symmetric matrices are developed. The power method for finding the dominant eigenvalue on a computer is introduced.

KEY DEFINITIONS

Eigenvalues and eigenvectors. *302*
Similarity, orthogonal similarity. *304, 316*
Eigenspaces. *307*
Symmetric matrix. *314*
Dominant eigenvalue. *331*

KEY FACTS

The matrix $P^{-1}AP$ is diagonal if, and only if, the columns of P are a linearly independent set of eigenvectors of A.

Eigenvectors, associated with distinct eigenvalues of A, are always linearly independent.

Complex eigenvalues of real matrices occur in conjugate pairs and the eigenvectors are also conjugates.

The eigenvalues of a symmetric matrix are always real.

For a symmetric matrix, the eigenvectors associated with distinct eigenvalues are orthogonal.

A real matrix, with all eigenvalues real, is orthogonally similar to a triangular matrix. (See Schur's theorem.)

A is orthogonally similar to a diagonal matrix if, and only if, A is symmetric.

Every matrix is a zero of its characteristic polynomial. (See Cayley-Hamilton theorem.)

COMPUTATIONAL PROCEDURES

Find the characteristic polynomial, the eigenvalues, and the eigenvectors of a matrix.
Find a matrix P so that $P^{-1}AP$ is a diagonal matrix.
Find an orthogonal matrix Q so that Q^TAQ is upper triangular.
If A is symmetric, find an orthogonal matrix Q so that Q^TAQ is diagonal.

APPLICATIONS

Uncouple a system of linear differential equations.
Quadratic forms (Section 7.3).

CHAPTER 5 REVIEW EXERCISES

1. Find the characteristic polynomial of $B = \begin{bmatrix} -1 & 1 & 3 \\ 2 & 1 & 1 \\ 4 & 2 & 3 \end{bmatrix}$.

2. For $A = \begin{bmatrix} 3 & -6 \\ -2 & 2 \end{bmatrix}$, find P such that $P^{-1}AP$ is diagonal.

3. Given that the characteristic polynomial of $A = \begin{bmatrix} 3 & -1 & 1 \\ -1 & 3 & -1 \\ 1 & -1 & 3 \end{bmatrix}$ is $(\lambda - 2)^2(\lambda - 5)$,

(a) find a basis for the two-dimensional subspace of all eigenvectors of A for $\lambda = 2$; and
(b) find an orthonormal basis for this subspace.

4. Let $\mathbf{T}$ be a linear operator on $\Re^2$ whose matrix representative with respect to the standard basis is $[\mathbf{T}]_{\mathcal{S}} = \begin{bmatrix} 1 & 1 \\ -2 & 4 \end{bmatrix}$.

(a) Find the matrix representative $[\mathbf{T}]_{\mathcal{B}}$ if

$$\mathcal{B} = \left\{ \mathbf{v}_1 = \begin{bmatrix} 1 \\ 1 \end{bmatrix}, \mathbf{v}_2 = \begin{bmatrix} 1 \\ 2 \end{bmatrix} \right\}.$$

(b) Find $\mathbf{T}(\mathbf{v}_1)$ and $\mathbf{T}(\mathbf{v}_2)$.

5. Given that the eigenvalues of $B = \begin{bmatrix} 2 & 1 & 1 \\ 2 & 3 & 2 \\ 1 & 1 & 2 \end{bmatrix}$ are 5, 1, and 1, find a matrix P so that $P^{-1}AP$ is a diagonal matrix.

6. Let $\mathbf{T}$ be a linear operator on $\Re^{1\times 3}$ defined by

$$\mathbf{T}[x_1 \quad x_2 \quad x_3] = [3x_1 - 5x_2 + 7x_3 \quad 3x_2 - 4x_3 \quad x_1 + x_2 + x_3].$$

(a) Find the matrix A that represents $\mathbf{T}$ with respect to the basis

$$\mathcal{B} = \{\mathbf{u}_1 = [1 \quad 3 \quad 2], \mathbf{u}_2 = [0 \quad 1 \quad 2], \mathbf{u}_3 = [0 \quad 2 \quad 3]\}.$$

(b) Use the matrix from part (a) to find $[\mathbf{T}(\mathbf{x})]_{\mathcal{B}}$ if $\mathbf{x} = \mathbf{u}_1 + 2\mathbf{u}_2 - 3\mathbf{u}_3$.
(c) Let $\mathcal{S}$ be the standard basis for $\mathfrak{R}^{1\times 3}$. Let A' be the matrix which represents $\mathbf{T}$ with respect to $\mathcal{S}$. Find a matrix P such that $A = P^{-1}A'P$.

7. Find the characteristic equation and the eigenvalues of

$$A = \begin{bmatrix} 3 & -4 & 3 \\ -4 & 3 & 0 \\ 3 & 0 & 3 \end{bmatrix}.$$

8. For $A = \begin{bmatrix} 2 & 6 \\ 2 & 1 \end{bmatrix}$, find P such that $P^{-1}AP$ is diagonal.

9. Let $A = \begin{bmatrix} 1 & 3 & 0 \\ 4 & 2 & 1 \\ 2 & 7 & -2 \\ 0 & -1 & 5 \end{bmatrix}$ be the matrix representative of the linear transformation $\mathbf{T}$ with respect to the basis

$$\mathcal{B} = \left\{ \mathbf{u}_1 = \begin{bmatrix} 1 \\ 0 \\ 1 \end{bmatrix}, \mathbf{u}_2 = \begin{bmatrix} 1 \\ 2 \\ 0 \end{bmatrix}, \mathbf{u}_3 = \begin{bmatrix} 1 \\ 1 \\ 1 \end{bmatrix} \right\} \text{ for } \mathfrak{R}^3$$

and the basis

$$\mathcal{B}' = \left\{ \begin{bmatrix} 1 & 1 \\ 1 & 1 \end{bmatrix}, \begin{bmatrix} 1 & 1 \\ 1 & 0 \end{bmatrix}, \begin{bmatrix} 1 & 1 \\ 0 & 0 \end{bmatrix}, \begin{bmatrix} 1 & 0 \\ 0 & 0 \end{bmatrix} \right\} \text{ for } \mathfrak{R}^{2\times 2}.$$

(a) Find $[\mathbf{T}(\mathbf{u}_1)]_{\mathcal{B}'}$.
(b) Find $\mathbf{T}(\mathbf{u}_2)$.
(c) Find $\mathbf{T}\begin{bmatrix} 6 \\ 7 \\ 4 \end{bmatrix}$.

10. Given that the characteristic equation of $D = \begin{bmatrix} 1 & 4 & -3 \\ 4 & 0 & -4 \\ -3 & -4 & 1 \end{bmatrix}$ is
$\lambda^3 - 2\lambda^2 - 40\lambda - 64$,
(a) find the eigenvalues of D;
(b) find a nonsingular matrix P such that $P^{-1}DP$ is a diagonal matrix; and
(c) find an orthogonal matrix Q such that Q^TDQ is a triangular matrix.

11. Find an orthogonal matrix R so that R^TAR is upper triangular if

$$A = \begin{bmatrix} 3 & 1 \\ -1 & 1 \end{bmatrix}.$$

12. Let $A = \begin{bmatrix} 5 & 2 & 3 \\ -4 & 6 & 2 \end{bmatrix}$ be the matrix representative of the linear transformation $\mathbf{T}$: $\mathfrak{R}^3 \to \mathfrak{R}^2$, with respect to the basis

$$\mathcal{B} = \left\{ \mathbf{u}_1 = \begin{bmatrix} 1 \\ 1 \\ 1 \end{bmatrix}, \mathbf{u}_2 = \begin{bmatrix} 1 \\ -1 \\ -1 \end{bmatrix}, \mathbf{u}_3 = \begin{bmatrix} 4 \\ -1 \\ 2 \end{bmatrix} \right\} \text{ for } \mathfrak{R}^3$$

and the basis $\mathscr{B}' = \left\{\mathbf{v}_1 = \begin{bmatrix} 2 \\ 3 \end{bmatrix}, \mathbf{v}_2 = \begin{bmatrix} 3 \\ 5 \end{bmatrix}\right\}$ for $\Re^2$.

(a) Find $[\mathbf{T}(\mathbf{u}_2)]_{\mathscr{B}'}$. **(b)** Find $\mathbf{T}(\mathbf{u}_3)$. **(c)** Find $\mathbf{T}\begin{bmatrix} 6 \\ -1 \\ 2 \end{bmatrix}$.

13. For $A = \begin{bmatrix} 3 & 4 & -3 \\ 4 & 2 & -4 \\ -3 & -4 & 3 \end{bmatrix}$, the characteristic polynomial is $\lambda^3 - 8\lambda^2 - 20\lambda$.
Find an orthogonal matrix Q so that Q^TAQ is a diagonal matrix.

14. (a) Show that if $B = P^{-1}AP$, then $B^k = P^{-1}A^kP$ for any positive integer k.
(b) Let $q(x) = a_0 + a_1x + a_2x^2 + \cdots + a_kx^k$ be any polynomial and define $q(A) = a_0I + a_1A + a_2A^2 + \cdots + a_kA^k$. Show that $q(B) = P^{-1}q(A)P$.

15. Find bases for the eigenspaces of the following triangular matrices.

$$A = \begin{bmatrix} 1 & 3 & -5 \\ 0 & 2 & -10 \\ 0 & 0 & 2 \end{bmatrix}, \quad B = \begin{bmatrix} 1 & 2 & 0 & 0 \\ 0 & 1 & 0 & 0 \\ 0 & 0 & 1 & 3 \\ 0 & 0 & 0 & 1 \end{bmatrix}.$$

16. Let A be an $n \times n$ matrix and let $c(\lambda) = \det(\lambda I - A)$ be the characteristic polynomial of A. Show that the coefficient of λ^{n-1} in $c(\lambda)$ is $-\text{tr}(A)$.

17. Find a nonsingular matrix P such that $P^{-1}AP$ is diagonal if

$$A = \begin{bmatrix} 7 & 4 & 16 \\ 2 & 5 & 8 \\ -2 & -2 & -5 \end{bmatrix}, \quad A = \begin{bmatrix} -1 & 4 & -2 \\ -3 & 4 & 0 \\ -3 & 1 & 3 \end{bmatrix}.$$

18. Show that if a matrix B is both diagonalizable and invertible, then so is B^{-1}.

19. Find the eigenvalues of the following matrices and decide if they are similar to a diagonal matrix:

$$A = \begin{bmatrix} 1 & 0 & 0 \\ 0 & 0 & -1 \\ 0 & 1 & 0 \end{bmatrix}, \quad B = \begin{bmatrix} 0 & 1 & 1 & 0 \\ 0 & 1 & 0 & -1 \\ 0 & 1 & 1 & 0 \\ 0 & 1 & 1 & 0 \end{bmatrix}, \quad C = \begin{bmatrix} 2 & 0 & 3 & 0 \\ 0 & 2 & 0 & 3 \\ 3 & 0 & 2 & 0 \\ 0 & 3 & 0 & 2 \end{bmatrix}.$$

20. Consider the matrix $A = \begin{bmatrix} 1 - x & x \\ x & 1 - x \end{bmatrix}$ where x is any real number. Show that

$$A^k = \frac{1}{2}\begin{bmatrix} 1 + (1 - 2x)^k & 1 - (1 - 2x)^k \\ 1 - (1 - 2x)^k & 1 + (1 - 2x)^k \end{bmatrix}.$$

21. Find the eigenvalues of the matrix $A = \begin{bmatrix} 3 & 0 & -4 \\ 0 & 2 & 0 \\ 2 & 0 & -1 \end{bmatrix}$, and find a complex matrix Q such that $Q^{-1}AQ$ is diagonal.

6 Linear Programming

In this chapter we will see how to use the row reduction techniques of Chapter 1 to solve a class of optimization problems that occur frequently in a wide variety of applications to business, economics, political, and social situations. We will first discuss a relatively simple example from a geometric point of view and then develop the simplex method for solving such problems. The simplex method was developed in the 1940s to solve supply and distribution problems for the military.

6.1 THE GEOMETRIC POINT OF VIEW

An office furniture manufacturer produces two items: file cabinets and desks. The profit on each file cabinet is \$30 and the profit on each desk is \$36. Manufacturing each product requires processing by three different departments: cutting, assembly, and finishing. Each file cabinet requires 3 hours in the cutting department, 1 hour in assembly, and 1 hour in the finishing department. Each desk requires 1 hour in the cutting department, 1 hour in the assembly department, and 2 hours in the finishing department. There is a total of 90 hours per week available in the cutting department, a total of 40 hours per week available in the assembly department, and a total of 70 hours per week available in the finishing department. Assuming that all products can be sold, how many of each product should be produced in order to maximize the profit?

We represent the data in the problem in Table 6.1. Suppose that we let x represent the number of file cabinets produced per week and let y be the number of desks produced per week. The profit for the week would then be $P = 30x + 36y$. The total time required in the cutting department is then $3x + y$ and the availability constraint on the cutting department means that x and y must

TABLE 6-1

	Cutting	*Assembly*	*Finishing*	*Profit*
File cabinets	3	1	1	30
Desks	1	1	2	36
	90	40	70	

satisfy the linear inequality

$$3x + y \leq 90.$$

Similarly, for the assembly department, we must have

$$x + y \leq 40,$$

and for the finishing department

$$x + 2y \leq 70.$$

We must also assume that x and y are greater than or equal to 0, since it makes no sense to produce a negative number of items. Thus we wish to choose x and y so that we maximize the objective function

$$P = 30x + 36y \tag{6.1}$$

subject to the constraints

$$\begin{aligned} x + 2y &\leq 70 \\ x + y &\leq 40 \\ 3x + y &\leq 90 \end{aligned} \qquad \begin{aligned} x &\geq 0 \\ y &\geq 0. \end{aligned} \tag{6.2}$$

This is an example of a linear programming problem; such problems occur frequently in a wide variety of business and scientific applications. We say that

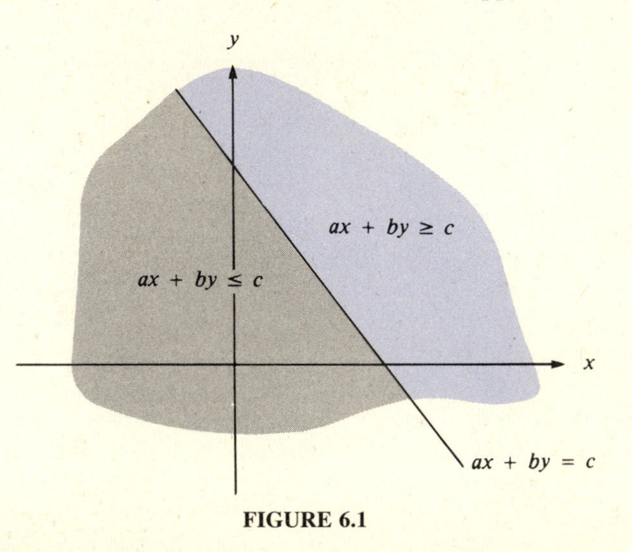

FIGURE 6.1

it is a linear problem because the objective function (Equation (6.1)) is a linear function. Here, "programming" refers to planning, not to computer programming. We will be able to solve this simple problem in a geometric way that will give insight into how we should proceed with more complicated problems of this type.

The set of all points (x, y) that satisfy the constraints (given in System (6.2)) is called the **feasible set** for the problem; its elements are called **feasible solutions.** In order to give a geometric description of the feasible set we first observe that a straight line $ax + by = c$ divides the plane into two half planes, as shown in Figure 6.1.

Thus, the set of all solutions of each of the linear inequalities in System (6.2) is represented by a half plane and the feasible set, determined by System (6.2), is the intersection of the five half planes shown in Figure 6.2. Thus the feasible

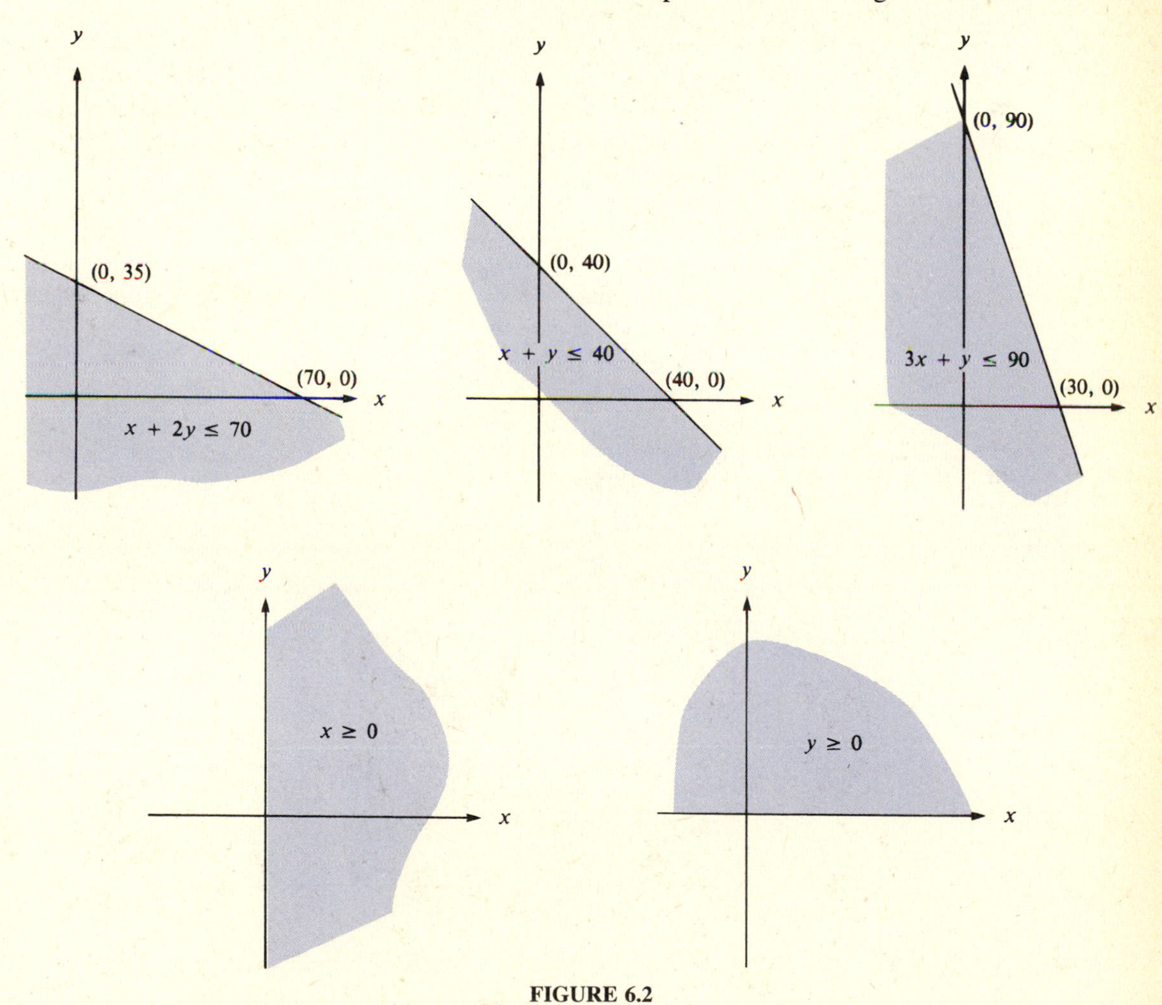

FIGURE 6.2

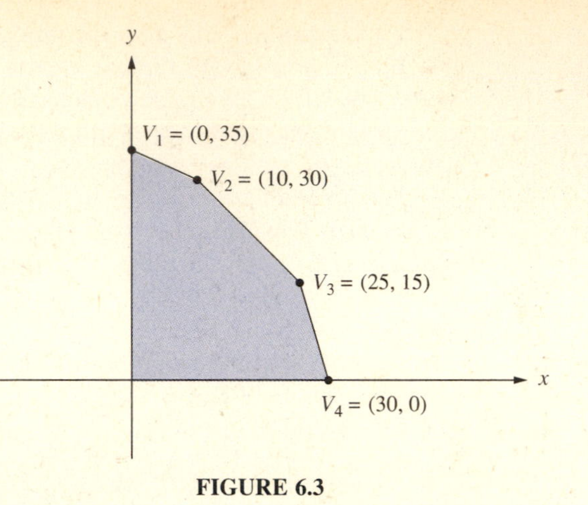

FIGURE 6.3

set determined by System (6.2) is the polygonal region shown in Figure 6.3. The vertices of the polygon are determined by finding the points of intersections of the lines, two at a time.

We want to find the maximum value of the objective function $P = 30x + 36y$ for (x, y) in the feasible set. Since the feasible set contains an infinite number of points, it is clearly impossible to check the value of P for each point in the feasible set. We can gain some further insight into where the maximum occurs by interpreting Equation (6.1) geometrically.

For any fixed value of P, the objective function (Equation (6.1)) represents a straight line with slope $-\frac{5}{6}$. Thus, we can interpret Equation (6.1) as a one-parameter family of parallel lines with P as the parameter. Several of these lines are shown in Figure 6.4. Note that the value of P increases as the lines move to the right of the origin.

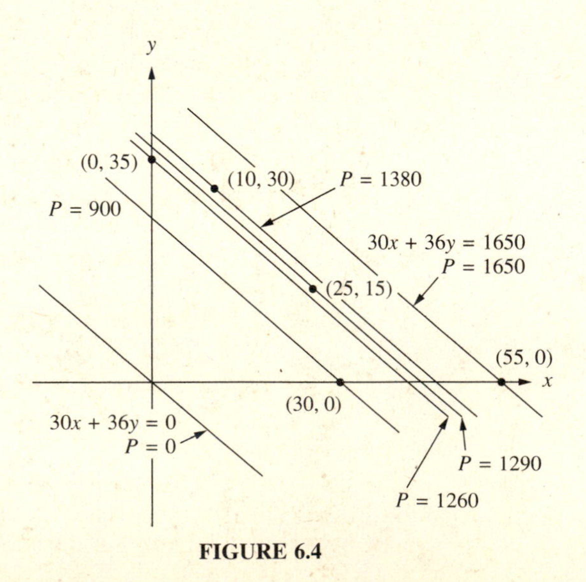

FIGURE 6.4

If we superimpose Figure 6.4 on Figure 6.3 we obtain the diagram in Figure 6.5. From Figure 6.5 we see that solving the optimization problem posed at the start of this section amounts to finding the right most line in Figure 6.4 that intersects the feasible set in Figure 6.3. Figure 6.5 shows that the maximum occurs for the line through the vertex $V_2 = (10, 30)$, for which the value of P is 1380. It should be clear geometrically that the maximum value must occur at a vertex of the feasible set; at such a vertex at least two of the inequalities of System (6.2) must actually be equalities.

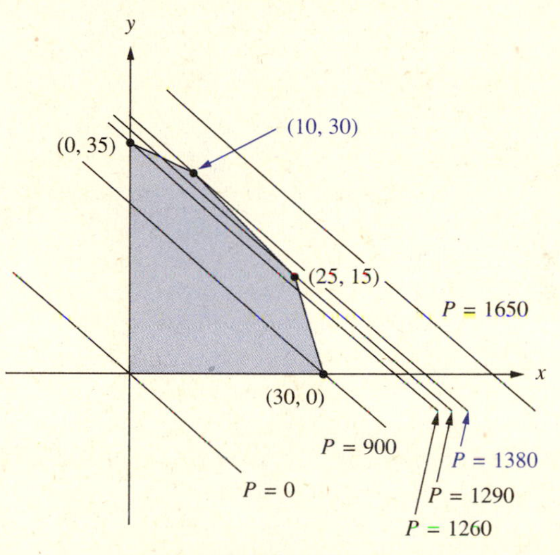

FIGURE 6.5

For more general problems of this type the preceding observation is still valid; that is, *the optimal solution occurs at one of the vertices of the feasible set.* If there are three unknowns in the problem, then the feasible set is a polyhedron that is the intersection of several half spaces of $\Re^3$; in this case each vertex is the intersection of three planes. The three-dimensional case is significantly more difficult to handle geometrically than the case $n = 2$. If there are n unknowns, then the feasible set will be the intersection of several half spaces of $\Re^n$. For $n > 3$ it will be impossible to proceed geometrically, so we must devise algebraic methods to handle the problem.

The key to success in the algebraic treatment of the general problem is to devise an efficient way of searching through the vertices of the feasible set for the optimal solution. This search involves finding the vertices of the feasible set and then deciding which of the vertices to check and in what order to check them. If the number of variables and the number of constraints are large, then the number of vertices of the feasible set may be so large that it is impractical to either find them all or to check the value of the objective function at all of the vertices of the feasible set.

It is easier to deal with linear equations than with linear inequalities, so we begin by changing the constraints in System (6.2) into equalities. We make this change by introducing new variables into the inequalities. Thus, the inequality

$x + 2y \leq 70$ is equivalent to the equation $x + 2y + u = 70$ with the restriction that $u \geq 0$. We will call variables like u **slack variables.** In many applications slack variables represent the amount of unused resources; in the problem we are discussing u is the amount of unused time in the finishing department. If we introduce slack variables into each of the inequalities of System (6.2), then we have the linear system

$$\begin{aligned} x + 2y + u \quad\quad\quad &= 70 \\ x + y \quad + v \quad &= 40 \\ 3x + y \quad\quad + w &= 90, \end{aligned} \tag{6.3}$$

in which all the variables are required to be positive.

The linear system (6.3) has three equations in five unknowns and is of rank 3, so any two of the unknowns can be assigned values arbitrarily and then the other three values will be uniquely determined. A solution obtained by setting any two of the variables equal to 0 and then computing the other values from the equation is called a **basic solution** of System (6.3). For example, if we set $x = y = 0$, then it follows at once that $u = 70$, $v = 40$, and $w = 90$. For this problem there are a total of ten basic solutions; they are listed in Table 6.2.

TABLE 6-2

	1	*2*	*3*	*4*	*5*	*6*	*7*	*8*	*9*	*10*
x	0	0	0	0	70	40	30	10	22	25
y	0	35	40	90	0	0	0	30	24	15
u	70	0	−10	−110	0	30	40	0	0	15
v	40	5	0	−50	−30	0	10	0	−6	0
w	90	55	50	0	−120	−30	0	30	0	0

Of the ten basic solutions listed in Table 6.2, only solutions *1, 2, 7, 8,* and *10* are feasible solutions; the others fail to be feasible because at least one variable has a negative value. These feasible solutions are called **basic feasible solutions;** in any basic feasible solution the **basic variables** are those that are not set equal to 0. For solution *8* the basic variables are $x = 10$, $y = 30$, and $w = 30$. There is a one-to-one correspondence between the basic feasible solutions of System (6.3) and the vertices of the feasible set shown in Figure 6.3; for each slack variable set equal to 0, the corresponding inequality in System (6.2) becomes an equality and the points that satisfy the equation are thus on the boundary of the feasible set. For example, in solution *10* the slack variables v and w are set equal to 0 so that the last two inequalities of System (6.2) become equalities and solution 10 corresponds to the intersection of the lines $x + y = 40$ and $3x + y = 90$. That intersection is the point $V_3 = (25, 15)$ of Figure 6.3. Similarly, solution *1* corresponds to the origin, solution *2* to $V_1 = (0, 35)$, solution *7* to $V_4 = (30, 0)$, and solution *8* to $V_2 = (10, 30)$.

The augmented matrix of System (6.3) is

$$\begin{array}{c} \begin{matrix} x & y & u & v & w & \end{matrix} \\ \left[\begin{array}{ccccc|c} 1 & 2 & 1 & 0 & 0 & 70 \\ 1 & 1 & 0 & 1 & 0 & 40 \\ 3 & 1 & 0 & 0 & 1 & 90 \end{array}\right]. \end{array} \tag{6.4}$$

Upon examination of Matrix (6.4) we see that the columns associated with the variables u, v, and w are columns of the identity matrix I. This situation makes it easy to solve for u, v, and w if x and y are set equal to 0; this leads to the basic feasible solution 1 in Table 6.2.

From the development in Chapter 1 we know that any matrix that is row equivalent to Matrix (6.4) is the augmented matrix of a linear system equivalent to System (6.3). We can find other basic feasible solutions by using elementary row operations to change Matrix (6.4) so that other columns become columns of I. For example, the row operations $R_2 \leftarrow -R_1 + R_2$ and $R_3 \leftarrow -3R_1 + R_3$ will change column 1 of Matrix (6.4) into the first column of I and will not change columns 4 and 5. These operations lead to the augmented matrix

$$\begin{array}{c} \begin{matrix} x & y & u & v & w & \end{matrix} \\ \left[\begin{array}{ccccc|c} 1 & 2 & 1 & 0 & 0 & 70 \\ 0 & -1 & -1 & 1 & 0 & -30 \\ 0 & -5 & -3 & 0 & 1 & -120 \end{array}\right]. \end{array}$$

If we now set $y = u = 0$, we find that $x = 70$, $v = -30$, and $z = -120$. This is solution 5 in Table 6.2, which is a basic solution of System (6.3) but not a basic feasible solution.

Similarly, the operations $R_1 \leftarrow \frac{1}{2}R_1$, $R_2 \leftarrow -R_1 + R_2$, and $R_3 \leftarrow -R_1 + R_3$ on Matrix (6.4) lead to the augmented matrix

$$\begin{array}{c} \begin{matrix} x & y & u & v & w & \end{matrix} \\ \left[\begin{array}{ccccc|c} \frac{1}{2} & 1 & \frac{1}{2} & 0 & 0 & 35 \\ \frac{1}{2} & 0 & -\frac{1}{2} & 1 & 0 & 5 \\ \frac{5}{2} & 0 & -\frac{1}{2} & 0 & 1 & 55 \end{array}\right]. \end{array} \tag{6.5}$$

If we set $x = u = 0$ we obtain, from the equations associated with Matrix (6.5), $y = 35$, $v = 5$, and $w = 55$. This is the second basic feasible solution listed in Table 6.2. It corresponds to the vertex $V_1 = (0, 35)$ of the feasible set shown in Figure 6.3.

For the original optimization problem {maximize (6.1) subject to (6.2)}, which is equivalent to the optimization problem {maximize (6.1) subject to (6.3)}, the easiest basic feasible solution to find is the first one in Table 6.2; that is, $[x \quad y \quad u \quad v \quad w] = [0 \quad 0 \quad 70 \quad 40 \quad 90]$. For this basic feasible solution the value of the objective function is $P = 30x + 36y = 0$, which is clearly not optimal. In order to increase P we need to increase one of the nonbasic variables; that is, we need to increase either x or y. Since a unit change in y will increase P more than a unit change in x, let us increase y while holding x fixed at $x = 0$. To see how much we can increase y, we look at the equations of System (6.3), with $x = 0$. These equations are

$$\begin{aligned} 2y + u &= 70 \\ y + v &= 40 \\ y + w &= 90 \end{aligned} \qquad \text{or} \qquad \begin{aligned} u &= 70 - 2y \\ v &= 40 - y \\ w &= 90 - y. \end{aligned}$$

The restriction that $u \geq 0$ means that $y \leq 35$; since $v \geq 0$ we must have $y \leq 40$; and since $w \geq 0$ we must have $y \leq 90$. If $y \leq 35$, then u, v, and w will all be nonnegative and the corresponding solution will remain in the feasible set. If we take $y = 35$, then $u = 0$, $v = 5$, and $w = 55$ are all nonnegative. This amounts to changing the set of basic variables from $\{u, v, w\}$ to $\{y, v, w\}$ and the new basic feasible solution is $[x \quad y \quad u \quad v \quad w] = [0 \quad 35 \quad 0 \quad 5 \quad 55]$, which corresponds to the vertex $V_1 = (0, 35)$ of the feasible set shown in Figure 6.3.

The value of the objective function for this feasible solution is

$$P = 30x + 36y = 30 \cdot 0 + 36 \cdot 35 = 1260.$$

We have already seen that System (6.3) is equivalent to the system whose augmented matrix is Matrix (6.5) and that this new system leads to the above basic feasible solution. The equations of that system are

$$\begin{aligned} \tfrac{1}{2}x + y + \tfrac{1}{2}u \qquad\qquad &= 35 \\ \tfrac{1}{2}x \qquad - \tfrac{1}{2}u + v \qquad &= 5 \\ \tfrac{5}{2}x \qquad - \tfrac{1}{2}u \qquad + w &= 55. \end{aligned} \tag{6.6}$$

Let us now use the first equation of System (6.6) to eliminate y from the objective function (6.1), thus expressing P in terms of the current nonbasic variables. This substitution yields

$$P = 30x + 36y = 30x + 36\left(35 - \frac{x}{2} - \frac{u}{2}\right) = 1260 + 12x - 18u.$$

In order to further increase P we need to increase x while holding $u = 0$. Note that increasing u would decrease P. To see how much we can increase x we look at the equations of System (6.6) which, with $u = 0$, reduce to

$$\begin{aligned} \tfrac{1}{2}x + y \qquad\qquad &= 35 \\ \tfrac{1}{2}x \qquad + v \qquad &= 5 \\ \tfrac{5}{2}x \qquad\qquad + w &= 55 \end{aligned} \qquad \text{or} \qquad \begin{aligned} y &= 35 - \tfrac{1}{2}x \\ v &= 5 - \tfrac{1}{2}x \\ w &= 55 - \tfrac{5}{2}x. \end{aligned}$$

Now $y \geq 0$ requires that $x \leq 70$, $v \geq 0$ requires that $x \leq 10$, and $w \geq 0$ requires that $x \leq 22$. If we take $x = 10$, then $y = 30$, $v = 0$, and $w = 30$ are all nonnegative. We have again changed the set of basic variables, this time from $\{y, v, w\}$ to $\{x, y, w\}$. The basic feasible solution associated with this new set of basic variables is $[x \quad y \quad u \quad v \quad w] = [10 \quad 30 \quad 0 \quad 0 \quad 30]$, which corresponds to the vertex $V_2 = (10, 30)$ of the feasible set shown in Figure 6.3. The value of the objective function for this feasible solution is

$$P = 1260 + 12x - 18u = 1380.$$

To see if any further increase is possible we need to express P in terms of the new nonbasic variables, and in order to do this we need to modify Matrix (6.5) so that the first column changes into the second column of I. The elementary row operations

$$R_2 \leftarrow 2R_2, \quad R_1 \leftarrow \frac{-1}{2}R_2 + R_1, \qquad \text{and} \qquad R_3 \leftarrow \frac{-5}{2}R_2 + R_3$$

change Matrix (6.5) to the matrix

$$\begin{bmatrix} 0 & 1 & 1 & -1 & 0 & 30 \\ 1 & 0 & -1 & 2 & 0 & 10 \\ 0 & 0 & 2 & -5 & 1 & 30 \end{bmatrix},$$

which is the augmented matrix of a system equivalent to System (6.3). The new system is

$$\begin{aligned} y + u - v &= 30 \\ x - u + 2v &= 10 \\ 2u - 5v + w &= 30. \end{aligned}$$

We can use the second equation of this system to eliminate x from the last expression for P and thus express P in terms of the current nonbasic variables:

$$P = 1260 + 12(10 - 2v + u) - 18u = 1380 - 24v - 6u.$$

Since any increase in the nonbasic variables u and v will result in a decrease in P we conclude that we have found the maximum P and the corresponding optimal production schedule for our original problem. Note that this is the same optimal solution that we obtained earlier by a geometric argument.

EXERCISES 6.1

1. Sketch the region in the xy-plane determined by the inequalities

$$\begin{aligned} 3x + 2y &\le 46 \qquad & x &\ge 0 \\ 3x - 4y &\le 16 \qquad & 0 \le y &\le 8. \end{aligned}$$

2. Sketch the region in the xy-plane determined by the inequalities

$$\begin{aligned} 2x + 15y &\le 300 & & \\ 16x + 5y &\le 330 \qquad & x &\ge 0 \\ 10x + 15y &\ge 150 \qquad & y &\ge 0. \\ 2x - 3y &\le 34 & & \end{aligned}$$

3. Sketch the region in $\Re^2$ determined by the inequalities

$$\begin{aligned} 5x_1 - 6x_2 &\le 38 & & \\ x_1 - x_2 &\ge -9 \qquad & x_1 &\ge 0 \\ 10x_1 + 7x_2 &\ge 114 \qquad & x_2 &\ge 0. \end{aligned}$$

4. Sketch the region in $\Re^2$ determined by the inequalities

$$\begin{aligned} x_1 + x_2 &\leq 8 \\ 3x_1 - x_2 &\leq 12 \\ -x_1 + 3x_2 &\leq 12. \end{aligned}$$

5. Sketch the region in $\Re^3$ determined by the inequalities

$$\begin{aligned} 6x + 5y \qquad &\leq 60 \\ 6x - 13y + 6z &\leq 60 \\ -78x + 25y + 30z &\leq 300 \\ 42x - 19y + 42z &\leq 636 \end{aligned} \qquad \begin{aligned} x &\geq 0 \\ y &\geq 0 \\ z &\geq 0. \end{aligned}$$

6. For the region described in Exercise 2, find the indicated values.
(a) Maximum value of $z_1 = 20x + 25y$
(b) Minimum value of z_1
(c) Maximum value of $z_2 = 20x - 15y$
(d) Minimum value of z_2

7. For the region described in Exercise 1, find the indicated values.
(a) Maximum value of $z_1 = 30x + 12y$
(b) Minimum value of z_1
(c) Maximum value of $z_2 = 24x + 16y$
(d) Minimum value of z_2

8. For the region described in Exercise 4, find the indicated values.
(a) Maximum value of $z = 3x_1 + 5x_2$
(b) Minimum value of z

9. For the region described in Exercise 3, find the indicated values.
(a) Maximum value of $z = 8x_1 + 6x_2$
(b) Minimum value of z

10. For the region described in Exercise 5, find the indicated values.
(a) Maximum value of $P = 10x + 12y + 15z$
(b) Maximum value of $R = 18x + 32y - 17z$

11. Change the system of linear inequalities in Exercise 1 into a system of linear equations by adding slack variables. List all the basic solutions of this system and then find all the basic feasible solutions. Match the basic feasible solutions with the vertices of the region in Exercise 1.

12. Change the system of linear inequalities in Exercise 5 into a system of linear equations by adding slack variables.
(a) How many basic solutions does this system have?
(b) Find all basic solutions for which $x = z = 0$. Are any of them basic feasible solutions?

13. A truck farmer plans to grow corn, green beans, and zucchini. He has 15 acres of land available and \$600 for seed and fertilizer. He has 80 hours per week of labor available to work the crops. The labor and capital requirements as well as the estimated profit for each crop are given in the following table.

	Labor (hours)	*Capital ($)*	*Profit ($)*
Corn/acre	5	30	45
Green beans/acre	8	20	60
Zucchini/acre	3	15	20

How much of each crop should be planted in order to maximize the profit to the farmer?

14. The farmer in Exercise 13 has just received word that corn prices are going down while bean and zucchini prices are going up. The revised estimates are that corn will produce $40 profit per acre and beans will produce $70 profit per acre, while zucchini will produce $30 profit per acre. How should the farmer change his planting plans in order to take advantage of this change in the market?

15. An investment analyst must make decisions about the investment of a $600,000 trust fund. He can invest in common stocks, which pay a 4 percent dividend and are expected to have 10 percent capital gain; he can invest in corporate bonds, which pay 8 percent interest and are expected to have a 1 percent capital loss; or he can invest in government bonds, which pay a 6 percent dividend but have no potential for capital growth. His company restricts trust fund investments so that no more than 60 percent can be invested in stocks, no more than 30 percent can be invested in corporate bonds, and at least 20 percent must be invested in government bonds.

(a) How should the funds be invested to generate the maximum amount of dividends and interest?

(b) The owner of the trust fund wants to use $30,000 of the income each year and will reinvest any additional income. How should the fund be invested to generate the maximum capital growth while meeting the income requirements?

16. Find the points in Figure 6.3 which correspond to the nonfeasible solutions in Table 6.2.

17. **(a)** Sketch the region in $\Re^3$ determined by the inequalities

$$\begin{aligned} x + 2y + 2z &\leq 50 \qquad & x &\geq 0 \\ x + y + 3z &\leq 40 \qquad & y &\geq 0 \\ x + y + z &\leq 30 \qquad & z &\geq 0. \end{aligned}$$

(b) How many vertices does this region have?

(c) Add slack variables to the inequalities to obtain an equivalent system of linear equations. How many basic solutions does this system have? How many of these basic solutions are feasible?

18. Find a sequence of elementary row operations which, when applied to

$$A = \begin{bmatrix} 6 & 5 & 0 & 1 & 0 & 0 & 0 & 60 \\ 6 & -13 & 6 & 0 & 1 & 0 & 0 & 60 \\ -78 & 25 & 30 & 0 & 0 & 1 & 0 & 300 \\ 42 & -19 & 42 & 0 & 0 & 0 & 1 & 636 \end{bmatrix}$$

(a) will reduce column 1 of A to column 2 of I?

(b) will reduce column 3 of A to column 3 of I?

(c) Use MATMAN to carry out the reductions in parts (a) and (b).

19. A certain factory makes three products; wingos, dingos, and dodas. Each product requires processing by 4 departments. Department 1 has 200 hours of time available per week, Departments 2 and 3 each have 100 hours, and Department 4 has 80 hours available. Each wingo requires 3 hours in Department 1, 1 hour in Department 2, 2 hours in Department 3, and 1 hour in Department 4. Each dingo requires 1 hour in Department 1, 2 hours in Department 2, 1 hour in Department 3, and 2 hours in Department 4. Each doda requires 4 hours in Department 1, 2 hours in Department 2, 1 hour in Department 3, and 2 hours in Department 4. The profit on each wingo is $20, on each dingo $16, and on each doda $24. Determine a production schedule for the week that will maximize the profit.

20. A retiree wishes to invest a portion of his $150,000 pension fund in money market funds, stocks, and real estate. He does not wish to invest more than 20 percent in money market funds and wants his investment in stocks to be at least 60 percent of the total. He estimates a 5 percent return from the money market funds, 12 percent from real estate and 14 percent from stocks. How should he invest his money in order to generate the maximum income?

21. A refinery can process both light crude and dark crude. It costs $30 per barrel to process light crude and $22 per barrel to process dark crude. Each barrel of light crude yields 0.21 barrels of fuel oil, 0.5 barrels of gasoline, and 0.25 barrels of jet fuel. Each barrel of dark crude yields 0.55 barrels of fuel oil, 0.3 barrels of gasoline, and 0.1 barrels of jet fuel. The refinery has orders for 4,000,000 barrels of fuel oil, 6,000,000 barrels of gasoline, and 2,500,000 barrels of jet fuel. Determine the amount of each type of crude that should be processed in order to fill the orders at minimum cost.

22. A dairy farmer knows that each of his animals should receive between 16,000 and 18,000 calories, at least 3 kg of protein, and at least 3 grams of vitamins daily. Three commercial feeds are available and the following table describes their nutritional content per kg.

Feed	*Cost*	*Calories*	*Protein*	*Vitamins*
1	$0.8	3600	0.25 kg	0.7 grams
2	0.6	2000	0.35	0.4
3	0.2	1600	0.15	0.25

How many kg of each food mix should be fed to each animal in order to minimize the feeding cost while meeting all the nutritional requirements?

23. A food store packages and sells three kinds of dried fruit mixture: regular, deluxe, and special. Each regular package contains 1 lb prunes, 1 lb apples, and 1 lb peaches. Each deluxe package contains 2 lb prunes, .5 lb apples, and 1 lb peaches. Each package of the special mix contains 1 lb prunes, 2 lb apples, and 2 lb peaches. The store has an inventory of 700 lb prunes, 500 lb apples, and 450 lb peaches. Suppose the profit on each package of the regular mix is $2, on each package of the deluxe mix is $2.50, and on each package of the special mix is $3. How many packages of each mix should be made in order to maximize the profit?

6.2 DIFFERENT TYPES OF LINEAR PROGRAMMING PROBLEMS

In the last section we considered a simple example of a linear programming problem. In this section we will describe the general linear programming problem, show how to change the general problem into standard form, and consider some additional examples. We will also show how to use matrix notation to describe the standard linear programming problem.

The **general linear programming problem,** in the n variables $x_1, x_2, \ldots, x_n$, requires that we either maximize or minimize the linear objective function

$$z = c_1x_1 + c_2x_2 + \cdots + c_nx_n \tag{6.7}$$

subject to certain linear constraints of the form

$$\begin{array}{llll} a_{11}x_1 + a_{12}x_2 + \cdots + a_{1n}x_n & (\leq, \geq, =) & b_1 \\ a_{21}x_1 + a_{22}x_2 + \cdots + a_{2n}x_n & (\leq, \geq, =) & b_2 \\ \vdots & & \vdots \\ a_{m1}x_1 + a_{m2}x_2 + \cdots + a_{mn}x_n & (\leq, \geq, =) & b_m \end{array} \tag{6.8}$$

where in each equation in System (6.8) one and only one of the symbols $(\leq)$, $(\geq)$, and $(=)$ appear. The variables may, or may not, be required to be nonnegative.

We will say that the above problem is a **standard linear programming problem** if it is a maximization problem in which each of the constraints is a $(\leq)$ inequality and the variables are all constrained to be nonnegative. The problem considered in Section 6.1 was an example of a standard linear programming problem. A general linear programming problem may fail to be in standard form for any of the following reasons:

(1) It is a minimization problem.
(2) Some of the constraints are either $(\geq)$ or $(=)$ constraints.
(3) Some variables are not required to be nonnegative.

EXAMPLE 1 **A Nonstandard Linear Programming Problem (LPP)**

The problem of finding the minimum value of $z = 3x_1 - 2x_2 + 4x_3$ subject to the constraints

$$\begin{array}{rl} 2x_1 + 3x_2 - 3x_3 \leq 4 & \quad x_1 \geq 0 \\ 3x_1 + 2x_2 + 2x_3 \leq 6 & \quad x_2 \geq 0 \\ -x_1 + 3x_2 + 2x_3 \geq -7 & \quad x_3 \geq 0 \end{array}$$

is not a standard linear programming problem for two reasons: it is a minimization problem and the third constraint is a $(\geq)$ inequality. ■

EXAMPLE 2 **Another Nonstandard LPP**

The problem of finding the maximum of $z = 5x_1 - 6x_2 + x_3$ subject to the constraints

$$\begin{aligned} x_1 + x_2 + x_3 &\geq 1 \\ x_1 + 3x_2 + 3x_3 &\leq 6 \\ x_1 + 3x_2 \quad\quad &= 7 \end{aligned} \qquad \begin{aligned} x_1 &\geq 0 \\ x_2 &\geq 0 \end{aligned}$$

fails to be in standard form because the first and third constraints are not $(\leq)$ inequalities and because x_3 is not constrained to be nonnegative. ■

We will now show how to change any linear programming problem into an equivalent standard linear programming problem.

If the given linear programming problem is a minimization problem, then we can change it into a maximization problem using the observation that

$$\min(z) = -\max(-z),$$

where z is any real-valued function. Thus, to find the minimum of the objective function we need only maximize its negative. Since the inequality $a \geq b$ is equivalent to the inequality $-a \leq -b$, changing a $(\geq)$ constraint to a $(\leq)$ constraint involves only changing the direction of the inequality by multiplying by -1.

EXAMPLE 3 **Change Example 1 to a Standard LPP**

We can change the linear programming problem in Example 1 to a standard linear programming problem by multiplying the third inequality by -1 and by changing the signs of the coefficients in the objective function. The equivalent standard linear programming problem is then

Maximize $w = -3x_1 + 2x_2 - 4x_3$ subject to the constraints

$$\begin{aligned} 2x_1 + 3x_2 - 3x_3 &\leq 4 \\ 3x_1 + 2x_2 + 2x_3 &\leq 6 \\ x_1 - 3x_2 - 2x_3 &\leq 7 \end{aligned} \qquad \begin{aligned} x_1 &\geq 0 \\ x_2 &\geq 0 \\ x_3 &\geq 0. \end{aligned}$$ ■

Since the equality $x = a$ is equivalent to the two inequalities $x \leq a$ and $-x \leq -a$ $(x \geq a)$, we can replace an $(=)$ constraint with two $(\leq)$ constraints.

Moreover, since any number, positive or negative, can be written as the difference of two positive numbers, we can replace any unconstrained variable x_j by the difference

$$x_j = x_{j^+} - x_{j^-}$$

where x_{j^+} and x_{j^-} are constrained to be nonnegative.

EXAMPLE 4 Change Example 2 to a Standard LPP

We can change the nonstandard linear programming problem of Example 2 to a standard linear programming problem by multiplying the first inequality by -1, replacing x_3 by the difference $x_{3^+} - x_{3^-}$, and replacing the third constraint with two inequalities. The equivalent standard linear programming problem is then

Maximize $z = 5x_1 - 6x_2 + x_{3^+} - x_{3^-}$ subject to the constraints

$$\begin{array}{rcl} -x_1 - x_2 - x_{3^+} + x_{3^-} \leq -1 & \quad & x_1 \geq 0 \\ x_1 + 3x_2 + 3x_{3^+} - 3x_{3^-} \leq 6 & & x_2 \geq 0 \\ x_1 + 3x_2 \leq 7 & & x_{3^+} \geq 0 \\ -x_1 - 3x_2 \leq -7 & & x_{3^-} \geq 0. \end{array}$$

■

In order to use matrix notation to describe linear programming problems we need to introduce inequalities for matrices.

DEFINITION 6.1 If A and B are matrices of the same size, say m by n, then $A \leq B$ means that $a_{ij} \leq b_{ij}$ for all relevant i and j. $A \geq B$, $A > B$, and $A < B$ are all defined in terms of their components in the obvious way. □

Using Definition 6.1 we can write the constraint system of Example 4 as the matrix inequality

$$\begin{bmatrix} -1 & -1 & -1 & 1 \\ 1 & 3 & 3 & -3 \\ 1 & 3 & 0 & 0 \\ -1 & -3 & 0 & 0 \end{bmatrix} \begin{bmatrix} x_1 \\ x_2 \\ x_{3^+} \\ x_{3^-} \end{bmatrix} \leq \begin{bmatrix} -1 \\ 6 \\ 7 \\ -7 \end{bmatrix}.$$

We can use Definition 6.1 to describe the standard linear programming problem as follows: given an $n \times 1$ matrix C, an $m \times 1$ matrix B, and an $m \times n$ matrix A, we seek an $n \times 1$ matrix X, an element of $\Re^n$, that maximizes $z = C^TX$ subject to the constraints $AX \leq B$ and $X \geq 0$.

The **feasible set** for this problem is the subset of $\Re^n$ satisfying the constraints; that is,

$$\mathcal{FS} = \{X \mid AX \leq B \text{ and } X \geq 0\}.$$

We will say that the feasible set $\mathcal{FS}$ is **bounded** if there exists a scalar r such that, for every X in $\mathcal{FS}$, $\|X\| = \sqrt{X^TX} \leq r$. In $\Re^2$ this means that the feasible set is contained in a circle of radius r with its center at the origin. If $\mathcal{FS}$ is not bounded we will say it is **unbounded.** Examples of both bounded and unbounded feasible sets were contained in the exercises of Section 6.1.

The basic theoretical result about the standard linear programming problem is contained in the following theorem. The proof will be omitted.

THEOREM 6.1 For the standard linear programming problem

$$\text{maximize } z = C^T X \text{ subject to } AX \leq B \text{ and } X \geq 0$$

either

(1) there is a unique solution;
(2) no solution exists; or
(3) there are an infinite number of solutions. □

The geometric interpretation of the problem, given in Section 6.1, makes this result clear if the number of variables is two or three. The no solution case can occur only if the feasible set is unbounded, but an unbounded feasible set does not necessarily mean that there is no optimal solution. If there are solutions, then they can be interpreted as vertices or extreme points of the feasible set. The case of an infinite number of solutions occurs when all the points on some "edge" or "face" of the feasible set are optimal solutions (recall part (c) of Exercise 6 of Section 6.1).

As we saw in Section 6.1, we can change the inequalities of a standard linear programming problem into equalities by introducing slack variables. For the standard linear programming problem,

$$\text{maximize } z = C^T X \text{ subject to } AX \leq B \text{ and } X \geq 0,$$

we introduce slack variables $s_1, s_2, \ldots, s_m$ satisfying

$$\begin{array}{llll} a_{11}x_1 + \cdots + a_{1n}x_n + s_1 & & & = b_1 \\ a_{21}x_1 + \cdots + a_{2n}x_n & + s_2 & & = b_2 \\ \vdots & & & \vdots \\ a_{m1}x_1 + \cdots + a_{mn}x_n & & + s_m & = b_m \end{array}$$

and the additional constraints $s_1 \geq 0, s_2 \geq 0, \ldots, s_m \geq 0$. This last system can be written in matrix form as

$$[A \mid I]\begin{bmatrix} X \\ S \end{bmatrix} = B; \quad X = \begin{bmatrix} x_1 \\ x_2 \\ \vdots \\ x_n \end{bmatrix} \geq 0 \quad \text{and} \quad S = \begin{bmatrix} s_1 \\ s_2 \\ \vdots \\ s_m \end{bmatrix} \geq 0.$$

The objective function can also be written in terms of the slack variables as

$$z = c_1x_1 + \cdots + c_nx_n + 0s_1 + \cdots + 0s_m = [C^T \quad 0]\begin{bmatrix} X \\ S \end{bmatrix}.$$

A $(\geq)$ constraint can also be changed into an $(=)$ constraint by subtracting a nonnegative slack variable. For example the inequality

$$3x_1 - 5x_2 + 7x_3 \geq 15$$

is equivalent to the equality

$$3x_1 - 5x_2 + 7x_3 - s_4 = 15$$

with the additional condition that the slack variable s_4 satisfies $s_4 \geq 0$. Slack variables with negative coefficients are used when it is important to keep the constants on the right-hand side of the the equation nonnegative.

A linear programming problem that requires the maximization of the objective function, in which all the constraints are (=) constraints and all the variables are constrained to be nonnegative, is said to be in **canonical form.** By adding slack variables we can turn any standard linear programming problem into a canonical linear programming problem. The solution techniques to be introduced in the next section require that the linear programming problem be in canonical form and that the matrix $B \geq 0$.

EXAMPLE 5 A Canonical LPP

The linear programming problem of Example 3 is equivalent to the following canonical problem.

Maximize $w = -3x_1 + 2x_2 - 4x_3$ subject to the constraints

$$\begin{array}{llllll} 2x_1 + 3x_2 - 3x_3 + s_1 & & & = 4 & x_1 \geq 0, & s_1 \geq 0 \\ 3x_1 + 2x_2 + 2x_3 & + s_2 & & = 6 & x_2 \geq 0, & s_2 \geq 0 \\ x_1 - 3x_2 - 2x_3 & & + s_3 & = 7 & x_3 \geq 0, & s_3 \geq 0. \end{array}$$

This problem can be written in matrix form:

$$\text{maximize } w = \begin{bmatrix} -3 & 2 & -4 & 0 & 0 & 0 \end{bmatrix} \begin{bmatrix} x_1 \\ x_2 \\ x_3 \\ s_1 \\ s_2 \\ s_3 \end{bmatrix} = \begin{bmatrix} C^T & 0 \end{bmatrix} \begin{bmatrix} X \\ S \end{bmatrix}$$

satisfying

$$[A \mid I] \begin{bmatrix} X \\ S \end{bmatrix} = \begin{bmatrix} 2 & 3 & -3 & 1 & 0 & 0 \\ 3 & 2 & 2 & 0 & 1 & 0 \\ 1 & -3 & -2 & 0 & 0 & 1 \end{bmatrix} \begin{bmatrix} X \\ S \end{bmatrix} = B = \begin{bmatrix} 4 \\ 6 \\ 7 \end{bmatrix} \text{ and } \begin{bmatrix} X \\ S \end{bmatrix} \geq 0.$$

■

Let us now consider the canonical linear programming problem:

$$\text{maximize } z = C^T X \text{ subject to } MX = B \text{ and } X \geq 0,$$

where M is m by $(m + n)$. We can find an element of the feasible set for this problem by assigning nonnegative values to n of the variables and then computing the values of the remaining m variables from the system $MX = B$. If the values turn out to be nonnegative, then we have found a feasible solution; if any of the values are negative, then we have a nonfeasible solution.

If n of the variables are set to 0, then we will call the resulting solution a **basic solution;** the variables set to 0 are called the **nonbasic variables,** and the variables that are computed are called the **basic variables.** If a basic solution is also a member of the feasible set, then it is called a **basic feasible solution.**

The basic feasible solutions are of importance because, as in the example of Section 6.1, they correspond to the vertices of the feasible set. The following theorem is the basis for the computational procedure (the simplex method) that will be introduced in the next section.

THEOREM 6.2 If the canonical linear programming problem

$$\text{maximize } z = C^T X \text{ subject to } MX = B \text{ and } X \geq 0$$

has an optimal solution, then there is a basic feasible solution that is optimal. □

EXERCISES 6.2

For Exercises 1–5, find an equivalent problem in standard form and rewrite the problem in matrix form.

1. Maximize $z = 2x + 5y$ subject to $\begin{cases} 2x + y \geq 2 \\ x + y \leq 8 \\ x + y \geq 3 \end{cases}$ $\quad x \geq 0,\ y \geq 0.$

2. Maximize $P = 2x + 3y + z$ subject to $\begin{cases} x + y + z \leq 12 \\ x + 3y + 3z = 24 \\ 3x + 6y + 4z \leq 90 \end{cases}$ $\quad x \geq 0,\ z \geq 0.$

3. Minimize $C = 3x - 4y + 6z$ subject to $\begin{cases} x + y + z = 1 \\ 2x + 3y + 5z \leq 4 \\ 4x + 3y + z \leq 2 \end{cases}$ $\quad x \geq 0.$

4. Minimize $P = 6x + y$ subject to $\begin{cases} x + y \leq 8 \\ 2x - y \geq -2 \\ 2x + y \leq 2 \end{cases}$ $\quad x \geq 0,\ y \geq 0.$

5. Minimize $C = x + 3y + 2u + 2v$ subject to $\begin{cases} x + y = 150 & x \geq 0 \\ u + v = 150 & y \geq 0 \\ x + u \leq 200 & u \geq 0 \\ y + v \leq 250 & v \geq 0. \end{cases}$

6. For each of the problems in Exercises 1–5, find an equivalent problem in canonical form with positive right-hand side.

7. Find all basic solutions and all basic feasible solutions of

$$\begin{aligned} 2x + y + s_1 \quad &= 6 \\ x + y \quad + s_2 &= 2. \end{aligned}$$

8. For the standard linear programming problem

$$\text{maximize } z = 24x + 16y \text{ subject to } \begin{cases} 3x + 2y \le 46 & x \ge 0 \\ 3x - 4y \le 16 & 0 \le y \le 10, \end{cases}$$

find the equivalent canonical problem. Find all basic solutions of the canonical restraint system and identify which of the basic solutions is also feasible.

9. Change the following linear programming problems to canonical form by adding slack variables:

(a) Maximize $w = x - 2y + 2z$ such that

$$\begin{aligned} 3x - 2y + 5z &\le 14 \\ x - y - 2z &\ge 1 \qquad x, y, z \ge 0. \\ y - z &= 5 \end{aligned}$$

(b) Minimize $P = 15x_1 + 15x_2 + 25x_3 + 25x_4$ such that

$$\begin{aligned} 2x_1 + 2x_2 + 3x_3 \qquad\quad &\ge 78 \\ 4x_1 + 2x_2 + 8x_3 + 8x_4 &\ge 214 \qquad \text{all } x_i \ge 0. \\ 4x_2 \qquad\qquad\quad + 12x_4 &\ge 198 \end{aligned}$$

10. Verify, or find a counter example, for each of the following assertions about matrix inequalities.

(a) $A \ge B$ implies $-A \le -B$.

(b) For any two matrices A and B, either $A > B$, $A < B$, or $A = B$.

(c) $A \ge 0$ and $B \ge C$ implies $AB \ge AC$.

(d) $A \ge B$ and $C \ge D$ implies $A + C \ge B + D$.

11. A subset $\mathcal{S}$ of $\Re^2$ is **convex** if whenever P and Q are any two points of $\mathcal{S}$, then the line segment joining P and Q lies entirely in $\mathcal{S}$.

(a) Which of the subsets of $\Re^2$ pictured below are convex?

(b) For a subset $\mathcal{S}$ of $\Re^2$ or $\Re^3$ show that $\mathcal{S}$ is convex if, and only if, $tx + (1 - t)y$ is in $\mathcal{S}$ for every x and y in $\mathcal{S}$ and every real number t such that $0 \le t \le 1$.

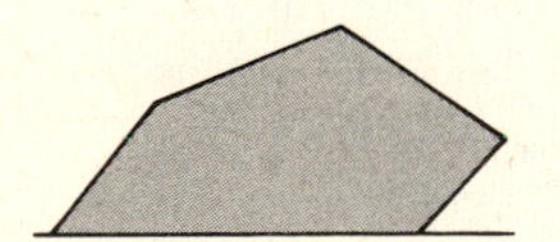

Define a convex set in $\mathfrak{R}^n$ by generalizing part (b); $\mathcal{S} \subseteq \mathfrak{R}^n$ is convex if, and only if, $t\mathbf{x} + (1 - t)\mathbf{y}$ is in $\mathcal{S}$ for every $\mathbf{x}$ and $\mathbf{y}$ in $\mathcal{S}$ and for every real number t such that $0 \leq t \leq 1$.

(c) Show that the half space of $\mathfrak{R}^n$ defined by

$$\mathcal{H} = \{\mathbf{x} \text{ in } \mathfrak{R}^n \mid a_1x_1 + a_2x_2 + \cdots + a_nx_n \leq b\}$$

is a convex set in $\mathfrak{R}^n$.

(d) Show that the intersection of convex sets is a convex set.

(e) Show that the feasible set of a linear programming problem is a convex set.

12. A truck-rental company has two models of refrigerated trucks available for rent. The old model has 2 cubic yards of refrigerated space, 4 cubic yards of unrefrigerated space, and costs \$.50 per mile to rent. The new model has 3 cubic yards of refrigerated space, 3 cubic yards of unrefrigerated space, and costs \$.65 per mile to rent. A shipper needs 90 cubic yards of refrigerated space and 120 cubic yards of unrefrigerated space. How many of each type of truck should he rent to keep his shipping costs to a minimum? Find a linear programming problem that models this situation.

13. A nutritionist wants to make an egg and milk mixture that contains at least 480 calories, at least 30 grams of protein, and at least 1600 units of vitamin A. Each large egg contains 80 calories, 6 grams of protein, 580 units of vitamin A, and costs \$.08. Each quart of whole milk contains 640 calories, 40 grams of protein, 1360 units of vitamin A, and costs \$.60. How should the two ingredients be combined so that the cost of the resulting mixture is minimized? Find a linear programming problem that models this situation. Solve the problem geometrically.

14. A manufacturer of heating and air conditioning units needs to have warehouse space available, but her needs vary with the seasons. She can rent space on either a monthly, quarterly, semiannual, or annual basis. The following table gives the rates per 1000 square feet.

Period	monthly	3 months	6 months	12 months
Rent	\$325	\$550	\$750	\$1200

The storage space required per month, in 1000 square foot units, is given in the following table.

Month	1	2	3	4	5	6	7	8	9	10	11	12
Space Needed	10	12	24	32	44	50	60	40	35	30	20	15

Set up a linear programming problem to determine how to meet the storage needs at minimum rental cost per year. Can you find some basic feasible solutions for this problem without any calculation? How does the problem change if we assume that annual leases are only available in January, semiannual leases are available only in January and July, and quarterly leases are available only in January, April, July, and October?

15. A local brewery makes four products: premium beer, premium ale, regular beer, and regular ale. The current stock of ingredients consists of 480 pounds of corn, 160 ounces of hops, and 1190 pounds of malt. There are 280 hours of labor available to process these ingredients. Each barrel of premium beer requires 5 pounds of corn, 4 ounces of hops, 40 pounds of malt, 14 hours of labor, and will generate a profit of $23. Each barrel of regular beer uses 10 pounds of corn, 4 ounces of hops, 35 pounds of malt, 7 hours of labor, and generates a profit of $15. Each barrel of premium ale uses 15 pounds of corn, 2 ounces of hops, 22 pounds of malt, 14 hours of labor, and generates a profit of $22. Each barrel of regular ale uses 17 pounds of corn, 2 ounces of hops, 22 pounds of malt, 14 hours of labor, and generates a profit of $14. How many barrels of each product should be made in order to maximize profit? Find a canonical linear programming problem to describe this situation.

6.3 THE SIMPLEX METHOD

In this section we will describe the simplex method for solving the canonical linear programming problem. The simplex method was developed by George Dantzig in 1947. This method, which uses relatively simple techniques, has been very successful in providing solutions to a wide class of linear programming problems arising in applications. The basic ideas of the method were illustrated in the algebraic solution of the production planning problem discussed in detail in Section 6.1.

We will describe the simplex method for solving a standard linear programming problem, or equivalently for solving the canonical linear programming problem that arises from a standard linear programming problem when slack variables are added to each of the constraints. The method starts from the basic feasible solution in which the slack variables are the basic variables, and at each step moves from the current basic feasible solution to another basic feasible solution for which the value of the objective function is larger. At each step the set of basic variables is modified by bringing in one new basic variable (the **entering variable**) and deleting one of the old basic variables (the **departing variable**). At each step the constraint system is modified, by elementary row operations, to make the computation of the new basic feasible solution easy and to express the objective function in terms of the new nonbasic variables.

To deal with the computations we will introduce the **simplex tableau;** it is essentially the augmented matrix of the constraint system with an additional row representing the objective function. Consider again the canonical form for the motivating problem of Section 6.1:

maximize $z = 30x + 36y$ subject to the constraints

$$\begin{aligned} x + 2y + u &= 70 \\ x + y + v &= 40 \\ 3x + y + w &= 90 \end{aligned} \qquad \begin{aligned} x &\geq 0 \\ y &\geq 0. \end{aligned}$$

The initial tableau for this problem is

	x	y	u	v	w	z	
u	1	2	1	0	0	0	70
v	1	1	0	1	0	0	40
w	3	1	0	0	1	0	90
obj	−30	−36	0	0	0	1	0 ← objective row

(6.9)

The first row and first column of the tableau contain labels; the labels in the first column specify the current basic variables u, v, and w. The center part of the tableau is the augmented matrix of the constraint system, with an extra column for the variable z. The last row represents the objective function $z = 30x + 36y$ written in the form

$$-30x - 36y - 0u - 0v - 0w + z = 0.$$

The standard linear programming problem, max $z = C^TX + u_1$ such that $AX \leq B$, $X \geq 0$, leads to a tableau of the form

Variables	*Slack Var.*	z	
A	I	0	B
$-C^T$	0	1	u_1

Here u_1 is a constant which may appear in the objective function. Initially u_1 is quite likely 0, but it will become nonzero as the solution proceeds.

In general, the canonical linear programming problem, maximize $z = C_1^TX + u_1$ subject to $A_1X = B_1$ and $X \geq 0$, would lead to a tableau of the form

Variables	z	
A_1	0	B_1
$-C_1^T$	1	u_1

If $B_1 \geq 0$ and if the matrix A_1 has the identity matrix I as an $m \times m$ submatrix, then we can easily obtain an initial basic feasible solution by taking as the basic variables those variables corresponding to the columns of A_1 which are columns of I. Note that Tableau (6.9) has these two properties; the simplex method is designed to preserve these two properties at each step, namely the constants on the right-hand sides of the constraint equations are all positive and the coefficient matrix has I as a submatrix. Moreover, if the objective row has zeros in the columns associated with the basic variables, as in (6.9), then the value of the objective function at the basic feasible solution is u_1.

For example, in the simplex tableau

	x_1	x_2	x_3	x_4	x_5	x_6	z	
x_3	0	5	1	−4	0	9	0	6
x_1	1	−2	0	3	0	−5	0	2
x_5	0	3	0	6	1	4	0	7
obj	0	−7	0	8	0	10	1	150

(6.10)

Columns 1, 3, and 5 are columns of the identity matrix. From Tableau (6.10) it is easy to compute the basic feasible solution associated with the basic variables $\{x_3, x_1, x_5\}$. That solution, which is obtained by setting $x_2 = x_4 = x_6 = 0$, is

$$X = [2 \quad 0 \quad 6 \quad 0 \quad 7 \quad 0]^T.$$

From the last row of Tableau (6.10) we see that the objective function, expressed in terms of the nonbasic variables, is $-7x_2 + 8x_4 + 10x_6 + z = 150$ or

$$z = 7x_2 - 8x_4 - 10x_6 + 150.$$

Thus the value of the objective function for the current basic feasible solution is $z = 150$.

Note that, for the problem associated with Tableau (6.10), we could increase the value of the objective function by increasing the value of the nonbasic variable x_2 (from its current value of 0 to a positive value), but increasing either x_4 or x_6 would decrease the value of the objective function. If all of the entries in the objective row of Tableau (6.10) had been positive, then the associated basic feasible solution would be optimal since there would be no way to increase the value of the objective function.

This argument extends easily to show that *a basic feasible solution is optimal if, and only if, there are no negative entries in the objective row of the associated simplex tableau.*

In going from one tableau to the next, the choice of the entering variable is made so as to increase the objective function as much as possible. *Thus we choose, as the entering variable, the variable with the most negative element in the objective row; in case of a tie we can choose either.* For our motivating problem we see, from Tableau (6.9), that y should be the entering variable for the first iteration.

The choice of the departing variable is dictated by the requirement that the new basic solution should be a basic feasible solution; that is, the value of all the variables must be nonnegative.

For Tableau (6.9), y is to become a basic variable, but x will still be nonbasic at the next step. Thus $x = 0$ for the new basic feasible solution and the other variables must satisfy the constraint equations

$$2y + u = 70, \quad y + v = 40, \qquad \text{and} \qquad y + w = 90.$$

It follows that $u = 70 - 2y \geq 0$ if, and only if, $y \leq \frac{70}{2} = 35$. Similarly, $v = 40 - y \geq 0$ if, and only if, $y \leq 40$, and $w = 90 - y \geq 0$ if, and only if, $y \leq 90$. All the current basic variables (u, v, and w) will be nonnegative if $y \leq 35$. Thus, the maximum possible value for the entering variable is $y = 35$. This value will force $u = 0$ so that u will be the departing basic variable.

For Tableau (6.10), x_2 is to become a basic variable but x_4 and x_6 will remain nonbasic. Thus, for the new basic feasible solution, $x_4 = x_6 = 0$. The equations of the constraint system then reduce to

$$5x_2 + x_3 = 6, \quad x_1 - 2x_2 = 2, \qquad \text{and} \qquad 3x_2 + x_5 = 7.$$

We therefore see that $x_3 \geq 0$ if, and only if, $x_2 \leq \frac{6}{5}$, $x_1 \geq 0$ if, and only if, $x_2 \geq -1$, and $x_5 \geq 0$ if, and only if, $x_2 \leq \frac{7}{3}$. All the current basic variables (x_3, x_1, and x_5) will be nonnegative if $x_2 \leq \frac{6}{5}$. Thus, the maximum possible value for the entering variable is $x_2 = \frac{6}{5}$; this forces $x_3 = 0$ so that x_3 is the departing variable. Note that the negative entry in the 2, 2 position of Tableau (6.10) leads to an equation, $x_2 \geq -1$, that imposes no additional restriction on x_2; it is already required to be nonnegative.

In general we divide the entries in the last column of the tableau by the corresponding positive entries in the column associated with the entering variable. We will call these ratios the **feasibility ratios** for the basic variables. *We choose as the departing variable the basic variable with the smallest nonnegative feasibility ratio; in case of a tie we can choose either.* It follows that the new basic solution will be nonnegative and hence be a basic feasible solution. When working by hand you may wish to compute the feasibility ratios in the column to the right of the tableau.

For Tableau (6.9) this modified tableau would be

		x	y	u	v	w	z		F. Ratios
←	u	1	2	1	0	0	0	70	$\frac{70}{2} = 35$
	v	1	1	0	1	0	0	40	$\frac{40}{1} = 40$
	w	3	1	0	0	1	0	90	$\frac{90}{1} = 90$
	obj	−30	−36	0	0	0	1	0	

and the feasibility ratios are

$$\text{for } u: \frac{70}{2} = 35, \quad \text{for } v: \frac{40}{1} = 40, \qquad \text{and} \qquad \text{for } w: \frac{90}{1} = 90.$$

Thus, u should be chosen as the departing variable.

For Tableau (6.10) the feasibility ratios are as follows:

$$\text{for } x_3: \frac{6}{5} = 1.2, \qquad \text{and} \qquad \text{for } x_5: \frac{7}{3} = 2.33.$$

Thus, x_3 should be chosen as the departing variable. Note that, in Tableau (6.10), we do not compute a feasibility ratio for x_1 since the entry, in the pivot column, corresponding to x_1 is negative.

Once the new set of basic variables has been determined, we need to modify the tableau in order to find the new basic feasible solution and a description of the objective function in terms of the new set of nonbasic variables. We do this modification by using elementary row operations to change the column associated with the entering variable (the **pivot column**) into the column of the identity matrix I that was associated with the departing variable in the old tableau. In the linear programming literature, this modification of the tableau is called **pivoting.**

THE SIMPLEX METHOD

1. Select the entering variable (pivot column) by choosing a column with the most negative entry in the objective row.
2. Select the departing variable (pivot row) by choosing a row that has a positive entry in the pivot column and for which the feasibility ratio is minimum.
3. Use elementary row operations to change the element at the intersection of the pivot row and the pivot column (the pivot) to 1 and to change all other entries in the pivot column to 0.
4. Stop if all the entries in the objective row are nonnegative; otherwise go to step 1.

EXAMPLE 1 The Simplex Method for Our Motivating Problem

Use the simplex method to solve the motivating example discussed in Section 6.1. The initial tableau for this problem was given in Tableau (6.9). It is

		x	$\downarrow$ y	u	v	w	z	
$\leftarrow$	u	1	2	1	0	0	0	70
	v	1	1	0	1	0	0	40
	w	3	1	0	0	1	0	90
	obj	−30	−36	0	0	0	1	0

We have already decided that the entering variable should be y and that the departing variable should be u. The pivot row and the pivot column have been highlighted and marked with small arrows. The pivot is the entry 2 in the 1, 2 position of the tableau. The reduction called for in step 3 of the simplex method, called pivoting, should change the pivot column into the first column of I. This reduction can be accomplished by the following sequence of elementary row operations:

$$R_1 \leftarrow \frac{1}{2}R_1, \quad R_2 \leftarrow -R_1 + R_2, \quad R_3 \leftarrow -R_1 + R_3, \quad \text{and} \quad R_4 \leftarrow 36R_1 + R_4.$$

The new tableau and the associated basic feasible solution are

	x ↓	y	u	v	w	z	
y	$\frac{1}{2}$	1	$\frac{1}{2}$	0	0	0	35
← v	$\frac{1}{2}$	0	$-\frac{1}{2}$	1	0	0	5
w	$\frac{5}{2}$	0	$-\frac{1}{2}$	0	1	0	55
obj	-12	0	18	0	0	1	1260

$$X = \begin{bmatrix} x \\ y \\ u \\ v \\ w \end{bmatrix} = \begin{bmatrix} 0 \\ 35 \\ 0 \\ 5 \\ 55 \end{bmatrix}.$$

This solution is not optimal because there is one negative entry (-12) in the objective row. The entering variable for the next iteration of the simplex method is x and the pivot column is the first column. The feasibility ratios are as follows:

$$y: \frac{35}{\frac{1}{2}} = 70, \qquad v: \frac{5}{\frac{1}{2}} = 10, \qquad w: \frac{55}{\frac{5}{2}} = 22.$$

The smallest feasibility ratio is associated with the variable v, so we choose v as the departing variable. Thus the pivot is the $\frac{1}{2}$ in the 2, 1 position of the tableau. We now need to modify the tableau so that the pivot column becomes the second column of I. This modification is accomplished by using the elementary row operations

$$R_2 \leftarrow 2R_2, \quad R_1 \leftarrow -\frac{1}{2}R_2 + R_1, \quad R_3 \leftarrow -\frac{5}{2}R_2 + R_3, \quad \text{and} \quad R_4 \leftarrow 12R_2 + R_4.$$

The modified tableau and the associated basic feasible solution are then

	x	y	u	v	w	z	
y	0	1	1	-1	0	0	30
x	1	0	-1	2	0	0	10
w	0	0	2	-5	1	0	30
obj	0	0	6	24	0	1	1380

$$X = \begin{bmatrix} x \\ y \\ u \\ v \\ w \end{bmatrix} = \begin{bmatrix} 10 \\ 30 \\ 0 \\ 0 \\ 30 \end{bmatrix}.$$

Since there are no negative elements in the objective row, we conclude that we have found the optimal solution. The maximum value of z is 1380. This is the same solution we found in Section 6.1 using the geometric approach. ■

■ EXAMPLE 2 Simplex Method for a Standard LPP

Use the simplex method to solve the following standard linear programming problem. Maximize $z = 8x_1 + 9x_2 + 5x_3 + 10$ subject to

$$\begin{aligned} x_1 + x_2 + 2x_3 &\le 2 \qquad & x_1 &\ge 0 \\ 3x_1 + 4x_2 + 6x_3 &\le 5 \qquad & x_2 &\ge 0 \\ 7x_1 + 7x_2 + 4x_3 &\le 10 \qquad & x_3 &\ge 0. \end{aligned}$$

If we add slack variables (x_4, x_5, x_6) to each of the inequalities to change the problem into canonical form, we obtain the initial tableau and initial basic feasible solution given below.

	x_1	↓ x_2	x_3	x_4	x_5	x_6	z	
x_4	1	1	2	1	0	0	0	2
← x_5	3	**4**	6	0	1	0	0	5
x_6	7	7	4	0	0	1	0	10
obj	−8	−9	−5	0	0	0	1	10

$$X = \begin{bmatrix} 0 \\ 0 \\ 0 \\ 2 \\ 5 \\ 10 \end{bmatrix}.$$

The most negative entry in the objective row is the -9 in the x_2 column, so we select x_2 as the entering variable and the second column as the pivot column. The feasibility ratios are $\frac{2}{1} = 2$ for x_4, $\frac{5}{4} = 1.25$ for x_5, and $\frac{10}{7} = 1.42$ for x_6, so we select x_5 as the departing variable. The pivot is thus the entry 4 in the 2, 2 position of the tableau. The pivoting operation is accomplished by the elementary row operations

$$R_2 \leftarrow \frac{1}{4}R_2, \quad R_1 \leftarrow -R_2 + R_1, \quad R_3 \leftarrow -7R_2 + R_3, \quad \text{and} \quad R_4 \leftarrow 9R_2 + R_4.$$

The modified tableau is then

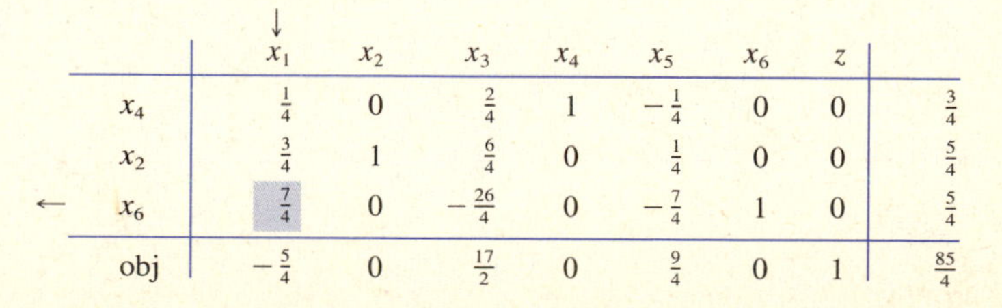

	↓ x_1	x_2	x_3	x_4	x_5	x_6	z	
x_4	$\frac{1}{4}$	0	$\frac{2}{4}$	1	$-\frac{1}{4}$	0	0	$\frac{3}{4}$
x_2	$\frac{3}{4}$	1	$\frac{6}{4}$	0	$\frac{1}{4}$	0	0	$\frac{5}{4}$
← x_6	$\frac{7}{4}$	0	$-\frac{26}{4}$	0	$-\frac{7}{4}$	1	0	$\frac{5}{4}$
obj	$-\frac{5}{4}$	0	$\frac{17}{2}$	0	$\frac{9}{4}$	0	1	$\frac{85}{4}$

from which we conclude that the value of z for the basic feasible solution $X = [0\ \ \frac{5}{4}\ \ 0\ \ \frac{3}{4}\ \ 0\ \ \frac{5}{4}]^T$ is $z = \frac{85}{4}$. Since there are still negative entries in the objective row, this solution is not optimal.

For the next step of the simplex method the entering variable will be x_1 since the only negative entry in the objective row is in column 1. The feasibility ratios at this step are then $(\frac{3}{4})/(\frac{1}{4}) = 3$ for x_4, $(\frac{5}{4})/(\frac{3}{4}) = 1.67$ for x_2, and $(\frac{5}{4})/(\frac{7}{4}) = .714$ for x_6, so we must select x_6 as the departing variable. The pivot is then the 3, 1 entry of the tableau, and the elementary row operations

$$R_3 \leftarrow \frac{4}{7}R_3,\quad R_2 \leftarrow -\frac{3}{4}R_3 + R_2,\quad R_1 \leftarrow -\frac{1}{4}R_3 + R_1,\quad \text{and}\quad R_4 \leftarrow \frac{5}{4}R_3 + R_4$$

lead to the modified tableau

	x_1	x_2	x_3	x_4	x_5	x_6	z	
x_4	0	0	$\frac{10}{7}$	1	0	$-\frac{1}{7}$	0	$\frac{4}{7}$
x_2	0	1	$\frac{30}{7}$	0	1	$-\frac{3}{7}$	0	$\frac{5}{7}$
x_1	1	0	$-\frac{26}{7}$	0	-1	$\frac{4}{7}$	0	$\frac{5}{7}$
obj	0	0	$\frac{27}{7}$	0	1	$\frac{5}{7}$	1	$\frac{155}{7}$

Since all entries in the objective row are positive, we can conclude that the optimal solution of the problem is $X = [\frac{5}{7}\ \ \frac{5}{7}\ \ 0\ \ \frac{4}{7}\ \ 0\ \ 0]^T$, for which the value of the objective function is $z = \frac{155}{7}$. ■

If there is a tie in the selection process for the entering variable (see step 1), then any one of the tied variables may be chosen. In the case of a tie let us agree to always choose the variable with the smallest subscript as the entering variable.

Ties in the choice of the departing variable require more attention. If two basic variables have the same minimum nonnegative feasibility ratio, then we can choose either of them as the departing variable, but, at the next step, the variable not chosen will be a basic variable with value zero. At the subsequent step this variable will have a feasibility ratio of zero and is the natural choice for the departing variable provided the corresponding entry in the pivot column is positive. In this case the next pivoting step will not increase the value of the objective function. For this reason such a problem is said to be **degenerate.** With degenerate problems there is a slim possibility that the algorithm will enter an infinite cycle and hence fail to find the optimal solution. We will leave a detailed discussion of the cycling phenomenon to more advanced texts, noting only that it rarely occurs in practical problems. In our solutions we will always break ties in the choice of the departing variable by choosing the variable which is first in the listing of the basic variables given in the first column of the simplex tableau.

■ EXAMPLE 3 A Degenerate LPP

Use the simplex method to find the maximum value of $P = 10x_1 + 12x_2 + 15x_3 + 100$ that satisfies the inequalities

$$\begin{aligned} 6x_1 + 5x_2 \qquad\qquad &\leq 60 \\ 6x_1 - 13x_2 + 6x_3 &\leq 60 \\ -78x_1 + 25x_2 + 30x_3 &\leq 300 \\ 42x_1 - 19x_2 + 42x_3 &\leq 636 \end{aligned} \qquad \begin{aligned} x_1 &\geq 0 \\ x_2 &\geq 0 \\ x_3 &\geq 0. \end{aligned}$$

If we denote the slack variables by x_4, x_5, x_6, and x_7, the initial tableau is

		x_1	x_2	↓ x_3	x_4	x_5	x_6	x_7	P	
	x_4	6	5	0	1	0	0	0	0	60
←	x_5	6	−13	**6**	0	1	0	0	0	60
	x_6	−78	25	30	0	0	1	0	0	300
	x_7	42	−19	42	0	0	0	1	0	636
	obj	−10	−12	−15	0	0	0	0	1	100

We choose x_3 as the entering variable and compute the feasibility ratios $\frac{60}{6} = 10$ for x_5, $\frac{300}{30} = 10$ for x_6, and $\frac{636}{42} = 15.14$ for x_7. Note that there is a tie in the choice of the departing variable. We choose x_5 and pivot on the 2, 3 position of the tableau. This step yields

		x_1	↓ x_2	x_3	x_4	x_5	x_6	x_7	P	
	x_4	6	5	0	1	0	0	0	0	60
	x_3	1	$-\frac{13}{6}$	1	0	$\frac{1}{6}$	0	0	0	10
←	x_6	−108	**90**	0	0	−5	1	0	0	0
	x_7	0	72	0	0	−7	0	1	0	216
	obj	5	$-\frac{89}{2}$	0	0	$\frac{5}{2}$	0	0	1	250

Note that the tie for departing variable at step 1 has resulted in a value of 0 for the basic variable x_6. Thus this is an example of a degenerate problem. We now choose x_2 as the entering variable and compute the feasibility ratios

$$\frac{60}{5} = 12 \text{ for } x_4, \quad \frac{0}{90} = 0 \text{ for } x_6, \quad \text{and} \quad \frac{216}{72} = 3 \text{ for } x_7.$$

We choose x_6 as the departing variable and pivot on the 3, 2 position of the tableau to obtain the new tableau

		$\downarrow$ x_1	x_2	x_3	x_4	x_5	x_6	x_7	P	
	x_4	12	0	0	1	$\frac{5}{18}$	$-\frac{1}{18}$	0	0	60
	x_3	$-\frac{8}{5}$	0	1	0	$\frac{5}{108}$	$\frac{13}{540}$	0	0	10
	x_2	$-\frac{6}{5}$	1	0	0	$-\frac{1}{18}$	$\frac{1}{90}$	0	0	0
$\leftarrow$	x_7	$\frac{432}{5}$	0	0	0	-3	$-\frac{4}{5}$	1	0	216
	obj	$-\frac{242}{5}$	0	0	0	$\frac{1}{36}$	$\frac{89}{180}$	0	1	250

Note that the current value of the objective function (250) did not increase at the last step. This is because the problem is degenerate and the departing basic variable had value 0. We now choose x_1 as the entering variable and compute the feasibility ratios

$$\frac{60}{12} = 5 \text{ for } x_4 \qquad \text{and} \qquad \frac{216}{\frac{432}{5}} = 2.5 \text{ for } x_7.$$

Note that we do not compute the feasibility ratio for x_3 and x_2 since these variables have negative entries in the pivot column. You should check that choosing either x_2 or x_3 as the departing variable would result in a nonfeasible basic solution. We choose x_7 as the departing variable and pivot on the 4, 1 position of the tableau to obtain

		x_1	x_2	x_3	x_4	$\downarrow$ x_5	x_6	x_7	P	
$\leftarrow$	x_4	0	0	0	1	$\frac{25}{36}$	$\frac{1}{18}$	$-\frac{5}{36}$	0	30
	x_3	0	0	1	0	$-\frac{1}{108}$	$\frac{1}{108}$	$\frac{1}{54}$	0	14
	x_2	0	1	0	0	$-\frac{7}{72}$	0	$\frac{1}{72}$	0	3
	x_1	1	0	0	0	$-\frac{5}{144}$	$-\frac{1}{108}$	$\frac{5}{432}$	0	$\frac{5}{2}$
	obj	0	0	0	0	$-\frac{119}{72}$	$\frac{5}{108}$	$\frac{121}{216}$	1	371

Note that the value of the objective function increased at the last step and that all of the current basic variables have positive values. This solution is still not optimal, so we choose x_5 as the entering variable and note that only x_4 has a positive value in the pivot column. Thus we choose x_4 as the departing variable and pivot on the 1, 5 position of the tableau to obtain

	x_1	x_2	x_3	x_4	x_5	x_6	x_7	P	
x_5	0	0	0	$\frac{36}{25}$	1	$\frac{2}{25}$	$-\frac{1}{5}$	0	$\frac{216}{5}$
x_3	0	0	1	$\frac{1}{75}$	0	$\frac{1}{100}$	$\frac{1}{60}$	0	$\frac{72}{5}$
x_2	0	1	0	$\frac{7}{50}$	0	$\frac{7}{900}$	$-\frac{1}{180}$	0	$\frac{36}{5}$
x_1	1	0	0	$\frac{1}{20}$	0	$-\frac{7}{1080}$	$\frac{1}{216}$	0	4
obj	0	0	0	$\frac{119}{50}$	0	$\frac{241}{1350}$	$\frac{31}{135}$	1	$\frac{2212}{5}$

Since all the entries in the objective row are now nonnegative, this tableau represents the optimal solution. The maximum value of 442.4 occurs at the point $(x_1, x_2, x_3) = (4, 7.2, 14.4)$. ■

The simplex method can be implemented using MATMAN, MATLAB, or the TI-85 graphics calculator by constructing the augmented matrix of the modified constraint system (the simplex tableau without the labels) and then using the machine to do the appropriate row operations. Both MATLAB and the TI-85 can be programmed to do the complete pivoting step.

MATMAN has a better way. The "L" command, on the main menu (type HELP to see this menu), changes the program into interactive linear programming mode. In this mode the problem can be entered in unmodified form, and the program will construct the appropriate tableau (even for the nonstandard cases considered in the next section). In linear programming mode, the MATMAN menu contains two new commands: the "O" command checks the current tableau for optimality, and the "P" command implements the complete pivoting procedure. When you give the "P" command, the program will lead you through the correct choice of entering and departing variable, rejecting any improper choices you make, and then does the row operations automatically and displays the modified tableau for your inspection. As in its primary mode, MATMAN can do arithmetic with fractions as well as decimals. For more details see Appendix B (page 431) which includes a MATMAN solution of Example 3.

EXERCISES 6.3

Given the following simplex tableaux identify the current basic variables, find the associated basic feasible solution and its objective value, choose the entering and departing variable, find the tableau associated with the new set of basic variables, and find the new basic feasible solution and its objective value.

1.

	x	y	u	v	w	z	
	3	2	2	1	0	0	10
	7	4	1	0	1	0	3
obj	−5	−8	−9	0	0	1	0

2.

	x_1	x_2	x_3	x_4	x_5	z	
	5	1	6	0	0	0	10
	3	0	3	1	0	0	15
	−10	0	2	0	1	0	20
obj	−9	0	2	0	0	1	80

3.

	x_1	x_2	x_3	x_4	x_5	x_6	x_7	z	
	1	4	1	0	1	0	3	0	12
	0	2	3	0	2	1	−2	0	6
	0	−5	5	1	3	0	4	0	10
obj	0	−8	−8	0	4	0	−6	1	120

4.

	x_1	x_2	x_3	x_4	x_5	x_6	x_7	z	
	0	−1	2	0	1	4	0	0	12
	0	2	5	1	0	0	2	0	10
	1	3	1	0	0	7	3	0	7
obj	0	−15	−20	0	0	8	3	1	325

In Exercises 5–16, use the simplex method to solve the given linear programming problem.

5. Maximize $z = 2x + 3y$ subject to $\begin{cases} x + y \leq 160 & x \geq 0 \\ 3x + 2y \leq 240 & y \geq 0. \end{cases}$

6. Maximize $z = 40x_1 + 60x_2$ subject to $\begin{bmatrix} 2 & 1 \\ 1 & 1 \\ 1 & 3 \end{bmatrix}\begin{bmatrix} x_1 \\ x_2 \end{bmatrix} \leq \begin{bmatrix} 70 \\ 40 \\ 90 \end{bmatrix}, \quad \begin{bmatrix} x_1 \\ x_2 \end{bmatrix} \geq 0.$

7. Maximize $w = x + 2y$ subject to $\begin{bmatrix} -2 & 1 \\ 1 & -1 \end{bmatrix}\begin{bmatrix} x \\ y \end{bmatrix} \leq \begin{bmatrix} 2 \\ 4 \end{bmatrix}, \quad \begin{bmatrix} x \\ y \end{bmatrix} \geq 0.$

8. Maximize $P = 200x + 300y + 100z$ subject to

$$\begin{aligned} x + 2y + 2z &\leq 50 & x &\geq 0 \\ x + y + 3z &\leq 40 & y &\geq 0 \\ x + y + z &\leq 30 & z &\geq 0. \end{aligned}$$

9. Maximize $w = 2x_1 + 5x_2 + 4x_3 - 15$ subject to

$$\begin{aligned} x_1 + 2x_2 + x_3 &\leq 4 \\ x_1 + 2x_2 + 2x_3 &\leq 6 \end{aligned} \qquad x_i \geq 0, \quad i = 1, 2, 3.$$

10. Maximize $M = 5x_1 + 4x_2$ subject to $\begin{cases} 4x_1 + 2x_2 \leq 80{,}000 & x_1 \geq 0 \\ x_1 + 3x_2 \leq 30{,}000 & x_2 \geq 0. \end{cases}$

11. Maximize $P = 23x_1 + 16x_2 + 19x_3 + 14x_4$ subject to

$$\begin{bmatrix} 1 & 1 & 1 & 1 \\ 4 & 3 & 4 & 3 \\ 1 & 0 & 1 & 0 \\ 0 & 1 & 0 & 1 \end{bmatrix}\begin{bmatrix} x_1 \\ x_2 \\ x_3 \\ x_4 \end{bmatrix} \leq \begin{bmatrix} 30 \\ 200 \\ 20 \\ 20 \end{bmatrix}, \quad \begin{bmatrix} x_1 \\ x_2 \\ x_3 \\ x_4 \end{bmatrix} \geq 0.$$

12. Maximize $P = 10x + 12y + 15z + 100$ subject to

$$\begin{aligned} 2x + 4y + 5z &\leq 10{,}000 \\ 3x + 3y + 4z &\leq 12{,}000 \end{aligned} \qquad x, y, z \geq 0.$$

13. Exercise 13 of Section 6.1 (see page 350).

14. Exercise 10(a) of Section 6.1 (see page 350).

15. Exercise 10(b) of Section 6.1 (see page 350).

16. Exercise 15 of Section 6.2 (see page 360).

†**17.** The following simplex tableau is not in a form for which a basic feasible solution is obvious. Show that $X = [0 \quad 6 \quad 3 \quad 0]^T$ is a basic feasible solution, make the necessary adjustments in the tableau, and then complete the solution.

	x_1	x_2	x_3	x_4	z	
	2	1	1	0	0	9
	1	1	0	−1	0	6
obj	−4	−5	0	0	1	0

†**18.** The following simplex tableau is not in a form for which a basic feasible solution is obvious. Show that $X = [2 \quad 2 \quad 5 \quad 0]^T$ is a basic feasible solution, make the necessary adjustments in the tableau, and then complete the solution.

x_1	x_2	x_3	x_4	z	
1	1	1	0	0	9
1	2	0	−1	0	6
0	1	0	0	0	2
−5	−4	0	0	1	0

19. Consider this canonical linear programming problem:

Maximize $z = C^TX$ subject to $AX = B$ and $X \geq 0$ where A is m by $m + n$.

Show that this problem is not degenerate if every m by m submatrix of $[A \mid B]$ is nonsingular.

20. Use the linear programming mode in MATMAN to solve:
(a) the example of Section 6.1,
(b) the problem of Example 2 of this section,
(c) the problem of Example 3 of this section,
(d) the problem of Exercise 13 of Section 6.2, and
(e) the problem of Exercise 23 of Section 6.1.

6.4 REFINEMENTS OF THE SIMPLEX METHOD

> What if the LPP is not in standard form?

The basic simplex method, as described in Section 6.3, will not work for all linear programming problems. Difficulties occur if an initial basic feasible solution is not readily available, thus making it difficult to get the algorithm started.

This is the case if the original problem has (=) constraints or (≥) constraints with positive right-hand sides.

Consider the following linear programming problem which is not in standard form: maximize $z = 40x_1 + 30x_2$ subject to the constraints

$$\begin{aligned} x_1 + 2x_2 &\geq 6 \\ 2x_1 + x_2 &\geq 4 \qquad x_1 \geq 0 \\ x_1 + x_2 &\leq 5 \qquad x_2 \geq 0. \\ 2x_1 + x_2 &\leq 8 \end{aligned} \tag{6.11}$$

Since this is a two-dimensional problem we can sketch the feasible set determined by the constraints in System (6.11). This set is shown in Figure 6.6. We could solve the given problem by simply evaluating the objective function at each of the five vertices of the feasible set, but we wish to explore an algebraic solution using the simplex algorithm.

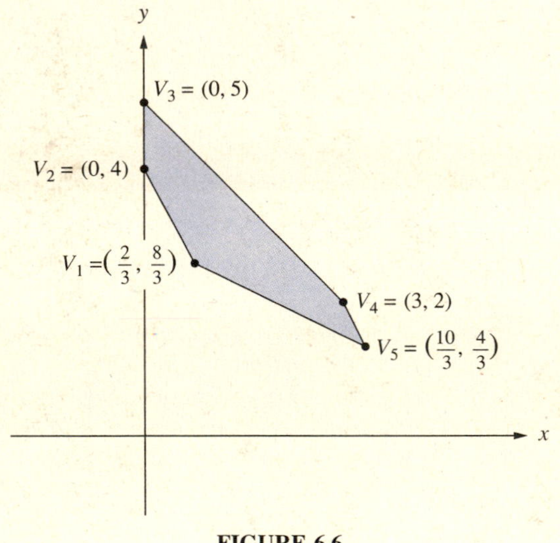

FIGURE 6.6

If we add nonnegative slack variables to the constraints of System (6.11) we obtain System (6.12).

$$\begin{aligned} x_1 + 2x_2 - x_3 \qquad\qquad\qquad &= 6 \\ 2x_1 + x_2 \qquad - x_4 \qquad\quad &= 4 \\ x_1 + x_2 \qquad\qquad + x_5 \quad &= 5 \\ 2x_1 + x_2 \qquad\qquad\qquad + x_6 &= 8 \end{aligned} \qquad \text{all } x_i \geq 0. \tag{6.12}$$

Note that the ($\geq$) inequalities in System (6.11) lead to slack variables with negative coefficients in System (6.12). A basic solution of System (6.12) is readily available:

$$x_1 = x_2 = 0, \quad x_3 = -6, \quad x_4 = -4, \quad x_5 = 5, \quad \text{and} \quad x_6 = 8.$$

This basic solution is not a feasible solution since x_3 and x_4 are negative. Thus this basic solution does not provide a suitable starting point for the simplex algorithm.

In order to get around this difficulty we modify System (6.12) as follows:

$$\begin{aligned} x_1 + 2x_2 - x_3 \qquad\qquad\qquad + x_7 \qquad &= 6 \\ 2x_1 + x_2 \qquad - x_4 \qquad\qquad\qquad + x_8 &= 4 \\ x_1 + x_2 \qquad\qquad + x_5 \qquad\qquad\qquad &= 5 \\ 2x_1 + x_2 \qquad\qquad\qquad + x_6 \qquad\qquad &= 8 \end{aligned} \qquad \text{all } x_i \geq 0 \qquad \textbf{(6.13)}$$

We will call x_7 and x_8 **artificial variables.** Note that a basic feasible solution of System (6.13) is readily available:

$$x_1 = x_2 = x_3 = x_4 = 0, \quad x_5 = 5, \quad x_6 = 8, \quad x_7 = 6, \quad \text{and} \quad x_8 = 4.$$

Note also that any basic feasible solution of System (6.13), for which $x_7 = x_8 = 0$, would yield a basic feasible solution of System (6.12); it would also yield a maximum value of 0 for the auxiliary objective function $P = -x_7 - x_8$. Thus we could try to find a basic feasible solution for System (6.12) by finding an optimal solution to the **auxiliary problem,**

$$\text{maximize } P = -x_7 - x_8 \quad \text{satisfying System (6.13).} \qquad \textbf{(6.14)}$$

The simplex tableau for the auxiliary problem (6.14) is

	x_1	x_2	x_3	x_4	x_5	x_6	x_7	x_8	P	
x_7	1	2	−1	0	0	0	1	0	0	6
x_8	2	1	0	−1	0	0	0	1	0	4
x_5	1	1	0	0	1	0	0	0	0	5
x_6	2	1	0	0	0	1	0	0	0	8
obj	0	0	0	0	0	0	1	1	1	0

which is not quite in suitable form for the simplex method—because the objective row has nonzero entries in the columns associated with the basic variables x_7 and x_8, that is the auxiliary objective function is not expressed in terms of the nonbasic variables. The elementary row operations

$$R_5 \leftarrow -R_1 + R_5 \qquad \text{and} \qquad R_5 \leftarrow -R_2 + R_5$$

will remove the troublesome entries in the objective row and lead to the tableau

	x_1	x_2	x_3	x_4	x_5	x_6	x_7	x_8	P	
x_7	1	2	−1	0	0	0	1	0	0	6
x_8	2	1	0	−1	0	0	0	1	0	4
x_5	1	1	0	0	1	0	0	0	0	5
x_6	2	1	0	0	0	1	0	0	0	8
obj	−3	−3	1	1	0	0	0	0	1	−10

(6.15)

which is suitable for use by the simplex algorithm because there are zeros, in the objective row, for each of the columns associated with the current basic variables.

The simplex method, applied to Tableau (6.15), leads, after 2 iterations, to Tableau (6.16).

	x_1	x_2	x_3	x_4	x_5	x_6	x_7	x_8	P	
x_2	0	1	$-\frac{2}{3}$	$\frac{1}{3}$	0	0	$\frac{2}{3}$	$-\frac{1}{3}$	0	$\frac{8}{3}$
x_1	1	0	$\frac{1}{3}$	$-\frac{2}{3}$	0	0	$-\frac{1}{3}$	$\frac{2}{3}$	0	$\frac{2}{3}$
x_5	0	0	$\frac{1}{3}$	$\frac{1}{3}$	1	0	$-\frac{1}{3}$	$-\frac{1}{3}$	0	$\frac{5}{3}$
x_6	0	0	0	1	0	1	0	−1	0	4
obj	0	0	0	0	0	0	1	1	1	0

(6.16)

From Tableau (6.16) we conclude that

$$x_1 = \frac{2}{3}, \quad x_2 = \frac{8}{3}, \quad x_3 = x_4 = 0, \quad x_5 = \frac{5}{3}, \quad x_6 = 4, \quad x_7 = x_8 = 0$$

is a basic feasible solution of System (6.13) and an optimal solution of the auxiliary problem (6.14). Since $x_7 = x_8 = 0$ we can also conclude that

$$x_1 = \frac{2}{3}, \quad x_2 = \frac{8}{3}, \quad x_3 = x_4 = 0, \quad x_5 = \frac{5}{3}, \quad x_6 = 4$$

is a basic feasible solution of System (6.12). Note that this solution corresponds to the vertex V_1 of the feasible set shown in Figure 6.6.

If we delete the columns of Tableau (6.16) associated with the artificial variables (x_7 and x_8), we obtain a tableau for a constraint system equivalent to System (6.12). This new tableau is associated with the basic feasible solution we have just found and we can use this basic feasible solution to start the simplex method for the original problem.

If we now change the objective row in Tableau (6.16), with columns 7 and 8 deleted, to correspond to the original objective function, we obtain Tableau (6.17).

	x_1	x_2	x_3	x_4	x_5	x_6	z	
x_2	0	1	$-\frac{2}{3}$	$\frac{1}{3}$	0	0	0	$\frac{8}{3}$
x_1	1	0	$\frac{1}{3}$	$-\frac{2}{3}$	0	0	0	$\frac{2}{3}$
x_5	0	0	$\frac{1}{3}$	$\frac{1}{3}$	1	0	0	$\frac{5}{3}$
x_6	0	0	0	1	0	1	0	4
obj	-40	-30	0	0	0	0	1	0

(6.17)

Note that Tableau (6.17) is not yet suitable for the simplex method—because the objective row has nonzero entries in the columns associated with the basic variables x_1 and x_2. The elementary row operations

$$R_5 \leftarrow 40R_2 + R_5 \qquad \text{and} \qquad R_5 \leftarrow 30R_1 + R_5$$

change Tableau (6.17) to

	x_1	x_2	x_3	x_4	x_5	x_6	z	
x_2	0	1	$-\frac{2}{3}$	$\frac{1}{3}$	0	0	0	$\frac{8}{3}$
x_1	1	0	$\frac{1}{3}$	$-\frac{2}{3}$	0	0	0	$\frac{2}{3}$
x_5	0	0	$\frac{1}{3}$	$\frac{1}{3}$	1	0	0	$\frac{5}{3}$
x_6	0	0	0	1	0	1	0	4
obj	0	0	$-\frac{20}{3}$	$-\frac{50}{3}$	0	0	1	$\frac{320}{3}$

(6.18)

If we apply the simplex method to Tableau (6.18) we are led, after two iterations, to Tableau (6.19)

	x_1	x_2	x_3	x_4	x_5	x_6	z	
x_2	0	1	0	0	2	-1	0	2
x_1	1	0	0	0	-1	1	0	3
x_3	0	0	1	0	3	-1	0	1
x_4	0	0	0	1	0	1	0	4
obj	0	0	0	2	20	10	1	180

(6.19)

from which we conclude that the maximum value of z is 180 for $x_1 = 3$ and $x_2 = 2$; this solution corresponds to the vertex V_4 shown in Figure 6.6.

The method we have just illustrated is called the **two-phase method.** It is widely used in computer software for solving linear programming problems.

■ TWO-PHASE METHOD ■

1. Change the original problem into a canonical linear programming problem by adding nonnegative slack variables to the inequality constraints. (The slack variables have a coefficient of -1 for a ($\geq$) inequality.)
2. Add artificial variables to those constraints that do not have a slack variable with a positive coefficient.
3. Set up the simplex tableau for the constraints in step 2 and the objective function $M = -$(sum of the artificial variables).
4. Use row operations to modify the tableau so that the objective row has zeros in the columns associated with the initial basic variables.
5. Use the simplex method to solve the auxiliary problem.
6. If the auxiliary problem has a solution with all the artificial variables equal to zero, then construct a basic feasible solution of the original problem by deleting the artificial variables from the solution of the auxiliary problem. If the solution of the auxiliary problem has nonzero artificial variables, then the original problem has no feasible solution.
7. Modify the final tableau of the auxiliary problem by
   (a) deleting the columns associated with the artificial variables; and
   (b) changing the objective row to correspond to the objective function of the original problem.
8. Use row operations to modify the tableau so that the objective row has zeros in the columns associated with the current basic variables.
9. Use the simplex method to solve the original problem.

The linear programming mode in MATMAN will handle the two-phase method interactively. After you enter the problem, in its natural form, the program adds both slack variables and artificial variables as needed and then sets up the initial tableau. If artificial variables were added, then the program will solve the auxiliary problem first and you will be prompted to modify the tableau as required in step 4. Note that the auxiliary objective function is in the row labeled "obj2" and the main objective function is in the row labeled "obj." After phase one (the solution of the auxiliary problem) is completed, the program will prompt you to remove the auxiliary variables and will delete the row containing the auxiliary objective function. The program will then prompt you to carry out the modifications required in step 8 before allowing you to proceed to the solution of the original problem (step 9). Appendix B (page 435) contains a MATMAN solution of the example of this section.

EXERCISES 6.4

1. Maximize $P = 2x + 3z$ subject to $\begin{cases} x + 3y + z = 5 \\ x + 2y + 7z \leq 4 \end{cases}$ $\quad x, y, z \geq 0.$

2. **(a)** Maximize $z = 40x + 80y$ subject to $\begin{cases} x + 2y \geq 4 \\ 3x + 5y \leq 30 \\ -2x + 3y \leq 9 \\ 3x - y \leq 12 \end{cases}$ $\quad \begin{matrix} x \geq 0 \\ y \geq 0. \end{matrix}$

(b) Minimize z subject to the constraints in part (a).

3. Maximize $P = 3x_1 + 2x_2 - x_3 + 4x_4$ subject to

$$\begin{aligned} 7x_2 + x_3 + 2x_4 &\geq 3 \\ 2x_1 + x_2 + 3x_3 - 4x_4 &\leq 7 \qquad \text{all } x_i \geq 0. \\ x_1 + x_2 + 2x_3 &= 9 \end{aligned}$$

4. Solve the minimization problem of Exercise 12 of Section 6.2.

5. Solve the minimization problem of Exercise 13 of Section 6.2.

6. Solve the brewery problem of Exercise 15 of Section 6.2 with the additional constraint that at least 5 barrels of premium ale must be produced.

7. Solve the transportation problem (recall Section 1.3) whose data graph is shown below.

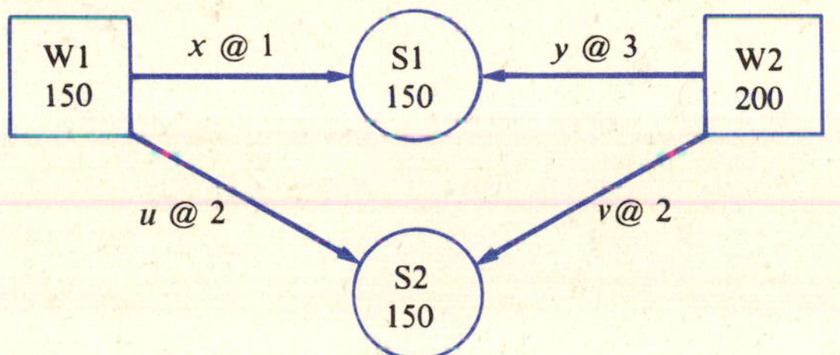

8. Solve the transportation problem whose data graph is shown below.

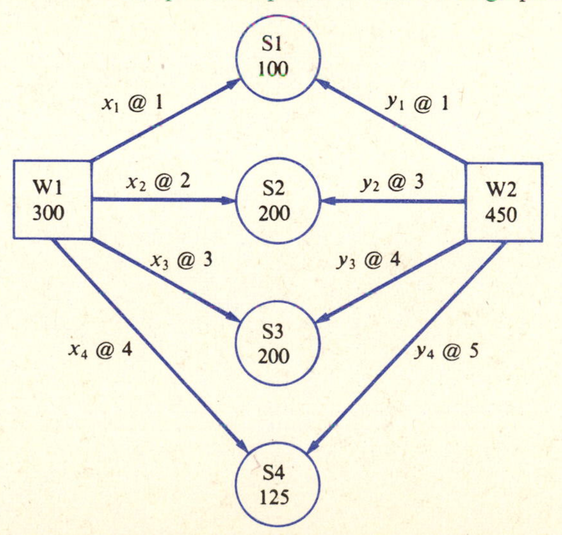

9. A small electronics firm has a contract to deliver 22,000 portable radios within the next four weeks. The customer will pay \$40 for each radio delivered by the end of the first week, \$37 for each radio delivered by the end of the second week, \$34 for those delivered by the end of the third week, and \$30 for those delivered by the end of the last week. The company currently has 40 workers who can each assemble 50 radios per week. Clearly the company must hire and train temporary workers. Any of the current workers can be taken off the assembly line to teach a class of three trainees. After a week of instruction each trainee can either be assigned to the assembly line or used to instruct additional trainees. The company has no other contracts, so some workers may become idle after the delivery is completed. All of the workers must be kept on the payroll for the full four weeks. The weekly wages of the assemblers is \$300 and the weekly wages of the trainees is \$200. The production costs, excluding labor, are \$7 per radio.

(a) Set up a linear programming problem to determine what hiring strategy the company should use in order to meet the order and maximize profit.

(b) Use MATMAN to solve the problem.

10. Use MATMAN to find the minimum value of

$$z = 30(x_1 + x_2 + x_3) + 50(x_4 + x_5 + x_6) + 80(x_7 + x_8 + x_9)$$

satisfying the constraints

$$\begin{aligned} x_1 + x_2 + x_3 &\leq 50 \\ x_4 + x_5 + x_6 &\leq 50 \\ x_7 + x_8 + x_9 &\leq 50 \\ 300x_1 + 600x_4 + 800x_7 &= 10{,}000 \\ 250x_2 + 400x_5 + 700x_8 &= 800 \\ 200x_3 + 350x_6 + 600x_9 &= 6000. \end{aligned}$$

7 Selected Applications

In this chapter we will present four applications of linear algebra. Each section of this chapter is essentially independent of the others. Those topics from Chapters 1 through 5 that are required for each section are indicated at the beginning of each section.

7.1 GRAPH THEORY

(Requires: Chapter 1)

Graph theory is a branch of mathematics that is used to analyze relationships among elements of a set. The set may consist of people, cities, objects, teams, or the like. Graph theory has important applications to many disciplines, including business, sociology, biology, engineering, medicine, and communications.

DEFINITION 7.1 A **graph** G is a finite set of vertices along with a finite set of edges, each of which connects a pair of vertices. □

EXAMPLE 1 Graph of a Highway System

Consider the graph in Figure 7.1 that represents the roads connecting seven different towns. Note that edges may cross where there is no vertex.

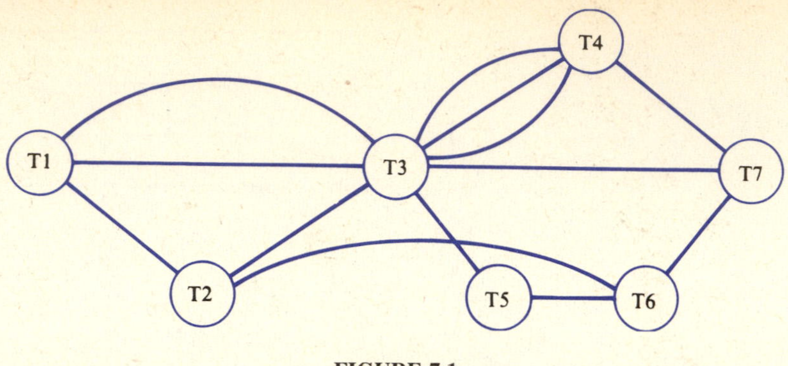

FIGURE 7.1

Note that there are two edges connecting T1 and T3, indicating that there are two roads connecting these two towns. Similarly, there are three roads connecting T3 and T4, one road connecting T5 and T6, and no direct route from T2 to T5. ■

■ EXAMPLE 2 An Interaction Graph

The graph in Figure 7.2 represents which members of a group of eight people are acquainted.

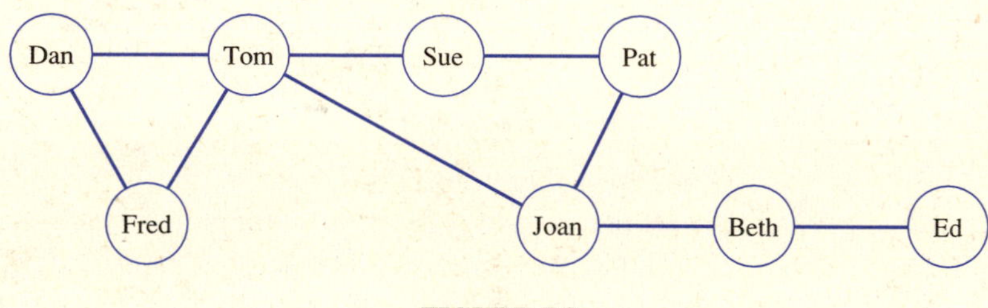

FIGURE 7.2

From this graph we see that Ed knows only Beth, Sue knows Tom and Pat, Fred and Dan do not know any of the women, and Pat does not know any of the men. ■

Many applications assign a direction to the edges of a graph; such graphs are called **directed graphs** or **digraphs.**

■ EXAMPLE 3 An Airlines Digraph

Consider the graph in Figure 7.3 that represents the routes of a small commuter airline serving five cities.

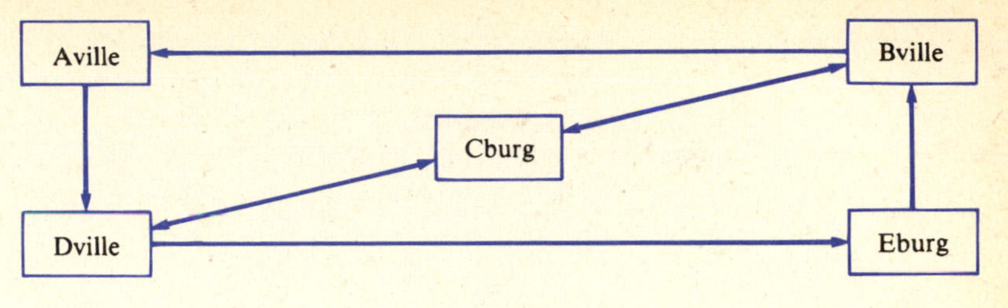

FIGURE 7.3

The arrows on the edges indicate the directions in which the planes fly. There is direct round-trip service from Cburg to Bville, but only one-way service from Bville to Aville. To get from Aville to Bville it is necessary to visit Dville and either Cburg or Eburg. ■

■ EXAMPLE 4 Incomplete Round-robin Tournament

The digraph in Figure 7.4 represents the results of an incomplete round-robin tennis tournament involving eight players.

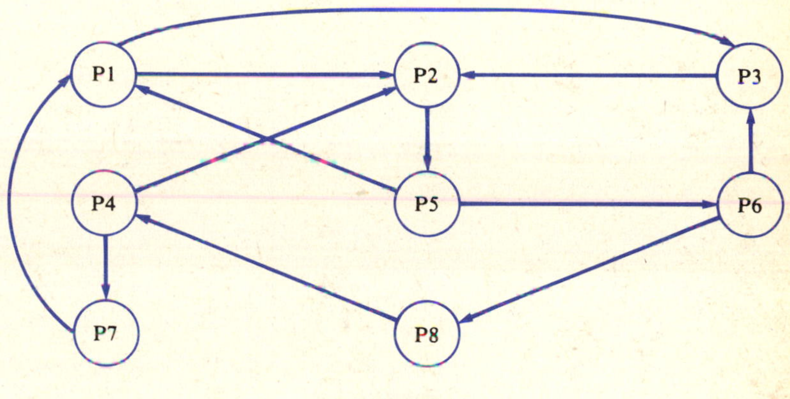

FIGURE 7.4

The existence of an edge between any two players means that those two players have competed. The arrow indicates the direction of dominance; that is, it points toward the loser. Thus, P1 has defeated P2 and P3, but has lost to P5 and P7, while P8 has defeated P4 but lost to P6. ■

In many applications of graph theory the graphs are very large and it is not easy to draw the desired conclusions by looking at the graph. Matrix theory can be useful in dealing with such graphs. We associate with a given graph two matrices that we now define formally.

DEFINITION 7.2 Let G be a graph with n vertices. The **matrix** of G is the $n \times n$ matrix $M(G)$ whose i, jth entry is the number of edges connecting vertex i to vertex j. The **adjacency matrix** of G is the $n \times n$ matrix $A(G)$ whose i, jth entry is 1 if, and only if, there is an edge containing vertex i and vertex j. All other entries of $A(G)$ are 0. If G is a digraph, then the adjacency matrix has a 1 in position i, j if, and only if, there is an edge from vertex i to vertex j. □

Note that for an undirected graph both $M(G)$ and $A(G)$ are always symmetric and $M(G) = A(G)$ if there are no multiple edges. The adjacency matrix $A(G)$ can be obtained from $M(G)$ by changing all nonzero entries to 1.

EXAMPLE 5 Matrices of Graphs

For the graph of Example 1,

$$M(G1) = \begin{bmatrix} 0 & 1 & 2 & 0 & 0 & 0 & 0 \\ 1 & 0 & 1 & 0 & 0 & 1 & 0 \\ 2 & 1 & 0 & 3 & 1 & 0 & 1 \\ 0 & 0 & 3 & 0 & 0 & 0 & 1 \\ 0 & 0 & 1 & 0 & 0 & 1 & 0 \\ 0 & 1 & 0 & 0 & 1 & 0 & 1 \\ 0 & 0 & 1 & 1 & 0 & 1 & 0 \end{bmatrix} \quad \text{and}$$

$$A(G1) = \begin{bmatrix} 0 & 1 & 1 & 0 & 0 & 0 & 0 \\ 1 & 0 & 1 & 0 & 0 & 1 & 0 \\ 1 & 1 & 0 & 1 & 1 & 0 & 1 \\ 0 & 0 & 1 & 0 & 0 & 0 & 1 \\ 0 & 0 & 1 & 0 & 0 & 1 & 0 \\ 0 & 1 & 0 & 0 & 1 & 0 & 1 \\ 0 & 0 & 1 & 1 & 0 & 1 & 0 \end{bmatrix}.$$

For the graph of Example 2,

$$M(G2) = A(G2) = \begin{array}{r} \\ \text{Dan} \\ \text{Tom} \\ \text{Fred} \\ \text{Sue} \\ \text{Joan} \\ \text{Beth} \\ \text{Ed} \\ \text{Pat} \end{array} \begin{array}{c} \begin{array}{cccccccc} \text{D} & \text{T} & \text{F} & \text{S} & \text{J} & \text{B} & \text{E} & \text{P} \end{array} \\ \begin{bmatrix} 0 & 1 & 1 & 0 & 0 & 0 & 0 & 0 \\ 1 & 0 & 1 & 1 & 1 & 0 & 0 & 0 \\ 1 & 1 & 0 & 0 & 0 & 0 & 0 & 0 \\ 0 & 1 & 0 & 0 & 0 & 0 & 0 & 1 \\ 0 & 1 & 0 & 0 & 0 & 1 & 0 & 1 \\ 0 & 0 & 0 & 0 & 1 & 0 & 1 & 0 \\ 0 & 0 & 0 & 0 & 0 & 1 & 0 & 0 \\ 0 & 0 & 0 & 1 & 1 & 0 & 0 & 0 \end{bmatrix} \end{array}.$$

For the digraph of Example 3 the adjacency matrix is

$$A(G3) = \begin{bmatrix} 0 & 0 & 0 & 1 & 0 \\ 1 & 0 & 1 & 0 & 0 \\ 0 & 1 & 0 & 1 & 0 \\ 0 & 0 & 1 & 0 & 1 \\ 0 & 1 & 0 & 0 & 0 \end{bmatrix}.$$

For the digraph of Example 4 the adjacency matrix is

$$A(G4) = \begin{bmatrix} 0 & 1 & 1 & 0 & 0 & 0 & 0 & 0 \\ 0 & 0 & 0 & 0 & 1 & 0 & 0 & 0 \\ 0 & 1 & 0 & 0 & 0 & 0 & 0 & 0 \\ 0 & 1 & 0 & 0 & 0 & 0 & 1 & 0 \\ 1 & 0 & 0 & 0 & 0 & 1 & 0 & 0 \\ 0 & 0 & 1 & 0 & 0 & 0 & 0 & 1 \\ 1 & 0 & 0 & 0 & 0 & 0 & 0 & 0 \\ 0 & 0 & 0 & 1 & 0 & 0 & 0 & 0 \end{bmatrix}. \quad \blacksquare$$

We will say that the vertices V_i and V_j of the graph G are **connected** if there is a sequence of edges connecting V_i and V_j. In the graph of Example 1 there is a two-step path (T1 $\to$ T3, T3 $\to$ T5) connecting T1 and T5. There is also a three-step path connecting these two vertices.

If A is the adjacency matrix of the graph G, then the entries of A^2 will tell us about the two-step paths in the graph G. To see how this works we suppose that the entry in the i, j position of A^2 is nonzero; that is,

$$[A^2]_{ij} = a_{i1}a_{1j} + a_{i2}a_{2j} + \cdots + a_{in}a_{nj} \neq 0. \tag{7.1}$$

There must be at least one term, say $a_{ik}a_{kj} = 1$, in this sum that is nonzero. In G this means that there is an edge from V_i to V_k and also an edge from V_k to V_j. Thus there is a two-step path from V_i to V_j. If $[A^2]_{ij} = r \neq 0$, then there must be r nonzero terms in Equation (7.1) and hence r two-step paths from V_i to V_j. This result extends easily to any power of A, so we have the following theorem.

THEOREM 7.1 If $A(G)$ is the adjacency matrix of the graph G and the i, j entry of $[A(G)]^s = r$, then there are r paths of length s connecting vertex i to vertex j. □

If the graph is undirected and has multiple edges, then a similar argument establishes the next theorem.

THEOREM 7.2 If the i, j entry of $[M(G)]^s = r$, then in G there are r paths of length s connecting vertex i to vertex j. □

EXAMPLE 6 Paths of Length 2 and 3 in $G1$

For the matrix $M = M(G1)$ from Examples 5 and 1 we use the MATALG program to compute

$$M^2 = \begin{bmatrix} 5 & 2 & 1 & 6 & 2 & 1 & \boxed{2} \\ 2 & 3 & 2 & 3 & 2 & 0 & 2 \\ 1 & 2 & \boxed{16} & 1 & 0 & 3 & 3 \\ 6 & 3 & 1 & 10 & 3 & 1 & 3 \\ 2 & 2 & 0 & 3 & 2 & 0 & 2 \\ 1 & 0 & 3 & 1 & 0 & 3 & 0 \\ 2 & 2 & 3 & 3 & 2 & 0 & 3 \end{bmatrix} \quad \text{and}$$

$$M^3 = \begin{bmatrix} 4 & 7 & 34 & 5 & 2 & 6 & \boxed{8} \\ 7 & 4 & 20 & 8 & 2 & 7 & 5 \\ 34 & 20 & 10 & 51 & 19 & 5 & 20 \\ 5 & 8 & 51 & 6 & \boxed{2} & 9 & 12 \\ 2 & 2 & 19 & 2 & 0 & 6 & 3 \\ 6 & 7 & 5 & 9 & 6 & 0 & 7 \\ 8 & 5 & 20 & 12 & 3 & 7 & 6 \end{bmatrix}.$$

Since the 1, 7 entries of M^2 and M^3 are 2 and 8 respectively, Theorem 7.2 tells us that there are only two paths of length 2 connecting T1 and T7 while there are eight paths of length 3 connecting T1 and T7. Can you find them? Note that all the diagonal entries of M are zero, but the diagonal entries of M^2 are nonzero. The "16" in the 3, 3 position of M^2 means that there are 16 paths of length 2 connecting T3 to T3. These closed paths are called cycles. Because the 4, 5 entry of M^3 is 2 there should be precisely two paths of length 3 connecting T4 and T5; this can be verified by direct inspection of the graph.

Because every *off-diagonal entry* (an entry not positioned on the diagonal of the matrix) of M^3 is nonzero it follows that every town is connected to every other town by at least one path of length 3. From M^2 we see that the only pairs not connected by a path of length 2 are (T2, T6), (T3, T5), (T5, T6), and (T6, T7); all of these are connected by paths of length 1. Thus, every pair of towns is connected by a path of length 1 or 2 and this is equivalent to the fact that the matrix $M + M^2$ has no zero entries. ■

DEFINITION 7.3 A graph is **strongly connected** if, and only if, there is a path connecting any pair of vertices. □

If the graph contained n vertices then it would never be necessary to consider a path of length greater than $n - 1$. Thus the graph would be strongly connected if, and only if, the matrix

$$A(G) + A(G)^2 + \cdots + A(G)^{n-1} = D$$

had no zero entries. The calculations in Example 6 show that the graph of Example 1 is strongly connected. If the edge from P2 to P5 is removed from the graph of Example 4, then the resulting graph is not strongly connected.

EXAMPLE 7 Breaking Ties in Incomplete Tournaments

Who should be declared the winner of the incomplete tournament described in Example 4, assuming that no further matches can be played?

At first glance it would appear that P4, P5, and P6 are tied with a record of two wins and one loss. Let us break the tie by considering the quality of the wins, that is, let us assume that a win over a strong player is more impressive than a win over a weak player. We will consider Pi to have second-order dominance over Pj if Pi beat someone who beat Pj, that is, if there is a path of length 2 from Pi to Pj in the graph of Example 4. The matrix of second-order dominance would be $A(G4)^2$ and the sum $A(G4) + A(G4)^2$ would represent both the first- and second-order dominance. This matrix is

$$A(G4) + A(G4)^2 = \begin{bmatrix} 0 & 2 & 1 & 0 & 1 & 0 & 0 & 0 \\ 1 & 0 & 0 & 0 & 1 & 1 & 0 & 0 \\ 0 & 1 & 0 & 0 & 1 & 0 & 0 & 0 \\ 1 & 1 & 0 & 0 & 1 & 0 & 1 & 0 \\ 1 & 1 & 2 & 0 & 0 & 1 & 0 & 1 \\ 0 & 1 & 1 & 1 & 0 & 0 & 0 & 1 \\ 1 & 1 & 1 & 0 & 0 & 0 & 0 & 0 \\ 0 & 1 & 0 & 1 & 0 & 0 & 1 & 0 \end{bmatrix}.$$

The sum of the entries in row i indicates the number of matches in which the ith player had either first- or second-order dominance. Using this evaluation method we conclude that P5 is the strongest player and should be declared the winner. ■

EXERCISES 7.1

1. Find the matrix $M(G)$ for the following graph. Find the number of paths of length 2 connecting T1 and T3. Find the number of paths of length 3 connecting T4 and T3.

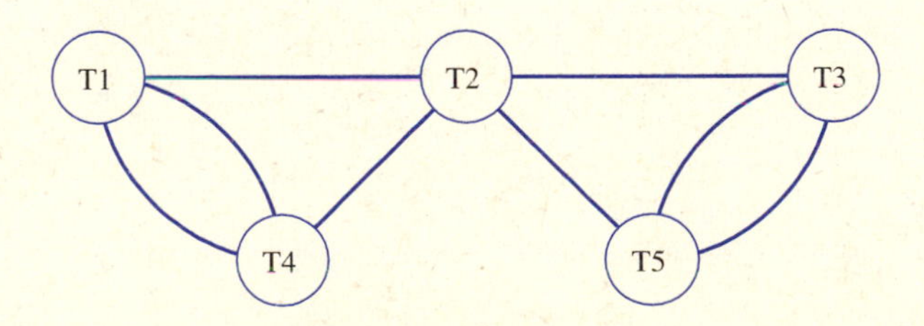

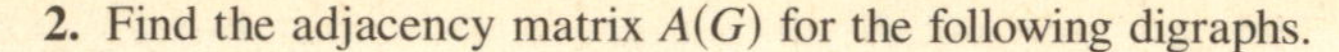

2. Find the adjacency matrix $A(G)$ for the following digraphs.

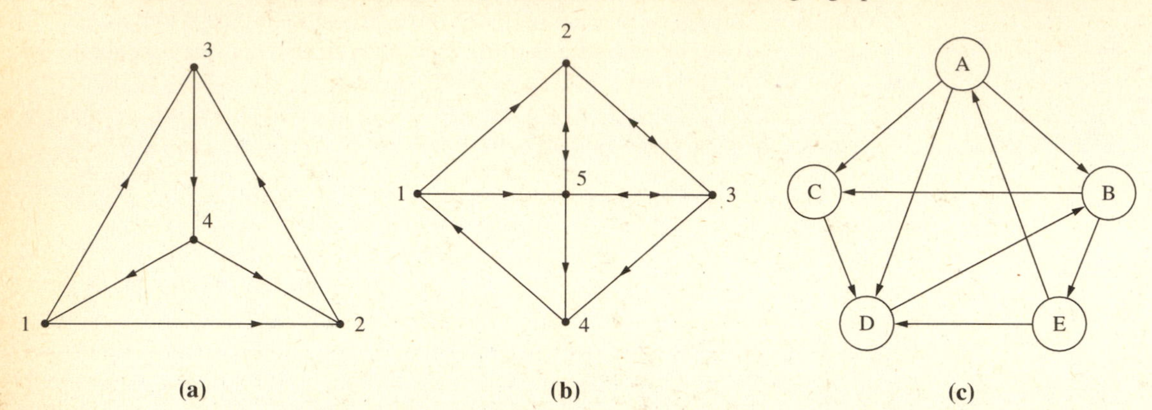

(a) (b) (c)

3. Find a directed graph whose adjacency matrix is as shown.

(a) $\begin{bmatrix} 0 & 0 & 1 & 1 & 0 \\ 0 & 0 & 0 & 1 & 1 \\ 0 & 1 & 0 & 0 & 0 \\ 1 & 0 & 0 & 0 & 0 \\ 0 & 0 & 1 & 0 & 0 \end{bmatrix}$ **(b)** $\begin{bmatrix} 0 & 1 & 0 & 0 & 0 \\ 0 & 0 & 1 & 0 & 0 \\ 0 & 0 & 0 & 1 & 0 \\ 0 & 0 & 0 & 0 & 1 \\ 1 & 0 & 0 & 0 & 0 \end{bmatrix}$ **(c)** $\begin{bmatrix} 0 & 0 & 0 & 1 & 0 & 0 \\ 1 & 0 & 1 & 1 & 0 & 0 \\ 0 & 0 & 0 & 1 & 0 & 1 \\ 0 & 0 & 1 & 0 & 0 & 1 \\ 0 & 0 & 0 & 0 & 0 & 0 \\ 1 & 1 & 0 & 0 & 1 & 0 \end{bmatrix}$

4. Which of the digraphs in Exercises 2 and 3 are strongly connected?

5. Make a reasonable definition of equivalence for two digraphs and then determine how the adjacency matrices of two equivalent graphs are related.

6. A **clique** in a digraph is a subset of the vertices of the graph satisfying the following three conditions:

(i) The subset contains at least three vertices.

(ii) For each pair of vertices in the subset there is a two-way path connecting them.

(iii) No other vertices can be added to the subset and still satisfy property 2.

(a) Find a clique in the graph of Exercise 2(b).

(b) Show that the ith vertex is a member of a clique if, and only if, the entry in the i, i position of S^3 is not 0, where S is defined by $s_{ij} = a_{ij}a_{ji}$, and $A = A(G)$.

7. A group of six individuals have been interacting for some time and the following influence pattern has been observed:

Adam influences Fred and Bill;
Dan is influenced by Ed;
Dan influences Charlie and Bill;
Bill and Fred influence Charlie;
Fred is influenced by Dan;
Charlie influences Adam and Ed; and
Ed is influenced by Adam, Fred, and Bill.

(a) Represent the above information by a digraph and find its adjacency matrix.
(b) Who influences the most people? Who is influenced by the most people?
(c) By considering second-order influence, determine who is the best choice for leader of the group.

8. Suppose that the Big Ten football season ended in a tie between Michigan and Iowa. Use graph theory to determine which team will be the conference's best representative in the Rose Bowl; consider first- and second-order dominance. The records of the teams are given in the table below, where + means a victory and 0 means the teams did not play.

	ILL	*IND*	*IOW*	*MIC*	*MSU*	*MIN*	*NWU*	*OSU*	*PUR*	*WIS*	*W-L*
Illinois (ILL)	0	+	−	−	+	0	+	+	−	+	5–3
Indiana (IND)	−	0	0	−	−	−	+	−	−	−	1–7
Iowa (IOW)	+	0	0	+	+	+	+	−	+	+	7–1
Michigan (MIC)	+	+	−	0	+	+	0	+	+	+	7–1
Michigan State (MSU)	−	+	−	−	0	+	+	0	+	+	5–3
Minnesota (MIN)	0	+	−	−	−	0	+	−	+	+	4–4
Northwestern (NWU)	−	−	−	0	−	−	0	−	−	−	0–8
Ohio State (OSU)	−	+	+	−	0	+	+	0	+	−	5–3
Purdue (PUR)	+	+	−	−	−	−	+	−	0	0	3–5
Wisconsin (WIS)	−	+	−	−	−	−	+	+	0	0	3–5

9. Find the adjacency matrices for the following graphs.

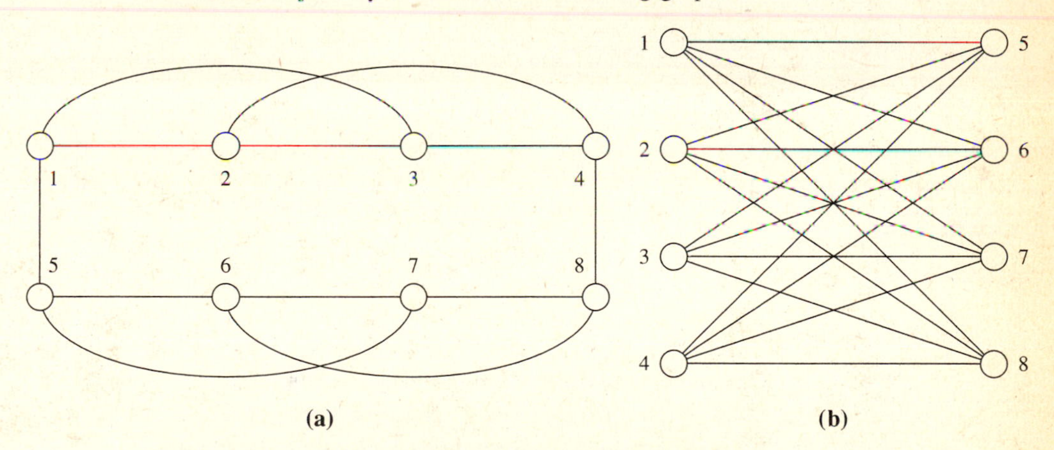

(a) (b)

10. Construct a graph whose vertices represent the students in this class. Connect two students with an edge if they have previously taken a course together. Find the adjacency matrix of this graph.

11. Let $A(G)$ be the adjacency matrix of a graph and let U be a column matrix with each entry 1. Show that the ith entry of $A(G)U$ tells how many nodes are connected directly to node i.

7.2 LEAST SQUARES APPROXIMATIONS

(Requires: Chapters 1 and 3)

A very common problem in applications is to find an analytic expression that represents a collection of experimental data. We know that two points determine a straight line, that three points determine a parabola, that four points determine a cubic polynomial, and so on. Suppose that we were given the six data points shown in Figure 7.5(a). The relationship between x and y seems to be almost linear, so it would not seem reasonable to fit a fifth-degree polynomial, as in Figure 7.5(b), to this data. A much more reasonable approach is to find a linear function $y = a + bx$ that best represents the given data; such a line is drawn in Figure 7.5(c).

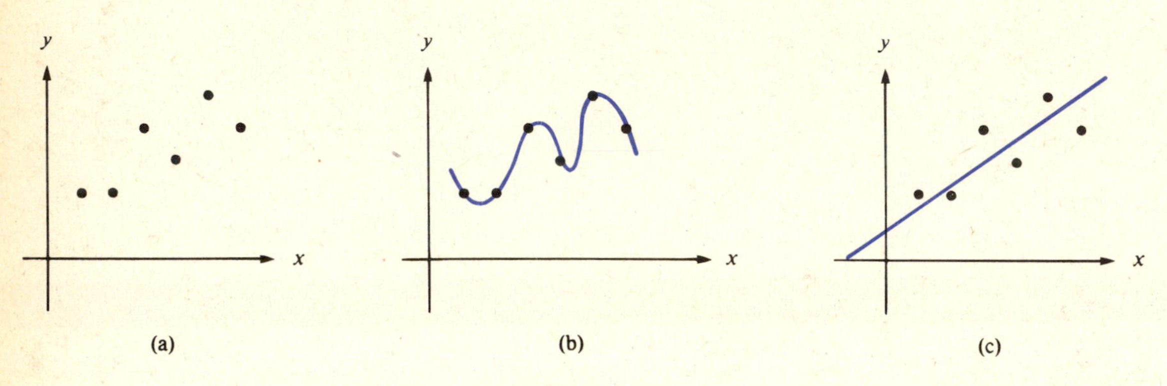

FIGURE 7.5

Let us try to find the straight line $y = a + bx$ that best fits the data points

$$(x_1, y_1), (x_2, y_2), \cdots, (x_k, y_k).$$

Ideally we would like each of the data points to lie on the line; that is, we would want a and b to satisfy each of the k equations

$$a + bx_i = y_i, \quad i = 1, 2, \cdots, k. \tag{7.2}$$

This system of k equations in 2 unknowns is overdetermined in the sense that there are more equations than unknowns. We have seen in Chapter 1 that such a system is almost certain to be inconsistent. Instead of insisting that each data point lie on the line, let us try to find a line that is close to each of the data points. We will measure the deviation of a given data point from the line $y = a + bx$ by the difference

$$d_i = y_i - (a + bx_i), \quad i = 1, 2, \cdots, k, \tag{7.3}$$

which represents the vertical distance from the point to the line, as shown in Figure 7.6.

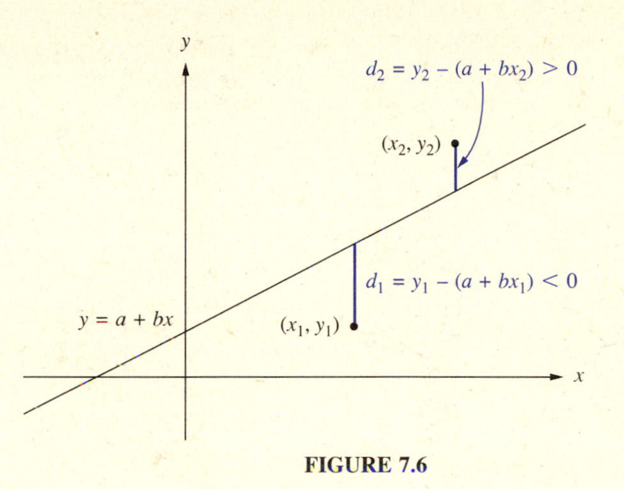

FIGURE 7.6

In order to see why we choose to measure the deviations parallel to the y-axis rather than by measuring the shortest distance from the point to the line we write the problem in matrix notation. If we define the matrices

$$A = \begin{bmatrix} 1 & x_1 \\ 1 & x_2 \\ \cdot & \cdot \\ \vdots & \vdots \\ 1 & x_k \end{bmatrix}, \quad X = \begin{bmatrix} a \\ b \end{bmatrix}, \quad Y = \begin{bmatrix} y_1 \\ y_2 \\ \cdot \\ \vdots \\ y_k \end{bmatrix}, \quad \text{and} \quad D = \begin{bmatrix} d_1 \\ d_2 \\ \cdot \\ \vdots \\ d_k \end{bmatrix},$$

then we can write (inconsistent) System (7.2) as $AX = Y$ and Equation (7.3) as the matrix equation

$$Y - AX = D.$$

The natural way to measure the deviation of the data points from the line is to use the norm of the vector D in $\Re^k$. Thus, we wish to choose X so that

$$\|Y - AX\|^2 = \|D\|^2 = \sum_{i=1}^{k} d_i^2$$

is as small as possible. The resulting vector X_0 in $\Re^2$ is called the **least squares solution** to the inconsistent system $AX = Y$.

EXAMPLE 1 **Two Lines Close to Four Points**

Consider the four data points: (1, 3), (2, 2), (3, 5), and (4, 4).

These points are plotted in Figure 7.7 along with the lines $y = 1 + x$ and $y = \frac{1}{3}(x + 8)$ which are reasonable approximations to the given data. The following table lists the values of these two approximations at the data points as well as the deviations.

Data		$y = \frac{1}{3}(x + 8)$			$y = 1 + x$		
x	y	y	d	d^2	y	d	d^2
1	3	3	0	0	2	1	1
2	2	$\frac{10}{3}$	$-\frac{4}{3}$	$\frac{16}{9}$	3	−1	1
3	5	$\frac{11}{3}$	$\frac{4}{3}$	$\frac{16}{9}$	4	1	1
4	4	4	0	0	5	−1	1
			$\Sigma d_i^2 = \frac{32}{9}$			$\Sigma d_i^2 = 4$	

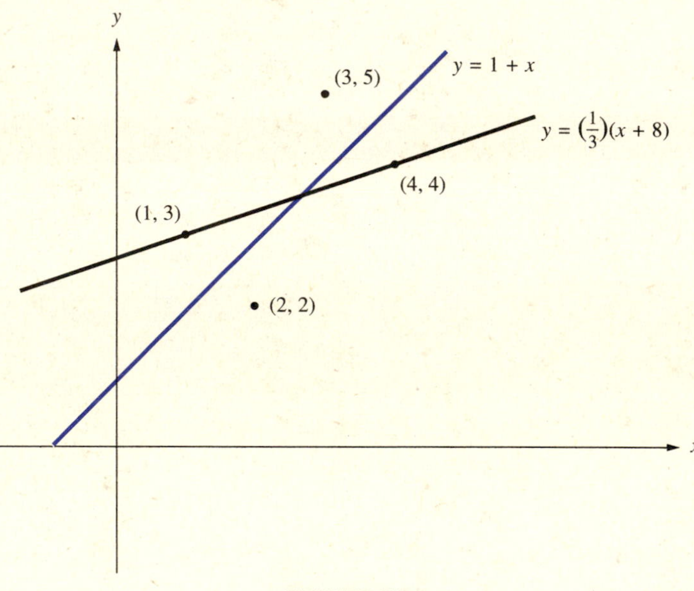

FIGURE 7.7

Note that for both of these approximations the sum of the deviations is zero. Clearly Σd_i is not a good measure of the quality of the fit. The sum of the squares of the deviations, Σd_i^2, provides a more reasonable measure of the quality of the fit since there will be no cancellation and the value of Σd_i^2 will be zero only if the fit is exact. Since $\frac{32}{9} < 4$, we see that $y = \frac{1}{3}(x + 8)$ is a better approximation (by our criterion) to the given data than $y = 1 + x$. It is not clear at this point that $y = \frac{1}{3}(x + 8)$ is the best approximation. ■

Note that the vector AX is an element of the column space of the matrix A. Thus, we wish to choose X_0 so that AX_0 is the element of $\mathscr{CS}(A)$ that is closest to Y. The diagram in Figure 7.8 illustrates the situation for the case $k = 3$. Note that we want to choose X_0 so that AX_0 is the orthogonal projection of Y onto $\mathscr{CS}(A)$ and $D_0 = Y - AX_0$ is orthogonal to the column space of A.

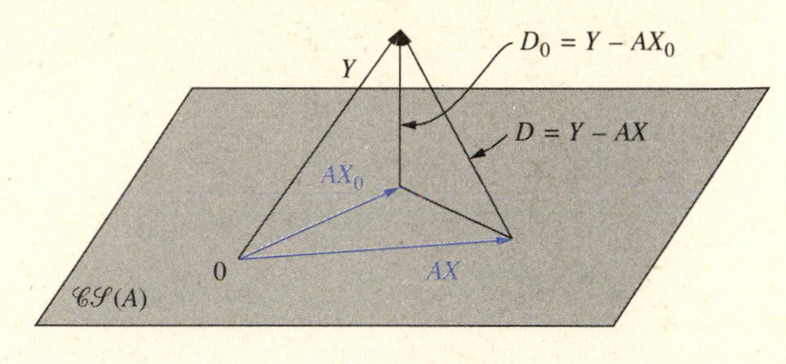

FIGURE 7.8

If $k > 3$, then the geometric argument is not valid, but it is still true that we need to choose X_0 so that $D_0 = Y - AX_0$ is orthogonal to $\mathscr{CS}(A)$. To see that this is true in general we let X in $\Re^k$ be any other candidate for a solution and consider

$$\begin{aligned} D = Y - AX &= Y - AX_0 + AX_0 - AX \\ &= (Y - AX_0) + A(X_0 - X). \end{aligned}$$

If $Y - AX_0$ is orthogonal to $\mathscr{CS}(A)$, then by the generalization of the Pythagorean theorem to $\Re^k$, given in Exercise 27 of Section 3.6, it follows that

$$\|D\|^2 = \|Y - AX_0\|^2 + \|AX_0 - AX\|^2 = \|D_0\|^2 + \|AX_0 - AX\|^2.$$

Thus $\|D_0\| \leq \|D\|$ for every choice of X.

If X is to be chosen so that $Y - AX$ is orthogonal to $\mathscr{CS}(A)$, then $Y - AX$ must be orthogonal to each column of the matrix A. This is equivalent to the matrix condition

$$A^T(Y - AX) = 0 \qquad \text{or} \qquad A^TAX = A^TY.$$

This last equation is called the **normal equation** for the least squares problem; for the case of finding the best fitting straight line it is a system of two linear equations in two unknowns.

Note that the above argument does not require that A have only two columns; it applies to any overdetermined system $AX = Y$. Note also that $\text{rank}(A^TA) = \text{rank}(A)$, by Exercise 9 of Section 3.6, so that A^TA will be nonsingular if, and only if, the columns of A are linearly independent. Even if A does not have full rank, this system will always be consistent (see Exercise 14). We summarize this discussion in the following theorem.

THEOREM 7.3 Let A be an $m \times n$ matrix with $m > n$. The least squares solution of the linear system $AX = Y$ is an exact solution of the normal equations

$$A^TAX = A^TY. \tag{7.4}$$

The normal equations are always consistent. If A has rank n, then the $n \times n$ matrix A^TA is nonsingular and the unique solution of the normal equations is

$$X = (A^TA)^{-1}A^TY. \quad \square$$

EXAMPLE 2 A Best Fitting Line

Find the straight line that best fits the following four data points: (1, 3), (2, 2), (3, 5), and (4, 4).

The matrices associated with this problem are

$$A = \begin{bmatrix} 1 & 1 \\ 1 & 2 \\ 1 & 3 \\ 1 & 4 \end{bmatrix}, \quad Y = \begin{bmatrix} 3 \\ 2 \\ 5 \\ 4 \end{bmatrix}, \quad X = \begin{bmatrix} a \\ b \end{bmatrix}, \quad A^TA = \begin{bmatrix} 4 & 10 \\ 10 & 30 \end{bmatrix}, \quad \text{and}$$

$$A^TY = \begin{bmatrix} 14 \\ 38 \end{bmatrix}.$$

Thus Normal Equations (7.4), in matrix form, are

$$\begin{bmatrix} 4 & 10 \\ 10 & 30 \end{bmatrix}\begin{bmatrix} a \\ b \end{bmatrix} = \begin{bmatrix} 14 \\ 38 \end{bmatrix}$$

and the solution, obtained by Gaussian elimination, is $a = 2$ and $b = \frac{3}{5}$. The best fitting line, $y = 2 + \frac{3}{5}x$, and the data points are shown in Figure 7.9.

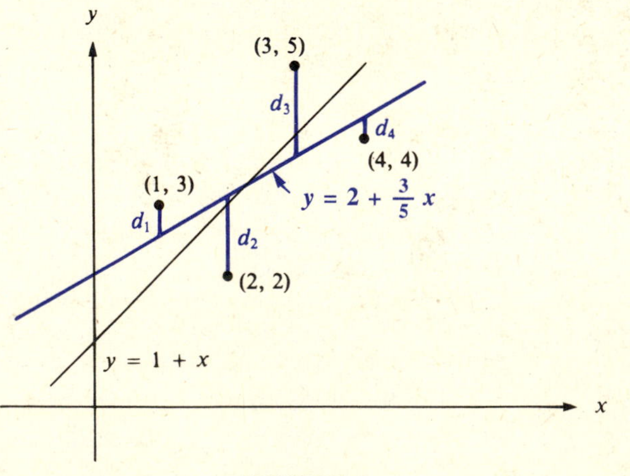

FIGURE 7.9

In the following table we show the deviations for the best fitting line and also for the line $y = 1 + x$, which also fits the data reasonably well.

Data		*Best Fitting Line*			$y = 1 + x$		
x	y	y	d	d^2	y	d	d^2
1	3	2.6	.4	.16	2	1	1
2	2	3.2	−1.2	1.44	3	−1	1
3	5	3.8	1.2	1.44	4	1	1
4	4	4.4	−.4	.16	5	−1	1
				$\Sigma\, d_i^2 = 3.20$			$\Sigma\, d_i^2 = 4$

Note that the sum of the squares of the deviations for the least squares line is 3.20, which is less than the sum of the squares of the deviations for both of the lines $y = 1 + x$ and $y = \frac{1}{3}(x + 8)$ considered in Example 1. ■

If we want to fit a higher degree polynomial to a given set of data points, the basic procedure is the same; the only thing that changes is the number of columns of the matrix A. Suppose we want to fit a polynomial of degree 3 to the data

$$(x_1, y_1), (x_2, y_2), \cdots, (x_k, y_k).$$

If we let $y = a + bx + cx^2 + dx^3$ be the desired cubic polynomial, we are led to the linear system

$$\begin{aligned} a + bx_1 + cx_1^2 + dx_1^3 &= y_1 \\ a + bx_2 + cx_2^2 + dx_2^3 &= y_2 \\ \vdots \quad\quad \vdots \quad\quad \vdots \quad\quad \vdots \quad &\quad \vdots \\ a + bx_k + cx_k^2 + dx_k^3 &= y_k. \end{aligned}$$

If $k > 4$, this system is overdetermined and we cannot expect Gaussian elimination to produce an exact solution. In this case we will seek a least squares solution. In matrix notation this system is

$$AX = \begin{bmatrix} 1 & x_1 & x_1^2 & x_1^3 \\ 1 & x_2 & x_2^2 & x_2^3 \\ \vdots & \vdots & \vdots & \vdots \\ 1 & x_k & x_k^2 & x_k^3 \end{bmatrix} \begin{bmatrix} a \\ b \\ c \\ d \end{bmatrix} = \begin{bmatrix} y_1 \\ y_2 \\ \vdots \\ y_k \end{bmatrix} = Y.$$

Note that every 4×4 submatrix of A is a Vandermonde matrix and is hence nonsingular by Review Exercise 14 of Chapter 2. Therefore A is of rank 4 provided only that the x_i are distinct. By Theorem 7.3 the least squares solution of this system is the exact solution of the normal equations

$$A^TAX = A^TY.$$

EXAMPLE 3 **A Least Squares Parabola**

Find the parabola $y = a + bx + cx^2$ that best fits the five data points (0, 4), (1, 1), (2, 4), (3, 6), (4, 8).

Attempting to fit $y = a + bx + cx^2$ to the given data leads to the overdetermined system

$$AX = \begin{bmatrix} 1 & 0 & 0 \\ 1 & 1 & 1 \\ 1 & 2 & 4 \\ 1 & 3 & 9 \\ 1 & 4 & 16 \end{bmatrix} \begin{bmatrix} a \\ b \\ c \end{bmatrix} = \begin{bmatrix} 4 \\ 1 \\ 4 \\ 6 \\ 8 \end{bmatrix} = Y,$$

for which the normal equations are

$$A^TAX = \begin{bmatrix} 5 & 10 & 30 \\ 10 & 30 & 100 \\ 30 & 100 & 354 \end{bmatrix} \begin{bmatrix} a \\ b \\ c \end{bmatrix} = \begin{bmatrix} 23 \\ 59 \\ 199 \end{bmatrix} = A^TY.$$

The solution of the normal equations is

$$X = \tfrac{1}{70}\begin{bmatrix} 230 \\ -89 \\ 45 \end{bmatrix} = \begin{bmatrix} 3.29 \\ -1.27 \\ 0.64 \end{bmatrix},$$

so the best fitting parabola is $y = \frac{1}{90}(230 - 89x + 45x^2)$. For this parabola the sum of the squares of the deviations is 4.64. The data points and the least squares parabola are shown in Figure 7.10. ■

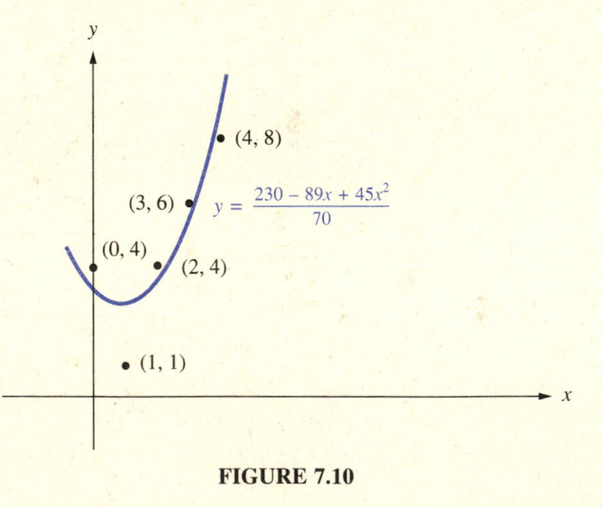

FIGURE 7.10

The basic procedure we have illustrated above can also be used to fit functions of several variables to given data. We illustrate this in our next example.

EXAMPLE 4 A Best Fitting Plane

Find the plane $z = a + bx + cy$ that best fits the data

$$(1, 1, 5), \quad (2, 1, 4), \quad (1, 2, 8), \quad (0, 1, 4), \quad (1, 0, 2), \quad \text{and} \quad (2, 2, 6).$$

If the given points are to lie on the plane $z = a + bx + cy$, then the following equations must be satisfied:

$$\begin{aligned} a + b + c &= 5 \\ a + 2b + c &= 4 \\ a + b + 2c &= 8 \\ a \quad + c &= 4 \\ a + b \quad &= 2 \\ a + 2b + 2c &= 6 \end{aligned} \quad \text{or} \quad \begin{bmatrix} 1 & 1 & 1 \\ 1 & 2 & 1 \\ 1 & 1 & 2 \\ 1 & 0 & 1 \\ 1 & 1 & 0 \\ 1 & 2 & 2 \end{bmatrix} \begin{bmatrix} a \\ b \\ c \end{bmatrix} = \begin{bmatrix} 5 \\ 4 \\ 8 \\ 4 \\ 2 \\ 6 \end{bmatrix}.$$

This system is overdetermined and does not have an exact solution. The best solution, in the least squares sense, is the solution of the normal equations

$$A^TAX = \begin{bmatrix} 6 & 7 & 7 \\ 7 & 11 & 9 \\ 7 & 9 & 11 \end{bmatrix} \begin{bmatrix} a \\ b \\ c \end{bmatrix} = \begin{bmatrix} 29 \\ 35 \\ 41 \end{bmatrix},$$

for which the exact solution is $a = \frac{96}{44}$, $b = -\frac{16}{44}$, and $c = \frac{116}{44}$. Thus the best fitting plane is $z = \frac{96 - 16x + 116y}{44}$, for which the sum of the squares of the errors is 2.362. The table on page 398 compares the value of this function with the values of z at the data points and with the plane $z = 2 + x + 2y$, which also fits the data reasonably well. Figure 7.11 is a MATLAB plot of the data points and the least squares plane. Note that the point (0, 1, 4) is below the plane and does not show. ■

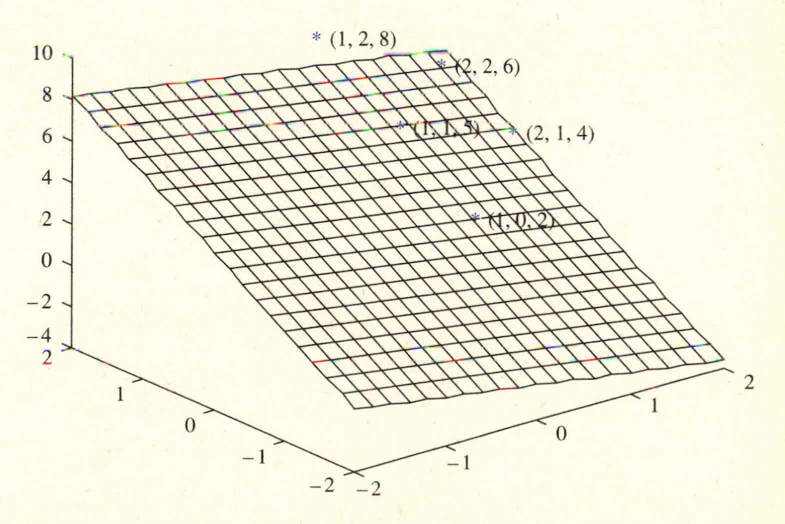

FIGURE 7.11

Data			*Least-Squares Plane*			$z = 2 + x + 2y$		
x	y	z	z	d	d^2	z	d	d^2
1	1	5	4.455	.545	.2970	5	0	0
2	1	4	4.091	−.091	.0083	6	−2	4
1	2	8	7.091	.909	.8263	7	1	1
0	1	4	4.818	−.818	.6691	4	0	0
1	0	2	1.818	.182	.0331	3	−1	1
2	2	6	6.727	−.727	.5285	8	−2	4
					$\Sigma\, d_i^2 = 2.362$			$\Sigma\, d_i^2 = 10$

The procedure just illustrated seems quite straightforward, but there is a computational problem that should be mentioned briefly. The difficulty is that the normal equations are notoriously ill-conditioned; that is, small changes in the data can lead to large changes in the solution. If the computations are being done on a computer this is certainly cause for concern and caution. A thorough discussion of ill-conditioning is beyond the scope of this book, but we should at least mention the problem and indicate how to improve the computational procedure.

The key ingredient is commonly called the QR factorization of the $m \times n$ coefficient matrix A. In this factorization the $m \times n$ matrix Q is to have orthonormal columns, so that $Q^TQ = I$, and the $n \times n$ matrix R is to be upper triangular and nonsingular. As we indicated in Section 3.6, finding the matrices Q and R is equivalent to the Gram-Schmidt process for the column space of A. In practice, Q is usually found as a product of plane rotations (see Exercises 17 and 20 of Section 4.4) or as a product of elementary reflectors (see Exercise 9 of Section 5.3 and Exercise 10 of this section). If we have the QR factorization of the matrix A, then we can use it to simplify Normal Equations (7.4) as follows:

$$A^TAX = A^TY \rightarrow (QR)^T(QR)X = (QR)^TY \rightarrow$$
$$R^T(Q^TQ)RX = R^TQ^TY \rightarrow R^TRX = R^TQ^TY \rightarrow RX = Q^TY.$$

Note that the final step used the fact that R, and hence R^T, is nonsingular if A has rank n. Note also that the final equation, $RX = Q^TY$, can be obtained directly from $AX = Y$ by substituting QR for A and then multiplying by Q^T.

EXAMPLE 5 Using the QR Factorization to Solve the Normal Equations

For the 6×3 matrix A of Example 4 we have the following MATALG generated QR factorization: $A = QR$ with

$$Q = \begin{bmatrix} .40825 & -.099015 & -.073127 \\ .40825 & .49507 & -.25595 \\ .40825 & -.099015 & .54845 \\ .40825 & -.69310 & .10969 \\ .40825 & .099015 & -.69471 \\ .40825 & .49507 & .36564 \end{bmatrix} \quad \text{and}$$

$$R = \begin{bmatrix} 2.4495 & 2.8577 & 2.8577 \\ 0 & 1.6833 & 0.49507 \\ 0 & 0 & 1.6088 \end{bmatrix}$$

and the system $RX = Q^T Y$ is

$$\begin{bmatrix} 2.4495 & 2.8577 & 2.8577 \\ 0 & 1.6833 & 0.49507 \\ 0 & 0 & 1.6088 \end{bmatrix} \begin{bmatrix} a \\ b \\ c \end{bmatrix} = \begin{bmatrix} 11.839 \\ .69310 \\ 4.2414 \end{bmatrix}.$$

The solution of this system is $a = 2.1818$, $b = -.36364$, and $c = 2.6364$, which agrees with the solution obtained in Example 4 by solving Normal Equations (7.4) exactly. ■

EXERCISES 7.2

1. Find the straight line that best fits the following sets of data points, and compare the sum of the squares of the deviations from this line and the line determined by the first two points.
(a) (1, 2), (2, 4), (3, 8)
(b) (1, 4), (2, 9), (3, 10), (4, 15), (5, 16), (6, 21)
(c) (1, 4.2), (2, 4.7), (3, 4.9), (4, 5.6)
(d) (40, 472), (45, 457), (50, 442), (55, 423), (60, 411)

2. Find the parabola that best fits the following sets of data points.
(a) (−2, 45), (0, 10), (2, −16), (4, −30)
(b) (1, 1), (2, 0), (3, 1), (4, −2), (5, −5)
(c) (3, 0), (4, −10), (6, −48), (7, −76)

3. Find the plane that best fits the following sets of data points.
(a) (1, 1, 1), (0, 2, 1), (1, 2, 0), (0, 4, 0)
(b) (1, 2, −1), (2, 3, −3), (3, 2, 5), (2, 1, 8), (2, 2, 3)
(c) Compare the sum of the squares of the deviations for the plane in part (b) and the plane $z = 3x - 5y + 6$.

4. Show that System (7.2) is consistent if, and only if, all the data points lie on a straight line. Show that A has rank 2 unless all the data points lie on a vertical line.

5. Solve each of the following 2×2 systems exactly

$$\begin{aligned} x + 3y &= 4 \\ 1.001x + 3y &= 4.001 \end{aligned} \qquad \begin{aligned} x + 3y &= 4 \\ 1.001x + 3y &= 4.004 \end{aligned}$$

Interpret each of the equations as a straight line and explain why the small change in the right-hand side generates such a large change in the answer. This is an example of an ill-conditioned system.

6. Find the QR factorization for each of the coefficient matrices in Exercises 1(a), 2(a), and 3(a) and use it to solve the normal equations.

7. In physics, Hooke's law asserts that there is a linear relationship, $s = a + hf$, between the length s of a spring and the force f applied to it. The following data were collected in an elementary physics laboratory; use it to estimate the natural length (a) of the spring and the Hooke's law constant (h) for the given spring.

f (pounds)	.5	1	2	4
s (feet)	4	8	12	24

8. Let $\mathcal{W}$ be an n-dimensional subspace of $\Re^m$ and let $\{\mathbf{w}_1, \ldots, \mathbf{w}_n\}$ be an orthonormal basis for $\mathcal{W}$. Extend this basis to an orthonormal basis $\{\mathbf{w}_1, \ldots, \mathbf{w}_n, \ldots, \mathbf{w}_m\}$ for $\Re^m$. Let

$$\mathbf{v} = a_1\mathbf{w}_1 + \cdots + a_n\mathbf{w}_n + \cdots + a_m\mathbf{w}_m$$

be an arbitrary vector in $\Re^m$. Consider any vector

$$x = b_1\mathbf{w}_1 + \cdots + b_n\mathbf{w}_n$$

in $\mathcal{W}$. Compute $\|\mathbf{v} - \mathbf{x}\|$ and determine when $\mathbf{x}$ is closest to $\mathbf{v}$.

9. Find a least squares solution for the following linear systems

(a) $$\begin{bmatrix} 1 & 6 & 11 \\ 2 & 7 & 12 \\ 3 & 8 & 12 \\ 3 & 8 & 13 \\ 4 & 9 & 14 \\ 5 & 10 & 15 \end{bmatrix} X = \begin{bmatrix} 1 \\ 2 \\ 3 \\ 4 \\ 5 \end{bmatrix}.$$

(b) $$\begin{bmatrix} 1 & 1 & 1 & 1 \\ 1 & 2 & 4 & 8 \\ 1 & -1 & 1 & -1 \\ 1 & -2 & 4 & -8 \\ 1 & 3 & 9 & 27 \\ 1 & -3 & 9 & -27 \end{bmatrix} X = \begin{bmatrix} -2 \\ -7 \\ 8 \\ 1 \\ -4 \\ -22 \end{bmatrix}.$$

10. Let $\mathbf{x}$ be a unit vector in $\Re^n$ and define $H = I - 2\mathbf{x}\mathbf{x}^T$.

(a) Show that H is both orthogonal and symmetric.

(b) Show that $H\mathbf{x} = -\mathbf{x}$ and that if $\mathbf{w} \cdot \mathbf{x} = 0$, then $H\mathbf{w} = \mathbf{w}$. Show that these results can be interpreted geometrically to mean that H is a reflection in the subspace orthogonal to $\mathbf{x}$. Hence, these matrices are called **elementary reflectors** or **Householder matrices.**

(c) Show that if $\|\mathbf{w}\| = \|\mathbf{v}\|$ and $\mathbf{x} = \frac{\mathbf{w} - \mathbf{v}}{\|\mathbf{w} - \mathbf{v}\|}$, then $H\mathbf{w} = \mathbf{v}$.

(d) For $\mathbf{w} = [3 \quad 4 \quad 12]^T$, find $\mathbf{x}$ such that $H\mathbf{w} = [13 \quad 0 \quad 0]^T$.

(e) If A is $m \times n$ with $m > n$, show that there exists elementary reflectors, $H_1, \ldots, H_{n-1}$ such that

$$H_{n-1}, \cdots H_2 H_1 A = \begin{bmatrix} R \\ 0 \end{bmatrix},$$

where R is $n \times n$ and triangular.

(f) Show how the result of part (e) can be used to find the QR factorization of the matrix A.

11. Determine scalars x_i such that $p(t) = x_1 \sin t + x_2 \cos t + x_3 \sin 2t + x_4 \cos 2t$ best fits the data $(-2, -4)$, $(-1, -4)$, $(0, -2)$, $(1, 2)$, $(2, 8)$, $(3, 16)$. Plot the resulting curve showing the data points on the same graph.

12. Use the least squares technique to estimate the population of the United States in 1990 from the following U.S. Census Bureau data by fitting a quadratic $p(t) = a + bt + ct^2$ to the given data. The population figures are rounded to the nearest thousand.

Year	*Population*
1900	75,995,000
1910	91,972,000
1920	105,711,000
1930	122,775,000
1940	131,669,000
1950	150,697,000
1960	179,323,000
1970	203,235,000
1980	226,546,000

13. Repeat the solution of Exercise 12 using $p(t) = a + b(t - 1900) + c(t - 1900)^2$.

14. Show that the normal equations $A^TAX = A^TY$ are consistent even if A has less than full rank.

7.3 QUADRATIC FORMS

(Requires: Sections 4.5, 4.6, and 5.2)

In this section we will show how matrix techniques are useful in the study of quadratic forms. Quadratic forms are of importance in geometry, statistics, physics, engineering, number theory, and other areas of mathematics.

In general, a **quadratic form** in the n variables $x_1, x_2, \ldots, x_n$ is a second-degree polynomial of the form

$$\begin{aligned} q &= a_{11}x_1^2 + a_{12}x_1x_2 + \cdots + a_{1n}x_1x_n \\ &\quad + a_{21}x_2x_1 + a_{22}x_2^2 + \cdots + a_{2n}x_2x_n \\ &\quad + \cdots \\ &\quad + a_{n1}x_nx_1 + a_{n2}x_2x_n + \cdots + a_{nn}x_n^2 \\ &= \sum_{j=1}^{n}\left(\sum_{i=1}^{n} a_{ij}x_ix_j\right). \end{aligned}$$

It is much easier to deal with quadratic forms in matrix notation than in the above double sum notation. If we assume, without loss of generality, that $a_{ij} =$

a_{ji} (if not, we can write $a_{ij}x_ix_j + a_{ji}x_jx_i = bx_ix_j + bx_jx_i$, where $b = [a_{ij} + a_{ji}]/2$), then we can write q as

$$q = [x_1\ x_2 \cdots x_n]A\begin{bmatrix} x_1 \\ x_2 \\ \cdot \\ \cdot \\ \cdot \\ x_n \end{bmatrix} = X^TAX,$$

where X is a vector from $\mathfrak{R}^n$ and the matrix A is symmetric. Thus we can view Q as defining a function from $\mathfrak{R}^n$ to $\mathfrak{R}$, that is, a real-valued function on $\mathfrak{R}^n$.

Some examples of quadratic forms and their matrix representations follow. Note carefully how the cross product terms are handled so that the associated matrix is symmetric.

$$\begin{aligned} q_1 &= 3x_1^2 + 2x_1x_2 + 3x_2^2 = 3x_1^2 + x_1x_2 + x_2x_1 + 3x_2^2 \\ &= x_1(3x_1 + x_2) + x_2(x_1 + 3x_2) \\ &= [x_1\quad x_2]\begin{bmatrix} 3x_1 + \ \ x_2 \\ x_1 + 3x_2 \end{bmatrix} = [x_1\quad x_2]\begin{bmatrix} 3 & 1 \\ 1 & 3 \end{bmatrix}\begin{bmatrix} x_1 \\ x_2 \end{bmatrix}; \end{aligned}$$

$$q_2 = x_1^2 - 5x_2^2 + 3x_3^2 = [x_1\quad x_2\quad x_3]\begin{bmatrix} 1 & 0 & 0 \\ 0 & -5 & 0 \\ 0 & 0 & 3 \end{bmatrix}\begin{bmatrix} x_1 \\ x_2 \\ x_3 \end{bmatrix}; \quad \text{and}$$

$$\begin{aligned} q_3 &= 3x_1^2 - 5x_1x_2 + 4x_2^2 + 3x_1x_3 + 8x_2x_3 \\ &\quad + x_3^2 + 10x_1x_4 + 12x_3x_4 \\ &= [x_1\quad x_2\quad x_3\quad x_4]\begin{bmatrix} 3 & -\frac{5}{2} & \frac{3}{2} & 5 \\ -\frac{5}{2} & 4 & 4 & 0 \\ \frac{3}{2} & 4 & 1 & 6 \\ 5 & 0 & 6 & 0 \end{bmatrix}\begin{bmatrix} x_1 \\ x_2 \\ x_3 \\ x_4 \end{bmatrix}. \end{aligned}$$

You have most likely encountered quadratic forms in two variables in your algebra and calculus studies. There you considered the problem of graphing the equation

$$ax^2 + 2bxy + cy^2 = k^2. \tag{7.5}$$

This problem is relatively easy if $b = 0$. Recall that the graph of

$$ax^2 + cy^2 = k^2$$

is an ellipse if a and c are both positive and a hyperbola if ac is negative (see Figure 7.12). If $k = 0$, the hyperbola degenerates to two intersecting lines and the ellipse shrinks to a single point. The easy case ($b = 0$) corresponds to the case where the matrix of the quadratic form is a diagonal matrix.

To identify and sketch the more general curve given in Equation (7.5) we will use the eigenvalue theory for symmetric matrices that was developed in Sections 5.2 and 5.3.

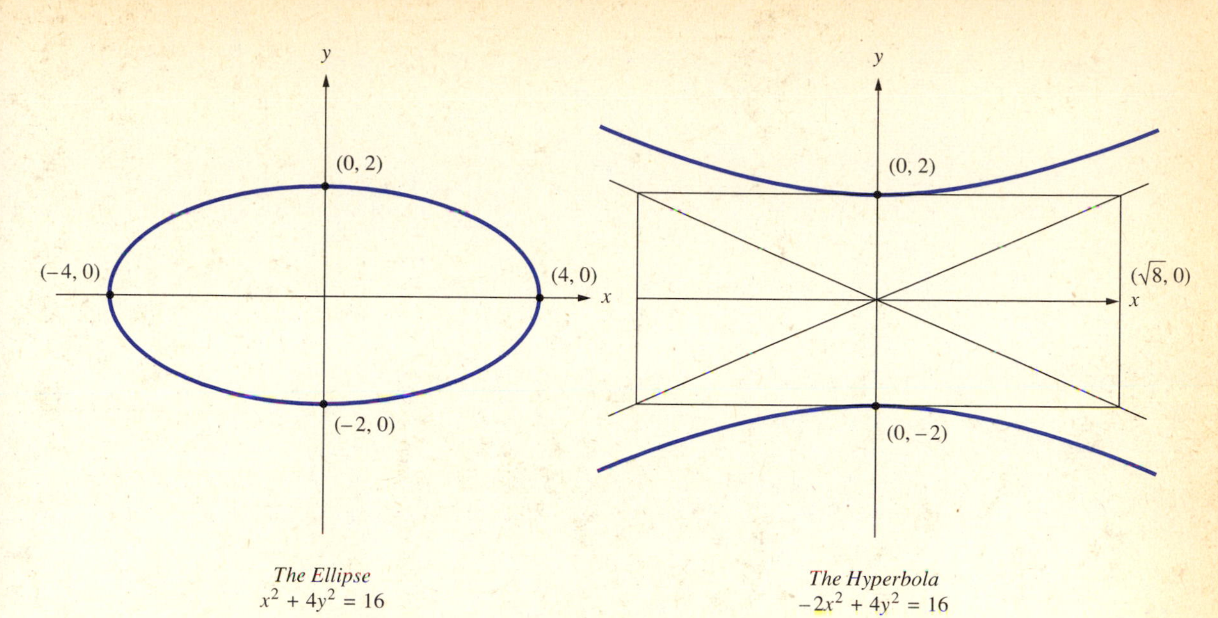

FIGURE 7.12

From Theorem 5.10 we know that for any 2×2 symmetric matrix A there exists an orthogonal matrix P such that P^TAP is a diagonal matrix with the eigenvalues of A on the diagonal;

$$P^TAP = \begin{bmatrix} \lambda_1 & 0 \\ 0 & \lambda_2 \end{bmatrix}. \tag{7.6}$$

Moreover, the columns of P are an orthonormal set of eigenvectors for the matrix A.

By Exercise 19 of Section 4.4 we know that every 2×2 orthogonal matrix is of one of the following forms:

$$P_1 = \begin{bmatrix} \cos\theta & -\sin\theta \\ \sin\theta & \cos\theta \end{bmatrix} \quad \text{or}$$

$$P_2 = \begin{bmatrix} \cos\theta & \sin\theta \\ \sin\theta & -\cos\theta \end{bmatrix} = \begin{bmatrix} \cos\theta & -\sin\theta \\ \sin\theta & \cos\theta \end{bmatrix}\begin{bmatrix} 1 & 0 \\ 0 & -1 \end{bmatrix}.$$

It was shown in Section 4.4 that P_1 represents a rotation through an angle θ and that P_2 represents a reflection in the x-axis followed by a rotation through an angle θ. Note that $\det(P_1) = 1$ and $\det(P_2) = -1$. It is easy to see that the matrix P of Equation (7.6) can always be chosen to be a rotation since if $P_2^TAP_2 = \text{Dg}\{\lambda_1, \lambda_2\}$, then

$$P_1^TAP_1 = \begin{bmatrix} 1 & 0 \\ 0 & -1 \end{bmatrix}\text{Dg}\{\lambda_1, \lambda_2\}\begin{bmatrix} 1 & 0 \\ 0 & -1 \end{bmatrix} = \text{Dg}\{\lambda_1, \lambda_2\}.$$

If we now make the substitution (coordinate change)

$$X = \begin{bmatrix} x \\ y \end{bmatrix} = PX' = P\begin{bmatrix} x' \\ y' \end{bmatrix}$$

where P is the matrix of Equation (7.6), we have

$$Q = X^TAX = (PX')^TA(PX) = X'T(P^TAP)X' = \lambda_1 x'^2 + \lambda_2 y'^2,$$

and the troublesome cross product term has been eliminated. Since

$$P = \begin{bmatrix} \cos\theta & -\sin\theta \\ \sin\theta & \sin\theta \end{bmatrix}$$

for some angle θ, the substitution $X = PX'$ amounts to a rotation of the coordinate axes through an angle θ. If we let $\mathbf{u}_1$ and $\mathbf{u}_2$ be the new basis vectors, then, using the development of Section 4.6, we see that P is the change-of-basis matrix from the new basis $\{\mathbf{u}_1, \mathbf{u}_2\}$ to the old basis $\mathcal{S} = \{\mathbf{e}_1, \mathbf{e}_2\}$. Thus, from Theorem 4.11

$$P = \begin{bmatrix} [\mathbf{u}_1]_{\mathcal{S}} & [\mathbf{u}_2]_{\mathcal{S}} \end{bmatrix} = [\mathbf{u}_1 \quad \mathbf{u}_2],$$

and we see that the eigenvectors of A are vectors in the direction of the new coordinate axes.

■ EXAMPLE 1 A Rotated Ellipse

Identify and sketch the graph of $3x^2 + 2xy + 3y^2 = 16$.

The matrix of this quadratic form is

$$A = \begin{bmatrix} 3 & 1 \\ 1 & 3 \end{bmatrix},$$

for which the characteristic polynomial is

$$\begin{aligned} \det(\lambda I - A) &= \det\begin{bmatrix} \lambda - 3 & -1 \\ -1 & \lambda - 3 \end{bmatrix} \\ &= \lambda^2 - 6\lambda + 8 = (\lambda - 2)(\lambda - 4). \end{aligned}$$

Thus the eigenvalues of A are $\lambda_1 = 2$ and $\lambda_2 = 4$ and the graph must be an ellipse.

For $\lambda_1 = 2$ the equation $(2I - A)X = 0$ leads to the augmented matrix

$$\begin{bmatrix} -1 & -1 & 0 \\ -1 & -1 & 0 \end{bmatrix} \rightarrow \begin{bmatrix} 1 & 1 & 0 \\ 0 & 0 & 0 \end{bmatrix}.$$

Thus $X_1 = \dfrac{1}{\sqrt{2}}\begin{bmatrix} 1 \\ -1 \end{bmatrix}$ is a unit eigenvector for $\lambda_1 = 2$.

Similarly, we find that $X_2 = \dfrac{1}{\sqrt{2}}\begin{bmatrix} 1 \\ 1 \end{bmatrix}$ is a unit eigenvector for $\lambda_2 = 4$.

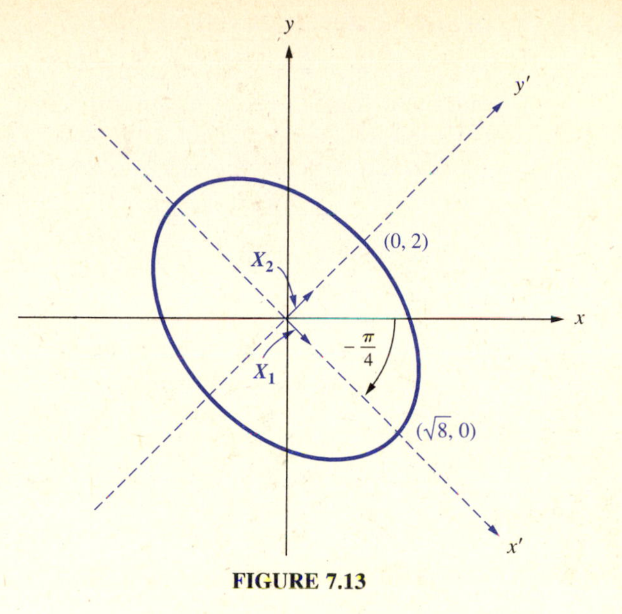

FIGURE 7.13

Thus

$$P = [X_1 \quad X_2] = \frac{1}{\sqrt{2}}\begin{bmatrix} 1 & 1 \\ -1 & 1 \end{bmatrix}$$

is an orthogonal matrix such that

$$P^TAP = \begin{bmatrix} \lambda_1 & 0 \\ 0 & \lambda_2 \end{bmatrix} = \begin{bmatrix} 2 & 0 \\ 0 & 4 \end{bmatrix}.$$

Note that det $(P) = +1$ so P represents a rotation. Since $\cos\theta = \frac{1}{\sqrt{2}}$ and $\sin\theta = -\frac{1}{\sqrt{2}}$, it follows that $\theta = -\frac{\pi}{4}$ and therefore P represents a rotation (clockwise) through an angle of $-\frac{\pi}{4}$ radians (-45 degrees). In Figure 7.13 we show the graph of the quadratic form and the two coordinate systems. Note that the x'-axis is in the direction of the eigenvector X_1 and that the y'-axis is in the direction of the eigenvector X_2. Relative to the $x'y'$ coordinate system the equation of the ellipse reduces to

$$2x'^2 + 4y'^2 = 16 \qquad \text{or} \qquad \frac{x'^2}{8} + \frac{y'^2}{4} = 1. \quad ■$$

The discussion preceding the example generalizes directly to the case of n variables. The result is called the Principal Axis theorem.

THEOREM 7.4 ***Principal Axis Theorem***

Let A be an $n \times n$ symmetric matrix and let $q(X) = X^TAX$ be the associated quadratic form. Let P be an orthogonal matrix, with $\det(P) = 1$, whose columns are eigenvectors of A that form an orthonormal basis for $\mathfrak{R}^n$. Then the substitution $X = PY$ reduces q to the equivalent form

$$q(Y) = Y^T(P^TAP)Y = \lambda_1 y_1^2 + \lambda_2 y_2^2 + \cdots + \lambda_n y_n^2,$$

where the λ_i are the eigenvalues of A. The coordinate change $X = PY$ amounts to a rotation in $\mathfrak{R}^n$. □

■ EXAMPLE 2 An Elliptic Hyperboloid

Identify and sketch the surface whose equation is

$$10x^2 + 8xy - 2y^2 - 20xz + 28yz + z^2 = 72.$$

The (symmetric) matrix of this quadratic form is

$$A = \begin{bmatrix} 10 & 4 & -10 \\ 4 & -2 & 14 \\ -10 & 14 & 1 \end{bmatrix},$$

for which the characteristic polynomial is

$$\begin{aligned} \det(\lambda I - A) &= \lambda^3 - 9\lambda^2 - 324\lambda + 2916 \\ &= (\lambda - 18)(\lambda + 18)(\lambda - 9). \end{aligned}$$

Routine calculations, like those in Section 5.2, show that the eigenvectors of A are

$$X_1 = \frac{1}{3}\begin{bmatrix} -2 \\ 1 \\ 2 \end{bmatrix} \quad \text{for the eigenvalue } \lambda_1 = 18,$$

$$X_2 = \frac{1}{3}\begin{bmatrix} 1 \\ -2 \\ 2 \end{bmatrix} \quad \text{for } \lambda_2 = -18, \qquad \text{and}$$

$$X_3 = \frac{1}{3}\begin{bmatrix} 2 \\ 2 \\ 1 \end{bmatrix} \quad \text{for } \lambda_3 = 9.$$

Now the matrix

$$P = \frac{1}{3}\begin{bmatrix} -2 & 1 & 2 \\ 1 & -2 & 2 \\ 2 & 2 & 1 \end{bmatrix}$$

is an orthogonal matrix such that

$$P^TAP = \mathrm{Dg}\{18, -18, 9\}.$$

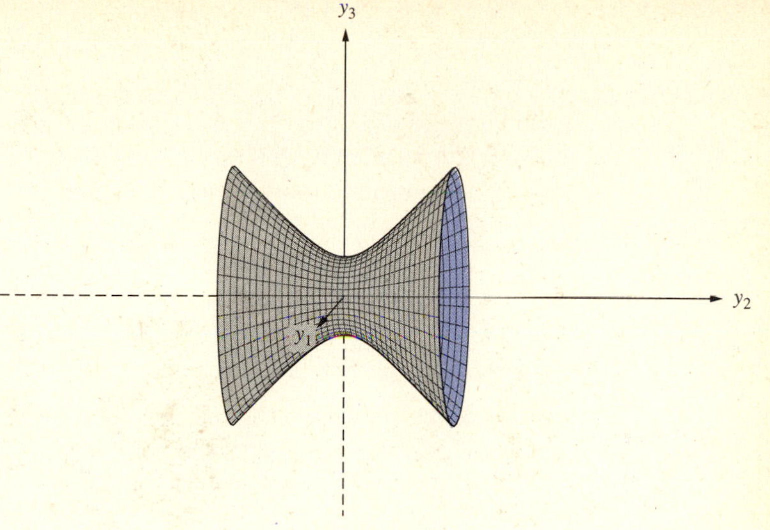

FIGURE 7.14

Note that $\det(P) = 1$, so P is the matrix of a rotation in $\Re^3$. The substitution $X = PY$ reduces the quadratic form $q = X^TAX$ to

$$q = Y^T(P^TAP)Y = 18y_1^2 - 18y_2^2 + 9y_3^2 = 72.$$

The y_1-coordinate axis is in the direction of $X_1 = \text{Col}_1(P)$, the y_2-coordinate axis is in the direction of $X_2 = \text{Col}_2(P)$, and the y_3-coordinate axis is in the direction of the eigenvector X_3. The surface is an elliptic hyperboloid of one sheet whose axis of symmetry is the y_2-axis; it is plotted in Figure 7.14 with respect to the y_i-coordinate axis and in Figure 7.15 with respect to the original coordinate system. ■

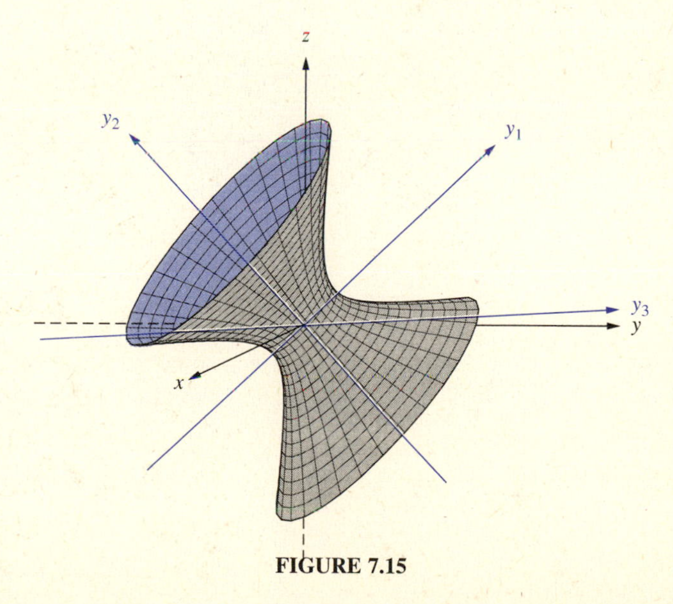

FIGURE 7.15

We now define a frequently cited property of quadratic forms and the associated symmetric matrices.

DEFINITION 7.4 A symmetric matrix A, or the associated quadratic form $q = X^TAX$, is said to be **positive definite** if $X^TAX > 0$ for every nonzero X in $\Re^n$. Similarly, A is **positive semidefinite** if $X^TAX \geq 0$ for all X, A is **negative definite** if $X^TAX < 0$ for all $X \neq 0$, A is **negative semidefinite** if $X^TAX \leq 0$ for all X, and A is **indefinite** otherwise. □

If λ is an eigenvalue of A with associated eigenvector X, then $AX = \lambda X$ and the value of the quadratic form at the eigenvector X is

$$X^TAX = X^T(\lambda X) = \lambda X^TX.$$

This value has the same algebraic sign as λ since X^TX is positive. Thus we see that the algebraic signs of the eigenvalues will be involved in deciding whether or not a given matrix is positive definite. The precise result is given in the next theorem.

THEOREM 7.5 A symmetric matrix A, with eigenvalues $\lambda_1, \lambda_2, \ldots, \lambda_n$ is

(1) positive definite if, and only if, all $\lambda_i > 0$;
(2) positive semidefinite if, and only if, all $\lambda_i \geq 0$;
(3) negative semidefinite if, and only if, all $\lambda_i \leq 0$;
(4) negative definite if, and only if, all $\lambda_i < 0$; and
(5) indefinite if A has both positive and negative eigenvalues.

Proof The above calculation shows that if A has a negative eigenvalue, then A is not positive definite. The contrapositive is that if A is positive definite then all eigenvalues of A must be positive. Conversely, suppose that all eigenvalues of A are positive and let $X_1, X_2, \ldots, X_n$ be an orthonormal basis for $\Re^n$ consisting of eigenvectors of A with $AX_i = \lambda_i X_i$. Then for any nonzero vector X in $\Re^n$ there exist unique scalars a_i such that

$$X = a_1X_1 + a_2X_2 + \cdots + a_nX_n.$$

The value of the quadratic form $q = X^TAX$ for this X is

$$\begin{aligned} X^TAX &= X^T(a_1\lambda_1X_1 + a_2\lambda_2X_2 + \cdots + a_n\lambda_nX_n) \\ &= (a_1X_1^T + \cdots + a_n X_n^T)(a_1\lambda_1X_1 + a_2\lambda_2X_2 + \cdots + a_n\lambda_nX_n) \\ &= a_1^2\lambda_1X_1^TX_1 + a_2^2\lambda_2X_2^TX_2 + \cdots + a_n^2\lambda_nX_n^TX > 0. \end{aligned}$$

Note that the last step used the fact that $X_i^TX_j = 0$ if $i \neq j$. This completes the proof of the first assertion. The other proofs are similar. □

It is sometimes possible to establish that a matrix is positive definite without finding all the eigenvalues. One such class of matrices for which this is possible is the class of *diagonally dominant* matrices. These matrices occur frequently in applications.

DEFINITION 7.5 An $n \times n$ matrix A is **diagonally dominant** if each diagonal entry of A is, in absolute value, greater than or equal to the sum of the absolute values of the off diagonal entries in that row. That is,

$$|a_{ii}| \geq \sum_{j \neq i} |a_{ij}| \quad \text{for } i = 1, 2, \ldots, n.$$

The matrix A is **strictly diagonally dominant** if

$$|a_{ii}| > \sum_{j \neq i} |a_{ij}| \quad \text{for } i = 1, 2, \ldots, n. \quad \square$$

The matrices

$$A = \begin{bmatrix} 5 & 2 & -2 \\ 2 & -4 & 1 \\ 0 & 0 & 1 \end{bmatrix} \quad \text{and} \quad B = \begin{bmatrix} 2 & 1 & 0 & 0 \\ 1 & 2 & 1 & 0 \\ 0 & 1 & 2 & 1 \\ 0 & 0 & 1 & 2 \end{bmatrix}$$

are both diagonally dominant, but only A is strictly diagonally dominant.

THEOREM 7.6 A strictly diagonally dominant symmetric matrix, with positive diagonal elements, must be positive definite.

Proof Let X be an eigenvector of A with $AX = \lambda X$ and suppose, without loss of generality, that X has been scaled so that all $|x_j| \leq 1$ and $x_i = 1$. Computing the ith component in the equation $AX = \lambda X$ yields

$$a_{i1}x_1 + \cdots + a_{ii}x_i + \cdots + a_{in}x_n = \lambda x_i.$$

Rearranging, using the fact that $x_i = 1$, and taking absolute values yields

$$|\lambda - a_{ii}| = \left| \sum_{j \neq i} a_{ij}x_j \right| \leq \sum_{j \neq i} |a_{ij}|\,|x_j|.$$

Since all $|x_j| \leq 1$ it follows that

$$|\lambda - a_{ii}| \leq \sum_{j \neq i} |a_{ij}| = r_i.$$

Since $a_{ii} > r_i$, by the assumed diagonal dominance of A, it follows that the eigenvalue λ must be positive. Thus A is positive definite by Theorem 7.5. $\square$

EXERCISES 7.3

Identify and sketch the graphs of the following curves.

1. $2x^2 + 6xy + 2y^2 = 10$
2. $2x^2 - 2xy + 2y^2 = 9$
3. $14x^2 + 16xy + 2y^2 = 36$
4. $16x^2 - 24xy + 9y^2 - 60x - 80y = 0$

Identify and sketch the graphs of the following surfaces.

5. $3x^2 - 2xy + 2xz + 3y^2 - 2yz + 3z^2 = 10$
6. $x^2 - 2y^2 + 4yz + z^2 = 12$
7. $3x^2 + 8xy + 4xz + 3y^2 + 4yz = 16$
8. $2x^2 + 2xy + 2xz + y^2 + z^2 = 10$
9. $-12xy - 12xz + 3y^2 - 3z^2 = 9$
10. Let A be an $n \times n$ symmetric matrix and define $r_i = \Sigma_{j \neq i} |a_{ij}|$. Extend the discussion in the proof of Theorem 7.6 to show that all the eigenvalues of A lie in the union of the intervals $|x - a_{ii}| \leq r_i$.
11. Let $A = \begin{bmatrix} a & b \\ b & c \end{bmatrix}$ and $A' = \begin{bmatrix} a' & b' \\ b' & c' \end{bmatrix} = P^TAP$, where P is orthogonal.
    (a) Show that $b^2 - ac = b'^2 - a'c'$; that is, the discriminant $(b^2 - ac)$ is invariant under rotation.
    (b) Show how the result in part (a) can be used to identify the graph of
    $$ax^2 + 2bxy + cy^2 + dx + ey + f = 0.$$
    (*Hint:* Consider the case $b' = 0$.)
12. Prove part 3 of Theorem 7.5.
13. Let A be any $n \times n$ matrix. Show that $X^TAX = X^TBX$ where $B = \frac{1}{2}(A + A^T)$ is a symmetric matrix.
14. Extend the result of Exercise 20 of Section 4.4 to show that every rotation in $\mathfrak{R}^3$ is a product of plane rotations. Does this result generalize to $\mathfrak{R}^n$?

7.4 LINEAR ECONOMIC MODELS

(Requires: Chapter 1 with minor references to Sections 2.4 and 7.3)

In this section we will take a brief look at the Leontief input-output model of an economy.* We begin with a simple example.

---

*Wassily Leontief received the 1973 Nobel prize in economics for his work on mathematical modeling.

EXAMPLE 1 **A Three-sector Closed Economy**

Consider a cooperative arrangement between an electrician, a plumber, and a painter who each agree to contribute twelve hours per week working for themselves and the other two. After some negotiations they agree on the following work schedule.

	Supplier		
Consumer	*Electrician*	*Plumber*	*Painter*
Electrician	2	5	3
Plumber	4	2	6
Painter	6	5	3

Thus the plumber will work 5 hours per week for the electrician, 5 hours per week for the painter, and 2 hours per week for himself.

For tax reasons the three workers must establish a value for their services. They wish to do this in such a way that they each come out even. What hourly rates should each worker set for his services?

If we let the rate matrix be

$$X = \begin{bmatrix} x \\ y \\ z \end{bmatrix} = \begin{bmatrix} \text{electrician's hourly rate} \\ \text{plumber's hourly rate} \\ \text{painter's hourly rate} \end{bmatrix},$$

then the electrician requires services worth $2x + 5y + 3z$ and provides services worth $12x$. Thus the equilibrium (break-even) condition for the electrician is

$$2x + 5y + 3z = 12x.$$

Similarly, the equilibrium conditions for the other two workers are $4x + 2y + 6z = 12y$ for the plumber, and $6x + 5y + 3z = 12z$ for the painter.

Thus, the unknown hourly wages must satisfy the linear system

$$\begin{bmatrix} 2 & 5 & 3 \\ 4 & 2 & 6 \\ 6 & 5 & 3 \end{bmatrix} \begin{bmatrix} x \\ y \\ z \end{bmatrix} = \begin{bmatrix} 12x \\ 12y \\ 12z \end{bmatrix}.$$

If we divide the last equation by 12 we obtain an equation of the form $EX = X$, where the coefficient matrix

$$E = \frac{1}{12} \begin{bmatrix} 2 & 5 & 3 \\ 4 & 2 & 6 \\ 6 & 5 & 3 \end{bmatrix}$$

has the properties that all $e_{ij} \geq 0$ and that each column sum is 1.

The system $EX = X$ is equivalent to the homogeneous system $(I - E)X = O$, for which the general solution can be obtained by Gaussian elimination. The solution is

$$X = \begin{bmatrix} x \\ y \\ z \end{bmatrix} = c \begin{bmatrix} \frac{3}{4} \\ \frac{9}{10} \\ 1 \end{bmatrix}, \quad \text{where } c \text{ is arbitrary.}$$

If we take $c = 20$, then $x = 15$, $y = 18$, and $z = 20$ are reasonable values for the desired hourly rates. ■

More generally, suppose that we have n interdependent industries, each of which produces a single product and uses a part of the output of the other industries in its operation. Let us denote the industries by $S_1, S_2, \ldots, S_n$ and the unit price for their products by $x_1, x_2, \ldots, x_n$. Let e_{ij} be the amount of product j required in the production of one unit of product i. Let us assume that each industry produces one unit of its product and that the economy is **closed** in the sense that no goods leave or enter the system. This means that the total consumption of each product must equal its total production.

It follows that the cost of producing one unit of product i is

$$e_{i1}x_1 + e_{i2}x_2 + \cdots + e_{in}x_n.$$

If we assume that each industry breaks even (income equals expenses), then we have the equations

$$e_{i1}x_1 + e_{i2}x_2 + \cdots + e_{in}x_n = x_i, \quad i = 1, 2, \ldots, n.$$

In matrix form this equation can be written as

$$EX = X \qquad \text{or} \qquad (I - E)X = 0,$$

and a solution of this system, provided it is positive, would provide equilibrium prices for the various products of the economy.

Note that all the elements of the matrix E are nonnegative and that, because the economy is closed, each of the columns of E has sum 1. Such matrices are called **exchange matrices** or **column stochastic matrices.** The fact that the column sums are all 1 can be interpreted to mean that each industry stores or uses any of its product that is not used by the other industries.

For an exchange matrix E it is easy to see that the sum of the rows of $I - E$ is the zero row. It follows that $I - E$ is a singular matrix and, from Theorem 1.9, that $(I - E)X = 0$ and hence $EX = X$ has a nonzero solution. Note that this is equivalent to saying that $\lambda = 1$ is an eigenvalue of E.

It is more difficult to show that $EX = X$ has a nonnegative solution. The details of this important result are included in *The Theory of Linear Economic Models* by David Gale (New York: McGraw Hill, 1960).

The following theorem states this important property of exchange matrices. Some insight into why the solution is positive is provided in Exercise 10.

THEOREM 7.7 If E is an $n \times n$ exchange matrix, then there exists a nonnegative vector X such that $EX = X$. □

We will now discuss a model of an **open** economy, where some of the industries may produce more of their product than is required by the other industries. The remainder of their product is then available for consumer use. In this model we will assume that the prices of the products are fixed and will show how to compute the outputs from each industry needed to meet a given set of demands from the marketplace.

Suppose that we have, as above, n interdependent industries, each of which produces a single product. Let y_i represent the value of the output from industry i and let c_{ij} be the value of product i needed to produce \$1.00 worth of product j. For example, if $c_{32} = .37$, then \$.37 worth of product 3 is required to make \$1.00 worth of product 2.

Note that the first subscript represents input and the second subscript represents output. The $n \times n$ matrix C is called the **consumption matrix** of the economy. A consumption matrix must have entries that satisfy $0 \leq c_{ij} \leq 1$.

For a consumption matrix C we observe that:

(1) the sum of the elements in column j represents the total value of inputs needed to produce \$1.00 worth of product j;
(2) industry j is profitable if the jth column sum is less than 1;
(3) $\text{Row}_i(C)Y = c_{i1}y_1 + \cdots + c_{in}y_n$ is the total value of the output of industry i used by the other industries; and
(4) $y_i - \text{row}_i(C)Y$ is the value of the output of industry i which is available to satisfy outside demand from consumers.

If D is a given demand vector, then the output vector Y must satisfy the linear system

$$Y - CY = D. \tag{7.7}$$

EXAMPLE 2 **A Six Industry Economy**

Suppose that the following consumption matrix defines the demands of six industries on each other:

$$\begin{array}{l} \\ \text{Automobile} \\ \text{Metal} \\ \text{Power} \\ \text{Coal} \\ \text{Plastics} \\ \text{Chemicals} \end{array} \begin{array}{c} \begin{array}{cccccc} A & M & P & Co & P & Ch \end{array} \\ \begin{bmatrix} .10 & .05 & .10 & .05 & .05 & .10 \\ .35 & .10 & .15 & .10 & .05 & .05 \\ .12 & .30 & .15 & .15 & .20 & .25 \\ .10 & .20 & .30 & .05 & .15 & .10 \\ .15 & .01 & .05 & .10 & .10 & .10 \\ .05 & .10 & .05 & .05 & .20 & .15 \end{bmatrix} \end{array} = C.$$

(a) What outputs are required to meet a demand of

$$D = [10 \quad 2 \quad 15 \quad 2 \quad 20 \quad 15]^T$$

million dollars?

(b) Suppose that the demand changes to $D_1 = [12 \quad 2 \quad 15 \quad 2 \quad 20 \quad 15]^T$ because of a good economic forecast. What output is required to meet demand D_1?

(c) Suppose that the price of power is increased 10 percent while all other prices remain constant. What output is required to meet demand D_1?

In order to answer the first question we need to solve the system

$$(I - C)Y = D.$$

The solution, obtained by Gaussian elimination, is

$$Y = [28.49 \quad 32.44 \quad 61.09 \quad 41.53 \quad 39.62 \quad 38.50]^T.*$$

Thus the total output of the automobile industry needs to be $28,490,000, while the total output of the power industry needs to be $61,090,000.

In order to answer the second question we need to solve the system

$$(I - C)Y_1 = D_1.$$

Note that the coefficient matrix is the same as for part (a). The solution, rounded to four significant figures, is

$$Y_1 = [31.15 \quad 33.94 \quad 62.60 \quad 42.80 \quad 40.38 \quad 39.17]^T.$$

Note that all sectors of the economy are affected by the increased demand for only one product and that the increase in the demand for automobiles is greater than the increased consumer demand.

In order to answer the third question we need to modify the consumption matrix C to reflect the change in power costs. Only the third row of C is changed; if the new matrix is called C', then

$$\text{Row}_3(C') = [.132 \quad .33 \quad .165 \quad .165 \quad .22 \quad .275].$$

The answer to part (c) is obtained by solving the linear system

$$(I - C')Y_2 = D_1,$$

for which the solution is

$$Y_2 = [32.68 \quad 36.53 \quad 71.13 \quad 46.56 \quad 41.72 \quad 40.61]^T.$$

Note that the price change in only one industry affected the production plans of all the other industries. ■

*The calculations were done using the MATMAN program.

A consumption matrix C is said to be **productive** if $(I - C)$ is nonsingular and all of the entries of $(I - C)^{-1}$ are nonnegative; that is, $(I - C)^{-1} \geq 0$. This of course implies that the solution of any system $Y - CY = D \geq 0$ will have a nonnegative (economically reasonable) solution.

If the consumption matrix C is productive, then the solution of (demand) Equation (7.7) can be written as $Y = (I - C)^{-1}D$. If one needs to make this calculation for many different values of D, not all of which are available at the same time, it might be worth the extra computational effort to find and store the inverse matrix $(I - C)^{-1}$.

Note that

$$(I - C)(I + C + C^2 + C^3 + \cdots + C^m) = I - C^{m+1}$$

so that, if $\lim_{m\to\infty} C^m = 0$, then

$$(I - C)^{-1} = I + C + C^2 + C^3 + \cdots \tag{7.8}$$

can be expressed as an infinite geometric series. Approximations for $(I - C)^{-1}$ can be obtained by truncating the above series. Thus,

$$(I + C), \qquad (I + C + C^2), \qquad \text{and} \qquad (I + C + C^2 + C^3)$$

are all approximations to $(I - C)^{-1}$. Since $C^k \geq 0$ for all k, the latter approximations will be better than the earlier ones, but will require more computation.

It can be shown that $\lim_{m\to\infty} C^m = 0$, if, and only if, the eigenvalues of C satisfy $|\lambda| < 1$. Since, by Exercise 10 of Section 7.3,

$$|\lambda| \leq \max_j \left\{ \sum_{i=1}^{n} |c_{ij}| \right\},$$

it follows that the consumption matrix C is productive if each column sum of C is less than 1, that is, if each industry is profitable. We summarize this discussion with Theorem 7.8.

THEOREM 7.8 A consumption matrix C is productive if each industry is profitable. □

EXERCISES 7.4

1. Show that each of the following is an exchange matrix and find a nonnegative price vector satisfying the equilibrium condition of Theorem 7.8.

(a) $\begin{bmatrix} \frac{1}{6} & \frac{1}{4} & \frac{1}{3} \\ \frac{2}{3} & \frac{1}{4} & \frac{2}{3} \\ \frac{1}{6} & \frac{1}{2} & 0 \end{bmatrix}$ **(b)** $\begin{bmatrix} .25 & .40 & .50 \\ .50 & .20 & .50 \\ .25 & .40 & .00 \end{bmatrix}$ **(c)** $\begin{bmatrix} \frac{7}{16} & \frac{1}{2} & \frac{3}{16} \\ \frac{5}{16} & \frac{1}{6} & \frac{5}{16} \\ \frac{1}{4} & \frac{1}{3} & \frac{1}{2} \end{bmatrix}$

2. Three doctors, an allergist, a dermatologist, and a surgeon, enter into a cooperative agreement of the type described in Example 1. They each agree to provide a total of six hours per month according to the following table.

	A	*D*	*S*
A	1	4	3
D	2	1	3
S	3	1	0

Determine reasonable rates for the three doctors so that each of the three doctors breaks even on the cooperative agreement.

3. Four neighbors, Tom, Paula, Bill, and Zeke, agree to share the produce from their backyard gardens. Tom will grow tomatoes, Paula will grow peas, Bill will grow beans, and Zeke will grow zucchini. The following table indicates how they will share the produce.

	Consumer			
Producer	*T*	*P*	*B*	*Z*
Tom	10	2	3	5
Paula	5	13	5	7
Bill	4	6	9	3
Zeke	5	3	7	9

What are reasonable prices for each of the four vegetables if each of the gardeners is to break even?

4. Suppose that the four gardeners in Exercise 3 decide to exchange their produce according to the following table.

	Consumer			
Producer	*T*	*P*	*B*	*Z*
Tom	10	2	3	5
Paula	5	3	5	7
Bill	4	1	2	4
Zeke	5	1	7	7

Is it still possible to assign prices to their products so that everyone breaks even?

5. The following consumption matrix defines the interdependence of four industries.

$$\begin{array}{c} \\ A \\ B \\ C \\ D \end{array} \begin{array}{c} \begin{array}{cccc} A & B & C & D \end{array} \\ \begin{bmatrix} .20 & .15 & .30 & .25 \\ .15 & .05 & .10 & .00 \\ .20 & .30 & .25 & .20 \\ .10 & .20 & .05 & .30 \end{bmatrix} \end{array}$$

(a) What is the total value of the raw materials needed to make \$1.00 worth of the product of industry *C*?

(b) What percentage of the output of industry *C* is used by the other industries?

(c) What percentage of the output of industry *B* is available to satisfy consumer demand?

6. **(a)** For the four-sector economy of Exercise 5 assume that prices are constant and find the production necessary to satisfy demands for \$100,000 of product A, \$75,000 of product B, \$10,000 of product C, and \$80,000 of product D.

(b) Suppose that the demand for product C increases to \$12,000 while the other demands remain constant. What production changes would be required to meet this new demand?

(c) Suppose the cost of product C decreased 10 percent while the other prices remained constant. What production changes would be necessary to meet the demands of part (b)?

7. Which of the following consumption matrices is productive?

(a) $\begin{bmatrix} .10 & .10 & .20 \\ .20 & .15 & .30 \\ .40 & .30 & .15 \end{bmatrix}$ **(b)** $\begin{bmatrix} .35 & .25 & .00 \\ .15 & .55 & .35 \\ .45 & .30 & .60 \end{bmatrix}$ **(c)** $\begin{bmatrix} .2 & .4 & .1 & .4 \\ .3 & .2 & .2 & .1 \\ .0 & .4 & .5 & .3 \\ .5 & .0 & .2 & .2 \end{bmatrix}$

8. Read "The World Economy of the Year 2000" by Wassily Leontief in *Scientific American,* November 1980, page 207. How large is the matrix that describes the world economy?

9. Let C be a consumption matrix. Compare the computational effort required to compute $(I - C)^{-1}$ using Gauss-Jordan elimination and Formula (7.8).

10. **(a)** Let E be an $n \times n$ exchange matrix. Show that, as a consequence of Exercise 10 of Section 7.3, that any eigenvalue of E must satisfy

$$|\lambda| \leq \max\left\{\sum_{j=1}^{n} |e_{ij}|\right\} = 1.$$

Thus $\lambda = 1$ is a dominant eigenvalue of E.

(b) Let us use the power method of Section 5.5 to compute an eigenvector of E for $\lambda = 1$. If we use an initial vector with nonnegative entries, then the eigenvector will have nonnegative entries. (What if there is a complex eigenvalue of modulus one?)

A Linear Algebra Software

There has been a great expenditure of scientific effort aimed at producing high-quality software for handling the basic problems of linear algebra. Any scientific or academic computing facility is certain to have such programs in its program library. There are three mainframe packages that deserve special mention.

LINPACK is a collection of FORTRAN subroutines for handling all aspects of the linear equations problem. The LINPACK user must write a FORTRAN driver which dimensions the arrays, inputs the data, calls the appropriate subroutines, and outputs the desired results. For example, calling the two subroutines

```
SGECO (A, LDA, N, IPVT, RCOND, Z)
SGESL (A, LDA, N, IPVT, B, JOB)
```

will result in computing the solution of the linear system $AX = B$, where A is an $N \times N$ matrix with $N \leq LDA$. The solution is returned in B and the original matrix A is destroyed. LINPACK should be available at most scientific computing installations. On-line documentation is frequently inadequate. Full documentation is contained in the LINPACK Users Guide available from SIAM (The Society for Industrial and Applied Mathematics).

EISPACK is a collection of FORTRAN subroutines for handling all aspects of the algebraic eigenvalue problem. The EISPACK user must write a FORTRAN driver which dimensions the arrays, inputs the data, calls the proper subroutines, and outputs the desired results. For example, calling the three subroutines

```
ORTHES(NM, N, LOW, IGH, A, ORT)
ORTRAN(NM, N, LOW, IGH, A, ORT, Z)
HQR2(NM, N, LOW, IGH, A, WR, WI, Z, ERROR)
```

will result in computing all the eigenvalues of the $N \times N$ matrix A ($N \leq NM$). The real parts of the eigenvalues are returned in the vector WR and the complex parts in the vector WI. The eigenvectors are contained in the matrix Z. The matrix A is destroyed. EISPACK should be available at most scientific computing installations; usually with adequate on-line documentation. Full documentation is contained in Volume 6 of the Springer-Verlag Lecture Notes in Computer Science.

LAPACK is an updated version of LINPACK and EISPACK. It uses algorithms which are faster, more accurate, more portable, and suitable for execution in parallel. It extends the power of LINPACK and EISPACK to include easy driver routines, equilibration, iterative refinement, and error bounds. The LAPACK users guide is available from SIAM.

An increasing amount of linear algebra software is becoming available for personal computers. The program PC-MATLAB, although expensive, is of high quality. There is an inexpensive student version of this program available.

The software available to users of this text consists of two main programs, MATMAN and MATALG. These programs are designed primarily to help teach linear algebra; they are not designed as production codes for large-scale problems. These programs were developed by

John Burkardt and Charles G. Cullen
Department of Mathematics
University of Pittsburgh
Pittsburgh, Pennsylvania, 15260
(E-mail to CULLEN@vms.cis.pitt.edu or burkardt@math.iastate.edu)

All rights to these programs are reserved by the University of Pittsburgh and the authors. They may not be reproduced in any manner without the written permission of the authors. This permission is automatically granted to schools using this textbook. These programs are not copy protected. They are available free of charge to users of this text.

Development of these programs was partially supported by a courseware development grant from the College of General Studies of the University of Pittsburgh.

These programs are menu driven and require no programming knowledge; all that is required is to know where to insert the program disk and how to turn on the machine. MATMAN (The MATrix MANipulator) is an interactive program to do row reductions; it can be used to solve linear systems, invert matrices, evaluate determinants, and solve linear programming problems using the tableau implementation of the simplex algorithm. In the program's primary mode the user specifies the row operations to be performed and the machine does the arithmetic quickly and accurately. The arithmetic can be done using either rational or real numbers. An automatic switch can be unlocked after the student has progressed to the desired point. MATALG (The MATrix ALGebraist) is an interactive program for automating basic matrix operations. It can be used to check hand calculation o save computation time with moderate-sized problems, to illustrate round-off error, and to provide experimental evidence in support of many of the theoretical observations in the text. These programs will run on most (all that we have tried) IBM-PCs and compatibles with a floating point chip; a version for the Apple Macintosh is also available. These programs were written in FORTRAN; the disk contains only the compiled versions of the programs. Full documentation for these programs is contained on the program disk.

Copies of these programs are available from the publisher either electronically or on disk. An electronic version of the programs can be obtained from "ftp://ftp.aw.com/math/authors/cullen/la2e"

If you wish to obtain the programs on disk, please write to:

Post-Calculus Editor
Addison Wesley Longman, Inc.
One Jacob Way
Reading, MA 01867

B The MATMAN Program

MATMAN is an interactive matrix manipulator which can carry out user-specified elementary row operations on a matrix. MATMAN is intended as a learning aid. As such, it requires interaction on the part of the user. The user may choose to use either real, rational (fractional) or decimal arithmetic with a fixed number of decimal places. Rational mode will most nearly duplicate normal hand calculations, and round-off inaccuracies will be largely avoided. The program will operate in rational arithmetic unless instructed to do otherwise. Using rational arithmetic for large problems may cause integer overflow; if this happens switch to real arithmetic.

Matrices are restricted to at most 16 rows and 30 columns.

Running the Program

IBM PC: The system must be booted and running already. Insert the MATMAN disk in the disk drive, open a DOS Window, and type "MATMAN." When not running the program, the command "DIR" will give a directory of all the files on the disk.

MACINTOSH: The system must be booted and running already. Insert the disk containing MATMAN. Move the mouse to the MATMAN disk and click it twice. All of the files on the disk will be displayed. Move the mouse to the "MATMAN" icon and click twice.

MATMAN Commands

The list of available commands follows. This list can be displayed at any time by issuing the "HELP" command. A short list of frequently used commands is available by typing "H."

In the following list, S, I, and J stand for numbers you must include in the command. S is typically a multiplier which may be in decimal or rational format, while I and J are row or column identifiers.

Command and Arguments		*Short Definition*
A	S, I, J (RJ <= S RI + RJ)	Add S times row I to row J.
B		Set up sample problem.
C	I,J,S	Change entry I,J to S.
D/M	I,S (RI <= S RI)	Divide/Multiply row I by S.
DET		Print the determinant of the matrix.
E	I,J	Enter a matrix with I rows and J columns.
EDET		Print ERO determinant.
F	F/R/D	Choose arithmetic (Fractional, Real, or Decimal).
G	+/−,R/C,K	Add/Delete a row or column to/of the matrix.
H		Display short menu.
HELP		Display full menu.
I	I,J (RI <=> RJ)	Interchange rows I and J.
K	filename	Open/close transcript file.
L		Linear programming.
N	S	Set the number of decimal digits to S.
Q		Quit.
S/R		Store/restore matrix.
T		Type the matrix.
TR		Transpose the matrix.
U		Undo last operation.
W/X	filename	Write/read matrix to/from file.
Z		Automatic operation (requires password).
?		Extensive help.
#		Begins a comment line.
$/%		Turn paging off/on.
<		Get input from a file

When entering matrix data in either real or rational arithmetic, you can use the decimal, integer, or fractional form of the number. In rational mode decimals will be interpreted as fractions, e.g., .33 will be interpreted as $\frac{33}{100}$. To multiply row 2 by $\frac{1}{3}$, you may type: D, 2, 3 or M 2, .33333 or M 2, $\frac{1}{3}$ or M 2, 1.0/3.0. In rational mode the first command would multiply by $\frac{33333}{100000}$, not by $\frac{1}{3}$.

If a command has several arguments, you may list them on the same line as the command letter, or not. If you do not list all needed arguments, the program will prompt you for them. If you have issued a command which requires several items of input, and you change your mind and want to cancel the entire command, you can type control-z (hold down the control key and type z). This should take you back to the main menu.

After most commands the new matrix will be printed on the screen for you to inspect.

A S, I, J (or RJ <= S RI + RJ) = ADD S TIMES ROW I TO ROW J

$(R_j \leftarrow S * R_i + R_j)$

will add S times row I to row J. This is one of the most frequently used commands in the program. Legal commands to add 2.5 times row 2 to row 6 would be A 2.5,2,6 or A 5/2, 2, 6.

B = SET UP SAMPLE PROBLEM

If you have not entered your own problem, this command will set up a sample problem which you can practice on.

C I, J, S = CHANGE ENTRY I, J TO S

The command to change entry 1,3 to 3.5 is C 1,3,3.5, or C 1,3,7/2, or C 1,3,7.0/2.0. This command is useful in correcting errors made in data entry as well as handling perturbations of the original problem.

D I,S = DIVIDE ROW I BY S

This command divides row I of the matrix by S. On a computer using real arithmetic the result may be slightly different than multiplying row I by 1/S.

DET = PRINT DETERMINANT OF THE CURRENT MATRIX

E I, J = ENTER MATRIX WITH I ROWS AND J COLUMNS

This command is used to set up a new problem. This command means you want to enter a matrix of I rows and J columns. The program will then request that you enter the elements of the matrix, one row at a time, separated by commas or spaces. To enter a 3 by 4 matrix, you must type:

```
E 3,4
1, 2, 3, 4
5, 6, 7, 8
9, 8, 7, 6
```

If you wish, you may enter several rows, or the entire matrix, at one time. Be careful not to lose track of where you are. If you do this, you could enter the same matrix this way:

```
E 3,4
1, 2, 3, 4, 5, 6, 7, 8, 9, 8, 7, 6
```

Each time you enter a matrix, it is automatically stored and typed. This means that, unless you store another matrix in the meantime, you can always recover your original input matrix with the "R" command if you regret the changes you have made to it. The program stores only one matrix. If additional storage is required, use the "W" command.

EDET = PRINT DETERMINANT OF THE ERO MATRIX

F = CHANGE ARITHMETIC

Change from rational to real arithmetic or decimal arithmetic, or vice versa. The program begins in rational arithmetic. Going from real to rational arithmetic, the program will want to know how many decimal places to save. For example, if you save two places, 35.678934 will become $\frac{3568}{100}$. Beware!! Using rational arithmetic for large problems may lead to integer overflow. The "N" command is used to specify the number of digits in the decimal mode.

G = ADD/DELETE A ROW OR COLUMN OF THE MATRIX

This command allows you to add or delete a row or column of the matrix.
Type "+" to add something and "−" to delete something.
Type "R" if you're interested in a row, or "C" for a column.
If you're adding a row or column, type the position the new object should occupy. For instance, you would type "1" if the new row or column was to be the first row or column in the matrix. All other rows or columns will automatically shift to make room for the new one.
If you're adding a row or column, you will then be requested to enter the values of its entries.
Here is the input that would insert a row at position 3 in the current matrix, and assign values to its 5 entries:

G +R 3

1.0, 2.0, 3.5, 4.7, 5.3

H = HELP

This command prints out the short list of frequently used commands. To see the full list type "HELP."

I K, J (or RK <=> RJ) = INTERCHANGE ROWS K AND J ($R_k \leftrightarrow R_j$)

This command interchanges rows K and J in the matrix. For example, I 2, 1 means switch rows 2 and 1.

K Filename

This command opens a file into which is put an exact copy of everything you or the program types. This is very useful for turning in homework, or checking your work. A second "K" command will close the file. The program will ask you to give the file a name such as MATMAN.LPT (in the default drive) or B:HOMEWORK (in drive B). On the Macintosh, you can write to the disk whose name is FRED by giving the file the name FRED:FILENAME. On the PC you can generate a hardcopy, as the file is being constructed, by pressing the CTL and PRINT SCREEN keys together. Once constructed, the file can be edited using any available editor or word processor. On the MAC you can convert the file to a MACWRITE or a WORD file and then print it.

L = SWITCH TO LINEAR PROGRAMMING MODE

M I, S = MULTIPLY ROW I BY S ($R_i \leftarrow S * R_i$)

For example, the command M 1, −4 will multiply row 1 by −4.

N = NUMBER OF DIGITS

This command allows you to select the number of digits in the decimal mode. If you select 4, then the program will simulate the calculations on a 4-digit decimal machine.

Q = QUIT

This command causes the program to terminate. The program will ask you to confirm this choice. Control is returned to the computer operating system.

R = RESTORE THE SAVED MATRIX

The current matrix is replaced by the matrix you stored earlier with the "S" command, or the last matrix you entered using the "E" or "X" command, whichever was most recent.

S = STORE THE CURRENT MATRIX

A copy of the current matrix is stored in memory, so that you can reuse it later. Only one matrix may be stored at any time, and storing a new matrix destroys the old information. The matrix is only stored in memory, so when you quit the program, this information is lost. See the "W" command for a more permanent storage method. Whenever a new matrix is entered with the "E" or "X" commands, an automatic store is done.

T = TYPE THE MATRIX

Since the program prints the matrix after most operations, you will not need this command often.

TR = TRANSPOSE THE CURRENT MATRIX

This command will transpose the current matrix if the transpose is within the size limits of the program. Use this command if you wish to use this program to manipulate the columns of a matrix.

U = UNDO LAST OPERATION

In most cases, this command will allow you to recover from a command you regret giving, or gave by mistake. The matrix will be restored to its form as it was just before your previous command. This only works one time; you cannot backtrack using several consecutive "U" commands.

W = WRITE AN EXAMPLE TO A FILE

This command allows you to specify the name of a file, such as MATMAN.DAT or B:DATA and a label, such as EXAMPLE 1. The current matrix will then be stored in the file. You can store several problems in the same file, and retrieve any of them later with the "X" command.

X = READ AN EXAMPLE FROM A FILE

This command allows you to specify the name of a file, such as MATMAN.DAT. It then searches through that file, and lists the labels of all the problems stored there. You will be asked to pick one problem to read in. Possibly, the example will require that the arithmetic mode be changed. If so, this will be done automatically. The file MATMAN.DAT comes with some examples in it already.

Z = AUTOMATIC ROW REDUCTION

The "Z" command will row reduce the active matrix using partial pivoting. Because MATMAN is a teaching program, use of the "Z" command should be discouraged until the students have had adequate opportunity to carry out reductions by hand. Use of the "Z" command requires a password which is included in the instructors manual. (See your instructor or E-mail the author.) The Z authorization lasts throughout the current MATMAN session. This command will not work if the password file MATKEY.DAT is missing. To change the password type the current password with a minus sign; the program will then ask you to enter a new password.

? = EXTENSIVE HELP

Sometimes the short one-line help offered by the "H" command is not enough. In those cases, using the "?" command will allow you, while running the program, to browse through this help file in search of information. This command will not work if the file MATMAN.HLP is not available on the same disk from which MATMAN was launched.

= COMMENT LINE

Lines that begin with a # contain user comments that will be part of the transcript file. They are not to be interpreted as commands to the program.

Linear Programming Mode

In linear programming mode MATMAN implements the tableau form of the simplex algorithm described in Chapter 6 of the text. In this mode several new commands will appear on the main menu in response to the "H" command. They are

L	Leave linear programming mode
O	Optimality check
P	Pivot, entering variable I, departing variable J
V	Remove artificial variables.

The "B" command, for setting up a sample problem, will now provide a sample linear programming problem and the "E" command will accept input in the form of inequalities. Read the prompts carefully. The workhorse command in this mode is the "P" command

P = PIVOT, ENTERING VARIABLE I, DEPARTING VARIABLE J

The program will guide you through one step of the pivoting operation. You will be shown the objective row, and asked to pick a proper entering variable, I. Then you will be shown the feasibility ratios, and asked to pick a proper departing variable, J. If the program accepts your choices, the pivoting operation will be carried out automatically. The program will list the operations used and display the new tableau. If you have made improper choices, the program will ask you to try again.

V = REMOVE ARTIFICIAL VARIABLES

If the problem required artificial variables, then, at the end of Phase I, you will need to issue this command to change to Phase II. The artificial variables will be deleted and the original objective function restored. If adjustments to the objective row are required before proceeding, the program will prompt you.

Sample MATMAN Transcript File

(User input in boldface, comments in parentheses or preceded by #.)
k sample.ero
Opening the transcript file "sample.ero".
Enter command? ("H" for short menu, "HELP" for full menu, ? for full help)
e 4 5
Enter entries 1 to 5 of row 1
5 11 7 3 1
Enter entries 1 to 5 of row 2
2, 1, 4, −5, 1 (entries can be separated by spaces or commas)
Enter entries 1 to 5 of row 3
3 −2 8 7 1 1 2 3 4 1 (more than one row can be entered at one time)
A copy of this matrix is being saved. The "R" command can bring it back.
The current matrix:

5	11	7	3	1
2	1	4	−5	1
3	−2	8	7	1
1	2	3	4	1

Enter command? ("H" for short menu, "HELP" for full menu, ? for full help)
c 4 5 2 (correcting data entry error)
Change entry 4, 5 to 2.

The current matrix:

5	11	7	3	1
2	1	4	−5	1
3	−2	8	7	1
1	2	3	4	2

Enter command? (“H” for short menu, “HELP” for full menu, ? for full help)
i 1,4 (or **R1 <=> R4**)
ERO: Row 4 <=> Row 1
The current matrix:

1	2	3	4	2
2	1	4	−5	1
3	−2	8	7	1
5	11	7	3	1

Enter command? (“H” for short menu, “HELP” for full menu, ? for full help)
a −2 1 2 (or **R2 <= −2 R1 + R2**)
ERO: Row 2 <= −2 Row 1 + Row 2
The current matrix:

1	2	3	4	2
0	−3	−2	−13	−3
3	−2	8	7	1
5	11	7	3	1

Enter command? (“H” for short menu, “HELP” for full menu, ? for full help)
R3 <= −3 R1 + R3 (or **a − 3 1 3**)
ERO: Row 3 <= −3 Row 1 + Row 3
The current matrix:

1	2	3	4	2
0	−3	−2	−13	−3
0	−8	−1	−5	−5
5	11	7	3	1

Enter command? (“H” for short menu, “HELP” for full menu, ? for full help)
a −5 1 4 (or **R4 <= −5 R1 + R4**)
ERO: Row 4 <= −5 Row 1 + Row 4
The current matrix:

1	2	3	4	2
0	−3	−2	−13	−3
0	−8	−1	−5	−5
0	1	−8	−17	−9

Enter command? (“H” for short menu, “HELP” for full menu, ? for full help)
i 2.4

Because of an error, your command was not completed. We return to the main menu. (The period in the command should be a comma.)
Enter command? ("H" for short menu, "HELP" for full menu, ? for full help)
i 2,4
ERO: Row 4 <=> Row 2
The current matrix:

$$\begin{matrix} 1 & 2 & 3 & 4 & 2 \\ 0 & 1 & -8 & -17 & -9 \\ 0 & -8 & -1 & -5 & -5 \\ 0 & -3 & -2 & -13 & -3 \end{matrix}$$

Enter command? ("H" for short menu, "HELP" for full menu, ? for full help)
a 8 2 3 (or **R3 <= 8 R2 + R3**)
ERO: Row 3 <= 8 Row 2 + Row 3
The current matrix:

$$\begin{matrix} 1 & 2 & 3 & 4 & 2 \\ 0 & 1 & -8 & -17 & -9 \\ 0 & 0 & -65 & -141 & -77 \\ 0 & -3 & -2 & -13 & -3 \end{matrix}$$

Enter command? ("H" for short menu, "HELP" for full menu, ? for full help)
a 3 2 4 (or **R4 <= 3 R2 + R4**)
ERO: Row 4 <= 3 Row 2 + Row 4
The current matrix:

$$\begin{matrix} 1 & 2 & 3 & 4 & 2 \\ 0 & 1 & -8 & -17 & -9 \\ 0 & 0 & -65 & -141 & -77 \\ 0 & 0 & -26 & -64 & -30 \end{matrix}$$

Enter command? ("H" for short menu, "HELP" for full menu, ? for full help)
a − 26/65 3 4 (or **R4 <= −26/65 R3 + R4**)
ERO: Row 4 <= −2/5 Row 3 + Row 4
The current matrix:

$$\begin{matrix} 1 & 2 & 3 & 4 & 2 \\ 0 & 1 & -8 & -17 & -9 \\ 0 & 0 & -65 & -141 & -77 \\ 0 & 0 & 0 & -38 & 4 \\ & & & 5 & 5 \end{matrix}$$

(Note how fractions appear in the output)

(Reduced far enough to solve the system by back substitution.)
Enter command? ("H" for short menu, "HELP" for full menu, ? for full help)
r
The saved matrix has been restored.

The current matrix:

5	11	7	3	1
2	1	4	−5	1
3	−2	8	7	1
1	2	3	4	1

(Note that the earlier change command did not affect this matrix.)
Enter command? ("H" for short menu, "HELP" for full menu, ? for full help)
z
Enter authorization key for "Z" command.
?????? (See the instructors manual for the password.)
Authorization confirmed.
ERO: Row 1 <= Row 1 / 5
ERO: Row 2 <= −2 Row 1 + Row 2
ERO: Row 3 <= −3 Row 1 + Row 3
ERO: Row 4 <= −1 Row 1 + Row 4
ERO: Row 2 <=> Row 3
ERO: Row 2 <= Row 2 / (−43/5)
ERO: Row 1 <= −11/5 Row 2 + Row 1
ERO: Row 3 <= 17/5 Row 2 + Row 3
ERO: Row 4 <= 1/5 Row 2 + Row 4
ERO: Row 3 <=> Row 4
ERO: Row 3 <= Row 3 / (65/43)
ERO: Row 1 <= −102/43 Row 3 + Row 1
ERO: Row 2 <= 19/43 Row 3 + Row 2
ERO: Row 4 <= 13/43 Row 3 + Row 4
ERO: Row 4 <= Row 4 / (−38/5)
ERO: Row 1 <= 209/65 Row 4 + Row 1
ERO: Row 2 <= −23/65 Row 4 + Row 2
ERO: Row 3 <= −141/65 Row 4 + Row 3
The current matrix:

1	0	0	0	$\frac{-31}{26}$
0	1	0	0	$\frac{105}{494}$
0	0	1	0	$\frac{343}{494}$
0	0	0	1	$\frac{-3}{38}$

Enter command? ("H" for short menu, "HELP" for full menu, ? for full help)
k
Closing the transcript file "sample.ero."
l
Switching to linear programming mode. (Chapter 6)
Enter "Y" to use current matrix in linear programming.
n
#
This is a solution of Example 3 of Section 6.3.
#
k sample.lp1
Opening the transcript file "sample.lp1"
Enter command? ("H" for short menu, "HELP" for full menu, ? for full help)
e
Enter number of constraints, number of variables.
4, 3
Enter sign <,> or = and coefficients and RHS of constraint 1.
< 6 5 0 60
Enter sign <,> or = and coefficients and RHS of constraint 2.
< 6 −13 6 60
Enter sign <,> or = and coefficients and RHS of constraint 3.
< −78 25 30 300
Enter sign <, > or = and coefficients and RHS of constraint 4.
< 42 −19 42 636
Enter coefficients and constant of objective function.
10 12 15 0
A copy of this matrix is being stored.
The "R" command can bring it back.
The linear programming tableau:

	1	2	3	4	5	6	7	P	C
$X4$	6	5	0	1	0	0	0	0	60
$X5$	6	−13	6	0	1	0	0	0	60
$X6$	−78	25	30	0	0	1	0	0	300
$X7$	42	−19	42	0	0	0	1	0	636
Obj	−10	−12	−15	0	0	0	0	1	0

Enter command? ("H" for short menu, "HELP" for full menu, ? for full help)
p (pivot)
Objective row

	1	2	3	4	5	6	7	P	C
Obj	−10	−12	−15	0	0	0	0	1	0

Variable with most negative objective coefficient?
Enter column (=variable number).
3
The entering variable is 3.
Variable with least nonnegative feasibility ratio?
Nonnegative feasibility ratios:
Row 2, variable 5, ratio = 10
Row 3, variable 6, ratio = 10
Row 4, variable 7, ratio = 15.1429
Enter the row of the departing variable.
2
Departing variable is 5 with feasibility ratio 10.0000.
ERO: Row 2 <= 1/6 Row 2
ERO: Row 3 <= −30 Row 2 + Row 3
ERO: Row 4 <= −42 Row 2 + Row 4
ERO: Row 5 <= 15 Row 2 + Row 5
Objective changed from 0 to 150.
The linear programming tableau:

	1	2	3	4	5	6	7	P	C
$X4$	6	5	0	1	0	0	0	0	60
$X3$	1	$-\frac{13}{6}$	1	0	$\frac{1}{6}$	0	0	0	10
$X6$	−108	90	0	0	−5	1	0	0	0
$X7$	0	72	0	0	−7	0	1	0	216
Obj	5	$-\frac{89}{2}$	0	0	$\frac{5}{2}$	0	0	1	150

Enter command? ("H" for short menu, "HELP" for full menu, ? for full help)
p
Objective row

	1	2	3	4	5	6	7	P	C
Obj	5	$-\frac{89}{2}$	0	0	$\frac{5}{2}$	0	0	1	150

Variable with most negative objective coefficient?
Enter column (=variable number).
2
Entering variable is 2.
Variable with least nonnegative feasibility ratio?
Nonnegative feasibility ratios:
Row 1, variable 4, ratio = 12
Row 3, variable 6, ratio = 0
Row 4, variable 7, ratio = 3

Enter the row of the departing variable.
3
Departing variable is 6 with feasibility ratio 0.000000.
ERO: Row 3 <= 1/90 Row 3
ERO: Row 1 <= −5 Row 3 + Row 1
ERO: Row 2 <= 13/6 Row 3 + Row 2
ERO: Row 4 <= −72 Row 3 + Row 4
ERO: Row 5 <= 89/2 Row 3 + Row 5
No change in objective.
The linear programming tableau:

	1	2	3	4	5	6	7	P	C
$X4$	12	0	0	1	$\frac{5}{18}$	$\frac{-1}{18}$	0	0	60
$X3$	$\frac{-8}{5}$	0	1	0	$\frac{5}{108}$	$\frac{13}{540}$	0	0	10
$X2$	$\frac{-6}{5}$	1	0	0	$\frac{-1}{18}$	$\frac{1}{90}$	0	0	0
$X7$	$\frac{432}{5}$	0	0	0	−3	$\frac{-4}{5}$	1	0	216
Obj	$\frac{-242}{5}$	0	0	0	$\frac{1}{36}$	$\frac{89}{180}$	0	1	150

Enter command? ("H" for short menu, "HELP" for full menu, ? for full help)
p (pivot again)
Objective row

	1	2	3	4	5	6	7	P	C
Obj	$-\frac{242}{5}$	0	0	0	$\frac{1}{36}$	$\frac{89}{180}$	0	1	150

Variable with most negative objective coefficient?
Enter column (=variable number).
1
Entering variable is 1.
Variable with least nonnegative feasibility ratio?
Nonnegative feasibility ratios:
Row 1, variable 4, ratio = 5
Row 4, variable 7, ratio = 2.50000.
Enter the row of the departing variable.
1
Not acceptable.
note that an incorrect choice was rejected by MATMAN.

Nonnegative feasibility ratios:
Row 1, variable 4, ratio = 5
Row 4, variable 7, ratio = 2.50000.
Enter the row of the departing variable.
4
Departing variable is 7 with feasibility ratio 2.50000.
ERO: Row 4 <= 5/432 Row 4
ERO: Row 1 <= −12 Row 4 + Row 1
ERO: Row 2 <= 8/5 Row 4 + Row 2
ERO: Row 3 <= 6/5 Row 4 + Row 3
ERO: Row 5 <= 242/5 Row 4 + Row 5
Objective changed from 150 to 271.
The linear programming tableau:

	1	2	3	4	5	6	7	P	C
$X4$	0	0	0	1	$\frac{25}{36}$	$\frac{1}{18}$	$-\frac{5}{36}$	0	30
$X3$	0	0	1	0	$-\frac{1}{108}$	$\frac{1}{108}$	$\frac{1}{54}$	0	14
$X2$	0	1	0	0	$-\frac{7}{72}$	0	$\frac{1}{72}$	0	3
$X1$	1	0	0	0	$-\frac{5}{144}$	$-\frac{1}{108}$	$\frac{5}{432}$	0	$\frac{5}{2}$
Obj	0	0	0	0	$-\frac{119}{72}$	$\frac{5}{108}$	$\frac{121}{216}$	1	271

Enter command? ("H" for short menu, "HELP" for full menu, ? for full help)
p
Objective row

	1	2	3	4	5	6	7	P	C
Obj	0	0	0	0	$-\frac{119}{72}$	$\frac{5}{108}$	$\frac{121}{216}$	1	271

Variable with most negative objective coefficient?
Enter column (=variable number).
5
Entering variable is 5.
Variable with least nonnegative feasibility ratio?
Nonnegative feasibility ratios:
Row 1, variable 4, ratio = 43.2000

Enter the row of the departing variable.
1
Departing variable is 4 with feasibility ratio 43.2000
ERO: Row 1 <= 36/25 Row 1
ERO: Row 2 <= 1/108 Row 1 + Row 2
ERO: Row 3 <= 7/72 Row 1 + Row 3
ERO: Row 4 <= 5/144 Row 1 + Row 4
ERO: Row 5 <= 119/72 Row 1 + Row 5
Objective changed from 271 to 1712/5 (= 342.4).
The linear programming tableau:

	1	2	3	4	5	6	7	P	C
*X*5	0	0	0	36	1	2	−1	0	216
				25		25	5		5
*X*3	0	0	1	1	0	1	1	0	72
				75		100	60		5
*X*2	0	1	0	7	0	7	−1	0	36
				50		900	180		5
*X*1	1	0	0	1	0	−7	1	0	4
				20		1080	216		
Obj	0	0	0	119	0	241	31	1	1712
				50		1350	135		5

We have found the optimal solution.
k
Closing the transcript file "sample.lp1"
#
#We will now solve the two-phase example of Section 6.4.
#
Enter command? ("H" for short menu, "HELP" for full menu, ? for full help)
k sample.lp2
Opening the transcript file "sample.lp2"
e 4 2
Enter sign <, > or = and coefficients and RHS of constraint 1.
> 1 2 6
Enter sign <, > or = and coefficients and RHS of constraint 2.
> 2 1 4
Enter sign <, > or = and coefficients and RHS of constraint 3.
< 1 1 5
Enter sign <, > or = and coefficients and RHS of constraint 4.
< 2 1 8
Enter coefficients and constant of objective function.
40 30 0

Because we have artificial variables, the objective function is also "artificial."
The true objective will be stored away until the artificial variables are gone.
A copy of this matrix is being stored.
The "R" command can bring it back.
The linear programming tableau:

	1	2	3	4	5	6	7	8	P	C
*X*7	1	2	−1	0	0	0	1	0	0	6
*X*8	2	1	0	−1	0	0	0	1	0	4
*X*5	1	1	0	0	1	0	0	0	0	5
*X*6	2	1	0	0	0	1	0	0	0	8
Obj2	0	0	0	0	0	0	1	1	1	0
Obj	−40	−30	0	0	0	0	0	0	1	0

Enter command? ("H" for short menu, "HELP" for full menu, ? for full help)
p
Nonzero objective entry in column 7 corresponding to a basic variable.
Use the "A" command to fix this! Because of an error, your command was not completed. We return to the main menu.
Enter command? ("H" for menu, "HELP" for full menu, "?" for full help)
a −1 1 5 (or **R5 <= −1 R1 + R5**)
ERO: Row 5 <= −1 Row 1 + Row 5
The linear programming tableau:

	1	2	3	4	5	6	7	8	P	C
*X*7	1	2	−1	0	0	0	1	0	0	6
*X*8	2	1	0	−1	0	0	0	1	0	4
*X*5	1	1	0	0	1	0	0	0	0	5
*X*6	2	1	0	0	0	1	0	0	0	8
Obj2	−1	−2	1	0	0	0	0	1	1	−6
Obj	−40	−30	0	0	0	0	0	0	1	0

Enter command? ("H" for short menu, "HELP" for full menu, ? for full help)
a −1 2 5 (to get rid of the entry in column 7 of the obj2 row)
ERO: Row 5 <= −1 Row 2 + Row 5
The linear programming tableau:

	1	2	3	4	5	6	7	8	P	C
*X*7	1	2	−1	0	0	0	1	0	0	6
*X*8	2	1	0	−1	0	0	0	1	0	4
*X*5	1	1	0	0	1	0	0	0	0	5
*X*6	2	1	0	0	0	1	0	0	0	8
Obj2	−3	−3	1	1	0	0	0	0	1	−10
Obj	−40	−30	0	0	0	0	0	0	1	0

Enter command? ("H" for short menu, "HELP" for full menu, ? for full help)
p
Objective row

	1	2	3	4	5	6	7	8	P	C
Obj2	−3	−3	1	1	0	0	0	0	1	−10

Variable with most negative objective coefficient?
Enter column (=variable number).
1
Entering variable is 1.
Variable with least nonnegative feasibility ratio?
Nonnegative feasibility ratios:
Row 1, variable 7, ratio = 6
Row 2, variable 8, ratio = 2
Row 3, variable 5, ratio = 5
Row 4, variable 6, ratio = 4
Enter the row of the departing variable.
2
Departing variable is 8 with feasibility ratio 2.00000.
ERO: Row 2 <= Row 2/2
ERO: Row 1 <= −1 Row 2 + Row 1
ERO: Row 3 <= −1 Row 2 + Row 3
ERO: Row 4 <= −2 Row 2 + Row 4
ERO: Row 5 <= 3 Row 2 + Row 5
Objective changed from −10 to −4.
The linear programming tableau:

	1	2	3	4	5	6	7	8	P	C
*X*7	0	$\frac{3}{2}$	−1	$\frac{1}{2}$	0	0	1	$\frac{-1}{2}$	0	4
*X*1	1	$\frac{1}{2}$	0	$\frac{-1}{2}$	0	0	0	$\frac{1}{2}$	0	2
*X*5	0	$\frac{1}{2}$	0	$\frac{1}{2}$	1	0	0	$\frac{-1}{2}$	0	3
*X*6	0	0	0	1	0	1	0	−1	0	4
Obj2	0	$\frac{-3}{2}$	1	$\frac{-1}{2}$	0	0	0	$\frac{3}{2}$	1	−4
Obj	−40	−30	0	0	0	0	0	0	1	0

Enter command? ("H" for short menu, "HELP" for full menu, ? for full help)
p (pivot)

Objective row

	1	2	3	4	5	6	7	8	P	C
Obj2	0	$-\frac{3}{2}$	1	$-\frac{1}{2}$	0	0	0	$\frac{3}{2}$	1	−4

Variable with most negative objective coefficient?
Enter column (=variable number).
2
Entering variable is 2.
Variable with least nonnegative feasibility ratio?
Nonnegative feasibility ratios:
Row 1, variable 7, ratio = 2.66667.
Row 2, variable 1, ratio = 4
Row 3, variable 5, ratio = 6
Enter the row of the departing variable.
1
Departing variable is 7 with feasibility ratio 2.66667.
ERO: Row 1 <= 2/3 Row 1
ERO: Row 2 <= −1/2 Row 1 + Row 2
ERO: Row 3 <= −1/2 Row 1 + Row 3
ERO: Row 5 <= 3/2 Row 1 + Row 5
Objective changed from −4 to 0.
The linear programming tableau:

	1	2	3	4	5	6	7	8	P	C
*X*2	0	1	−2	1	0	0	2	−1	0	8
			3	3			3	3		3
*X*1	1	0	1	−2	0	0	−1	2	0	2
			3	3			3	3		3
*X*5	0	0	1	1	1	0	−1	−1	0	5
			3	3			3	3		3
*X*6	0	0	0	1	0	1	0	−1	0	4
Obj2	0	0	0	0	0	0	1	1	1	0
Obj	−40	−30	0	0	0	0	0	0	1	0

Enter command? ("H" for short menu, "HELP" for full menu, ? for full help)
o
Optimality test
Are all objective entries nonnegative?
Yes. The current solution is optimal.

The linear programming solution

1	2	3	4	5	6	7	8
$\frac{2}{3}$	$\frac{8}{3}$	0	0	$\frac{5}{3}$	4	0	0

Objective = 0.
Problem has artificial variables. Use the "V" command to remove them.
Enter command? ("H" for short menu, "HELP" for full menu, "?" for full help)
v
All the artificial variables were deleted. The original objective function is restored. Examine the tableau, to verify that the objective row entries for all basic variables are 0.
If not, use the "A" command to zero them out.
The linear programming tableau:

	1	2	3	4	5	6	P	C
$X2$	0	1	$\frac{-2}{3}$	$\frac{1}{3}$	0	0	0	$\frac{8}{3}$
$X1$	1	0	$\frac{1}{3}$	$\frac{-2}{3}$	0	0	0	$\frac{2}{3}$
$X5$	0	0	$\frac{1}{3}$	$\frac{1}{3}$	1	0	0	$\frac{5}{3}$
$X6$	0	0	0	1	0	1	0	4
Obj	−40	−30	0	0	0	0	1	0

Enter command? ("H" for short menu, "HELP" for full menu, ? for full help)
a 40 2 5 (or **R5 <= 40 R2 + R5**)
ERO: Row 5 <= 40 Row 2 + Row 5
The linear programming tableau:

	1	2	3	4	5	6	P	C
$X2$	0	1	$\frac{-2}{3}$	$\frac{1}{3}$	0	0	0	$\frac{8}{3}$
$X1$	1	0	$\frac{1}{3}$	$\frac{-2}{3}$	0	0	0	$\frac{2}{3}$
$X5$	0	0	$\frac{1}{3}$	$\frac{1}{3}$	1	0	0	$\frac{5}{3}$
$X6$	0	0	0	1	0	1	0	4
Obj	0	−30	$\frac{40}{3}$	$\frac{-80}{3}$	0	0	1	$\frac{80}{3}$

Enter command? ("H" for short menu, "HELP" for full menu, ? for full help)
a 30 1 5
ERO: Row 5 <= 30 Row 1 + Row 5
The linear programming tableau

	1	2	3	4	5	6	P	C
*X*2	0	1	−2	1	0	0	0	8
			3	3				3
*X*1	1	0	1	−2	0	0	0	2
			3	3				3
*X*5	0	0	1	1	1	0	0	5
			3	3				3
*X*6	0	0	0	1	0	1	0	4
Obj	0	0	−20	−50	0	0	1	320
			3	3				3

Enter command? ("H" for short menu, "HELP" for full menu, ? for full help)
p (pivot)
Objective row

	1	2	3	4	5	6	P	C
Obj	0	0	$-\frac{20}{3}$	$-\frac{50}{3}$	0	0	1	$\frac{320}{3}$

Variable with most negative objective coefficient?
Enter column (=variable number).
4
Entering variable is 4.
Variable with least nonnegative feasibility ratio?
Nonnegative feasibility ratios:
Row 1, variable 2, ratio = 8
Row 3, variable 5, ratio = 5
Row 4, variable 6, ratio = 4
Enter the row of the departing variable.
4
Departing variable is 6 with feasibility ratio 4.00000.
ERO: Row 1 <= − 1/3 Row 4 + Row 1
ERO: Row 2 <= 2/3 Row 4 + Row 2
ERO: Row 3 <= −1/3 Row 4 + Row 3
ERO: Row 5 <= 50/3 Row 4 + Row 5
Objective changed from 320/3 = 106.667 to 520/3 = 173.333

The linear programming tableau:

	1	2	3	4	5	6	P	C
$X2$	0	1	$\frac{-2}{3}$	0	0	$\frac{-1}{3}$	0	$\frac{4}{3}$
$X1$	1	0	$\frac{1}{3}$	0	0	$\frac{2}{3}$	0	$\frac{10}{3}$
$X5$	0	0	$\frac{1}{3}$	0	1	$\frac{-1}{3}$	0	$\frac{1}{3}$
$X4$	0	0	0	1	0	1	0	4
Obj	0	0	$\frac{-20}{3}$	0	0	$\frac{50}{3}$	1	$\frac{520}{3}$

Enter command? ("H" for short menu, "HELP" for full menu, ? for full help)
p
Objective row

	1	2	3	4	5	6	P	C
Obj	0	0	$-\frac{20}{3}$	0	0	$\frac{50}{3}$	1	$\frac{520}{3}$

Variable with most negative objective coefficient?
Enter column (=variable number).
3
Entering variable is 3.
Variable with least nonnegative feasibility ratio?
Nonnegative feasibility ratios:
Row 2, variable 1, ratio = 10
Row 3, variable 5, ratio = 1
Enter the row of the departing variable.
3
Departing variable is 5 with feasibility ratio 1.00000.
ERO: Row 3 <= 3 Row 3
ERO: Row 1 <= 2/3 Row 3 + Row 1
ERO: Row 2 <= −1/3 Row 3 + Row 2
ERO: Row 5 <= 20/3 Row 3 + Row 5
Objective changed from 520/3 to 180.
The linear programming tableau:

	1	2	3	4	5	6	P	C
$X2$	0	1	0	0	2	−1	0	2
$X1$	1	0	0	0	−1	1	0	3
$X3$	0	0	1	0	3	−1	0	1
$X4$	0	0	0	1	0	1	0	4
Obj	0	0	0	0	20	10	1	180

Enter command? (“H” for short menu, “HELP” for full menu, “?” for full help)
o
Optimality test
Are all objective entries nonnegative?
Yes. The current solution is optimal.
The linear programming solution

1	2	3	4	5	6
3	2	1	4	0	0

Objective = 180.
Enter command? (“H” for short menu, “HELP” for full menu, ? for full help)
k
Closing the transcript file “sample.lp2”
Enter command? (“H” for short menu, “HELP” for full menu, ? for full help)
q
Enter “Y” to confirm you want to quit.
y
MATMAN is stopping now.

C The MATALG Program

This appendix describes MATALG, an interactive matrix algebra program. This appendix contains information on the background and purpose of MATALG, how to run the program, what the commands are, what the legal formulas and symbols are, and instructions for running the program on an IBM PC or compatible, or on a MACINTOSH.

This program can perform most common calculations with real scalars, vectors, and matrices. You name and dimension the variables, give them values, and define formulas involving them; the program interprets and evaluates those formulas. The program can be used as a simple programming language for carrying out matrix calculations. For example, the following commands declare A to be a 3 by 3 matrix, give it a value, and request the determinant. Then the value of $A(3, 3)$ is changed, the determinant is reevaluated, and the matrix is inverted.

E A,3,3	(Enter the 3×3 matrix A.)
1,2,3	(Entries are entered one row at a time
4,5,6	separated by either commas or spaces.)
7 8 9	
=DET(A)	(Computes $\det(A) = 0$.)
C A,3,3, 0.0	(Changes $A(3,3)$ to 0.)
=DET(A)	(New value of the determinant is 27.)
B = INV(A)	(Inverts the new matrix and names it B.)

The program will print out prompts, ask for missing input, help you decide what to do next, and show the intermediate results as you go along. These messages were not included in the example above. Formulas can be more complicated, involving several functions. In order to compute the eigenvalues of A^TA in the above example, you could type the single line "= EVAL(TRANS(A)*A)".

Running the Program

IBM PC: The system must be booted and running already. Insert the MATALG disk in the disk drive, open a DOS Window and type "MATALG." When not running the program, the command "DIR" will give a directory of all the files on the disk.

MACINTOSH: The system must be booted and running already. Insert the disk containing MATALG. Move the mouse to the MATALG disk and click it twice. All of the files on the disk will be displayed. Move the mouse to the "MATALG" icon and click twice.

Commands

A command to MATALG is usually a single letter. The list of available commands follows. Most must be followed by one or more arguments. These arguments are listed following the command letter. You can give the full command on a single line. You do not have to give the full command on a single line, but can specify part of the command. In that case, MATALG will prompt you for the rest of the information it needs.

Command and Arguments		*Short Definitions*
C	name, row, column, value	Change an entry of a matrix.
D	filename	Open/close a transcript file.
E	name, nrow, ncolumn	Enter a variable of the given size.
H		Help; types this list.
I		Initialize.
K	name(s)	Clears specified variables.
L	name, lo, hi, nstep	Tabulate a formula.
M	LU/QR/SVD, name	Factor matrix: *LU*, *QR*, or *SVD*.
P	name	Characteristic polynomial.
Q		Quit.
R	steps	Repeated evaluation of $X = F(X)$.
S	name	Save formula value in a variable.
T	name/ALL/blank	Type out the given variables.
U	name	Partition a matrix.
V		Reevaluate the last formula.
W	filename, name	Write variable to file.
X	filename	Read variable from file.
variable = formula		Set variable to value of formula.
?		Display extensive help.
#		Comment line.

A command and its arguments may be entered on the same line, separated by commas or spaces, or each argument may be given on its own line. The program will prompt you for the next argument it needs if you do not supply it. Thus, "E A 3, 4" is the concise way to prepare to enter a matrix named A which has 3 rows and 4 columns. But if you just type "E" the program will ask you for the name of the variable, and then the number of rows and columns. Variable names may be up to 10 characters long, and may only consist of letters and numbers. The program is not case sensitive, that is $A = a$ and so on.

All variables are stored as real numbers inside the program. All arithmetic operations are performed in single precision and the usual round-off problems can occur. When you are inputting a number to the program, you may use integer, rational, or real form. The following are all legal input values for a variable: 2, 3/4, 17.23

C (name, row, column, value) = CHANGE A VARIABLE

Changes the given entry of the variable to the new value. If the variable is a scalar or vector, you still must specify both row and column. For a scalar, the row and column should be 1. "C A 3,4, 5.0" will change entry $A(3,4)$ to 5. The program will echo print the variable after its value has been changed using the "C" command. The command "A(3, 4) = 5" will do the same thing, but the new value will not be printed.

D (filename) = OPEN OR CLOSE TRANSCRIPT FILE

Requests that a disk file be opened with the given name, in which a transcript of the session will be stored. A second "D" command (with no argument) will close the file. File names may be up to 30 characters long. Transcript files cannot be reopened from the program.

"D B:RECORD.DAT" would store the transcript file in the file RECORD.DAT on the disk in drive B (or on the MAC disk named B). "D RECORD.DAT" would store the transcript in the file RECORD.DAT on the disk in the currently active drive.

E (variable, rows, columns) = ENTER A VARIABLE

Sets up a variable of the given name and size. The number of rows and columns must be no more than 12. You must use this command first, to set up variables, before entering their values and using them in a formula. The program will immediately request that you set the value of the variable. If the variable is a scalar (1 row and 1 column), then its value may be included on the same line. The program will echo print the variable when you have finished entering it. Complex numbers cannot be entered.

"E SQUARE 5,4" would declare a matrix named SQUARE of five rows and four columns. "E FRED 1,1, 17" or "FRED = 17" would declare a scalar named FRED whose value was 17.

"=" INDICATES FORMULA EVALUATION

An equal sign requests that the formula be interpreted and evaluated. Formulas such as "=1+2" need no variables, but a formula such as "=X+Y" requires that you have declared and defined the variables X and Y earlier. "= A*B+A**2+INV(TRANS(A)*A)" is an example of formula input. Because nothing appears on the left-hand side, the formula is evaluated, the value is printed, but nothing is done with the value. You can, if you like, assign the value of the formula to a variable using the "S" command. Thus you can say "=1+2" in which case the result will be printed, or "A=1+2" in which case the result will be printed and stored in A. The following are all legal commands: "A(1,2)=2", "X=4+SIN(Y)", "=X+Y". The program expects the formula to be in standard FORTRAN and will reject incorrect syntax as well as incompatible matrix operations.

H = HELP

Requests the display of the list of all the commands, with one line explanations, given at the beginning of this section.

I = INITIALIZE

This command initializes or reinitializes the program. If you issue this command you will be asked to affirm that you really mean it. It clears out a fixed amount of workspace to be used by the program. This is done automatically when the program begins, but it may be necessary to issue this command during program execution if all the workspace has been used up. WARNING: the "I" command erases the definitions and values of all user variables; to delete only part of the data, use the "K" command.

K (variable) = KILL (DELETE) THE LISTED VARIABLES

MATALG has a finite amount of space to store variables. To clear out all the space used so far, the "I" command will do. But the "I" command destroys all user variables and formulas. To delete just some variables, the "K" command can be used. It will delete the variables named, and return the space they were occupying to the free space available to the program. "K X,Y" will kill the variables named X and Y and return the memory allocated to X and Y to the free space.

L (Variable, start value, stop value, number of steps) = LIST A TABLE

This command allows you to list the successive values of a formula for equally spaced values of one of its arguments. The most recent formula you entered is the one that will be tabulated. For example, the two commands

"=DET(x-A)"

"L X, 0.0, 10, 100"

will list the value of the characteristic polynomial of A for every tenth of a unit from 0 to 10.

M (LU/QR/SVD, variable name) = MATRIX FACTORIZATION

This command allows you to carry out either the LU, QR, or SVD factorization of a rectangular matrix. Type either "LU", "QR", or "SVD" and then give the name of the matrix to be factored.

"M LU A" would carry out the LU factorization of the matrix A.

The LU factorization of an $m \times n$ matrix A returns an $m \times n$ unit lower triangular matrix L, an $m \times m$ permutation matrix P, and an $m \times n$ upper triangular matrix U, such that $PA = LU$. This factorization is useful in solving linear systems.

"M QR B" would carry out the QR factorization of the matrix B.

The QR factorization of an $m \times n$ matrix B returns an $m \times m$ orthogonal matrix Q, and an $m \times n$ upper triangular matrix R, such that $B = QR$. This factorization is useful in solving overdetermined systems in the least squares sense and is the essential ingredient in the best algorithms for finding the eigenvalues of a matrix.

"M SVD C" would carry out the singular value decomposition of the matrix C.

The SVD (singular value decomposition) of the $m \times n$ matrix returns an $m \times m$ orthogonal matrix U, an $m \times n$ diagonal matrix S, and an $n \times n$ orthogonal matrix V such that $U^TCV = S$. The singular value decomposition is useful in statistics and in solving least squares problems. It provides the best method of computing the rank of a matrix on a computer.

Once a factorization is carried out, you will be asked if you want to save the factors in variables. If so, you must provide names of the variables in which to store the factors.

P (Variable) = POLYNOMIAL

If the variable is a square matrix (of order n), this command computes the characteristic polynomial of the matrix. The result is a list of $n+1$ coefficients. The program will offer to store this information in a vector variable if you like. Then you can evaluate the polynomial using formulas involving the "POLY" function.

"P A", assuming A is a square matrix, produces the coefficients of the characteristic polynomial of A. If we store these coefficients in a vector, say PCOEF, then the command "=POLY(PCOEF, A)" would evaluate the polynomial with the matrix A as its argument. In exact arithmetic, the result should be the zero matrix (Cayley-Hamilton theorem).

If the variable is a vector, then it is assumed that the n roots of an nth degree polynomial have been stored in the vector. The command computes the $n+1$ coefficients of the corresponding polynomial.

"P V" where V is the vector [1 2 3] would result in the computation of the coefficients of the polynomial $(X - 1)*(X - 2)*(X - 3)$, which equals $X^3 - 6X^2 + 11X - 6$.

Q = QUIT

This stops the program. Just in case you did not mean to type the letter Q, the program requires that you affirm your decision by asking you to enter "Y" or "YES" to quit.

R (steps) = REPEATED EVALUATION

This command only makes sense if you have typed in a formula of the form $X = F(X)$. Here, we understand this equation to mean "Take the current value of X, evaluate $F(X)$, and update X to have that value. The commands

"X = (A*X)/NORM2(A*X)"

"R 15"

where A is a matrix and X a vector, implements 15 steps of the power method for finding the dominant eigenvalue of A.

S (Variable) = SAVE A VARIABLE

If a formula is not of the form "variable=expression", the value returned by the formula is not going to be saved. If you decide you want to save the value, the S command will do it.

"=SQRT(16)"

"S X"

will save the value 4 in X.

T (Variable or 'All' or 'Debug' or blank) = TYPE A VARIABLE

Requests that the program print the value of one or more variables. If a variable name follows, as in "T X", its value will be typed out. "All" requests that all the symbols used by the program be displayed. "Debug" prints out all sorts of stuff, primarily for debugging.

U (Variable) = UNPACK (PARTITION) A MATRIX

This command lets you construct new matrices by partitioning an existing matrix. You will be asked: the rows to keep, followed by a zero; the columns to keep, followed by a zero. Type the name of a matrix in which to store the result, or RETURN if you don't want to save it.

You actually define four new matrices by this process, and you can see and save the other three if you want. The second matrix is made by deleting the rows you listed, keeping the columns you deleted. The third and fourth matrices are made similarly. The program will let you know that you can examine them.

For example, if A is the matrix $\begin{bmatrix} 11 & 12 & 13 \\ 21 & 22 & 23 \\ 31 & 32 & 33 \end{bmatrix}$, then if your row list is 1,0 and your column list is 1,3,0, the (first) submatrix will be $[11 \quad 13]$ and the others will be $\begin{bmatrix} 21 & 23 \\ 31 & 33 \end{bmatrix}$, $\begin{bmatrix} 22 \\ 32 \end{bmatrix}$, and $[12]$.

V = VALUE OF LAST FORMULA

Evaluate the current formula (presumably with a new value for at least one of the variables). For example, the sequence of commands

"E X, 1,1, 2"

"=X*3"

"C X, 1,1, 3"

"V"

will produce a value of 6 on the first evaluation, and a value of 9 on the second. This command can be used in place of the "R" command if you wish to monitor the calculation closely, perhaps to determine convergence of a sequence of approximations.

W (Filename, variable) = WRITE A VARIABLE TO A FILE

Writes the value of a variable into a file, which may later be read in by this program, or by MATMAN.

X (Filename) = READ A VARIABLE FROM A FILE

Requests that the given file be opened, and the names of the variables stored there be listed. Presumably, the file was made by this program with the "W" command, or by MATMAN. Once the program has listed the variables, you may pick one to read in.

The file "MATALG.DAT" contains some examples you can use.

? = DETAILED HELP

While running the program, you can always use the "H" command for quick help. Sometimes, though, that isn't enough. If you really want to see some detailed explanation of a particular command, you can use "?" instead of "H". Effectively, this causes the program to read the help file, MATALG.HLP, and to display the main topics. At this point your options are:

Type a topic, or enough of the topic to be unique. In this case you will get information on the given topic, plus possibly a list of subtopics.

Hit the RETURN key. This will back you up one "level." If you've just entered the help program, you will back up into the main program. If you're down one subtopic, you'll back up to the main topic, and so on.

Type CTRL-Z, "Control-Z", by holding down the "CONTROL" key on most terminals or the "Clover-leaf" key on a Macintosh, and typing "Z", possibly followed by RETURN. This gets you out of the help program and back into the main program immediately, no matter how many levels down you are.

Type "?" to have information about the current topic repeated. You might give this command several times in a row, just to browse through the same information over and over again. But you might also issue it when backing up, because when backing up to a higher topic, that topic's information is not printed out again unless you specially request it with the "?".

= COMMENT LINE

Lines that begin with a # contain user comments that will be part of the transcript file. They are not to be interpreted as commands to the program.

Formulas

In order to use the formula option correctly, you must use the proper abbreviations for mathematical functions, and the right symbols for operations like multiplication and exponentiation. Remember that every formula must contain an "=".

Symbols and constants: The operators allowed are +, −, *, / ,**, ^ , and =.

* is standard scalar or matrix multiplication if both arguments are scalars or both are matrices. If one argument is a scalar, then each entry of the other argument will be multiplied by the scalar. If both arguments are row or column vectors, the dot product is taken. A row vector times a column vector yields a scalar. A column times a row vector yields a matrix.

/ is standard scalar division. You may "divide" a matrix by a scalar, in which case each entry of the matrix is divided by the scalar, e.g., $A/2 = (\frac{1}{2})*A$.

+ and − may be used for scalars, vectors, or matrices of the same order. Also, a **nonstandard** shorthand allows you to add a scalar to a matrix. In this case, the scalar is added only to the diagonal. Thus, if B is a square matrix, $B+2$ is a legal formula, which really means $B+2*I$. This allows you to save storage space since the identity matrix need not be stored.

** (scalar exponentiation) is legal for scalar base and power, such as $2**3(= 2^3)$, or for a square matrix base and integer power, such as $A**2 = A^\wedge 2$.

Functions

S, S1, S2: Arguments may ONLY be scalar.
M : Arguments may be scalar or square matrix.
* : Arguments may be scalar, vector, or matrix.
I : Arguments may only be a positive integer.

ABS(*) = Absolute value of *.
ACOS(S) = The arc cosine of S. $1 < S < 1$

ALOG(S) = Natural logarithm of S. $S > 0$

ALOG10(S) = Logarithm base 10 of S. $S > 0$

ASIN(S) = Arc sine of S. $1 < S < 1$

ATAN(S) = Arc tangent of S.

ATAN2(S1,S2) = Arc tangent of $(S1/S2)$. Correctly computes ATAN2(0,0) = PI/2.

COS(S) = Cosine of S.

COSH(S) = Hyperbolic cosine of S.

DET(M) = Determinant of matrix M.

DIAG(M) = Diagonal of matrix M, stored in a column vector.

EPS = Machine epsilon. This is the smallest power of two such that 1+EPS.GT.1.0

EVAL(M) = Real and imaginary parts of eigenvalues of matrix, stored in an array of m rows and 2 columns, real parts in first column, imaginary in second.

EVEC(M) = Eigenvectors of matrix, stored in square matrix of same size as M. (Eigenvectors corresponding to a complex pair: if eigenvalues j and $j+1$ are a complex pair, then the eigenvector for eigenvalue j is column j + i*column $j+1$, and the eigenvector for eigenvalue $j+1$ is column j − i*column $j+1$.)

EXP(S) = Exponential of S.

ID(N) = Matrix identity of order N.

INT(*) = Replace real values by nearest integer.

INV(M) = Inverse of matrix M.

IVAL(M) = Imaginary parts of eigenvalues of matrix, stored in a column vector.

LN(S) = Natural logarithm of S. $0 < S$

LOG(S) = Natural logarithm of S. $0 < S$

LOG10(S) = Logarithm base 10 of S. $0 < S$

MAX(S1,S2) = Maximum of $S1$, $S2$.

MIN(S1,S2) = Minimum of $S1$, $S2$.

NEG(*) = Changes sign of *.

NORM0(V) = Infinity or Max-norm of a vector. NORM0(V) returns the maximum of the absolute values of the entries of V.

NORM1(V) = 1-norm of a vector. NORM1(V) returns the sum of the absolute values of the entries of V.

NORM2(V) = 2-norm. Euclidean norm, or root-mean-square norm of the vector V. Returns the square root of the sum of the squares of the entries of V.

PI = 3.14159265 . . .

POLY(V,M)	= Polynomial evaluation. V contains the coefficients of the polynomial, with $V(1)$ the coefficient of the highest order, and the last entry of V being the constant term. M is the square matrix argument of the polynomial.
RAN(*)	= Fills * with random numbers between 0 and 1.
RVAL(M)	= Real parts of eigenvalues of matrix, stored in a column vector.
SIN(S)	= Sine of S.
SINE(S)	= Sine of S.
SINH(S)	= Hyperbolic sine of S.
SQRT(S)	= Square root of S. $0 \leq S$
STEP(S)	= Heavyside step function. STEP=0 if $S < 0$, STEP=1 if $S > 0$.
TAN(S)	= Tangent of S.
TANH(S)	= Hyperbolic tangent of S.
TRACE(M)	= Trace (sum of diagonal elements) of matrix.
TRANS(*)	= The transpose of *. A=TRANS(A) is a legal formula if A is square.
ZERO(I)	= The zero matrix of order I.
I!	= I factorial. Valid for I=0 to 25. $I! = I * (I - 1) * (I - 2) * \cdots * 2 * 1$.

Sample MATALG Transcript File

(User input in boldface, comments in parentheses or preceded by #.)
Enter command (H for menu or ? for full help)
D
Enter file name (default=matalg.lpt)
sample.dat
MATALG is opening sample.dat
E A 4 4
Added variable A.
Enter A (1 1), A (1 2), A (1 3), A (1 4), . . .
17 −8 −12 14
Enter A (2 1), A (2 2), A (2 3), A (2 4), . . .
46,−22,−35, 41
Enter A (3 1), A (3 2), A (3 3), A (3 4), . . .
−2 1 4 −4 4 −2 −2 3
A =

$$\begin{matrix} 17 & -8 & -12 & 14 \\ 46 & -22 & -35 & 41 \\ -2 & 1 & 4 & -4 \\ 4 & -2 & -2 & 3 \end{matrix}$$

Enter command (H for menu or ? for full help)
= DET(A)
FORMULA = −1.00000
Enter command (H for menu or ? for full help)
B=INV(A)
Implicit definition of B
B =

17.0000	−7.99999	−12.0000	14.0000	
41.9999	−20.0000	−31.0000	36.0000	$(B = A^{-1})$
9.99998	−4.99999	−8.00000	11.0000	
12.0000	−5.99999	−10.00000	13.0000	

Enter command (H for menu or ? for full help)
E K 4 1
Added variable K.
Enter K(1),K(2),K(3),K(4),
1 2 3 4
K TRANSPOSED
1 2 3 4
Enter command (H for menu or ? for full help)
=B*K
FORMULA= Transposed 21.0001 53.0002 20.0000 22.000 $(=A^{-1}K)$
Enter command (H for menu or ? for full help)
P A
X**4−2.00000*X**3+2.00000*X−1.00000
(CHARACTERISTIC POLYNOMIAL OF A)
Enter variable to save polynomial in, or RETURN.
COEF
Added variable COEF.
Coefficients saved in COEF.
Enter command (H for menu or ? for full help)
=POLY(COEF,A)
FORMULA =

0	0	0	0
0	0	0	0
0	0	0	0
0	0	0	0

(The Cayley-Hamilton theorem holds for this matrix.)
Enter command (H for menu or ? for full help)
=A3 − A**2 − A + 1**

FORMULA =

0	0	0	0
0	0	0	0
0	0	0	0
0	0	0	0

(The minimum polynomial of A is $x^3 - x^2 - x + 1$.)
Enter command (H for menu or ? for full help)
C A 3 3 5
A =

17	−8	−12	14
46	−22	−35	41
−2	1	5	−4
4	−2	−2	3

Enter command (H for menu or ? for full help)
=INV(A)
FORMULA =

−.142856	.571428	1.71429	−4.85714
−2.28571	2.14285	4.42857	−12.7143
−1.42857	.714285	1.14286	−1.57143
−2.28571	1.14286	1.42857	−2.71429

Enter command (H for menu or ? for full help)
=EVAL(A)
FORMULA =

3.63196	0	
−.815968	1.12321	(TWO COMPLEX EIGENVALUES)
−.815968	−1.12321	
.999966	0	

Enter command (H for menu or ? for full help)
P = EVEC(A)
Implicit definition of *P*
FORMULA =

0.297464	−0.213015	0.256110	0.447214
0.905275	−0.571984	0.730475	0.894427
−0.302205	−0.67449E−01	0.151782E−02	0.147742E−05
−0.257684E−01	−0.134127	0.757088E−01	0.177456E−05

Enter command (H for menu or ? for full help)
=INV(P)*A*P

FORMULA =

3.63198	0	0.26226E−05	−0.725591E−06
0.148565E−04	−0.816006	1.12315	0.21901E−04
0.483394E−04	−1.12317	−0.815976	0.485050E−04
−0.700168E−05	0.578165E−05	−0.579655E−05	1.00000

This matrix is block diagonal, except for round-off error.
Enter command (H for menu or ? for full help)
E X 4 1
Added variable *X*.
Enter X(1),X(2),X(3),X(4),
1 1 1 1
X TRANSPOSED

1 1 1 1

Enter command (H for menu or ? for full help)
X=A*X/NORM2(A*X)
FORMULA = Transposed

.342747 .934766 0 0.934765E−01

Enter command (H for menu or ? for full help)
R 10 (POWER METHOD)

K = 1Transposed	−.324796	−.915334	−.118107	−.206688
K = 2 Transposed	.349957	.922615	−.318143E−01	.159069
K = 3 Transposed	.320037	.934164	−.155697	.259475E−01
K = 4 Transposed	.210639	.723498	−.622750	−.210634
K = 5 Transposed	.306069	.915789	−.260099	.120969E−02
K = 6 Transposed	.298564	.907435	−.294650	−.245543E−01
K = 7 Transposed	.295681	.902280	−.312275	−.307889E−01
K = 8 Transposed	.298128	.906226	−.298847	−.237672E−01
K = 9 Transposed	.297431	.905262	−.302261	−.259600E−01
K = 10 Transposed	.297385	.905141	−.302666	−.259771E−01

(DOMINANT EIGENVECTOR)
Enter command (H for menu or ? for full help)
S=1
Implicit definition of *S*.
S = 1.
Enter command (H for menu, or ? for full help)
V = DET(S-A)
Implicit definition of *V*.
V = 0. (1 IS AN EIGENVALUE OF A)

Enter command (H for menu or ? for full help)
L S, 0, 10, 11
Table of X, $F(X)$

.000000	7.00000	
.909091	1.04880	
1.81818	−12.1695	
2.72727	−21.5902	
3.63636	.243662	(THERE IS AN EIGENVALUE BETWEEN 2.727 AND 3.636)
4.54545	97.1816	
5.45455	329.466	
6.36364	773.730	
7.27273	1523.00	
8.18182	2686.70	
9.09091	4390.63	
10.0000	6777.00	

D The MATLAB Program

MATLAB, the MATrix LABoratory, is an interactive computing environment for scientific calculation. It supports matrix calculation, complex arithmetic, a programming command set, two-and three-dimensional graphics, and is expandable to meet the needs of various users. The basic data element in MATLAB is a matrix that does not require dimensioning. The program is easy to learn and to use and has extensive on-line help and demonstrations. The professional version of the program is very powerful and quite expensive. It may be available to you on the mainframe at your base of operations. The inexpensive student version, which is available for both Macintosh and IBM PC's, is more than adequate for the needs of this text.

Any MATLAB command must be followed by a return or enter in order to activate it. Several commands can be typed on the same line separated by commas or semicolons.

Data Entry and Display

(1) To see a list of MATLAB commands type "help" followed by a return or enter.
(2) To get help on a particular command type "help command".
(3) To run a MATLAB demonstration type "demo".
(4) To exit MATLAB type "exit" or "quit".

The simplest way to enter a matrix is to type its entries inside of square brackets with the entries separated by spaces or commas and the rows separated by semicolons or returns. Note that MATLAB is case sensitive ($A \neq a$), so you need to be consistent in the names you give to variables. Typing either

```
A = [1  2  3
 4  5  6
 7  8  9]
```

or A = [1 2 3; 4 5 6; 7 8 9] will enter

$$A = \begin{bmatrix} 1 & 2 & 3 \\ 4 & 5 & 6 \\ 7 & 8 & 9 \end{bmatrix}.$$

Once the matrix A has been entered into the program it, or its parts, may be displayed:

(1) Type "A" to see the entire matrix.
(2) Type "A(i, j)" to see the entry in row i and column j.
(3) Type "A(i, :)" to see row i of A. "v = A(1, :)" would construct a row vector **v** from the first row of A.
(4) Type "A(:, j)" to see column j of A. "w = A(:, 3)" constructs a column vector **w** from the third column of A.
(5) Type "A(2:3, [1 3])" to see the submatrix obtained by deleting row 1 and column 2 of A.

If the data to be entered contains some complex numbers, then they are to be entered in the form "a + b*i" where "i = "sqrt(−1)". If complex numbers are involved, do not use i as an index in a loop.

If the column vectors **u**, **v**, and **w** have already been defined, then the command "A = [u v w]" builds a matrix whose columns are the given vectors. Similarly, the command "B = [u'; v'; w']" builds a matrix whose rows are $u' = \mathbf{u}^T, \mathbf{v}^T$, and $\mathbf{w}^T$. The augmented matrix of the system $AX = B$ is constructed by the command "M = [A, B]".

Some special matrices are easily constructed:

(1) "eye(n)" is the $n \times n$ identity matrix and "eye(A)" is an identity matrix of the same size as A.
(2) "zeros(n)" is the $n \times n$ zero matrix.
(3) "ones(n)" is the $n \times n$ matrix of all ones.
(4) "rand(n)" is an $n \times n$ matrix with random entries.
(5) "hilb(n)" is the $n \times n$ Hilbert matrix.
(6) "x = [a:h:b]" = $[a \quad a + h \quad a + 2h \quad \cdots \quad a + nh]$ where n is as large as possible so that $a + nh \leq b$.

The display format can be changed to suit the user. The choice of display format does not affect the computations, all of which are done in double precision. The choices are:

(1) "format short" (the default) displays 4 decimal places.
(2) "format long" displays 16 decimal places.
(3) "format short e" uses floating point notation with 5 significant figures.
(4) "format long e" uses floating point notation with 16 significant digits.

If you do not wish to see the result of a computation displayed, then you should end the command line with a semicolon. This is frequently convenient when all you are interested in is the final result of a sequence of calculations.

If you wish to keep a record of your MATLAB session on your disk, then type "diary filename," where "filename" is a name you choose for the diary file. This file can be edited, annotated, and printed after you exit MATLAB.

The currently active variables can be saved to a file, for later use by MATLAB, using the command "save filename." The command "load filename" will enter the variables saved from the earlier session. Do not try to use the "save" command to construct a diary file.

Matrix Operations

MATLAB will perform all of the standard matrix operations. If a requested operation is undefined, because of size incompatibilities, the program will give appropriate explanations.

The basic matrix operations are:
+, −, *, ^, ′ for addition, subtraction, multiplication, exponentiation, and transpose. In addition the following element-wise operations are available:
A.*B = $(a_{ij}*b_{ij})$, A./B = (a_{ij}/b_{ij}), and A.^ k = (a_{ij}^k).

These vectorized operations can be very useful in avoiding loops and in preparing data for graphs.

Note that the operation A + 2 is treated as A.+2, that is 2 is added to every element of the matrix A. In order to compute $3A^2+5A-6I$ you should type "B = 3*A^2 + 5*A − 6*eye(A)".

Matrix Functions

A full list of the available matrix functions can be obtained by typing "help". A short list of those functions which will be useful in this text follows. For additional information type help followed by the name of the command you are interested in.

(1) "inv(A)" = inverse of the matrix A.
(2) "rref(A)" = row reduced echelon form of the matrix A.
(3) "A\B" generates the solution of the linear system $AX = B$. The results may be inaccurate if A is singular.
(4) "det(A)" = determinant of A.
(5) "trace(A)" = trace of A.
(6) "rank(A)" = rank of A.
(7) "null(A)" generates a basis for the null space of A.
(8) "norm(A)" = norm(A, 2) = $\|A\|$. (A may be a matrix or a vector.)
(9) "lu(A)" generates the LU factorization of the matrix A. "[L, U, P] = lu(A)" generates a permutation matrix P, an upper triangular matrix U, and a unit lower triangular matrix L such that $PA = LU$.

(10) "chol(A)" computes the Cholesky factorization of the positive definite matrix A. If A is positive definite, then "R = chol(A)" generates an upper triangular matrix R such that $A = R^T R$.
(11) "v = poly(A)" computes the coefficients of the characteristic polynomial of A and stores them in the vector **v** with the coefficient of the highest power first.
(12) "polyval(v, s)" is the value of the polynomial, with coefficients in **v**, at the scalar s. "polyval(v, A)" evaluates the polynomial at each entry of the matrix A.
(13) "polyvalm(v, A)" is the value of the polynomial, with coefficients in **v**, at the matrix A.
(14) "roots(v)" generates the roots of the polynomial whose coefficients are in the vector **v**. "roots(poly(A))" would generate the eigenvalues of A.
(15) "eig(A)" = column vector of the eigenvalues of A. "[P, D] = eig(A)" generates a matrix P whose columns are the eigenvectors of A and a diagonal matrix D whose entries are the eigenvalues of A; $P^{-1}AP = D$. The results may be inaccurate if cond(P) is large.
(16) "qr(A)" generates the QR factorization of A. "[Q, R] = qr(A)" produces an upper triangular matrix R and an orthogonal (unitary) matrix Q such that $A = QR$. Note that the diagonal entries of R may not be positive.

Scalar Functions

The usual scalar functions are available. Among them are: sqrt, abs, conj, sin, cos, tan, asin, acos, atan, exp, log, log10, sinh, cosh, tanh, asinh, acosh, atanh, gamma, erf. Note carefully, that if these functions are applied to a matrix or a vector, then the function is evaluated at every entry of the array. This can be very useful in preparing data for graphs.

Graphs

MATLAB has several ways of displaying information graphically. It will be helpful to run demos 2 and 9 to see the various possibilities. If you are running MATLAB on a mainframe, using a remote terminal, you should type "terminal" and identify your terminal type for MATLAB.

The basic Cartesian plot of a function $y = f(x)$ is obtained using the "plot" command. The first step is to construct two data vectors; the first contains the x coordinates of the points to be plotted and the second contains the y coordinates (the functional values).

As an example let us plot $y = e^{-3x} \sin 2x$ for $0 \leq x \leq 5$. The command "x = 0:.1:5;" constructs the vector $x = [0\ .1\ .2\ .3 \cdots 4.9\ 5]$. Note that the semicolon suppresses the display of this long uninteresting vector. The command "y = (exp(−3*x)).*sin(2*x)" will construct and display a vector whose entries are the values of the function at the entries of the vector x; again the semicolon could be used to suppress the display. The command "plot(x,y,x, zeros(x))" will now cause the graph to be drawn and displayed. A title and axis labels can be added if desired using the commands, "title", "xlabel", and "ylabel". For various options regarding line type and point markings type "help plot." Polar plots, parametric plots, and various log plots are also available. It is possible to plot several curves on the same graph using different line types for each graph.

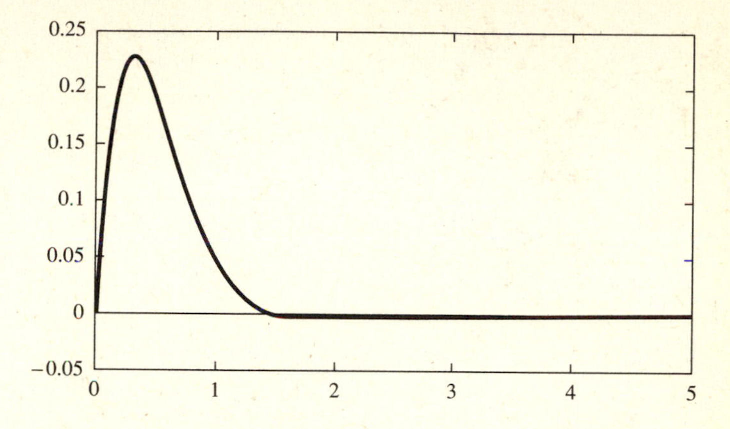

MATLAB can also plot surfaces representing functions $z = f(x, y)$. The procedure is to define a rectangle in the xy plane that will be the domain of the function. As an example let us plot the function $z = \sin(x)\ \cos(y)$ for $-7 < x, y < 8$. The commands

```
x = -7 :.5:8;   y = x;   [x1, y1] = meshgrid (x, y);
```

will construct the mesh points for the domain and then the command "z = sin(x1).* cos(y1);" will generate a matrix of functional values for the function. Finally, the command "mesh(z)" will generate the surface. The graphics will not be part of the diary file. You can use the "print" command or do a "print screen" on the PC or use the MAC clipboard to get a hard copy.

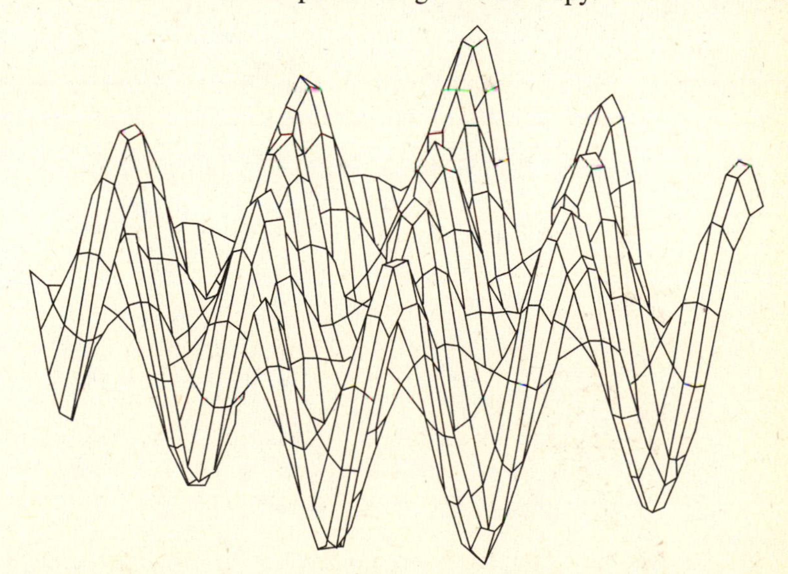

Programming

MATLAB has significant programming capabilities and can be used instead of FORTRAN or PASCAL to carry out complicated calculations, especially those which are iterative in nature. Looping is accomplished using "for" or "while" loops. Some simple examples will suffice to illustrate this capability. It is best to avoid using i as an index.

(1)
```
for j = 1:10
      w(j) =  j^3 + 3*j
      end
```
This code will construct the vector

$$\mathbf{w} = [4\ \ 14\ \ 36\ \ 76\ \ 140\ \ 234\ \ 364\ \ 536\ \ 756\ \ 1030].$$

(2)
```
v = [-1 2 3 5 7],
     A = zeros(5);
     for j = 1: 5
        for k = 1: 5
           A(j,k) = v(j)^(5-k);
        end
     end
     A
```
This code will generate a 5 × 5 Vandermonde matrix. Note that loops can be nested, but that there must be an "end" for each "for" statement. The semicolon at the end of line 5 prevents the printing of 25 5 × 5 matrices. The last command prints the final matrix.

(3)
```
E = zeros(A);
      F = eye(A);
      k = 1;
while norm((E + F) - E) > 0
      E = E + F;
      F = A*F/k;
      k = k+1;
end
k, E
```
This code will compute the exponential of the matrix A; $e^A = \sum_{k=0}^{\infty} \frac{Ak}{k!} = I + A + \frac{A^2}{2!} + \frac{A^3}{3!} + \cdots$. The summation continues until the term added makes no difference on the machine.

M-Files

MATLAB is normally used in a command-driven mode; when single-line commands are entered, MATLAB processes them immediately and displays the results. MATLAB is also capable of executing sequences of commands that are stored in files. These files are called m-files because they have a file type of ".m". These files can be constructed and edited using an editor or a word processor; on a MAC they can be constructed without leaving MATLAB. Once they have been stored, they are available as a command. If, for example, the commands in the last example are stored in a file named exponA.m, then the command "exponA" will cause the program to compute the exponential of the matrix called A. Use of m-files makes it much easier to debug long programs and takes the burden off typing skill.

Sample MATLAB Diary

(User input in boldface, comments in parentheses or after %)

diary sample.dat (opens transcript or diary file named "sample.dat")

A=[17 −8 −12 14; 46 −22 −35 41; −2 1 4 −4; 4 −2 −2 3]

```
A =

    17    -8   -12    14
    46   -22   -35    41
    -2     1     4    -4
     4    -2    -2     3
```

det(A)

```
ans =

   -1
```

B = inv(A)

```
B =

   17.0000    -8.0000   -12.0000   14.0000
   42.0000   -20.0000   -31.0000   36.0000
   10.0000    -5.0000    -8.0000   11.0000
   12.0000    -6.0000   -10.0000   13.0000
```

K = [1 2 3 4]′

```
K =

   1
   2
   3
   4
```

X = A \ K

```
X =
    21.0000
    53.0000
    20.0000
    22.0000
```

rref([A K]) (Row reduced echelon form of the augmented matrix; this will give better results if A is singular, or close to singular.)

```
ans =
    1  0  0  0  21
    0  1  0  0  53
    0  0  1  0  20
    0  0  0  1  22
```

v = poly(A) (computes the characteristic polynomial of A and stores coefficients in **v**.)

```
v =
    1.0000  -2.0000  -0.0000  2.0000  -1.0000
```

($c(x) = \det(xI - A) = x^4 - 2x^3 + 2x - 1$)

polyvalm(v, A)

```
ans =
  1.0e - 11*
 -0.0442  -0.0711  -0.1080  0.1414
 -0.1421  -0.1891  -0.2728  0.3627
  0.0902  -0.0338  -0.0712  0.0739
  0.0469  -0.0462  -0.0831  0.0961
```

(This is the zero matrix!! To see why it is not exactly *0* change to long format and look at **v** more closely.)

$A^3 - A^2 - A$ + eye(A)

```
ans =
    0  0  0  0
    0  0  0  0
    0  0  0  0
    0  0  0  0
```

(The minimum polynomial of A is $x^3 - x^2 - x + 1$.)

[P, D] = eig(A)

```
P =
    0.3357   0.0000  -0.0000  -0.4586
    0.9231   0.2673   0.2673  -0.8762
    0.0839  -0.8018  -0.8018  -0.1234
    0.1678  -0.5345  -0.5345  -0.0823
```

```
D =

   -1.0000        0        0        0
         0   1.0000        0        0
         0        0   1.0000        0
         0        0        0   1.0000
```

inv(P)*A*P

```
ans =

   -1.0000   0.0000   0.0000  -0.0000
   -0.0000   1.0000  -0.0000   0.0000
    0.0000   0.0000   1.0000  -0.0000
   -0.0000  -0.0000  -0.0000   1.0000
```

A(3,3) = 5 % change the (3,3) entry of A to 5.

```
A =

    17   -8  -12   14
    46  -22  -35   41
    -2    1    5   -4
     4   -2   -2    3
```

eig(A)

```
ans =

  -0.8160 + 1.1232i
  -0.8160 - 1.1232i   (note that the new A has two complex eigenvalues.)
   3.6320
   1.0000
```

x=[1 1 1 1]′

```
x =

     1
     1
     1
     1
```

```
for j = 1:10
    z=A*x;
    x = z/norm(z);
end
x
```

```
x =

    0.3286
    1.0000
   -0.3339
   -0.0287
```

(10 iterations of the power method)

A*x./x

```
ans =

   3.6241
   3.6242
   3.6295
   3.6270
```

% **x** is an approximate eigenvector for the eigenvalue 3.62.

E = zeros(A); F = eye(A); k=1;
while norm ((E+F)−E)>0

E = E + F;
F = A*F/k;
k = k + 1;

end, k, E

```
k =

   31

E =

    41.2433   -19.2625    -58.3108    47.0301
   116.1248   -55.3441   -176.9186   143.5393
   -30.2550    15.1275     57.1689   -45.1076
    -1.0996     0.5498      3.9726    -2.1794
```

(This is the sum of the first 31 terms of the series defining e^A.)

v = poly(A)

```
v =

   1.0000   -3.0000   -2.0000   -7.0000
```

x=[−3:.1:6]; y=polyval(v,x); plot(x,y,x,zeros(x))

(Plots the characteristic polynomial and the x-axis.)

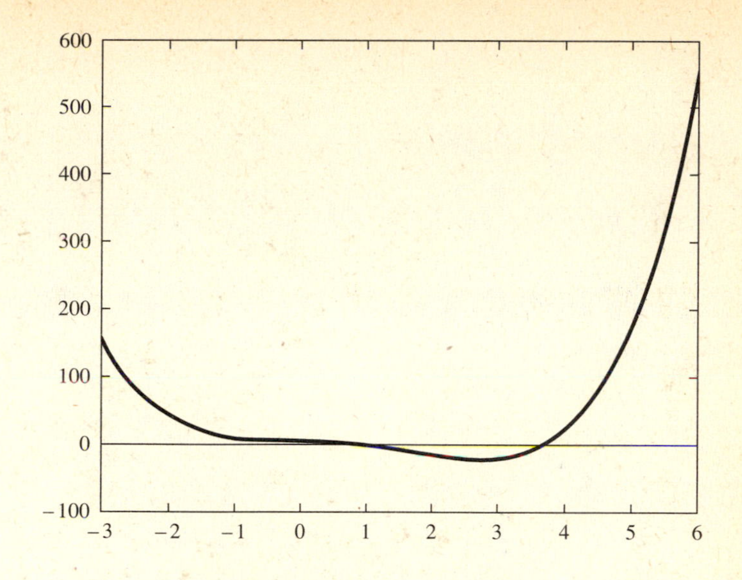

x=[−2: .1 : 4]; y = polyval(v,x); plot(x,y,x,zeros(x))
(Decrease the range of x so that the range of y is smaller.)

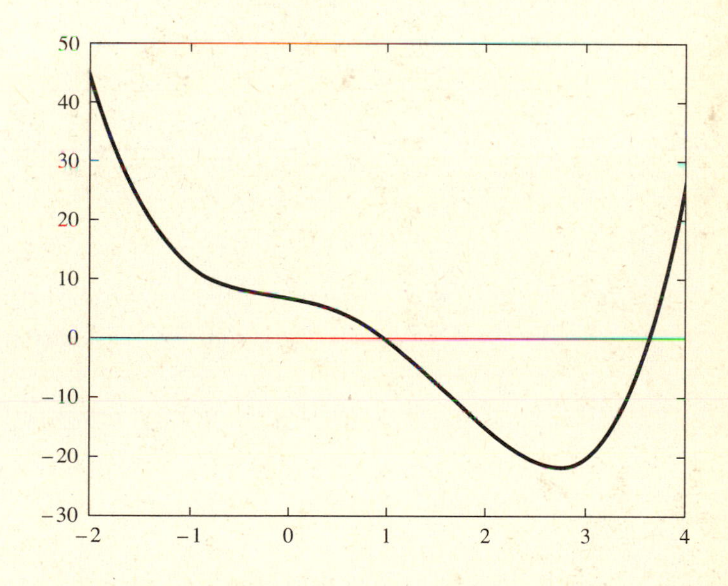

Answers to Odd-numbered Exercises

CHAPTER 1

EXERCISES 1.1 *(page 10)*

1. **(a)** $x = 4, y = 0$ **(b)** No solution. **(c)** $x = 6, y = -2$ **3.** **(a)** and **(b)**

5. **(a)** $\begin{bmatrix} 2 & 0 & -3 \\ 4 & 5 & 1 \\ 0 & 4 & 12 \end{bmatrix}, \begin{bmatrix} 2 & 0 & -3 & 11 \\ 4 & 5 & 1 & 6 \\ 0 & 4 & 12 & 0 \end{bmatrix}$ **(c)** $\begin{bmatrix} 8 & 8 & 8 & 0 \\ 9 & -15 & -18 & 0 \\ -5 & 11 & 14 & 0 \end{bmatrix}$

(e) $\begin{bmatrix} 6 & -5 & 0 & 0 & 0 & 11 \\ 4 & 6 & -5 & 0 & 0 & 12 \\ 0 & 4 & 6 & -5 & 0 & 13 \\ 0 & 0 & 4 & 6 & -5 & 14 \\ 0 & 0 & 0 & 4 & 6 & 15 \end{bmatrix}$ **(g)** $\begin{bmatrix} 1 & \frac{1}{2} & \frac{1}{3} & \frac{1}{4} & 1 \\ \frac{1}{2} & \frac{1}{3} & \frac{1}{4} & \frac{1}{5} & 2 \\ \frac{1}{3} & \frac{1}{4} & \frac{1}{5} & \frac{1}{6} & 3 \end{bmatrix}$

7. **(a)** $x = \frac{17}{2}, y = -6, z = 2$ **(d)** $(-4, 13, -4)$ **9.**

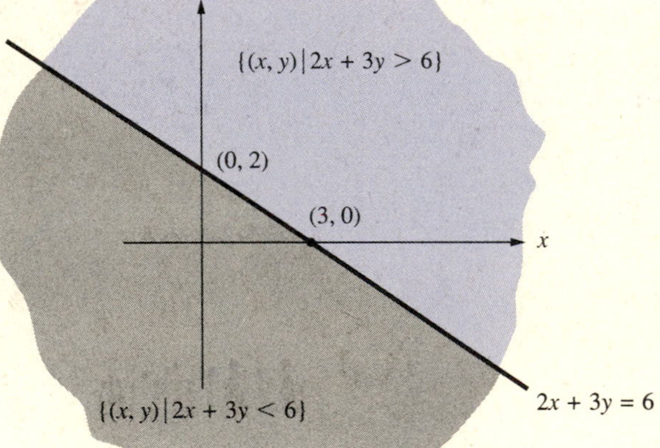

11. 40 cows and 20 chickens **13.** 20 lb peanuts, 10 lb cashews, $\frac{10}{3}$ lb almonds, and $66\frac{2}{3}$ lb sunflower seeds

15. $y = 2x^2 + 5x - 6$ **17.** $\frac{14}{13}, \frac{16}{13}, \frac{8}{13}$ **19.** $z = \frac{1}{3}(-8x - 2y + 19)$ **21.** $y = -x^2 + 2x + 4$

EXERCISES 1.2 *(page 21)*

1. **(a)** $-193, 57, -11$ **(c)** $5, -\frac{10}{3}, \frac{16}{3}$ **(e)** Inconsistent **(f)** $c, \frac{5 - c}{2}, 3 - 2c, -8 + 7c$

3. $x_1 = -1 - \frac{3}{5}k, x_2 = 1 + \frac{1}{5}k, x_3 = k$ (k is arbitrary) **5.** $x_1 = -66, x_2 = 27, x_3 = 6, x_4 = 4$ **7.** Inconsistent

9. $x_1 = c_2 - 1, x_2 = 2c_1, x_3 = c_1, x_4 = c_2$ (c_i are arbitrary) **13.** $h = \frac{5}{2}$ **15.** If $t = 4$ there are many solutions, if $t = -4$ there are no solutions. If $|t| \neq 4$, there is a unique solution for each value of t. **17.** $a_{11}a_{22} - a_{12}a_{21} \neq 0$
19. $x = \pm\sqrt{\frac{11}{7}}, y = \pm\sqrt{\frac{1}{7}}, z = \pm\sqrt{\frac{2}{7}}$ **21.** **(a)** $k_2 - 2k_1 = 0$ **(b)** $k_4 - 2k_1 = 0$ and $-k_1 + k_2 + k_3 = 0$.
23. $R = 23, D = 6, L = 4$ **25.** $t_1 = 69.6, t_2 = 107.2, t_3 = 144.8, t_4 = 182.4$

EXERCISES 1.3 *(page 34)*

1. **(a)** $\begin{bmatrix} 1 & -3 & 0 & 0 & 0 & 1 \\ 0 & 1 & 1 & 1 & -1 & -1 \\ 0 & 0 & 1 & \frac{7}{6} & -\frac{4}{3} & -\frac{5}{6} \end{bmatrix}$ (not unique) **(c)** $\begin{bmatrix} 1 & 1 & 2 & 6 \\ 0 & 1 & -2 & -3 \\ 0 & 0 & 1 & 2 \\ 0 & 0 & 0 & 0 \end{bmatrix}$ (not unique)

3. **(a)** $x_1 = 0, x_2 = 1, x_3 = 3$ **(c)** Inconsistent **5.** **(a)** $x = 0, y = -1, z = 1$ **(b)** $x = 14, y = -14, z = 5$
(c) $x = -9, y = 5, z = 0$
7. $(R_1 \leftarrow 2R_2 + R_1), (R_2 \leftarrow -R_3 + R_2), (R_1 \leftarrow 3R_3 + R_1), (R_3 \leftarrow 5R_1 + R_3), (R_2 \leftarrow 2R_1 + R_2)$;
$R = \begin{bmatrix} 1 & 0 & 0 & 41 \\ 0 & 1 & 0 & -11 \\ 0 & 0 & 1 & -6 \end{bmatrix}$ **9.** $R_i \leftarrow -\left(\frac{b}{a}\right)R_j + \left(\frac{1}{a}\right)R_i$ **11.** $x = 40, y = 60, z = 60$ **13.** $y = 2x^3 - x^2 - x + 1$

15. $x = 100, y = 50, u = 0, v = 150; C = 550$ **17.** $C = 1300; x_3 = 200, y_1 = 100, y_2 = 200, y_4 = 100$

EXERCISES 1.4 *(page 42)*

1. **(a)** $x_1 = c_2 - 2c_1, x_2 = -4c_2 - 5c_1, x_3 = c_1, x_4 = c_2$ (c_1, c_2 arbitrary)
(c) $x_1 = 46c, x_2 = 40c, x_3 = -35c, x_4 = 9c$ (c arbitrary) **(e)** $x_1 = -2c, x_2 = -2c, x_3 = c$ (c arbitrary)
3. **(a)** $\lambda = 4, 9$ **(c)** $\lambda = 1, 5$ **5.** For $B_1: x = 1, y = -1, z = -1$; for $B_2: x = -1, y = 2, z = 1$; for B_3: inconsistent **7.** $a_{11}a_{22} - a_{12}a_{21} = 0$ **9.** **(a)** 15, 30, 45, 60 **(c)** 4, 28, 52, 76 **11.** **(a)** $\frac{225}{4}, \frac{275}{4}, \frac{425}{4}, \frac{475}{4}$
(b) 25, 25, 75, 75 **(c)** $\frac{385}{6}, \frac{110}{6}, \frac{110}{6}, \frac{55}{6}$ **13.** $-29 - 2x - 4y + (x^2 + y^2) = 0$ or $(x - 1)^2 + (y - 2)^2 = 34$.

EXERCISES 1.5 *(page 54)*

1. $AB = \begin{bmatrix} 5 & 18 \\ 14 & 39 \end{bmatrix}, BA = \begin{bmatrix} 1 & 2 & 3 \\ 14 & 19 & 24 \\ 16 & 20 & 24 \end{bmatrix}$ **3.** **(a)** $AB = BA = \begin{bmatrix} 1 & 0 & 0 \\ 0 & 1 & 0 \\ 0 & 0 & 1 \end{bmatrix}$ **(c)** $\begin{bmatrix} -9 & 17 & 22 \\ -5 & 12 & 6 \\ 5 & -7 & -1 \end{bmatrix}$

5. **(a)** $(AB)C = A(BC) = A$ **(b)** $A(B + C) = AB + AC = \begin{bmatrix} 6 & -8 & 14 \\ -5 & 16 & 28 \\ -28 & 54 & 64 \end{bmatrix}$

7. **(a)** $x = 2, y = \frac{29}{11}, u = 4, v = \frac{2}{11}$ **(b)** $\frac{1}{3}\begin{bmatrix} 11 & 2 \\ 8 & -1 \end{bmatrix}$ **9.** $\text{ent}_{ij}(AD) = a_{ij}d_j$; $\text{ent}_{ij}(DA) = d_i a_{ij}$

11. $AB = BA = \begin{bmatrix} 1 & 0 \\ 0 & 1 \end{bmatrix}$ **13.** **(a)** $\begin{bmatrix} a & b & c \\ g & h & i \\ d & e & f \end{bmatrix}$ **(c)** $\begin{bmatrix} a & b & c \\ d & e & f \\ ka + g & kb + h & kc + i \end{bmatrix}$ **(e)** $\begin{bmatrix} a & kb & c \\ d & ke & f \\ g & kh & i \end{bmatrix}$

15. $\begin{bmatrix} a & b \\ 0 & a \end{bmatrix}$ **17.** $\begin{bmatrix} 24 & 45 & 55 \\ 16 & 16 & -38 \\ 31 & 71 & 77 \end{bmatrix}\begin{bmatrix} u \\ v \\ w \end{bmatrix} = \begin{bmatrix} 12 \\ 0 \\ 13 \end{bmatrix}$ **19.** **(a)** $a_{ij} + (b_{ij} + c_{ij})$ **(c)** $(h + k)a_{ij}$

(e) $\Sigma_{k=1}^{n} a_{ik}(b_{kj} + c_{kj})$ **(g)** $\Sigma_{k=1}^{n} \Sigma_{s=1}^{n} a_{ik}(b_{ks}c_{sj})$ **23. (a)** $\begin{bmatrix} -4.3889 & -7.7847 & 60.1663 \\ 16.1048 & 66.7516 & -10.4901 \\ 13.8449 & 79.2839 & -59.3043 \end{bmatrix}$

(b) $\begin{bmatrix} 149.164 & 294.830 & 239.038 \\ -76.1603 & 1998.38 & 1874.13 \\ 1357.39 & 3397.91 & -4190.01 \end{bmatrix}$ **25. (a)** $4x^2 - 2xy + 4xz + 2yz$ **(b)** $[x \quad y \quad z]\begin{bmatrix} 3 & -4 & 3 \\ -4 & 2 & 0 \\ 3 & 0 & -3 \end{bmatrix}\begin{bmatrix} x \\ y \\ z \end{bmatrix}$
is the "best" answer. There are other answers where A is not symmetric. **29.** 1107, 1032, 861

EXERCISES 1.6 *(page 63)*

1. $A^{-1} = \begin{bmatrix} 1 & -1 \\ 0 & 1 \end{bmatrix}$, $B^{-1} = \begin{bmatrix} 3 & -1 \\ -5 & 2 \end{bmatrix}$, $A^{-1} + B^{-1} = \begin{bmatrix} 4 & -2 \\ -5 & 3 \end{bmatrix}$, $(A + B)^{-1} = \begin{bmatrix} 2 & -1 \\ -\frac{5}{2} & \frac{3}{2} \end{bmatrix} \neq A^{-1} + B^{-1}$.

3. $B = \frac{1}{2}\begin{bmatrix} -6 & 4 \\ 5 & -3 \end{bmatrix}$.

5. $A^3 + A^2B + ABA + BA^2 + AB^2 + BAB + B^2A + B^3$; $A^3 + 3A^2B + 3AB^2 + B^3$(if $AB = BA$).

7. $A = \frac{1}{10}\begin{bmatrix} 3 & -5 \\ -2 & 4 \end{bmatrix}$. **9.** $J^{-1} = \begin{bmatrix} \frac{1}{3} & -\frac{1}{9} & 0 \\ 0 & \frac{1}{3} & 0 \\ 0 & 0 & -1 \end{bmatrix}$. **11.** If $\text{Row}_i(A) = 0$, then $\text{Row}_i(AB) = \text{Row}_i(A)B = 0$.

13. If A and B are lower triangular, then $a_{ij} = b_{ij} = 0$ if $i > j$. Let $AB = C$, then for $i > j$ $c_{ij} = \Sigma_{k=1}^{n} a_{ik}b_{kj} = \Sigma_{k=1}^{j} a_{ik}b_{kj} + \Sigma_{k=j+1}^{n} a_{ik}b_{kj}$. In the first sum all $a_{ik} = 0$ while in the second sum all $b_{kj} = 0$. Therefore $c_{ij} = 0$ and C is lower triangular.

15. $\begin{bmatrix} 13 & -2 & -3 & 4 \\ 47 & 4 & -7 & 16 \\ 29 & -4 & -10 & 7 \\ 30 & 10 & -9 & 10 \end{bmatrix}$ **17.** $X = A^{-1}K$. **19.** $A^{-1} = \begin{bmatrix} 1 & -1 \\ -1 & \frac{4}{3} \end{bmatrix}$, B^{-1} does not exist, $C^{-1} = \frac{1}{4}\begin{bmatrix} 3 & 1 & -1 \\ 1 & 3 & 1 \\ -1 & 1 & 3 \end{bmatrix}$.

21. $A^T = \begin{bmatrix} 1 & 2 & -1 \\ 2 & 0 & 1 \end{bmatrix}$, $B^T = \begin{bmatrix} 1 & 2 \\ -1 & 1 \\ 2 & 0 \end{bmatrix}$, $(AB)^T = B^TA^T = \begin{bmatrix} 5 & 2 & 1 \\ 1 & -2 & 2 \\ 2 & 4 & -2 \end{bmatrix}$, $A^TB^T = \begin{bmatrix} -3 & 4 \\ 4 & 4 \end{bmatrix}$, $A^TA = \begin{bmatrix} 6 & 1 \\ 1 & 5 \end{bmatrix}$.

23. $(A^n)^T = (AA \cdots A)^T = (A^TA^T \cdots A^T) = (A^T)^n$. **25.** $d_{ii} = \pm 1$ **27.** $X = \begin{bmatrix} 0 & 1 \\ 1 & 1 \end{bmatrix}$.

EXERCISES 1.7 *(page 75)*

1. $X = \begin{bmatrix} -9 & -6 & -6 \\ 7 & 5 & \frac{9}{2} \end{bmatrix}$. **3.** $X = \begin{bmatrix} 2-a & 1-b & 2-c \\ a & b & c \\ 0 & 1 & 1 \end{bmatrix}$. **5.** $A^{-1} = \begin{bmatrix} 2 & -1 & -1 \\ 1 & 0 & -1 \\ -2 & 1 & 2 \end{bmatrix}$.

7. $B^{-1} = \begin{bmatrix} 0 & 1 & 2 \\ -1 & 3 & 0 \\ 1 & -2 & 1 \end{bmatrix}$. **9.** $T^{-1} = \begin{bmatrix} 1 & 2 & -\frac{14}{3} \\ 0 & -1 & \frac{4}{3} \\ 0 & 0 & \frac{1}{3} \end{bmatrix}$. **11.** $B^{-1} = \frac{1}{18}\begin{bmatrix} 2 & 5 & -7 & 1 \\ 5 & -1 & 5 & -2 \\ -7 & 5 & 11 & 10 \\ 1 & -2 & 10 & 5 \end{bmatrix}$.

13. $R^{-1} = \begin{bmatrix} \cos\theta & \sin\theta \\ -\sin\theta & \cos\theta \end{bmatrix}$. **15.** $L^{-1} = \begin{bmatrix} 1 & 0 & 0 & 0 \\ -2 & 1 & 0 & 0 \\ 13 & -5 & 1 & 0 \\ -39 & 13 & -3 & 1 \end{bmatrix}$. **17.** $P = \begin{bmatrix} 5 & -1 & -5 \\ 3 & -\frac{6}{5} & -\frac{11}{5} \\ -8 & \frac{13}{5} & \frac{38}{5} \end{bmatrix}$.

19. $[PA \mid P] \to [P_1PA \mid P_1P] = [I \mid A^{-1}]$. **21. (a)** Keep left arm straight. **(c)** Tack on the header.
23. (a) 93, 113, 98, 36, 37, 41, 77, 96, 77, 126, 149, 141, 87, 87, 112
(b) 73, 91, 74, 95, 95, 120, 61, 61, 81, 124, 138, 145, 94, 114, 99

25. **(a)** $X = \begin{bmatrix} 1 + 2c + d & 8 + 2c \\ c & d \end{bmatrix}$. **(b)** $X = \frac{1}{11}\begin{bmatrix} -158 & -208 \\ 56 & 52 \end{bmatrix}$. **27.** $A = \frac{1}{16}\begin{bmatrix} 4 & -5 & 1 \\ 4 & -3 & -1 \\ 4 & -11 & -9 \end{bmatrix}$.

EXERCISES 1.8 *(page 86)*

1. $\begin{bmatrix} 1 & 0 & 0 \\ 4 & 1 & 0 \\ 7 & 2 & 1 \end{bmatrix}\begin{bmatrix} 1 & 2 & 3 \\ 0 & -3 & -6 \\ 0 & 0 & -9 \end{bmatrix}$ **3.** $\begin{bmatrix} 1 & 0 & 0 \\ 2 & 1 & 0 \\ 1 & -2 & 1 \end{bmatrix}\begin{bmatrix} 1 & 2 & 4 \\ 0 & -1 & -1 \\ 0 & 0 & 1 \end{bmatrix}$

5. $\begin{bmatrix} 1 & 0 & 0 & 0 & 0 \\ \frac{1}{4} & 1 & 0 & 0 & 0 \\ \frac{1}{4} & \frac{1}{5} & 1 & 0 & 0 \\ 0 & \frac{4}{15} & \frac{2}{9} & 1 & 0 \\ 0 & 0 & \frac{5}{18} & \frac{7}{32} & 1 \end{bmatrix}\begin{bmatrix} 4 & 1 & 1 & 0 & 0 \\ 0 & \frac{15}{4} & \frac{3}{4} & 1 & 0 \\ 0 & 0 & \frac{18}{5} & \frac{4}{5} & 1 \\ 0 & 0 & 0 & \frac{32}{9} & \frac{7}{9} \\ 0 & 0 & 0 & 0 & \frac{341}{96} \end{bmatrix}$ **7.** $B = \begin{bmatrix} 1 & 0 & 0 & 0 \\ 1 & 1 & 0 & 0 \\ \frac{3}{2} & 0 & 1 & 0 \\ 2 & -3 & 10 & 1 \end{bmatrix}\begin{bmatrix} 2 & 4 & 3 & 2 \\ 0 & 1 & -1 & -5 \\ 0 & 0 & \frac{1}{2} & -1 \\ 0 & 0 & 0 & 5 \end{bmatrix}$

9. $x = \frac{1}{209}\begin{bmatrix} 389 \\ 268 \\ 113 \\ -46 \end{bmatrix}$. **11.** $\begin{bmatrix} 1 & 0 & 0 \\ 0 & 0 & 1 \\ 0 & 1 & 0 \end{bmatrix} A = \begin{bmatrix} 1 & 0 & 0 \\ 3 & 1 & 0 \\ 2 & 0 & 1 \end{bmatrix}\begin{bmatrix} 1 & 2 & 3 \\ 0 & -1 & -3 \\ 0 & 0 & -1 \end{bmatrix}$.

13. $\begin{bmatrix} 1 & 0 & 0 & 0 \\ 0 & 0 & 1 & 0 \\ 0 & 1 & 0 & 0 \\ 0 & 0 & 0 & 1 \end{bmatrix} A = \begin{bmatrix} 1 & 0 & 0 & 0 \\ 1 & 1 & 0 & 0 \\ \frac{3}{2} & 0 & 1 & 0 \\ 2 & -3 & 10 & 1 \end{bmatrix}\begin{bmatrix} 2 & 4 & 3 & 2 \\ 0 & 1 & -1 & -5 \\ 0 & 0 & \frac{1}{2} & -1 \\ 0 & 0 & 0 & 6 \end{bmatrix}$.

15. If $A = LU$, then $U = DU'$ where $D = \text{Dg}\{u_{11}, u_{22}, \ldots, u_{nn}\}$ and $u'_{ij} = \frac{u_{ij}}{u_{ii}}$. If $A = LDU = L_1 D_1 U_1$, then $L_1^{-1}L = (D_1 U_1)(DU)^{-1} = I$ since $L_1^{-1}L$ is unit lower triangular and $(D_1 U_1)(DU)^{-1}$ is upper triangular. Thus $L_1^{-1}L = I \Rightarrow L_1 = L$. Now $(D_1 U_1)(DU)^{-1} = I \Rightarrow D_1 U_1 = DU \Rightarrow D^{-1}D_1 = UU_1^{-1} = I$ since $D^{-1}D_1$ is diagonal and UU_1^{-1} is unit upper triangular. Thus $D = D_1$ and $U = U_1$.

EXERCISES 1.9 *(page 92)*

1. **(a)** $X = \frac{1}{7}\begin{bmatrix} 16 \\ 27 \\ 20 \end{bmatrix}$. **(c)** $X = \begin{bmatrix} \frac{20}{21} \\ -\frac{1}{2} \\ \frac{58}{21} \\ \frac{1}{42} \end{bmatrix}$. **(e)** $X = \begin{bmatrix} 86.71 \\ 27.54 \\ -83.96 \\ -6.205 \end{bmatrix}$. **3.** $\begin{bmatrix} 0 & 0 & 1 \\ 0 & 1 & 0 \\ 1 & 0 & 0 \end{bmatrix} A = \begin{bmatrix} 1 & 0 & 0 \\ \frac{2}{3} & 1 & 0 \\ \frac{1}{3} & \frac{1}{2} & 1 \end{bmatrix}\begin{bmatrix} 3 & 5 & 6 \\ 0 & \frac{2}{3} & 1 \\ 0 & 0 & \frac{1}{2} \end{bmatrix}$.

5. $\begin{bmatrix} 0 & 0 & 0 & 1 \\ 0 & 0 & 1 & 0 \\ 0 & 1 & 0 & 0 \\ 1 & 0 & 0 & 0 \end{bmatrix} A = \begin{bmatrix} 1 & 0 & 0 & 0 \\ \frac{1}{2} & 1 & 0 & 0 \\ \frac{3}{4} & \frac{9}{10} & 1 & 0 \\ \frac{1}{2} & \frac{3}{5} & 1 & 1 \end{bmatrix}\begin{bmatrix} 4 & 5 & 14 & 14 \\ 0 & \frac{5}{2} & -5 & -10 \\ 0 & 0 & -1 & \frac{1}{2} \\ 0 & 0 & 0 & \frac{1}{2} \end{bmatrix}$. **7.** $X = \begin{bmatrix} -17.8 \\ 7.2 \\ 2.2 \\ .6 \end{bmatrix}$.

9. $A^{-1} = \frac{1}{44}\begin{bmatrix} 24 & -4 & -8 \\ 3 & 5 & -12 \\ 10 & 2 & 4 \end{bmatrix}$, $C^{-1} = \begin{bmatrix} .435 & -.135 & -.0156 & -.0451 \\ -1.97 & .310 & .160 & .444 \\ 1.90 & -.251 & -.193 & -.390 \\ .481 & -.0381 & -.00252 & -.113 \end{bmatrix}$.

(Correct to three significant figures.)

CHAPTER 1 REVIEW EXERCISES *(page 95)*

1. **(a)** $a_1x_1 + a_2x_2 + \cdots + a_nx_n = b$. **(b)** B can be obtained from A by a finite sequence of elementary row operations. **(c)** A single elementary row operation **(d)** $AB = BA = I$. **(e)** I **(f)** Elementary matrices **(g)** $X = 0$. **(h)** Unique

3. $I = \begin{bmatrix} 1 & 0 & 0 \\ 0 & 1 & 0 \\ 0 & 0 & 1 \end{bmatrix}$. **5.** $-\frac{1}{3}(B^2 - 7B + 7I)$ **7.** $M^{-1} = \begin{bmatrix} 1 & -3 & 2 \\ -3 & 3 & -1 \\ 2 & -1 & 0 \end{bmatrix}$. **9.** $x = z$, $y = -2z$, z arbitrary

11. $\begin{bmatrix} -31 & -17 & -20 \\ 19 & 10 & 12 \end{bmatrix}$ **13. (a)** $\begin{bmatrix} 3 & -16 \\ -24 & 43 \end{bmatrix}$ **(b)** $d_{57} = \Sigma_{k=1}^{10} (2a_{5k} + 3b_{5k})c_{k7}$ **(c)** $x = -2$.

15. (a) $\left[\begin{array}{cccc|c|c} 1 & -1 & 2 & -1 & -1 & k_1 \\ 0 & 1 & -2 & 0 & 0 & k_1 + k_3 \\ 0 & 0 & 0 & 0 & 0 & -5k_1 + k_2 + -3k_3 \\ 0 & 0 & 0 & 0 & 0 & -6k_1 - 4k_3 + k_4 \end{array}\right]$ **(b)** $\begin{bmatrix} -1 + c_2 \\ 2c_1 \\ c_1 \\ c_2 \end{bmatrix}$

(c) $k_1 = -\frac{2}{3}k_3 + \frac{1}{6}k_4$ and $k_2 = -\frac{1}{3}k_3 + \frac{5}{6}k_4$. **(d)** $\begin{bmatrix} c_2 \\ 2c_1 \\ c_1 \\ c_2 \end{bmatrix}$ **17.** $\frac{1}{3}\begin{bmatrix} -5 \\ 2 \\ 0 \end{bmatrix} + c_1 \begin{bmatrix} -5 \\ -1 \\ 3 \end{bmatrix}$ **19.** $\lambda = 4, 9$.

21. $J^{-1} = \frac{1}{12}(J^2 - 7J + 16I) = \begin{bmatrix} \frac{1}{2} & 0 & 0 \\ -\frac{1}{4} & \frac{1}{2} & 0 \\ 0 & 0 & \frac{1}{3} \end{bmatrix}$.

CHAPTER 2

EXERCISES 2.1 *(page 105)*

1. $-2, 17, -34, -34, -34, -\frac{1}{2}, -2$ **3. (a)** 0 **(b)** $x^3 - 9x^2 + 8x + 5$ **(c)** $x^3 + 2x^2 - x - 2$ **5. (a)** 27 **(c)** -117 **7. (a)** $ad - bc$ **(b)** $cb - da = -\det(A)$, **(c)** $kad - kbc = k\det(A)$ **(d)** $\det(A)$ **9.** $-\det(A)$, $k\det(A)$, $\det(A)$ **11. (a)** $\text{tr}(A) = 9$, $\text{tr}(B) = 2$, $\text{tr}(AB) = \text{tr}(BA) = 52$ **(b)** $\text{tr}(A) = \Sigma_{i=1}^{n} a_{ii}$.

13. 1.37×10^{53} years **15.** $\begin{bmatrix} a & b & c \\ b & e & f \\ c & f & i \end{bmatrix}$ **17.** $X^TX = 29$, $XX^T = \begin{bmatrix} 4 & -6 & -8 \\ -6 & 9 & 12 \\ -8 & 12 & 16 \end{bmatrix}$.

19. For the matrices of Exercise 1, $\det(A + B) = 22$, $\det(A) + \det(B) = 15$. **21.** $\det(S) = -216$, $\det(T) = 4\pi$ **23.** $AA^T = I$.

EXERCISES 2.2 *(page 113)*

1. -12 **3.** -18 **5.** 132 **7.** 6 **9.** $\det(A) = -12$, $\det(B) = -4$ **11.** $-24, 40$ **13.** 0, 1, 1 **15.** $\det(AB) = \det(B) = 0$. $\det(A + B) = -246 \neq \det(A) + \det(B) = -12 + 0$. **17.** If $\text{Row}_i(A) = \Sigma_{j=1,\neq i}^{n} k_j \text{Row}_j(A)$, then the sequence of elementary row operations $\{R_i \leftarrow -k_jR_j + R_i\}_{j=1,\neq i}^{n}$ yields a matrix B with a zero row. It follows that $\det(A) = \det(B) = 0$. **19.** The matrix kA can be computed by multiplying each of the n rows of A by k. Each of these row multiplications multiplies the determinant by k. Thus $\det(kA) = k^n \det(A)$. **21.** $R_i \leftrightarrow R_j$ **23.** Expanding by the first row yields an equation of the form $rx + sy + t = 0$ which is the equation of a straight line. Moreover, the equation is satisfied if $(x, y) = (a, b)$ since in that case the determinant has two equal rows and hence has value 0. Similarly, the point $(x, y) = (c, d)$ is also on the line.

25. Area $= \frac{|\det(A)|}{2} = \frac{9}{2}$. **27. (a)** -13 **(b)** $\frac{29}{6}$ **(c)** No real solutions **29.** Use the column operation $C_5 \leftarrow 10C_4 + C_5$, $C_5 \leftarrow 10^2C_3 + C_5$, $C_5 \leftarrow 10^3C_2 + C_5$, $C_5 \leftarrow 10^4C_1 + C_5$. The last column of the resulting matrix has a factor of 17.

EXERCISES 2.3 *(page 122)*

1. $A^{-1} = -\frac{1}{12}\begin{bmatrix} 4 & 4 & -4 \\ -8 & -14 & 5 \\ 0 & 6 & -3 \end{bmatrix}$, $D^{-1} = \frac{1}{117}\begin{bmatrix} 39 & 14 & 1 & -65 \\ 39 & -61 & 4 & -26 \\ 0 & 24 & -15 & 39 \\ 0 & 9 & 9 & 0 \end{bmatrix}$, $G^{-1} = \frac{1}{6}\begin{bmatrix} -2 & 2 & 6 & -2 \\ -1 & 1 & 0 & 2 \\ 5 & -5 & -6 & 2 \\ -1 & 7 & 6 & -4 \end{bmatrix}$.

3. $x_3 = -\frac{1}{6}$ **5.** $z = 2$. **7.** The result follows from the given factorization using Property 2.9 and Exercise 6.

9. $A = \pm\begin{bmatrix} 5 & -0.5 & -10.5 \\ -4 & 1 & 9 \\ 1 & -0.5 & -0.5 \end{bmatrix}$. **11.** For A, 4×4. For N, 3×3. **13. (a)** -5 **(b)** 25 **(c)** 40 **(d)** $\frac{125}{27^3}$

15. (a) $\frac{3}{2}$ **(b)** 9 **(c)** $2^{15} \times 3^{10} = 1934917632$ **(d)** $\frac{1}{2}$ **17. (a)** 10.5 **(b)** $\frac{27}{2}$ **(c)** 16.5.

19. $A\,\text{Adj}(A) = \det(A)I \Rightarrow \det(A)\det(\text{Adj}(A)) = \det(A)^n \Rightarrow \det(\text{Adj}(A)) = \det(A)^{n-1}$.

$\text{Adj}(\text{Adj}(A))\text{Adj}(A) = \det(\text{Adj}(A))I = (\det(A))^{n-1}I \Rightarrow \text{Adj}(A)^{-1} = \left(\frac{1}{(\det(A))^{n-1}}\text{Adj}(\text{Adj}(A))\right)$.

Also $A\,\text{Adj}(A) = \det(A)I \Rightarrow \text{Adj}(A)^{-1} = \left(\frac{1}{\det(A)}\right)A$. The desired result follows by comparing the two different forms for the inverse.

21. $\lambda^4 + a_3\lambda^3 + a_2\lambda^2 + a_1\lambda + a_0$ **23. (a)** 8 **(b)** 4, -4 **(c)** -1 **(d)** -2, 1

EXERCISES 2.4 *(page 131)*

1. (a) X is an eigenvector for $\lambda = 2$ since $AX = 2X$. Y is not an eigenvector. **(b)** Both X and Y are eigenvectors for $\lambda = 1$. **3.** $\lambda^2 - 25\lambda - 52$; $\left(\frac{25 \pm \sqrt{833}}{2}\right)$; $X_1 = \begin{bmatrix} 28 \\ 7 - \sqrt{833} \end{bmatrix}$, $X_2 = \begin{bmatrix} 28 \\ 7 + \sqrt{833} \end{bmatrix}$.

5. $\lambda^3 - 9\lambda^2 - 9\lambda + 81$; $\lambda_1 = 3$, $\lambda_2 = -3$, $\lambda_3 = 9$; $X_1 = \begin{bmatrix} -2 \\ -2 \\ 1 \end{bmatrix}$, $X_2 = \begin{bmatrix} 2 \\ -1 \\ 2 \end{bmatrix}$, $X_3 = \begin{bmatrix} -1 \\ 2 \\ 2 \end{bmatrix}$.

7. $\lambda^3 - 4\lambda^2 + 3\lambda$; 0, 3, 1; $\begin{bmatrix} 1 \\ -2 \\ 4 \end{bmatrix}$, $\begin{bmatrix} 1 \\ 1 \\ 1 \end{bmatrix}$, $\begin{bmatrix} 1 \\ -1 \\ 1 \end{bmatrix}$

9. $\lambda^3 + \lambda^2 - 12\lambda$; -4, 3, 0; $\begin{bmatrix} 1 \\ 0 \\ 1 \end{bmatrix}$, $\begin{bmatrix} 1 \\ -1 \\ -1 \end{bmatrix}$, $\begin{bmatrix} -1 \\ -2 \\ 1 \end{bmatrix}$

11. (8) Dg{5, -5, -4}; **(10)** Dg{7, -4, -1} **13.** The eigenvalues are the diagonal elements.

15. $\det(\lambda I - B) = \det(\lambda P^{-1}IP - P^{-1}AP) = \det(P^{-1}(\lambda I - A)P) = \det(P^{-1})\det(\lambda I - A)\det(P) = \det(\lambda I - A)$.

17. The eigenvalues of A^2 are the squares of the eigenvalues of A. A^2 and A have the same eigenvectors. The eigenvalues of A^3 are the cubes of the eigenvalues of A. **19.** $(\lambda - 2)^2(\lambda^2 - 5\lambda + 6)$

CHAPTER 2 REVIEW EXERCISES *(page 134)*

1. T, F, F, T, T, T, F, F, T, F **3.** $\lambda^3 - 4\lambda^2 + 4\lambda - 1$ **5.** $x = -\frac{20}{33}$, $y = -\frac{4}{33}$, $z = \frac{15}{33}$ **7.** $\lambda^2 - 9\lambda + 18$; $\lambda = 6, 3$; $X_1 = \begin{bmatrix} 4 \\ -1 \end{bmatrix}$, $X_2 = \begin{bmatrix} -1 \\ 1 \end{bmatrix}$. **9.** $\begin{bmatrix} x_1 \\ y_1 \end{bmatrix} = \begin{bmatrix} \cos\theta & \sin\theta \\ -\sin\theta & \cos\theta \end{bmatrix}\begin{bmatrix} x \\ y \end{bmatrix}$. **11.** $\det(B) = 0$. **(b)** -20

13. (a) $\lambda^3 - 7\lambda^2 + 6\lambda$ **(b)** $\begin{bmatrix} 1 \\ 2 \\ 5 \end{bmatrix}$

CHAPTER 3

EXERCISES 3.1 *(page 148)*

1.

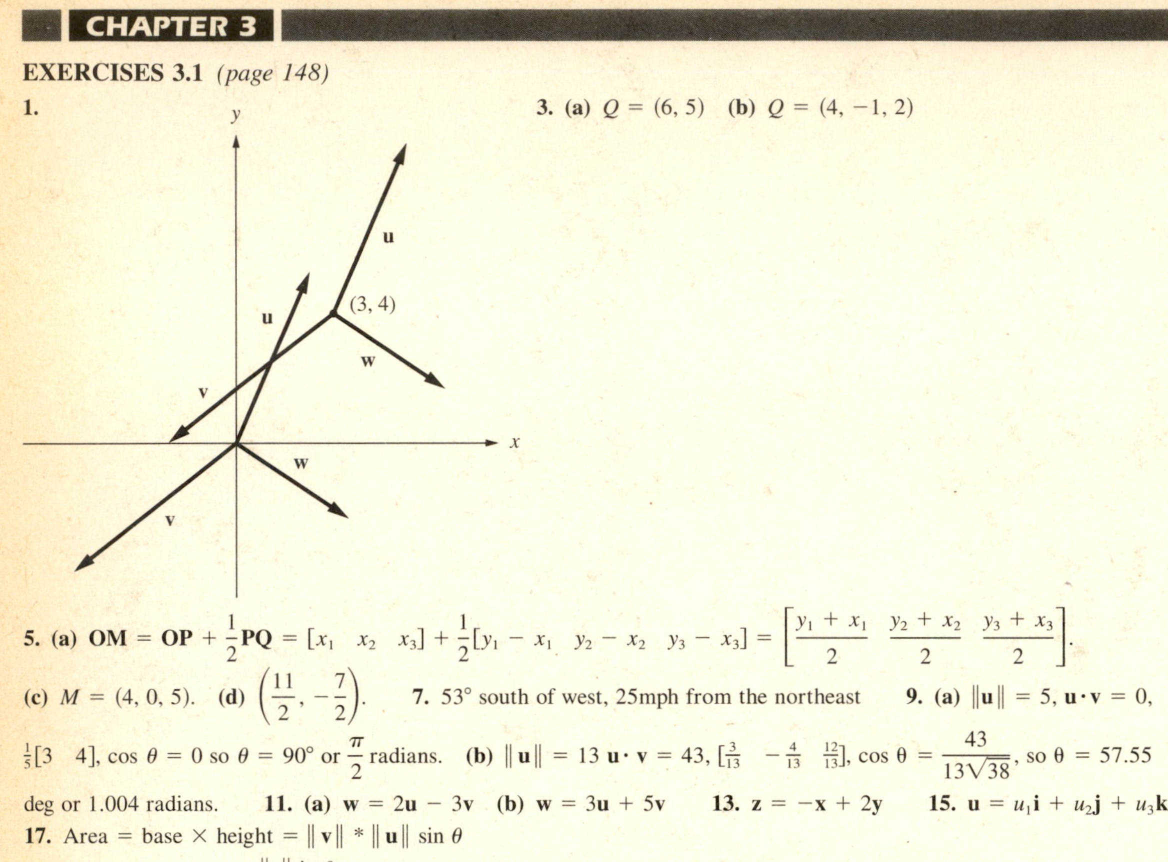

3. (a) $Q = (6, 5)$ **(b)** $Q = (4, -1, 2)$

5. (a) $\mathbf{OM} = \mathbf{OP} + \frac{1}{2}\mathbf{PQ} = [x_1 \quad x_2 \quad x_3] + \frac{1}{2}[y_1 - x_1 \quad y_2 - x_2 \quad y_3 - x_3] = \left[\frac{y_1 + x_1}{2} \quad \frac{y_2 + x_2}{2} \quad \frac{y_3 + x_3}{2}\right]$.
(c) $M = (4, 0, 5)$. **(d)** $\left(\frac{11}{2}, -\frac{7}{2}\right)$. **7.** 53° south of west, 25mph from the northeast **9. (a)** $\|\mathbf{u}\| = 5$, $\mathbf{u}\cdot\mathbf{v} = 0$, $\frac{1}{5}[3 \quad 4]$, $\cos\theta = 0$ so $\theta = 90°$ or $\frac{\pi}{2}$ radians. **(b)** $\|\mathbf{u}\| = 13$ $\mathbf{u}\cdot\mathbf{v} = 43$, $[\frac{3}{13} \quad -\frac{4}{13} \quad \frac{12}{13}]$, $\cos\theta = \frac{43}{13\sqrt{38}}$, so $\theta = 57.55$ deg or 1.004 radians. **11. (a)** $\mathbf{w} = 2\mathbf{u} - 3\mathbf{v}$ **(b)** $\mathbf{w} = 3\mathbf{u} + 5\mathbf{v}$ **13.** $\mathbf{z} = -\mathbf{x} + 2\mathbf{y}$ **15.** $\mathbf{u} = u_1\mathbf{i} + u_2\mathbf{j} + u_3\mathbf{k}$
17. Area = base × height = $\|\mathbf{v}\| * \|\mathbf{u}\| \sin\theta$

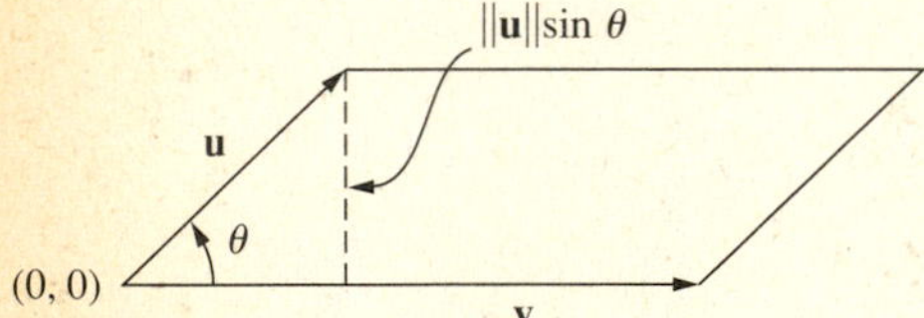

19. **b**, **h**, **d**, 2**c**, **0**, **d**, 2**f**, **b**, **0** **21.** (b) and (c) are correct statements. **23.** Let $A = (x_2, y_2)$ and $B = (x_1, y_1)$ be on the line. Then, $\mathbf{AB} = \left(x_1 - x_2, -\frac{a}{b}(x_1 - x_2)\right)$ and $\mathbf{v}\cdot\mathbf{AB} = 0$.

EXERCISES 3.2 *(page 157)*

1. $\sqrt{14}, 9, \sqrt{6}, \sqrt{38}$
3. (a) $a\mathbf{0} = a[0 \quad 0 \quad \cdots \quad 0]^T = [a0 \quad a0 \quad \cdots \quad a0]^T = [0 \quad 0 \quad \cdots \quad 0]^T = \mathbf{0}$.
$0\mathbf{u} = 0[u_1 \quad u_2 \quad \cdots \quad u_n]^T = [0u_1 \quad 0u_2 \quad \cdots \quad 0u_n]^T = \mathbf{0}$. **(c)** $a\mathbf{u} = b\mathbf{u} \Leftrightarrow (a - b)\mathbf{u} = \mathbf{0} \Leftrightarrow a - b = 0$ or $\mathbf{u} = \mathbf{0}$ by part (b).
5. (a) If $\mathbf{u}$, $\mathbf{v}$ are in $\mathcal{W}$ then $A\mathbf{u} = 2\mathbf{u}$ and $A\mathbf{v} = 2\mathbf{v}$. Since $A(\mathbf{u} + \mathbf{v}) = A\mathbf{u} + A\mathbf{v} = 2\mathbf{u} + 2\mathbf{v} = 2(\mathbf{u} + \mathbf{v})$ it follows that $\mathbf{u} + \mathbf{v}$ is in $\mathcal{W}$. Since $A(k\mathbf{u}) = k(A\mathbf{u}) = k(2\mathbf{u}) = 2(k\mathbf{u})$, $k\mathbf{u}$ is in $\mathcal{W}$. It follows from Definition 3.7 that $\mathcal{W}$ is a subspace of $\mathfrak{R}^4$.
7. (a), (b) are subspaces. **9. (a)** $\frac{1}{6}[5 \quad 2 \quad -1 \quad 0]^T$ **(b)** $\frac{5}{3}$ **13.** Subset (c) is a subspace.

EXERCISES 3.3 *(page 166)*

1. $\mathbf{w} = 3\mathbf{u} - \mathbf{v}$. **3.** The system $\begin{bmatrix} 1 & 1 & 0 \\ 0 & 1 & 1 \\ 1 & 1 & 1 \end{bmatrix} \mathbf{X} = \begin{bmatrix} a \\ b \\ c \end{bmatrix}$ has a unique solution because the coefficient matrix is nonsingular. **5.** $\mathbf{v}_2 = \mathbf{u}_1 + 2\mathbf{u}_2$, $\mathbf{v}_3 = 3\mathbf{u}_1 - \mathbf{u}_2$. $\mathbf{v}_1$ is not in SPAN$\{\mathbf{u}_1, \mathbf{u}_2\}$.
7. $\mathbf{v}_1 = 2\mathbf{u}_1 - \mathbf{u}_2$, $\mathbf{v}_2 = -3\mathbf{u}_1 + 2\mathbf{u}_2$, $\mathbf{u}_1 = 2\mathbf{v}_1 + \mathbf{v}_2$, $\mathbf{u}_2 = 3\mathbf{v}_1 + 2\mathbf{v}_2$. **9. (a)** $B = [1 \quad -2 \quad 1]$.
(b) $B = [1 \quad -2 \quad -1 \quad -0.5]$. **(c)** $B = \begin{bmatrix} 1 & 0 & -3 & 2 \\ 0 & 1 & -2 & 1 \end{bmatrix}$. **11. (a)** Yes **(b)** Yes **(c)** Yes

13. $z = x + y$. **15.** $\mathcal{W}$ is not a subspace because it does not contain the zero vector. **17.** $\begin{bmatrix} 1 \\ 2 \\ 0 \\ 0 \end{bmatrix}, \begin{bmatrix} 1 \\ -3 \\ 1 \\ 0 \end{bmatrix}, \begin{bmatrix} -1 \\ 4 \\ -3 \\ 4 \end{bmatrix}$

19. (a) No **(b)** Yes ($\mathbf{v}_3 = -\mathbf{v}_1 + 2\mathbf{v}_2$) **(c)** Yes ($\mathbf{v}_3 = 2\mathbf{v}_1 + \mathbf{v}_2$)

EXERCISES 3.4 *(page 173)*

1. Independent because the system $\begin{bmatrix} 1 & 3 \\ 2 & 5 \end{bmatrix} \mathbf{X} = \mathbf{0}$ has only the zero solution. **3.** Independent **5.** Independent

7. Independent **9.** Dependent because $\text{Col}_4(C) = \text{Col}_1(C) + \text{Col}_2(C) - \frac{1}{2}\text{Col}_3(C)$. **11.** Independent

13. (a) Yes **(b)** Not coplanar **(c)** Not coplanar **15.** $2\mathbf{u}_3 = 7\mathbf{u}_1 + 3\mathbf{u}_2$.
17. Suppose $\mathbf{v}_4 = a_1\mathbf{v}_1 + a_2\mathbf{v}_2 + a_3\mathbf{v}_3$. Then a typical element of SPAN$\{\mathbf{v}_1, \mathbf{v}_2, \mathbf{v}_3, \mathbf{v}_4\}$ is $\mathbf{w} = b_1\mathbf{v}_1 + b_2\mathbf{v}_2 + b_3\mathbf{v}_3 + b_4\mathbf{v}_4 = b_1\mathbf{v}_1 + b_2\mathbf{v}_2 + b_3\mathbf{v}_3 + b_4(a_1\mathbf{v}_1 + a_2\mathbf{v}_2 + a_3\mathbf{v}_3) = (b_1 + b_4a_1)\mathbf{v}_1 + (b_2 + b_4a_2)\mathbf{b}_2 + (b_3 + b_4a_3)\mathbf{v}_3$. Therefore $\mathbf{w}$ is in SPAN$\{\mathbf{v}_1, \mathbf{v}_2, \mathbf{v}_3\}$ and SPAN$\{\mathbf{v}_1, \mathbf{v}_2, \mathbf{v}_3, \mathbf{v}_4\} \subseteq$ SPAN$\{\mathbf{v}_1, \mathbf{v}_2, \mathbf{v}_3\}$. The inclusion in the other order is obvious, so the two sets are equal.

EXERCISES 3.5 *(page 181)*

1. (a) No, the vectors are dependent. **(b)** No, the vectors do not span $\mathfrak{R}^3$. **(c)** Yes **(d)** No, the vectors are dependent. **(e)** No, the vectors are dependent. **(f)** No, the vectors are dependent.

3. Add the vector $\begin{bmatrix} 1 \\ 0 \\ 0 \end{bmatrix}$. **5.** $\mathbf{e}_1 = \mathbf{u}_1 - 2\mathbf{u}_2 + 7\mathbf{u}_3 - 25\mathbf{u}_4$, $\mathbf{e}_2 = \mathbf{u}_2 - 2\mathbf{u}_3 + 6\mathbf{u}_4$, $\mathbf{e}_3 = \mathbf{u}_3 - 3\mathbf{u}_4$, $\mathbf{e}_4 = \mathbf{u}_4$.

7. Rank$(A) = 2$, rank$(B) = 3$, rank$(C) = 4$.

9. $\mathcal{NS}(A) = \text{BASIS}\left\{\begin{bmatrix} -4 \\ -5 \\ 1 \\ 0 \end{bmatrix}, \begin{bmatrix} -2 \\ -3 \\ 0 \\ 1 \end{bmatrix}\right\}$. $\mathcal{CS}(A) = \text{BASIS}\left\{\begin{bmatrix} 1 \\ 0 \\ 0 \end{bmatrix}, \begin{bmatrix} 0 \\ 1 \\ 0 \end{bmatrix}\right\}$.

11. $\mathcal{NS}(A) = \text{BASIS}\left\{\begin{bmatrix} -2 \\ 1 \\ 0 \\ 0 \\ 0 \end{bmatrix}, \begin{bmatrix} -4 \\ 0 \\ 1.4 \\ 1 \\ 0 \end{bmatrix}\right\}$. $\mathcal{CS}(A) = \text{BASIS}\{\text{Col}_1(A), \text{Col}_3(A), \text{Col}_5(A)\}$.

EXERCISES 3.6 *(page 194)*

1. $\mathbf{u} \cdot \mathbf{v} = 9$, $\|\mathbf{u}\| = \sqrt{31}$, $\|\mathbf{v}\| = 3\sqrt{6}$, $\|\mathbf{u} + \mathbf{v}\| = \sqrt{103} \leq \|\mathbf{u}\| + \|\mathbf{v}\| = 12.916$, $a = \pm\frac{1}{\sqrt{31}}$, $h = -\frac{5}{4}$, $a = \frac{34}{13}$, $b = -\frac{11}{13}$, $c = -6$. **3.** $X_1 = \begin{bmatrix} 2 \\ -1 \\ 0 \end{bmatrix}$ for $\lambda = 1$, $X_2 = \begin{bmatrix} 1 \\ 2 \\ -1 \end{bmatrix}$ for $\lambda = 0$, $X_3 = \begin{bmatrix} 1 \\ 2 \\ 5 \end{bmatrix}$ for $\lambda = 6$.

5. **(a)** $\frac{1}{2}, \frac{1}{2}, \frac{1}{2}, \frac{1}{2}$ **(b)** 5, 0, −1, −2 **(c)** 7, 2, 2, −9.

7. $\|\mathbf{u} + \mathbf{v}\| = \mathbf{u} \cdot \mathbf{u} + 2\mathbf{u} \cdot \mathbf{v} + \mathbf{v} \cdot \mathbf{v} = \|\mathbf{u} - \mathbf{v}\| = \mathbf{u} \cdot \mathbf{u} - 2\mathbf{u} \cdot \mathbf{v} + \mathbf{v} \cdot \mathbf{v} \Leftrightarrow \mathbf{u} \cdot \mathbf{v} = 0$. The diagonals of a square are of equal length. **9.** X in $\mathcal{NS}(A) \Rightarrow AX = 0 \Rightarrow A^TAX = 0 \Rightarrow X$ in $\mathcal{NS}(A^TA)$. Therefore $\mathcal{NS}(A) \subseteq \mathcal{NS}(A^TA)$. Y in $\mathcal{NS}(A^TA) \Rightarrow A^TAY = 0 \Rightarrow Y^TA^TAY = 0 \Rightarrow \|AY\| = 0 \Rightarrow AY = 0$. Therefore $\mathcal{NS}(A^TA) \subseteq \mathcal{NS}(A)$ and $\mathcal{NS}(A^TA) = \mathcal{NS}(A)$. Both A and A^TA have n columns, so by Theorem 3.11 $\text{rank}(A) = n - \dim(\mathcal{NS}(A)) = n - \dim(\mathcal{NS}(A^TA)) = \text{rank}(A^TA)$. **11.** $\frac{1}{3}\mathbf{u}_1, \frac{1}{\sqrt{41}}[6 \quad -1 \quad -2]^T$

13. **(a)** $\frac{1}{\sqrt{3}}\begin{bmatrix}1\\1\\1\end{bmatrix}, \frac{1}{\sqrt{6}}\begin{bmatrix}-2\\1\\1\end{bmatrix}, \frac{1}{\sqrt{2}}\begin{bmatrix}0\\-1\\1\end{bmatrix}$ **(b)** $\frac{1}{5}\begin{bmatrix}3\\0\\4\end{bmatrix}, \frac{1}{\sqrt{850}}\begin{bmatrix}-12\\25\\9\end{bmatrix}, \frac{1}{\sqrt{34}}\begin{bmatrix}4\\3\\-3\end{bmatrix}$

15. The answer is not unique. Applying Gram-Schmidt to the basis $\{\mathbf{u}_1, \mathbf{u}_2, \mathbf{e}_1, \mathbf{e}_2\}$ yields the orthonormal basis $\left\{\frac{1}{\sqrt{7}}\mathbf{u}_1, \frac{1}{\sqrt{6}}\mathbf{u}_2, [-0.8309 \quad 0.3438 \quad -0.2292 \quad 0.3725]^T, [0 \quad -0.5571 \quad 0.3714 \quad 0.7428]^T\right\}$.

17. $\left\{\frac{1}{\sqrt{30}}\begin{bmatrix}-5\\2\\1\\0\\0\end{bmatrix}, \begin{bmatrix}-0.3131\\-0.8767\\0.1879\\0.3131\\0\end{bmatrix}, \begin{bmatrix}0.0443\\0.1240\\-0.0266\\0.4074\\-0.9033\end{bmatrix}\right\}$ **19.** **(a)** $\begin{bmatrix}\frac{3}{5} & 0 & \frac{4}{5} & 0\\0 & 1 & 0 & 0\\\frac{4}{5} & 0 & -\frac{3}{5} & 0\\0 & 0 & 0 & 1\end{bmatrix}$ **(c)** $\frac{1}{2}\begin{bmatrix}1 & -1 & 1 & -1\\-1 & 1 & 1 & -1\\1 & 1 & 1 & 1\\-1 & -1 & 1 & 1\end{bmatrix}$

21. $H^TH = I - 4\mathbf{w}\mathbf{w}^T + 4\mathbf{w}(\mathbf{w}^T\mathbf{w})\mathbf{w}^T = I$ since $\mathbf{w}^T\mathbf{w} = 1$. $H = \frac{1}{169}\begin{bmatrix}151 & -24 & -72\\-24 & 137 & -96\\-72 & -96 & -119\end{bmatrix}$. **23.** It does not make a difference because eventually the vectors have to be scaled to make them unit vectors.

25. **(a)** $A = \begin{bmatrix}1 & 0 & 0\\1 & 1 & 0\\1 & 1 & 1\end{bmatrix} = \begin{bmatrix}-0.5774 & 0.8165 & -0.0000\\-0.5774 & -0.4082 & -0.7071\\-0.5774 & -0.4082 & 0.7071\end{bmatrix}\begin{bmatrix}-1.7321 & -1.1547 & -0.5774\\0 & -0.8165 & -0.4082\\0 & 0 & 0.7071\end{bmatrix}$.

(b) $B = \begin{bmatrix}3 & 0 & 5\\0 & 1 & -2\\4 & 1 & 1\end{bmatrix} = \begin{bmatrix}-0.6000 & 0.4116 & -0.6860\\0 & -0.8575 & -0.5145\\-0.8000 & -0.3087 & 0.5145\end{bmatrix}\begin{bmatrix}-5.0000 & -0.8000 & -3.8000\\0 & -1.1662 & 3.4643\\0 & 0 & -1.8865\end{bmatrix}$.

27. $\|X - Y\|^2 = (X - Y) \cdot (X - Y) = X \cdot X - 2X \cdot Y + Y \cdot Y = \|X\|^2 + \|Y\|^2 - 2X \cdot Y = \|X\|^2 + \|Y\|^2 \Leftrightarrow X \cdot Y = 0$.

EXERCISES 3.7 *(page 203)*

1. No (10 fails) **3.** No (9 and 10 fail) **5.** Yes **7.** Yes **9.** No (1 fails) **11.** Closed **13.** Not closed **15.** Not closed **17.** Closed **19.** $(f + g)(x) = f(x) + g(x) = g(x) + f(x) = (g + f)(x)$, so $f + g = g + f$. $[a(f + g)](x) = a(f + g)(x) = a(f(x) + g(x)) = af(x) + ag(x) = (af + bg)(x)$, so $a(f + g) = af + ag$. **21.** The single element $\mathbf{u}$ must be the additive identity, so define $\mathbf{u} + \mathbf{u} = \mathbf{u}$, and $a\mathbf{u} = \mathbf{u}$. **23.** $(\mathbf{u} + \mathbf{v}) + -\mathbf{v} = (\mathbf{w} + \mathbf{v}) + -\mathbf{v} \Rightarrow \mathbf{u} + (\mathbf{v} + -\mathbf{v}) = \mathbf{w} + (\mathbf{v} + -\mathbf{v}) \Rightarrow \mathbf{u} + \mathbf{0} = \mathbf{w} + \mathbf{0} \Rightarrow \mathbf{u} = \mathbf{w}$. **25.** $\mathbf{u} + (-\mathbf{u}) = \mathbf{0}$ and $-(-\mathbf{u}) + -\mathbf{u} = \mathbf{0}$. By the uniqueness of the additive inverse of $-\mathbf{u}$, it follows that $-(-\mathbf{u}) = \mathbf{u}$. **27.** Yes

EXERCISES 3.8 *(page 212)*

1. Subsets (a), (c), and (f) are subspaces; (a) and (c) are planes through the origin, (b) is the unit sphere, (d) is a plane not containing the origin, (e) is a half space (all points above the plane $x_1 + x_2 = x_3$), and (f) is just the origin. **3.** Subsets (a) and (c) are subspaces. **5.** Only $\mathbf{d} = \mathbf{u}_1 + \mathbf{u}_2$ is in $\mathcal{W}$. **7.** Subsets (a) and (f) **9.** Subsets (a) and (c) are subspaces. Subset (b) is not closed under addition.

11. $\mathcal{NS}(A) = \text{SPAN}\left\{\begin{bmatrix}2\\1\\0\\0\end{bmatrix}, \begin{bmatrix}-4\\0\\3\\1\end{bmatrix}\right\}$, $\mathcal{NS}(B) = \text{SPAN}\left\{\begin{bmatrix}-5\\2\\0\\0\end{bmatrix}, \begin{bmatrix}-3\\0\\1\\0\end{bmatrix}, \begin{bmatrix}-7\\0\\0\\2\end{bmatrix}\right\}$, $\mathcal{NS}(C) = \text{SPAN}\left\{\begin{bmatrix}-1\\1\\0\\0\end{bmatrix}, \begin{bmatrix}-1\\0\\1\\0\end{bmatrix}\right\}$.

13. $\frac{1}{\sqrt{2}}[1 \quad 0 \quad 1 \quad 0], \frac{1}{2}[1 \quad -1 \quad -1 \quad 1], \frac{1}{2}[1 \quad 1 \quad -1 \quad -1]$

EXERCISES 3.9 *(page 223)*

1. Does the system $\begin{bmatrix}1 & 1 & 4\\-1 & 1 & 1\\2 & 2 & 2\\0 & 0 & 1\end{bmatrix}\begin{bmatrix}a\\b\\c\end{bmatrix} = \begin{bmatrix}0\\0\\0\\0\end{bmatrix}$ have a nonzero solution? The vectors are independent.

3. Does the system $\begin{bmatrix}1 & 0 & 0 & 0\\0 & 3 & 0 & 0\\2 & 0 & 6 & 0\\5 & 2 & 0 & 2\end{bmatrix}\begin{bmatrix}a\\b\\c\\d\end{bmatrix} = \begin{bmatrix}0\\0\\0\\0\end{bmatrix}$ have a nonzero solution? The answer is no, so the vectors are independent.

5. Does the system $BX = 0$ have a nonzero solution? The vectors are dependent.

7. Does the system $\begin{bmatrix}1 & 1 & 1\\2 & 16 & 4\\3 & 81 & 9\end{bmatrix}\begin{bmatrix}a\\b\\c\end{bmatrix} = \begin{bmatrix}0\\0\\0\end{bmatrix}$ have a nonzero solution? The vectors are independent.

9. Does the system $\begin{bmatrix}0 & 0 & 0 & -12\\0 & -4 & 3 & -1\\1 & 1 & 1 & 1\end{bmatrix}\begin{bmatrix}a\\b\\c\\d\end{bmatrix} = \begin{bmatrix}0\\0\\0\end{bmatrix}$ have a nonzero solution? The vectors are dependent.

11. Does the system $\begin{bmatrix}1 & 0 & 0 & 1\\2 & 1 & 2 & 0\\1 & 1 & 3 & -1\\-2 & 0 & 1 & 2\end{bmatrix}\begin{bmatrix}a\\b\\c\\d\end{bmatrix} = \begin{bmatrix}0\\0\\0\\0\end{bmatrix}$ have a nonzero solution? The vectors are independent.

13. Does the system $\begin{bmatrix}1 & 1 & 5 & -1\\0 & 2 & -1 & 7\\1 & 0 & 4 & 6\\0 & 0 & 0 & 0\end{bmatrix}\begin{bmatrix}a\\b\\c\\d\end{bmatrix} = 0$ have a nonzero solution? The vectors are dependent.

15. $a\mathbf{x}_1 + b\mathbf{x}_2 + c\mathbf{x}_3 = \mathbf{0} \Rightarrow a(\mathbf{u} + \mathbf{v}) + b(\mathbf{v} + \mathbf{w}) + c(\mathbf{w} + \mathbf{u}) = \mathbf{0} \Rightarrow (a + c)\mathbf{u} + (a + b)\mathbf{v} + (b + c)\mathbf{w} = \mathbf{0} \Rightarrow$ (because $\mathbf{u}$, $\mathbf{v}$, $\mathbf{w}$ are independent) $\begin{cases}a + c = 0\\a + b = 0\\b + c = 0\end{cases} \Rightarrow a = b = c = 0$. Therefore the vectors $\mathbf{x}_1$, $\mathbf{x}_2$, and $\mathbf{x}_3$ are linearly independent.

17. $\det\begin{bmatrix}\cos^2 t & \sin^2 t & \sin 2t\\-\sin 2t & \sin 2t & 2\cos 2t\\-2\cos 2t & 2\cos 2t & -4\sin 2t\end{bmatrix} = -8 \neq 0$. The functions are linearly independent.

19. Any 3×4 homogeneous system has a nontrivial solution.

21. $\mathcal{S}_1 = \{[1 \quad 2 \quad 4 \quad 1], [2 \quad 3 \quad 7 \quad 1]\}$.

23. $\mathcal{S}_1 = \{\sin^2 x, \cos^2 x, e^x\}$.

25. Sets (a) and (d) are bases.

27. $\begin{bmatrix}10.5\\-13\\3.5\end{bmatrix}, \begin{bmatrix}2\\-3\\1\end{bmatrix}, \begin{bmatrix}.5\\-1\\.5\end{bmatrix}$

29. $\mathscr{E} = \{[1 \quad 0 \quad 2 \quad -1], [0 \quad 1 \quad 1 \quad 1]\}$. $[\mathbf{u}_1]_{\mathscr{E}} = \begin{bmatrix} 3 \\ -2 \end{bmatrix}$, $[\mathbf{u}_2]_{\mathscr{E}} = \begin{bmatrix} 1 \\ -1 \end{bmatrix}$.

31. $A = \begin{bmatrix} 1 & 1 & 2 & 4 \\ 1 & -1 & -4 & 0 \\ 2 & 1 & 1 & 6 \\ 2 & -1 & -5 & 2 \end{bmatrix} \underset{R}{\sim} \begin{bmatrix} 1 & 0 & -1 & 0 \\ 0 & 1 & 3 & 2 \\ 0 & 0 & 0 & 0 \\ 0 & 0 & 0 & 0 \end{bmatrix}$. Also $\{\mathbf{w}_1, \mathbf{w}_2\}$ is a basis. **33. (a)** Use as a basis the set of matrices which have one entry 1 and all other entries 0. **35.** Each matrix has rank 2.

37. (a) $\left\{ \begin{bmatrix} -.4 \\ 1.2 \\ 1 \\ 0 \\ 0 \end{bmatrix}, \begin{bmatrix} -1.2 \\ .6 \\ 0 \\ 1 \\ 0 \end{bmatrix}, \begin{bmatrix} .2 \\ -.6 \\ 0 \\ 0 \\ 1 \end{bmatrix} \right\}$, 3, 2 **(b)** $\left\{ \begin{bmatrix} 3 \\ -4 \\ 1 \\ 0 \end{bmatrix} \right\}$, 1, 3 **39.** $\begin{bmatrix} -3 \\ 1 \\ 6 \\ 2 \end{bmatrix}$

41. $\left\{ \begin{bmatrix} -4 \\ -5 \\ 1 \\ 0 \end{bmatrix}, \begin{bmatrix} -2 \\ -3 \\ 0 \\ 1 \end{bmatrix} \right\}$, $\{[1 \quad 0 \quad 4 \quad 2], [0 \quad 1 \quad 5 \quad 3]\}$, $\left\{ \begin{bmatrix} 1 \\ 0 \\ 0 \end{bmatrix}, \begin{bmatrix} 0 \\ 1 \\ 0 \end{bmatrix} \right\}$

43. $\left\{ \begin{bmatrix} \frac{85}{18} \\ -\frac{13}{9} \\ \frac{5}{18} \\ 1 \\ 0 \end{bmatrix}, \begin{bmatrix} -\frac{149}{6} \\ \frac{14}{3} \\ -\frac{7}{6} \\ 0 \\ 1 \end{bmatrix} \right\}$, $\{[1 \quad 0 \quad 0 \quad -\frac{85}{18} \quad \frac{149}{6}], [0 \quad 1 \quad 0 \quad \frac{13}{9} \quad -\frac{14}{3}], [0 \quad 0 \quad 1 \quad -\frac{5}{18} \quad \frac{7}{6}]\}$, $\left\{ \begin{bmatrix} 1 \\ -2 \\ -1 \\ 1 \end{bmatrix}, \begin{bmatrix} 3 \\ -5 \\ -2 \\ 7 \end{bmatrix}, \begin{bmatrix} -5 \\ 8 \\ 3 \\ 5 \end{bmatrix} \right\}$

CHAPTER 3 REVIEW EXERCISES *(page 228)*

1. T, F, F, F, T, T, F, T, T **3.** The zero vector is not in $\mathscr{W}$. $\mathscr{W}$ is not closed under addition.

5. (a) $\mathbf{u}_1 \perp \mathbf{u}_2$ **(b)** $\left\{ \frac{1}{\sqrt{17}}\mathbf{u}_1, \frac{1}{\sqrt{6}}\mathbf{u}_2, \frac{1}{\sqrt{21}} \begin{bmatrix} 2 \\ -3 \\ 2 \\ 2 \end{bmatrix} \right\}$ **7.** $\left\{ \begin{bmatrix} -2 \\ 1 \\ 0 \\ 0 \end{bmatrix}, \begin{bmatrix} 1 \\ 0 \\ -1 \\ 1 \end{bmatrix} \right\}$, $\text{rank}(A) = 4 - \dim(\mathscr{NS})(A)) = 2$.

9. (a) 13 **(b)** $\sqrt{35}$ **(c)** $\frac{1}{\sqrt{10}} \begin{bmatrix} 1 \\ 0 \\ 3 \end{bmatrix}$ **(d)** $\frac{13}{\sqrt{29}}$ **11. (a)** $\mathbf{u}_1 \cdot \mathbf{u}_2 = 0$. **(b)** $\left\{ \frac{1}{2}\mathbf{u}_1, \frac{1}{6}\mathbf{u}_2, \frac{1}{\sqrt{18}} \begin{bmatrix} 3 \\ -2 \\ -2 \\ 1 \end{bmatrix} \right\}$ **13.** Each matrix has rank 3 and each matrix has dependent columns. **15.** A basis for $\mathscr{RS}(A)$ is $\{\text{Row}_1(R), \text{Row}_2(R), \text{Row}_3(R)\}$. A basis for $\mathscr{NS}(A)$ is $\left\{ \begin{bmatrix} -3 \\ 1 \\ 0 \\ 0 \\ 0 \end{bmatrix}, \begin{bmatrix} -2 \\ 0 \\ -1 \\ 1 \\ 0 \end{bmatrix} \right\}$. $\mathscr{CS}(A) = \text{BASIS}\{\text{Col}_1(A), \text{Col}_3(A), \text{Col}_5(A)\}$. **17. (a)** $\dim(\mathscr{CS}(A)) = 3$.

(b) $\dim(\mathscr{NS}(A)) = 2$. **(c)** $\left\{ \begin{bmatrix} -2 \\ 1 \\ 0 \\ 0 \\ 0 \end{bmatrix}, \begin{bmatrix} -\frac{2}{7} \\ 0 \\ -\frac{2}{7} \\ 0 \\ 1 \end{bmatrix} \right\}$ is a basis for $\mathscr{NS}(A)$. **(d)** $\{[1 \quad 2 \quad 0 \quad 0 \quad \frac{2}{7}], [0 \quad 0 \quad 1 \quad 0 \quad \frac{2}{7}], [0 \quad 0 \quad 0 \quad 1 \quad 0]\}$ is a basis for $\mathscr{RS}(A)$. **(e)** $\{\text{Col}_1(A), \text{Col}_3(A), \text{Col}_4(A)\}$ **(f)** $\text{Col}_5(A) = \frac{2}{7}\text{Col}_1(A) + \frac{2}{7}\text{Col}_3(A)$.
(g) $[0 \quad 1 \quad 0 \quad 0 \quad 0]$ **(h)** $\dim(\mathscr{NS}(A^T)) = 1$.

CHAPTER 4

EXERCISES 4.1 *(page 238)*

1. (a) T is linear; the domain is $\Re^3$; $\mathcal{W} = \Re^3$. **(b)** T is nonlinear; the domain is $\Re^3$; $\mathcal{W} = \Re^3$. **(c)** T is nonlinear; the domain is $\Re^3$; $\mathcal{W} = \Re^3$. **(d)** T is linear; the domain is $\mathcal{P}_3$; $\mathcal{W} = \mathcal{P}_2$. **(e)** T is nonlinear; the domain is $\mathcal{C}^2[a \quad b]$; $\mathcal{W} = \mathcal{C}^0[a \quad b]$. **(f)** T is linear; the domain is $\Re^{2\times 2}$; $\mathcal{W} = \Re^{2\times 2}$. **3. (a)** T is linear; $\mathcal{R}(\mathbf{T}) = \{p(x) \text{ in } \mathcal{P} \mid p(0) = 0\}$. **(b)** T is not linear. **(c)** T is linear; $\mathcal{R}(\mathbf{T}) = \text{SPAN}\{x^2\}$. **(d)** T is linear; $\mathcal{R}(\mathbf{T}) = \mathcal{P}$. **5.** No, since $\mathbf{T}(2x) \neq 2\mathbf{T}(x)$ if $b \neq 0$. **7.** $\begin{bmatrix} 3 \\ 0 \\ 6 \end{bmatrix}, \begin{bmatrix} -2 \\ -4 \\ -10 \end{bmatrix}, \begin{bmatrix} -5 \\ -14 \\ -31 \end{bmatrix}, \begin{bmatrix} 5 \\ 4 \\ 16 \end{bmatrix}$ **9.** $[1 \quad 14 \quad 11], [1 \quad 2 \quad 2], [-9 \quad 12 \quad 0]$ **11.** U is linear, but is neither one-to-one nor onto $\Re^3$.

13. $\mathbf{T}(\mathbf{u}) = \begin{bmatrix} -1 \\ 12 \\ 6 \end{bmatrix}$, $\mathbf{T}(\mathbf{T}(\mathbf{u})) = \begin{bmatrix} -13 \\ 43 \\ 36 \end{bmatrix}$, $\mathbf{T}(\mathbf{T}(\mathbf{u}) + 3\mathbf{u}) = \begin{bmatrix} -16 \\ 79 \\ 54 \end{bmatrix} = (A^2 + 3A)\mathbf{u}$.

15. (a) $\mathbf{I}(a\mathbf{u} + b\mathbf{v}) = a\mathbf{u} + b\mathbf{v} = a\mathbf{I}(\mathbf{u}) + b\mathbf{I}(\mathbf{v})$. If $\mathcal{V} = \Re^n$, then **I** is multiplication by the identity matrix I.

17. $\begin{bmatrix} -45 \\ 90 \end{bmatrix}, \begin{bmatrix} 10 \\ 15 \end{bmatrix}$

19. $a_1[\mathbf{v}_1]_{\mathcal{B}} + a_2[\mathbf{v}_2]_{\mathcal{B}} + \cdots + a_k[\mathbf{v}_k]_{\mathcal{B}} = 0 \Rightarrow [a_1\mathbf{v}_1 + a_2\mathbf{v}_2 + \cdots + a_k\mathbf{v}_k]_{\mathcal{B}} = 0 \Rightarrow a_1\mathbf{v}_1 + a_2\mathbf{v}_2 + \cdots + a_k\mathbf{v}_k = 0$. Since the $\mathbf{v}_i$ are independent, all $a_i = 0$ and the $[\mathbf{v}_i]_{\mathcal{B}}$ are also independent. **21.** As shown below, **R** maps the parallelogram determined by the vectors **u** and **v** into a congruent parallelogram. Thus $\mathbf{R}(\mathbf{u} + \mathbf{v}) = \mathbf{R}(\mathbf{u}) + \mathbf{R}(\mathbf{v})$.

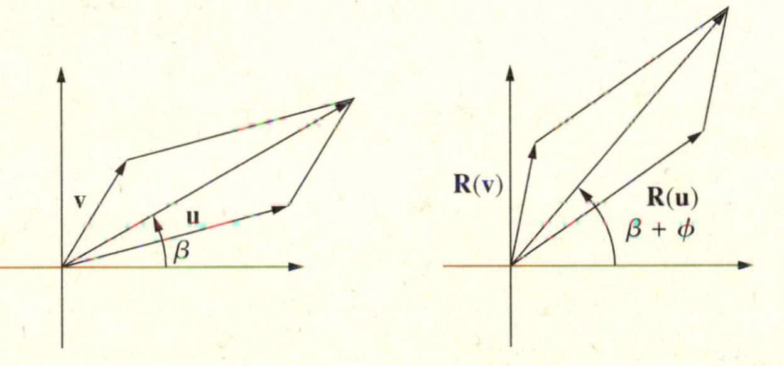

23. D is not one-to-one since $\mathbf{D}(x^2 + 1) = \mathbf{D}(x^2 + 2) = 2x$. D is onto $\mathcal{P}$ since $\mathbf{D}(\int_0^x p(x)\,dx) = p(x)$ for any $p(x)$ in $\mathcal{P}$. The integral operator $(p(x) \to \int_0^x p(x)\,dx)$ is one-to-one, but not onto since no nonzero constant is in the range of this operator. **25.** Let $\mathbf{v} = a_1\mathbf{u}_1 + a_2\mathbf{u}_2 + \cdots + a_n\mathbf{u}_n$ be such that $\mathbf{T}(\mathbf{v}) = \mathbf{0}$. Then, $\mathbf{T}(\mathbf{v}) = a_1\mathbf{T}(\mathbf{u}_1) + a_2\mathbf{T}(\mathbf{u}_2) + \cdots + a_n\mathbf{T}(\mathbf{u}_n) = \mathbf{0} \Rightarrow$ all $a_i = 0 \Rightarrow \mathbf{v} = \mathbf{0}$. Therefore T is one-to-one by part 5 of Theorem 4.2. To see that T is onto let **w** be arbitrary in $\mathcal{W}$. Since the $\mathbf{T}(\mathbf{u}_i)$ form a basis there exists scalars b_i such that $\mathbf{w} = b_1\mathbf{T}(\mathbf{u}_1) + b_2\mathbf{T}(\mathbf{u}_2) + \cdots + b_n\mathbf{T}(\mathbf{u}_n) = \mathbf{T}(b_1\mathbf{u}_1 + b_2\mathbf{u}_2 + \cdots + b_n\mathbf{u}_n)$. It follows that T is onto $\mathcal{W}$.

EXERCISES 4.2 *(page 250)*

1. (a) $\mathcal{NS}(\mathbf{T}) = \text{SPAN}\left\{\begin{bmatrix} -2 \\ 1 \end{bmatrix}\right\}$, $\mathcal{R}(\mathbf{T}) = \text{SPAN}\left\{\begin{bmatrix} 1 \\ 2 \end{bmatrix}\right\}$, rank(T) = 1, nullity(T) = 1.
(b) $\mathcal{NS}(\mathbf{D}^2) = \text{SPAN}\{1, x\}$, $\mathcal{R}(\mathbf{D}^2) = \text{SPAN}\{1, x\}$, $\text{rank}(\mathbf{D}^2) = 2$, $\text{nullity}(\mathbf{D}^2) = 2$.
(c) $\mathcal{NS}(\mathbf{T}) = \left\{\begin{bmatrix} a & b \\ c & d \end{bmatrix} \middle| \, a + b = 0, c + d = 0\right\}$, $\mathcal{R}(\mathbf{T}) = \left\{\begin{bmatrix} x & 0 \\ 0 & y \end{bmatrix}\right\}$, rank(T) = 2, nullity(T) = 2.
(d) $\mathcal{NS}(\mathbf{T}) = \{\mathbf{0}\}$, $\mathcal{R}(\mathbf{T}) = \Re^{1\times 3}$, rank(T) = 3, nullity(T) = 0. **3. (a)** $\{e^x\}$ **(b)** $\{1, x\}$ **(c)** $\{\sin 2x, \cos 2x\}$
5. $\dim(\mathcal{V}) < \dim(\mathcal{W})$, $\dim(\mathcal{R}(\mathbf{T})) < \min\{\dim(\mathcal{V}), \dim(\mathcal{W})\}$. **7. (a)** 0 **(b)** 4 **(c)** nullity = 2, $\mathcal{R}(\mathbf{T}) = \mathcal{W}$.
(d) rank(T) = 1, nullity(T) = 4. **9.** $(\mathcal{RS}(A))^T = \mathcal{R}(A^T)$.

11. **(a)** $\begin{bmatrix} -1 \\ 0 \\ 1 \\ 2 \\ 0 \end{bmatrix} + c_1 \begin{bmatrix} -5 \\ 1 \\ 0 \\ 0 \\ 0 \end{bmatrix} + c_2 \begin{bmatrix} -5 \\ 0 \\ -3 \\ -4 \\ 1 \end{bmatrix}$ **(b)** $\begin{bmatrix} -\frac{29}{3} \\ 2 \\ 0 \\ \frac{5}{3} \end{bmatrix} + c_1 \begin{bmatrix} \frac{13}{2} \\ -\frac{3}{2} \\ 1 \\ 0 \end{bmatrix}$ **13.** If **T** is onto, then $m \geq n$. If **T** is one-to-one, then $m \leq n$. **15.** No, yes, yes, yes **17.** **(a)** Columns 1, 3, and 5 of A form a basis for $\mathcal{CS}(A)$.

(b) Rows 1, 2, and 3 of R form a basis for $\mathcal{RS}(A)$. **(c)** $\left\{ \begin{bmatrix} -3 \\ 1 \\ 0 \\ 0 \\ 0 \end{bmatrix}, \begin{bmatrix} -2 \\ 0 \\ 1 \\ 1 \\ 0 \end{bmatrix} \right\}$ **(d)** $\mathcal{R}(\mathbf{T}) = \mathcal{CS}(A)$.

19. $\mathcal{R}(\mathbf{U}) = \text{SPAN}\{a + bx, ax + bx^2, ax^2 + bx^3, ax^3 + bx^4\} \subset \mathcal{P}_5$. **21.** 4, 3, 2, 3, 3 **23.** **(a)** The first and third vectors are in $\mathcal{R}(\mathbf{T})$. **(b)** The first and third vectors are in $\mathcal{NS}(\mathbf{T})$.

EXERCISES 4.3 *(page 256)*

1. **(a)** [2 1] **(b)** [−24 −5] **(c)** [6 5] **3.** $(\mathbf{T}^3 - 7\mathbf{T}^2 + 11\mathbf{T})(\mathbf{u}) = (A^3 - 7A^2 + 11A)\mathbf{u} = (5I)\mathbf{u} = 5\mathbf{u}$.
5. **(a)** $\mathbf{D}^2 - 1$ **(b)** $\mathbf{D}^3 + 7\mathbf{D}^2 - \mathbf{D} - 22$ **(c)** $\mathbf{x}\mathbf{D}^3 + (3\mathbf{x} + 3)\mathbf{D}^2 + (9 + \mathbf{x})\mathbf{D} + 3$
(d) $\mathbf{x}\mathbf{e}^{\mathbf{x}}\mathbf{D}^2 + (\mathbf{x}\mathbf{e}^{\mathbf{x}} + \mathbf{x} \sin 3\mathbf{x} + 3\mathbf{e}^{\mathbf{x}})\mathbf{D} + 3(\sin 3\mathbf{x} + \mathbf{x} \cos 3\mathbf{x})$
7. $(k\mathbf{T})(a\mathbf{u} + b\mathbf{v}) = k(\mathbf{T}(a\mathbf{u} + b\mathbf{v})) = k(a\mathbf{T}(\mathbf{u}) + b\mathbf{T}(\mathbf{v})) = a(k\mathbf{T})(\mathbf{u}) + b(k\mathbf{T})(\mathbf{v})$. **9.** A is nonsingular.
11. If $\mathbf{T}(\mathbf{u}) = \mathbf{v}$, then define $\mathbf{T}^{-1}(\mathbf{v}) = \mathbf{u}$. If $\mathbf{T}(\mathbf{u}_1) = \mathbf{v}_1$ and $\mathbf{T}(\mathbf{u}_2) = \mathbf{v}_2$, then
$\mathbf{T}^{-1}(a\mathbf{v}_1 + b\mathbf{v}_2) = \mathbf{T}^{-1}(a\mathbf{T}(\mathbf{u}_1) + b\mathbf{T}(\mathbf{u}_2)) = \mathbf{T}^{-1}(\mathbf{T}(a\mathbf{u}_1 + b\mathbf{u}_2)) = a\mathbf{u}_1 + b\mathbf{u}_2 = a\mathbf{T}^{-1}(\mathbf{v}_1) + b\mathbf{T}^{-1}(\mathbf{v}_2)$.
13. $\mathbf{T}_1\mathbf{T}_2(A) = \text{tr}(A)I = \text{tr}(A^T)I = \mathbf{T}_2\mathbf{T}_1(A)$. **15.** Let $\{\mathbf{u}_1, \ldots, \mathbf{u}_n\}$ be a basis for $\mathcal{V}$. Then $\{\mathbf{T}(\mathbf{u}_1), \ldots, \mathbf{T}(\mathbf{u}_n)\}$ is a basis for $\mathcal{W}$. $(\mathbf{TU})(\mathbf{T}(\mathbf{u}_i)) = \mathbf{T}(\mathbf{UT}(\mathbf{u}_i)) = \mathbf{T}(\mathbf{u}_i)$ since $\mathbf{UT} = \mathbf{I}_{\mathcal{V}}$. Therefore $\mathbf{TU} = \mathbf{I}_{\mathcal{W}}$.

EXERCISES 4.4 *(page 268)*

1. **(a)** $\begin{bmatrix} \frac{5}{2} & -\frac{1}{6} \\ -\frac{3}{2} & \frac{15}{6} \end{bmatrix}$ **(b)** $\begin{bmatrix} -2 & 3 \\ -2 & 3 \end{bmatrix}$ **(c)** $\begin{bmatrix} -7 & 5 \\ -10 & 8 \end{bmatrix}$

3. $\begin{bmatrix} \cos \alpha & -\sin \alpha \\ \sin \alpha & \cos \alpha \end{bmatrix} \begin{bmatrix} \cos \beta & -\sin \beta \\ \sin \beta & \cos \beta \end{bmatrix} = \begin{bmatrix} \cos(\alpha + \beta) & -\sin(\alpha + \beta) \\ \sin(\alpha + \beta) & \cos(\alpha + \beta) \end{bmatrix}$.

5. $\begin{bmatrix} \frac{1}{2} & 1 \\ -\frac{5}{4} & -3 \end{bmatrix}$, $\mathbf{T}(1, 0) = \left(\frac{1}{2}, -\frac{5}{4}\right)$, $\mathbf{T}(0, 1) = (1, -3)$, $\mathbf{T}(1, 1) = \left(\frac{3}{2}, -\frac{17}{4}\right)$.

7. $\begin{bmatrix} \frac{2}{3} & -\frac{5}{3} \\ \frac{1}{3} & -1 \end{bmatrix}$, $\mathbf{T}(1, 0) = \left(\frac{2}{3}, \frac{1}{3}\right)$, $\mathbf{T}(0, 1) = \left(-\frac{5}{3}, -1\right)$, $\mathbf{T}(1, 1) = \left(-1, -\frac{2}{3}\right)$. The answers to 9 and 11 are not unique; one answer is given. **9.** **(a)** Shear in the x-direction with factor −1 **(b)** Shear in the y-direction with factor 3 **(c)** Shear in the x-direction with factor 1 **11.** **(a)** Shear in the x-direction with factor $-\tan \theta$ **(b)** Expansion in the y-direction with factor $\sec \theta$ **(c)** Shear in the y-direction with factor $\sin \theta$ **(d)** Expansion in the x-direction with factor $\cos \theta$

13. $\begin{bmatrix} 0 & 0 & 0 \\ 1 & 0 & 0 \\ 3 & 0 & 2 \end{bmatrix}$, $\mathbf{T}(1,0,0) = (0,1,3)$, $\mathbf{T}(0,1,0) = (0,0,0)$, $\mathbf{T}(0,0,1) = (0,0,2)$. **15.** **(a)** Shear in the x-direction of the xy-plane, factor −1 **(b)** Expansion in the y-direction by a factor of 3 **(c)** Reflection in the yz-plane **(d)** Shear in the x-direction of the xz-plane, factor 3 **(e)** Shear in the y-direction of the yz-plane, factor 7 **(f)** Shear in the z-direction of the yz-plane, factor 2 **(g)** Shear in the z-direction of the xz-plane, factor −4 **(h)** Shear in the y-direction of the xz-plane, factor −2. **17.** $\tan \theta = -\dfrac{c}{a}$.

19. This follows easily from Exercise 18 of Section 3.6 since $\begin{bmatrix} \cos \theta & \sin \theta \\ \sin \theta & -\cos \theta \end{bmatrix} = \begin{bmatrix} \cos \theta & -\sin \theta \\ \sin \theta & \cos \theta \end{bmatrix} \begin{bmatrix} 1 & 0 \\ 0 & -1 \end{bmatrix}$.
21. **(d)** The eigenvectors determine lines through the origin which are invariant under the transformation.

23. $\frac{17}{30}\mathbf{v}_1 + \frac{2}{51}\mathbf{v}_2$ **25. (a)** **(b)**

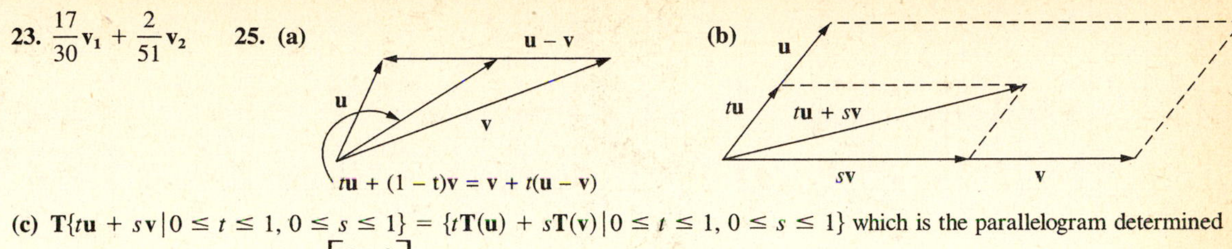

(c) $\mathbf{T}\{t\mathbf{u} + s\mathbf{v} \mid 0 \le t \le 1, 0 \le s \le 1\} = \{t\mathbf{T}(\mathbf{u}) + s\mathbf{T}(\mathbf{v}) \mid 0 \le t \le 1, 0 \le s \le 1\}$ which is the parallelogram determined by $\mathbf{T}(\mathbf{u})$ and $\mathbf{T}(\mathbf{v})$. **27. (a)** $\begin{bmatrix} 1 & 3 \\ 2 & 5 \\ 5 & 1 \end{bmatrix}$ **(b)** Yes **(c)** No

EXERCISES 4.5 *(page 279)*

1. (a) $\begin{bmatrix} 3 \\ 1 \\ -3 \end{bmatrix}, \begin{bmatrix} -2 \\ 6 \\ 0 \end{bmatrix}, \begin{bmatrix} 1 \\ 2 \\ 7 \end{bmatrix}, \begin{bmatrix} 0 \\ 1 \\ 1 \end{bmatrix}$ **(b)** $\begin{bmatrix} 11 \\ 5 \\ 22 \end{bmatrix}, \begin{bmatrix} -42 \\ 32 \\ -10 \end{bmatrix}, \begin{bmatrix} -56 \\ 87 \\ 17 \end{bmatrix}, \begin{bmatrix} -13 \\ 17 \\ 2 \end{bmatrix}$ **(c)** $\begin{bmatrix} -31 \\ 37 \\ 12 \end{bmatrix}$ **3.** $\begin{bmatrix} 2 & 2 & 1 \\ 0 & 0 & 0 \\ 2 & 1 & 2 \\ 0 & 0 & 0 \\ 2 & 1 & 1 \end{bmatrix}$

5. $\begin{bmatrix} 6 & 1 & -7 & 12 \\ 8 & 1 & 9 & 9 \\ 6 & 1 & -7 & 9 \\ 0 & 0 & 0 & 0 \end{bmatrix}$ **7.** If $\mathcal{B} = \{\mathbf{u}_1, \ldots, \mathbf{u}_n\}$ is any basis for $\mathcal{V}$, then the ith column of $[\mathbf{I}]_{\mathcal{B}}$ is $[\mathbf{I}(\mathbf{u}_i)]_{\mathcal{B}} = [\mathbf{u}_i]_{\mathcal{B}} = \mathbf{e}_i = \text{Col}_i(I)$. **9.** $\frac{1}{9}\begin{bmatrix} 8 & -1 & -5 \\ 2 & 11 & 10 \\ 7 & 7 & 44 \end{bmatrix}$ **11.** $\begin{bmatrix} 1 & 0 & 0 & 0 \\ 0 & 0 & 1 & 0 \\ 0 & 1 & 0 & 0 \\ 0 & 0 & 0 & 1 \end{bmatrix}$ **13.** $\begin{bmatrix} 5 & 0 & 0 \\ 0 & 3 & 0 \\ 0 & 0 & 1 \end{bmatrix}$

15. The correspondence is one-to-one and onto by Theorem 4.2. The linearity follows from the calculation: $[(a\mathbf{T} + b\mathbf{U})(\mathbf{x})]_{\mathcal{B}'} = [a\mathbf{T}(\mathbf{x}) + b\mathbf{U}(x)]_{\mathcal{B}'} = a[\mathbf{T}(\mathbf{x})]_{\mathcal{B}'} + b[\mathbf{U}(x)]_{\mathcal{B}'} = a[\mathbf{T}]_{\mathcal{B}'\mathcal{B}}[\mathbf{x}]_{\mathcal{B}} + b[\mathbf{U}]_{\mathcal{B}'\mathcal{B}} = (a[\mathbf{T}]_{\mathcal{B}'\mathcal{B}} + b[\mathbf{U}]_{\mathcal{B}'\mathcal{B}})[\mathbf{x}]_{\mathcal{B}}$.

EXERCISES 4.6 *(page 287)*

1. (a) $P = \begin{bmatrix} -1 & -3 \\ 2 & 5 \end{bmatrix}$ **(b)** $\frac{1}{2}\begin{bmatrix} -1 & 5 & 7 \\ 1 & -9 & 5 \\ 0 & 0 & -24 \end{bmatrix}$ **(c)** $\frac{1}{6}\begin{bmatrix} 0 & 2 & -2 \\ -3 & -10 & 4 \\ 15 & 18 & 6 \end{bmatrix}$ **(d)** $\begin{bmatrix} 1 & 1 & 5 & -1 \\ 0 & 2 & -1 & 7 \\ 1 & 0 & 4 & 0 \\ 0 & 0 & 0 & 6 \end{bmatrix}$

(e) $\frac{1}{\sin 1}\begin{bmatrix} \sin 3 & -\cos 3 \\ -\sin 2 & \cos 2 \end{bmatrix}$ **3.** $\frac{1}{2}\begin{bmatrix} 7 \\ -3 \\ -16 \end{bmatrix}$ **5.** $\mathcal{S}$ to $\mathcal{N}$: $\begin{bmatrix} \cos\theta & \sin\theta \\ -\sin\theta & \cos\theta \end{bmatrix}$; $\mathcal{N}$ to $\mathcal{S}$: $\begin{bmatrix} \cos\theta & -\sin\theta \\ \sin\theta & \cos\theta \end{bmatrix}$

7. $\mathcal{B}$ to $\mathcal{S}: M = [\mathbf{u}_1 \ \ \mathbf{u}_2 \ \ \cdots \ \ \mathbf{u}_n]$; $\mathcal{S}$ to $\mathcal{B}: M^{-1} = M^T$. **9.** The $\mathcal{B}$ to $\mathcal{B}'$ matrix is $\begin{bmatrix} 0 & 0 & 1 & 0 \\ 1 & 0 & 0 & 0 \\ 0 & 0 & 0 & 1 \\ 0 & 1 & 0 & 0 \end{bmatrix}$.

EXERCISES 4.7 *(page 294)*

1. (a) $[\mathbf{T}]_{\mathcal{B}} = \begin{bmatrix} 2 & 4 & 1 & 5 \\ -3 & 0 & 0 & 1 \\ 1 & 0 & 4 & 0 \\ 0 & -5 & 0 & -1 \end{bmatrix}$ **(b)** $P = \begin{bmatrix} 1 & 0 & 0 & 0 \\ -2 & 1 & 0 & 0 \\ 1 & 1 & 1 & 0 \\ 1 & -3 & 5 & 1 \end{bmatrix}$ **(c)** $[\mathbf{T}]_{\mathcal{B}'} = \begin{bmatrix} 0 & -10 & 26 & 5 \\ -2 & -23 & 57 & 11 \\ 7 & 37 & -79 & -16 \\ -32 & -246 & 535 & 107 \end{bmatrix}$

(d) $\begin{cases} T(\mathbf{v}_1) = -2\mathbf{v}_2 + 7\mathbf{v}_3 - 32\mathbf{v}_4 \\ T(\mathbf{v}_2) = -10\mathbf{v}_1 - 23\mathbf{v}_2 + 37\mathbf{v}_3 - 246\mathbf{v}_4 \\ T(\mathbf{v}_3) = 26\mathbf{v}_1 + 57\mathbf{v}_2 - 79\mathbf{v}_3 + 535\mathbf{v}_4 \\ T(\mathbf{v}_4) = 5\mathbf{v}_1 + 11\mathbf{v}_2 - 16\mathbf{v}_3 + 107\mathbf{v}_4 \end{cases}$

3. (a) $\begin{bmatrix} 65 & 46 \\ -85 & -60 \end{bmatrix} = \begin{bmatrix} 4 & 3 \\ 3 & 2 \end{bmatrix}^{-1} \begin{bmatrix} 2 & -1 \\ 4 & 3 \end{bmatrix} \begin{bmatrix} 4 & 3 \\ 3 & 2 \end{bmatrix}$.

5. $[\mathbf{T}]_{\mathcal{B}'} = \begin{bmatrix} 8 & 4 \\ -9 & -4 \end{bmatrix} = \begin{bmatrix} -4 & 3 \\ 3 & 2 \end{bmatrix}^{-1} \begin{bmatrix} 2 & 1 \\ 0 & 2 \end{bmatrix} \begin{bmatrix} -4 & -3 \\ 3 & 2 \end{bmatrix}$.

7. $[\mathbf{T}]_{\mathcal{B}} = \begin{bmatrix} 5 & 0 & 0 \\ 0 & 1 & 0 \\ 0 & 0 & 1 \end{bmatrix}$, **T** is expansion by a factor of 5 in the $\mathbf{u}_1$ direction.

9. $\begin{bmatrix} 3 & 0 & 0 & 0 \\ 0 & 2 & 1 & 0 \\ 0 & 0 & 2 & 0 \\ 0 & 0 & 0 & 1 \end{bmatrix}$ shear by a factor of $\frac{1}{2}$ in the x_2 direction of the x_2x_3 plane, expansions by a factor of 2 in the x_2 and x_3 directions, expansion by a factor of 3 in the x_1 direction.

11. $\det(\lambda I - P^{-1}AP) = \det(\lambda P^{-1}IP - P^{-1}AP) = \det(P^{-1}(\lambda I - A)P) = \det(\lambda I - A)$.

13. (a) $B^k = (P^{-1}AP)^k = (P^{-1}AP)(P^{-1}AP) \cdots (P^{-1}AP) = P^{-1}A^kP$.

(b) $p(b) = a_0I + a_1B + a_2B^2 + \cdots + a_nB^n = a_0P^{-1}IP + a_1(P^{-1}AP) + a_2(P^{-1}AP)^2 + \cdots + a_n(P^{-1}AP)^n = a_0P^{-1}IP + a_1(P^{-1}AP) + a_2(P^{-1}A^2P) + \cdots + a_n(P^{-1}A^nP) = P^{-1}(a_0I + a_1A + a_2A^2 + \cdots + a_nA^n)P = P^{-1}p(A)P$.

15. Let $\mathbf{u}_i$ be the ith basis vector of $\mathcal{B}$. Then the ith column of the $\mathcal{B}$ to $\mathcal{B}'$ change-of-basis matrix is $[\mathbf{u}_i]_{\mathcal{B}'} = [\mathbf{I}(\mathbf{u}_i)]_{\mathcal{B}'}$ which is the ith column of $[\mathbf{I}]_{\mathcal{B}'\mathcal{B}}$.

17. $[\mathbf{T}]_{\mathcal{C}'\mathcal{B}'} = [\mathbf{I}]_{\mathcal{C}'\mathcal{C}}[\mathbf{T}]_{\mathcal{C}\mathcal{B}}[\mathbf{I}]_{\mathcal{B}\mathcal{B}'}$; that is, the two matrix representatives are equivalent matrices.

CHAPTER 4 REVIEW EXERCISES *(page 296)*

1. (a) 2 **(b)** 8 **(c)** 7 **(d)** 10 **(e)** 9 or 11 **3.** **T** is not linear because $\mathbf{T}(2A) \neq 4\mathbf{T}(A)$. **5.** **T** is not linear because $\mathbf{T}(\mathbf{0}) \neq \mathbf{0}$. **7.** $\mathcal{NS}(\mathbf{Q}) = \{a + dx^3\}$, $\mathcal{R}(\mathbf{Q}) = \{b + cx^2\}$, $\text{rank}(\mathbf{Q}) = 2$, $\text{nullity}(\mathbf{Q}) = 2$.

9. $\begin{bmatrix} \frac{2}{3} & -5 \\ \frac{1}{3} & -3 \end{bmatrix} = \begin{bmatrix} 0 & 1 \\ 1 & 0 \end{bmatrix} \begin{bmatrix} 1 & 0 \\ 2 & 1 \end{bmatrix} \begin{bmatrix} 1 & -3 \\ 0 & 1 \end{bmatrix} \begin{bmatrix} \frac{1}{3} & 0 \\ 0 & 1 \end{bmatrix}$.

11. $-16x^3 - 5x^2 + 12x + 12$, $a_1(x^2 + x + 1) + (-a_1 + 2a_2 - a_3)(2x^3 + x^2) + (-a_2 + a_3)(x^3 + x + 1)$.

13. $\begin{bmatrix} -23 & -9 & 6 \\ 8 & 3 & -2 \\ -3 & -1 & 1 \end{bmatrix}$, $\begin{bmatrix} -14 \\ 5 \\ -1 \end{bmatrix}$

CHAPTER 5

EXERCISES 5.1 *(page 310)*

In Exercises 1–15, the diagonalizing matrix P is not unique.

1. $\lambda^2 - 13\lambda + 36$; $\lambda = 4, 9$; $P = \begin{bmatrix} 1 & 1 \\ -1 & 4 \end{bmatrix}$. **3.** $\lambda^2 - 6\lambda + 8$; $\lambda = 2, 4$; $P = \begin{bmatrix} 3 & 1 \\ 1 & 1 \end{bmatrix}$.

5. $\lambda^3 - 9\lambda^2 + 24\lambda - 20$; $\lambda = 2, 2, 5$; $P = \begin{bmatrix} 1 & -1 & 1 \\ 1 & 0 & -1 \\ 0 & 1 & 1 \end{bmatrix}$.

7. $\lambda^3 - 9\lambda^2 + 24\lambda - 20$; $\lambda = 2, 2, 5$; not diagonalizable.

9. $\lambda^3 + 4\lambda^2 - \lambda - 4$; $\lambda = 1, -4, -1$; $P = \begin{bmatrix} 4 & 1 & 4 \\ 5 & 0 & -3 \\ 1 & -1 & -1 \end{bmatrix}$.

11. $\lambda^4 - 13\lambda^2 + 36$; $\lambda = 3, -3, 2, -2$; $P = \begin{bmatrix} 1 & 1 & 1 & 1 \\ 3 & -3 & 2 & -2 \\ 9 & 9 & 4 & 4 \\ 27 & -27 & 8 & -8 \end{bmatrix}$.

13. $(\lambda - 2)^4 - 3(\lambda - 2)^2 + 1$; $\lambda = 2 \pm \sqrt{\dfrac{3 \pm \sqrt{5}}{2}}$; $P = \begin{bmatrix} -.6180 & .6180 & -1 & -1 \\ 1 & 1 & -.6180 & .6180 \\ -1 & 1 & .6180 & .6180 \\ .6180 & .6180 & 1 & -1 \end{bmatrix}$.

15. $\lambda^3 - 3\lambda^2 + 4\lambda - 12$; $\lambda = 3, \pm 2i$; $P = \begin{bmatrix} 1 & 1 & 1 \\ 3 & 2i & -2i \\ 9 & -4 & -4 \end{bmatrix}$.

17. If the eigenvalues of A are $\lambda_1, \lambda_2, \ldots, \lambda_n$, then the eigenvalues of $p(A)$ are $p(\lambda_1), p(\lambda_2), \ldots, p(\lambda_n)$. The eigenvectors of A and $p(A)$ are the same. **19.** $P^{-1}AP = D \Rightarrow (P^{-1}AP)^T = D^T \Rightarrow P^TA^T(P^T)^{-1} = D$. Thus A and A^T are both similar to D and hence, by the transitivity of similarity, are similar to each other.

21. The operations suggested in the hint, none of which change the determinant, reduce

$(\lambda I - C)$ to $\begin{bmatrix} 0 & -1 & 0 & \cdots & 0 \\ 0 & \lambda & -1 & \cdots & 0 \\ 0 & 0 & \lambda & \cdots & 0 \\ \vdots & & & & \\ 0 & 0 & 0 & \cdots & -1 \\ p(\lambda) & -a_1 & -a_2 & \cdots & -a_{n-1} \end{bmatrix}$. Expanding by the first column yields

$\det(\lambda I - C) = (-1)^{n+1}p(\lambda)(-1)^{n-1} = p(\lambda)$. **23.** $\det(\lambda I - A) = \lambda^2 - (a + d)\lambda + (ad - b^2)$. The discriminant of this quadratic is $(a + d)^2 - 4(ad - b^2) = (a - d)^2 + 4b^2 \geq 0$, so the roots must be real.

25. $Y(t) = -\dfrac{3}{2}\begin{bmatrix} 3 \\ 1 \end{bmatrix} e^{2t} + \dfrac{11}{2}\begin{bmatrix} 1 \\ 1 \end{bmatrix} e^{4t}$. **27.** $A = I^{-1}AI \Rightarrow A$ similar to A. If A is similar to B with $B = P^{-1}AP$ then $A = PBP^{-1}$ so B is similar to A. If $B = P^{-1}AP$ and $C = Q^{-1}BQ$, then $C = Q^{-1}P^{-1}APQ = (PQ)^{-1}A(PQ)$ so that C is similar to A. **29.** $\Lambda(A) = \{4, 9\}$, $\Lambda(B) = \{2, 4\}$, $\Lambda(A + B) = \{4.57557, 14.4244\}$, $\Lambda(AB) = \{5.72894, 50.2711\}$.

31. $\det(\lambda I - A) = \det(\lambda I - A)^T = \det(\lambda I - A^T)$. **33.** $\begin{bmatrix} 1 & 0 & 0 \\ -2 & 2.5 & .5 \\ -2 & .5 & 2.5 \end{bmatrix}$

35. 1, 0.2, 0.5; $\begin{bmatrix} 3 \\ 6 \\ 1 \end{bmatrix}, \begin{bmatrix} -1 \\ 0 \\ 1 \end{bmatrix}, \begin{bmatrix} 1 \\ -3 \\ 2 \end{bmatrix}$ **37.** $\Lambda\left(\begin{bmatrix} A & O \\ O & B \end{bmatrix}\right) = \Lambda(A) \cup \Lambda(B)$

EXERCISES 5.2 *(page 319)*

1. $7 - 3i$, $-6 + 19i$, $7 + 7i$, $7 + 3i$, $22 + 7i$, $22 + 7i$, 13, 54 **3. (a)** Set the i, j entry of $A_1(A_2)$ equal to the real (imaginary) part of a_{ij}. **(b)** The entry in the i, j position of $\overline{AB}$ is $\overline{\Sigma\, a_{ik}b_{kj}} = \Sigma\, \overline{a_{ik}}\,\overline{b_{kj}}$ which is the entry in the i, j position of $\bar{A}\bar{B}$. Now $(AB)^H = (\overline{AB})^T = (\bar{A}\bar{B})^T = \bar{B}^T\bar{A}^T = B^HA^H$. **(c)** $[\bar{A}^T]_{ij} = [\bar{A}]_{ji} = \overline{a_{ji}} = \overline{[A^T]_{ij}} = [\overline{A^T}]_{ij}$.

5. (a) $\lambda_1 = 2$, $\lambda_2 = 4$, $X_1 = \begin{bmatrix} 1 \\ -1 \end{bmatrix}$, $X_2 = \begin{bmatrix} 1 \\ 1 \end{bmatrix}$, $X_1 \cdot X_2 = 0$. **(b)** $\lambda_1 = 1$, $\lambda_2 = 7$, $\lambda_3 = 4$; $X_1 = \begin{bmatrix} -2 \\ 2 \\ 1 \end{bmatrix}$, $X_2 = \begin{bmatrix} 2 \\ 1 \\ 2 \end{bmatrix}$,

$X_3 = \begin{bmatrix} -1 \\ -2 \\ 2 \end{bmatrix}$, $X_1 \cdot X_2 = X_1 \cdot X_3 = X_2 \cdot X_3 = 0$. **7.** $\dfrac{1}{5}\begin{bmatrix} 3 & -4 \\ 4 & 3 \end{bmatrix}$ **9.** $\dfrac{1}{3}\begin{bmatrix} 2 & 1 & 2 \\ -2 & 2 & 1 \\ -1 & -2 & 2 \end{bmatrix}$ **11.** $\dfrac{1}{\sqrt{5}}\begin{bmatrix} \sqrt{5} & 0 & 0 \\ 0 & 1 & -2 \\ 0 & 2 & 1 \end{bmatrix}$

13. $\begin{bmatrix} \frac{1}{\sqrt{2}} & \frac{1}{\sqrt{2}} & 0 & 0 \\ -\frac{1}{\sqrt{2}} & \frac{1}{\sqrt{2}} & 0 & 0 \\ 0 & 0 & \frac{3}{5} & -\frac{4}{5} \\ 0 & 0 & \frac{4}{5} & \frac{3}{5} \end{bmatrix}$ **15.** $A^T = A \Rightarrow (P^TAP)^T = P^TA^TP^{T^T} = P^TAP$. **17.** $\det(\lambda I - A) = \lambda^2 + b^2$, so the roots are $\pm bi$. **19.** $A = \frac{1}{2}(A + A^T) + \frac{1}{2}(A - A^T)$ is the unique solution. **21.** Let $X = \Sigma_{i=1}^n a_1 X_1$ where $AX_i = \lambda_i X_i$ and the X_i are orthonormal. Then $X^TAX = (\Sigma\, a_i X_i^T)(\Sigma\, a_j\lambda_j X_j) = \Sigma\, a_i^2\lambda_i > 0$. **23.** Only A is orthogonal.

25. $\begin{bmatrix} b \\ \lambda_1 - a \end{bmatrix}, \begin{bmatrix} b \\ \lambda_2 - a \end{bmatrix}$ **27.** $\Lambda(A) = \{4a, 0, 0, 0\}, \begin{bmatrix} 1 \\ 1 \\ 1 \\ 1 \end{bmatrix}, \begin{bmatrix} -1 \\ 1 \\ 0 \\ 0 \end{bmatrix}, \begin{bmatrix} -1 \\ 0 \\ 1 \\ 0 \end{bmatrix}, \begin{bmatrix} -1 \\ 0 \\ 0 \\ 1 \end{bmatrix}$

29. $A^{12} = \begin{bmatrix} 2 & 3 \\ 1 & 2 \end{bmatrix}\begin{bmatrix} (-3)^{12} & 0 \\ 0 & 2^{12} \end{bmatrix}\begin{bmatrix} 2 & -3 \\ -1 & 2 \end{bmatrix} = (-3)^{12}\begin{bmatrix} 4 & -6 \\ 2 & -3 \end{bmatrix} + 2^{12}\begin{bmatrix} -3 & 6 \\ -2 & 4 \end{bmatrix} = \begin{bmatrix} 2113476 & -3164070 \\ 1054690 & -1577939 \end{bmatrix}$.

31. $\begin{bmatrix} \frac{1}{2} & -\frac{1}{\sqrt{2}} & -\frac{1}{\sqrt{2}} & -\frac{1}{\sqrt{2}} \\ \frac{1}{2} & \frac{1}{\sqrt{2}} & 0 & 0 \\ \frac{1}{2} & 0 & \frac{1}{\sqrt{2}} & 0 \\ \frac{1}{2} & 0 & 0 & \frac{1}{\sqrt{2}} \end{bmatrix}$

EXERCISES 5.3 *(page 327)*

1. $P = \frac{1}{\sqrt{17}}\begin{bmatrix} 1 & 4 \\ 4 & -1 \end{bmatrix}, P^TAP = \begin{bmatrix} 9 & 3 \\ 0 & 4 \end{bmatrix}$. **3.** $\det(\lambda I - A) = \lambda^3 - 3\lambda - 14$ has complex roots, so the given matrix is not orthogonally similar to a triangular matrix.

5. $\Lambda(A) = \{1, 1, 5\}, P = \frac{1}{\sqrt{6}}\begin{bmatrix} -\sqrt{3} & 1 & -\sqrt{2} \\ \sqrt{3} & 1 & -\sqrt{2} \\ 0 & -2 & -\sqrt{2} \end{bmatrix}, P^TAP = \begin{bmatrix} 1 & 0 & -1.2247 \\ 0 & 1 & -.70711 \\ 0 & 0 & 5 \end{bmatrix}$. The matrix P is not unique.

7. $(AB)^T(AB) = B^TA^TAB = B^TIB = I \Rightarrow AB$ is orthogonal.
$\begin{bmatrix} A & 0 \\ 0 & B \end{bmatrix}^T\begin{bmatrix} A & 0 \\ 0 & B \end{bmatrix} = \begin{bmatrix} A^T & 0 \\ 0 & B^T \end{bmatrix}\begin{bmatrix} A & 0 \\ 0 & B \end{bmatrix} = \begin{bmatrix} A^TA & 0 \\ 0 & B^TB \end{bmatrix} = \begin{bmatrix} I & 0 \\ 0 & I \end{bmatrix} = I.$

9. **(a)** $H^T = (I - 2\mathbf{w}\mathbf{w}^T)^T = I^T - 2\mathbf{w}^{T^T}\mathbf{w}^T = I - 2\mathbf{w}\mathbf{w}^T = H$. $H^TH = I - 4\mathbf{w}\mathbf{w}^T + 4\mathbf{w}(\mathbf{w}^T\mathbf{w})\mathbf{w}^T = I$.
(b) $H\mathbf{w} = -\mathbf{w}$. **(c)** $H\mathbf{u} = \mathbf{u} - 2\mathbf{w}\mathbf{w}^T\mathbf{u} = \mathbf{u}$ since $\mathbf{w}^T\mathbf{u} = 0$.

11. **(a)** $(P^TAP)^T(P^TAP) = P^TA^TPP^TAP = P^TA^TAP$. $(P^TAP)(P^TAP)^T = P^TAPP^TA^TP = P^TAA^TP = P^TA^TAP$.
(b) If $\begin{bmatrix} a_{11} & A_{12} \\ 0 & A_{22} \end{bmatrix}$ is $n \times n$, normal and upper triangular, then $AA^T = \begin{bmatrix} a_{11}^2 + A_{12}A_{12}^T & * \\ * & * \end{bmatrix}$ and $A^TA = \begin{bmatrix} a_{11}^2 & * \\ * & * \end{bmatrix}$. It follows that $A_{12} = 0$ and that A_{22} is $(n - 1) \times (n - 1)$, normal, and upper triangular. The result now follows by induction on n. **(c)** Since such a matrix is orthogonally similar to a normal upper triangular matrix T, it follows from (b) that T is diagonal and that A is symmetric by Theorem 5.10. **(d)** $A^TA = I = A^TA \Rightarrow A$ is normal.

13. **(1)** $A^2 - 13A + 36I = 0$ **(3)** $A^3 - 3A - 14I = 0$ **(5)** $A^3 - 7A^2 + 11A - 5I = 0$.

15. $p(A) = q(A)c(A) + r(A) = r(A)$ since $c(A) = 0$. **17.** By Theorem 5.12, A^{-1} is a polynomial in A. If A is symmetric, then so is $p(A)$ for any polynomial $p(x)$. Thus A^{-1} is also symmetric.

19. $\begin{bmatrix} \lambda^k & k\lambda^{k-1} & k(k-1)\lambda^{k-2}/2 \\ 0 & \lambda^k & k\lambda^{k-1} \\ 0 & 0 & \lambda^k \end{bmatrix} = \lambda^k I + k\lambda^{k-1}N + \frac{k(k-1)}{2}\lambda^{k-2}N^2$ where $N = \begin{bmatrix} 0 & 1 & 0 \\ 0 & 0 & 1 \\ 0 & 0 & 0 \end{bmatrix}$.

21. $D_1 = \frac{1}{7}I, D_2 = \text{Dg}\left\{\frac{1}{\sqrt{12}}, \frac{1}{\sqrt{18}}, \frac{1}{\sqrt{6}}\right\}, \quad D_3 = \text{Dg}\left\{\frac{1}{2}, \frac{1}{6}, \frac{1}{6}, \frac{1}{6}\right\}$.

23. $\sqrt{A} = \begin{bmatrix} 1 & 1 & 1 \\ -1 & 0 & 2 \\ 0 & -1 & 1 \end{bmatrix} \begin{bmatrix} 1 & 0 & 0 \\ 0 & 1 & 0 \\ 0 & 0 & \sqrt{5} \end{bmatrix} \begin{bmatrix} 1 & 1 & 1 \\ -1 & 0 & 2 \\ 0 & -1 & 1 \end{bmatrix}^{-1} = \frac{1}{4} \begin{bmatrix} 3+\sqrt{5} & -1+\sqrt{5} & -1+\sqrt{5} \\ -2+2\sqrt{5} & 2+2\sqrt{5} & -2+2\sqrt{5} \\ -1+\sqrt{5} & -1+\sqrt{5} & 3+\sqrt{5} \end{bmatrix}$

There are five other correct answers.

25. $Q = \begin{bmatrix} 1 & -1 & -1 & 2 \\ 1 & 3 & 0 & -1 \\ 1 & -3 & -1 & 0 \\ 1 & 1 & 2 & -1 \end{bmatrix}$, $Q^{-1}AQ = \text{Dg}\{34, 0, -8, 8\}$.

EXERCISES 5.4 *(page 335)*

1. $Y_0 = \begin{bmatrix} 1 \\ 1 \end{bmatrix} \Rightarrow \lambda_1 = -4.41421$, $X_1 = \begin{bmatrix} -.225440 \\ 1.0 \end{bmatrix}$. **3.** $Y_0 = \begin{bmatrix} 1 \\ 1 \\ 1 \end{bmatrix} \Rightarrow \lambda = 2$, $X = \begin{bmatrix} -1 \\ -1 \\ 0 \end{bmatrix}$, $Y_0 = \begin{bmatrix} 1 \\ 2 \\ 3 \end{bmatrix}$, $\Rightarrow \lambda = 3$, $X = \begin{bmatrix} .5 \\ 1.0 \\ 1.0 \end{bmatrix}$. **5. (a)** $\begin{bmatrix} 1 \\ 1 \end{bmatrix}, \begin{bmatrix} 1 \\ 2 \end{bmatrix}$ **(b)** $Y_1 = Y_3 = Y_5 = \begin{bmatrix} .8 \\ 1.0 \end{bmatrix}$, $Y_2 = Y_4 = Y_6 = \begin{bmatrix} 1.0 \\ .5 \end{bmatrix}$. **(c)** Each subsequence converges to an eigenvector of A^2 for $\lambda = 4$. **(e)** $\begin{bmatrix} 1 \\ 3 \\ 2 \end{bmatrix}$ and $\begin{bmatrix} 3 \\ 1 \\ 2 \end{bmatrix}$ are eigenvectors of A^2 for $\lambda = 1$.

7. (a) $\lambda^2 - 5\lambda + 22$, $\lambda = \left(\frac{5 \pm i\sqrt{63}}{2}\right)$ **(b)** No apparent convergence
(c) $Z_8 - 5Z_7 + 22Z_6 = (A^2 - 5A + 22I)Y_6 = 0$.
9. $(A - \lambda X_1 X_1^T)\, X_1 = AX_1 - \lambda_1 X_1(X_1^T X_1) = \lambda_1 X_1 - \lambda_1 X_1(1) = 0 = 0X_1$.
For $i \neq 1$, $(A - \lambda X_1 X_1^T)\, X_i = AX_i - \lambda_1 X_1(X_1^T X_i) = \lambda_i X_i - \lambda_1 X_1(0) = \lambda_i X_i$.

11. For A, $\lambda = 30.2887$, $X = \begin{bmatrix} 0.5286 \\ 0.3803 \\ 0.5520 \\ 0.5209 \end{bmatrix}$, for B, $\lambda = 3.6180$, $X = \begin{bmatrix} -0.3697 \\ 0.6002 \\ -.6028 \\ 0.3738 \end{bmatrix}$, for C, $\lambda = 8.7667$, $X = \begin{bmatrix} 0.0786 \\ 0.1492 \\ 0.5779 \\ 0.7718 \\ 0.2047 \end{bmatrix}$.

13. The power method, applied to A^{-1} should yield the reciprocal of the smallest eigenvalue of A and its eigenvector. The power method applied to $(A - kI)^{-1}$ should yield the reciprocal of the eigenvalue closest to k and its eigenvector.
15. The sequence initially is headed towards 9, but slowly shifts away from 9 and eventually approaches 10. With the second choice of starting vector, the sequence converges quickly to 10. The original starting vector had a zero component in the direction of the dominant eigenvector. Round off introduced a small component in the direction of X_1 which eventually took over and produced the correct result.

17. (b) $A_1 = \begin{bmatrix} 14.3000 & -2.9850 & -2.6579 & 3.5827 \\ -2.9850 & -2.6636 & 1.2899 & 2.1604 \\ -2.6579 & 1.2899 & -2.1030 & -0.3819 \\ 3.5827 & 2.1604 & -0.3819 & -2.5333 \end{bmatrix}$, $A = QR \Rightarrow Q^TA = R \Rightarrow A_1 = RQ = (Q^TA)Q$.

(e) $A_6 = \begin{bmatrix} 15.8328 & .0331 & 0.0006 & 0.000 \\ 0.0331 & -5.9209 & -0.0082 & -0.0000 \\ 0.0006 & -0.0082 & -2.6854 & -0.0000 \\ 0.0000 & -0.0000 & -0.0000 & -0.2264 \end{bmatrix}$ is almost upper triangular and the diagonal entries are good approximations for the eigenvalues of A.

CHAPTER 5 REVIEW EXERCISES *(page 338)*

1. $\lambda^3 - 3\lambda^2 - 17\lambda + 3$

3. (a) $\begin{bmatrix}1\\1\\0\end{bmatrix}, \begin{bmatrix}-1\\0\\1\end{bmatrix}$ **(b)** $\frac{1}{\sqrt{2}}\begin{bmatrix}1\\1\\0\end{bmatrix}, \frac{1}{\sqrt{6}}\begin{bmatrix}-1\\1\\2\end{bmatrix}$ **5.** $P = \begin{bmatrix}1 & -1 & -1\\2 & 1 & 0\\1 & 0 & 1\end{bmatrix}$, $P^{-1}AP = \text{Dg}\{5, 1, 1\}$.

7. $\lambda^3 - 9\lambda^2 + 2\lambda + 48$; $8, -2, 3$ **9. (a)** $[1 \quad 4 \quad 2 \quad 0]^T$ **(b)** $\mathbf{T}(\mathbf{u}_2) = \begin{bmatrix}11 & 12\\5 & 3\end{bmatrix}$ **(c)** $\begin{bmatrix}41 & 28\\18 & 7\end{bmatrix}$

11. For $R = \frac{1}{\sqrt{2}}\begin{bmatrix}1 & 1\\-1 & 1\end{bmatrix}$, $R^TAR = \begin{bmatrix}2 & 2\\0 & 2\end{bmatrix}$. **13.** For $Q = \begin{bmatrix}-\frac{1}{\sqrt{6}} & \frac{1}{\sqrt{2}} & -\frac{1}{\sqrt{3}}\\ \frac{2}{\sqrt{6}} & 0 & -\frac{1}{\sqrt{3}}\\ \frac{1}{\sqrt{6}} & \frac{1}{\sqrt{2}} & \frac{1}{\sqrt{3}}\end{bmatrix}$, $Q^TAQ = \text{Dg}\{-2, 0, 10\}$.

15. (a) For $\lambda = 1$, $\begin{bmatrix}1\\0\\0\end{bmatrix}$; for $\lambda = 2$, $\begin{bmatrix}3\\1\\0\end{bmatrix}$. **(b)** for $\lambda = 1$, $\begin{bmatrix}1\\0\\0\\0\end{bmatrix}$.

17. (a) $P = \begin{bmatrix}-1 & 4 & -2\\1 & 0 & -1\\0 & 1 & 1\end{bmatrix}$. **(b)** $P = \begin{bmatrix}1 & 2 & 1\\1 & 3 & 3\\1 & 3 & 4\end{bmatrix}$.

19. $\Lambda(A) = \{\pm i, 1\}$; A is diagonalizable because the roots are distinct. $\Lambda(B) = \{0, 0, 1 \pm i\}$; B is diagonalizable because $\text{rank}(0I - B) = 2$. $\Lambda(C) = \{5, 5, -1, -1\}$; C is diagonalizable because it is symmetric.

21. For $Q = \begin{bmatrix}1+i & 1-i & 0\\0 & 0 & 1\\1 & 1 & 0\end{bmatrix}$, $Q^{-1}AQ = \text{Dg}\{1 + 2i, 1 - 2i, 2\}$.

CHAPTER 6

EXERCISES 6.1 *(page 349)*

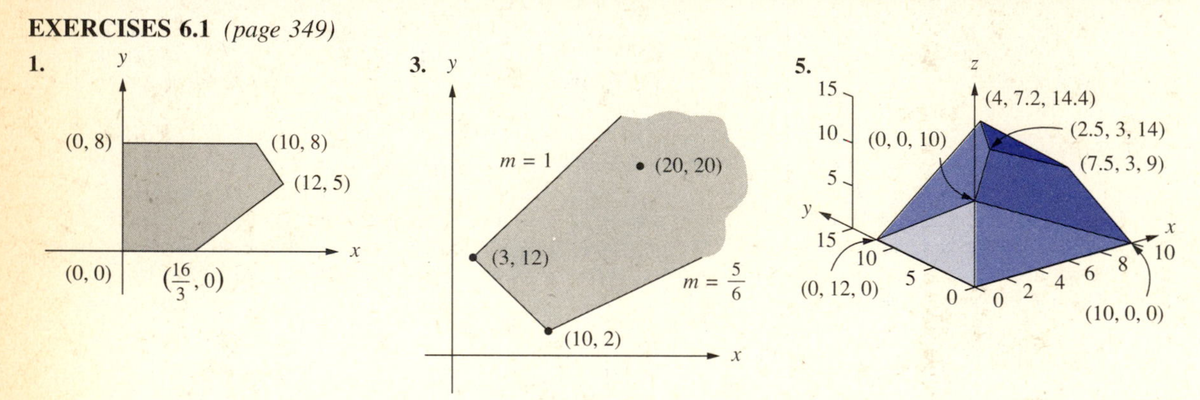

Unbounded feasible set

7. (a) max $z_1 = 420$ at (12,5), **(b)** min $z_1 = 0$ at (0, 0), **(c)** max $z_2 = 368$ at (12, 5). **9.** max $z = \infty$ since z is not bounded above on this set, min $z = 92$ at (10, 2).

11.

	1	2	3	4	5	6	7	8	9	10
x	0	0	0	0	$\frac{46}{3}$	$\frac{16}{3}$		12	10	16
y	0	23	−4	8	0	0	0	5	8	8
u	46	0	54	30	0	30		0	0	−18
v	16	108	0	48	−30	0		0	18	0
w	8	−15	12	0	8	8	0	3	0	0

There are 10 possible basic solutions. System 7 is inconsistent, so there are only 9 basic solutions. The basic feasible solutions are 1, 4, 6, 8, and 9 as shown in the above table. **13.** Plant $\frac{40}{3}$ acres of corn and $\frac{5}{3}$ acres of beans for a profit of \$700. **15. (a)** \$180,000 in corporate bonds, \$420,000 in government bonds; \$39,600. **(b)** \$360,000 in stocks, \$120,000 in corporate bonds, and \$120,000 in government bonds; \$36,000.
17. (b) There are 8 vertices: $(0, 0, 0)$, $\left(0, 0, \frac{40}{3}\right)$, $(0, 25, 0)$, $\left(0, \frac{35}{2}, \frac{15}{2}\right)$, $(30, 0, 0)$, $(15, 0, 5)$, $(10, 20, 0)$, $(10, 15, 4)$.
(c) There are 20 possibilities for basic solutions, but one system is inconsistent, so there are only 19 basic solutions, 8 of which are feasible. **19.** Make 40 wingos and 20 dodas for a profit of \$1280. **21.** 9,905,660 barrels of light crude, 3,490,566 barrels of dark crude for a total production cost of \$373,962,252. **23.** 200 boxes of regular mix and 250 boxes of deluxe mix for a profit of 1025.

EXERCISES 6.2 *(page 358)*

1. $\max[2 \quad 5]\begin{bmatrix} x \\ y \end{bmatrix}$ such that $\begin{bmatrix} -2 & -1 \\ 1 & 1 \\ -1 & -1 \end{bmatrix}\begin{bmatrix} x \\ y \end{bmatrix} \le \begin{bmatrix} -2 \\ 8 \\ -3 \end{bmatrix}$, $\begin{bmatrix} x \\ y \end{bmatrix} \ge 0.$

3. $\max[-3 \quad 4 \quad -4 \quad -6 \quad 6]\begin{bmatrix} x \\ y_+ \\ y_- \\ z_+ \\ z_- \end{bmatrix}$ such that $\begin{bmatrix} 1 & 1 & -1 & 1 & -1 \\ -1 & -1 & 1 & -1 & 1 \\ 2 & 3 & -3 & 5 & -5 \\ 4 & 3 & -3 & 1 & -1 \end{bmatrix}\begin{bmatrix} x \\ y_+ \\ y_- \\ z_+ \\ z_- \end{bmatrix} \le \begin{bmatrix} 1 \\ -1 \\ 4 \\ 2 \end{bmatrix}$, $\begin{bmatrix} x \\ y_+ \\ y_- \\ z_+ \\ z_- \end{bmatrix} \ge 0.$

5. $\max[-1 \quad -3 \quad -2 \quad -2]\begin{bmatrix} x \\ y \\ u \\ v \end{bmatrix}$ such that $\begin{bmatrix} 1 & 1 & 0 & 0 \\ -1 & -1 & 0 & 0 \\ 0 & 0 & 1 & 1 \\ 0 & 0 & -1 & -1 \\ 1 & 0 & 1 & 0 \\ 0 & 1 & 0 & 1 \end{bmatrix}\begin{bmatrix} x \\ y \\ u \\ v \end{bmatrix} \le \begin{bmatrix} 150 \\ -150 \\ 150 \\ -150 \\ 200 \\ 250 \end{bmatrix}$, $\begin{bmatrix} x \\ y \\ u \\ v \end{bmatrix} \ge 0.$

7.

	1	2	3	4	5	6
x	0	0	0	3	2	4
y	0	6	2	0	0	−2
s_1	6	0	4	0	2	0
s_2	2	−4	0	−1	0	0

Solutions 1, 3, and 5 are feasible.

9. (a) Maximize $w = x - 2y + 2z$ such that $\begin{bmatrix} 3 & -2 & 5 & 1 & 0 \\ 1 & -1 & -2 & 0 & -1 \\ 0 & 1 & -1 & 0 & 0 \end{bmatrix}\begin{bmatrix} x \\ y \\ z \\ u \\ v \end{bmatrix} = \begin{bmatrix} 14 \\ 1 \\ 5 \end{bmatrix}$, $\begin{bmatrix} x \\ y \\ z \\ u \\ v \end{bmatrix} \ge \mathbf{0}.$

(b) Maximize $Q = -15x_1 - 15x_2 - 25x_3 - 25x_4$ such that

$$\begin{bmatrix} 2 & 2 & 3 & 0 & -1 & 0 & 0 \\ 4 & 2 & 8 & 8 & 0 & -1 & 0 \\ 0 & 4 & 0 & 12 & 0 & 0 & -1 \end{bmatrix}\begin{bmatrix} x_1 \\ x_2 \\ x_3 \\ x_4 \\ x_5 \\ x_6 \\ x_7 \end{bmatrix} = \begin{bmatrix} 78 \\ 214 \\ 198 \end{bmatrix}, \quad \text{all } x_i \ge 0.$$

11. **(a)** The first and fourth figures are convex. **(b)** For $0 \le t \le 1$, $t\mathbf{x} + (1 - t)\mathbf{y}$ is a point between $\mathbf{x}$ and $\mathbf{y}$. Recall Exercise 4.4.25. **(c)** For $\mathbf{x}$ and $\mathbf{y}$ in $\mathcal{H}$ and $0 \le t \le 1$, $t\mathbf{x} + (1 - t)\mathbf{y} = [tx_1 + (1 - t)y_1 \ldots tx_n + (1 - t)y_n]$ and $a_1[tx_1 + (1 - t)y_1] + a_2[tx_2 + (1 - t)y_2] + \cdots + a_n[tx_n + (1 - t)y_n] = t(a_1x_1 + a_2x_2 + \cdots + a_nx_n) + (1 - t)(a_1y_1 + a_2y_2 + \cdots + a_ny_n) \le tb + (1 - t)b = b$. Therefore $t\mathbf{x} + (1 - t)\mathbf{y}$ is in $\mathcal{H}$ and $\mathcal{H}$ is convex. **(d)** Let $\mathcal{C}_1$ and $\mathcal{C}_2$ be convex. If $\mathbf{x}$ and $\mathbf{y}$ are in $\mathcal{C}_1 \cap \mathcal{C}_2$, then $\mathbf{x}$ and $\mathbf{y}$ in $\mathcal{C}_1 \Rightarrow t\mathbf{x} + (1 - t)\mathbf{y}$ in $\mathcal{C}_1$ and $\mathbf{x}$ and $\mathbf{y}$ in $\mathcal{C}_2 \Rightarrow t\mathbf{x} + (1 - t)\mathbf{y}$ in $\mathcal{C}_2$. Thus $t\mathbf{x} + (1 - t)\mathbf{y}$ is in $\mathcal{C}_1 \cap \mathcal{C}_2$ and $\mathcal{C}_1 \cap \mathcal{C}_2$ is convex.

13. Minimize $C = .08E + .60M$ such that $\begin{cases} 80E + 640M \ge 480 \\ 6E + 40M \ge 30 \\ 580E + 1360M \ge 1600 \end{cases}$ $E = \dfrac{58}{41}, M = \dfrac{47}{82}, C = \dfrac{937}{2050} = \dfrac{\$.457}{\text{quart}}$.

15. max $P = 23PB + 22PA + 15RB + 14RA$ such that

$$\begin{bmatrix} 5 & 15 & 10 & 17 & 1 & 0 & 0 & 0 \\ 4 & 2 & 4 & 2 & 0 & 1 & 0 & 0 \\ 40 & 22 & 35 & 22 & 0 & 0 & 1 & 0 \\ 14 & 14 & 7 & 14 & 0 & 0 & 0 & 1 \end{bmatrix} \begin{bmatrix} PB \\ PA \\ RB \\ RA \\ s_1 \\ s_2 \\ s_3 \end{bmatrix} = \begin{bmatrix} 480 \\ 160 \\ 1190 \\ 280 \end{bmatrix}.$$

EXERCISES 6.3 *(page 371)*

1. Basic variables: v, w; basic feasible solution: $[0 \ \ 0 \ \ 0 \ \ 10 \ \ 3]^T$; objective value: 0; entering variable: u; departing variable: w; new tableau:

	x	y	u	v	w	z	
v	−11	−6	0	1	−2	0	4
u	7	4	1	0	1	0	3
obj	58	28	0	0	9	1	27

New basic feasible solution: $[0 \ \ 0 \ \ 3 \ \ 4 \ \ 0]^T$; new objective value: 27.

3. Basic variables: x_1, x_6, x_4; basic feasible solution: $[12 \ \ 0 \ \ 0 \ \ 10 \ \ 0 \ \ 6 \ \ 0]^T$; objective value: 120; entering variable; x_2; departing variable: x_1 (not unique); new tableau:

	x_1	x_2	x_3	x_4	x_5	x_6	x_7	z	
x_2	$\frac{1}{4}$	1	$\frac{1}{4}$	0	$\frac{1}{4}$	0	$\frac{3}{4}$	0	3
x_6	$-\frac{1}{2}$	0	$\frac{5}{2}$	0	$\frac{3}{2}$	1	$-\frac{7}{2}$	0	0
x_4	$\frac{5}{4}$	0	$\frac{25}{4}$	1	$\frac{17}{4}$	0	$\frac{31}{4}$	0	25
obj	2	0	−6	0	6	0	0	1	144

New basic feasible solution: $[0 \ \ 3 \ \ 0 \ \ 25 \ \ 0 \ \ 0 \ \ 0]^T$; new objective value: 144.

5. $x = 0, y = 120$; $z = 360$. **7.** No optimal solution since the feasible set is unbounded. **9.** $w = -2$ for $x_1 = 0$, $x_2 = 1, x_3 = 2$. **11.** $P = 620$ for $x_1 = 20, x_2 = 10, x_3 = x_4 = 0$. **13.** $P = 700$ for $C = \dfrac{40}{3}, B = \dfrac{5}{3}, Z = 0$.

15. $R = 384$ at $[0 \ \ 12 \ \ 0]^T$. **17.** $R_1 \leftarrow R_2 + R_1$ and $R_3 \leftarrow 5R_2 + R_3$ yield

	x_1	x_2	x_3	x_4	z	
x_3	1	0	1	1	0	3
x_2	1	1	0	−1	0	6
obj	1	0	0	−5	1	30

The solution is then $z = 45$ for $X = [0 \ \ 9 \ \ 0 \ \ 3]^T$.

19. If the problem is degenerate, then B is a linear combination of $m - 1$ columns of A. These columns of A and B yield a singular submatrix.

EXERCISES 6.4 *(page 379)*

1. $P = 4$ for $x = 2, y = 1, z = 0$. **3.** No optimal solution because the feasible set is unbounded. **5.** $C = .45707$ for $E = 1.4146$ and $M = .5732$. **7.** Cost $= \$450$ for $x = 150, u = 0, y = 0, v = 150$. **9.** Week one: 5 assemblers and 35 trainers. Weeks 2, 3, 4: 145 assemblers. Profit will be $424,750.

CHAPTER 7

EXERCISES 7.1 *(page 387)*

1. $M(G) = \begin{bmatrix} 0 & 1 & 0 & 2 & 0 \\ 1 & 0 & 1 & 1 & 1 \\ 0 & 1 & 0 & 0 & 2 \\ 2 & 1 & 0 & 0 & 0 \\ 0 & 1 & 2 & 0 & 0 \end{bmatrix}$. There is one path of length 2 connecting T1 and T3. There are 4 paths of length 3 connecting T4 and T3.

3.

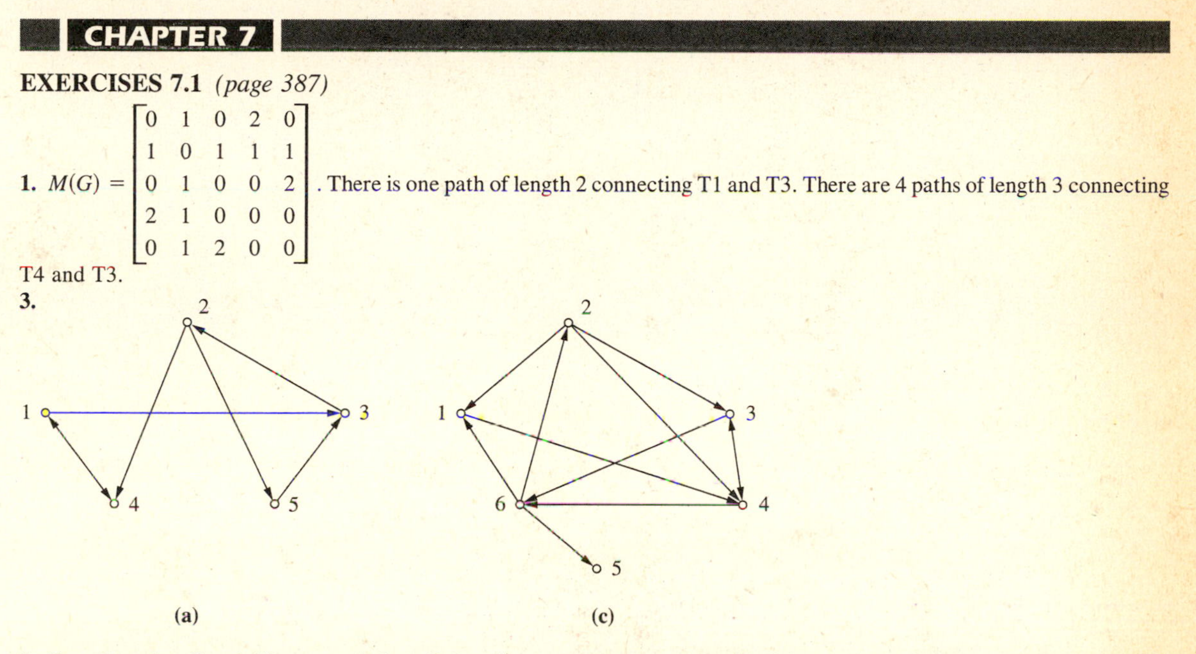

(a) (c)

5. Two digraphs, G and G', are equivalent if there is a one-to-one mapping **T** of the vertices of G onto the vertices of G', which preserves the edges in the sense that there is an edge connecting $\mathbf{T}(Vi)$ and $\mathbf{T}(Vj)$ in G' if, and only if, there is an edge connecting Vi and Vj in G. If A is the matrix of G and A' is the matrix of G', then there is a permutation matrix P such that $A' = PAP^T$.

7. $A(G) =$

	A	B	C	D	E	F
A	0	1	0	0	1	1
B	0	0	1	0	1	0
C	1	0	0	0	1	0
D	0	1	1	0	0	1
E	0	0	0	1	0	0
F	0	0	1	0	1	0

Adam and Dan influence the most people. Ed is influenced by the most people. Dan is the best choice for leader.

9. (a) $\begin{bmatrix} 0 & 1 & 1 & 0 & 1 & 0 & 0 & 0 \\ 1 & 0 & 1 & 1 & 0 & 0 & 0 & 0 \\ 1 & 1 & 0 & 1 & 0 & 0 & 0 & 0 \\ 0 & 1 & 1 & 0 & 0 & 0 & 0 & 1 \\ 1 & 0 & 0 & 0 & 0 & 1 & 1 & 0 \\ 0 & 0 & 0 & 0 & 1 & 0 & 1 & 1 \\ 0 & 0 & 0 & 0 & 1 & 1 & 0 & 1 \\ 0 & 0 & 0 & 1 & 0 & 1 & 1 & 0 \end{bmatrix}$ **(b)** $\begin{bmatrix} 0 & 0 & 0 & 0 & 1 & 1 & 1 & 1 \\ 0 & 0 & 0 & 0 & 1 & 1 & 1 & 1 \\ 0 & 0 & 0 & 0 & 1 & 1 & 1 & 1 \\ 0 & 0 & 0 & 0 & 1 & 1 & 1 & 1 \\ 1 & 1 & 1 & 1 & 0 & 0 & 0 & 0 \\ 1 & 1 & 1 & 1 & 0 & 0 & 0 & 0 \\ 1 & 1 & 1 & 1 & 0 & 0 & 0 & 0 \\ 1 & 1 & 1 & 1 & 0 & 0 & 0 & 0 \end{bmatrix}$

EXERCISES 7.2 *(page 399)*

1. (a) $y = -\frac{4}{3} + 3x$, $\Sigma\, d^2 = .666$; for $y = 2x$, $\Sigma\, d^2 = 4$. **(c)** $y = 3.75 + .44x$, $\Sigma\, d^2 = .042$; for $y = 3.7 + .5x$, $\Sigma\, d^2 = .10$. **(d)** $y = 597 - 3.12x$, $\Sigma\, d^2 = 8.4$; for $y = 592 - 3x$, $\Sigma\, d^2 = 17$. **3. (a)** $z = 2.3 - .9x - .6y$. **(c)** For $z = 7.4 + 3x - 5.5y$, $\Sigma\, d^2 = 0.7$. For $z = 6 + 3x - 5y$, $\Sigma\, d^2 = 2$. **5.** (1, 1) and (4, 0); the lines are nearly parallel. **7.** $a = 1.565$, $h = 5.565$. **9. (a)** $X = \begin{bmatrix} 1.96 \\ -1.26 \\ .600 \end{bmatrix}$. **(b)** $X = \begin{bmatrix} 5 \\ -6 \\ -2 \\ 1 \end{bmatrix}$, $\Sigma\, d^2 = 0$.

11. $X = [5.507 \quad -6.793 \quad -2.044 \quad 3.607]^T$.

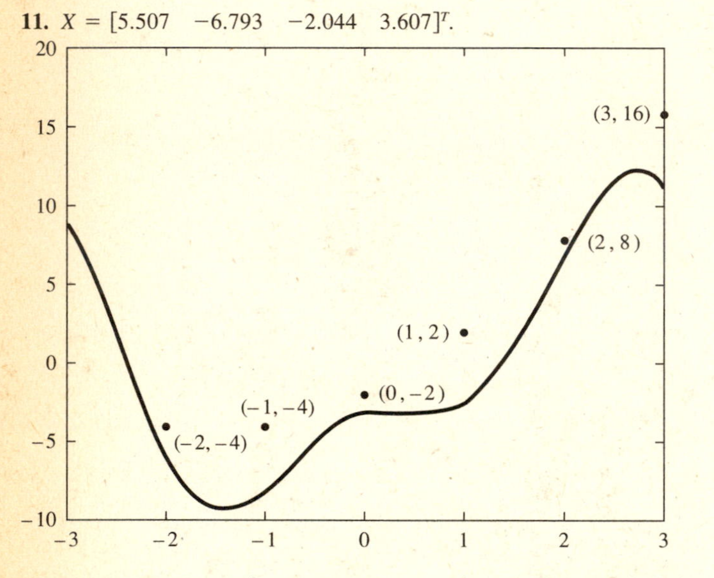

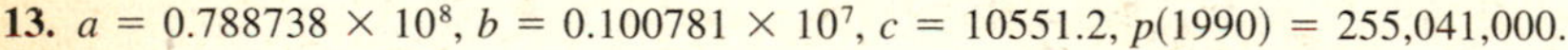

13. $a = 0.788738 \times 10^8$, $b = 0.100781 \times 10^7$, $c = 10551.2$, $p(1990) = 255{,}041{,}000$.

EXERCISES 7.3 *(page 410)*

1. Hyperbola; $5x'^2 - y'^2 = 10$; $P = \frac{1}{\sqrt{2}}\begin{bmatrix} 1 & -1 \\ 1 & 1 \end{bmatrix}$. **3.** Hyperbola; $18x'^2 - 2y'^2 = 36$; $P = \frac{1}{\sqrt{5}}\begin{bmatrix} 2 & -1 \\ 1 & 2 \end{bmatrix}$.

5. Ellipsoid; $2x'^2 + 5y'^2 + 2z'^2 = 10$; $P = \frac{1}{\sqrt{6}}\begin{bmatrix} -2 & -\sqrt{2} & 0 \\ -1 & \sqrt{2} & \sqrt{3} \\ 1 & -\sqrt{2} & \sqrt{3} \end{bmatrix}$.

7. Hyperboloid of two sheets; $-x'^2 + 8y'^2 - z'^2 = 16$; $P = \begin{bmatrix} -\frac{5}{\sqrt{45}} & \frac{2}{3} & 0 \\ \frac{4}{\sqrt{45}} & \frac{2}{3} & \frac{1}{\sqrt{5}} \\ \frac{2}{\sqrt{45}} & \frac{1}{3} & -\frac{2}{\sqrt{5}} \end{bmatrix}$.

9. Hyperbolic cylinder along the z'-axis; $x'^2 - y'^2 = 1$; $P = \frac{1}{3}\begin{bmatrix} -2 & 2 & -1 \\ 2 & 1 & -2 \\ 1 & 2 & 2 \end{bmatrix}$.

11. (a) $b'^2 - a'c' = -\det A' = -\det(P^TAP) = -\det P^T \det A \det P = -(1)\det A(1) = -\det A = b^2 - ac$.
(b) $b^2 - ac > 0 \Rightarrow$ Ellipse, $b^2 - ac < 0 \Rightarrow$ Hyperbola, $b^2 - ac = 0 \Rightarrow$ Parabola.

13. $X^TBX = \frac{1}{2}x^T(A + A^T)X = \frac{1}{2}X^TAX + \frac{1}{2}X^TA^TX = \frac{1}{2}X^TAX + \frac{1}{2}X^TAX = X^TAX$.

EXERCISES 7.4 *(page 415)*

1. (a) $\begin{bmatrix} 30 \\ 56 \\ 33 \end{bmatrix}$ **(b)** $\begin{bmatrix} 24 \\ 25 \\ 16 \end{bmatrix}$ **(c)** $\begin{bmatrix} 4 \\ 3 \\ 4 \end{bmatrix}$ **3. (c)** $\begin{bmatrix} 570 \\ 1058 \\ 723 \\ 739 \end{bmatrix}$ **5. (a)** \$.70 **(b)** 70% **(c)** 70% **7.** (a) is productive, (b) and (c) are not productive. **9.** Computing $(I - C)^{-1}$, by Gauss-Jordan elimination requires n^3 multiplications. C^2 requires n^3 multiplications, C^3 requires n^3 additional multiplications, and so on. Thus approximating $(I - C)^{-1}$ by $I + C + C^2 + C^3 + \cdots + C^k$ requires $(k - 1)n^3$ multiplications.

Index